A CUSTOM EDITION OF

STATISTICS FOR MANAGERS

USING MICROSOFT EXCEL

SECOND EDITION   BY LEVINE, BERENSON, AND STEPHAN

WITH LECTURE NOTES FOR E370

MARY ELIZABETH CAMP
INDIANA UNIVERSITY

Pearson
Custom
Publishing

Cover photo courtesy of IU Home Pages

Excerpts taken from:

Statistics for Managers, Second Edition
by David M. Levine, Mark L. Berenson, and David Stephan
Copyright © 1999, 1997 by Prentice-Hall, Inc.
A Pearson Education Company
Upper Saddle River, New Jersey 07458

Students Solutions Manual for Statistics for Managers, Second Edition,
by David M. Levine, Mark L. Berenson, and David Stephan
Copyright © 1999 by Prentice-Hall, Inc.

Microsoft and Windows are registered trademarks of the Microsoft Corporation in the
U.S.A. and other countries. This book is not sponsored or endorsed by or affiliated with
Microsoft Corporation.

The information, illustrations, and/or software contained in this book, and regarding the
above-mentioned programs, are provided "As Is," without warranty of any kind, express or
implied, including without limitation any warranty concerning the accuracy, adequacy, or
completeness of such information. Neither the publisher, the authors, nor the copyright
holders shall be responsible for any claims attributable to errors, omissions, or other
inaccuracies contained in this book. Nor shall they be liable for direct, indirect, special,
incidental, or consequential damages arising out of the use of such information or material.

This special edition published in cooperation with Pearson Custom Publishing

This publication has been printed using selections as they appeared in their original format.
Layout and appearance will vary accordingly.

Printed in the United States of America

10 9 8 7 6 5 4 3 2 1

Please visit our web site at www.pearsoncustom.com

ISBN 0–536–61632–9

BA 992553

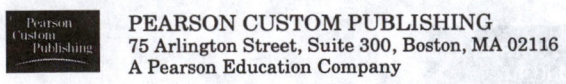
PEARSON CUSTOM PUBLISHING
75 Arlington Street, Suite 300, Boston, MA 02116
A Pearson Education Company

Contents

Contents

Chapter 1
Introduction
and Data Collection

CHAPTER OBJECTIVES

✓ *To present a broad overview of the subject of statistics and its applications*
✓ *To distinguish between descriptive and inferential statistics*
✓ *To discuss sources of data*
✓ *To discuss types of data*
✓ *To introduce methods of sample selection*
✓ *To study how to evaluate survey worthiness*

1.1 WHY A MANAGER NEEDS TO KNOW ABOUT STATISTICS

A century ago H. G. Wells commented, "statistical thinking will one day be as necessary as the ability to read and write." As we approach the next millennium, the issue facing managers is not a shortage of information but how to use the available information to make better decisions.

It is from this perspective of informed decision making that we consider why a manager needs to know about statistics. Managers need an understanding of statistics for the following four key reasons:

1. To know how to properly present and describe information
2. To know how to draw conclusions about large populations based only on information obtained from samples
3. To know how to improve processes
4. To know how to obtain reliable forecasts

On the next page is a road map of this text from the perspective of these four reasons for learning statistics. From this road map we observe that the first three chapters include coverage of methods involved in the collection, presentation, and description of information. Chapters 4–6 provide coverage of the basic concepts of probability; the binomial, normal, and other distributions; decision making; and sampling distributions so that in chapters 7–11 the reader will learn how to draw conclusions about large populations based only on information obtained from samples. Chapter 12 contains coverage of statistical applications in quality and productivity management that is essential for process improvement. Chapters 13–15 focus on regression, multiple regression, modeling, and time-series analysis that provide methods for obtaining forecasts.

We may apply statistical methods in the functional areas of business, accounting, finance, management, and marketing: Accounting uses statistical methods to select samples for auditing purposes and to understand the cost drivers in cost accounting. Finance uses statistical methods to choose between alternative portfolio investments and to track trends in financial measures over time. Management uses statistical methods to improve the quality of the products manufactured or the services delivered by an organization. Marketing uses statistical methods to estimate the proportion of customers who prefer one product over another and why they do and to draw conclusions about what advertising strategy might be most useful in increasing sales of a product.

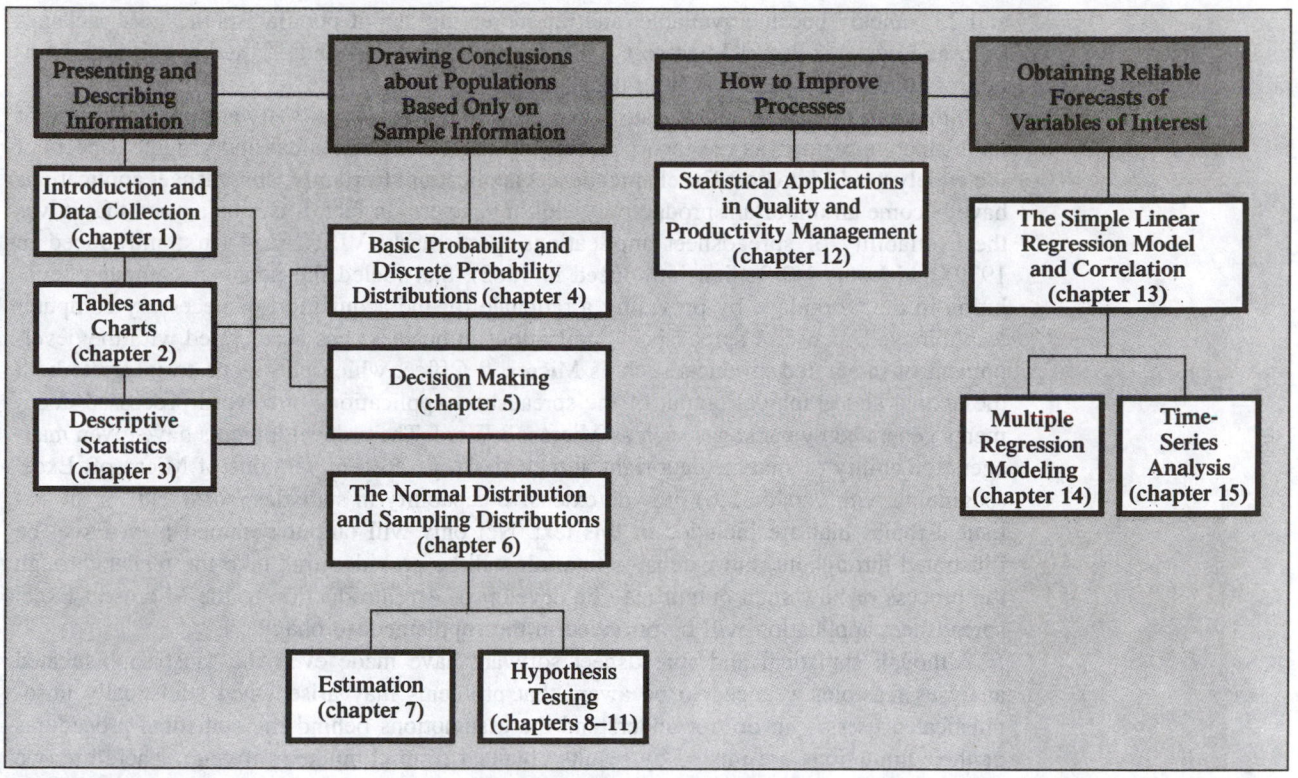

Road map

1.2 THE GROWTH AND DEVELOPMENT OF MODERN STATISTICS

Historically, the growth and development of modern statistics can be traced to three separate phenomena: the needs of government to collect data on its citizenry (see references 12, 19, 20, 24, and 25), the development of the mathematics of probability theory, and the advent of the computer.

Data have been collected throughout recorded history. During the Egyptian, Greek, and Roman civilizations, data were obtained primarily for the purposes of taxation and military conscription. In the Middle Ages, church institutions often kept records concerning births, deaths, and marriages. In America, various records were kept during colonial times (see reference 25), and beginning in 1790 the federal Constitution required the taking of a census every 10 years. In fact, the expanding needs of the census helped spark the development of tabulating machines at the beginning of the twentieth century. This achievement led to the development of large-scale mainframe computers and eventually to the personal computer.

This infusion of computer technology has profoundly changed the field of statistics in the last 30 years. Mainframe packages such as SAS and SPSS became popular during the 1960s and 1970s. During the 1980s, statistical software experienced a vast technological revolution. Besides the usual improvements made in periodic updates, the availability of personal computers led to the development of new packages. In addition, personal computer versions of existing packages such as SAS, SPSS, and Minitab (see references 16, 18, and 23) quickly became available, and the increasing use of popular spreadsheet packages such as Lotus 1-2-3 and Microsoft Excel (see references 14 and 15) led to the incorporation of statistical features in these packages.

Although this text is appropriate for those who use statistical software packages, one of its distinctive features is the incorporation of many of the statistical and graphic aspects of the widely used Microsoft Excel spreadsheet application. Certainly, spreadsheet applications have become an important productivity tool in business. In fact, it can be argued that it was the availability of spreadsheet applications, particularly VISICALC (first introduced in 1979) and Lotus 1-2-3 (first introduced in 1983), that fueled the personal computer revolution in the workplace by providing a rationale for the acquisition of necessary computer technology. This use of spreadsheet applications in business has accelerated with the development of integrated products such as Microsoft Office, which allow one to integrate both the textual and graphical output of the spreadsheet applications into word processed documents generated by packages such as Microsoft Word. These developments have given managers the ability to analyze data right at their desktops. Recent versions of Microsoft Excel (beginning with version 5.0) provide extensive capability in statistics, sufficient for all statistical topics that are included in this text. Not only will output obtained from Excel be illustrated throughout, but extensive tutorials will be provided that take the reader through the process of how such output may be developed. An introduction to the Microsoft Excel spreadsheet application will be provided in the supplement to chapter 1.

Although statistical and spreadsheet software have made even the most sophisticated analyses feasible, we need to be aware that problems may arise when statistically unsophisticated users who do not understand the assumptions behind the statistical procedures or their limitations are misled by results obtained from computer software. Therefore, we believe that for pedagogical reasons it is important that the applications of the methods covered in the text be illustrated through worked-out examples.

◆ 1.3 STATISTICAL THINKING AND MODERN MANAGEMENT

In the past decade the emergence of a global economy has led to an increasing focus on the quality of products manufactured and services delivered. In fact, more than that of any other individual, it was the work of a statistician, W. Edwards Deming, that led to this changed business environment. An integral part of the managerial approach that contains this increased focus on quality (often referred to as **total quality management**) is the application of certain statistical methods and the use of statistical thinking on the part of managers throughout a company.

Statistical thinking can be defined as thought processes that focus on ways to understand, manage, and reduce variation.

Statistical thinking includes the recognition that data are inherently variable (no two things or people will be exactly alike in all ways) and that the identification, measurement, control, and reduction of variation provide opportunities for quality improvement. Statistical methods can provide the vehicle for taking advantage of these opportunities.

The role of statistical methods in the context of quality improvement can be better understood if we refer to a model of quality improvement as presented in Figure 1.1. We may observe from Figure 1.1 that the triangle consists of three parts: at the top we have management philosophy, and at the two lower corners we have statistical methods and behavioral

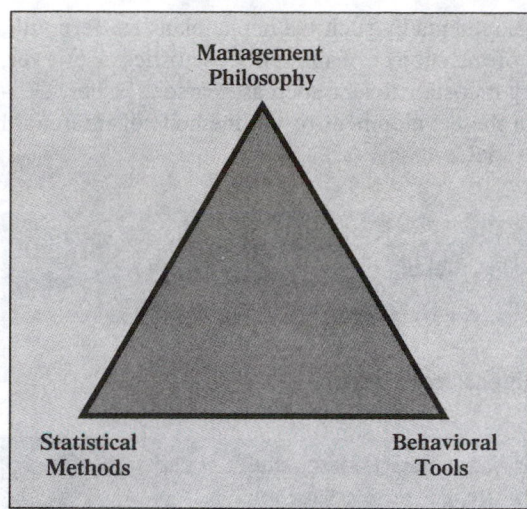

FIGURE 1.1
A model of the quality improvement process

tools. Each of these three aspects is indispensable for long-term quality improvement of either the products manufactured or the services provided by an organization. A management philosophy provides a constant foundation for quality improvement efforts. Among the approaches available are those advocated by W. Edwards Deming (see references 4 and 5 and section 12.2) and Joseph Juran (see references 10 and 11).

In order to implement a quality improvement approach in an organization, one needs to use both behavioral tools and statistical methods. Each of these aids in understanding and improving processes. Among the useful behavioral tools are process flow and fishbone diagrams, brainstorming, nominal group decision making, and team building. (For further discussion, see references 4 and 22.) Among the most useful statistical methods for quality improvement are the numerous tables and charts and descriptive statistics discussed in chapters 2 and 3 and the control charts developed in chapter 12.

 **DESCRIPTIVE VERSUS INFERENTIAL STATISTICS**

The need for data on a nationwide basis was closely intertwined with the development of descriptive statistics.

> **Descriptive statistics** can be defined as those methods involving the collection, presentation, and characterization of a set of data in order to describe the various features of that set of data properly.

Although descriptive statistical methods are important for presenting and characterizing data (see chapters 2 and 3), it has been the development of inferential statistical methods as an outgrowth of probability theory that has led to the wide application of statistics in all fields of research today.

The initial impetus for the formulation of the mathematics of probability theory came from the investigation of games of chance during the Renaissance. The foundations of the subject of probability can be traced back to the middle of the seventeenth century in the correspondence between the mathematician Pascal and the gambler Chevalier de Mere (see references 12 and 13). These and other developments by such mathematicians as Bernoulli, DeMoivre, and Gauss were the forerunners of the subject of inferential statistics. However, it has only been since the turn of this century that statisticians such as Pearson, Fisher, Gosset, Neyman, Wald, and Tukey pioneered in the development of the methods of inferential statistics that are widely applied in so many fields today.

> **Inferential statistics** can be defined as those methods that make possible the estimation of a characteristic of a population or the making of a decision concerning a population based only on sample results.

To clarify this definition, a few more definitions are necessary.

> A **population** (or **universe**) is the totality of items or things under consideration.
> A **sample** is the portion of the population that is selected for analysis.
> A **parameter** is a summary measure that is computed to describe a characteristic of an entire population.
> A **statistic** is a summary measure that is computed to describe a characteristic from only a sample of the population.

Suppose that the president of your college wanted to conduct a survey to learn about student perceptions concerning the quality of life on campus. The population, or universe, in this instance would be all currently enrolled students, whereas the sample would consist only of those students who had been selected to participate in the survey. The goal of the survey would be to describe various attitudes or characteristics of the entire population (the parameters). This would be achieved by using the statistics obtained from the sample of students to estimate various attitudes or characteristics of interest in the population. Thus, one major aspect of inferential statistics is the process of using sample statistics to draw conclusions about the population parameters.

The need for inferential statistical methods derives from the need for sampling. As a population becomes large, it is usually too costly, too time consuming, and too cumbersome to obtain our information from the entire population. Decisions about the population's characteristics have to be based on the information contained in a sample of that population. Probability theory provides the link by determining the likelihood that the results from the sample reflect the results from the population.

◆1.5◆ WHY DO WE NEED DATA?

◆ USING STATISTICS: *Exchange International Resort Evaluation*

Exchange International offers a vacation exchange network for time-share resort owners throughout the world. Members of Exchange International may use its services to arrange for vacation exchanges with other time-share resort owners. Exchange International recognizes the importance of offering its members high-quality resort destinations. To monitor the quality of its affiliated resorts, Exchange International provides a survey to all members prior to their vacation departures and offers the possibility of a free vacation exchange as an incentive to respond. Exchange International seeks to improve member services and to affiliate with only the finest vacation resorts. This survey is partially reproduced here.

- How would you rate the following aspects of your vacation?

	EXCELLENT	GOOD	FAIR	DISAPPOINTING	UNACCEPTABLE
Quality of area	5	4	3	2	1
Quality of unit	5	4	3	2	1
Quality of resort	5	4	3	2	1
Quality of services	5	4	3	2	1

- Would you be likely to return to this resort again?   Yes ☐   No ☐
- About how much money (in U.S. dollars) did you pay for your time-share?_____ ◆

Obtaining appropriate information is essential to conducting business. We may think of **data** as the information needed to help us make a more informed decision in a particular situation. There are many instances in which data are needed.

- A market researcher needs to assess product characteristics to distinguish one product from another.
- A pharmaceutical manufacturer needs to determine whether a new drug is more effective than those currently in use.
- A manager wants to monitor a process on a regular basis to find out whether the quality of service being provided or products being manufactured are conforming to company standards.
- An auditor wants to review the financial transactions of a company in order to ascertain whether or not it is in compliance with generally accepted accounting principles.
- A potential investor wants to determine which firms within which industries are likely to have accelerated growth in a period of economic recovery.
- A student wants to get data on classmates' favorite rock groups to satisfy a curiosity.

There are six main reasons for data collection, as illustrated in Exhibit 1.1.

Exhibit 1.1   Reasons for Obtaining Data

✓ **1.** Data are needed to provide the necessary input to a survey.

✓ **2.** Data are needed to provide the necessary input to a study.

✓ **3.** Data are needed to measure performance of an ongoing service or production process.

✓ **4.** Data are needed to evaluate conformance to standards.

✓ **5.** Data are needed to assist in formulating alternative courses of action in a decision-making process.

✓ **6.** Data are needed to satisfy our curiosity.

The Exchange International survey in our Using Statistics example illustrates reasons 1, 3, 4, and 5. For example, Exchange International compiles data as the result of a survey. It then analyzes the data to measure performance, to evaluate standards, and to help formulate alternative courses of action should this be required.

It is extremely important that we begin our statistical analysis by identifying the most appropriate data collection sources. If the data are flawed by biases, ambiguities, or other types of errors, even the fanciest and most sophisticated statistical methodologies would not likely be enough to compensate for such deficiencies.

 ## 1.6 SOURCES OF DATA

There are four key data collection sources, as illustrated in Exhibit 1.2.

> ### Exhibit 1.2   Key Data Collection Sources
> ✓ **1.** We may obtain data already published by governmental, industrial, or individual sources.
> ✓ **2.** We may design an experiment to obtain the necessary data.
> ✓ **3.** We may conduct a survey.
> ✓ **4.** We may make observations through an observational study.

Data collectors are labeled **primary sources;** data compilers are called **secondary sources.** As illustrated in Exhibit 1.2, the first method of obtaining data is via governmental, industrial, or individual sources. Of these three, the federal government is the major collector and compiler of data for both public and private purposes.

Many governmental agencies facilitate this work. The Bureau of Labor Statistics is responsible for collecting data on employment as well as for establishing the well-known monthly *Consumer Price Index.* In addition to its constitutional requirement for conducting a decennial census, the Bureau of the Census oversees a variety of ongoing surveys regarding population, housing, and manufacturing. Additionally, it undertakes special studies on topics such as crime, travel, and health care.

In addition to the federal government, various trade publications present data pertaining to specific industrial groups. Investment services such as Moody's display financial data on a company basis. Syndicated services such as A. C. Nielsen provide clients with information enabling the comparison of client products with their competitors. Daily newspapers are filled with numerical information regarding stock prices, weather conditions, and sports statistics.

The second data collection source is through experimentation. In an experiment, strict control is exercised over the treatments given to participants. For example, in a study testing the effectiveness of toothpaste, the researcher would determine which participants in the study would use the new brand and which would not, instead of leaving the choice to the subjects. Proper experimental designs are usually the subject matter of more advanced texts, because they often involve sophisticated statistical procedures. However, in order to develop a feeling for testing and experimentation, the fundamental experimental design concepts will be considered in chapters 8–11.

The third data collection source is obtained by conducting a survey. Here no control is exercised over the behavior of the people being surveyed. They are merely asked questions about their beliefs, attitudes, behaviors, and other characteristics. Responses are then edited, coded, and tabulated for analysis.

The fourth method for obtaining data is through an observational study. A researcher observes the behavior directly, usually in its natural setting. Most knowledge of animal behavior is developed in this way, as is our scientific knowledge in many fields, such as astronomy and geology, in which experimentation and surveys are impractical if not impossible.

Observational study has many formats in business, all of which are intended to collect information in a group setting to assist in the decision-making process. As one example, the **focus group** is a popular marketing research tool that is used for eliciting unstructured responses to open-ended questions. A moderator leads the discussion and all the participants respond to the questions asked. Other, more structured formats involving group dynamics for obtaining information (and consensus building) include various organizational behavior/industrial psychology tools such as brainstorming, the Delphi technique, and the nominal-group method (see reference 22). These tools have become more popular in recent years owing to the impact of the total quality management (TQM) philosophy on business, because TQM emphasizes the importance of teamwork and employee empowerment in an attempt to improve every product and service.

COMMENT: Data Sources and the Age of Technology

Thanks to the widespread use of information technology, never before have so much timely and accurate data and information been so readily available—from so many sources. Bar codes automatically record inventory information as products are purchased from supermarkets, department stores, and other outlets. ATMs enable banking transactions in which information is immediately recorded on account balances. Airline ticketing offices and travel agents have up-to-the-minute information regarding space availability on flights and at hotels. Transactions that took hours, or even days, a decade ago are now accomplished in a matter of seconds.

Use of the library for research has taken on a new meaning. No longer need one be confined to using print media such as books, journals, magazines, pamphlets, and newspapers. With computer-based information systems, we can search for and retrieve data electronically by using CD-ROM databases, surfing the World Wide Web, or exchanging e-mail messages with other Internet users. The phrase "library visit" has come to include electronic visits from our home or office using a properly equipped personal computer system.

In order to design an experiment, conduct a survey, or perform an observational study, one must understand the different types of data and measurement levels. To demonstrate some of the issues involved in obtaining data, we will present them in the context of a survey, although most of the same issues will arise in other types of research.

 **1.7** **TYPES OF DATA**

Statisticians develop surveys to deal with a variety of phenomena or characteristics. These phenomena or characteristics are called **random variables.** The data, which are the observed outcomes of these random variables, will undoubtedly differ from response to response.

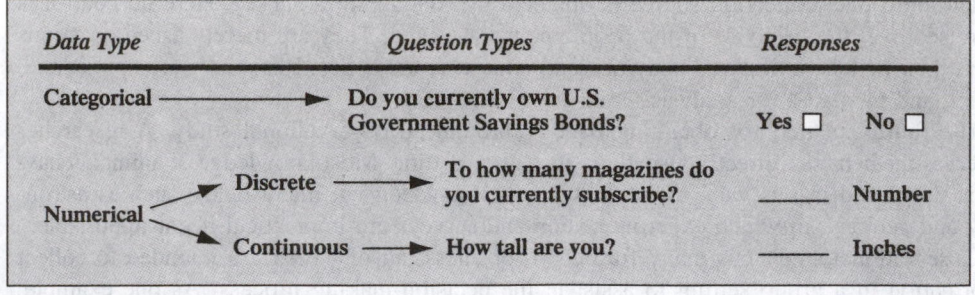

Data Type	Question Types	Responses
Categorical ⟶	Do you currently own U.S. Government Savings Bonds?	Yes ☐ No ☐
Numerical ⟶ Discrete ⟶	To how many magazines do you currently subscribe?	_____ Number
Continuous ⟶	How tall are you?	_____ Inches

FIGURE 1.2 Types of data

As illustrated in Figure 1.2, there are two types of *random variables* that yield the observed outcomes or data: categorical and numerical.

Categorical random variables yield categorical responses, such as yes or no answers. An example is the response to the question "Do you currently own U.S. Government Savings Bonds?" because it is limited to a simple yes or no answer. Another example is the response to the question on the Exchange International survey "Would you be likely to return to this resort again?"

Numerical random variables yield numerical responses such as your height in inches. Other examples are how much money you paid for your time-share from the Exchange International survey or the response to the question "To how many magazines do you currently subscribe?" There are two types of numerical variables: discrete and continuous.

Discrete random variables produce numerical responses that arise from a counting process. "The number of magazines subscribed to" is an example of a discrete numerical variable, because the response is one of a finite number of integers. You subscribe to zero, one, two, and so on, magazines.

Continuous random variables produce numerical responses that arise from a measuring process. Your height is an example of a continuous numerical variable, because the response takes on any value within a continuum or interval, depending on the precision of the measuring instrument. For example, your height may be 67 inches, 67¼ inches, 67⁷⁄₃₂ inches, or 67⁵⁸⁄₂₅₀ inches depending on the precision of the available instruments.

Theoretically, no two persons could have exactly the same height, because the finer the measuring device used, the greater the likelihood of detecting differences among them. However, most measuring devices are not sophisticated enough to detect small differences. Hence, *tied observations* are often found in experimental or survey data even though the random variable is truly continuous.

Problems for Section 1.7

Learning the Basics

- **1.1** Suppose that three different beverages are sold at a fast-food restaurant—soft drinks, tea, and coffee. Explain why the type of beverage sold is an example of a categorical variable.

  1.2 Suppose that soft drinks are sold in three sizes in a fast-food restaurant—small, medium, and large. Explain why the size of the soft drink is a categorical variable.

  1.3 Suppose that we measure the time of airplane flight from New York to Los Angeles from takeoff to landing. Explain why the time of airplane flight is a numerical variable.

Applying the Concepts

● **1.4** For each of the following random variables determine whether the variable is categorical or numerical. If the variable is numerical, determine whether the phenomenon of interest is discrete or continuous.

(a) Number of telephones per household
(b) Type of telephone primarily used
(c) Number of long-distance calls made per month
(d) Length (in minutes) of longest long-distance call made per month
(e) Color of telephone primarily used
(f) Monthly charge (in dollars and cents) for long-distance calls made
(g) Ownership of a cellular phone
(h) Number of local calls made per month
(i) Length (in minutes) of longest local call per month
(j) Whether there is a telephone line connected to a computer modem in the household
(k) Whether there is a fax machine in the household

1.5 Suppose that the following information is obtained from students upon exiting from the campus bookstore during the first week of classes:

(a) Amount of money spent on books
(b) Number of textbooks purchased
(c) Amount of time spent shopping in the bookstore
(d) Academic major
(e) Gender
(f) Ownership of a personal computer
(g) Ownership of a videocassette recorder
(h) Number of credits registered for in the current semester
(i) Whether or not any clothing items were currently purchased at the bookstore
(j) Method of payment

Classify each of these variables as categorical or numerical. If the variable is numerical, determine whether the variable is discrete or continuous.

1.6 For each of the following random variables, determine whether the variable is categorical or numerical. If the variable is numerical, determine whether the phenomenon of interest is discrete or continuous.

(a) Brand of personal computer primarily used
(b) Cost of personal computer system
(c) Amount of time the personal computer is used per week
(d) Primary use for the personal computer
(e) Number of persons in the household who use the personal computer
(f) Number of computer magazine subscriptions
(g) Word processing package primarily used
(h) Whether the personal computer is connected to the Internet

1.7 For each of the following random variables, determine whether the variable is categorical or numerical. If the variable is numerical, determine whether the phenomenon of interest is discrete or continuous.

(a) Amount of money spent on clothing in the last month
(b) Number of winter coats owned
(c) Favorite department store
(d) Amount of time spent shopping for clothing in the last month
(e) Most likely time period during which shopping for clothing takes place (weekday, weeknight, or weekend)
(f) Number of pairs of women's gloves owned
(g) Primary type of transportation used when shopping for clothing

1.8 Suppose the following information is obtained from Robert Keeler on his application for a home mortgage loan at the Metro County Savings and Loan Association:

(a) Place of Residence: Stony Brook, New York
(b) Type of Residence: Single-family home
(c) Date of Birth: April 9, 1962
(d) Monthly Payments: $1,427
(e) Occupation: Newspaper reporter/author
(f) Employer: Daily newspaper
(g) Number of Years at Job: 14
(h) Number of Jobs in Past 10 Years: 1
(i) Annual Family Salary Income: $66,000
(j) Other Income: $16,000
(k) Marital Status: Married

(l) Number of Children: 2

(m) Mortgage Requested: $120,000

(n) Term of Mortgage: 30 years

(o) Other Loans: Car

(p) Amount of Other Loans: $8,000

Classify each of the responses by type of data.

1.9 One of the variables most often included in surveys is income. Sometimes the question is phrased "What is your income (in thousands of dollars)?" In other surveys, the respondent is asked to "Place an X in the circle corresponding to your income level."

   ○ Under $20,000?   ○ $20,000–$39,999   ○ $40,000 or more

(a) In the first format, explain why income might be considered either discrete or continuous.

(b) Which of these two formats would you prefer to use if you were conducting a survey? Why?

(c) Which of these two formats would likely bring you a greater rate of response? Why?

1.10 If two students score a 90 on the same examination, what arguments could be used to show that the underlying random variable—test score—is continuous?

1.11 Suppose that the director of market research at a large department store chain wanted to conduct a survey throughout a metropolitan area to determine the amount of time working women spend shopping for clothing in a typical month.

(a) Describe both the population and the sample of interest and indicate the type of data the director might wish to collect.

(b) Develop a first draft of the questionnaire needed in (a) by writing a series of three categorical questions and three numerical questions that you feel would be appropriate for this survey.

◆1.8 ▸ TYPES OF SAMPLING METHODS

As mentioned in section 1.4, a sample is the portion of the population that has been selected for analysis. Rather than taking a complete census of the whole population, statistical sampling procedures focus on a small representative group of the larger population. The resulting sample provides information that can be used to estimate characteristics of the entire population.

The sampling process begins by locating appropriate data sources, such as population lists, directories, maps, and other sources, which are called **frames.** Samples are drawn from these frames. If the frame is inadequate because certain groups of individuals or items in the targeted population were not properly included, then the samples will be inaccurate and biased. Using different frames to generate data can lead to opposite conclusions, as illustrated in Example 1.1.

Example 1.1 *A Case of Opposing Conclusions*

Consider the following headline that appeared in a suburban New York newspaper a decade ago: "Off, With a Head Count: Is Suffolk More Populous than Nassau? LILCO and the Census Bureau Disagree" (*Newsday,* April 25, 1988). Given Suffolk's survey data, the Suffolk county executive felt it was more populous, whereas the Nassau county executive disagreed, citing Nassau's own survey data. Who was right?

SOLUTION

The differences in the two estimates come from the fact that the Census Bureau and the Long Island Lighting Company (LILCO) used different frames and formulas to estimate population in the two counties. The Census Bureau used birth and death rates, migration patterns as shown on income tax returns, and a demographic formula that estimates that the average number of people per household had been shrinking in the past several years. For its definition, LILCO used the number of year-round electric and gas meters, building permits, and a factor for the number of people in each house.

There are three main reasons for drawing a sample, as depicted in Exhibit 1.3.

Exhibit 1.3 Reasons for Drawing a Sample

✓ **1.** A sample is less time consuming than a census.

✓ **2.** A sample is less costly to administer than a census.

✓ **3.** A sample is less cumbersome and more practical to administer than a census of the targeted population.

As depicted in Figure 1.3, there are basically two kinds of samples: the nonprobability sample and the probability sample.

A **nonprobability sample** is one in which the items or individuals included are chosen without regard to their probability of occurrence.

Because nonprobability samples have chosen participants without knowing their probabilities of selection (and in some cases participants have self-selected), the theory that has been developed for probability sampling cannot be applied. For many studies, only a nonprobability sample such as a judgment sample is available. In these instances, the opinion of an expert in the subject matter of a study is crucial to being able to use the results obtained to make changes in a process. Some other common procedures of nonprobability sampling are quota sampling and chunk sampling; these are discussed in detail in specialized books on sampling methods (see references 1, 3, and 9).

Nonprobability samples can have certain advantages such as convenience, speed, and lower cost. On the other hand, two major disadvantages—a lack of accuracy due to selection

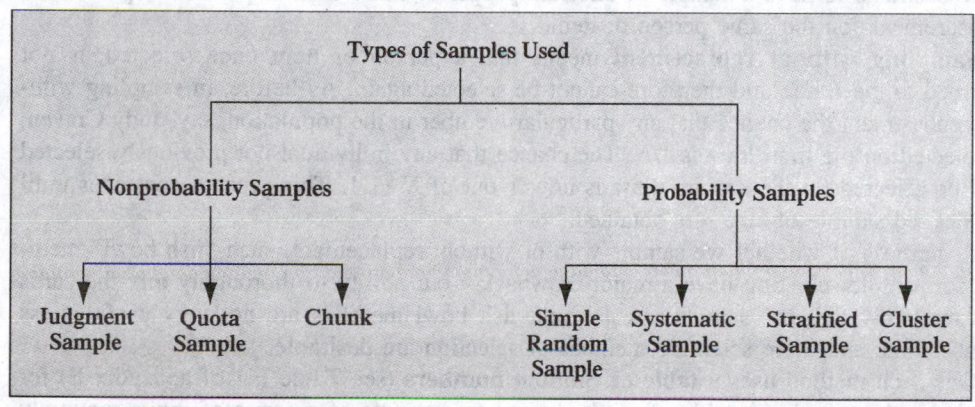

FIGURE 1.3 Types of samples

13

bias and a lack of generalizability of the results—more than offset the advantages. Therefore, we should restrict our use of nonprobability sampling methods to situations in which we want to obtain rough approximations at low cost in order to satisfy our curiosity about a particular subject or to small-scale initial or pilot studies that will later be followed up by more rigorous investigations.

Probability sampling should be used whenever possible because it is the only method by which correct statistical inferences can be made from a sample.

> **A probability sample** is one in which the subjects of the sample are chosen on the basis of known probabilities.

The four types of probability samples most commonly used are simple random, systematic, stratified, and cluster. These sampling methods vary from one another in their cost, accuracy, and complexity. A discussion of these types of samples follows.

Simple Random Sample

A **simple random sample** is one in which every individual or item from a population has the same chance of selection as every other individual or item. In addition, every sample of a fixed size has the same chance of selection as every other sample of that size. Simple random sampling is the most elementary random sampling technique and as such forms the basis for the other random sampling techniques.

With simple random sampling, we use n to represent the sample size and N to represent the population size. Every item or person in the frame is numbered from 1 to N. The chance that any particular member of the population is selected on the first draw is $1/N$.

There are two basic methods by which samples are selected: with replacement or without replacement.

Sampling with replacement means that once a person or item is selected, it is returned to the frame where it has the same probability of being selected again. Imagine a fish bowl with 100 business name cards. On the first selection, suppose the name Judy Craven is selected. Pertinent information is recorded, and the business card is replaced in the bowl. The cards in the bowl are then well shuffled and the second card is drawn. On the second selection, Judy Craven has the same probability of being selected again, $1/N$. The process is repeated until the desired sample size n is obtained. However, it is generally considered more desirable to have a sample of different people or items than to permit a repetition of measurements on the same person or item.

Sampling without replacement means that a person or item once selected, is not returned to the frame and therefore cannot be selected again. As before, in sampling without replacement the chance that any particular member in the population, say, Judy Craven, is selected on the first draw is $1/N$. The chance that any individual not previously selected will be selected on the second draw is now 1 out of $N - 1$. This process continues until the desired sample of size n is obtained.

Regardless of whether we sample with or without replacement, such "fish bowl" methods for sample selection have a major drawback—our ability to thoroughly mix the cards and randomly pull the sample. As a result, fish bowl methods are not very useful. Less cumbersome and more scientific methods of selection are desirable.

One such method uses a **table of random numbers** (see Table E.1 of appendix E) for obtaining the sample. A table of random numbers consists of a series of digits randomly generated and listed in the sequence in which the digits were generated (see references 9

Table 1.1 *Using a table of random numbers*

		COLUMN							
	Row	00000 12345	00001 67890	11111 12345	11112 67890	22222 12345	22223 67890	33333 12345	33334 67890
	01	49280	88924	35779	00283	81163	07275	89863	02348
	02	61870	41657	07468	08612	98083	97349	20775	45091
	03	43898	65923	25078	86129	78496	97653	91550	08078
	04	62993	93912	30454	84598	56095	20664	12872	64647
	05	33850	58555	51438	85507	71865	79488	76783	31708
Begin	06	97340	03364	88472	04334	63919	36394	11095	92470
selection	07	70543	29776	10087	10072	55980	64688	68239	20461
(row 06, column 05)	08	89382	93809	00796	95945	34101	81277	66090	88872
	09	37818	72142	67140	50785	22380	16703	53362	44940
	10	60430	22834	14130	96593	23298	56203	92671	15925
	11	82975	66158	84731	19436	55790	69229	28661	13675
	12	39087	71938	40355	54324	08401	26299	49420	59208
	13	55700	24586	93247	32596	11865	63397	44251	43189
	14	14756	23997	78643	75912	83832	32768	18928	57070
	15	32166	53251	70654	92827	63491	04233	33825	69662
	16	23236	73751	31888	81718	06546	83246	47651	04877
	17	45794	26926	15130	82455	78305	55058	52551	47182
	18	09893	20505	14225	68514	46427	56788	96297	78822
	19	54382	74598	91499	14523	68479	27686	46162	83554
	20	94750	89923	37089	20048	80336	94598	26940	36858
	21	70297	34135	53140	33340	42050	82341	44104	82949
	22	85157	47954	32979	26575	57600	40881	12250	73742
	23	11100	02340	12860	74697	96644	89439	28707	25815
	24	36871	50775	30592	57143	17381	68856	25853	35041
	25	23913	48357	63308	16090	51690	54607	72407	55538
	⋮	⋮	⋮	⋮	⋮	⋮	⋮	⋮	⋮

Source: Partially extracted from The Rand Corporation, A Million Random Digits with 100,000 Normal Deviates *(Glencoe, IL: The Free Press, 1955) and displayed in Table E.1 in appendix E at the back of this text.*

and 21). Because our numeric system uses 10 digits (0, 1, 2, ... , 9), the chance of randomly generating any particular digit is equal to the probability of generating any other digit. This probability is 1 out of 10. Hence, if a sequence of 800 digits were generated, we would expect about 80 of them to be the digit 0, 80 to be the digit 1, and so on. In fact, researchers who use tables of random numbers usually test out such generated digits for randomness before employing them. Table E.1 has met all such criteria for randomness. Because every digit or sequence of digits in the table is random, we may use the table by reading either horizontally or vertically. The margins of the table designate row numbers and column numbers. The digits themselves are grouped into sequences of five to make reading the table easier.

To use such a table instead of a fish bowl for selecting the sample, it is first necessary to assign code numbers to the individual members of the population. We then obtain our random sample by reading the table of random numbers and selecting those individuals from the population frame whose assigned code numbers match the digits found in the table. To better understand the process of sample selection from its inception, consider Example 1.2.

Example 1.2 *Selecting a Simple Random Sample Using a Table of Random Numbers*

Suppose that a company wants to select a sample size of 32 full-time workers out of a population of 800 full-time employees in order to obtain information on expenditures from a company-sponsored dental plan. We assume that not everyone will be willing to respond to the survey, so we reason that we must have a larger mailing than 32 to get our desired 32 responses. If we assume that 8 out of 10 full-time workers are expected to respond to such a survey (that is, a rate of return of 80%), we calculate that a total of 40 such employees must be contacted to obtain the desired 32 responses. Therefore, our survey will be distributed to 40 full-time employees drawn from the personnel files of the company. How will the simple random sample actually be drawn?

SOLUTION

To select the random sample, we use a table of random numbers. The population frame consists of a listing of the names and company mailbox numbers of all $N = 800$ full-time employees obtained from the company personnel files. Because the population size (800) is a three-digit number, each assigned code number must also be three digits so that every full-time worker has an equal chance for selection. Thus, a code of 001 is given to the first full-time employee in the population listing, a code of 002 is given to the second full-time employee in the population listing, and so on, until a code of 800 is given to the Nth full-time worker in the listing. Because $N = 800$ is the largest possible coded value, all three-digit code sequences greater than N (i.e., 801 through 999 and 000) are discarded.

To select the simple random sample, a random starting point for the table of random numbers is chosen. One such method is to close one's eyes and strike the table of random numbers with a pencil. Suppose we use such a procedure and thereby select row 06, column 05, of Table 1.1 on page 15 (which is a replica of Table E.1) as the starting point. Although we can go in any direction in the table, suppose we read from left to right in sequences of three digits without skipping.

The individual with code number 003 is the first full-time employee in the sample (row 06 and columns 05–07), the second individual has code number 364 (row 06 and columns 08–10), and the third individual has code number 884. Because the highest code for any employee is 800, this number is discarded. Individuals with code numbers 720, 433, 463, 363, 109, 592, 470, and 705 are selected third through tenth, respectively.

The selection process continues in a similar manner until the needed sample size of 40 full-time employees is obtained. During the selection process, if any three-digit coded sequence repeats, the employee corresponding to that coded sequence is included again as part of the sample if we are sampling with replacement; however, the repeating coded sequence is discarded if we are sampling without replacement.

◆ *Systematic Sample* In a **systematic sample,** the N individuals or items in the population frame are partitioned into k groups by dividing the size of the population frame N by the desired sample size n. That is,

$$k = \frac{N}{n}$$

where k is rounded to the nearest integer. To obtain a systematic sample, the first individual or item to be selected is chosen at random from the k individuals or items in the first partitioned group in the population frame, and the rest of the sample is obtained by selecting every kth individual or item thereafter from the entire population frame listing.

If the population frame consists of a listing of prenumbered checks, sales receipts, or invoices or if the population frame pertains to club membership listings, student registration listings, or perhaps a preset number of consecutive items coming off an assembly line, a systematic sample is faster and easier to obtain than a simple random sample. In such situations, the systematic sample would be a convenient mechanism for obtaining the desired data.

Although they are simpler to use, simple random sampling methods and systematic sampling methods are generally less efficient than other, more sophisticated probability sampling methods. That is, for any one sample obtained by either simple random sampling or systematic sampling, the data obtained may or may not be a good representation of the population's underlying characteristics (parameters). Although most simple random samples are representative of their underlying population, it is not possible to know if the particular sample taken is in fact representative.

Even greater possibilities for selection bias and lack of representation of the population characteristics occur from systematic samples. If a pattern were to exist in the population frame listing, severe selection biases could result. To overcome the potential problem of disproportionate representation of specific groups in a sample, we can use either stratified sampling methods or cluster sampling methods.

◆ **Stratified Sample**   In a **stratified sample,** the N individuals or items in the population are first subdivided into separate subpopulations, or **strata,** according to some common characteristic. A simple random sample is conducted within each of the strata and the results from the separate simple random samples are then combined. Such sampling methods are more efficient than either simple random sampling or systematic sampling because they ensure representation of individuals or items across the entire population, which ensures a greater precision in the estimates of underlying population parameters. It is the homogeneity of individuals or items within each stratum that, when combined across strata, provides the precision.

◆ **The Cluster Sample**   In a **cluster sample,** the N individuals or items in the population are divided into several *clusters* so that each cluster is representative of the entire population. A random sampling of clusters is then taken and all individuals or items in each selected cluster are then studied. Clusters can be naturally occurring designations, such as counties, election districts, city blocks, apartment buildings, or families.

Cluster sampling methods can be more cost effective than simple random sampling methods, particularly if the underlying population is spread over a wide geographic region. However, cluster sampling methods tend to be less efficient than either simple random sampling methods or stratified sampling methods and would require a larger overall sample size to obtain results as precise as those that would be obtained from the more efficient procedures.

A detailed discussion of systematic sampling, stratified sampling, and cluster sampling procedures can be found in references 3 and 9.

Problems for Section 1.8

Learning the Basics

1.12 For a population list containing $N = 902$ individuals, what code number would you assign for
  (a) the first person on the list?
  (b) the fortieth person on the list?
  (c) the last person on the list?

1.13 For a population of $N = 902$, verify that by starting in row 05 of the table of random numbers (Table E.1), only six rows are needed to draw a sample of size $n = 60$ *without* replacement.

• **1.14** Given a population of $N = 93$, starting in row 29 of the table of random numbers (Table E.1), and reading across the row, draw a sample of size $n = 15$.
  (a) *without* replacement.
  (b) *with* replacement.

Applying the Concepts

1.15 For a study that would involve doing personal interviews with participants (rather than mail or phone surveys), tell why a simple random sample might be less practical than some other methods.

1.16 Suppose that I want to select a random sample of size 1 from a population of three items (which we can call A, B, and C). My rule for drawing the sample is: Flip a coin; if it is heads, pick item A; if it is tails, flip the coin again; this time, if it is heads, choose B; if tails, choose C. Explain why this is a random sample but not a simple random sample.

1.17 Suppose that a population has four members (call them A, B, C, and D). I would like to draw a random sample of size 2, which I decide to do in the following way: Flip a coin; if it is heads, my sample will be items A and B; if it is tails, the sample will be items C and D. Although this is a random sample, it is not a simple random sample. Explain why. (If you did Problem 1.16, compare the procedure described there with the procedure described in this problem.)

• **1.18** Suppose that the registrar of a college with a population of $N = 4{,}000$ full-time students is asked by the president to conduct a survey to measure satisfaction with the quality of life on campus. The following table contains a breakdown of the 4,000 registered full-time students by gender and class designation:

GENDER	CLASS DESIGNATION				
	FR.	SO.	JR.	SR.	TOTAL
Female	700	520	500	480	2,200
Male	560	460	400	380	1,800
Total	1,260	980	900	860	4,000

The registrar intends to take a probability sample of $n = 200$ students and project the results from the sample to the entire population of full-time students.
  (a) If the population frame available from the registrar's files is an alphabetical listing of the names of all $N = 4{,}000$ registered full-time students, what type of samples could be taken? Discuss.
  (b) What would be the advantage of selecting a simple random sample in (a)?
  (c) What would be the advantage of selecting a systematic sample in (a)?

(d) If the population frame available from the registrar's files is a listing of the names of all $N = 4,000$ registered full-time students compiled from eight separate alphabetical lists based on the gender and class designation breakdowns shown in the above table, what type of sample should be taken? Discuss.

(e) Suppose that all $N = 4,000$ registered full-time students lived in one of the 20 campus dormitories. Each dormitory contains four floors with 50 beds per floor, thereby accommodating 200 students. It is college policy to fully integrate students by gender and class designation in each floor of each dormitory. If the registrar was able to compile a population frame through a listing of all student occupants on each floor within each dormitory, what type of sample should be taken? Discuss.

1.19 Prenumbered sales invoices are kept in a sales journal. The invoices are numbered from 0001 to 5000.

(a) Beginning in row 16, column 1, and proceeding horizontally in Table E.1, select a simple random sample of 50 invoice numbers.

(b) Select a systematic sample of 50 invoice numbers. Use the random numbers in row 20, columns 5–7 as the starting point for your selection.

(c) Are the invoices selected in (a) the same as those selected in (b)? Why or why not?

1.20 Suppose that 5,000 sales invoices are separated into four strata. Stratum 1 contains 50 invoices, stratum 2 contains 500 invoices, stratum 3 contains 1,000 invoices, and stratum 4 contains 3,450 invoices. All 50 invoices in stratum 1 are to be selected, and 50 invoices from each of the other strata are to be selected.

(a) What type of sampling should be done? Why?

(b) Explain how you would carry out the sampling according to the method stated in (a).

(c) Why is the type of sampling in (a) not a simple random sample?

◆ 1.9 EVALUATING SURVEY WORTHINESS

Nearly every day, we read or hear about survey or opinion poll results in our newspapers or on radio or television. Clearly, advances in information technology have led to a proliferation of survey research. Not all this research is good, meaningful, or important (reference 2), however.

To avoid those surveys lacking in objectivity or credibility, we must critically evaluate what we read and hear by examining the worthiness of the survey. First, we must evaluate the purpose of the survey, why it was conducted, and for whom. An opinion poll or survey conducted to satisfy curiosity is mainly for entertainment. Its result is an end in itself rather than a means to an end. We should be more skeptical of such a survey because the result should not be put to further use.

The second step in evaluating the worthiness of a survey is to determine whether it was based on a probability or a nonprobability sample (as discussed in section 1.8). You may recall that the only way for us to make correct statistical inferences from a sample to a population is through the use of a probability sample. Surveys employing nonprobability sampling methods are subject to serious, perhaps unintentional, interview biases that may render the results meaningless, as illustrated in the comment box on the following page.

Survey Errors

Even when surveys employ random probability sampling methods, they are subject to potential errors. As illustrated in Exhibit 1.4, there are four types of survey errors (reference 8). Good survey research design attempts to reduce or minimize these various survey errors, often at considerable cost.

COMMENT: *A Nonprobability Sampling Disaster*

In 1948, major pollsters predicted the outcome of the American presidential election between Harry S. Truman, the incumbent president, and Thomas E. Dewey, then governor of New York, as going to Dewey. The *Chicago Tribune* was so confident of the polls' predictions that it printed its early edition based on the predictions rather than waiting for the ballots to be counted.

An embarrassed newspaper and the pollsters they had relied on had a lot of explaining to do. How had the pollsters been so wrong? Intent on discovering the source of the error, the pollsters found that their use of a nonprobability sampling method was the culprit (see reference 17). As a result, polling organizations adopted probability sampling methods for future elections.

Exhibit 1.4 Survey Errors

✓ **1.** Coverage error or selection bias
✓ **2.** Nonresponse error or nonresponse bias
✓ **3.** Sampling error
✓ **4.** Measurement error

◆ *Coverage Error* The key to proper sample selection is an adequate population frame or up-to-date list of all the subjects from which the sample will be drawn. **Coverage error** occurs if we exclude certain groups of subjects from this population listing so that they have no chance of being selected in the sample. Coverage error results in a **selection bias.** If the listing is inadequate because certain groups of subjects in the population were not properly included, any random probability sample selected will provide an estimate of the characteristics of the *target* population, not the *actual* population. For a presentation of a famous case of selection bias we turn to the comment box on page 21.

◆ *Nonresponse Error* Not everyone will be willing to respond to a survey. In fact, research has indicated that individuals in the upper and lower economic classes tend to respond less frequently to surveys than do people in the middle class. **Nonresponse error** arises from the failure to collect data on all subjects in the sample and results in a **nonresponse bias**. Because it cannot be generally assumed that persons who do not respond to surveys are similar to those who do, it is extremely important to follow up on the nonresponses after a specified period of time. Several attempts should be made, either by mail or by telephone, to convince such individuals to change their minds. Based on these results, the estimates obtained from the initial respondents are subsequently tied to those obtained from the follow-ups so that the inferences made from the survey are valid (reference 1).

The mode of response affects the rate of response. The personal interview and the telephone interview usually produce a higher response rate than does the mail survey—but at a higher cost. The comment on the 1936 *Literary Digest* poll (page 21) also addresses nonresponse bias.

◆ *Sampling Error* There are three main reasons for drawing a sample rather than taking a complete census: It is more expedient, less costly, and more efficient. However, chance dictates who in the population frame will or will not be included. **Sampling error** reflects the heterogeneity, or "chance differences," from sample to sample based on the probability of particular individuals or items being selected in the particular samples.

When we read about the results of surveys or polls in newspapers or magazines, there is often a statement regarding margin of error or precision; for example, "the results of this poll are expected to be within ±4 percentage points of the actual value." This margin of error is our sampling error. Sampling error can be reduced by taking larger sample sizes, although this will increase the cost of conducting the survey.

◆ **Measurement Error**   In the practice of good survey research, a questionnaire is designed with the intent that it will allow meaningful information to be gathered. The obtained data must be *valid;* that is, the "right" responses must be assessed, and in a manner that will elicit meaningful measurements.

But there is a dilemma here—obtaining meaningful measurements is often easier said than done. Consider the following proverb:

A man with one watch always knows what time it is;
A man with two watches always searches to identify the correct one;
A man with ten watches is always reminded of the difficulty in measuring time.

Unfortunately, the process of obtaining a measurement is often governed by what is convenient, not what is needed. And the measurements obtained are often only a proxy for the ones really desired.

> **Measurement error** refers to inaccuracies in the recorded responses that occur because of a weakness in question wording, an interviewer's effect on the respondent, or the effort made by the respondent.

Much attention has been given to measurement error that occurs because of a weakness in question wording (reference 7). A question should be clear, not ambiguous. It should be objectively presented in a neutral manner; "leading questions" must be avoided.

There are three sources of measurement error: ambiguous wording of questions, the halo effect, and respondent error. As an example of ambiguous wording, in November 1993 the Labor Department reported that the unemployment rate in the United States had been underestimated for more than a decade because of poor questionnaire wording in the Current Population Survey. In particular, the wording led to a significant undercount of women in the labor force. Because unemployment rates are tied to benefit programs such as state unemployment compensation systems, it was imperative that government survey researchers rectify the situation by adjusting the questionnaire wording.

The "halo effect" occurs when the respondent feels obligated to please the interviewer. This type of error can be minimized by proper interviewer training.

Respondent error occurs as a result of overzealous or underzealous effort by the respondent. We can minimize this type of error in two ways: (1) by carefully scrutinizing the data and calling back those individuals whose responses seem unusual and (2) by establishing a program of random callbacks in order to ascertain the reliability of the responses.

Ethical Issues

 With respect to the proliferation of survey research (reference 2), Eric Miller, editor of the newsletter *Research Alert,* stated that "There's been a slow sliding in ethics. The scary part is that people make decisions based on this stuff. It may be an invisible crime, but it's not a victimless one." Not all survey research is good, meaningful, or important and not all survey research is ethical. We must try to distinguish between poor survey design and unethical survey design.

Ethical considerations arise with respect to the four types of potential errors that may occur when designing surveys that use random probability samples: coverage error or selection bias, nonresponse error or nonresponse bias, sampling error, and measurement error. Coverage error or selection bias becomes an ethical issue only if particular groups or individuals are *purposely* excluded from the population frame so that the survey results are skewed, indicating a position more favorable to that of the survey's sponsor.

In a similar vein, nonresponse error or nonresponse bias becomes an ethical issue only if particular groups or individuals are less likely to respond to a particular survey format and the sponsor knowingly designs the survey in a manner aimed at excluding such groups or individuals. Sampling error becomes an ethical issue only if the findings are purposely presented without reference to sample size and margin of error so that the sponsor can promote a viewpoint that might otherwise be truly insignificant. Measurement error becomes an ethical issue in one of three ways. (1) A survey sponsor may purposely choose loaded, lead-in questions that would guide the responses in a particular direction. (2) An interviewer, through mannerisms and tone, may purposely create a halo effect or otherwise guide the responses in a particular direction. (3) A respondent having a disdain for the survey process may willfully provide false information.

Problems for Section 1.9

Applying the Concepts

1.21 "A survey indicates that Americans overwhelmingly preferred a Chrysler to a Toyota after test-driving both." What information would you want to know before you accept the results of this survey?

1.22 "A survey indicates that the vast majority of college students picked Gap jeans as the most 'in' clothing." What information would you want to know before you accept the results of this survey?

1.23 A simple random sample of $n = 300$ full-time employees is drawn from a company list containing the names of all $N = 5,000$ full-time employees in order to evaluate job satisfaction.
(a) Give an example of possible coverage error.
(b) Give an example of possible nonresponse error.
(c) Give an example of possible sampling error.
(d) Give an example of possible measurement error.

 SUMMARY

As you can see in the summary chart on the following page, this chapter provided an introduction to statistics and discussed data collection. We have studied different modes of response to a survey, various types of data, and different ways of selecting random samples. In addition, we examined several aspects of survey worthiness.

Once they have been collected, the data must be organized and prepared in order to assist us in making various analyses. In the next two chapters, methods of tabular and chart presentation will be demonstrated, various "exploratory data analysis" techniques will be described, and a variety of descriptive summary measures useful for data analysis and interpretation will be developed.

Key Terms

categorical random variables 10	nonresponse bias 20	selection bias 20
cluster sample 17	nonresponse error 20	simple random sample 14
continuous random variables 10	numerical random variables 10	statistic 6
coverage error 20	parameter 6	statistical thinking 4
data 7	population 6	strata 17
descriptive statistics 5	primary and secondary sources 8	stratified sample 17
discrete random variables 10	probability sample 14	systematic sample 16
focus group 9	random variables 9	table of random numbers 14
frames 12	sample 6	total quality management 4
inferential statistics 6	sampling with replacement 14	universe 6
measurement error 22	sampling without replacement 14	
nonprobability sample 13	sampling error 21	

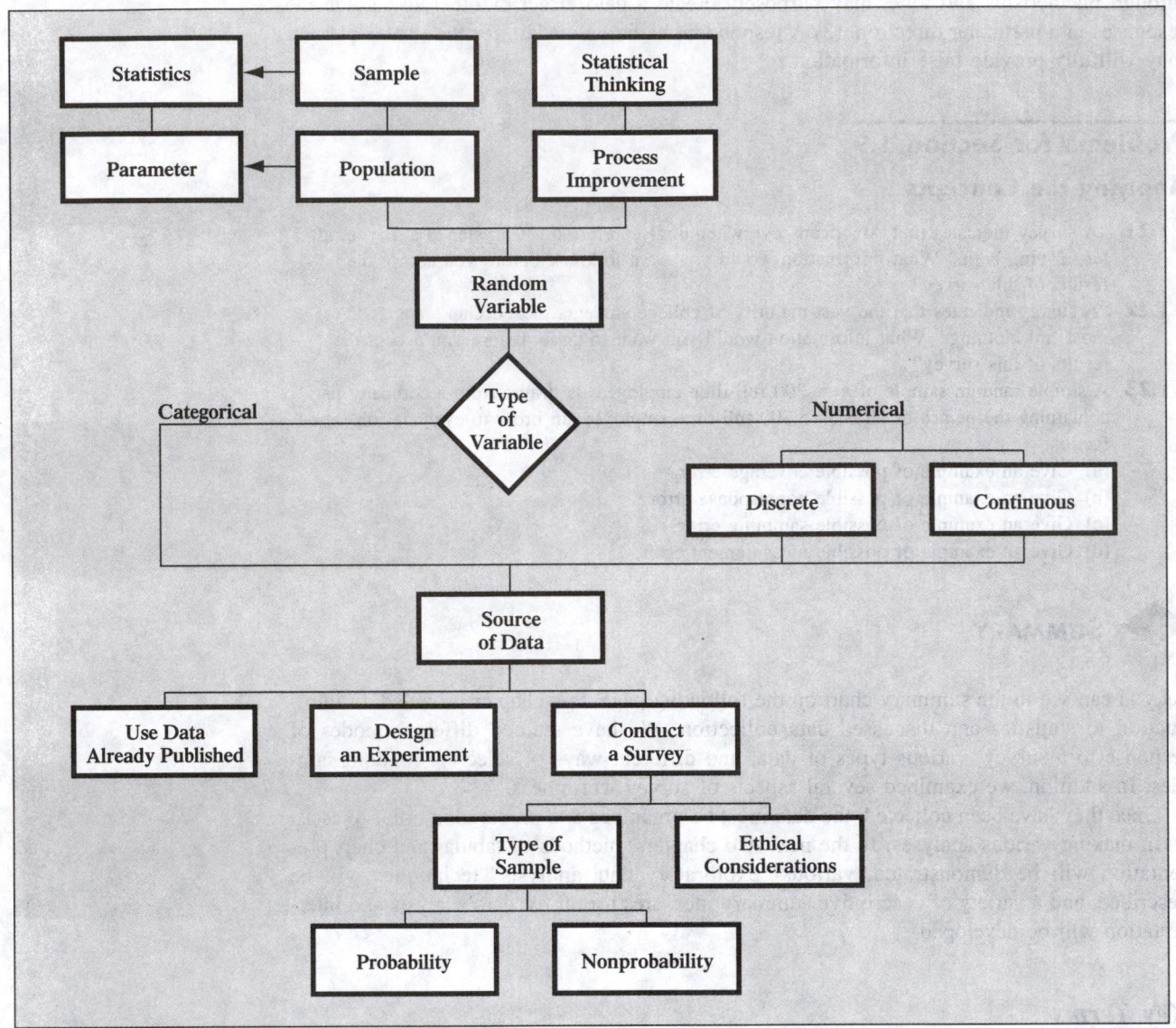

Chapter 1 summary chart

Checking Your Understanding

1.24 What is the difference between a sample and a population?

1.25 What is the difference between a statistic and a parameter?

1.26 What is the difference between descriptive and inferential statistics?

1.27 How can statistical methods be useful to a manager?

1.28 How has the field of statistics been changed by the development of computer technology?

1.29 How was the development of the field of statistics intertwined with the needs of the census?

1.30 How have statistical software packages changed in the past 30 years?

1.31 What are the three aspects of quality improvement?

1.32 What is the difference between a categorical and a numerical random variable?

1.33 What is the difference between discrete and continuous data?

1.34 What are the main reasons for obtaining data?

1.35 What is the difference between probability and nonprobability sampling?

1.36 What are some potential problems with using "fish bowl" methods to draw a simple random sample?

1.37 What is the difference between sampling *with* replacement versus *without* replacement?

1.38 What is the difference between a simple random sample and a systematic sample?

1.39 What is the difference between a stratified sample and a systematic sample?

1.40 What is the difference between a stratified sample and a cluster sample?

1.41 What distinguishes the four potential sources of error when dealing with surveys designed using probability sampling?

Chapter Review Problems

1.42 The Data and Story Library (DASL) (**http://lib.stat.cmu.edu/DASL**) is an on-line library of data files and stories that illustrate the use of basic statistical methods. Each data set has one or more associated stories. The stories are classified by method and by subject. Access this World Wide Web site and after reading a story, summarize how statistics have been used in one of the subject areas.

1.43 Access one of the following two World Wide Web sites provided by Microsoft Corporation for Microsoft Excel (**http://www.microsoft.com/msexcel** and **http://www.microsoft.com/msexcelsupport**). Explain how you think Microsoft Excel could be useful in the field of statistics.

1.44 The AT&T World Net Service provides a weekly poll at the following Internet address:

http://www.worldnet.att.net/poll/survey.poll

Access this site.

(a) Indicate whether the current poll question concerns a categorical or a numerical variable.

(b) Develop three questions with categorical variables and three questions with numerical variables that could be used in the weekly poll.

(c) Is the sample of respondents a random sample? Explain.

(d) What difficulties can you foresee in trying to apply the results of the survey to a population?

1.45 The AT&T World Net Service also provides periodic surveys of interest to its subscribers at its home address:

http://www.worldnet.att.net/

Access this site, and determine if a survey is currently being conducted. If it is,

(a) indicate which of the poll questions are categorical variables.

(b) indicate which of the poll questions are numerical variables.

(c) develop three questions with categorical variables and three questions with numerical variables that could be used in the survey.

1.46 The British Airways Internet site (**http://www.britishairways.com/feedback/feedback.shtml**) provides a questionnaire instrument that can be answered electronically.

Among the questions that have been listed in the past are:

1. How did you first hear about the British Airways Internet site?
2. What was your age at your last birthday?
3. What is the speed of your Internet connection?

(a) Does the survey on the British Airways Internet site represent a random sample? Explain.

(b) How can a random sample be obtained?

(c) Define the target population for British Airways.

(d) If the survey has not already done so, write a question that presents age as a numerical variable.

(e) Is "How did you first hear about the British Airways Internet site?" a categorical or a numerical variable? Explain.

(f) List three questions not currently part of the British Airways survey that you believe should be included.

● **1.47** Suppose that the manager of the customer service division of a consumer electronics company was interested in determining whether customers who had purchased a videocassette recorder over the past 12 months were satisfied with their products. Using the warranty cards submitted after the purchase, the manager was planning to survey these customers.
(a) Describe the target population.
(b) Describe the frame.
(c) What differences are there between the target population and the frame? How might these differences affect the results?
(d) Develop three categorical questions that you feel would be appropriate for this survey.
(e) Develop three numerical questions that you feel would be appropriate for this survey.
(f) How could a simple random sample of warranty cards be selected?
(g) If the manager wanted to select a sample of warranty cards for each brand of videocassette recorder sold, how should the sample be selected? Explain.

1.48 Political polls are taken to try to predict the outcome of an election. The results of such polls are routinely reported in newspapers and on television in the weeks and months prior to an election. For a specific election such as for president of the United States,
(a) what is the population to which we usually want to generalize?
(b) how might we get a random sample from that population?
(c) from what you know about how such polls are actually conducted, what might be some problems with the sampling in these polls?

1.49 The following questionnaire is placed in each room of a well-known and widely respected hotel. The intent of management is to evaluate guest satisfaction.

How well did we serve you?	😊	🙂	☹	😐
Reservations				
Doorman				
Front desk				
Room				
Cleanliness				
Restaurant				

(a) Do you think that this four-category device with "faces" provides management with enough information on each service or facility being rated? Discuss.
(b) Would the addition of another "frown face" to balance the two "smile faces" improve this questionnaire instrument? Discuss.
(c) Of what value are such self-selecting surveys? Can they be used to project the opinions of all guests during a particular interval of time (say, a week or a month)? Discuss.
(d) How could a random sample of guests be selected?
(e) What categorical question would you add to the survey?
(f) What numerical question would you add to the survey?

1.50 Suppose that a manufacturer of cat food was planning to survey households in the United States to determine purchase habits of cat owners. Among the questions to be included are those that relate to
1. where cat food is primarily purchased.
2. whether dry or moist cat food is purchased.
3. the number of cats living in the household.
4. whether or not the cat is pedigreed.

(a) Describe the target population.
(b) Define the frame.
(c) Indicate the type of sampling that you would use and why you would select it.
(d) For each of the four questions listed, indicate whether the variable is categorical or numerical.
(e) Develop five categorical questions for the survey.
(f) Develop five numerical questions for the survey.

 TEAM PROJECT

TP1.1 Suppose the following information is obtained for F. Jay Mori upon his admittance to the Brandwein College infirmary:

(a) Gender: Male
(b) Residence or Dorm: Mogelever Hall
(c) Class: Sophomore
(d) Temperature: 102.2°F (oral)
(e) Pulse: 70 beats per minute

(f) Blood Pressure: 130/80 mg/mm(g)
(g) Blood Type: B positive
(h) Known Allergies to Medicines: None
(i) Preliminary Diagnosis: Influenza
(j) Estimated Length of Stay: 3 days

Classify each of the 10 responses by type of data. (*Hint*: Be careful with blood pressure; it's tricky.)

 Case Study — **ALUMNI ASSOCIATION SURVEY**

Suppose that the president of the alumni association of a state university wishes to take a survey of its membership from the classes of 1988 and 1989 to determine their past achievements, current activities, and future aspirations. Toward this end, information pertaining to the following areas is desired: gender of the alumnus; major area of study; undergraduate grade-point index; further educational pursuits (i.e., master's degree or doctorate); current employment status; current annual salary; number of full-time positions held since graduation; annual salary anticipated in 5 years; political party affiliation; and marital status.

As director of institutional research you are asked to write a proposal demonstrating how you plan to conduct the survey. Included in this proposal must be

1. a statement of objectives (i.e., what you want to find out and why).
2. a discussion of *how* and *when* the survey will be conducted (i.e., how you plan to sample 300 alumni from the list of 3,000 alumni association members in the two classes).

3. a first draft of the questionnaire instrument (containing an organized sequence of both numerical and categorical questions).
4. a first draft of the cover letter to be used with the questionnaire.
5. a first draft of any special instructions to respondents to aid them in filling out the questionnaire.
6. a discussion of *how* you plan to test the questionnaire for validity and/or ambiguity.
7. a discussion of the type of sampling to be used in the survey.
8. a statement that you have taken into consideration such things as the costs involved in conducting the survey, personnel needs, and the amount of time required for implementation and completion.
9. a statement regarding the target population of alumni association members versus the actual population of graduates from the two classes, 1988 and 1989, and whether the survey results can be projected to all graduates from these two classes.

27

References

1. Cochran, W. G., *Sampling Techniques*, 3d ed. (New York: Wiley, 1977).

2. Crossen, C., "Margin of Error: Studies and Surveys Proliferate, but Poor Methodology Makes Many Unreliable," *The Wall Street Journal*, November 14, 1991, A1 and A9.

3. Deming, W. E., *Sample Design in Business Research* (New York: Wiley, 1960).

4. Deming, W. E., *Out of the Crisis* (Cambridge, MA: Massachusetts Institute of Technology Center for Advanced Engineering Study, 1986).

5. Deming, W. E., *The New Economics for Industry, Government, Education* (Cambridge, MA: Massachusetts Institute of Technology Center for Advanced Engineering Study, 1993).

6. Gallup, G. H., *The Sophisticated Poll-Watcher's Guide* (Princeton, NJ: Princeton Opinion Press, 1972).

7. Goleman, D., "Pollsters Enlist Psychologists in Quest for Unbiased Results," *The New York Times*, September 7, 1993, C1 and C11.

8. Groves, R. M., *Survey Errors and Survey Costs* (New York: Wiley, 1989).

9. Hansen, M. H., W. N. Hurwitz, and W. G. Madow, *Sample Survey Methods and Theory*, vols. 1 and 2 (New York: Wiley, 1953).

10. Juran, J. M., *Juran on Leadership for Quality* (New York: The Free Press, 1989).

11. Juran, J. M., and F. M. Gryna, *Quality Planning and Analysis,* 2d ed. (New York: McGraw-Hill, 1980).

12. Kendall, M. G., and R. L. Plackett, eds., *Studies in the History of Statistics and Probability,* vol. 2 (London: Charles W. Griffin, 1977).

13. Kirk, R. E., ed., *Statistical Issues: A Reader for the Behavioral Sciences* (Monterey, CA: Brooks/Cole, 1972).

14. *Lotus 1-2-3 Release 5* (Cambridge, MA: Lotus Development Corporation, 1994).

15. *Microsoft Excel 97* (Redmond, WA: Microsoft Corporation, 1997).

16. *Minitab Version 12* (State College, PA: Minitab, Inc., 1998).

17. Mosteller, F., et al., *The Pre-Election Polls of 1948* (New York: Social Science Research Council, 1949).

18. Norusis, M., *SPSS Guide to Data Analysis for SPSS-X: With Additional Instructions for SPSS/PC+* (Chicago: SPSS Inc., 1986).

19. Pearson, E. S., ed., *The History of Statistics in the Seventeenth and Eighteenth Centuries* (New York: Macmillan, 1978).

20. Pearson, E. S., and M. G. Kendall, eds., *Studies in the History of Statistics and Probability* (Darien, CT: Hafner, 1970).

21. Rand Corporation, *A Million Random Digits with 100,000 Normal Deviates* (New York: The Free Press, 1955).

22. Robbins, S. P., *Management*, 5th ed. (Upper Saddle River, NJ: Prentice Hall, 1997).

23. *SAS Language and Procedures Usage, Version 6* (Raleigh, NC: SAS Institute, 1988).

24. Walker, H. M., *Studies in the History of the Statistical Method* (Baltimore: Williams & Wilkins, 1929).

25. Wattenberg, B. E., ed., *Statistical History of the United States: From Colonial Times to the Present* (New York: Basic Books, 1976).

Chapter 1

Student Solutions Manual

1.2   Three sizes of soft drink are classified into distinct categories—small, medium, and large—in which order is implied.

•1.4   (a)   discrete numerical   (g)   categorical
       (b)   categorical   (h)   discrete numerical
       (c)   discrete numerical   (i)   continuous numerical
       (d)   continuous numerical   (j)   categorical
       (e)   categorical   (k)   categorical
       (f)   continuous numerical

1.6   (a)   categorical   (e)   discrete numerical
       (b)   continuous numerical   (f)   discrete numerical
       (c)   continuous numerical   (g)   categorical
       (d)   categorical   (h)   categorical

1.8   (a)   categorical   (i)   continuous numerical *
       (b)   categorical   (j)   continuous numerical *
       (c)   continuous numerical   (k)   categorical
       (d)   continuous numerical *   (l)   discrete numerical
       (e)   categorical   (m)   continuous numerical *
       (f)   categorical   (n)   continuous numerical
       (g)   discrete numerical **   (o)   categorical
       (h)   discrete numerical   (p)   continuous numerical *

       *Some researchers consider money as a discrete numerical variable because it can be "counted."
       **Some researchers would "measure" the time since starting the job and consider this a
       continuous numerical variable.

1.10   While it is theoretically true that ties cannot occur with continuous data, the grossness of the
       measuring instruments used often leads to the reporting of ties in practical applications. Hence
       two students may both score 90 on an exam—not because they possess identical ability but
       rather because the grossness of the scoring method used failed to detect a difference between
       them.

1.12   (a) 001   (b) 040   (c) 902

•1.14 (a)   Row 29: 12 47 83 76 22 ~~99~~ 65 93 10 ~~65 83~~ 61 36 ~~98~~ 89 58 86 92 71
              Note: All sequences above 93 and all repeating sequences are discarded.
       (b)   Row 29: 12 47 83 76 22 ~~99~~ 65 93 10 65 83 61 36 ~~98~~ 89 58 86
              Note: All sequences above 93 are discarded. Elements 65 and 83 are repeated.

1.16   This is a random sample because the selection is based on chance. It is not a simple random sample because A is more likely to be selected than B or C.

•1.18   (a)   Since a complete roster of full-time students exists, a simple random sample of 200 students could be taken. If student satisfaction with the quality of campus life randomly fluctuates across the student body, a systematic 1-in-20 sample could also be taken from the population frame. If student satisfaction with the quality of life may differ by gender and by experience/class level, a stratified sample using eight strata, female freshmen through female seniors and male freshmen through male seniors, could be selected. If student satisfaction with the quality of life is thought to fluctuate as much within clusters as between them, a cluster sample could be taken.

        (b)   A simple random sample is one of the simplest to select. The population frame is the registrar's file of 4,000 student names.

        (c)   A systematic sample is easier to select by hand from the registrar's records than a simple random sample, since an initial person at random is selected and then every 20th person thereafter would be sampled. The systematic sample would have the additional benefit that the alphabetic distribution of sampled students' names would be more comparable to the alphabetic distribution of student names in the campus population.

        (d)   If rosters by gender and class designations are readily available, a stratified sample should be taken. Since student satisfaction with the quality of life may indeed differ by gender and class level, the use of a stratified sampling design will not only ensure all strata are represented in the sample, it will generate a more representative sample and produce estimates of the population parameter that have greater precision.

        (e)   If all 4,000 full-time students reside in one of 20 on-campus residence halls which fully integrate students by gender and by class, a cluster sample should be taken. A cluster could be defined as an entire residence hall, and the students of a single randomly selected residence hall could be sampled. Since the dormitories are fully integrated by floor, a cluster could alternatively be defined as one floor of one of the 20 dormitories. Four floors could be randomly sampled to produce the required 200 student sample. Selection of an entire dormitory may make distribution and collection of the survey easier to accomplish. In contrast, if there is some variable other than gender or class that differs across dormitories, sampling by floor may produce a more representative sample.

1.20   (a)   The proposed sample design is a nonprobability quota sample. Since the invoices are already separated into strata, a stratified sample should be used to reduce selection bias and improve generalizability of results.

        (b)   Sampling 4% of the invoices in each of the four strata would produce a sample with the same number of units.

        (c)   The proposed sample design is not a simple random sample because all invoices do not have an equal chance of being selected.

1.22   Before accepting the results of a survey of college students, you might want to know, for example:
        Who funded the survey? Why was it conducted?
        What was the population from which the sample was selected?
        What sampling design was used?
        What mode of response was used: a personal interview, a telephone interview, or a mail
            survey? Were interviewers trained? Were survey questions field-tested?
        What questions were asked? Were they clear, accurate, unbiased, valid?
        What operational definition of "the most 'in' clothing" was used?
        What was the response rate?

1.24   A population contains all the items whereas a sample contains only a portion of the items in the population.

1.26   Descriptive methods deal with the collection, presentation, summarization, and analysis of data whereas inferential methods deal with decisions arising from the projection of sample information to the characteristics of a population.

1.48   (a)   Population: Actual voters
        (b)   Sample: "Exit" poll enables an estimate based on actual voters
        (c)   This is superior to a prior telephone poll of registered voters because not all registered voters will actually vote.

1.50   (a)   Population: Cat owners
        (b)   Sample frame: Households in the United States
        (d)   (1)   categorical   (3)   numerical
               (2)   categorical   (4)   categorical

Chapter 2
Presenting Data
in Tables and Charts

CHAPTER OBJECTIVES

✓ *To demonstrate how to organize numerical data*
✓ *To develop tables and charts for numerical data*
✓ *To develop tables and charts for categorical data*
✓ *To develop tables and charts for bivariate categorical data*
✓ *To demonstrate the principles of proper graphical presentation*

Introduction

In chapter 1 we learned about data collection. As a general rule, whenever a set of data contains about 20 or more observations, it is best to examine it in summary form by constructing appropriate tables and charts. We can then extract the important features of the data from these tables and charts. In this chapter we demonstrate how large sets of data can be organized and most effectively presented in the form of tables and charts in order to enhance data analysis and interpretation—two key aspects of the decision-making process. We begin with an example concerning equity mutual funds.

◆ USING STATISTICS: *Comparing the Performance of Equity Mutual Funds*

In recent years millions of individuals have invested billions of dollars in a variety of mutual funds. These investments have been made for a variety of reasons involving short-term objectives and long-term objectives. Suppose that you were employed by a financial investment service that was evaluating currently traded domestic general stock funds so that it could make purchase recommendations to potential investors. If you were to study the financial performance measures of these domestic general stock funds based on various features such as fund objective (growth versus blend), fee structure (no load versus fee payment), and capitalization size (large, mid, small) of companies making up a fund's portfolio, how might this help in pinpointing funds for possible investment as part of a long-term financial plan?

2.1 ORGANIZING NUMERICAL DATA

How can we go about answering the question raised in our Using Statistics example? One way is by obtaining recent data on the performance of a sample of 194 domestic general stock funds with high Morningstar Inc. dual ratings of 4 or 5. These data are described in detail in appendix D. The data relating to these 194 mutual funds are stored in an Excel workbook named MUTUAL on the CD-ROM that accompanies this text.

One way we may wish to compare performance, as measured by the 1-year return percentage, is based on fund objective (growth versus blend). There are 59 growth funds and

135 blend funds in our sample. The data contained in the file are in **raw form;** they are listed alphabetically by name of the mutual fund. In addition to the name of the mutual fund, information is provided on a variety of variables (see appendix D).

As the number of observations gets large, it becomes more and more difficult to focus on the major features in a set of data. We need ways to organize the observations so that we can better understand what information the data are conveying. Two commonly used methods for accomplishing this are the *ordered array* and the *stem-and-leaf display*.

The Ordered Array

If we place the raw data in rank order, from the smallest to the largest observation, the ordered sequence obtained is called an **ordered array.** As we begin our analysis with the growth funds, Table 2.1 indicates the 1-year total returns achieved by the 59 sampled growth funds. When data are sorted into an ordered array, it becomes easier to pick out extremes, typical values, and concentrations of values.

Table 2.1 *Ordered array of 1-year total percentage returns achieved by the 59 growth funds*

20.4	23.8	25.6	26.2	27.6	27.7	28.3	28.6	28.8	28.9
28.9	29.3	29.3	29.5	29.9	30.1	31.5	31.6	31.6	31.8
31.9	32.1	32.3	32.3	32.4	32.8	32.9	32.9	33.0	33.3
33.4	33.7	33.8	34.0	34.0	34.3	34.7	34.7	34.8	35.0
38.2	39.0	39.4	40.7	41.1	42.8	42.9	43.3	43.4	43.5
43.6	43.7	44.6	44.7	45.4	45.7	46.6	48.0	48.6	

Although it is useful to place the raw data into an ordered array prior to developing summary tables and charts or computing descriptive summary measures (see chapter 3), the greater the number of observations present in a data set, the more useful it is to organize the data set into a stem-and-leaf display in order to study its characteristics (references 1, 13, and 14).

The Stem-and-Leaf Display

The **stem-and-leaf display** is a valuable and versatile tool for organizing a set of data and understanding how the values distribute and cluster over the range of the observations in the set of data. A stem-and-leaf display separates data entries into leading digits, or stems, and trailing digits, or leaves. For example, because the 1-year total returns in the growth fund data set all have two-digit integer numbers, the 10s and units columns are the leading digits, and the remaining column (the 10ths column) is the trailing digit. Thus, an entry of 32.3 (corresponding to a 1-year total percentage return of 32.3) has a stem of 32 and a trailing digit, or leaf, of 3.

Figure 2.1 depicts the stem-and-leaf display of the 1-year total returns achieved by the 59 growth funds obtained from Microsoft Excel. The first column of numbers is the stem, or leading digits, of the data and the leaves, or trailing digits, branch out to the right of these numbers.

Figure 2.1

Stem-and-leaf display of the 1-year total returns (in percentages) achieved by the 59 growth funds obtained from the PHStat add-in of Microsoft Excel

Source: Data are taken from Table 2.1.

	A	B	C	D	E	F
1				Stem-and-Leaf Display for		
2				One-Year Total Percentage Returns		
3				Stem unit: 1		
4						
5	Statistics			20	4	
6	n	59		21		
7	Mean	34.964407		22		
8	Median	33.3		23	8	
9	Std. dev.	6.6534607		24		
10	Minimum	20.4		25	6	
11	Maximum	48.599998		26	2	
12				27	6 7	
13				28	3 6 8 9 9	
14				29	3 3 5 9	
15				30	1	
16				31	5 6 6 8 9	
17				32	1 3 3 4 8 9 9	
18				33	0 3 4 7 8	
19				34	0 0 3 7 7 8	
20				35	0	
21				36		
22				37		
23				38	2	
24				39	0 4	
25				40	7	
26				41	1	
27				42	8 9	
28				43	3 4 5 6 7	
29				44	6 7	
30				45	4 7	
31				46	6	
32				47		
33				48	0 6	

An examination of Figure 2.1 allows us to begin drawing conclusions about the 1-year percentage returns of the growth funds. Among the conclusions we can reach from the stem-and-leaf display are:

1. The lowest 1-year percentage return is 20.4.

2. The highest 1-year percentage return is 48.6.

3. The returns of the 59 growth funds are spread out between the lowest and highest returns with some concentration of percentage returns between 28 and 34.

4. There seem to be more growth funds that have a high percentage return above 40 than a low percentage return below 25.

5. Some mutual funds have the same percentage return; for example, there are two mutual funds for each percentage return of 28.9, 31.6, 32.3, 32.9, 34.0, and 34.7.

To understand how the stem-and-leaf display is constructed, refer to Example 2.1.

Example 2.1 *Constructing a Stem-and-Leaf Display*

The following raw data represent the weekly salary checks earned by a sample of eight secretaries in a large law firm:

$555 \quad $490 \quad $648 \quad $832 \quad $710 \quad $590 \quad $576 \quad $627

Construct the stem-and-leaf display.

SOLUTION

Because all the values are three-digit integers, to form the stem-and-leaf display, two approaches are demonstrated.

First, we may use the 100s column as the stems and the 10s column as the leaves and ignore the units column:

$555 \quad $490 \quad $648 \quad $832 \quad $710 \quad $590 \quad $576 \quad $627

```
4|9
5|597
6|42
7|1
8|3
```

or, second, we may use the 100s column as the stems and the 10s column as the leaves after rounding the units column:

$555 \quad $490 \quad $648 \quad $832 \quad $710 \quad $590 \quad $576 \quad $627

```
4|9
5|698
6|53
7|1
8|3
```

In the first approach, the values listed in the second row indicate that these weekly salaries are in the 550s, 590s, and 570s. In the second approach, the values listed in the second row show that these weekly salaries are rounded to $560, $590, and $580.

 **ORGANIZING NUMERICAL DATA USING MICROSOFT EXCEL**

Overview

◆ *For Quick Results Users and Developers* Quick results users and developers should both use the Data | Sort command to order numerical data and use the PHStat add-in to generate stem-and-leaf displays.

The 2-1E.XLS workbook file contains an ordered array of data and a stem-and-leaf display for the one-year total percentage return data for 59 growth funds presented in Table 2.1 on page 65.

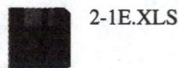
2-1E.XLS

Details for All Users

◆ *Ordering Data Using the Sort Command* Any worksheet cell range containing raw data values can be sorted in either ascending or descending order in Microsoft Excel. To order raw data values, do the following:

❶ Enter the raw data in a column range of contiguous cells. This range must contain no cells that are blank and include a column heading in the first cell of the range. (Or select the worksheet in which the raw data were previously entered.)

❷ Select a cell in the column range.

❸ Select Data | Sort.

❹ In the Sort dialog box:
  a. Verify that the Sort by drop-down list box contains the appropriate column (heading).
  b. Select either the Ascending or Descending option button as appropriate.
  c. Select the Header row option button.
  d. Click the OK button.

Example: Generating an ordered array using the Sort command   We can use the Sort command to reorder the data in the Growth Funds Sample workbook file by their one-year total percentage return values. (As stored, these data are ordered by fund name.) To reorder these data, do the following:

❶ Open the Growth Funds Sample workbook (Growth Funds Sample.XLS) and click the Data sheet tab.

❷ Select any cell in column B, the column containing the one-year return values.

❸ Select Data | Sort.

❹ In the Sort dialog box (see Figure 2E.1):
  a. Verify that the Sort by list box contains the 1-Yr Return column heading.
  b. Select the Ascending and Header row option buttons.
  c. Click the OK button.

The data are re-sorted by Microsoft Excel.

FIGURE 2E.1
Sort dialog box

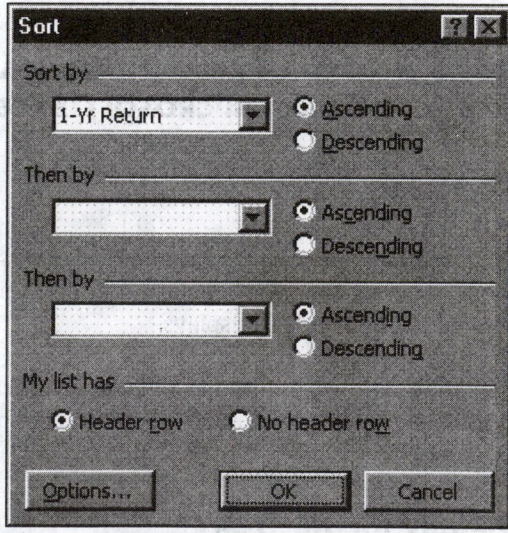

◆ ***Generating a Stem-and-Leaf Display for the Growth Funds Sample Data***   To generate a stem-and-leaf display based on the growth funds sample data of Table 2.1, do the following.

❶ If the PHStat add-in has not been previously loaded, load the add-in using the instructions of Section S4.2 on page 59.

❷ Open the Growth Funds Sample workbook (Growth Funds Sample.XLS) and click the Data sheet tab.

❸ Select PHStat | Stem-and-Leaf Display.

❹ In the Stem-and-Leaf dialog box (see Figure 2E.2):
  a. Enter B1:B60 in the Variable Cell Range: edit box.
  b. Select First cell contains label check box.
  c. Select the Set stem unit as: option button and enter 1 in its edit box.
  d. Enter 1-Year Total Percentage Returns in the Output Title: edit box.
  e. Select the Summary Statistics check box.
  f. Click the OK button.

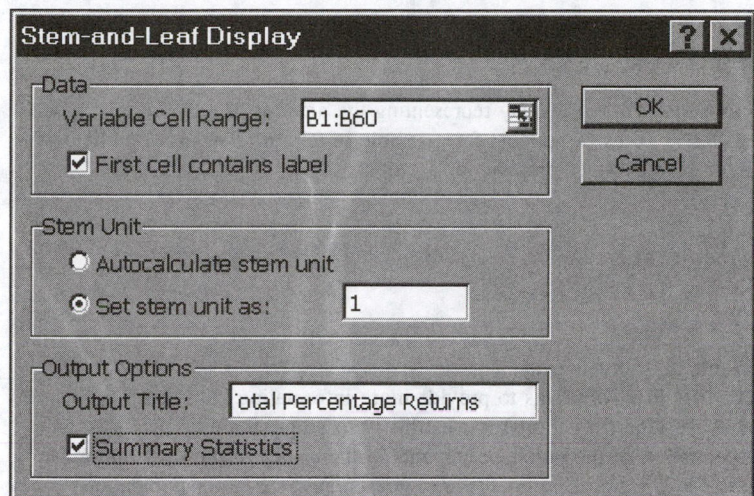

FIGURE 2E.2
The PHStat Stem-and-Leaf dialog box

The add-in produces a stem-and-leaf display similar to the one shown in Figure 2.1 on page 66. This chart is *not* dynamically changeable, so changes made to the underlying data would require repeating the procedure in order to produce a new chart.

Problems for Section 2.1

Learning the Basics

● 2.1 Form the ordered array given the following raw data from a sample of $n = 7$ midterm exam scores in accounting:

68 94 63 75 71 88 64

● 2.2 Form the stem-and-leaf display given the following raw data from a sample of $n = 7$ midterm exam scores in finance:

80 54 69 98 93 53 74

2.3 Form the ordered array given the following raw data from a sample of $n = 7$ midterm exam scores in marketing:

88 78 78 73 91 78 85

2.4 Form the stem-and-leaf display given the following raw data from a sample of $n = 7$ midterm exam scores in organizational behavior:

76 68 76 87 95 63 87

2.5 Form the stem-and-leaf display given the following ordered array from a sample of $n = 7$ midterm exam scores in economics:

46 58 69 76 82 82 96

2.6 Form the ordered array given the following stem-and-leaf display from a sample of $n = 7$ midterm exam scores in information systems:

```
5 | 0
6 |
7 | 464
8 | 91
9 | 2
```

Applying the Concepts

2.7 Given the following stem-and-leaf display representing the amount of gasoline purchased in gallons (with leaves in 10ths of gallons) for a sample of 25 cars that use a particular service station on the New Jersey Turnpike:

```
 9 | 714
10 | 82230
11 | 561776735
12 | 394282
13 | 20
```

(a) Place the data into an ordered array.
(b) Which of these two displays seems to provide more information? Discuss.
(c) What amount of gasoline (in gallons) is most likely to be purchased?
(d) Is there a concentration of the purchase amounts in the center of the distribution?
(e) Do you think these 25 purchase amounts are representative of larger population amounts? Explain.

2.8 Upon examining the monthly billing records of a mail-order CD and cassette company, the auditor takes a sample of 20 of its unpaid accounts. The amounts owed the company are

DATA FILE
MAILORDER.XLS

$4 $18 $11 $7 $ 7 $10 $ 5 $33 $ 9 $12
$3 $11 $10 $6 $26 $37 $15 $18 $10 $21

(a) Develop the ordered array.
(b) Form the stem-and-leaf display.
(c) What conclusions can you reach about the amounts owed on the unpaid accounts?

2.9 The following data represent the retail price (in dollars) of a sample of 29 different types of attaché cases that were being sold in department stores:

DATA FILE
ATTACHE.XLS

395 395 215 40 75 140 250 240 450 245 410 130 120 200 65 130
258 485 75 55 220 30 60 80 70 135 150 100 70

Source: "Attaché Cases," Copyright © 1996 by Consumers Union of U.S., Inc., Yonkers, NY 10703-1057. Adapted from CONSUMER REPORTS, *December 1996, 30–32, by permission of Consumers Union of U.S., Inc. Although these data sets originally appeared in* CONSUMER REPORTS, *the selective adaptation and resulting conclusions presented are those of the authors and are not sanctioned or endorsed in any way by Consumers Union, the publisher of* CONSUMER REPORTS.

(a) Develop the ordered array.
(b) Form the stem-and-leaf display.

(c) Are you more likely to encounter an expensive attaché case or an inexpensive attaché case? Explain.

(d) Do you think that if you are interested in purchasing an attaché case for under $100, you will be able to do so in a department store? Explain.

•2.10 The following data are the book values (in dollars, i.e., net worth divided by number of outstanding shares) for a random sample of 50 stocks from the New York Stock Exchange:

7	9	8	6	12	6	9	15	9	16
8	5	14	8	7	6	10	8	11	4
10	6	16	5	10	12	7	10	15	7
10	8	8	10	18	8	10	11	7	10
7	8	15	23	13	9	8	9	9	13

DATA FILE
STOCK.XLS

(a) Develop the ordered array.
(b) Form the stem-and-leaf display.
(c) On the basis of these data, are the book values on the New York Stock Exchange likely to be high or low? Explain.
(d) Are you more likely to find a stock with a book value below $10 or above $20? Explain.

2.11 The following data represent the annual family premium rates (in thousands of dollars) charged by 36 randomly selected HMOs throughout the United States:

3.8	4.1	4.7	5.2	2.8	5.6	4.9	6.7	9.2
4.9	4.9	4.9	5.2	5.9	5.2	4.8	4.8	9.1
4.6	8.0	4.9	4.2	4.1	5.3	5.5	8.0	7.2
7.2	4.1	4.5	8.0	4.4	4.2	4.6	4.2	4.8

DATA FILE
HMOPREMIUM.
XLS

Source: "HMO Annual Family Premium Rates," Copyright © 1996 by Consumers Union of U.S., Inc., Yonkers, NY 10703-1057. Adapted from CONSUMER REPORTS, *October 1996, 35. Although these data sets originally appeared in* CONSUMER REPORTS, *the selective adaptation and resulting conclusions presented are those of the authors and are not sanctioned or endorsed in any way by Consumers Union, the publisher of* CONSUMER REPORTS.

(a) Develop the ordered array.
(b) Form the stem-and-leaf display.
(c) Does there appear to be a concentration of premium rates in the center of the distribution?
(d) Your friend Kathy Rae said that her family has been considering whether or not to join an HMO. Based on your findings in parts (a) and (b), what would you tell her?

2.12 The following data are the retail prices for a random sample of 22 VCR models:

350	300	340	220	320	450	270	265
210	250	180	300	190	170	190	
170	170	200	180	220	200	250	

DATA FILE
VCR.XLS

Source: "VCRs," Copyright © 1996 by Consumers Union of U.S., Inc., Yonkers, NY 10703-1057. Adapted from CONSUMER REPORTS, *November 1996, 36–38. Although these data sets originally appeared in* CONSUMER REPORTS, *the selective adaptation and resulting conclusions presented are those of the authors and are not sanctioned or endorsed in any way by Consumers Union, the publisher of* CONSUMER REPORTS.

(a) Develop the ordered array.
(b) Form the stem-and-leaf display.
(c) Are you more likely to find a VCR for under $200 or over $300?
(d) From the stem-and-leaf display, does there seem to be a concentration of prices around or near any specific dollar amount? Explain.

The Frequency Distribution

Regardless of whether an ordered array (see Table 2.1) or a stem-and-leaf display (see Figure 2.1) is selected for organizing the data, as the number of observations obtained gets large, it becomes necessary to further condense the data into appropriate summary tables in order to properly present, analyze, and interpret the findings. Thus, we may wish to arrange the data into **class groupings** (i.e., categories) according to conveniently established divisions of the range of the observations. Such an arrangement of data in tabular form is called a frequency distribution.

> A **frequency distribution** is a summary table in which the data are arranged into conveniently established, numerically ordered class groupings or categories.

When the observations are grouped or condensed into frequency distribution tables, the process of data analysis and interpretation is made much more manageable and meaningful. The major data characteristics can be approximated, which compensates for the fact that when the data are so grouped, the initial information pertaining to individual observations that was previously available is lost through the grouping process.

In constructing the frequency distribution table, attention must be given to selecting the appropriate *number* of class groupings for the table, obtaining a suitable *class interval*, or *width* of each class grouping, and establishing the *boundaries* of each class grouping to avoid overlapping.

◆ *Selecting the Number of Classes*   The number of class groupings to be used is primarily dependent on the number of observations in the data. Larger numbers of observations require a larger number of class groups. In general, however, the frequency distribution should have at least 5 class groupings but no more than 15. If there are not enough class groupings or if there are too many, little new information would be learned.

◆ *Obtaining the Class Intervals*   When developing the frequency distribution table, it is desirable that each class grouping has the same width. To determine the **width of each class interval**, the *range* of the data is divided by the number of class groupings desired:

> ### Determining the Width of a Class Interval
>
> $$\text{Width of interval} > \frac{\text{range}}{\text{number of desired class groupings}} \qquad (2.1)$$

Because there were only 59 growth funds sampled, six class groupings are sufficient. From the ordered array in Table 2.1 (page 65), the range is computed as $48.6 - 20.4 = 28.2$. Using equation (2.1), the width of the class interval is approximated by

$$\text{Width of interval} \cong \frac{28.2}{6} = 4.7$$

For convenience and ease of reading, the selected interval or width of each class grouping is rounded to 5.0.

◆ **Establishing the Boundaries of the Classes** To construct the frequency distribution table, it is necessary to establish clearly defined **class boundaries** for each class grouping so that the observations can be properly tallied into the classes. Overlapping of classes must be avoided.

Because the width of each class interval for the 1-year total return data has been set at 5.0%, the boundaries of the various class groupings must be established so as to include the entire range of observations. Whenever possible, these boundaries should be chosen to facilitate the reading and interpreting of data. Thus, the first class interval ranges from 20.0 to under 25.0%, the second from 25.0 to under 30.0%, and so on, until they have been tallied into six classes, each having an interval width of 5.0%, without overlapping. By establishing these boundaries, all 59 observations can be tallied into each class as shown in Table 2.2.

Table 2.2 *Frequency distribution of 1-year total percentage returns achieved by 59 growth funds*

1-YEAR TOTAL PERCENTAGE RETURN	NUMBER OF FUNDS
20.0 but less than 25.0	2
25.0 but less than 30.0	13
30.0 but less than 35.0	24
35.0 but less than 40.0	4
40.0 but less than 45.0	11
45.0 but less than 50.0	5
Total	59

Source: Data are taken from Table 2.1 on page 65.

The main advantage of using this summary table is that the major data characteristics should become immediately clear to the reader. For example, we see from Table 2.2 that the *approximate range* in the 1-year total percentage returns achieved by these 59 sampled growth funds is from 20.0 to 50.0 and, typically, the 1-year total percentage returns tend to cluster between 30.0 and 35.0%.

On the other hand, the major disadvantage of this summary table is that we cannot know how the individual values are distributed within a particular class interval without access to the original data. Thus, for the four funds whose 1-year total returns are between 35.0 and 40.0%, it is not clear from Table 2.2 whether the values are distributed throughout the interval, cluster near 35.0%, or cluster near 40.0%. The class *midpoint* (37.5%), however, is the value used to represent the 1-year total percentage returns for all four funds contained in the particular interval.

> The **class midpoint** is the point halfway between the boundaries of each class and is representative of the data within that class.

The class midpoint for the interval "20.0 but less than 25.0" is 22.5%. (The other class midpoints are, respectively, 27.5, 32.5, 37.5, 42.5, and 47.5%.)

◆ *Subjectivity in Selecting Class Boundaries*   The selection of class boundaries for frequency distribution tables is highly subjective. For data sets that do not contain many observations the choice of a particular set of class boundaries over another may yield a different picture to the reader. For example, for the 1-year total percentage return data, using a class-interval width of 6.0 instead of 5.0 (as was used in Table 2.2) may cause shifts in the way in which the observations distribute among the classes. This is particularly true if the number of observations in the data set is not very large.

Such shifts in data concentration do not occur only because the width of the class interval is altered. We may keep the interval width at 5.0% but choose different lower and upper class boundaries. Such manipulation may also cause shifts in the way in which the data distribute—especially if the size of the data set is not very large. Fortunately, as the number of observations in a data set increases, alterations in the selection of class boundaries affect the concentration of data less and less.

The Relative Frequency Distribution and the Percentage Distribution

The frequency distribution is a summary table into which the original data are grouped to facilitate data analysis. To enhance the analysis, however, it is almost always desirable to form either the relative frequency distribution or the percentage distribution, depending on whether we prefer proportions or percentages. These two equivalent distributions are shown in Table 2.3.

Table 2.3   *Relative frequency distribution and percentage distribution of 1-year total percentage returns achieved by 59 growth funds*

1-YEAR TOTAL PERCENTAGE RETURN	PROPORTION OF FUNDS	PERCENTAGE OF FUNDS
20.0 but less than 25.0	.034	3.4
25.0 but less than 30.0	.220	22.0
30.0 but less than 35.0	.407	40.7
35.0 but less than 40.0	.068	6.8
40.0 but less than 45.0	.186	18.6
45.0 but less than 50.0	.085	8.5
Total	1.000	100.0

Source: Data are taken from Table 2.2 on page 73.

The **relative frequency distribution** is formed by dividing the frequencies in each class of the frequency distribution (Table 2.2 on page 73) by the total number of observations. A **percentage distribution** may then be formed by multiplying each relative frequency or proportion by 100.0. Thus, the proportion of growth funds that achieved 1-year total percentage returns of 35.0 to under 40.0 is .068; it can be seen that 6.8% of the funds have achieved such performance results.

Working with a base of 1 for proportions or 100.0 for percentages is usually more meaningful than using the frequencies themselves. Indeed, the use of the relative frequency distribution or percentage distribution becomes essential whenever one set of data is being compared with other sets of data, especially if the number of observations in each set is different.

Now suppose that in order to make purchase recommendations to potential investors we want to compare the 1-year total percentage returns achieved by the 59 sampled growth funds with those from the 135 sampled blend funds. To compare the 1-year total percentage returns achieved by the 59 growth funds with those from the 135 blend funds, we develop a percentage distribution for the latter group. This new table will then be compared with Table 2.3.

Table 2.4 depicts both the frequency distribution and the percentage distribution of the 1-year total returns achieved by the 135 blend funds. Note that the class groupings selected in Table 2.4 match, where possible, those selected in Table 2.3 for the growth funds. The boundaries of the classes should match or be multiples of each other in order to facilitate comparisons.

Table 2.4 *Frequency distribution and percentage distribution of 1-year total percentage returns achieved by 135 blend funds*

1-YEAR TOTAL PERCENTAGE RETURN	NUMBER OF FUNDS	PERCENTAGE OF FUNDS
10.0 but less than 15.0	1	0.7
15.0 but less than 20.0	3	2.2
20.0 but less than 25.0	9	6.7
25.0 but less than 30.0	41	30.4
30.0 but less than 35.0	67	49.6
35.0 but less than 40.0	14	10.4
Total	135	100.0

With the use of the percentage distributions of Tables 2.3 and 2.4, it is now meaningful to compare the differences in the 1-year total returns achieved by growth funds versus blend funds. Even though the 1-year total returns achieved by growth funds and by blend funds are typically clustering between 30 and 35% and, secondarily, are clustering between 25 and 30%, there is a vast difference in the manner in which the two sets of data distribute. The 1-year total percentage returns achieved by growth funds are generally at a substantially higher level than those attained by blend funds. We can observe this if we evaluate the *ranges* in 1-year total returns for both types of funds. In growth funds, the range in 1-year total percentage returns is approximated to be 30.0 (i.e., the difference between 50.0, the upper boundary of the last class, and 20.0, the lower boundary of the first class). In blend funds, however, even though the range in 1-year total percentage returns also is approximated as 30.0, the difference in the extreme boundaries here is 40.0 − 10.0. Other descriptive summary measures that would enhance a comparative analysis of the 1-year total returns achieved by growth funds versus blend funds will be discussed in chapter 3.

The Cumulative Distribution

Another useful method of data presentation that facilitates analysis and interpretation is the **cumulative distribution** table. This may be formed from the frequency distribution, the relative frequency distribution, or the percentage distribution. In this text, we will focus on the percentage distribution.

Because we already have the percentage distributions of 1-year total returns achieved by 59 growth funds and 135 blend funds (see Tables 2.3 and 2.4), we can use these tables to construct the respective cumulative percentage distributions as depicted in Tables 2.5 and 2.6.

Table 2.5 *Cumulative percentage distribution of 1-year total percentage returns achieved by 59 growth funds*

1-YEAR TOTAL PERCENTAGE RETURN	PERCENTAGE OF FUNDS "LESS THAN" INDICATED VALUE
20.0	0.0
25.0	3.4
30.0	25.4
35.0	66.1
40.0	72.9
45.0	91.5
50.0	100.0

Source: Data are taken from Table 2.3 on page 74.

Table 2.6 *Cumulative percentage distribution of 1-year total percentage returns achieved by 135 blend funds*

1-YEAR TOTAL PERCENTAGE RETURN	PERCENTAGE OF FUNDS "LESS THAN" INDICATED VALUE
10.0	0.0
15.0	0.7
20.0	2.9
25.0	9.6
30.0	40.0
35.0	89.6
40.0	100.0

Source: Data are taken from Table 2.4 on page 75.

A comparison of these two tables demonstrates that the growth funds have achieved a higher level of 1-year total returns. For example, we see from Table 2.5 that only 25.4% of the growth funds have achieved a 1-year total percentage return of less than 30.0, whereas from Table 2.6 we observe that 40.0% of the blend funds have attained a 1-year total percentage return of less than 30.0. In addition, only 72.9% of the growth funds have achieved a 1-year total percentage return of less than 40.0, whereas all the blend funds have performance records less than that level.

A cumulative percentage distribution table is constructed by first recording the lower boundaries of each class from the percentage distribution and then inserting an extra boundary at the end. We compute the cumulative percentages in the "less than" column by determining the percentage of observations less than each of the stated boundary values.

In Example 2.2 we demonstrate how the cumulative percentage distribution shown in Table 2.5 is constructed by using the percentage distribution of 1-year total returns achieved by 59 growth funds displayed in Table 2.3 on page 74.

Example 2.2 *Forming the Cumulative Percentage Distribution*

Using the percentage distribution of Table 2.3 on page 74, form the cumulative percentage distribution.

SOLUTION

From Table 2.5 we see that 0.0% of the 1-year total returns in growth funds are less than 20.0%, 3.4% of the 1-year total returns in growth funds are less than 25.0%, 25.4% of the 1-year total returns in growth funds are less than 30.0%, and so on, until all (100.0%) of the 1-year total returns in growth funds are less than 50.0%. This cumulating process is observed in the following table.

1-YEAR TOTAL RETURN (IN %)	PERCENTAGE OF FUNDS IN CLASS INTERVAL	PERCENTAGE OF FUNDS "LESS THAN" LOWER BOUNDARY OF CLASS INTERVAL
20.0 but less than 25.0	3.4	0.0
25.0 but less than 30.0	22.0	3.4
30.0 but less than 35.0	40.7	25.4 = 3.4 + 22.0
35.0 but less than 40.0	6.8	66.1 = 3.4 + 22.0 + 40.7
40.0 but less than 45.0	18.6	72.9 = 3.4 + 22.0 + 40.7 + 6.8
45.0 but less than 50.0	8.5	91.5 = 3.4 + 22.0 + 40.7 + 6.8 + 18.6
50.0 but less than 55.0	0.0	100.0 = 3.4 + 22.0 + 40.7 + 6.8 + 18.6 + 8.5

The Histogram

A saying sometimes mistakenly attributed to Confucius is "one picture is worth a thousand words." Indeed, statisticians often employ graphic techniques to more vividly describe sets of data. In particular, a histogram is used to describe numerical data that have been grouped into frequency, relative frequency, or percentage distributions.

A **histogram** is a vertical bar chart in which the rectangular bars are constructed at the boundaries of each class.

When plotting a histogram, we display the random variable of interest along the horizontal axis; the vertical axis represents the number, proportion, or percentage of observations per class interval.

A percentage histogram of the 1-year total percentage returns achieved by just the 59 growth funds is presented in Figure 2.2, which has been obtained from Microsoft Excel. We can observe that there is a strong concentration of funds with a return between 30 and 35%, a smaller concentration of funds with returns between 25 and 30% and between 40 and 45%, and little concentration of funds in the other class groupings.

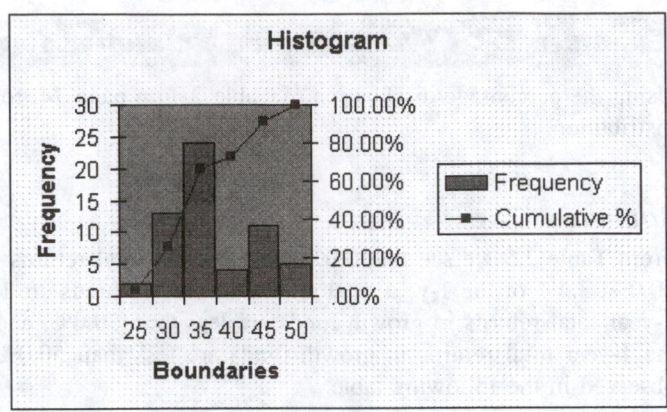

FIGURE 2.2 Percentage histogram of 1-year total percentage returns achieved by 59 growth funds obtained from Microsoft Excel

When comparing two or more sets of data, we cannot construct stem-and-leaf displays or histograms on the same graph. For example, superimposing the vertical bars of one histogram on another would cause difficulty in interpretation. For such cases it is necessary to construct relative frequency or percentage polygons.

The Polygon

As with histograms, when plotting polygons, we display the phenomenon of interest along the horizontal axis, and the vertical axis represents the number, proportion, or percentage of observations per class interval. In this text, we will concern ourselves with the latter.

> The **percentage polygon** is formed by having the midpoint of each class represent the data in that class and then connecting the sequence of midpoints at their respective class percentages.

Because consecutive midpoints are connected by a series of straight lines, the **polygon** is sometimes jagged in appearance. However, when dealing with a very large set of data, were we to make the boundaries of the classes in its frequency distribution closer together (and thereby increase the number of classes in that distribution), the jagged lines of the polygon would "smooth out."

Figure 2.3 shows the percentage polygons for the 1-year total returns achieved by the 59 growth funds and the 135 blend funds obtained from Excel. The differences in the structure of the two distributions, previously discussed when comparing Tables 2.3 and 2.4, are clearly indicated here.

◆ *Polygon Construction* Notice that the polygon in Figure 2.3 by Microsoft Excel has points whose values on the X axis represent the upper limit of the class interval. For example, observe the points plotted on the X axis at 34.99. The value for the blend funds (the higher value) represents the fact that 49.6% of these funds have 1-year percentage returns of between 30 and 34.99. The value for the growth funds (the lower value) represents the fact that 40.7% of these funds have 1-year percentage returns of between 30 and 34.99. Notice, too, that when polygons or histograms are constructed, the vertical axis should show the true zero or "origin" so as not to distort or otherwise misrepresent the character of the

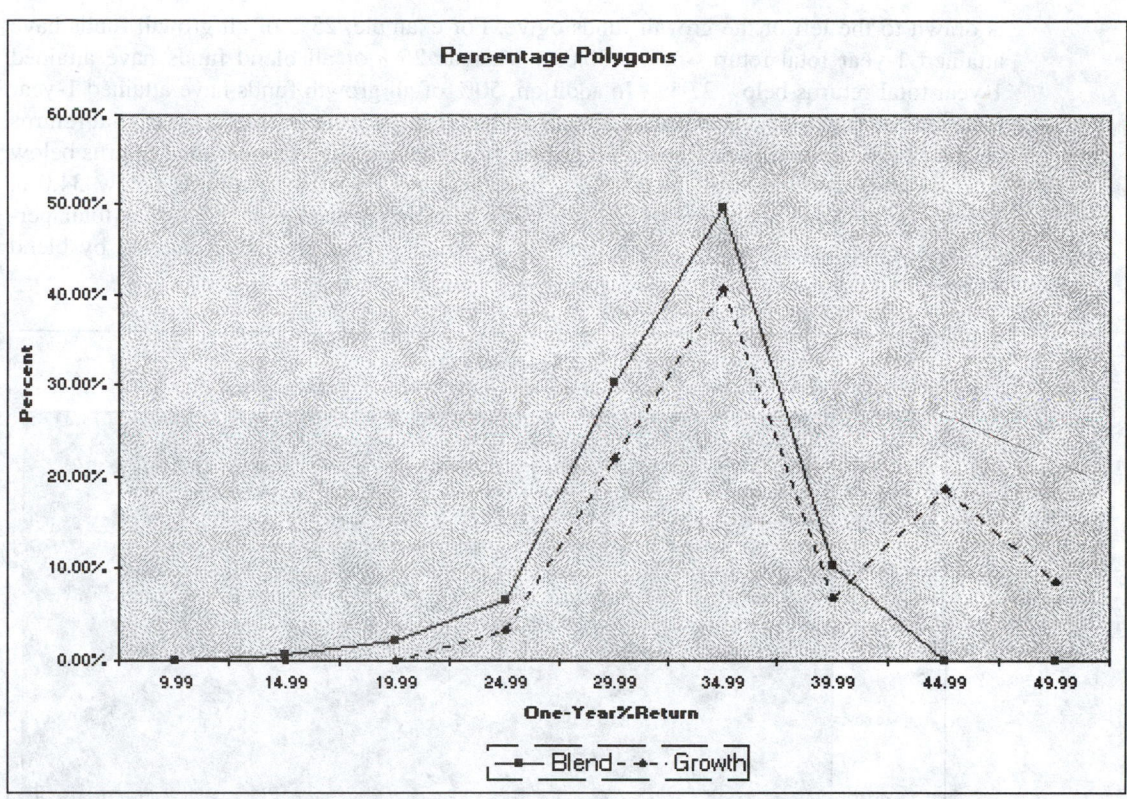

FIGURE 2.3 Percentage polygons of 1-year total percentage returns achieved by 59 growth funds and 135 blend funds obtained from Microsoft Excel

data. The horizontal axis, however, does not need to specify the zero point for the phenomenon of interest. For aesthetic reasons, the range of the random variable should constitute the major portion of the chart.

The Cumulative Polygon (Ogive)

The **cumulative percentage polygon,** or **ogive,** is a graphic representation of a cumulative distribution table. As with histograms and polygons, when plotting cumulative polygons, we display the phenomenon of interest along the horizontal axis, and the vertical axis represents the number, proportion, or percentage of cumulated observations. Again, we will concern ourselves here with the latter.

To construct a cumulative percentage polygon (also known as a percentage ogive), we note that our random variable of interest—1-year total percentage returns—is again plotted on the horizontal axis, whereas the cumulative percentages (from the "less than" column) are plotted on the vertical axis.

Figure 2.4 illustrates the cumulative percentage polygons of the 1-year total percentage returns achieved by the 59 growth companies and the 135 blend funds obtained from Microsoft Excel. As was the case with the percentage polygons obtained from Excel, the upper limits of the classes are noted on the *X* axis. Thus, for example, for the blend funds 9.6% had a 1-year percentage return of less than or equal to 24.99, whereas for the growth funds the percentage was 3.4. From Figure 2.4 we note that in general the blend funds ogive is drawn to the left of the growth funds ogive. For example, 25% of all growth funds have attained 1-year total returns below 30.0%, whereas 25% of all blend funds have attained 1-year total returns below 27.5%. In addition, 50% of all growth funds have attained 1-year total returns below 33.0%, whereas 50% of all blend funds have attained 1-year total returns below 31.5%. Furthermore, 75% of all growth funds have attained 1-year total returns below 41.0%, whereas 75% of all blend funds have attained 1-year total returns below 34.0%. These comparisons enable us to confirm our earlier impression that the 1-year total percentage return achieved by growth funds is at a higher level than that attained by blend funds.

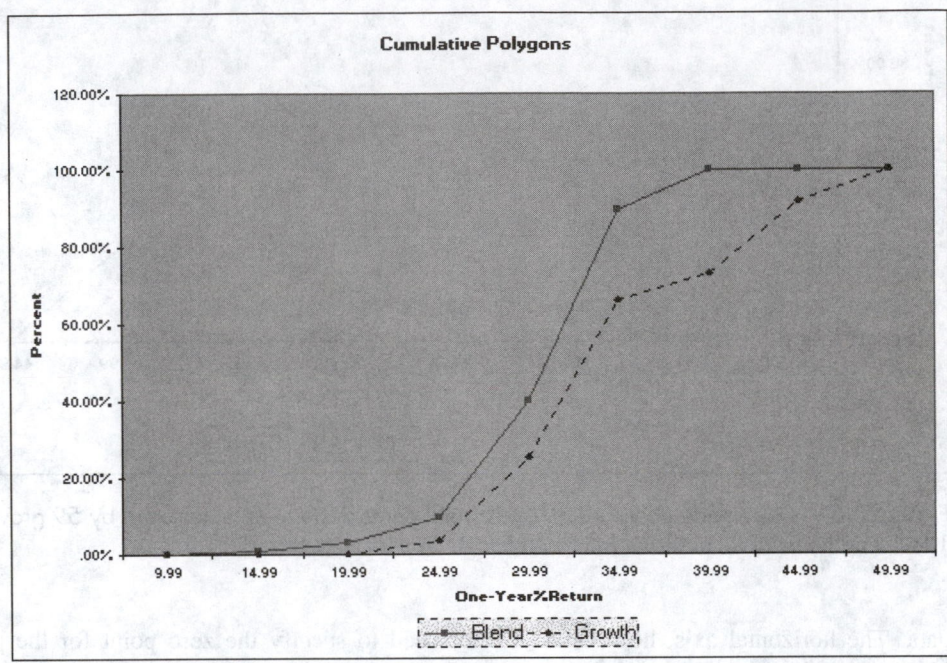

FIGURE 2.4 Cumulative percentage polygons of 1-year total percentage returns achieved by 59 growth funds and 135 blend funds obtained from Microsoft Excel

2.2E Generating Tables and Charts for Numerical Data Using Microsoft Excel

Overview

◆ *For Quick Results Users*   Use this two-step procedure:

A. Use the Data Analysis Histogram tool to generate a distribution table and histogram.

B. Modify the Histogram tool output to correct errors of presentation, if desired.

◆ *For Developers*   If only one-time results are needed, the Quick Results procedure can be used. For cases that involve examining the effects of changing data values or class boundaries or in cases that examine many distributions, the following two-step process can be used:

A. Implement a distribution table incorporating the FREQUENCY Excel function.

B. Use the Chart Wizard to generate charts based on this table.

The 2-2E.XLS workbook file contains distribution tables and charts for the one-year total percentage return data for 59 growth funds presented in Table 2.1 on page 65.

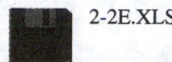
2-2E.XLS

Quick Results Approach

◆ *A. Generating Tables and Charts Using the Data Analysis Histogram Tool* The Data Analysis Histogram tool can be used to generate both a frequency distribution and charts such as a histogram and cumulative percentage polygon. For optimal results, the class intervals (called bins by Microsoft Excel) need to be specified. Class intervals in Excel are specified as an ordered list of class maximum values and not as ranges. Table 2E.1 shows how the ranges of Table 2.2 on page 73 can be approximated as an ordered list of values.

Table 2E.1 *Approximating class intervals*

ORIGINAL CLASS INTERVAL RANGE	APPROXIMATED AS:
20.0 but less than 25.0	24.99
25.0 but less than 30.0	29.99
30.0 but less than 35.0	34.99
35.0 but less than 40.0	39.99
40.0 but less than 45.0	44.99
45.0 but less than 50.0	49.99

We must enter the ordered list of class maximum values into a cell range on the worksheet containing the data to be analyzed. The Starting Point for Section 2.2E workbook (STARTING POINT 2-2E.XLS) is a modified version of the Growth Funds Sample workbook in which the class maximum values of Table 2E.1 have been entered into column D of the Data sheet. With this workbook, we can use the Data Analysis Histogram tool as follows:

❶ Open the Starting Point for Section 2.2E workbook and click the Data sheet tab. Verify that the upper-class maximum values have been entered in cell range D3:D8.

❷ Select Tools | Data Analysis. Select Histogram from the Analysis Tools list box in the Data Analysis dialog box. Click the OK button.

❸ In the Histogram dialog box (see Figure 2E.3):
a. Enter B1:B60 in the Input Range: edit box.
b. Enter D2:D8 in the Bin Range: edit box.
c. Select the Labels check box.
d. Select the New Worksheet Ply option button and enter Histogram as the name of the new sheet.
e. Select the Cumulative Percentage check box.
f. Select the Chart Output check box. Leave unchecked (deselected) the Pareto check box.
g. Click the OK button.

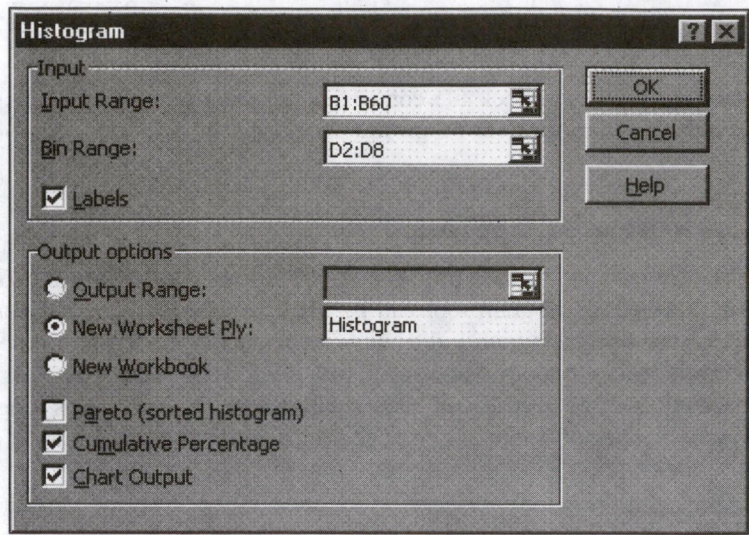

Microsoft Excel generates both a frequency distribution and cumulative percentage distribution and superimposes the cumulative percentage polygon onto the histogram (see Figure 2E.4).

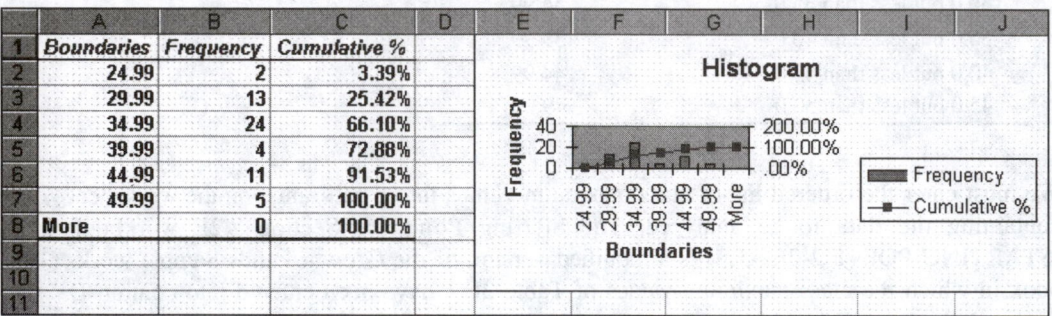

FIGURE 2E.4 Original frequency distribution table and histogram produced the Data Analysis Histogram tool

◆ *B. Modifying the Data Analysis Histogram Output* Note that Figure 2E.4 contains several errors: there is an additional class, labeled More by Excel, in both the frequency distribution and the histogram, there are gaps between the bars that correspond to the class intervals in the histogram, and the secondary *Y* axis scale exceeds 100%. To correct these errors, do the following:

❶ Click the Histogram sheet tab.

❷ Select the cell range A8:C8 (containing the More row of the frequency distribution table).

❸ Select Edit | Delete.

❹ In the Delete dialog box, select the Shift cells up option button and then click OK.

The "More" class is deleted from both the frequency distribution table and the histogram. Continue by removing the gaps between the bars.

⑤ Right-click on one of the histogram bars. (A tool tip that includes the words "Series 'Frequency'" appears when the mouse pointer is over a bar.)

⑥ Select Format Data Series from the shortcut menu.

⑦ In the Format Data Series dialog box, click the Options tab (see Figure 2E.5) and change the value in the Gap width: edit box to 0. Click the OK button.

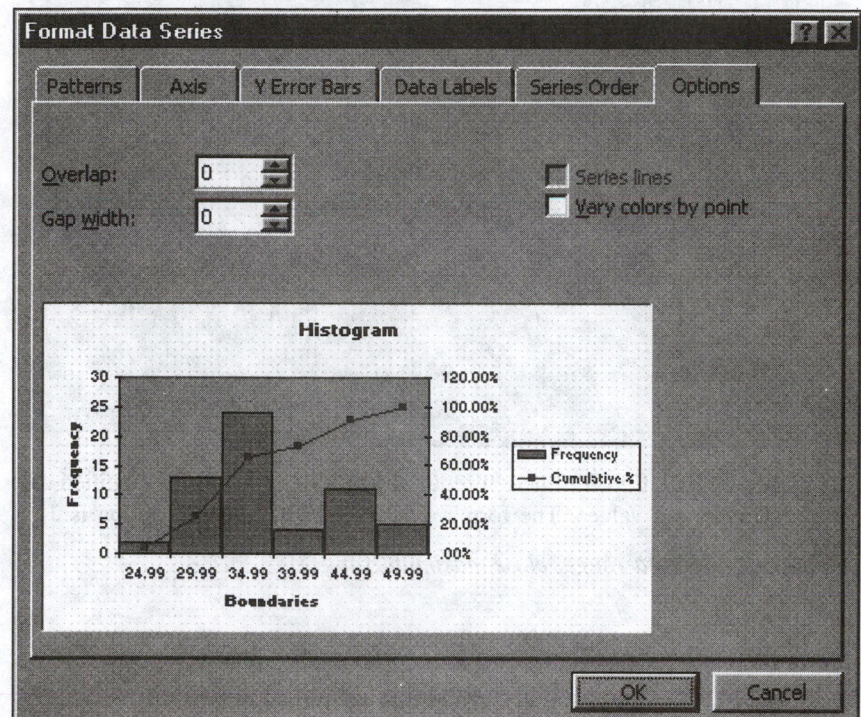

FIGURE 2E.5
Format Data Series
dialog box

The gaps are eliminated. Continue by changing the secondary Y axis.

⑧ Right-click the secondary (right) Y axis. (The tool tip "Secondary Value Axis" appears when the mouse pointer is over this axis.)

⑨ Select Format Axis from the shortcut menu.

⑩ In the Scale tab of the Format Axis dialog box, change the value in the Maximum: edit box to 1 and the maximum check box automatically deselects (see Figure 2E.6 on page 84). Click the OK button.

All errors are now eliminated and the resulting histogram will be similar to the one shown in Figure 2.2 on page 78.

Developer Details

◆ *A. Implementing Distribution Tables Incorporating the **FREQUENCY** Function*
We can use the FREQUENCY array function (a special type of function) to generate a list of frequencies around which a distribution table can be built. As was the case using the

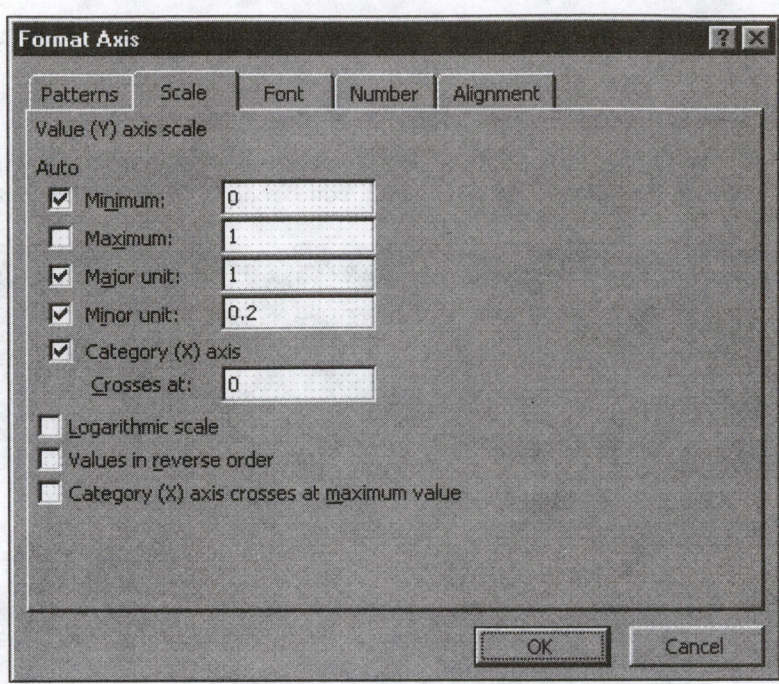

FIGURE 2E.6
Format Axis dialog box

Histogram tool, using the FREQUENCY function requires class intervals to be specified as an ordered list of class maximum values. The format of the FREQUENCY function is:

FREQUENCY(*data range, class maximum range*)

where

> *data range* is the cell range of the data to be analyzed
>
> *class maximum range* is the cell range that contains the ordered list of class maximum values being used to approximate class intervals

Table 2E.2 presents a Calculations sheet design that uses this function as the basis for developing a distribution table containing frequencies, relative frequencies, and percentage frequencies. Note that this design assumes that the data are in the range B2:B60 of a Data sheet, as is the case for the one-year total percentage return data of the Growth Funds Sample workbook. To implement this design, do the following:

Table 2E.2 Calculations sheet design for the growth funds sample analysis

	A	B	C	D
1	Distribution Table for One-Year Total Percentage Return Data			
2				
3	Class	Frequency	Relative Frequency	Percentage
4	24.99	{=FREQUENCY(DATA!B2:B60,A4:A9)}	=B4/$B$10	=C4
5	29.99	{=FREQUENCY(DATA!B2:B60,A4:A9)}	=B5/$B$10	=C5
6	34.99	{=FREQUENCY(DATA!B2:B60,A4:A9)}	=B6/$B$10	=C6
7	39.99	{=FREQUENCY(DATA!B2:B60,A4:A9)}	=B7/$B$10	=C7
8	44.99	{=FREQUENCY(DATA!B2:B60,A4:A9)}	=B8/$B$10	=C8
9	49.99	{=FREQUENCY(DATA!B2:B60,A4:A9)}	=B9/$B$10	=C9
10	Total:	=SUM(B4:B9)		

56

① Open the Growth Funds Sample workbook (GROWTH FUNDS SAMPLE.XLS) and click the Data sheet tab and verify the data.

② Select Insert | Worksheet and rename the new sheet Calculations.

③ Enter the title in row 1, the column headings in row 3, and the upper-class maximum values in the cell range A4:A9. Enter the Total: label in A10.

④ Enter the formulas in column B containing the FREQUENCY function. Because the function is an "array function", use the following special procedure to enter these formulas:
   a. Select the cell range (B4:B9) that will contain these formulas.
   b. Type =FREQUENCY(Data!B2:B60,A4:A9) but do *not* press the Enter key.
   c. While holding down the Control and Shift keys, press the Enter key.

A set of identical formulas now appear in the range B4:B9. Note that Microsoft Excel places a pair of curly braces around each of these formulas to indicate that they cannot be edited individually. Continue the design implementation of Table 2E.2 by doing the following:

⑤ Enter the formula =SUM(B4:B9) in cell B10 to calculate the total frequency.

⑥ Enter the formula =B4/$B$10 in cell C4 and copy the formula down through cell C9. Note that the mixed use of relative and absolute cell references (discussed and defined in section S3.8) is necessary.

⑦ Enter the formula =C4 in cell D4 and copy the formula down through cell C9. (This duplicates the column C values in column D.)

⑧ Select the cell range D4:D9 and click the Percentage format button on the Formatting toolbar (see section S3.11 on page 53).

⑨ Adjust the number of decimal places to two decimal places by clicking the Increase Decimal button on the Formatting toolbar twice.

A Calculations sheet, similar to the one illustrated in Figure 2E.7, will be produced.

	A	B	C	D	E
1	Distribution Table for One-Year Total Percentage Return Data				
2					
3	Class	Frequency	Relative Frequency	Percentage	
4	24.99	2	0.033898305	3.39%	
5	29.99	13	0.220338983	22.03%	
6	34.99	24	0.406779661	40.68%	
7	39.99	4	0.06779661	6.78%	
8	44.99	11	0.186440678	18.64%	
9	49.99	5	0.084745763	8.47%	
10	Total:	59			

FIGURE 2E.7
Implemented Calculations sheet based on the design of Table 2E.2

◆ **B. Generating Charts Using the Chart Wizard** Once implemented, distribution tables can serve as the source data for generating both polygons and histograms. Such charts are most easily generated by the Microsoft Excel Chart Wizard, a series of four linked dialog boxes. To use the Chart Wizard, follow this general procedure:

① Click on the tab of the worksheet containing the source data for the charts.

② Select Insert | Chart to begin the wizard.

③ Make selections and enter information, as necessary, in the Steps 1, 2, and 3 dialog boxes, clicking the Next button to advance to the next dialog box.

④ In the Step 4 dialog box, select the As new sheet: option button and type a self-descriptive name for the new chart sheet. (The authors recommend always placing charts on their own sheets.)

⑤ Click the Finish button.

Example: Generating a histogram using the Chart Wizard   To generate a histogram from the original Calculations sheet distribution table implemented in Step A on page 83, do the following:

❶ Click the Calculations sheet tab.

❷ Select Insert | Chart.

❸ In the Step 1 dialog box:
   a. Select the Standard Types tab and then select Column from the Chart type: list box. Select the first choice in the first row of choices under the Chart sub-type: heading. When this choice is selected, the phrase " Clustered Column" is displayed in the description box below the sub-types. (See Figure 2E.8, panel A.)
   b. Click the Next button.

FIGURE 2E.8

Panel A: Standard Types tab of the Chart Wizard Step 1 dialog box

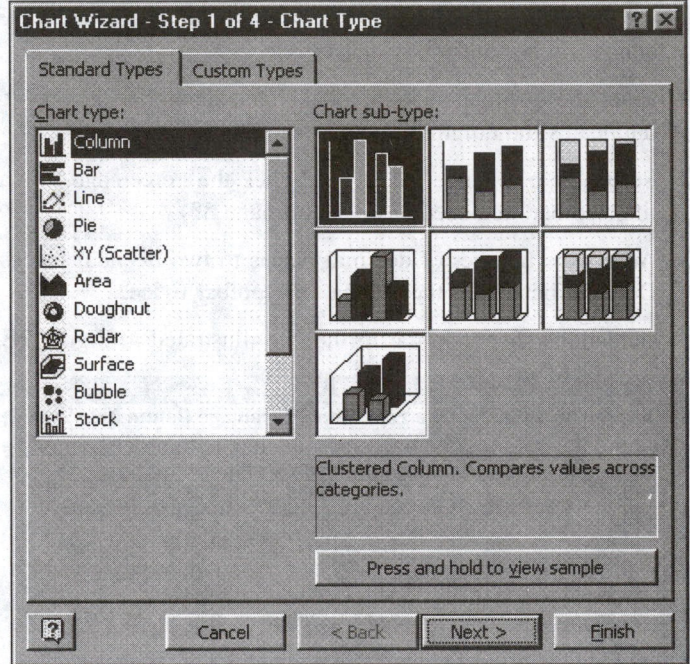

❹ In the Step 2 dialog box:
   a. Select the Data Range tab. Enter Calculations!B4:B9 in the Data range: edit box and select the Columns option button in the Series in: group. (See Figure 2E.8, panel B.)
   b. Select the Series tab (Figure 2E.8, panel C). Enter =Calculations!A4:A9 in the Category (X) axis labels: edit box as is shown in Figure 2E.8, panel C. Note that this entry must include the leading equal sign (=), one of the rare times in a Microsoft Excel dialog box that this must be done.
   c. Click the Next button.

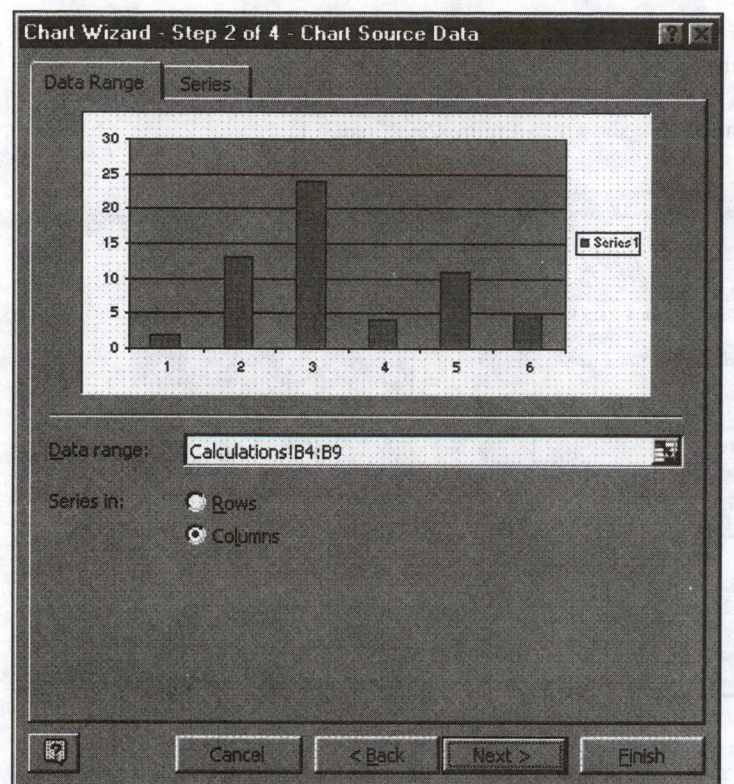

FIGURE 2E.8
Panel B: Data Range tab of the Chart Wizard Step 2 dialog box

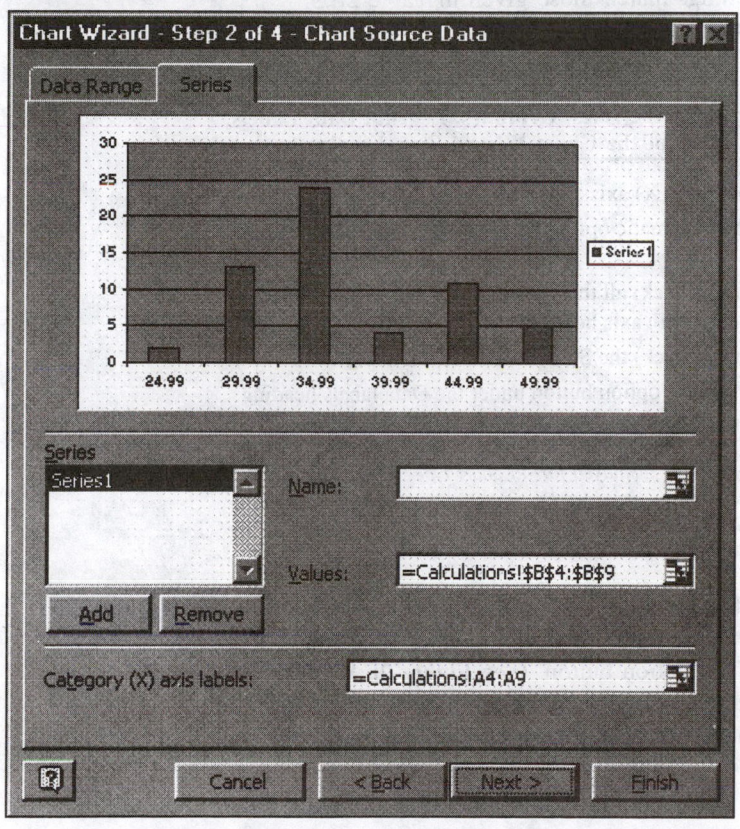

FIGURE 2E.8
Panel C: Series tab of the Chart Wizard Step 2 dialog box

❺ In the Step 3 dialog box:
   a. Select the Titles tab. Enter Histogram for 1-Year Returns in the Chart title: edit box, enter Class in the Category (X) axis: edit box, and enter Frequency in the Value (Y) axis: edit box as shown in Figure 2E.8, panel D.

FIGURE 2E.8

Panel D: Titles tab of the Chart Wizard Step 3 dialog box

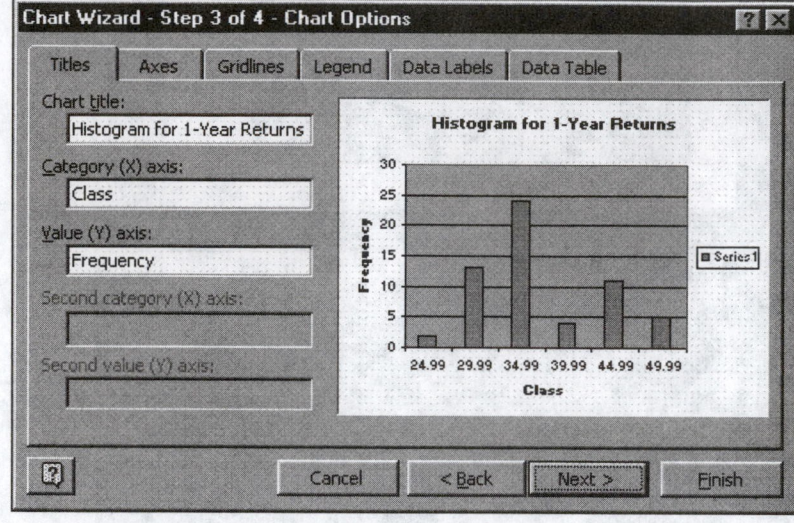

   b. Select, in turn, the Axes, Gridlines, Legend, Data Labels, and Data Table tabs and verify that their settings match those given in Table 2E.3.
   c. Click the Next button.

Table 2E.3 Common settings for the Axes, Gridlines, Legend, Data Labels, and Data Table tabs of the Chart Wizard Step 3 dialog box

Axes tab	Select both the (X) axis and (Y) axis check boxes.
	Select the Automatic option button under the (X) axis check box.
Gridlines tab	Deselect (uncheck) all the choices under the (X) axis heading and under the (Y) axis heading.
Legend tab	Deselect (uncheck) the Show legend check box.
Data Labels tab	Select the None option button under the Data labels heading.
Data Table tab	Deselect (uncheck) the Show data table check box.

Note: Not all settings will be enabled or displayed for all chart types. Depending on the chart type, (X) and (Y) axis labels will be preceded by either the word Category or Value, as in Category (X) axis or Value (Y) axis.

❻ In the Step 4 dialog box:
   a. Select the As new sheet: option button and enter Frequency Histogram in the edit box to the right of the option button. (See Figure 2E.8, panel E.)
   b. Click the Finish button to create the chart. (The floating Chart toolbar that may appear after the chart is created can be closed or dragged to the side of the application window.)

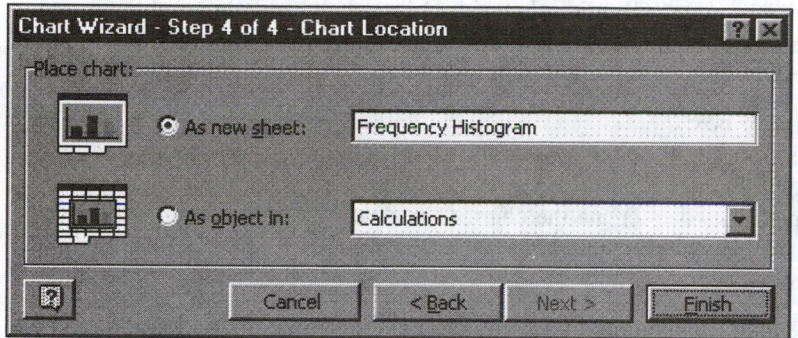

As was the case with the histogram created by the Data Analysis Histogram tool discussed earlier, there are gaps between the bars that need to be removed. To correct this error, do the following:

7 Right-click on one of the histogram bars. (The mouse pointer is over a bar when a tool tip that includes the word "Series" is displayed.)

8 Select Format Data Series from the shortcut menu.

9 In the Format Data Series dialog box (see Figure 2E.5 on page 83), click the Options tab and change the value in the Gap width: edit box to 0. Click OK.

The gaps between the bars of the histogram are now eliminated.

Example: Generating a frequency polygon using the Chart Wizard Many kinds of charts for numerical data can be generated using the Chart Wizard. For example, to generate a frequency polygon from the Calculations sheet distribution table implemented in Step 1, do the following:

1 Click the Calculations sheet tab.

2 Select Insert | Chart.

3 In the Step 1 dialog box:
  a. Select the Standard Types tab and then select Line from the Chart type: list box. Select the first choice in the second row of choices under the Chart sub-type: heading. When this choice is selected, the phrase "Line with markers displayed at each data value" is displayed in the description box below the sub-types.
  b. Click the Next button.

4 In the Step 2 dialog box:
  a. Select the Data Range tab. Enter Calculations!B4:B9 in the Data range: edit box and select the Columns option button in the Series in: group.
  b. Select the Series tab. Enter =Calculations!A4:A9 in the Category (X) axis labels: edit box. Note that this entry must include the leading equal sign (=).
  c. Click the Next button.

5 In the Step 3 dialog box:
  a. Select the Titles tab. Enter Growth Funds in the Chart title: edit box, enter 1-Year Return in the Category (X) axis: edit box, and enter Frequency in the Value (Y) axis: edit box.

b. Select, in turn, the Axes, Gridlines, Legend, Data Labels, and Data Table tabs and verify that their settings match those given in Table 2E.3 on page 88.

c. Click the Next button.

6 In the Step 4 dialog box:

a. Select the As new sheet: option button and enter Frequency Polygon in the edit box to the right of the option button.

b. Click the Finish button to create the chart. (The floating Chart toolbar that may appear after the chart is created can be closed or dragged to the side of the application window.)

(To generate a percentage polygon or a cumulative percentage polygon, we enter either Calculations!C4:C9 or Calculations!D4:D9 of the Data Range edit box in the Step 2 dialog box.)

The frequency polygon that is generated fails to connect to the horizontal X axis. To create a proper frequency polygon, we need to add two fictitious classes to the distribution table as explained in the Polygon Construction subsection on page 79. To add these fictitious classes, do the following:

7 Click on the Calculations sheet tab.

8 Select cell A4. Select Insert | Rows. An extra row is inserted above the first real class row.

9 Select cell A11 (containing the Total: label). Select Insert | Rows. An extra row is inserted after the last real class row.

10 Enter 0 (zero) in cells B4 and B11. (Cells A4 and A11 can be left blank.)

11 Click on the Frequency Polygon sheet tab.

12 Right-click in the white area surrounding the chart. (The mouse pointer is over this area when the tool tip "Chart Area" is displayed.)

13 Select Source Data from the shortcut menu.

14 In the Source Data dialog box:

a. Select the Series tab.

b. Change the contents of the Values: edit box to =Calculations!B4:B11.

c. Change the contents of the Category (X) axis labels: edit box to =Calculations!A4:A11.

d. Click the OK button.

The Frequency Polygon chart now properly connects with the X axis. A precise person might note that the category markings on the X axis refer to the upper limits of the classes, not the class midpoints as they should. To change these markings, do the following:

15 Enter the new class midpoint labels (22.5, 27.5, 32.5, 37.5, 42.5, 47.5) off to the side of the distribution table in the cell range F5:F10 of the Calculations sheet.

16 Right-click over the X axis. (The mouse pointer is over the axis when the tool tip phrase "Category Axis" is displayed.)

17 Select Format Axis from the shortcut menu.

18 In the Format Axis dialog box, select the Scale tab and select the Value (Y) Axis crosses between categories check box. Leave the other check boxes unselected. Click the OK button.

⑲ Right-click over the plot area. (The mouse pointer is over the plot area when the tool tip "Plot Area" appears.)

⑳ Select Source Data from the shortcut menu.

㉑ In the Source Data dialog box, select the Series tab and change the contents of the Category (X) axis labels edit box to =Calculations!F4:F11. Click the OK button.

The frequency polygon on the Frequency Polygon chart sheet changes. As before, if we wanted a percentage histogram, we would use the column of percentages instead of the column of frequencies.

WHAT IF ANALYSES and the SCENARIO MANAGER

In this analysis, the class maximum values are used as parameters that influence the statistical results produced. With the Calculations sheet completed, we might wonder how the distributions would look if we used an alternate set of class maximums such as 22.49, 27.49, 32.49, 37.49, 42.49, and 47.49. To discover the effects of these values on the distribution, we could enter the values in the cell range A4:A9 and note the changes in the distribution table results. Observing changes to results while changing the source data for a problem is commonly known as a "what if?" analysis.

The Scenario Manager feature of Microsoft Excel facilitates this type of analysis by allowing us to switch between alternate sets of data previously entered. To use the Scenario Manager for this problem, first complete the Calculations sheet, entering the original set of class maximums, and then do the following:

❶ Select Tools | Scenario.

❷ In the Scenario Manager dialog box
   a. Click the Add button. In the Add Scenario dialog box (see Figure 2E.9):
      i. Enter "Original set of classes" in the Scenario name: edit box.
      ii. Enter A4:A9 in the Changing cells: edit box. Enter A5:A10 if you followed the steps to generate a frequency polygon. Select the Prevent changes check box.
      iii. Click the OK button. The Scenario Values dialog box appears.
      iv. Immediately click the OK button. The original set of class maximum values is now defined as the scenario named "Original set of classes" and the Scenario Manager dialog box reappears.
   b. Click the Add button in the Scenario Manager dialog box a second time. In the Add Scenario dialog box:
      i. Enter "Alternate set of classes" in the Scenario name: edit box.
      ii. Enter A4:A9 in the Changing cells: edit box. (Or A5:A10 as noted above.)
      iii. Click the OK button.
      iv. In the Scenario Values dialog box, enter the alternate set of values in the edit boxes. (Use the dialog box scroll bars, if necessary to view all edit boxes.)
      v. Click OK. The alternate set of upper-class maximums is now defined as the scenario "Alternate set of classes" and the Scenario Manager dialog box reappears.
   c. Click OK (in the Scenario Manager dialog box).

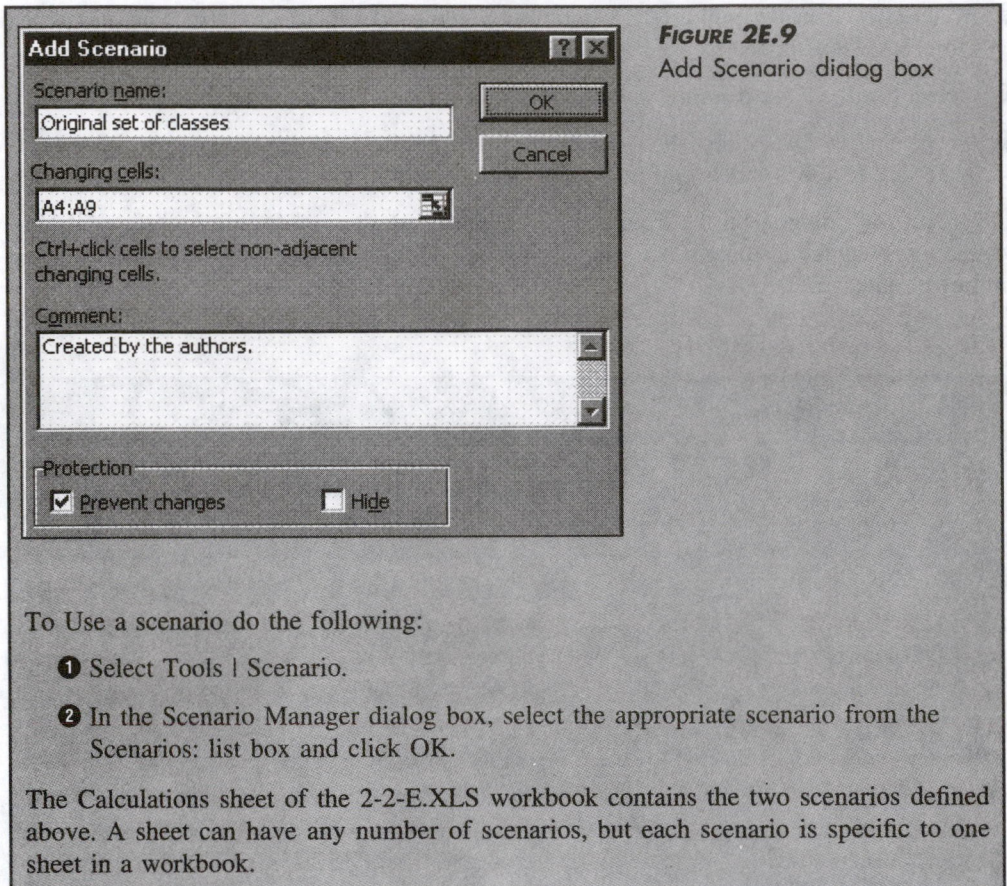

FIGURE 2E.9
Add Scenario dialog box

To Use a scenario do the following:

❶ Select Tools | Scenario.

❷ In the Scenario Manager dialog box, select the appropriate scenario from the Scenarios: list box and click OK.

The Calculations sheet of the 2-2-E.XLS workbook contains the two scenarios defined above. A sheet can have any number of scenarios, but each scenario is specific to one sheet in a workbook.

Problems for Section 2.2

Learning the Basics

2.13 The price-to-earnings ratios from a sample of 50 companies whose stocks are traded on the New York Stock Exchange range from 5.2 to 63.4.
  (a) What should be the width of each of your class intervals if you wanted to construct a frequency distribution having six class groupings?
  (b) For convenience, and ease of reading the table, how would you round off the width of the interval obtained in part (a), and what value would you assign to the lowest class boundary in the table you would be constructing?
  (c) On the basis of part (b), what would the class midpoints for each of the class groupings be?

2.14 A random sample of 50 executive vice presidents is selected from various public universities in the United States, and the annual salaries of these officials are obtained. The salaries range from $62,000 to $247,000. Set up the class boundaries for a frequency distribution
  (a) if five class intervals are desired.
  (b) if six class intervals are desired.
  (c) if seven class intervals are desired.
  (d) if eight class intervals are desired.

• **2.15** The asking price of one-bedroom cooperative and condominium apartments in Queens, a borough of New York City, varies from $103,000 to $295,000.
  (a) Indicate the class boundaries of 10 classes into which these values can be grouped.
  (b) What class-interval width did you choose?
  (c) What are the 10 class midpoints?

2.16 A frequency distribution is constructed on the basis of a sample of 40 observations. The most typical, or *modal*, class grouping contains 12 observations.
  (a) When the corresponding relative frequency distribution is constructed, what value will appear in the table for this class grouping?
  (b) When the corresponding percentage distribution is constructed, what value will appear in the table for this class grouping?

2.17 A percentage distribution is constructed on the basis of a sample of 60 observations. The value that appears in the table for the first class grouping is 20%.
  (a) When the corresponding relative frequency distribution is constructed, what value will appear in the table for this class grouping?
  (b) When the corresponding frequency distribution is constructed, what value will appear in the table for this class grouping?

2.18 In differentiating the histogram from the polygon:
  (a) Which of the two diagrams is plotted at the class midpoints from a frequency distribution?
  (b) Which of the two diagrams is plotted at the class boundaries from a frequency distribution?
  (c) Which of the two diagrams contains a series of vertical rectangular bars?
  (d) Which of the two diagrams results from connecting a set of consecutive plotted points?
  (e) Which of the two diagrams can be used for comparing two or more sets of data that have been tallied into corresponding frequency distributions?

• **2.19** In constructing either a histogram or polygon, on which axis (vertical or horizontal) must the true zero, or "origin," be shown so as not to visually distort or otherwise misrepresent the characteristics of the data?

• **2.20** In constructing an ogive, on which axis (vertical or horizontal) must the true zero, or "origin," be shown so as not to visually distort or otherwise misrepresent the characteristics of the data?

2.21 In constructing a percentage ogive (i.e., a cumulative percentage polygon):
  (a) What must be the largest value denoted on the scale along the vertical axis?
  (b) What value does the first plotted point on the horizontal axis represent?

2.22 In constructing a percentage ogive (i.e., a cumulative percentage polygon) pertaining to the GMAT scores from a sample of 50 applicants to an MBA program, it was noted from a previously obtained frequency distribution that none of the applicants scored below 450 and that the frequency distribution was formed by choosing class intervals $450 < 500$, $500 < 550$, and so on, with the last class grouping $700 < 750$. If two applicants scored in the interval $450 < 500$ and 16 applicants scored in the interval $500 < 550$:
  (a) What percentage of applicants scored below 500?
  (b) What percentage of applicants scored below 550?
  (c) What percentage of applicants scored between 500 and 550?
  (d) How many applicants scored between 500 and 550?
  (e) What percentage of applicants scored below 750?

Applying the Concepts

• **2.23** The data displayed here represent the electricity cost during the month of July 1997 for a random sample of 50 two-bedroom apartments in a large city.

Raw Data on Utility Charges ($)

96	171	202	178	147	102	153	197	127	82
157	185	90	116	172	111	148	213	130	165
141	149	206	175	123	128	144	168	109	167
95	163	150	154	130	143	187	166	139	149
108	119	183	151	114	135	191	137	129	158

(a) Form a frequency distribution
  (1) having five class intervals.
  (2) having six class intervals.
  (3) having seven class intervals.
[*Hint:* To help you decide how best to set up the class boundaries, you should first place the raw data either in a stem-and-leaf display (by letting the leaves be the trailing digits) or in an ordered array.]

(b) Form a frequency distribution having seven class intervals with the following upper class limits: $79, $99, $119, and so on.

(c) Form the percentage distribution from the frequency distribution developed in part (b).

(d) Plot the percentage histogram.

(e) Plot the percentage polygon.

(f) Form the cumulative frequency distribution.

(g) Form the cumulative percentage distribution.

(h) Plot the ogive (cumulative percentage polygon).

(i) Around what amount does the monthly electricity cost seem to be concentrated?

(j) Which of the graphs do you think is best in presenting the distribution of electricity cost? Explain.

2.24 Given the ordered arrays in the accompanying table dealing with the lengths of life (in hours) of a sample of forty 100-watt light bulbs produced by manufacturer A and a sample of forty 100-watt light bulbs produced by manufacturer B:

Manufacturer A					Manufacturer B				
684	697	720	773	821	819	836	888	897	903
831	835	848	852	852	907	912	918	942	943
859	860	868	870	876	952	959	962	986	992
893	899	905	909	911	994	1,004	1,005	1,007	1,015
922	924	926	926	938	1,016	1,018	1,020	1,022	1,034
939	943	946	954	971	1,038	1,072	1,077	1,077	1,082
972	977	984	1,005	1,014	1,096	1,100	1,113	1,113	1,116
1,016	1,041	1,052	1,080	1,093	1,153	1,154	1,174	1,188	1,230

(a) Form the frequency distribution for each brand. (*Hint:* For purposes of comparison, choose class-interval widths of 100 hours for each distribution.)

(b) For purposes of answering part (d), form the frequency distribution for each brand according to the following schema [if you have not already done so in part (a) of this problem]:
  (1) Manufacturer A: 650 but less than 750, 750 but less than 850, and so on.
  (2) Manufacturer B: 750 but less than 850, 850 but less than 950, and so on.

(c) Change the class-interval width in (b) to 50 so that you have intervals from 650 to under 700, 700 to under 750, 750 to under 800, and so on. Comment on the results of these changes.

(d) Form the percentage distributions from the frequency distributions developed in (b).

(e) Plot the percentage histograms on separate graphs.

(f) Plot the percentage polygons on one graph.

(g) Form the cumulative frequency distributions.

(h) Form the cumulative percentage distributions.

(i) Plot the ogives (cumulative percentage polygons) on one graph.

(j) Which manufacturer has bulbs with a longer life, manufacturer A or manufacturer B? Explain.

•2.25 Given the following stem-and-leaf display representing the amount of gasoline purchased in gallons (with leaves in 10ths of gallons) for a sample of 25 cars that use a particular service station on the New Jersey Turnpike:

```
 9 | 714
10 | 82230
11 | 561776735
12 | 394282
13 | 20
```

(a) Construct a frequency distribution and a percentage distribution.
(b) Form the cumulative frequency distribution and the cumulative percentage distribution.
(c) Plot the percentage histogram.
(d) Plot the percentage polygon.
(e) Plot the ogive (cumulative percentage polygon).
(f) Around what amount of gasoline do most of the purchases seem to be concentrated?

2.26 The following data are the book values (in dollars, i.e., net worth divided by number of outstanding shares) for a random sample of 50 stocks from the New York Stock Exchange:

7	9	8	6	12	6	9	15	9	16
8	5	14	8	7	6	10	8	11	4
10	6	16	5	10	12	7	10	15	7
10	8	8	10	18	8	10	11	7	10
7	8	15	23	13	9	8	9	9	13

DATA FILE
STOCK.XLS

(a) Construct a frequency distribution and a percentage distribution.
(b) Form the cumulative frequency distribution and the cumulative percentage distribution.
(c) Plot the percentage histogram.
(d) Plot the percentage polygon.
(e) Plot the ogive (cumulative percentage polygon).
(f) Are there any book values that seem to be more likely to occur than others? Explain.

2.27 The following data represent the number of cases of salad dressing purchased per week by a local supermarket chain over a period of 30 weeks:

WEEK	CASES PURCHASED	WEEK	CASES PURCHASED	WEEK	CASES PURCHASED
1	81	11	86	21	91
2	61	12	133	22	99
3	77	13	91	23	89
4	71	14	111	24	96
5	69	15	86	25	108
6	81	16	84	26	86
7	66	17	131	27	84
8	111	18	71	28	76
9	56	19	118	29	83
10	81	20	88	30	76

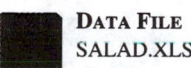

DATA FILE
SALAD.XLS

(a) Construct a stem-and-leaf display.
(b) Construct the frequency distribution and the percentage distribution.
(c) Plot the percentage histogram.

(d) Plot the percentage polygon.

(e) Form the cumulative percentage distribution.

(f) Plot the cumulative percentage polygon.

(g) On the basis of the results of (a)–(f), does there appear to be any concentration of the number of cases of salad dressing ordered by the supermarket chain around specific values?

(h) If you had to make a prediction of the number of cases of salad dressing that would be ordered next week, how many cases would you predict? Why?

2.28 The following data represent the amount of soft drink filled in a sample of 50 consecutive 2-liter bottles. The results, listed horizontally in the order of being filled, were:

2.109	2.086	2.066	2.075	2.065	2.057	2.052	2.044	2.036	2.038
2.031	2.029	2.025	2.029	2.023	2.020	2.015	2.014	2.013	2.014
2.012	2.012	2.012	2.010	2.005	2.003	1.999	1.996	1.997	1.992
1.994	1.986	1.984	1.981	1.973	1.975	1.971	1.969	1.966	1.967
1.963	1.957	1.951	1.951	1.947	1.941	1.941	1.938	1.908	1.894

(a) Construct a stem-and-leaf display.

(b) Construct the frequency distribution and the percentage distribution.

(c) Plot the frequency histogram.

(d) Plot the percentage polygon.

(e) Form the cumulative percentage distribution.

(f) Plot the cumulative percentage polygon.

(g) On the basis of the results of (a)–(f), does there appear to be any concentration of the amount of soft drink filled in the bottles around specific values?

(h) If you had to make a prediction of the amount of soft drink filled in the next bottle, what would you predict? Why?

2.3 TABLES AND CHARTS FOR CATEGORICAL DATA

Thus far in this chapter we have learned that when collecting a large set of numerical data, the best way to examine it is first to organize and present it in appropriate tabular and chart format. Often, however, the data we collect are categorical, not numerical. Thus, in this and the following section we will demonstrate how categorical data can be organized and presented in the form of tables and charts.

In order to do this, let us suppose once again that we want to evaluate various features pertaining to domestic general stock funds. In addition to net asset value and the two performance measures (1-year total percentage return and 3-year annualized total return) and fund objective (growth versus blend), the data set in appendix D displays information on fee schedule and fund size. We note that net asset value (in dollars) and the two performance measure variables (in percentages) are *numerical,* whereas the fund objective variable along with fee schedule and fund size are *categorical.* Earlier in this chapter we were concerned only with the tabular and chart presentation of numerical data. A detailed study of the responses to the categorical variables will be undertaken here.

When dealing with categorical phenomena, we may tally the observations into summary tables and then graphically display them as bar charts, pie charts, or Pareto diagrams.

The Summary Table

A **summary table** for categorical data is similar in format to the frequency distribution table for numerical data that we studied in section 2.2. To illustrate the development of a summary table, let us consider the data obtained on fee schedule. By tallying the observations from Special Data Set 1 in appendix D, we see that of the 194 sampled domestic general stock funds, 17 have fees covering marketing costs paid from fund assets, 5 have deferred sales charges or redemption fees, 19 have "front-load" sales charges, 46 are classified as charging multiple fees, and 107 are categorized as "no-load" funds (i.e., there are no fee charges). This information is presented in Table 2.7.

Table 2.7 *Frequency and percentage summary table pertaining to fee schedule for 194 domestic general stock funds*

FEE SCHEDULE	NUMBER OF FUNDS	PERCENTAGE OF FUNDS
Fees from fund assets	17	8.8
Deferred fees	5	2.6
Front-load fees	19	9.8
Multiple fees	46	23.7
No-load funds	107	55.2
Total	194	100.1[a]

[a] Error due to rounding.
Source: Data are taken from Special Data Set 1 in appendix D.

DATA FILE
2E.5.XLS

The Bar Chart

To express the information provided in Table 2.7 graphically, a frequency bar chart can be displayed. Figure 2.5 depicts a frequency bar chart obtained from Microsoft Excel for the fee schedule data presented in Table 2.7. In **bar charts,** each category is depicted by a bar, the length of which represents the frequency or percentage of observations falling into a category. From Figure 2.5 we observe that the bar chart allows us to directly compare the percentage of funds in terms of the type of fee schedule. More than half the funds are no-load funds, and almost 25% of the funds are multiple-fee funds. Very few funds receive fees from fund assets, deferred fees, or front-load fees.

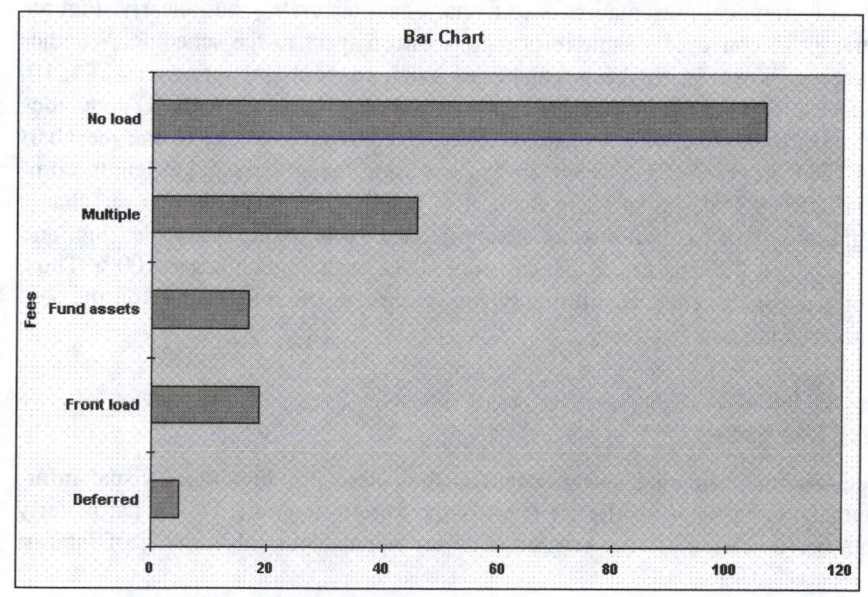

FIGURE 2.5
Frequency bar chart pertaining to fee schedule for 194 sampled domestic general stock funds obtained from Microsoft Excel
Source: Data are taken from Table 2.7.

The Pie Chart

Another widely used graphical display to visually express categorical data from a summary table is the **pie chart.** Figure 2.6 depicts a percentage pie chart for the fee schedule data presented in Table 2.7.

FIGURE 2.6
Percentage pie chart pertaining to fee schedule for 194 sampled domestic general stocks obtained from Microsoft Excel

Source: Data are taken from Table 2.7 on page 97.

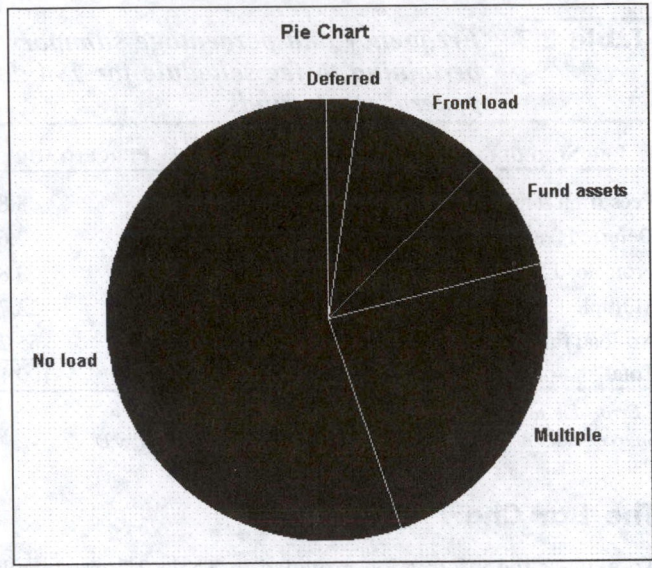

The pie chart is based on the fact that the circle has 360°. The pie is divided into slices according to the percentage in each category. As an example, in Table 2.7, 23.7% of the domestic general stock funds sampled are classified as charging multiple fees. Thus, in constructing the pie chart, 360° is multiplied by .237, resulting in a sector that takes up 85° of the 360° of the circle. From Figure 2.6 we observe that the pie chart lets us visualize the portion of the entire pie that is in each category. We can see that no-load funds take up more than half the pie and multiple-fee funds comprise almost one-quarter.

The purpose of graphical presentation is to display data accurately and clearly. Figures 2.5 and 2.6 attempt to convey the same information with respect to fee schedule. Whether these charts succeed, however, has been a matter of much concern (see references 2–4, 10, 11). In particular, some research in the human perception of graphs (reference 4) concludes that the pie chart presents the weaker display. The bar chart is preferred to the pie chart because it has been observed that the human eye can more accurately judge length comparisons against a fixed scale (as in a bar chart) than angular measures (as in a pie chart). Nevertheless, the pie chart has two distinct advantages: (1) it is aesthetically pleasing, and (2) it clearly shows that the total for all categories or slices of the pie comes to 100%. Thus, the selection of a particular chart is still highly subjective and often dependent on the aesthetic preferences of the user.

The Pareto Diagram

A graphical device for portraying categorical data that often provides more visual information than either the bar chart or the pie chart is the Pareto diagram. This is particularly true as the number of classifications or groupings for our categorical variable of interest increases. The **Pareto diagram** is a special type of vertical bar chart in which the catego-rized responses are plotted in the descending rank order of their frequencies and combined with a cumulative polygon on the same scale. The main principle behind this graphical device is its ability to separate the "vital few" from the "trivial many," enabling us to focus on the important responses. Hence, the chart achieves its greatest utility when the categor-ical variable of interest contains many categories. The Pareto diagram is widely used in the statistical control of process and product quality (see chapter 12).

To illustrate the Pareto diagram, we observe that in Figure 2.5 the bar chart pertaining to fee schedule presents the categories as fees taken from fund assets, deferred fees, front-load fees, multiple fees, and no fees (i.e., no-load funds). Because no-load funds dominate, a Pareto diagram may be formed by changing the ordering. Such a plot obtained from the PHStat add-in for Excel is depicted in Figure 2.7. From the lengths of the vertical bars, we observe that five out of every nine of these funds do not charge fees and are classified as no-load funds (code 5). From the cumulative polygon we note that 78.9% of these funds are classified either as no-load or as charging multiple fees (code 4).

	A	B	C	D	E	F
1	One-Year Total Percentage Returns					
2						
3	Count of Fees					
4	Fees	Total	Fees	Frequency	Percentage	Cumulative Pct.
5	Deferred	5	No load	107	55.15%	55.15%
6	Front load	19	Multiple	46	23.71%	78.87%
7	Fund assets	17	Front load	19	9.79%	88.66%
8	Multiple	46	Fund assets	17	8.76%	97.42%
9	No load	107	Deferred	5	2.58%	100.00%
10	Grand Total	194				

FIGURE 2.7

Pareto diagram depicting fee schedule for 194 sampled domestic general stock funds obtained from the PHStat add-in for Microsoft Excel

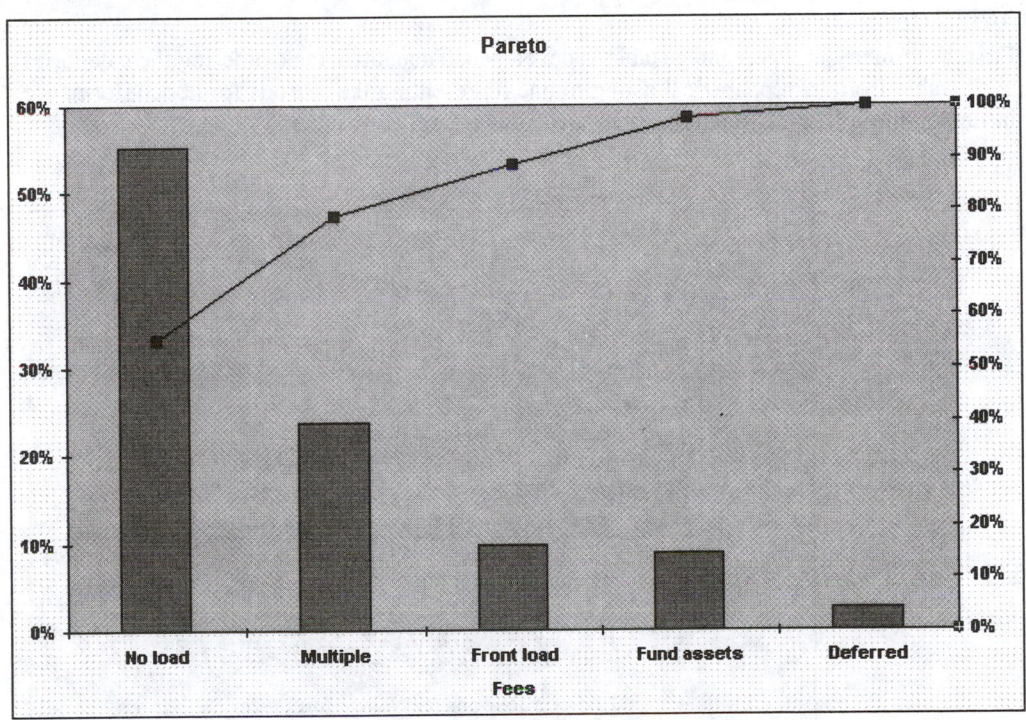

71

In the construction of the Pareto diagram, the vertical axis on the left contains the frequencies or percentages and the vertical axis on the right contains the cumulative percentages (from 100 on top to 0 on bottom) and the horizontal axis contains the categories of interest. The equally spaced bars are of equal width. The point on the cumulative percentage polygon for each category is centered at the midpoint of each respective bar. Hence, when studying a Pareto diagram, we should be focusing on two things: the magnitudes of the differences in bar lengths corresponding to adjacent descending categories and the cumulative percentages of these adjacent categories.

The Pareto diagram is a very useful tool for presenting categorical data, particularly when the number of classifications or groupings increases. To further demonstrate its value in these situations, we turn to Example 2.3, which is an application in operations management.

Example 2.3   *The Pareto Diagram*

The operations manager at a cereal packaging plant said that in her experience, typically there are nine reasons that result in the production of unacceptable cereal cartons at the end of the packaging process: broken carton (R), bulging carton (G), cracked carton (C), dirty carton (D), hole in carton (H), improper package weight (I), printing error (P), unreadable label (U), and unsealed box top (S).

The raw data below represent a sample of 50 unacceptable cereal cartons taken from the past week's production, and the reasons for nonconformance are indicated:

U G U S H D D R I U S U S U G C S U D R S U D U S
S D P R S I S U D G S S U S D G S C U D D S S S U

Using these data, construct the Pareto diagram.

SOLUTION

First, a summary and percentage table is developed. Then, after rank-ordering the categories in the table from its alphabetical listing to one representing the reasons for nonconformance in descending order of importance, a Pareto diagram is constructed.

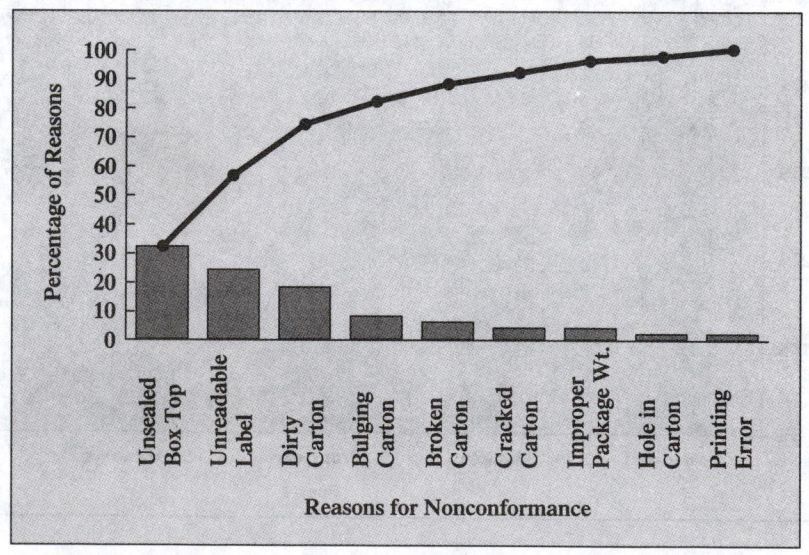

Summary and percentage table of reasons for nonconformance

REASON FOR NONCONFORMANCE	No.	%
Broken carton (R)	3	6.0
Bulging carton (G)	4	8.0
Cracked carton (C)	2	4.0
Dirty carton (D)	9	18.0
Hole in carton (H)	1	2.0
Improper package weight (I)	2	4.0
Printing error (P)	1	2.0
Unreadable label (U)	12	24.0
Unsealed box top (S)	16	32.0
Total	50	100.0

Summary and percentage table of ranked reasons for nonconformance

REASON FOR NONCONFORMANCE	No.	%
Unsealed box top (S)	16	32.0
Unreadable label (U)	12	24.0
Dirty carton (D)	9	18.0
Bulging carton (G)	4	8.0
Broken carton (R)	3	6.0
Cracked carton (C)	2	4.0
Improper package weight (I)	2	4.0
Hole in carton (H)	1	2.0
Printing error (P)	1	2.0
Total	50	100.0

Separating the "vital few" from the "trivial many," we may determine that unsealed box tops (32.0%), unreadable labels (24.0%), and dirty cartons (18.0%) together account for 74.0% of the reasons for nonconformance. The remaining six reasons account for 26.0%.

 **2.3E** **GENERATING TABLES AND CHARTS FOR CATEGORICAL DATA USING MICROSOFT EXCEL**

Overview

◆ *For Quick Results Users* Use the One-Way Tables & Charts choice of the PHStat add-in to tabulate and graph categorical data.

◆ *For Developers*   Use the following two-step process to tabulate and graph categorical data:

    A. Use the PivotTable Wizard to generate a summary table of the categorical data.

    B. Use the Chart Wizard to generate charts based on this table.

The 2E-3.XLS workbook file contains frequency tables and several charts based on the fee schedule data for 194 domestic general stock funds sample used in section 2E.2.

Quick Results

To generate a frequency table, bar chart, and pie chart from the mutual funds fee schedule data (Table 2.7 on page 97), do the following:

❶ If the PHStat add-in has not been previously loaded, load the add-in using the instructions of section S4.2 on page 59.

❷ Open the Mutual Funds Sample workbook (MUTUAL FUNDS SAMPLE.XLS) and click the Data sheet tab.

❸ Select PHStat | One-Way Tables & Charts.

❹ In the One-Way Tables & Charts dialog box (see Figure 2E.10):
    a. Enter E1:E195 in the Cell Range: edit box.
    b. Select the First cell contains label check box.
    c. Enter Fee Schedules in the Output Title: edit box.
    d. Select the Bar Chart, Pie Chart, and Pareto Diagram check boxes.
    e. Click the OK button

On separate sheets the add-in produces a frequency table and a bar chart, pie chart, and Pareto diagram similar to those shown in Figures 2.5, 2.6, and 2.7, respectively.

Developer Details

◆ *A. Generating Tables Using the PivotTable Wizard*   We can use the Microsoft Excel PivotTable Wizard to tabulate categorical variables in summary tables called PivotTables. Unlike the contents of manually entered summary tables, the contents of a PivotTable dynamically changes as the data on which it is based changes. To use the PivotTable Wizard, follow this general procedure:

❶ Click the tab of the worksheet containing the source data for the PivotTable.

❷ Select Data | PivotTable Report.

❸ In the Step 1 dialog box, select the Microsoft Excel list or database option button. Click the Next button.

❹ In the Step 2 dialog box, enter the cell range containing the data to be tabulated.

❺ In the Step 3 dialog box, specify the design of the summary table.

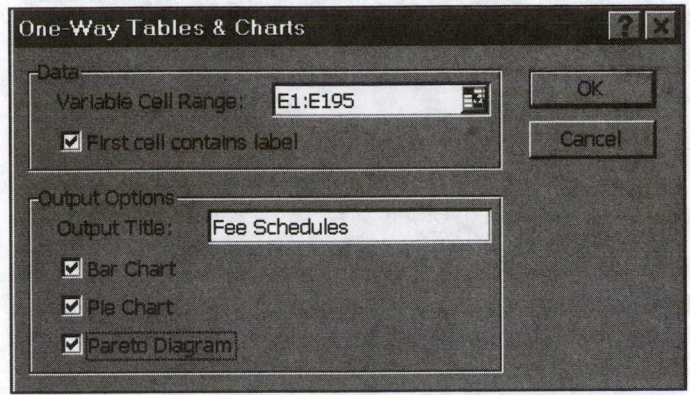

FIGURE 2E.10
One-Way Tables & Charts dialog box

6 In the Step 4 dialog box, select the New worksheet option button and click the Options button to display the PivotTable Options dialog box.

7 In the PivotTable Options dialog box, make selections and enter information, including a worksheet name in the Name: edit box, as necessary. Click the OK button to return to the Step 4 dialog box.

8 Click the Finish button.

Example: Generating a fee schedule summary table To produce a fee schedule summary table similar to Table 2.7 on page 97 from the sample of 194 domestic general stock funds, do the following:

1 Open the Mutual Funds Sample workbook (MUTUAL FUNDS SAMPLE.XLS) and click the Data sheet tab.

2 Select Data | PivotTable Report.

3 In the Step 1 dialog box, select the Microsoft Excel list or database option button. (See Figure 2E.11, panel A on page 104.) Click the Next button.

4 In the Step 2 dialog box, enter A1:G195 in the Range: edit box. (See Figure 2E.11, panel B on page 104.) Click Next.

5 In the Step 3 dialog box:
  a. Verify that the label boxes on the right of the dialog box correspond to the column headings on the Data sheet (Name, Type, NAV, etc.). If they do not, click Back and verify that the proper range has been entered in the Step 2 dialog box.
  b. Drag the Fees label and drop it in the Row box. A copy of the Fees label snaps into the Row box.
  c. Drag the Name label box and drop it into the Data box. Verify that the wording on the label has changed to Count of Name. (See Figure 2E.11, panel C on page 105.)
  d. If the label reads something other than Count of Name, double-click on the label to display the PivotTable Field dialog box. In this dialog box:
    i. Select Count from the Summarize by: list box.
    ii. Click the Options button.

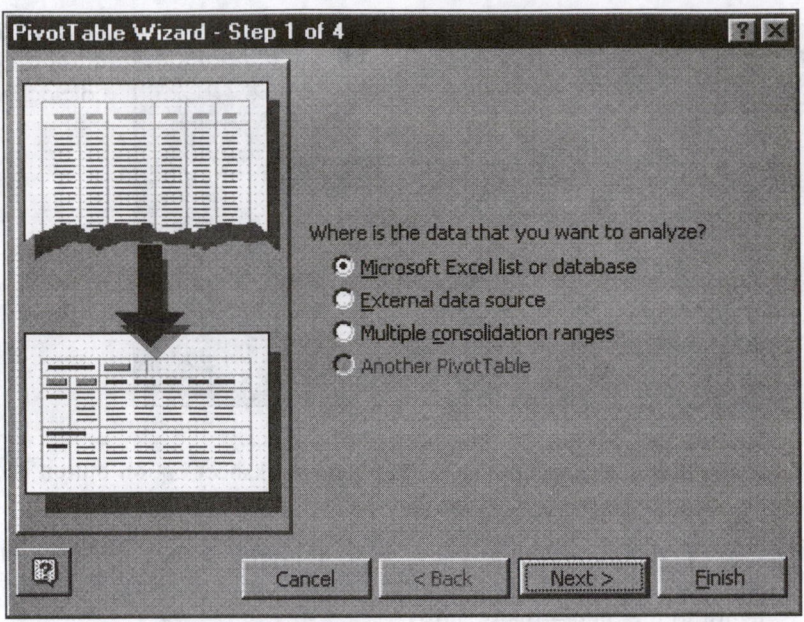

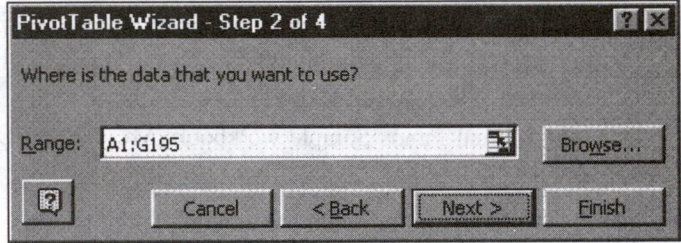

   iii. Select Normal in the Show data as: drop-down list box. (See Figure 2E.11, panel D.)

   iv. Click the OK button to redisplay the Step 3 dialog box.

  e. Click the Next button.

6 In the Step 4 dialog box (see Figure 2E.11, panel E on page 106):

  a. Select the New worksheet option button.

  b. Click the Options button.

  c. In the PivotTable Options dialog box (see Figure 2E.11, panel F on page 106):

    i. Enter OneWayTable in Name: edit box.

    ii. Select the Grand totals for columns and AutoFormat table check boxes of the Format options group. Deselect the Grand totals for rows and Preserve formatting check boxes, if selected.

    iii. Select the For empty cells, show: check box and enter 0 (zero) in its edit box.

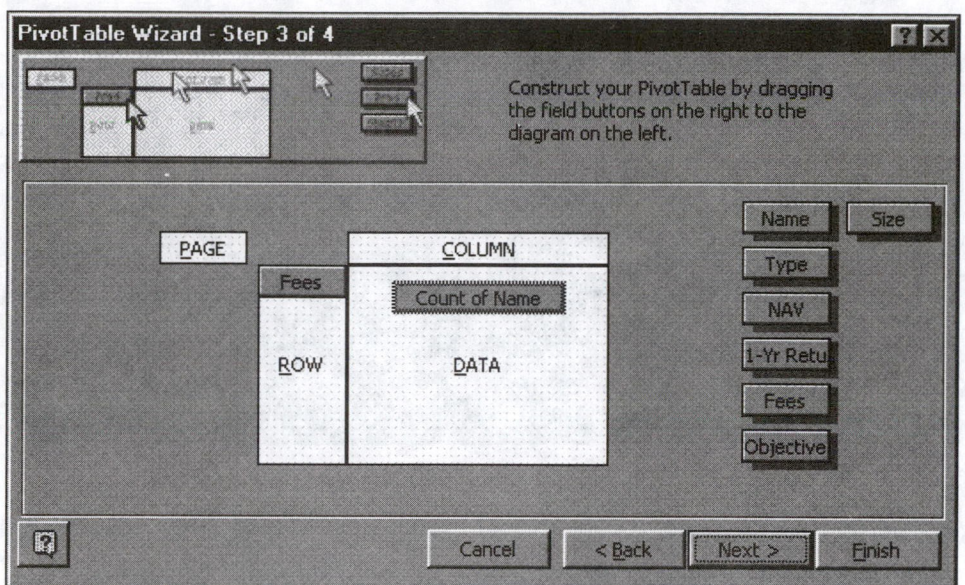

FIGURE 2E.11 Panel C: PivotTable Wizard Step 3 dialog box

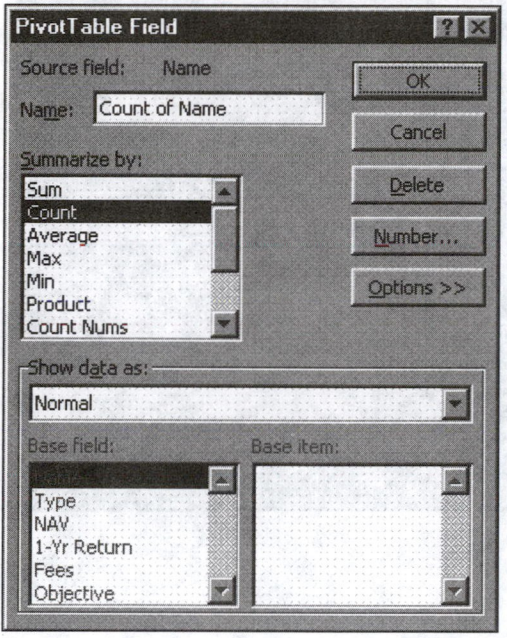

FIGURE 2E.11
Panel D: PivotTable Field
dialog box

    iv. Select the Save data with table layout and Enable drilldown option buttons
       under the Data source options: heading.
    v. Click the OK button to return to the Step 4 dialog box.
  d. Click the Finish button.

FIGURE 2E.11
Panel E: PivotTable Wizard
Step 4 dialog box

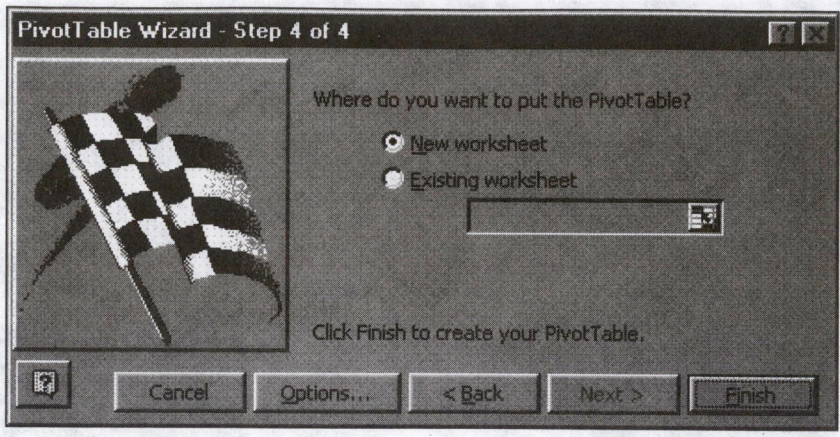

FIGURE 2E.11
Panel F: PivotTable Options
dialog box

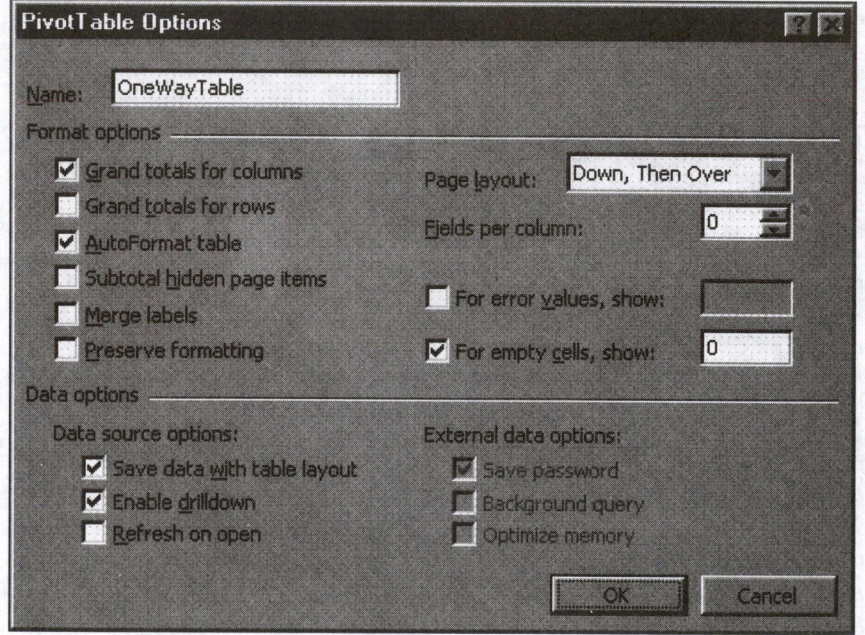

❼ Rename the new worksheet OneWayTable. (The floating PivotTable toolbar that may appear after the table is created can be closed or dragged to the side of the application window.)

❽ Select cell C1. Select Insert | Rows two times to insert two empty rows above the PivotTable. Enter a title in the new row 1.

The PivotTable produced will be similar to the one shown in Figure 2.7 on page 99.

◆ **B. Generating Charts from a Summary Table**   Having generated the PivotTable that tabulates the domestic growth funds by fund objective, we can use the Chart Wizard to generate bar and pie charts. To generate a bar chart from the PivotTable, do the following:

❶ Click the OneWayTable sheet tab.

❷ Select Insert | Chart.

❸ In the Step 1 dialog box:
   a. Select the Standard Types tab and then select Bar from the Chart type: list box. Select the first choice in the first row of choices under the Chart sub-type: heading. When this choice is selected, "Clustered Bar" is displayed in the description box below the sub-types.
   b. Click the Next button.

❹ In the Step 2 dialog box:
   a. Select the Data Range tab and enter OneWayTable!B5:B9 in the Data range: edit box and select the Columns option button in the Series in: group.
   b. Select the Series tab. Enter =OneWayTable!A5:A9 In the Category (X) axis labels: edit box. Note that this entry must include the leading equal sign (=).
   c. Click the Next button.

❺ In the Step 3 dialog box:
   a. Select the Titles tab. Enter Mutual Funds in the Chart title: edit box, enter Fee Schedule in the Category (X) axis: edit box, and enter Frequency in the Value (Y) axis: edit box.
   b. Select, in turn, the Axes, Gridlines, Legend, Data Labels, and Data Table tabs and verify that their settings match those given in Table 2E.3 on page 88.
   c. Click the Next button.

❻ In the Step 4 dialog box:
   a. Select the As new sheet: option button and enter Bar Chart in the edit box to the right of the option button.
   b. Click the Finish button to create the chart. (The floating Chart toolbar that may appear after the chart is created can be closed or dragged to the side of the application window.)

To generate a pie chart from the PivotTable, do the following:

❶ Click the OneWayTable sheet tab.

❷ Select Insert | Chart.

❸ In the Step 1 dialog box:
   a. Select the Standard Types tab and then select Pie from the Chart type: list box. Select the first choice in the first row of choices under the Chart sub-type: heading. When this choice is selected, "Pie" is displayed in the description box below the sub-types.
   b. Click the Next button.

❹ In the Step 2 dialog box:
   a. Select the Data Range tab and enter OneWayTable!B5:B9 in the Data range: edit box and select the Columns option button in the Series in: group.

b. Select the Series tab. Enter =OneWayTable!A5:A9 in the Category labels: edit box. Note that this entry must include the leading equal sign (=).

c. Click the Next button.

❺ In the Step 3 dialog box:

a. Select the Titles tab. Enter Mutual Funds Sample in the Chart title: edit box.

b. Select the Legend tab and verify that the Show legend check box is deselected (unchecked).

c. Select the Data Labels tab. Select the Show label and percent option button and the Show leader lines check box.

d. Click the Next button.

❻ In the Step 4 dialog box:

a. Select the As new sheet: option button and enter Pie Chart in the edit box to the right of the option button.

b. Click the Finish button to create the chart. (The floating Chart toolbar that appears after the chart is created can be closed or dragged to the side of the application window.)

Special Case: Generating a Pareto Diagram

Generating a Pareto diagram first requires that the frequency percentages and cumulative percentages be added to the distribution table created earlier. Table 2E.4 shows the design of a modified OneWayTable sheet containing these additions:

Table 2E.4 Additions to the OneWayTable sheet for generating a Pareto diagram

	A	B	C	D	E	F
1	One-Way Summary Table for Fee Schedule Data					
2						
3						
4	Range		Range		Percentage	Cumulative Percentage
5	of the		of the		=D5/$B$10	=E5
6					=D6/$B$10	=F5+E6
7	original		copy of the		=D7/$B$10	=F6+E7
8					=D8/$B$10	=F7+E8
9	PivotTable		original		=D9/$B$10	=F8+E9
10			PivotTable			

To implement these additions, do the following:

❶ Click the OneWayTable sheet tab.

❷ Select cell A3. This selects the entire PivotTable.

❸ Select Edit | Copy. Select cell C3. Select Edit | Paste. A copy of the PivotTable now occupies the range C3:D10.

❹ Select cell C3. Change the value of this cell to Count of Name Sorted.

❺ Right-click cell C3. Select Refresh Data from the shortcut menu.

❻ Select cell C4 and then right-click this cell.

❼ Select Field from the shortcut menu.

❽ In the PivotTable Field dialog box (see Figure 2E.12, panel A):
  a. Click the Advanced button. In the PivotTable Field Advanced Option dialog box (see Figure 2E.12, panel B):
     i. Select the Descending option button of the AutoSort options group.
     ii. In the Using field: list box of the AutoSort options group, select Count of Name Sorted.
     iii. Click the OK button to return to the PivotTable Field dialog box.
  b. Click the OK button (of the PivotTable Field dialog box).

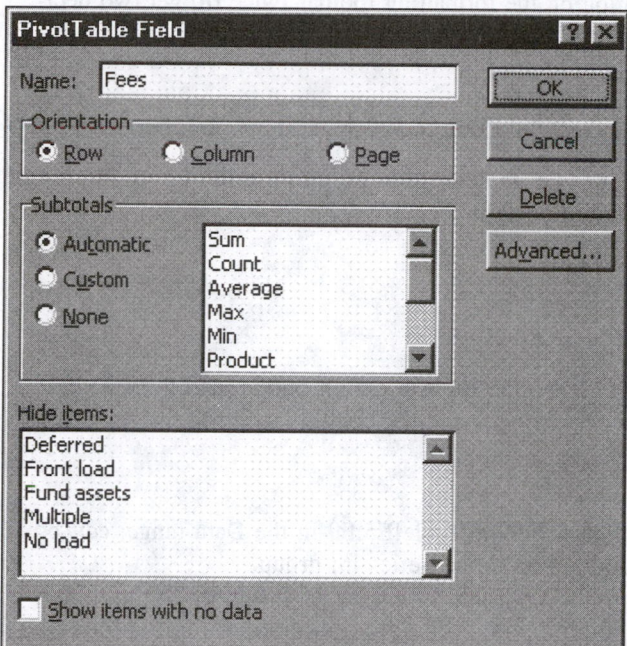

FIGURE 2E.12
Panel A: PivotTable Field dialog box

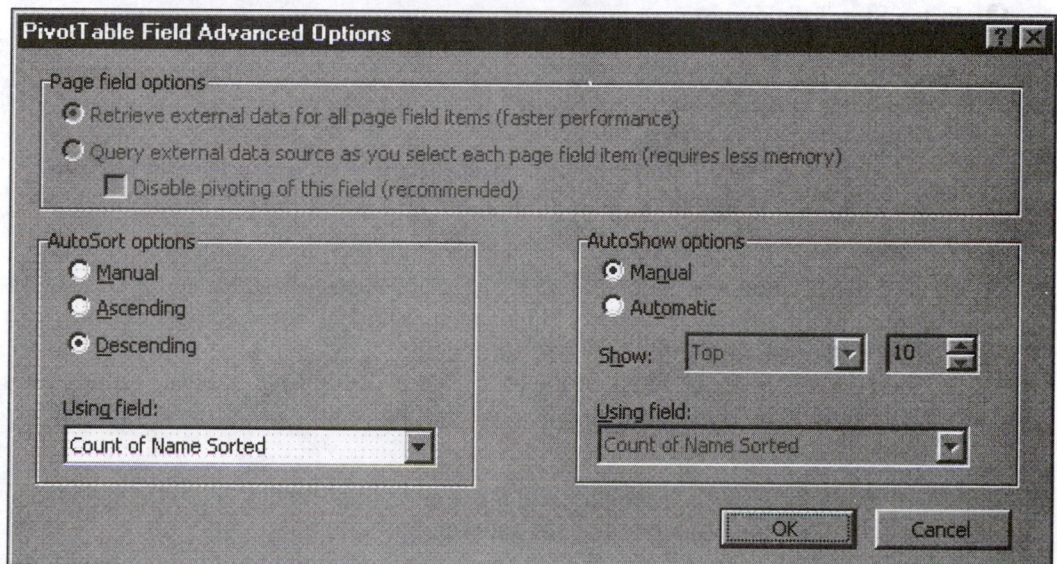

FIGURE 2E.12 Panel B: PivotTable Field Advanced Options dialog box

The frequencies in column D are now sorted in descending order. Continue by entering the formulas for the percentages and cumulative percentages.

⑨ Enter the column headings for cells E4 and F4.

⑩ Enter the formula =D5/$D$10 in cell E5 and copy this formula down through cell E9.

⑪ Select the cell range E5:E9.

⑫ Click the Percent button on the formatting toolbar.

⑬ Click the Increase Decimal button on the formatting toolbar twice (to get two decimal places).

⑭ Enter the formula =E5 in cell F5.

⑮ Enter the formula =F5+E6 into cell F6. Copy this formula down through cell F9.

With these additions to the OneWayTable implemented, we can use the Chart Wizard to produce a Pareto diagram, using the following procedure:

❶ Click the OneWayTable sheet tab.

❷ Select Insert | Chart.

❸ In the Step 1 dialog box:
   a. Select the Custom Types tab. Select the Built-in option button, and then select Line-Column on 2 Axes from the Chart type: list box.
   b. Click the Next button.

❹ In the Step 2 dialog box:
   a. Select the Data Range tab. Enter OneWayTable!E5:F9 in the Data range: edit box and select the Columns option button in the Series in: group.
   b. Select the Series tab. Enter =OneWayTable!C5:C9 in the Category (X) axis labels: edit box.
   c. Click the Next button.

❺ In the Step 3 dialog box:
   a. Select the Titles tab. Enter Pareto Diagram in the Chart title: edit box, and enter Fee Schedule in the Category (X) axis: edit box.
   b. Select the Axes tab. Select the Primary axis Category (X) axis check box and both Value (Y) axis check boxes.
   c. Select, in turn, the Gridlines, Legend, Data Labels, and Data Table tabs and verify that their settings match those given in Table 2E.3 on page 88.
   d. Click the Next button.

❻ In the Step 4 dialog box:
   a. Select the As new sheet: option button and enter Pareto in the edit box to the right of the option button.
   b. Click the Finish button to create the chart. (The floating Chart toolbar that appears after the chart is created can be closed or dragged to the side of the application window.)

The Chart Wizard produces a Pareto diagram that contains a secondary Y axis that wrongly exceeds 100%. To correct this error, do the following:

❼ Right-click when the mouse pointer is over the secondary Y axis. The tool tip phrase "Secondary Value Axis" appears when the mouse pointer is over the axis.

❽ Select Format Axis from the shortcut menu.

❾ In the Format Axis dialog box:
   a. Select the Scale tab and change the value in the Maximum: edit box to 1. (Excel will automatically deselect the Auto Maximum check box.)
   b. Click the OK button.

The Pareto diagram produced will be similar to the one shown in Figure 2.7 on page 99.

Problems for Section 2.3

Learning the Basics

• **2.29** Suppose that a categorical variable had three categories with the following frequency of occurrence:

CATEGORY	FREQUENCY
A	13
B	28
C	9

(a) Compute the percentage of values in each category.
(b) Construct a bar chart.
(c) Construct a pie chart.
(d) Form a Pareto diagram.

2.30 Suppose that a categorical variable had four categories with the following percentage of occurrence:

CATEGORY	PERCENTAGE	CATEGORY	PERCENTAGE
A	12	C	35
B	29	D	24

(a) Construct a bar chart.
(b) Construct a pie chart.
(c) Form a Pareto diagram.

Applying the Concepts

2.31 If you were going to make a presentation in front of an audience of 100 people and needed to use a visual display for a categorical variable with four classifications, which would you use—the bar chart, the pie chart, or the Pareto diagram? Explain.

2.32 The following data represent the breakdown of sales (in millions of dollars) of organic foods purchased from natural food stores in the United States during 1995:

FOOD PRODUCT	SALES (IN MILLIONS OF DOLLARS)
Grocery, soy foods, and dairy	512
Herbs	228
Food service and bakery	208
Produce	402
Bulk	231
Frozen foods	98
Miscellaneous	192
Total	1,871

Sources: Natural Foods Merchandiser; New Product News Magazine; The New York Times.

(a) Construct a bar chart.

(b) Construct a pie chart.

(c) Form a Pareto diagram.

(d) Which of these charts do you prefer to use here? Why?

(e) On the basis of the results of (a)–(c), in what categories does the major portion of the sales of organic foods purchased from natural food stores occur? Explain.

2.33 The following data represent the percentage of sales by various companies in the music industry during 1995:

COMPANY	% OF SALES
Sony	13.6
Warner, Elektra, Atlantic	22.1
Polygram	13.8
Bertelsmann Music	12.0
MCA	10.1
Capital/EMI	9.2
Independents	19.1
Total	99.9[a]

[a] Due to rounding.
Sources: Soundscan; Nielsen Media Research;
Exhibitor Relations Company; The New York
Times, December 6, 1995.

(a) Construct a bar chart.

(b) Construct a pie chart.

(c) Form a Pareto diagram.

(d) Which of these charts do you prefer to use here? Why?

(e) Would you conclude that any one company dominates sales in the music industry? Explain.

• **2.34** During the year 1995, oil consumption in the United States was 17.7 million barrels per day. The following data represent the percentage breakdown of the sources of that consumption:

SOURCE OF CONSUMPTION	% USAGE
Electric utilities	1.4
Highway transportation	53.4
Jet fuel	8.5
Plastics and fertilizers	10.2
Railroad, boat, and some construction equipment	4.8
Other uses for homes, industry, and businesses	21.7
Total	100.0

Source: U.S. Department of Energy.

(a) Construct a bar chart.

(b) Construct a pie chart.

(c) Form a Pareto diagram.

(d) Which of these charts do you prefer to use here? Why?

(e) What sources account for the bulk of oil consumption in the United States? Explain.

2.35 The following data represent the number of accidental deaths in the United States due to various causes during a recent year:

CAUSE OF DEATH	NUMBER
Agricultural machines	553
Airplanes	39
Buses	23
Caught in or between objects	119
Dog bites	13
Domestic wiring and appliances	66
Drowning (other than bathtub)	4,186
Drowning in bathtub	345
Falling objects	712
Falls	12,646
Falls into holes and openings	99
Fire and burns	3,958
Inhalation and ingestion of food	1,196
Lightning	53
Passenger cars and taxis	21,257
Side effects of therapeutic drug use	156
Suffocation by falling earth	58
Venomous plants and animals	68
Total	45,547

Source: National Safety Council.

(a) Form a Pareto diagram.
(b) What causes accounted for most of the accidental deaths? Explain.

2.36 The following data represent daily water consumption per household in a suburban water district during a recent summer.

REASON FOR WATER USAGE	GALLONS PER DAY
Bathing and showering	99
Dish washing	13
Drinking and cooking	11
Laundering	33
Lawn watering	150
Toilet	88
Miscellaneous	20
Total	414

(a) Form a Pareto diagram.
(b) If the water district wanted to develop a water reduction plan, which reasons for water usage should be focused on?

• **2.37** A patient-satisfaction survey conducted for a sample of 210 individuals discharged from a large urban hospital during the month of June led to the following list of 384 complaints:

REASON FOR COMPLAINT	NUMBER
Anger with other patients/visitors	13
Failure to respond to buzzer	71
Inadequate answers to questions	38
Lateness for tests	34
Noise	28
Poor food service	117
Rudeness of staff	62
All others	21
Total	384

(a) Form a Pareto diagram.

(b) Which reasons for complaint do you think the hospital should focus on if it wishes to reduce the number of complaints? Explain.

2.38 From a study of commercial properties conducted in March 1996 the following data represent the mix of retail establishments in the 32-block Times Square area of New York City:

TYPE OF RETAIL ESTABLISHMENT	NUMBER
Accessories	13
Adult use	25
Bar, lounge, or nightclub	15
Beauty/barber	15
Books/news	6
Clothing	18
Electronics	27
Entertainment	8
Fabric	5
Gifts	49
Grocery/convenience	34
Hardware	7
Health and fitness	4
Health, beauty, and housewares	10
Music	18
Quick service (food)	83
Restaurant	149
Services	45
Theater-movies	6
Wines and liquor	5
Other	14
Vacant store	69
Total	625

Source: The New York Times, *March 10, 1996, R11.*
Reprinted by permission of The New York Times.

(a) Form a Pareto diagram.

(b) What type of retail establishments make up the major portion of commercial properties in the Times Square area of New York City? Explain.

(c) If you were employed by a public relations firm whose goal was to promote business opportunities in New York City for tourists, what would you say about this renovated Times Square district?

 2.4 TABULATING AND GRAPHING BIVARIATE CATEGORICAL DATA

Often we need to simultaneously examine the responses to two categorical variables. For example, we might be interested in examining whether or not there is any pattern or relationship between fund objective (i.e., growth or blend) and the fee schedule. In this section we will examine some tabular and graphical methods of cross-classifying and presenting such data. In particular, we will develop the contingency table and the side-by-side bar chart.

The Contingency Table

In order to simultaneously study the responses to two categorical variables, we first form a two-way table of cross-classification known as a **contingency** or **cross-classification table.** For example, using the data set in appendix D, we can cross-classify the responses to the two categorical variables fund objective and fee schedule in order to determine if there is a pattern or relationship between them. Table 2.8 depicts this information for all 194 sampled domestic general stock funds.

To construct this contingency table, the joint responses for each of the 194 domestic general stock funds with respect to fund objective and fee schedule are tallied into one of the 10 possible "cells" of the table. Thus, from the data set in appendix D, the first fund listed (AARP Investment GrowInc) is classified as a blend fund for which there is no fee (i.e., it is a no-load fund). These joint responses are tallied into the cell composed of the second row and fifth column. The second institution (AIM BlueCh A) is a growth fund charging multiple fees. These joint responses are tallied into the cell composed of the first row and fourth column. The remaining 192 joint responses are recorded in a similar manner.

Table 2.8 *Contingency table displaying fund objective and fee schedule*

FUND OBJECTIVE	FEE SCHEDULE					
	FUND ASSETS	DEFERRED FEES	FRONT LOAD	MULTIPLE FEES	NO LOAD	TOTAL
Growth	4	0	7	16	32	59
Blend	13	5	12	30	75	135
Total	17	5	19	46	107	194

Source: Data are taken from Special Data Set 1 in appendix D.

In order to explore any possible pattern or relationship between fund objective and fee schedule, it is useful to first convert these results into percentages based on the following three totals:

1. The overall total (i.e., the 194 sampled domestic general stock funds)

2. The row totals (i.e., growth or blend funds)

3. The column totals (i.e., the five fee schedule classifications)

This is accomplished in Tables 2.9, 2.10, and 2.11, respectively.

Table 2.9 *Contingency table displaying fund objective and fee schedule (percentages based on overall total)*

FUND OBJECTIVE	FUND ASSETS	DEFERRED FEES	FRONT LOAD	MULTIPLE FEES	NO LOAD	TOTAL
Growth	2.1	0.0	3.6	8.2	16.5	30.4
Blend	6.7	2.6	6.2	15.5	38.7	69.6
Total	8.8	2.6	9.8	23.7	55.2	100.0

Source: Data are taken from Table 2.8.

Table 2.10 *Contingency table displaying fund objective and fee schedule (percentages based on row totals)*

FUND OBJECTIVE	FUND ASSETS	DEFERRED FEES	FRONT LOAD	MULTIPLE FEES	NO LOAD	TOTAL
Growth	6.8	0.0	11.9	27.1	54.2	100.0
Blend	9.6	3.7	8.9	22.2	55.6	100.0
Total	8.8	2.6	9.8	23.7	55.2	100.0

Source: Data are taken from Table 2.9.

Table 2.11 *Contingency table displaying fund objective and fee schedule (percentages based on column totals)*

FUND OBJECTIVE	FUND ASSETS	DEFERRED FEES	FRONT LOAD	MULTIPLE FEES	NO LOAD	TOTAL
Growth	23.5	0.0	36.8	34.8	29.9	30.4
Blend	76.5	100.0	63.2	65.2	70.1	69.6
Total	100.0	100.0	100.0	100.0	100.0	100.0

Let us examine some of the many findings present in these tables. From Table 2.9 we note that 30.4% of the domestic general stock funds sampled are growth funds, 55.2% are no-load funds, and 16.5% are growth funds that are no-load. From Table 2.10 we note that 0.0% of the growth funds have a deferred-fee schedule and 22.2% of the blend funds have a multiple-fee schedule. From Table 2.11 we note that 36.8% of the funds with a front-load fee schedule are growth funds and 76.5% of the funds whose fees are paid from fund assets are blend funds. The tables, therefore, do not reveal any major patterns: blend funds predominate over growth funds, regardless of the fee schedule. In addition, no-load funds and those with multiple-fee schedules predominate over other fee schedules, regardless of the objective of the fund (i.e., growth versus blend).

We may now wish to condense these 2×5 contingency tables (i.e., 2 rows by 5 columns) into a set of 2×2 tables of cross-classification by simplifying the fee schedule variable into only two categories—funds that charge fees versus those that do not (i.e., the no-load funds).

The Side-by-Side Bar Chart

A useful way to visually display bivariate categorical data when looking for patterns or relationships is by constructing a **side-by-side bar chart.** This graphic form is best suited when primary interest is in demonstrating differences in magnitude rather than differences in percentages. Thus, for example, using the data from Table 2.8, Figure 2.8 is a side-by-side bar chart obtained from Excel that enables a comparison of the two fund objectives based on the various fee schedules. From Figure 2.8 and Table 2.11 we observe that although about 70% of all the mutual funds have a blend fund objective, a slightly larger proportion of mutual funds whose fees are paid from fund assets have a blend objective, whereas a slightly smaller proportion of funds whose fees are front-loaded or are from multiple sources have a blend fund objective. In addition, all five funds with a deferred fee schedule have a blend fund objective.

	A	B	C
1	Summary Table for Fee Schedule		
2			
3	Count of Name		
4	Fees	Total	
5	Deferred	5	
6	Front load	19	
7	Fund assets	17	
8	Multiple	46	
9	No load	107	
10	Grand Total	194	

FIGURE 2.8

A side-by-side bar chart of fund objective based on fee schedule obtained from Microsoft Excel

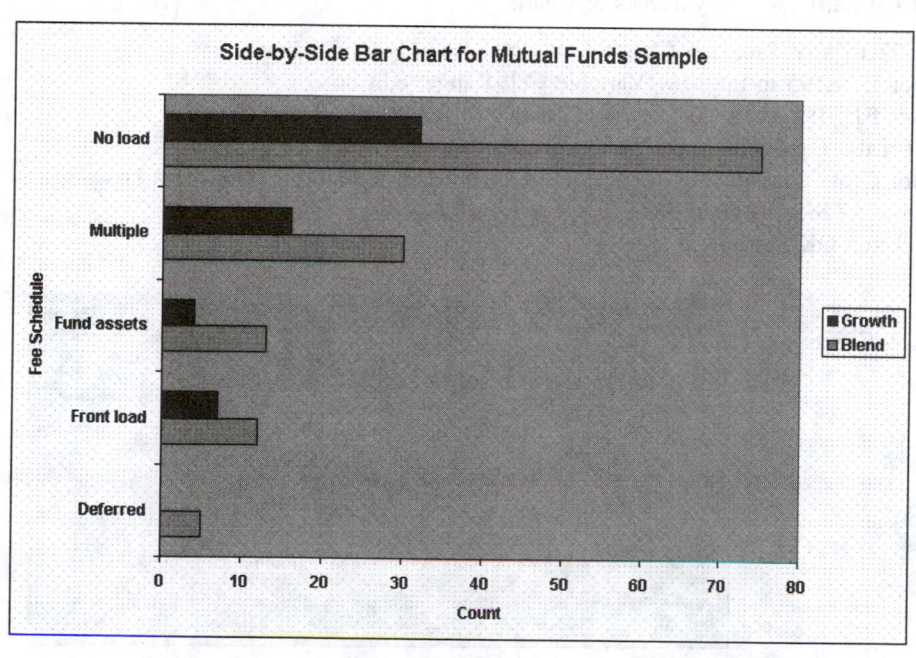

2.4E ◆ GENERATING TABLES AND CHARTS FOR BIVARIATE CATEGORICAL DATA USING MICROSOFT EXCEL

Overview

◆ *For Quick Results Users*   Select PHStat | Two-Way Tables & Charts to tabulate and graph bivariate categorical data.

◆ *For Developers*   Use the following two-step process:

    A. Use the PivotTable Wizard to create a summary table containing the bivariate categorical data.

    B. Use the Chart Wizard to generate a side-by-side chart based on this table.

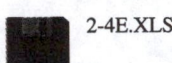
2-4E.XLS

The 2-4E.XLS workbook file contains a PivotTable and a side-by-side bar chart based on the mutual funds sample data used in this chapter.

Quick Results

To generate a two-way table that cross-classifies fee schedule and fund objective as well as a side-by-side bar chart from the mutual funds fee schedule data (Table 2.8 on page 115), do the following:

❶ If the PHStat add-in has not been previously loaded, load the add-in using the instructions of section S4.2 on page 59.

❷ Open the Mutual Funds Sample workbook (MUTUAL FUNDS SAMPLE.XLS) and click the Data sheet tab.

❸ Select PHStat | Two-Way Tables & Charts.

❹ In the Two-Way Tables & Charts dialog box (see Figure 2E.13):
    a. Enter E1:E195 in the Row Variable Cell Range: edit box.
    b. Enter F1:F195 in the Column Variable Cell Range: edit box.
    c. Select the First cells in both ranges contain label check box.
    d. Enter Cross Classification of Fee and Objective in the Output Title: edit box.
    e. Select the Side-by-Side Bar Chart check box.
    f. Click the OK button.

FIGURE 2E.13
Two-Way Tables & Charts
dialog box

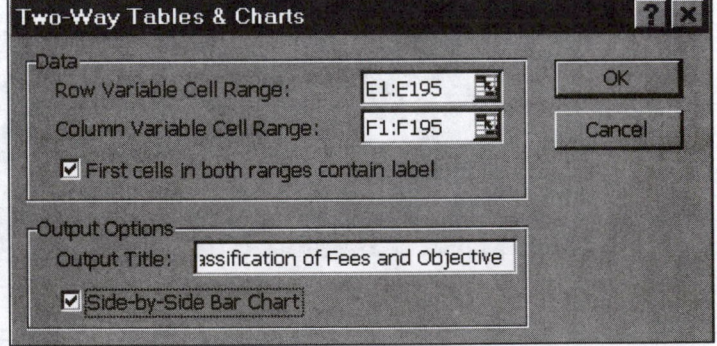

On separate sheets, the add-in produces a two-way summary table and side-by-side bar chart similar to those shown in Figure 2.8 on page 117.

Developer Details

◆ *A. Generating a Contingency Table Using the PivotTable Wizard*   We can create a contingency table using the same general procedure for using the PivotTable Wizard first presented in section 2.3E. To create a contingency table similar to Table 2.8 on page 115 that cross-classifies the 194 domestic general stock funds by fund objective and fee schedule, do the following:

❶ Open the Mutual Funds Sample workbook and click the Data sheet tab.

❷ Select Data | PivotTable Report.

❸ In the Step 1 dialog box, select the Microsoft Excel List or Database option button. Click the Next button.

❹ In the Step 2 dialog box, enter A1:G195 in the Range: edit box. Click Next.

❺ In the Step 3 dialog box:
    a. Verify that the label boxes displayed on the right side of the dialog box correspond to the column headings on the Data sheet (Name, Type, NAV, etc.). If they do not, click Back and verify that the proper range has been entered in the Step 2 dialog box.
    b. Drag the Fees label and drop it in the Row box. A copy of the Fees label snaps into the Row box.
    c. Drag the Objective label and drop it in the Column box.
    d. Drag the Name label and drop it into the Data box. Verify that the wording on the label has changed to Count of Name.
    e. If the label reads something other than Count of Name, double-click on the label to display the PivotTable Field dialog box. In this dialog box:
        i. Select Count from the Summarize by: list box.
        ii. Click the Options button.
        iii. Select Normal in the Show data as: drop-down list box.
        iv. Click the OK button to redisplay the Step 3 dialog box.
    f. Click the Next button.

❻ In the Step 4 dialog box:
    a. Select the New worksheet option button and click the Options button.
    b. In the PivotTable Options dialog box:
        i. Enter TwoWayTable in Name: edit box.
        ii. Select the Grand totals for columns, Grand totals for rows, AutoFormat table, and Preserve formatting check boxes of the Format options group.
        iii. Select the For empty cells, show: check box and enter 0 (zero) in its edit box.
        iv. Select the Save data with table layout and Enable drilldown option buttons under the Data source options: heading.
        v. Click the OK button to return to the Step 4 dialog box.
    c. Click the Finish button.

❼ Rename the new worksheet TwoWayTable. (The floating PivotTable toolbar that may appear after the table is created can be closed or dragged to the side of the application window.)

❽ Select cell E1. Select Insert | Rows two times to insert two empty rows above the PivotTable. Enter a title in the new row 1.

The contingency table produced will be similar to the one shown in Figure 2.8 on page 117.

◆ *Generating a Side-by-Side Bar Chart from a Contingency Table* Having generated this PivotTable that cross-classifies fund objective and fee schedule, we can use the Chart Wizard to generate a side-by-side bar chart. To generate this chart from the PivotTable just created, do the following:

❶ Click the TwoWayTable sheet tab.

❷ Select Insert | Chart.

❸ In the Step 1 dialog box:
   a. Select the Standard Types tab and then select Bar from the Chart type: list box. Select the first choice in the first row of choices under the Chart sub-type: heading. When this choice is selected, the phrase "Clustered Bar" is displayed in the description box below the sub-types.
   b. Click the Next button.

❹ In the Step 2 dialog box:
   a. Select the Data Range tab. Enter =TwoWayTable!B5:C9 in the Data range: edit box and select the Columns option button in the Series in: group.
   b. Select the Series tab.
      i. Select the first item (Series1) in the Series list box. Enter =TwoWayTable!B4 in the Name: edit box.
      ii. Select the second item (Series2) in the Series list box. Enter =TwoWayTable!C4 in the Name: edit box.
      iii. Enter =TwoWayTable!A5:A9 in the Category (X) axis labels: edit box.
   c. Click the Next button.

❺ In the Step 3 dialog box:
   a. Select the Titles tab. Enter Side-by-Side Bar Chart for Mutual Funds Sample in the Chart title: edit box, enter Fee Schedule in the Category (X) axis: edit box, and enter Count in the Value (Y) axis edit box.
   b. Select the Axes tab. Select the Primary axis Category (X) axis and the Value (Y) axis check boxes.
   c. Select the Legend tab. Select the Show legend check box.
   d. Select, in turn, the Gridlines, Data Labels, and Data Table tabs and verify that their settings match those given in Table 2E.3 on page 88.

❻ In the Step 4 dialog box:
   a. Select the As new sheet: option button and enter SidebySide in the edit box to the right of the option button.
   b. Click the Finish button to create a chart similar to Figure 2.8. (The floating Chart toolbar that appears after the chart is created can be closed or dragged to the side of the application window.)

Problems for Section 2.4

Learning the Basics

• **2.39** The following data represent the bivariate responses to two questions asked in a survey of 40 college students majoring in business—Gender (Male = M; Female = F) and Major (Accountancy = A; Computer Information Systems = C; Retailing = R):

Gender: M M M F M F F M F M F M F M M M M F F M F F
Major: A C C R A C A A C C A A A R C R A A A C

Gender: M M M M F M F F M M F M M M M F M F M M
Major: C C A A R R C A A A C C A A A A C C A C

(a) Tally the data into a 2 × 3 contingency table where the two rows represent the gender categories and the three columns represent the student-major categories.
(b) Form a contingency table based on percentages of all 40 student responses.
(c) Form a contingency table based on row percentages.
(d) Form a contingency table based on column percentages.
(e) Using the results from (a), construct a side-by-side bar chart of gender based on student major.

2.40 Given the following two-way cross-classification table, construct a side-by-side bar chart comparing A and B for each of the three column categories on the vertical axis.

	1	2	3	TOTAL
A	20	40	40	100
B	80	80	40	200

Applying the Concepts

• **2.41** An employee survey was conducted by the Human Resources Department at Leonel Industries. Responses to a questionnaire from 400 full-time employees yielded the following breakdown with respect to gender and occupational title:

					OCCUPATIONAL TITLE			
GENDER	MGT.	PROF.	SALES	ADM.	SUPPORT SERVICE	PRODUCTION	LABORER	TOTAL
Male	36	33	34	14	18	51	47	233
Female	29	33	23	51	11	03	17	167
Total	65	66	57	65	29	54	64	400

(a) Construct a table of row percentages.
(b) Construct a side-by-side bar chart for the results in (a).
(c) Does there seem to be an overrepresentation of males in some categories and an underrepresentation in others? Explain.

•**2.42** The victory of the incumbent, Bill Clinton, in the 1996 presidential election was attributed to improved economic conditions and low unemployment. Suppose that a survey of 800 adults taken soon after the election resulted in the following cross-classification of financial condition with education level:

| | EDUCATION LEVEL | | | |
FINANCIAL CONDITIONS	H.S. DEGREE OR LOWER	SOME COLLEGE	COLLEGE DEGREE OR HIGHER	TOTAL
Worse off now than before	91	39	18	148
No difference	104	73	31	208
Better off now than before	235	48	161	444
Total	430	160	210	800

(a) Construct a table of column percentages.
(b) Construct a side-by-side bar chart to visually highlight the results in (a).
(c) On the basis of the results of (a) and (b), do you think there is a clear difference in the current financial condition as compared with before based on level of education?

2.43 Using Special Data Set 1 from appendix D.
(a) Set up a 2 × 3 table of cross-classifications of fund objective (growth versus blend) and size of fund capitalization (large, medium, or small).
(b) Construct a table based on total percentages.
(c) Construct a table based on row percentages.
(d) Construct a table based on column percentages.
(e) Construct a side-by-side bar chart of fund objective based on size of fund capitalization.
(f) On the basis of your results in (e), does there appear to be a pattern or relationship between fund objective and size of fund capitalization? Discuss.

2.44 Each day at a large hospital several hundred laboratory tests are performed. The rate at which these tests are improperly done for a variety of reasons (and therefore need to be redone) seems steady at about 4%. In an effort to get to the root cause of these nonconformances (tests that need to be redone), the director of the lab decides to keep records for a period of 1 week of the nonconformances subdivided by the shift of workers who performed the lab tests. The results were as follows:

| | SHIFT | | |
LAB TESTS PERFORMED	DAY	EVENING	TOTAL
Nonconforming	16	24	40
Conforming	654	306	960
Total	670	330	1,000

(a) Construct a table of row percentages.
(b) Construct a table of column percentages.
(c) Construct a table of total percentages.
(d) Which type of percentage—row, column, or total—do you think is most informative for these data? Explain.
(e) What conclusions concerning the pattern of nonconforming laboratory tests can the laboratory director reach?

2.45 A savings bank conducted a customer satisfaction survey on a monthly basis to measure satisfaction with several areas of services offered by a branch office. The results from a sample of 200 customers were as follows:

AREA OF SERVICE	NUMBER SATISFIED	NUMBER DISSATISFIED
Waiting time for tellers	123	65
Automatic teller machine (ATM)	73	7
Investment advisement	43	6
Traveler's-check service	25	11
Safe deposits	24	5
Account maintenance services	46	4

Note: Because all customers did not use each service, the number of responses for each area of service is different.

(a) Construct a table of row percentages.
(b) Construct a table of column percentages.
(c) Construct a table of total percentages.
(d) Which type of percentage—row, column, or total—do you think is most helpful in understanding these data. Why?
(e) Construct a side-by-side bar chart of customer satisfaction by area of service.
(f) Do customers seem equally satisfied with all areas of service? Which areas seem to need improvement more than others? Discuss.

 **2.5** **GRAPHICAL EXCELLENCE**

To this point we have studied how a collected set of data is presented in tabular and chart form. Among the methods for describing and communicating statistical information, well-designed graphical displays are usually the simplest and the most powerful. Good graphical displays reveal what the data are conveying. If our analysis is to be enhanced by visual displays of data, it is essential that the tables and charts be presented clearly and carefully. Tabular frills and other "chartjunk" must be eliminated so as not to cloud the message given by the data with unnecessary adornments (references 7, 10, 11, 12, and 16).

The widespread use of spreadsheet applications and graphics software has led to a proliferation of graphics in recent years. Although much of the graphics presented have served as useful representations of the data, unfortunately the inappropriate and improper nature of many presentations has often hindered understanding and analysis.

Principles of Graphical Excellence

Perhaps the most well-known proponent of the proper presentation of data in graphs is Professor Edward R. Tufte, who has written a series of books devoted to proper methods of graphical design (see references 10, 11, and 12). It is the work of Tufte that we will focus on in this section. Exhibit 2.1 lists the essential features of graphing data.

Exhibit 2.1   Features of Graphing Data

The basic features of a proper graph include the following:

✓ **1.** Showing the data

✓ **2.** Getting the viewer to focus on the substance of the graph, not on how the graph was developed

✓ **3.** Avoiding distortion

✓ **4.** Encouraging comparisons of data

✓ **5.** Serving a clear purpose

✓ **6.** Being integrated with the statistical and verbal descriptions of the graph

In *The Visual Display of Quantitative Information* (reference 10), Tufte has suggested five principles of **graphical excellence.** These are listed in Exhibit 2.2.

Exhibit 2.2   Principles of Graphical Excellence

✓ **1.** Graphical excellence is a well-designed presentation of data that provides substance, statistics, and design.

✓ **2.** Graphical excellence communicates complex ideas with clarity, precision, and efficiency.

✓ **3.** Graphical excellence gives the viewer the largest number of ideas in the shortest time with the least ink.

✓ **4.** Graphical excellence almost always involves several dimensions.

✓ **5.** Graphical excellence requires telling the truth about the data.

There are several ways in which the excellence of a graph can be evaluated. One important measure is the **data-ink ratio.**

Data-Ink Ratio

The data-ink ratio is the proportion of the graphic's ink that is devoted to nonredundant display of data information.

$$\text{Data-ink ratio} = \frac{\text{data-ink}}{\text{total ink used to print the graphic}}$$

The objective is to maximize the proportion of the ink used in the graph that is devoted to the data. Within reasonable limits, non-data-ink and redundant data-ink should be eliminated. Non-data-ink includes aspects of the graph that do not relate to the substantive features of the data as well as grid lines that may be imposed on the graph. We refer to such adornments as chartjunk, which can be defined as follows.

Chartjunk is decoration that is non-data-ink or redundant data-ink.

In its extreme form, chartjunk represents self-promoting graphics that focuses the viewer on the style of the graph, not the data presented in the graph. Tufte has referred to this type of graph as "The Duck."

A central feature of graphical excellence is the importance of not using a graph to distort the data that it represents. A graph does not distort if its visual representation is consistent with its numerical representation. The amount of distortion can be measured by the lie factor.

Lie Factor

The **lie factor** is the ratio of the size of the effect shown in the graph to the size of the effects shown in the data.

One principle involved here is that any variation in the design of a graph must be consistent with the variation that exists in the data. Often changes in the graph are not consistent with variations in the data, producing a distortion between what the data represent and what the graph is showing.

In order to better understand these principles, it is useful to study several examples of graphs that are deficient in graphical excellence.

Figure 2.9 on page 126 is a graph printed in *The New York Times* of the annual oyster catch in Chesapeake Bay in millions of bushels for a time period that stretches more than a century, from the 1890s until 1992. The icon representing the estimated 20 million bushels of oysters caught in the 1890s does not appear to be five times the size of the icon representing the estimated 4 million bushels caught in 1962. Such an illustration may catch the eye, but it usually doesn't show anything that could not be presented better in a summary table or a plot of the data over time.

Exaggerated icons and symbols are chartjunk. They result in a distortion of the visual impact. Let's examine Figure 2.10 on page 127, a display of jobs supported through U.S. agriculture and agricultural exports. Note that in this chart the magnitude of the 10.8 million transportation, trade, and retail jobs is underrepresented by a truck icon that is smaller than the icon representing the 5.6 million jobs for farm family members. Also, the can icon representing the 1.3 million food processing jobs is far too large compared with the icons representing the 2.1 million farm worker jobs, the 2.6 million manufacturing jobs, and the 4.1 million jobs in other areas of agriculture and exporting. A simple summary table or a bar chart, a pie chart, or a Pareto diagram would have been more effective in accurately portraying the data.

FIGURE 2.9
"Improper" display of estimated oyster catch (in millions of bushels) in the Chesapeake Bay over various time periods

Source: Reprinted by permission of The New York Times. The New York Times, October 17, 1993, 26.

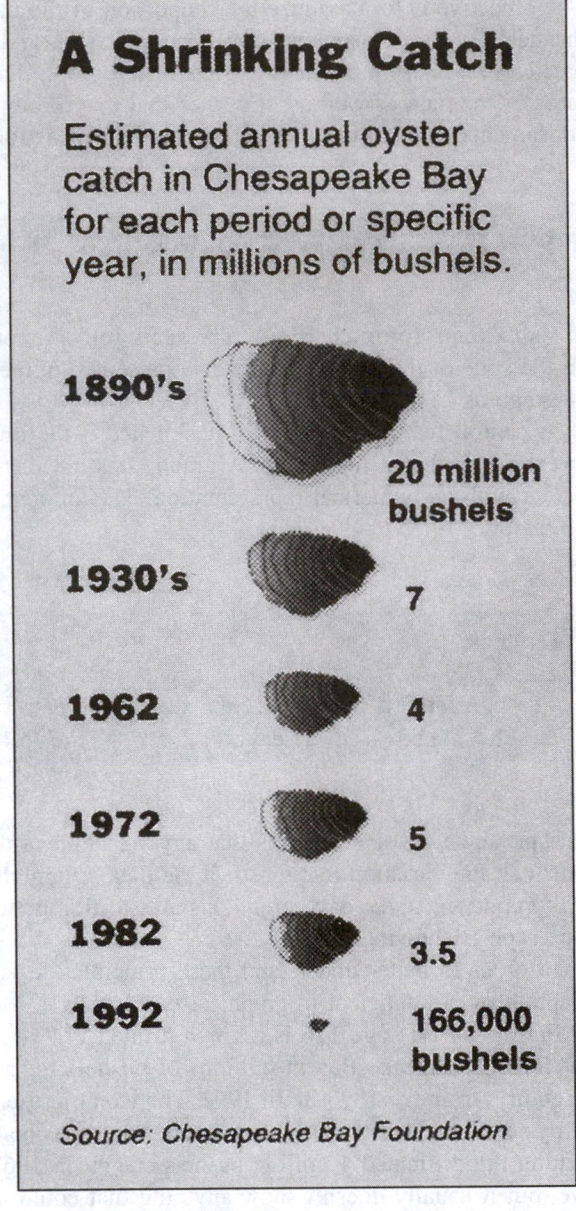

A Shrinking Catch

Estimated annual oyster catch in Chesapeake Bay for each period or specific year, in millions of bushels.

1890's — 20 million bushels

1930's — 7

1962 — 4

1972 — 5

1982 — 3.5

1992 — 166,000 bushels

Source: Chesapeake Bay Foundation

If we do not display the zero point on the vertical axis it becomes easy to visually distort a set of data. In what at first glance appears to be an attractive chart, Figure 2.11 provides a clear demonstration of such distortion. The five graphs, taken from a South American newspaper, are intended to depict the negative impact of what is referred to as the "caipirinha effect" on various Latin American stock markets from July 7 through July 15, 1997. The chart is so poorly drawn, however, that the same information could have been better presented in a small table highlighting the downward trends from July 7 through July 15.

There are numerous problems with this graphical display. First, the eye is automatically drawn to the São Paulo, Brazil, graph, and there is no explanation as to why the graph of

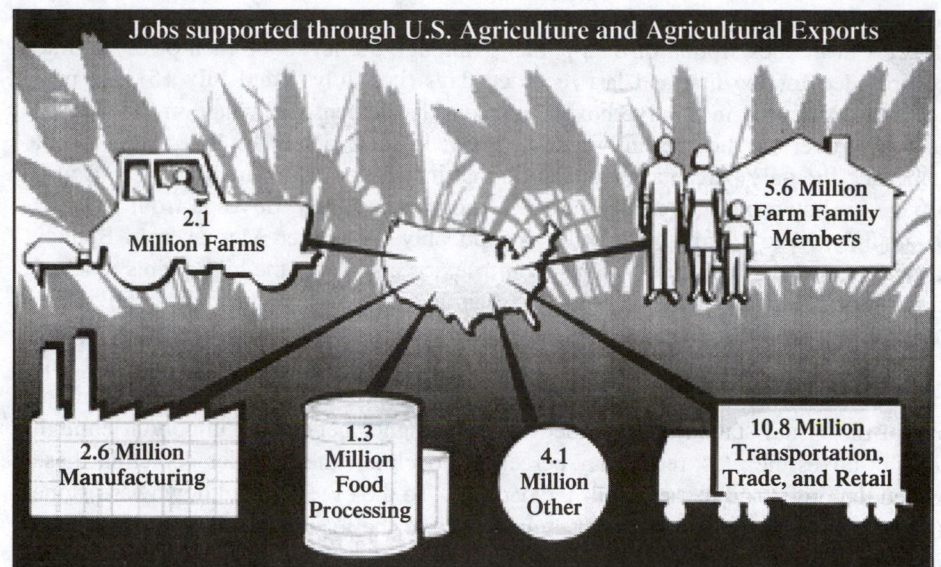

Jobs supported through U.S. Agriculture and Agricultural Exports

2.1 Million Farms

5.6 Million Farm Family Members

2.6 Million Manufacturing

1.3 Million Food Processing

4.1 Million Other

10.8 Million Transportation, Trade, and Retail

FIGURE 2.10

"Improper" display of jobs supported through United States agriculture and agricultural exports

Source: Reprinted by permission of The New York Times. The New York Times, *October 19, 1993, Advertising Supplement, D18.*

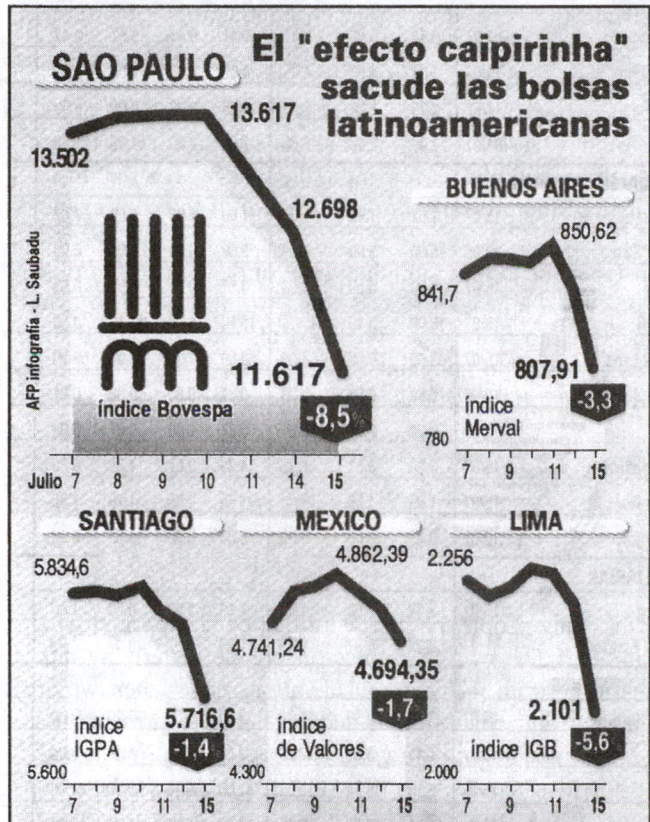

SAO PAULO

El "efecto caipirinha" sacude las bolsas latinoamericanas

13.617

13.502

12.698

índice Bovespa

11.617

-8,5%

AFP infografía - L. Saubadu

Julio 7 8 9 10 11 14 15

BUENOS AIRES

850,62

841,7

807,91

índice Merval

-3,3

780

7 9 11 15

SANTIAGO

5.834,6

índice IGPA

5.716,6

-1,4

5.600

7 9 11 15

MEXICO

4.862,39

4.741,24

4.694,35

-1,7

índice de Valores

4.300

7 9 11 15

LIMA

2.256

índice IGB

2.101

-5,6

2.000

7 9 11 15

FIGURE 2.11

"Improper" displays of time-series data

the Indice Bovespa is drawn to dwarf the other four graphed indices that appear comparable in size. Moreover, we have no explanation as to why an icon was drawn under this graph but not under any of the others.

As we examine the five graphs more closely and try to make comparisons, we note some inconsistencies in the description of data points. Although the levels of each of these time series are provided for the first and last recorded days (i.e., July 7 and July 15), the percentage declines indicated in the five boxes represent the percentage change in the respective indices only over the most recently recorded day, July 14 to July 15. The reader is not told that. In fact, the only percentage calculation that the reader can actually make and verify from what is shown in the boxes is the 8.5% drop in the Indice Bovespa from its level of 12.698 on July 14 to 11.617 on July 15. Inexplicably, the Indice Merval from Buenos Aires, Argentina, and the Indice de Valores from Mexico also show the high points reached in this time period, but the Indice IGPA from Santiago, Chile, and the Indice IGB from Lima, Peru, do not give these values.

Even if none of the above-mentioned deficiencies had been observed, the chart still suffers from a very major flaw—improper compression of the vertical axes that present major distortions in the visual statements provided. If the designer believed it important enough to highlight in boxes the most recent day percentage declines, the graphics depicting these declines need to correspond to the size of the declines. To the eye, the final downward slope segments in each of the five graphs do not correspond to the magnitudes of the percentage changes displayed in the boxes. Because of the vertical compression of the axes, to the eye the steepest negative slope appears to have occurred in Lima (a 5.6% drop), then in Buenos Aires (a 3.3% drop), then in São Paulo (an 8.5% drop), then in Santiago (a 1.4% drop), and, finally, in Mexico (a 1.7% drop).

Other types of eye-catching displays that we typically see in magazines and newspapers often include information that is not necessary and just adds excessive data-ink. Figure 2.12 is an example of one such display. From Figure 2.12, the partial male-female icons do not adequately portray the gender composition of top management at the nine Wall Street firms. The same information could be conveyed more clearly through the use of the contingency table or the side-by-side bar chart.

In summation, we are active consumers of information that we hear or see daily through the various media. Because much of what we hear or read is junk, we must learn to evaluate critically and discard that which has no real value. We must also keep in mind that sometimes the junk we are provided is based on ignorance; other times, it is planned and malicious. The bottom line—be critical and be skeptical of information provided.

Ethical Issues

According to Tufte (reference 10), for many people the first word that comes to mind when they think about statistical charts is "lie." Too many graphics distort the underlying data, making it hard for the reader to learn the truth. Ethical considerations arise when we are deciding what data to present in tabular and chart format and what not to present. It is vitally important when presenting data to document both good and bad results. When making oral presentations and presenting written reports, it is essential that the results be given in a fair, objective, and neutral manner. Thus, we must try to distinguish between poor data presentation and unethical presentation. Again, as in our discussion of ethical considerations in data collection (section 1.9), the key is *intent*. Often, when fancy tables and chartjunk are presented or pertinent information is omitted, it is simply done out of ignorance. However, unethical behavior occurs when an individual willfully hides the facts by distorting a table or chart or by failing to report pertinent findings.

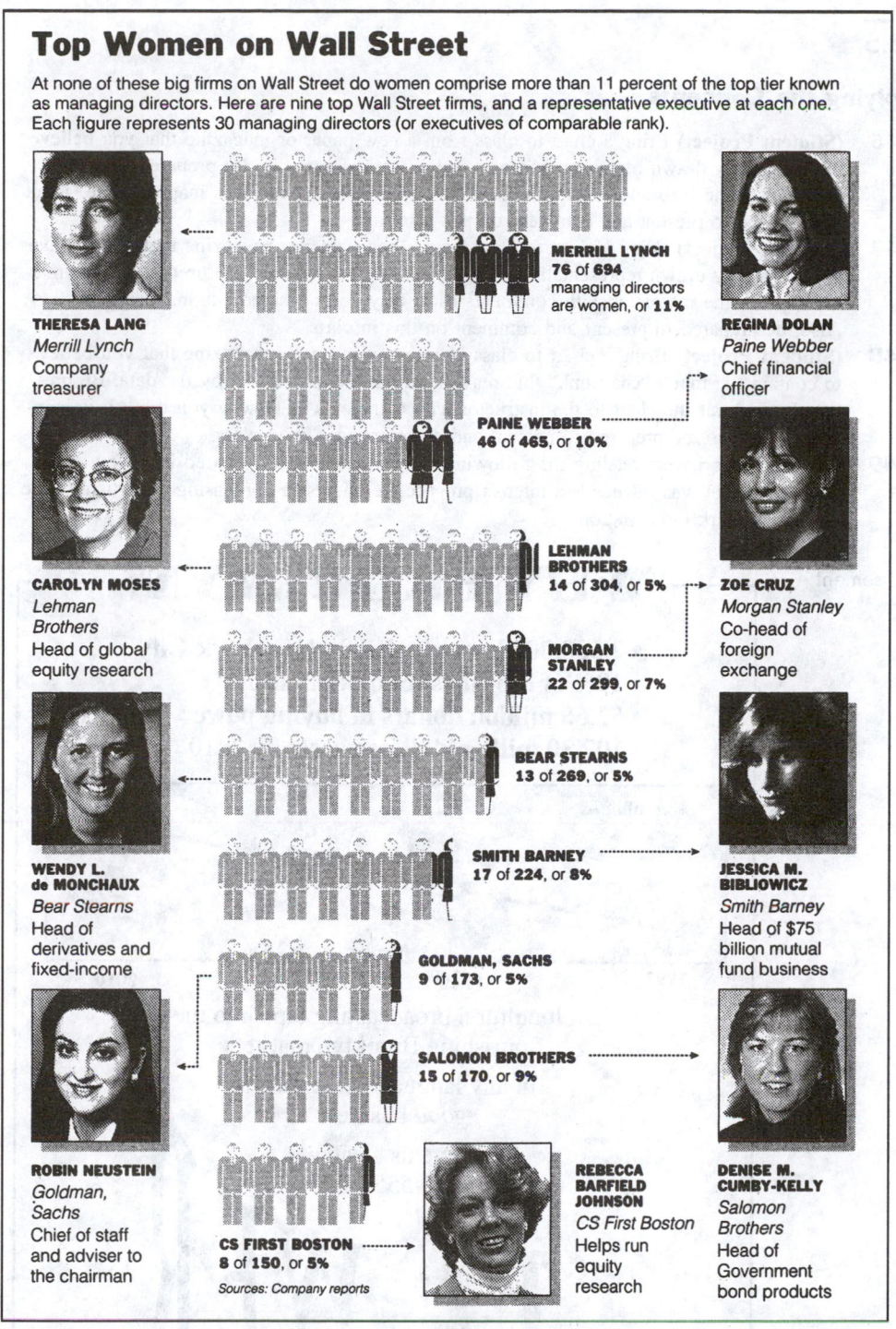

Top Women on Wall Street

At none of these big firms on Wall Street do women comprise more than 11 percent of the top tier known as managing directors. Here are nine top Wall Street firms, and a representative executive at each one. Each figure represents 30 managing directors (or executives of comparable rank).

THERESA LANG
Merrill Lynch
Company treasurer

MERRILL LYNCH
76 of **694** managing directors are women, or **11%**

REGINA DOLAN
Paine Webber
Chief financial officer

PAINE WEBBER
46 of **465**, or **10%**

CAROLYN MOSES
Lehman Brothers
Head of global equity research

LEHMAN BROTHERS
14 of **304**, or **5%**

ZOE CRUZ
Morgan Stanley
Co-head of foreign exchange

MORGAN STANLEY
22 of **299**, or **7%**

WENDY L. de MONCHAUX
Bear Stearns
Head of derivatives and fixed-income

BEAR STEARNS
13 of **269**, or **5%**

SMITH BARNEY
17 of **224**, or **8%**

JESSICA M. BIBLIOWICZ
Smith Barney
Head of $75 billion mutual fund business

GOLDMAN, SACHS
9 of **173**, or **5%**

ROBIN NEUSTEIN
Goldman, Sachs
Chief of staff and adviser to the chairman

SALOMON BROTHERS
15 of **170**, or **9%**

CS FIRST BOSTON
8 of **150**, or **5%**
Sources: Company reports

REBECCA BARFIELD JOHNSON
CS First Boston
Helps run equity research

DENISE M. CUMBY-KELLY
Salomon Brothers
Head of Government bond products

FIGURE 2.12 "Improper" display of workforce size for each of 30 managing directors at nine large Wall Street firms

Source: Reprinted by permission of The New York Times. The New York Times, July 2, 1996, D1 and D4.

101

Problems for Section 2.5

Applying the Concepts

2.46 (**Student Project**) Bring a chart to class from a newspaper or magazine that you believe to be a poorly drawn representation of some numerical variable. Be prepared to submit the chart to the instructor with comments as to why you believe it is inappropriate. Also, be prepared to present and comment on this in class.

2.47 (**Student Project**) Bring a chart to class from a newspaper or magazine that you believe to be a poorly drawn representation of some categorical variable. Be prepared to submit the chart to the instructor with comments as to why you consider it is inappropriate. Also, be prepared to present and comment on this in class.

2.48 (**Student Project**) Bring a chart to class from a newspaper or magazine that you believe to contain too much "chartjunk" that may cloud the message given by the data. Be prepared to submit the chart to the instructor with comments as to why you think it is inappropriate. Also, be prepared to present and comment on this in class.

2.49 Cassie and Lori were reading the following advertisement that appeared in a New York newspaper that was intended to interest prospective clients in purchasing advertising time on a New York radio station.

A newspaper advertisement

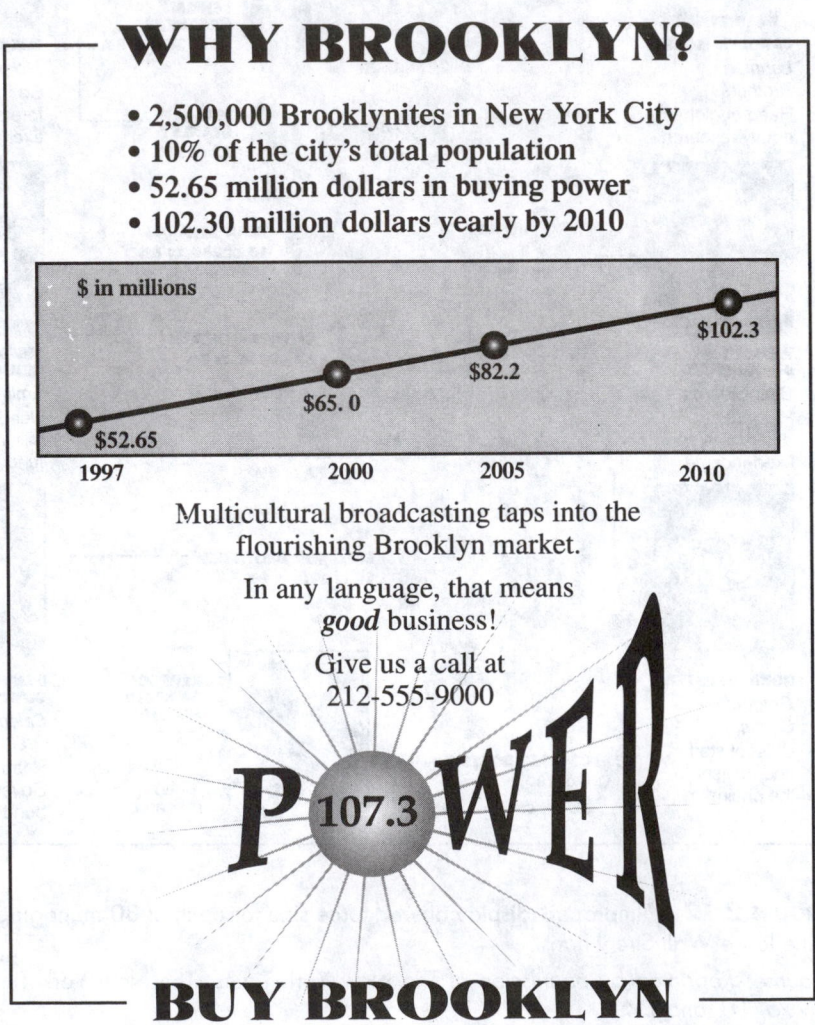

WHY BROOKLYN?

- 2,500,000 Brooklynites in New York City
- 10% of the city's total population
- 52.65 million dollars in buying power
- 102.30 million dollars yearly by 2010

$ in millions

$102.3
$82.2
$65. 0
$52.65

1997   2000   2005   2010

Multicultural broadcasting taps into the flourishing Brooklyn market.

In any language, that means *good* business!

Give us a call at
212-555-9000

P WER
107.3

BUY BROOKLYN

Cassie looks at the graph and says, "Lori, this graph does not accurately present the intended information to its readers." Lori replies, "Aside from being inaccurate, the graph has failed to demonstrate the projected growth in buying power that would make the radio station even more attractive to potential purchasers of advertising time." Cassie says, "I agree with you."

(a) Redraw this graph by properly plotting the time series.

(b) Do you agree with the conclusions reached by Cassie and Lori? Discuss.

2.50 The following visual display contains an overembellished chart that appeared as part of an article in *The New York Times* dealing with the decline of market share for Crest toothpaste.

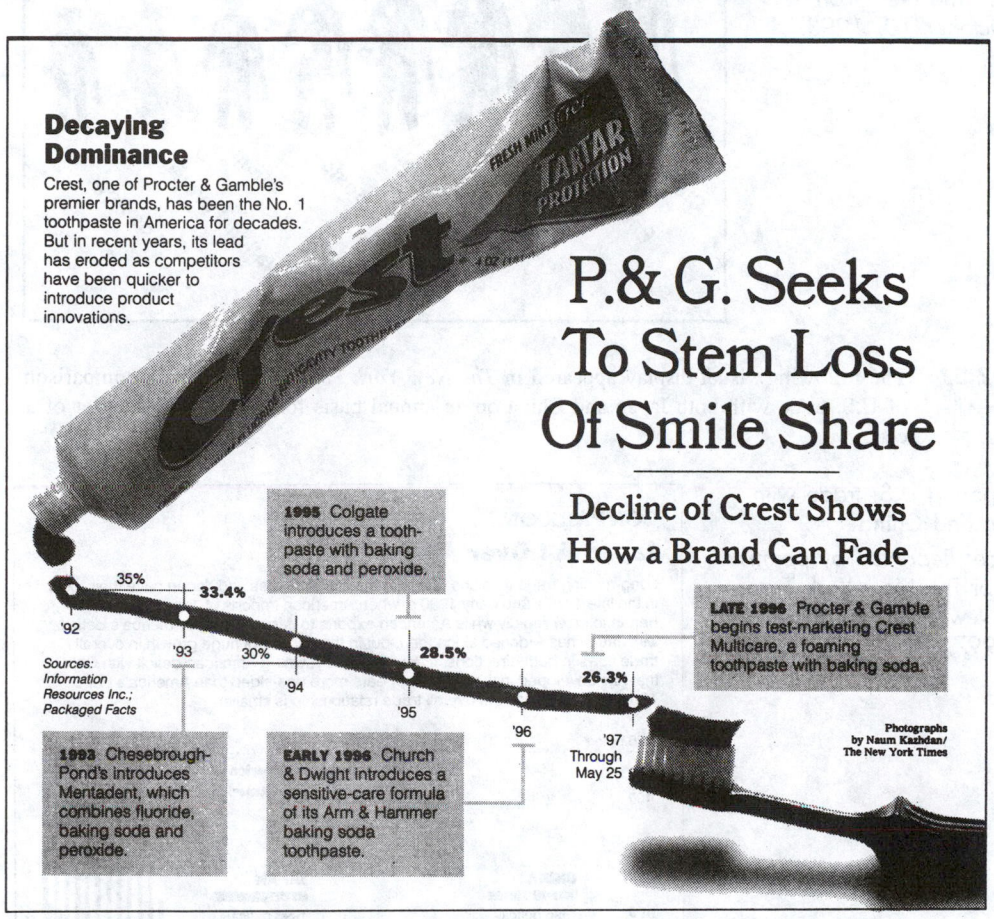

Source: Reprinted by permission of The New York Times. The New York Times, *June 20, 1997, D1.*

(a) Describe at least one good feature of this visual display.

(b) Describe at least one bad feature of this visual display.

(c) Locate the errors in this visual display. Discuss.

(d) Redraw the graph using the principles of graphical excellence.

2.51 The following visual display appeared in *The New York Times* and permits a comparison of the relative size of police departments in major U.S. cities.
(a) Indicate a feature of this chart that violates Tufte's principles of graphical excellence.
(b) Set up an alternative graph for the data provided in this figure.

"Improper" display of police department size for each 1,000 residents of major cities in the United States

Source: Reprinted by permission of The New York Times. *Extracted from R. Powell, "A Statistical Portrait of the N.Y.P.D.,"* The New York Times, *October 10, 1993, 35.*

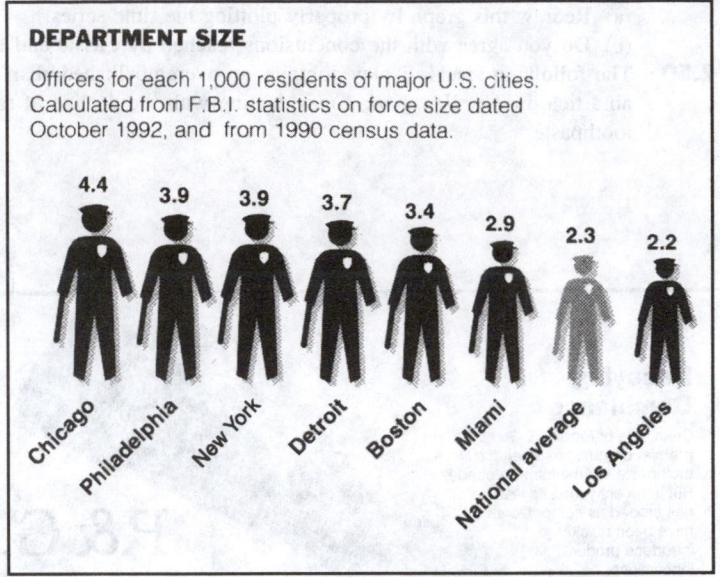

2.52 The following visual display appeared in *The New York Times* and permits a comparison of U.S. trade with both Japan and China on an annual basis for more than a quarter of a century.

Comparing U.S. trade with Japan and China

Source: Reprinted by permission of The New York Times. The New York Times, *March 2, 1997, 14.*

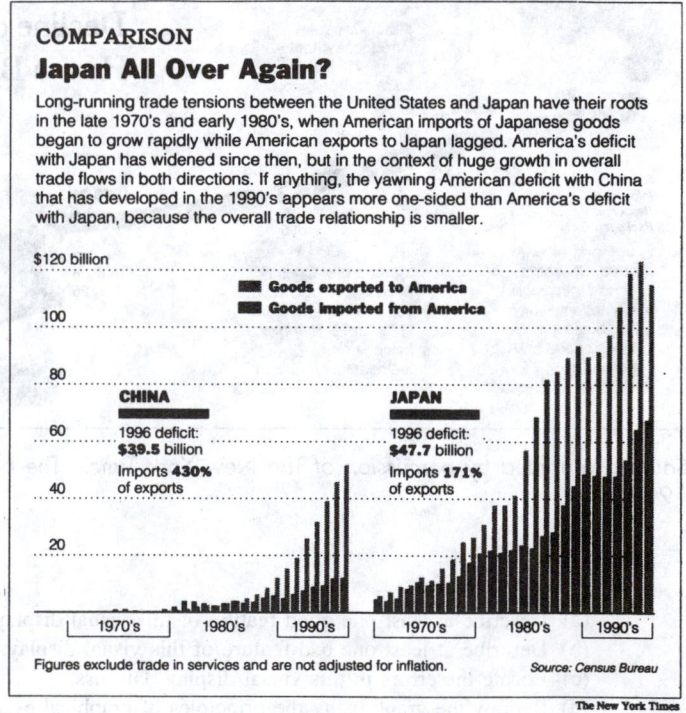

(a) Defend or refute the following statement: "This is an attractive, well-designed visual display that permits direct comparisons of trade over time between the United States and China and between the United States and Japan."

(b) Defend or refute the following statement: "The graphic 'keys' for both China and Japan adequately assist the reader in interpreting the data."

◆ SUMMARY

As you can see in the following summary chart, this chapter was about data presentation. Once the collected data have been presented in tabular and chart format, we are ready to make the analyses. In the following chapter, a variety of descriptive summary measures useful for data analysis and interpretation will be developed.

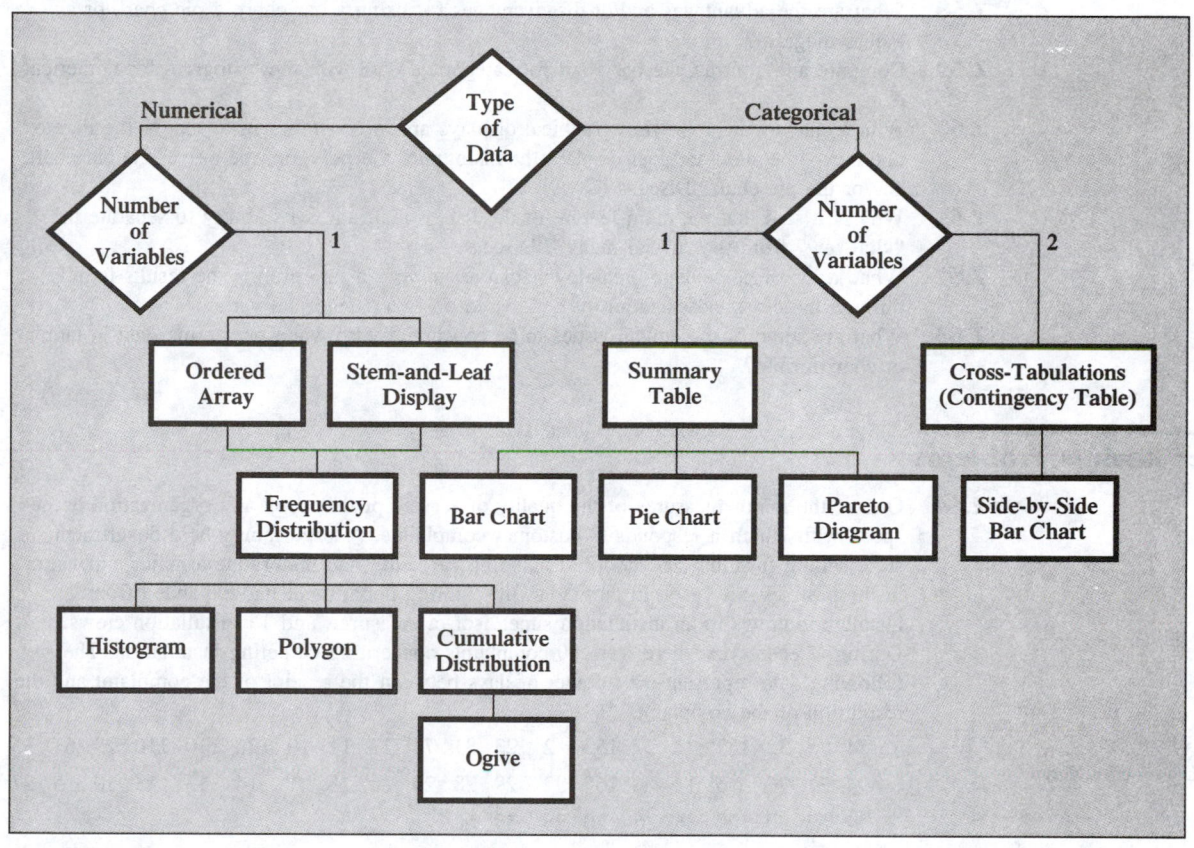

Chapter 2 Summary Chart

Key Terms

bar chart 97	contingency table or cross-classification table 115	data-ink ratio 124
chartjunk 125		frequency distribution 72
class boundaries 73	cumulative percentage distribution table 75	graphical excellence 124
class groupings 72		histogram 77
class midpoint 73	cumulative percentage polygon 79	lie factor 125

Checking Your Understanding

2.53 Why is it necessary to organize a set of numerical data that we collect?

2.54 What are the main differences between an ordered array and a stem-and-leaf display?

2.55 How do histograms and polygons differ with respect to their construction and use?

2.56 Why is the percentage ogive such a useful tool?

2.57 Why would you construct a frequency and percentage summary table?

2.58 What are the advantages and/or disadvantages for using a bar chart, a pie chart, or a Pareto diagram?

2.59 Compare and contrast the bar chart for categorical data with the histogram for numerical data.

2.60 Which ones of the following graphical displays are most similar in format to the Pareto diagram—the stem-and-leaf display, the histogram, the polygon, the ogive, the bar chart, and/or the pie chart? Discuss.

2.61 Why is it said that the main feature of the Pareto diagram is its ability to separate the "vital few" from the "trivial many"? Discuss.

2.62 What kinds of percentage breakdowns can assist you in interpreting the results found through the cross-classification of data based on two categorical variables?

2.63 What are some of the ethical issues to be concerned with when presenting data in tabular or chart format?

Chapter Review Problems

• **2.64** One of the major measures of the quality of service provided by any organization is the speed with which it responds to customer complaints. A large family-held department store selling furniture and flooring including carpeting had undergone a major expansion in the past several years. In particular, the flooring department had expanded from 2 installation crews to an installation supervisor, a measurer, and 15 installation crews. During a recent year there were 50 complaints concerning carpeting installation. The following data represent the number of days between the receipt of the complaint and the resolution of the complaint.

DATA FILE
FURNITURE.XLS

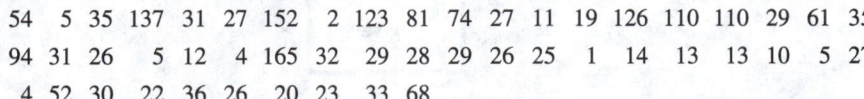

54   5 35 137 31 27 152   2 123 81 74 27 11 19 126 110 110 29 61 35
94 31 26   5 12   4 165 32 29 28 29 26 25   1 14 13 13 10   5 27
 4 52 30 22 36 26 20 23 33 68

(a) Form the frequency distribution and percentage distribution.

(b) Plot the histogram.

(c) Plot the percentage polygon.

(d) Form the cumulative percentage distribution.

(e) Plot the ogive (cumulative percentage polygon).

(f) On the basis of the results of (a)–(e), does there appear to be a great deal of variation in the time it takes to resolve complaints? Explain.

(g) On the basis of the results of (a)–(e), if you had to tell the president of the company how long a customer should expect to wait to have a complaint resolved, what would you say? Explain.

• 2.65 The following sets of data are based on closing stock price for random samples of 25 issues traded on the American Exchange and 50 issues traded on the New York Exchange:

AMERICAN EXCHANGE (25 ISSUES)		NEW YORK EXCHANGE (50 ISSUES)			
$ 6.88	$ 4.88	$36.50	$26.00	$ 8.75	$29.38
.75	6.38	23.50	19.00	8.62	3.75
3.88	33.62	8.25	46.00	5.75	64.75
4.12	4.88	57.50	23.50	21.88	14.25
11.88	9.00	27.12	22.62	6.12	46.38
15.88	2.00	3.75	12.88	25.00	4.75
16.50	20.00	25.00	5.50	15.88	25.00
8.75	14.25	15.50	37.50	24.00	35.00
9.25	4.00	36.12	9.88	10.88	9.00
7.50	15.25	6.00	59.12	18.75	12.38
5.38	2.38	9.12	35.25	53.88	31.00
14.38	49.50	33.38	20.62	20.38	
2.50		22.50	24.00	80.50	

DATA FILE
NYSEAX.XLS

(a) Using interval widths of $10, form the frequency distribution and percentage distribution for each exchange.

(b) Plot the frequency histogram for each exchange.

(c) On one graph, plot the percentage polygon for each exchange.

(d) Form the cumulative percentage distribution for each exchange.

(e) On one graph, plot the ogive (cumulative percentage polygon) for each exchange.

(f) On the basis of parts (a)–(e), can you conclude that there is a difference in the closing prices of stocks traded on both exchanges? Explain.

2.66 For a sample of 40 pizza products, the following data represent pie weight in ounces (PWt), cost of a slice in dollars (SCost), amount of calories per slice (SCal), and amount of fat in grams (SFat) per slice for three types of products—pizza-chain cheese (type-1), supermarket cheese (type-2), and supermarket pepperoni (type-3):

(a) For each of the four numerical variables in the data set (weight, cost, calories, and fat)
  (1) develop the ordered array.
  (2) form the stem-and-leaf display.

(b) On the basis of whether the pizza product is classified as either chain or supermarket cheese versus supermarket pepperoni, for each of the four numerical variables in the data set (weight, cost, calories, and fat)
  (1) develop the ordered array.
  (2) form the stem-and-leaf display.

(c) Construct separate frequency distributions and percentage distributions for weight, cost, and calories.

(d) Form the respective cumulative frequency distributions and cumulative percentage distributions for weight, cost, and calories.

(e) Plot the respective percentage polygons.

(f) Plot the respective ogives (cumulative percentage polygons).

(g) Construct scatter diagrams of weight and cost, weight and calories, weight and fat, cost and calories, cost and fat, and calories and fat.

PRODUCT	PWT	SCOST	SCAL	SFAT	TYPE
Pizza Hut Hand Tossed	31	1.51	305	9	1
Domino's Deep Dish	29	1.53	382	16	1
Pizza Hut Pan Pizza	31	1.51	338	14	1
Domino's Hand Tossed	21	1.90	327	9	1
Little Caesars Pan! Pan!	26	1.23	309	10	1
Little Caesars Pizza! Pizza!	25	1.28	313	11	1
Pizza Hut Stuffed Crust	47	1.23	349	13	1
DiGiorno Rising Crust Four Cheese	29	0.90	332	10	2
Tombstone Special Order Four Cheese	26	0.85	364	17	2
Red Baron Premium 4-Cheese	22	0.80	393	19	2
Boboli crust with Boboli sauce	29	1.00	347	12	2
Jack's Super Cheese	17	0.69	350	17	2
Pappalo's Three Cheese	19	0.75	353	12	2
Tombstone Original Extra Cheese	21	0.81	357	16	2
Master Choice Gourmet Four Cheese	17	0.90	296	13	2
Celeste Pizza For One	7	0.92	358	16	2
Totino's Party	10	0.64	322	14	2
The New Weight Watchers Extra Cheese	6	1.54	337	10	2
Jeno's Crisp 'N Tasty	7	0.72	323	14	2
Stouffer's French Bread2—Cheese	10	1.15	333	13	2
Ellio's 9-slice	24	0.52	299	9	2
Kroger	7	0.72	316	7	2
Healthy Choice French Bread	6	1.50	275	4	2
Lean Cuisine French Bread	6	1.49	288	7	2
DiGiorno Rising Crust	30	0.87	360	15	3
Tombstone Special Order	27	0.81	394	21	3
Pappalo's	20	0.73	390	17	3
Jack's New More Cheese!	18	0.64	372	21	3
Tombstone Original	22	0.77	387	20	3
Red Baron Premium	22	0.80	409	22	3
Tony's Italian Style Pastry Crust	15	0.83	436	25	3
Red Baron Deep Dish Singles	12	1.13	442	26	3
Totino's Party	10	0.62	367	20	3
The New Weight Watchers	6	1.52	348	11	3
Jeno's Crisp 'N Tasty	7	0.71	365	19	3
Stouffer's French Bread	11	1.14	370	18	3
Celeste Pizza For One	7	1.11	381	20	3
Tombstone For One French Bread	12	1.11	361	15	3
Healthy Choice French Bread	6	1.46	264	3	3
Lean Cuisine French Bread	5	1.71	312	7	3

DATA FILE
PIZZA.XLS

Source: "Pizza," Copyright © 1997 by Consumers Union of U.S. Adapted from CONSUMER REPORTS, *January 1997, 26–28, by permission of Consumers Union of U.S., Inc., Yonkers, NY 10703-1057. Although these data sets originally appeared in* CONSUMER REPORTS, *the selective adaptation and resulting conclusions presented are those of the authors and are not sanctioned or endorsed in any way by Consumers Union, the publisher of* CONSUMER REPORTS.

(h) What is the typical pie weight for these pizzas? Explain.

(i) What is the typical cost per slice for these pizzas? Explain.

(j) What is the typical amount of calories per slice for these pizzas? Explain.

(k) On the basis of the results of (g), do any of the variables of weight, cost, calories, and fat seem to be related? Explain.

2.67 The following data are intended to show the gap between families with the highest income and families with the lowest income in each of the 50 states and the District of Columbia as measured by the average of the bottom fifth and the top fifth of families with children during 1994–1996. The results classified by states were as follows:

STATE	BOTTOM FIFTH ($000)	TOP FIFTH ($000)	STATE	BOTTOM FIFTH ($000)	TOP FIFTH ($000)
New York	6.787	132.390	Kansas	10.790	110.341
Louisiana	6.430	102.339	Oregon	9.627	97.589
New Mexico	6.408	91.741	New Jersey	14.211	143.010
Arizona	7.273	103.392	Indiana	11.115	110.876
Connecticut	10.415	147.594	Montana	9.051	89.902
California	9.033	127.719	South Dakota	9.474	93.822
Florida	7.705	107.811	Idaho	10.721	104.725
Kentucky	7.364	99.210	Delaware	12.041	116.965
Alabama	7.531	99.062	Arkansas	8.995	83.434
West Virginia	6.439	84.479	Colorado	14.326	131.368
Tennessee	8.156	106.966	Hawaii	12.735	116.060
Texas	8.642	113.149	Missouri	11.090	100.837
Mississippi	6.257	80.980	Alaska	14.868	129.065
Michigan	9.257	117.107	Wyoming	11.174	94.845
Oklahoma	7.483	94.380	Minnesota	14.655	120.344
Massachusetts	10.694	132.962	Nebraska	12.546	102.992
Georgia	9.978	123.837	Maine	11.275	92.457
Illinois	10.002	123.233	New Hampshire	14.299	116.018
Ohio	9.346	111.894	Nevada	12.276	98.693
South Carolina	8.146	96.712	Iowa	13.148	104.253
Pennsylvania	10.512	124.537	Wisconsin	13.398	103.551
North Carolina	9.363	107.490	Vermont	13.107	97.898
Rhode Island	9.914	111.015	North Dakota	12.424	91.041
Washington	10.116	112.501	Utah	15.709	110.938
Maryland	13.346	147.971	District of		
Virginia	10.816	116.202	Columbia	5.293	149.508

Source: United States Census Bureau.

DATA FILE
STATEINC.XLS

(a) For each of the numerical variables
  (1) develop the ordered array.
  (2) form the stem-and-leaf display.

(b) Construct separate frequency distributions and percentage distributions for income of the bottom fifth and top fifth of families.

(c) Form the respective cumulative frequency distributions and cumulative percentage distributions for income of the bottom fifth and top fifth of families.

(d) Plot the respective percentage polygons.

(e) Plot the respective ogives (cumulative percentage polygons).

(f) What conclusions can you reach concerning the average income in the bottom fifth of families?

(g) What conclusions can you reach concerning the average income in the top fifth of families?

(h) For each state, obtain the ratio of the income of the top fifth divided by the income of the bottom fifth. Set up a stem-and-leaf display of this ratio.

(i) What conclusions can you reach concerning the ratio of the average income in the top fifth of families divided by the average income in the bottom fifth of families?

2.68 The following data represent the number of daily calls received at a toll-free telephone number of a large European airline over a period of 30 consecutive nonholiday workdays (Monday to Friday):

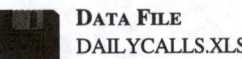

DATA FILE
DAILYCALLS.XLS

DAY	NUMBER OF CALLS	DAY	NUMBER OF CALLS	DAY	NUMBER OF CALLS	DAY	NUMBER OF CALLS
1	3,060	9	3,235	17	2,685	25	3,252
2	3,370	10	3,174	18	3,618	26	3,161
3	3,087	11	3,603	19	3,369	27	3,186
4	3,135	12	3,256	20	3,353	28	3,347
5	3,805	13	3,075	21	3,277	29	3,275
6	3,234	14	3,187	22	3,066	30	3,129
7	3,105	15	3,060	23	3,341		
8	3,168	16	3,004	24	3,181		

(a) Form the frequency distribution and percentage distribution.

(b) Plot the percentage histogram.

(c) Plot the percentage polygon.

(d) Form the cumulative percentage distribution.

(e) Plot the ogive (cumulative percentage polygon).

(f) On the basis of the results of (a)–(e), does there appear to be a great deal of variation in the daily number of calls?

2.69 The following data indicate fat and cholesterol information concerning popular protein foods (fresh red meats, poultry, and fish):

FOOD	CALORIES (G)	PROTEIN	FAT	% CALORIES FROM SATURATED FAT	CHOLESTEROL (MG)
Beef, ground, extra lean	250	25	58	23	82
Beef, ground, regular	287	23	66	26	87
Beef, round	184	28	24	12	82
Brisket	263	28	54	21	91
Flank steak	244	28	51	22	71

(continued)

FOOD	CALORIES (G)	PROTEIN	FAT	% CALORIES FROM SATURATED FAT	CHOLESTEROL (MG)
Lamb leg roast	191	28	38	16	89
Lamb loin chop, broiled	215	30	42	17	94
Liver, fried	217	27	36	12	482
Pork loin roast	240	27	52	18	90
Sirloin	208	30	37	15	89
Spareribs	397	29	67	27	121
Veal cutlet, fried	183	33	42	20	127
Veal rib roast	175	26	37	15	131
Chicken, with skin, roasted	239	27	51	14	88
Chicken, no skin, roasted	190	29	37	10	89
Turkey, light meat, no skin	157	30	18	6	69
Clams	98	16	6	0	39
Cod	98	22	8	1	74
Flounder	99	21	12	2	54
Mackerel	199	27	77	20	100
Ocean perch	110	23	13	3	53
Salmon	182	27	24	5	93
Scallops	112	23	8	1	56
Shrimp	116	24	15	2	156
Tuna	181	32	41	10	48

Source: United States Department of Agriculture.

DATA FILE
PROTEIN.XLS

For the data relating to the amount of calories, protein, the percentage of calories from fat and from saturated fat, and the cholesterol for the popular protein foods:

(a) Construct the stem-and-leaf display.
(b) Construct the frequency distribution and the percentage distribution.
(c) Plot the percentage histogram.
(d) Plot the percentage polygon.
(e) Form the cumulative percentage distribution.
(f) Plot the cumulative percentage polygon.
(g) What conclusions can you reach concerning the amount of calories in these foods?

(h) What conclusions can you reach concerning the amount of protein in these foods?
(i) What conclusions can you reach concerning the percentage of calories from fat and saturated fat in these foods?
(j) Are there any foods that seem very different from the others in terms of these variables? Explain.

2.70 A wholesale appliance-distributing firm wished to study its accounts receivable for two successive months. Two independent samples of 50 accounts were selected for each of the two months. The results are summarized in the following table:

Frequency distributions for accounts receivable

AMOUNT	MARCH FREQUENCY	APRIL FREQUENCY
$0 to under $2,000	6	10
$2,000 to under $4,000	13	14
$4,000 to under $6,000	17	13
$6,000 to under $8,000	10	10
$8,000 to under $10,000	4	0
$10,000 to under $12,000	0	3
Total	50	50

(a) Plot the frequency histogram for each month.
(b) On one graph, plot the percentage polygon for each month.
(c) Form the cumulative percentage distribution for each month.
(d) On one graph, plot the ogive (cumulative percentage polygon) for each month.
(e) On the basis of parts (a)–(d), do you think the distribution of the accounts receivable has changed from March to April? Explain.

2.71 The following table contains the cumulative distributions and cumulative percentage distributions of braking distance (in feet) at 80 miles per hour for a sample of 25 U.S.-manufactured automobile models and for a sample of 72 foreign-made automobile models obtained in a recent year:

Cumulative frequency and percentage distributions for the braking distance (in feet) at 80 mph for U.S.-manufactured and foreign-made automobile models

BRACKING DISTANCE (IN FT)	U.S.-MADE AUTOMOBILE MODELS "LESS THAN" INDICATED VALUES		FOREIGN-MADE AUTOMOBILE MODELS "LESS THAN" INDICATED VALUES	
	NUMBER	PERCENTAGE	NUMBER	PERCENTAGE
210	0	0.0	0	0.0
220	1	4.0	1	1.4
230	2	8.0	4	5.6
240	3	12.0	19	26.4
250	4	16.0	32	44.4
260	8	32.0	54	75.0
270	11	44.0	61	84.7
280	17	68.0	68	94.4
290	21	84.0	68	94.4
300	23	92.0	70	97.2
310	25	100.0	71	98.6
320	25	100.0	72	100.0

Based on these data, answer the following questions:
(a) How many models of U.S.-made automobiles have braking distances of 240 feet or more?
(b) What is the percentage of U.S.-made automobiles with braking distances less than 260 feet?
(c) Which group of car models—U.S.-made or foreign-made—have the wider range in braking distance?

(d) How many foreign-made automobile models have braking distances between 260 feet and 269.9 feet (inclusive)?

(e) Use the cumulative distributions to construct the frequency distributions and percentage distributions for each group of car models.

(f) On one graph, plot the two percentage ogives.

(g) Do you think there is a difference in the braking distance between the two groups of car models? Explain.

2.72 At the start of the 1995 National Football League season, Jerry Jones, owner of the Dallas Cowboys, argued that because his team was more popular than other teams and generated the most revenues from sales of team products, it deserved a proportionally higher share in earnings that, under league policy, are distributed equally to all 30 teams by NFL properties. The following summary tables respectively display the percentage market share of sales of licensed products by the various teams, the percentage of sales attributed to the specific kinds of products sold, and, last, the amount of revenues generated (in millions of dollars) from various sources:

TEAM	% SALES OF LICENSED PRODUCTS
Carolina	5.0
Dallas	20.8
Green Bay	5.1
Kansas City	4.8
Miami	6.6
New England	3.6
New York Giants	3.2
Oakland	4.7
Pittsburgh	4.7
San Francisco	11.3
Other 20 teams	30.2
Total	100.0

TYPE OF PRODUCT	% SALES
Clothing/hats	50.0
Home products	7.0
Toys/sporting goods	11.0
Trading cards	14.0
Other	18.0
Total	100.0

SOURCE	REVENUES (IN MILLIONS OF $)
International marketing	14.9
Publishing	10.6
Retail licensing	94.3
Special events	3.8
Sponsorships	58.8
Trading cards	15.3
Other operations	17.9
Total	215.6

Sources: Reprinted by permission of The New York Times. *NFL Properties;* The New York Times, *September 24, 1995, D1.*

(a) For the data on percentage of sales of licensed products, construct
   (1) a bar chart.
   (2) a pie chart.
   (3) a Pareto diagram.

(b) Which of these graphical displays do you prefer for purposes of presentation? Why?

(c) For the data on percentage of sales by type of product, develop the appropriate graph to pinpoint the "vital few" from the "trivial many."

(d) For the data on the amount of revenues generated (in millions of dollars) from various sources, develop the appropriate graph to pinpoint the "vital few" from the "trivial many."

(e) Analyze the data and summarize your findings.

• **2.73** The following data represent the trade relationships (in billions of dollars) between the United States and China in the year 1996:

Chinese and U.S. trade in 1996

CHINESE GOODS EXPORTED TO U.S.	BILLIONS OF $
Apparel, fabrics, and fibers	7.38
Electrical components, equipment, and parts	9.22
Food, agriculture, and forest products	1.77
Footwear	6.70
Furniture, bedding, and lighting	2.62
Leather goods (other than footwear)	2.76
Machinery, equipment, and parts (incl. computers)	4.64
Mineral, chemical, plastic, and rubber goods	3.99
Toys, games, and sporting goods	7.99
All other goods	7.33
Total	54.40

U.S. GOODS EXPORTED TO CHINA	BILLIONS OF $
Aircraft, spacecraft, and parts	1.71
Electronic components, equipment, and parts	1.43
Food and other farm produce	1.34
Leather, fur, tobacco, wood, cork, and paper	0.60
Machinery, equipment, and parts (incl. computers)	2.31
Mineral, chemical, plastic, and rubber goods	1.96
Textiles, apparel, and fibers	0.95
All other goods	1.69
Total	11.99

Sources: Reprinted by permission of The New York Times. *National Trade Data Bank,* The New York Times, *March 2, 1997, 14.*

(a) For the data on U.S. goods exported to China, construct
   (1) a bar chart.
   (2) a pie chart.
   (3) a Pareto diagram.
(b) For the data on Chinese goods exported to the U.S., construct
   (1) a bar chart.
   (2) a pie chart.
   (3) a Pareto diagram.
(c) Which pair of graphs do you prefer for purposes of presentation? Why?
(d) What differences are there between Chinese goods exported to the United States and U.S. goods exported to China? Explain.

2.74 It has been noted by consumer researchers that where people buy computers depends on whether they are buying one for the first time. First-time buyers tend to buy at big retail stores, whereas repeat buyers favor catalogues, small stores that assemble computers, and manufacturers that market directly.

The table at the top of page 143 provides a percentage breakdown of where first-time buyers and repeat buyers purchased personal computers in 1996:

WHERE PURCHASED	FIRST-TIME BUYERS	REPEAT BUYERS
Assembly stores	15%	23%
Catalogues/resellers	5	10
Computer superstores	23	21
Consumer electronics stores	28	11
Manufacturers	8	20
Mass merchants	11	4
Office superstores	4	5
Other	6	6
Total	100	100

In addition, the following table displays the percentage of sales of personal computers that were purchased in various types of stores in 1996:

STORE CATEGORY	% 1996 SALES
Assembly stores	20
Catalogues/resellers	8
Computer superstores	22
Consumer electronics stores	18
Manufacturers	15
Mass merchants	7
Office superstores	5
Other	5
Total	100

Sources: Reprinted by permission of The New York Times. *Computer Intelligence and* The New York Times, *January 8, 1997, D1–D2.*

(a) For each table, construct an appropriate graph and analyze the data.
(b) Discuss the implications of these shifting trends in repeat purchases and how this may affect product promotion and advertising. What marketing strategy might be useful for selling to people buying computers for the first time? Explain.

2.75 The owner of a restaurant serving Continental-style entrées was interested in studying patterns of demand by patrons for the Friday to Sunday weekend time period. Records were maintained that indicated the number of entrées ordered for each type. The following data were obtained:

TYPE OF ENTRÉE	NUMBER SERVED
Beef	187
Chicken	103
Duck	25
Fish	122
Pasta	63
Shellfish	74
Veal	26

(a) Construct a bar chart for the types of entrées ordered.

(b) Construct a Pareto diagram for the types of entrées ordered.

(c) Construct a pie chart for the types of entrées ordered.

(d) Do you prefer a Pareto diagram or a pie chart for these data? Why?

(e) What conclusions can the owner draw concerning demand for different types of entrées?

Suppose that the owner was also interested in studying the demand for dessert during the same time period. She decided that two other variables were to be studied along with whether a dessert was ordered, the gender of the individual, and whether a beef entrée was ordered. The results were as follows:

DESSERT ORDERED	GENDER		TOTAL
	MALE	FEMALE	
Yes	96	40	136
No	224	240	464
Total	320	280	600

DESSERT ORDERED	BEEF ENTRÉE		TOTAL
	YES	NO	
Yes	71	65	136
No	116	348	464
Total	187	413	600

Concerning each of the two cross-classification tables:

(f) Construct a table of row percentages.

(g) Construct a table of column percentages.

(h) Construct a table of total percentages.

(i) Which type of percentage (row, column, or total) do you think is most informative for each table? Explain.

(j) What conclusions concerning the pattern of dessert ordering can the owner of the restaurant reach?

2.76 (**Class Project**) Let each student in the class respond to the question "Which type of carbonated soft drink do you most prefer?" so that the teacher may tally the results into a summary table on the blackboard.

1. Convert the data to percentages and construct a Pareto diagram.

2. Analyze the findings.

2.77 (**Class Project**) Let each student in the class be cross-classified on the basis of gender (male, female) and current employment status (yes, no) so that the results are tallied on the blackboard.

(a) Construct a table with either row or column percentages, depending on which you think is more informative.

(b) What would you conclude from this study?

(c) What other variables would you want to know regarding employment in order to enhance your findings?

 ## TEAM PROJECT

TP2.1 A financial investment service is evaluating the list of domestic general stock funds so that it can make purchase recommendations to potential investors. The vice president for research has hired your group, the _____ Corporation, to study the financial characteristics of currently traded

domestic general stock funds. The vice president is interested in a comparison of some features of these funds based on fee structure (no-load versus fee-payment), objective (growth fund versus blend fund), and capitalization size of companies making up a fund's portfolio (large, mid, or small). Having access to Special Data Set 1 in appendix D pertaining to various characteristics from a sample of 194 domestic general stock funds with high Morningstar Inc. dual ratings of 4 or 5, the _____ Corporation is ready to

DATA FILE
MUTUAL.XLS

(a) outline how the group members will proceed with their tasks.

(b) form the respective frequency and percentage distributions of net asset value (in dollars) and 3-year annualized total returns (in percentage rates) for the 107 no-load funds versus the 87 fee-payment funds.

(c) form the respective frequency and percentage distributions of net asset value (in dollars) and 3-year annualized total returns (in percentage rates) for the 59 growth funds versus the 135 blend funds.

(d) form the respective frequency and percentage distributions of net asset value (in dollars) and 3-year annualized total returns (in percentage rates) for the 119 large-size funds, the 44-mid size funds, and the 31 small-size funds.

(e) on the basis of parts (b), (c), and (d), plot the various percentage polygons.

(f) form the needed cumulative percentage distributions.

(g) plot the various percentage ogives.

(h) additionally, form a contingency table cross-classifying type of fee schedule (fees from fund assets, deferred fee, front-load fee, multiple fees, or no load) with type of fund (large-growth, mid-growth, small-growth, large-blend, mid-blend, or small-blend). Discuss the results.

(i) form a Pareto diagram for the type of fund (large-growth, mid-growth, small-growth, large-blend, mid-blend, or small-blend). Discuss.

(j) write and submit a summary of your descriptive analysis, attaching all tables and charts.

(k) prepare and deliver a 15-minute oral presentation to the vice president for research at this financial investment service.

Note: Additional team projects can be found on the World Wide Web site for this text
http://www.prenhall.com/levine

These team projects deal with characteristics of 80 universities and colleges (see the UNIV&COL file) and the features of 89 automobiles (see the AUTO96 file).

THE SPRINGVILLE HERALD CASE

BACKGROUND

Springville represents a suburban area that is about 50 miles outside a large city in the western United States. The area was heavily agricultural before World War II and experienced a great expansion in population and industry between 1950 and 1980, with little growth in the years since 1980. The *Herald* was originally a family-owned newspaper that published only a daily and Sunday edition since 1957. Current circulation is 250,000 on weekdays (Monday–Saturday) and 300,000 on Sunday, growing only moderately since 1980. The financial status of the company is healthy, but senior management has become more conscious of costs and the need to improve the efficiency of operations.

PHASE 1

A task force consisting of corporate-level officers and department heads was formed to consider how to go about the quality improvement effort. There was agreement that the first step was the development of a mission statement for the newspaper that could succinctly communicate its mission to both customers and employees.

Once the mission statement was developed with the aid of both customers and employees at all levels of the organization, the task force turned to a discussion of which areas of operations should be examined for improvement opportunities. After much brainstorming and discussion, the task force decided by consensus that one critical area for improvement was represented by errors that were made in the process of filling advertising orders (which represented a critical revenue source for the newspaper) from the time the advertisement was ordered to when it actually appeared in the newspaper. Unfortunately, under certain circumstances, errors had been made and incorrect ads were displayed in the newspaper or may have been printed on the wrong day. Such occurrences called for an immediate effort to satisfy the customer by a variety of sometimes costly devices including refunds and rerunning the ads on other days. Members of the task force realized that data relating to the occurrence of these errors were already available from periodic reports that were routinely generated by the advertising production department. One such report containing the number of occurrences for each type of error in the most recent calendar year is presented in Table SH2.1.

Table SH2.1 *Summary table of errors for advertising production, composing room, policy, and sales for last calendar year*

TYPE OF ERROR	TALLY	TOTAL
Copy error	‖‖ ‖‖	54
Layout	‖‖ ‖	7
Omits	‖‖ ‖‖ ‖	13
Paste-up	‖‖ ‖‖ ‖	11
Poor reproduction	‖‖ ‖	8
Ran in error	‖‖ ‖‖ ‖‖ ‖‖ ‖‖ ‖‖	30
Rate quote	‖‖ ‖‖ ‖	13
Space not ordered	‖‖ ‖	7
Typesetting	‖‖ ‖‖ ‖‖ ‖‖ ‖‖ ‖‖ ‖‖ ‖‖ ‖‖ ‖‖ ‖‖ ‖	53
Velox	‖‖ ‖‖ ‖‖ ‖‖ ‖‖ ‖	28
Wrong ad	‖‖ ‖‖ ‖‖ ‖‖ ‖‖	25
Wrong date	‖‖ ‖‖ ‖	14
Wrong position	‖‖ ‖‖ ‖‖ ‖‖ ‖‖ ‖‖ ‖‖ ‖‖ ‖‖	45
Wrong manual paste-up	‖‖	5
Wrong size	‖‖ ‖	6
Total		319

Exercises

2.1 Develop the appropriate table for the data presented in Table SH2.1.

2.2 (a) Construct the graphical presentation you think is most appropriate and useful in gaining insights from the data of Table SH2.1.

(b) Explain in detail why you chose the graph in (a) instead of other alternative graphical presentations.

(c) On the basis of results of the frequency and percentage distribution and the graph that was constructed in (a), write a report to management concerning the frequency of different types of advertising errors.

2.3 At this stage of the analysis, what other information concerning the different types of errors would be useful to obtain?

 Do not continue until the Phase 1 exercises have been completed.

Phase 2

At the first meeting of the task force after the data of Table SH2.1 were made available, Bob Tatum, the head of advertising production, suggested that this was not the most appropriate way to examine the problem. He stated that the frequency of the errors was not the only issue. He argued that certain types of errors, although perhaps less frequent, might involve much greater costs than other errors. Fortunately, data concerning the cost of each type of error were also available. The data for the most recent year are summarized in Table SH2.2.

Table SH2.2 *Cost of advertising errors for the last calendar year*

TYPE OF ERROR	AMOUNT ($000)	TYPE OF ERROR	AMOUNT ($000)
Copy error	32.6	Typesetting	53.1
Layout	3.0	Velox	23.3
Omits	36.5	Wrong ad	53.6
Paste-up	59.4	Wrong date	35.9
Poor reproduction	13.0	Wrong position	74.9
Ran in error	108.2	Wrong manual paste-up	16.5
Rate quote	5.3	Wrong size	5.3
Space not ordered	12.9	Total	533.5

Exercises

2.4 Prepare the appropriate table for the data presented in Table SH2.2.

2.5 (a) Construct the graphical presentation you think is most appropriate for the data presented in Table SH2.2.

(b) Explain in detail why you chose the graph in (a) instead of other alternative graphical presentations.

(c) On the basis of results of the percentage distribution and the graph that was constructed in (a), write a report to management concerning the dollar amount of the different types of advertising errors.

2.6 On the basis of your analysis in Exercise 2.5 (c), what action would you recommend be taken next to study the reasons for advertising errors?

 Do not continue until the Phase 2 exercises have been completed.

Phase 3

Once the data of Table SH2.2 had been analyzed, it became evident that the most costly error involved the ran-in-error category, which accounted for over $100,000, or more than 20% of the total cost of the errors for the year. Further investigation of the errors in this category subdivided errors into different types. This is presented in Table SH2.3 on page 148.

Table SH2.3 — *Frequency and dollar amount of different types of ran-in-error problems in most recent calendar year*

TYPE	FREQUENCY	AMOUNT ($000)
Composing room	10	12.8
Policy	16	88.7
Sales	4	6.6
Total	30	108.1

Exercises

2.7 Prepare a percentage summary table for the data of Table SH2.3.

2.8 (a) Construct the graphical presentation you think is most appropriate for the data of Table SH2.3.

(b) Explain in detail why you chose the graph in (a) instead of other alternative graphical presentations.

(c) If the graph chosen in (a) was not the same one selected in Exercises 2.2(a) and 2.5(a), explain why you chose it.

(d) On the basis of results of the percentage distribution and the graph that was constructed in (a), write a report to management concerning the dollar amount of the different types of ran-in-error problems.

(e) What course of action would you recommend be taken next to reduce these types of errors in the future?

 Do not continue until the Phase 3 exercises have been completed.

Phase 4

One of the functions of the computer systems department of the newspaper relates to the reporting of the activities of the mainframe computer system. Typically, in any given day more than 100 different jobs need to be processed on the system. These jobs vary in requirements, from very small jobs that require a minimum of access to data cartridge storage devices, to large complex jobs that need to access in excess of 200 different data cartridges. The data presented in Table SH2.4 consist of an ordered array of the number of data cartridges that needed to be accessed by 111 jobs on a recent day.

Table SH2.4 — *Ordered array of number of data cartridges accessed per job on a recent day*

1	1	1	1	1	1	2	2	2	2	2	3	3	3	3	4	4	4	4
4	4	5	5	5	5	5	5	5	6	6	6	7	7	7	7	8	8	8
8	9	10	10	10	10	10	11	12	12	13	14	14	15	17	18	18	18	18
19	20	20	20	20	21	22	23	24	28	28	29	30	30	30	30	31	32	33
35	37	40	40	42	43	50	52	55	56	59	60	60	67	74	80	86	91	94
96	100	111	126	127	131	137	140	144	147	164	166	170	182	212	237			

DATA FILE
SH2.XLS

120

Exercises

2.9 (a) Set up all appropriate tables and charts for the number of data cartridges accessed by jobs on a recent day.

(b) Write a report to management that summarizes the results obtained from the tables and charts developed in (a).

References

1. Chambers, J. M., W. S. Cleveland, B. Kleiner, and P. A. Tukey, *Graphical Methods for Data Analysis* (Boston, MA: Duxbury Press, 1983).

2. Cleveland, W. S., "Graphs in Scientific Publications," *The American Statistician* 38 (November 1984), 261–269.

3. Cleveland, W. S., "Graphical Methods for Data Presentation: Full Scale Breaks, Dot Charts, and Multibased Logging," *The American Statistician* 38 (November 1984), 270–280.

4. Cleveland, W. S., and R. McGill, "Graphical Perception: Theory, Experimentation, and Application to the Development of Graphical Methods," *Journal of the American Statistical Association* 79 (September 1984), 531–554.

5. Croxton, F., D. Cowden, and S. Klein, *Applied General Statistics*, 3d ed. (Englewood Cliffs, NJ: Prentice Hall, 1967).

6. Ehrenberg, A. S. C., "Rudiments of Numeracy," *Journal of the Royal Statistical Society,* series A, vol. 140 (1977), 277–297.

7. Huff, D., *How to Lie with Statistics* (New York: Norton, 1954).

8. Kimble, G. A., *How to Use (and Misuse) Statistics* (Englewood Cliffs, NJ: Prentice Hall, 1978).

9. *Microsoft Excel 97* (Redmond, WA: Microsoft Corporation, 1997).

10. Tufte, E. R., *The Visual Display of Quantitative Information* (Cheshire, CT: Graphics Press, 1983).

11. Tufte, E. R., *Envisioning Information* (Cheshire, CT: Graphics Press, 1990).

12. Tufte, E. R., *Visual Explanations* (Cheshire, CT: Graphics Press, 1997).

13. Tukey, J., *Exploratory Data Analysis* (Reading, MA: Addison-Wesley, 1977).

14. Velleman, P. F., and D. C. Hoaglin, *Applications, Basics, and Computing of Exploratory Data Analysis* (Boston, MA: Duxbury Press, 1981).

15. Wainer, H., "How to Display Data Badly," *The American Statistician* 38 (May 1984), 137–147.

16. Wainer, H., *Visual Revelations: Graphical Tales of Fate and Deception from Napoleon Bonaparte to Ross Perot* (New York: Copernicus/Springer-Verlag, 1997).

Chapter 2

Student Solutions Manual

•2.2   Stem-and-leaf of Finance Scores
        5   34
        6   9
        7   4
        8   0
        9   38
             $n = 7$

2.4   Stem-and-leaf of Organizational Behavior Scores
        6   38
        7   66
        8   77
        9   5
             $n = 7$

2.6   Ordered array: 50 74 74 76 81 89 92

2.8   (a) Ordered array: 3 4 5 6 7 7 9 10 10 10
                             11 11 12 15 18 18 21 26 33 37
        (b) Stem-and-leaf of Monthly Billing Records
              0   3456778
             10   000112588
             20   16
             30   37
                   $n = 20$
        (c) Amounts owed on monthly billing records are concentrated between $10 and $19.

•2.10   (a) Ordered array: 4 5 5 6 6 6 6 7 7 7 7 7 7 8 8
                              8 8 8 8 8 8 8 9 9 9 9 9 9 10 10
                             10 10 10 10 10 10 11 11 12 12 13 13 14 15 15
                             15 16 16 18 23
        (b) Stem-and-leaf of Book Values
             0   45566667777778888888888999999
             1   000000001122334555668
             2   3
                  $n = 50$
        (c) Book values on the New York Stock Exchange are more likely to be low, since they are concentrated below $10. Better than half of the stocks sampled had book values below $10.
        (d) You are much more likely to find a New York stock with a book value below $10 than above $20. In fact, 28 of the 50 stocks sampled had book values below $10, compared to one stock with a book value above $20.

2.12 (a) Ordered array: 170 170 170 180 180 190 190 200 200 210 220
                                     220 250 250 265 270 300 300 320 340 350 450
      (b) Stem-and-leaf of VCR Model Prices

	A	B	C	D	E	F
1				Stem-and-Leaf Display		
2				for Price ($)		
3				Stem unit: 100		
4						
5	Statistics			1	7 7 7 8 8 9 9	
6	Sample Size	22		2	0 0 1 2 2 5 5 6 7	
7	Mean	245.2273		3	0 0 2 4 5	
8	Median	220		4	5	
9	Std. Deviation	73.26529				
10	Minimum	170				
11	Maximum	450				

      (c) While you can find VCR models priced from $170 to $450, you are more
           likely to find models priced under $200 than over $300.
      (d) Nine of the 22 models are priced from $200 through $290, which is the greatest
           concentration of VCR model prices.

2.14 (a) Width of interval $\cong \dfrac{247,000 - 62,000}{5} = 37,000 \cong 40,000$

Annual Salaries	Midpoint
$60,000 up to $100,000	$80,000
$100,000 up to $140,000	$120,000
$140,000 up to $180,000	$160,000
$180,000 up to $220,000	$200,000
$220,000 up to $260,000	$240,000

      (b) Width of interval $\cong \dfrac{247,000 - 62,000}{6} = 30,833.3\overline{3} \cong 35,000$

Annual Salaries	Midpoint
$60,000 up to $95,000	$77,500
$95,000 up to $130,000	$112,500
$130,000 up to $165,000	$147,500
$165,000 up to $200,000	$182,500
$200,000 up to $235,000	$217,500
$235,000 up to $270,000	$252,500

2.14
Cont.

(c) Width of interval $\cong \dfrac{247,000 - 62,000}{7} \cong 26,430 \cong 30,000$

Annual Salaries	Midpoint
$50,000 up to $80,000	$65,000
$80,000 up to $110,000	$95,000
$110,000 up to $140,000	$125,000
$140,000 up to $170,000	$155,000
$170,000 up to $200,000	$185,000
$200,000 up to $230,000	$215,000
$230,000 up to $260,000	$245,000

(d) Width of interval $\cong \dfrac{247,000 - 62,000}{8} \cong 23,125 \cong 25,000$

Annual Salaries	Midpoint
$50,000 up to $75,000	$62,500
$75,000 up to $100,000	$87,500
$100,000 up to $125,000	$112,500
$125,000 up to $150,000	$137,500
$150,000 up to $175,000	$162,500
$175,000 up to $200,000	$187,500
$200,000 up to $225,000	$212,500
$225,000 up to $250,000	$237,500

2.16 (a) The value appearing in a relative frequency distribution for a class containing 12 out of a total of 40 observations is 12/40 or 0.3.
(b) The value appearing in a percentage distribution for a class containing 12 out of a total of 40 observations is 12/40 • 100% or 30%.

2.18 (a) The percentage polygon is plotted at the class midpoints from a frequency distribution.
(b) The histogram is plotted at the class boundaries from a frequency distribution.
(c) The histogram contains a series of vertical rectangular bars.
(d) The percentage polygon is formed by connecting a set of consecutive plotted points.
(e) The percentage polygon is used for comparing two or more sets of data that have been tallied into corresponding frequency distributions.

•2.20 In constructing an ogive, the vertical axis must show the true zero or "origin" so as not to distort or otherwise misrepresent the character of the data.

2.22 (a) 4% (b) 36% (c) 32% (d) 16 (e) 100%

2.24   (a)

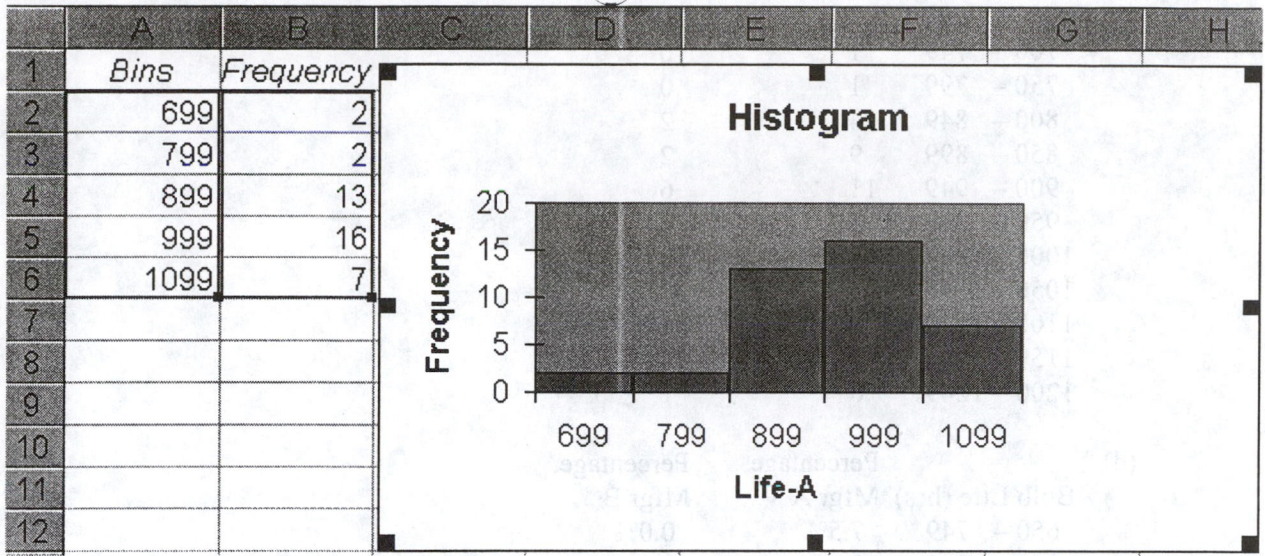

	A	B	C	D	E	F	G	H
1	*Bins*	*Frequency*						
2	699	2						
3	799	2						
4	899	13						
5	999	16						
6	1099	7						
7								
8								
9								
10								
11								
12								

(b)

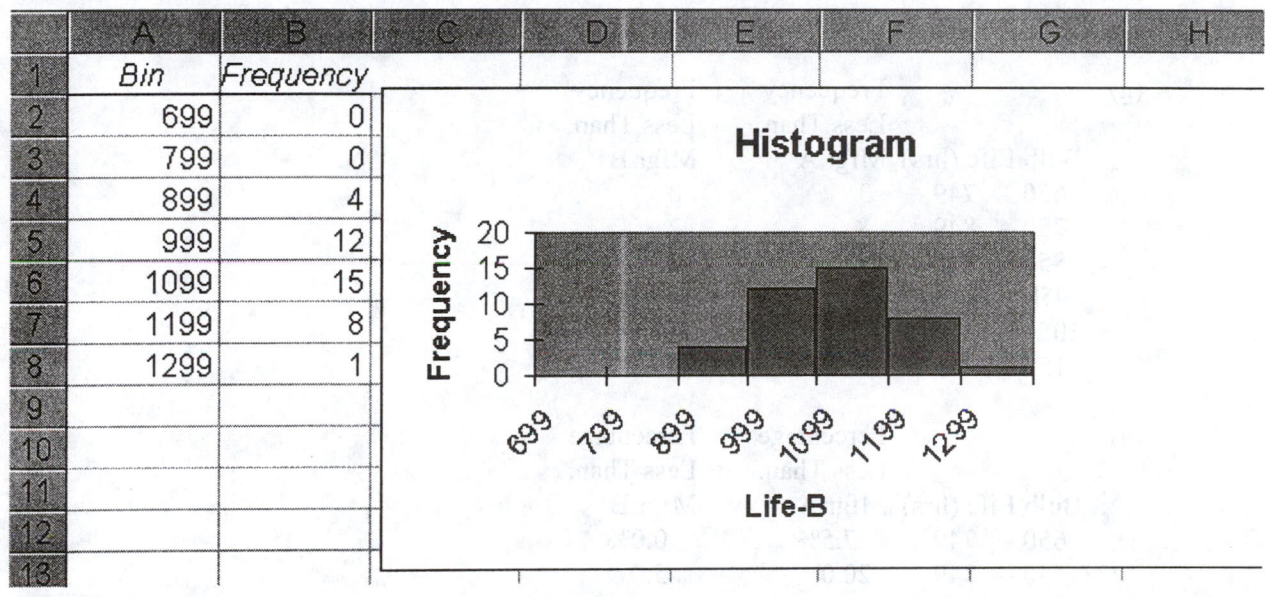

	A	B	C	D	E	F	G	H
1	*Bin*	*Frequency*						
2	699	0						
3	799	0						
4	899	4						
5	999	12						
6	1099	15						
7	1199	8						
8	1299	1						
9								
10								
11								
12								
13								

2.24
cont.

(c)

Bulb Life (hrs)	Frequency, Mfgr A	Frequency, Mfgr B
650 – 699	2	0
700 – 749	1	0
750 – 799	1	0
800 – 849	4	2
850 – 899	9	2
900 – 949	11	6
950 – 999	5	6
1000 – 1049	4	10
1050 – 1099	3	5
1100 – 1149	0	4
1150 – 1199	0	4
1200 – 1249	0	1

(d)

Bulb Life (hrs)	Percentage, Mfgr A	Percentage, Mfgr B
650 – 749	7.5%	0.0%
750 – 849	12.5	5.0
850 – 949	50.0	20.0
950 – 1049	22.5	40.0
1050 – 1149	7.5	22.5
1150 – 1249	0.0	12.5

(g)

Bulb Life (hrs)	Frequency Less Than, Mfgr A	Frequency Less Than, Mfgr B
650 – 749	3	0
750 – 849	8	2
850 – 949	28	10
950 – 1049	37	26
1050 – 1149	40	35
1150 – 1249	40	40

(h)

Bulb Life (hrs)	Percentage Less Than, Mfgr A	Percentage Less Than, Mfgr B
650 – 749	7.5%	0.0%
750 – 849	20.0	5.0
850 – 949	70.0	25.0
950 – 1049	92.5	65.0
1050 – 1149	100.0	87.5
1150 – 1249	100.0	100.0

(j)   Manufacturer B produces bulbs with longer lives than Manufacturer A. The cumulative percentage for Manufacturer B shows 65% of their bulbs lasted 1049 hours or less contrasted with 70% of Manufacturer A's bulbs which lasted 949 hours or less. None of Manufacturer A's bulbs lasted more than 1149 hours, but 12.5% of Manufacturer B's bulbs lasted between 1150 and 1249 hours. At the same time, 7.5% of Manufacturer A's bulbs lasted less than 750 hours, while all of Manufacturer B's bulbs lasted at least 750 hours.

2.26   (a)

Book Values	Frequency	Percentage
0 – 4	1	2%
5 – 9	27	54
10 – 14	15	30
15 – 19	6	12
20 – 24	1	2

(b)

Book Values	Frequency Less Than	Percentage Less Than
0 – 4	1	2%
5 – 9	28	56
10 – 14	43	86
15 – 19	49	98
20 – 24	50	100

(f)   Better than half of the book values for the sampled stocks on the New York Stock Exchange had book values between $5 and $9.

2.28   (a)   Stem-and-leaf of Amount of Soft Drink

1.8H	9
1.9L	03444
1.9H	555666677788899999
2.0L	00111111111222223334
2.0H	556678
2.1L	0

$n = 50$

Note: 1.8H are the "high 1.8s" such as 1.85, 1.86, 1.87, 1.88, or 1.89. 1.9L are the "low 1.9s" such as 1.90, 1.91, 1.92, 1.93, or 1.94. 1.9H are the "high 1.9s" such as 1.95, 1.96, 1.97, 1.98, or 1.99.

(b)

Amount of Soft Drink	Frequency	Percentage
1.85 – 1.89	1	2%
1.90 – 1.94	5	10
1.95 – 1.99	18	36
2.00 – 2.04	19	38
2.05 – 2.09	6	12
2.10 – 2.14	1	2

(c)

Amount of Soft Drink	Frequency Less Than	Percentage Less Than
1.85 – 1.89	1	2%
1.90 – 1.94	6	12
1.95 – 1.99	24	48
2.00 – 2.04	43	86
2.05 – 2.09	49	98
2.10 – 2.14	50	100

2.28
cont.

(g)   The amount of soft drink filled in the two liter bottles is most concentrated in two intervals on either side of the two-liter mark, from 1.95 to 1.99 and from 2.00 to 2.04 liters. Almost three-fourths of the 50 bottles sampled contained between 1.95 liters and 2.04 liters.

(h)   You would predict that the amount of soft drink filled in the next bottle will be between 1.95 liters and 2.04 liters because 74% of the bottles sampled fell within those bounds. If the prediction is for a specific value, you would predict 2.00 liters because it is the midpoint of the combined interval.

2.32

(d)   The bar chart does not facilitate comparison of the size of the various categories to the whole. The multiple divisions present in the pie chart are in some instances narrow and difficult to discern. The Pareto diagram is preferable here because it builds on the strength of the bar chart and conveys the relative sense of importance in sales of various food product groups through the cumulative polygon.

(e)   Grocery, soy foods, and dairy represent 27.4% of the sales of organic foods purchased from natural food stores in the U.S. during 1995, while Produce represented an additional 21.5% of sales. Together they represent nearly 50% of the sales for that year.

•2.34

(e)   Highway transportation accounted for better than half of the oil consumption in the United States in 1995.

2.36

(b)   The number one area for residential water consumption in a suburban area during a recent summer was lawn watering, accounting for better than one-third of all water consumed by households that summer. When bathing/showering and toilet usage were added, better than 80% of all water consumed was accounted for. The suburban water district should target those three areas in developing a water reduction plan.

2.38

(b)   Restaurants make up almost 24% of the commercial properties in the Times Square area of New York City. Quick-service (food) outlets and vacant stores made up additional 13.3% and 11% of the commercial properties, respectively.

(c)   Restaurant-goers will find ample choice among the 232 food establishments serving this 32-block region of New York City, 149 of which afford a full dining experience and the remaining 83 provide quick food service.

•2.42

(a)   Table based on column percentages

Financial Conditions	H.S. Degree or Lower	Education Level Some College	College Degree or Higher	Totals
Worse off now	21.2%	24.4%	8.6%	18.5%
No difference	24.2%	45.6%	14.8%	26.0%
Better off now	54.7%	30.0%	76.7%	55.5%
Totals	100%	100%	100%	100%

(c)   Financial conditions were rated as better now than before by a majority of the groups with the lowest and highest education levels. But the largest segment of the group with some college rated their financial conditions as no different now than before.

130

2.44   (d)   The row percentages allow us to block the effect of disproportionate group size and show us that the pattern for day and evening tests among the nonconforming group is very different from the pattern for day and evening tests among the conforming group. Where 40% of the nonconforming group was tested during the day, 68% of the conforming group was tested during the day.

    (e)   The director of the lab may be able to cut the number of nonconforming tests by reducing the number of tests run in the evening, when there is a higher percent of tests run improperly.

•2.64   (f)   There is a great deal of variation in the number of days it took to resolve customer complaints. Better than one-quarter of all complaints were resolved in less than 20 days, over half within 30 days. But the amount of time to resolution ranged from one day to 165 days, with over 40% of the complaints requiring more than 30 days to settle and 20% requiring more than 60 days to settle.

    (g)   You should tell the president of the company that over half of the complaints are resolved within a month, but point out that some complaints take as long as three or four months to settle.

2.66   (a)   (1)   Ordered arrays:

Weight:	5	6	6	6	6	6	7	7	7	7	7	10
	10	10	11	12	12	15	17	17	18	19	20	21
	21	22	22	22	24	25	26	26	27	29	29	29
	30	31	31	47								

Cost:	0.52	0.62	0.64	0.64	0.69	0.71	0.72	0.72	0.73	0.75	0.77
	0.80	0.80	0.81	0.81	0.83	0.85	0.87	0.90	0.90	0.92	1.00
	1.11	1.11	1.13	1.14	1.15	1.23	1.23	1.28	1.46	1.49	1.50
	1.51	1.51	1.52	1.53	1.54	1.71	1.90				

Calories:	264	275	288	296	299	305	309	312	313	316	322
	323	327	332	333	337	338	347	348	349	350	353
	357	358	360	361	364	365	367	370	372	381	382
	387	390	393	394	409	436	442				

Fat:	3	4	7	7	7	9	9	9	10	10	10	11
	11	12	12	13	13	13	14	14	14	15	15	16
	16	16	17	17	17	18	19	19	20	20	20	21
	21	22	25	26								

(2)

Stem-and-leaf for Weight		Stem-and-leaf for Calories	
0	56666677777	Hundreds	Tens
1	00012257789	2H	67899
2	01122245667999	3L	001112223333444
3	011	3H	55556666677888999
4	7	4L	034
	$n = 40$		$n = 40$

Stem-and-leaf for Cost		Stem-and-leaf for Fat	
0.5	2	0L	34
0.6	2449	0H	777999
0.7	122357	1L	0001122333444
0.8	0011357	1H	55666777899
0.9	002	2L	000112
1.0	0	2H	56
1.1	11345		$n = 40$
1.2	338	*Note*:	L is "low" and H is "high"
1.3			to create more stems.
1.4	69		
1.5	011234		
1.6			
1.7	1		
1.8			
1.9	0		
	$n = 40$		

(b) Ordered arrays by pizza type:

Weight:
Type 1:	21	25	26	29	31	31	47		
Type 2:	6	6	6	7	7	7	10	10	17	17
	19	21	22	24	26	29	29		
Type 3:	5	6	6	7	7	10	11	12	12	15
	18	20	22	22	27	30			

Cost:
Type 1:	1.23	1.23	1.28	1.51	1.51	1.53	1.90		
Type 2:	0.52	0.64	0.69	0.72	0.72	0.75	0.80	0.81	0.85
	0.90	0.90	0.92	1.00	1.15	1.49	1.50	1.54	
Type 3:	0.62	0.64	0.71	0.73	0.77	0.80	0.81	0.83	0.87
	1.11	1.11	1.13	1.14	1.46	1.52	1.71		

Calories:
Type 1:	305	309	313	327	338	349	382		
Type 2:	275	288	296	299	316	322	323	332	333
	337	347	350	353	357	358	364	393	
Type 3:	264	312	348	360	361	365	367	370	372
	381	387	390	394	409	436	442		

Fat:
Type 1:	9	9	10	11	13	14	16		
Type 2:	4	7	7	9	10	10	12	12	13	13
	14	14	16	16	17	17	19		
Type 3:	3	7	11	15	15	17	18	19	20	20
	20	21	21	22	25	26			

2.66
cont. (c)-(d)

Weight	Frequency	Percentage	Frequency Less Than	Percentage Less Than
0 up to 10	11	27.5%	11	27.5%
10 up to 20	11	27.5	22	55.0
20 up to 30	14	35.0	36	90.0
30 up to 40	3	7.5	39	97.5
40 up to 50	1	2.5	40	100.0

Cost	Frequency	Percentage	Frequency Less Than	Percentage Less Than
0.50 up to 0.75	9	22.5%	9	22.5%
0.75 up to 1.00	12	30.0	21	52.5
1.00 up to 1.25	8	20.0	29	72.5
1.25 up to 1.50	3	7.5	32	80.0
1.50 up to 1.75	7	17.5	39	97.5
1.75 up to 2.00	1	2.5	40	100.0

Calories	Frequency	Percentage	Frequency Less Than	Percentage Less Than
250 up to 280	2	5.0%	2	5.0%
280 up to 310	5	12.5	7	17.5
310 up to 340	10	25.0	17	42.5
340 up to 370	12	30.0	29	72.5
370 up to 400	8	20.0	37	92.5
400 up to 430	1	2.5	38	95.0
430 up to 460	2	5.0	40	100.0

(h)   The typical weight for a slice of pizza is between 10 and 20 ounces, where the cumulative percentage shows 55% of the sample is less than or equal to 20 ounces.

(i)   The typical cost for a slice of pizza is between $.75 and $1.00, since that is both the most frequently occurring interval and better than 50% of the sample is less than or equal to $1.00.

(j)   The typical caloric content for a slice of pizza is between 340 and 370 calories, since that is the most frequently occurring interval and better than 70% of the sample is less than or equal to 370 calories.

(k)   Based on the results of the pairwise scatter diagrams prepared for part (g), calories and fat seem to be related. Other variables do not show any particular pattern in the scatter plot, but the graph of calories and fat has a positive slope because it rises from left to right, showing that as one variable increases, the other tends to also increase. The data points are tightly distributed within the cloud of points, indicating that the relationship between calories and fat content is strong.

2.68   (a)

Number of Calls	Frequency	Percentage
2650-2850	1	3.33%
2850-3050	1	3.33
3050-3250	16	53.33
3250-3450	9	30.00
3450-3650	2	6.67
3650-3850	1	3.33

(f)   There is very little variation in the daily number of calls over the 30 day period sampled.

2.70   (c)

Amount	Percentage Less Than, March	Percentage Less Than, April
$0 up to $2,000	12%	20%
$2,000 up to $4,000	38	48
$4,000 up to $6,000	72	74
$6,000 up to $8,000	92	94
$8,000 up to $10,000	100	94
$10,000 up to $12,000	100	100

(d)

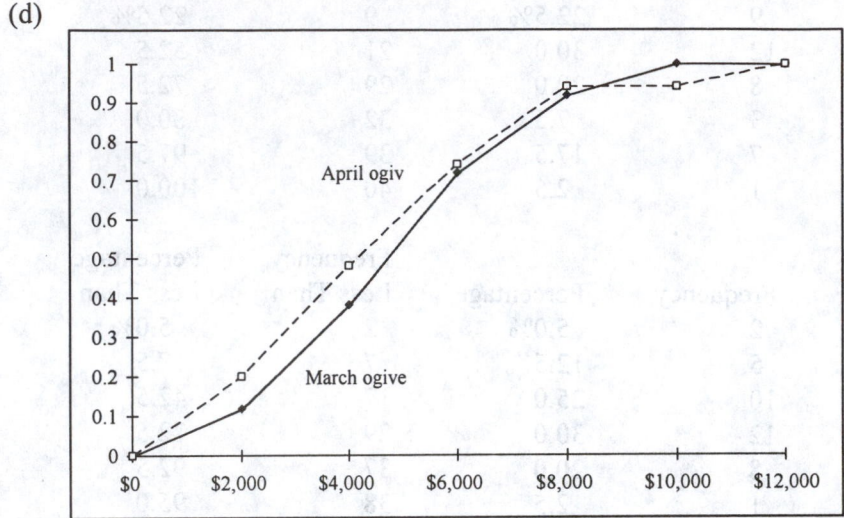

(e)   The distribution of the accounts receivable does not change much from March to April.

2.72   (e)   Two of the 30 NFL teams account for over 30% of the sales of licensed products, and six of the 30 NFL teams account for over 50% of the sales of licensed products. So a relatively few number of NFL teams account for a majority of the sales of licensed products. Clothing and hats account for 50% of the sales of licensed products. Retail licensing and sponsorships account for more than 70% of the reported revenues.

2.74   (b)   A first-time buyer of computer equipment is more than twice as likely to make a purchase from a consumer electronics store or mass merchant as is a repeat buyer. A repeat buyer is at least

Chapter 3
Summarizing and Describing Numerical Data

CHAPTER OBJECTIVES

✓ *To describe central tendency in numerical data*
✓ *To describe variation in numerical data*
✓ *To describe the shape of a distribution*
✓ *To introduce the box-and-whisker plot as a graphical tool for describing the characteristics of numerical data*
✓ *To calculate descriptive summary measures from a population*

Introduction

In the preceding chapter we learned how to present numerical and categorical data in both tabular and chart format. Now, how do we make sense out of such information? Although presenting data is an essential component of descriptive statistics, it does not tell the whole story. When dealing with numerical information, good **data analysis** involves not only *presenting* the data and *observing* what the data are trying to convey, but also *computing* and *summarizing* the key features and *analyzing* the findings.

In this chapter we explore numerical data and their properties. Next we discuss measures of central tendency, variation, and shape. Then we explore data analysis and how to calculate descriptive summary measures from a population. Finally we address how to recognize and practice proper descriptive summarization and deal with pertinent ethical issues.

 **EXPLORING NUMERICAL DATA AND THEIR PROPERTIES**

◆ USING STATISTICS: *Evaluating the Performance of Equity Mutual Funds*

To introduce the relevant ideas of this chapter, let us return to our study of the performance of currently traded domestic general stock funds. In determining an investment strategy, it would be useful to compare the 1-year total percentage return of the stock funds based on their fee structures. To illustrate such an evaluation, we will first study the 17 funds whose fee structure consists of marketing fees paid from fund assets and subsequently compare the returns for all five fee structures. ◆

The 1-year total percentage returns are given in Table 3.1. We note that the 17 funds are recorded in alphabetical order along with their attained 1-year total returns (in percents). What can be learned from such data that will assist us in an evaluation for future investment purposes? On the basis of this sample, we can make the following three observations:

1. Two of the funds achieved 1-year total percentage returns of 28.6 and two other funds attained 1-year total percentage returns of 30.5. These are the *most typical*, or *modal*, values and this data set is considered to be *bimodal*. The 1-year total returns achieved by each of the other funds differ from one another.

Table 3.1 *1-year total percentage returns for stocks funds whose fee structure consists of marketing fees paid from funds assets*

FUND	1-YEAR TOTAL PERCENTAGE RETURNS
Amcore Vintage Equity	32.2
Baron Funds Asset	29.5
Berger SmCoGrow	29.9
Chicago Trust GrowInc	32.4
Dodge & Cox DominiSo	30.5
Federated Institut MaxCapSvc	30.1
First Funds GroInc III	32.1
Harris Insight Inst Haven	35.2
Mentor Merger	10.0
Rainler Reich Tang	20.6
Robertson Stephens ValGrow	28.6
SSgA S&P500Idx	30.5
SSgA SmallCap	38.0
1784 GrowInc	33.0
Stagecoach CorpStk	29.4
Westwood Eq R	37.1
Wright Yacktman	28.6

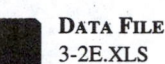

DATA FILE
3-2E.XLS

2. The *spread* in the 1-year total returns ranges from 10.0% to 38.0%.

3. There appears to be at least one unusual or extraordinarily low 1-year total return in this data set—that attained by Mentor Merger. Arranged in an ordered array, the 1-year total percentage returns attained by these domestic general stock funds (in percents) are

10.0 20.6 28.6 28.6 29.4 29.5 29.9 30.1 30.5 30.5 32.1 32.2 32.4 33.0 35.2 37.1 38.0

Here the 10.0 and, perhaps, the 20.6 would be considered **outliers,** or **extreme values.**

We might want to explore the reasons why the 1-year total percentage return attained by Mentor Merger was so much lower than that achieved by any of the other funds whose marketing fees were paid from fund assets. A comparison of the prospectus for this fund against those with higher-achieving performance records might provide some useful insight for future investment purposes.

Nevertheless, had we been asked to examine the data and present a short summary of our findings, statements similar to the three above are basically all that we could be expected to make without more knowledge of the subject of statistics.

We can add to our understanding of what the data are telling us by more formally examining the three major properties that describe a set of numerical data: *central tendency, variation,* and *shape.*

3.2 MEASURES OF CENTRAL TENDENCY, VARIATION, AND SHAPE

In any analysis and/or interpretation, a variety of descriptive measures representing the properties of central tendency, variation, and shape may be used to summarize the major features of the data set. If these descriptive summary measures are computed from a sample of data, they are called *statistics;* if they are computed from an entire population of data, they are called *parameters.* Because statisticians usually take samples rather than use entire populations, our primary emphasis in this text when describing the properties of central tendency, variation, and shape is on statistics, rather than on parameters.

Measures of Central Tendency

Most sets of data show a distinct tendency to group or cluster about a certain central point. Thus, for any particular set of data, it usually becomes possible to select some typical value, or **average,** to describe the entire set. Such a descriptive typical value is a measure of **central tendency,** or **location.**

Five types of averages often used as measures of central tendency are the *arithmetic mean,* the *median,* the *mode,* the *midrange,* and the *midhinge.*

◆ *The Arithmetic Mean*   The **arithmetic mean** (also called the **mean**) is the most commonly used *average*[1] or measure of central tendency. It is calculated by summing all the observations in a set of data and then dividing the total by the number of items involved. Thus, for a sample containing a set of n observations $X_1, X_2, X_3, \ldots , X_n$, the arithmetic mean (given by the symbol $\overline{X}$—called "X bar") can be written as

$$\overline{X} = \frac{X_1 + X_2 + X_3 + \cdots + X_n}{n}$$

To simplify the notation, the term

$$\sum_{i=1}^{n} X_i$$

(meaning the *summation* of all the X_i values) is used whenever we wish to add together a series of observations. That is,

$$\sum_{i=1}^{n} X_i = X_1 + X_2 + X_3 + \cdots + X_n$$

Rules pertaining to summation notation are presented in appendix B. Using this summation notation, the arithmetic mean of the sample can be expressed as follows.

[1] *Although the word* average *refers to any summary measure of central tendency, it is most often used as a synonym for the mean.*

138

Arithmetic Mean

The arithmetic mean is the sum of the values divided by the number of values.

$$\bar{X} = \frac{\sum_{i=1}^{n} X_i}{n} \qquad\qquad (3.1)$$

where

$\bar{X}$ = sample arithmetic mean

n = sample size

X_i = ith observation of the random variable X

$\sum_{i=1}^{n} X_i$ = summation of all X_i values in the sample (see appendix B)

The computation of the arithmetic mean is illustrated in Example 3.1.

Example 3.1   *Computing the Arithmetic Mean*

Compute the arithmetic mean of the 1-year total percentage returns of the 17 domestic general stock funds presented in Table 3.1 on page 153.

SOLUTION

We begin by assigning X_i values to the 17 funds.

X_1 = 32.2 at Amcore Vintage Equity

X_2 = 29.5 at Baron Funds Asset

X_3 = 29.9 at Berger SmCoGrow

X_4 = 32.4 at Chicago Trust GrowInc

X_5 = 30.5 at Dodge & Cox DominiSo

X_6 = 30.1 at Federated Institut MaxCapSvc

X_7 = 32.1 at First Funds GroInc III

X_8 = 35.2 at Harris Insight Inst Haven

X_9 = 10.0 at Mentor Merger

X_{10} = 20.6 at Rainler Reich Tang

X_{11} = 28.6 at Robertson Stephens ValGrow

X_{12} = 30.5 at SSgA S&P500Idx

X_{13} = 38.0 at SSgA SmallCap

X_{14} = 33.0 at 1784 GrowInc

X_{15} = 29.4 at Stagecoach CorpStk

X_{16} = 37.1 at Westwood Eq R

X_{17} = 28.6 at Wright Yacktman

The arithmetic mean for this sample is then calculated as follows.

$$\overline{X} = \frac{\sum_{i=1}^{n} X_i}{n} = \frac{32.2 + 29.5 + 29.9 + \cdots + 28.6}{17} = 29.86$$

We observe here that the mean 1-year total percentage return is computed as 29.86 even though not one particular fund in the sample actually had that value. In addition, we see from the **dot scale** of Figure 3.1 that for this set of data, 6 observations are smaller than the mean and 11 are larger. The mean acts as a *balancing point* so that smaller observations balance out larger ones. Note that the calculation of the mean is based on all the observations $(X_1, X_2, X_3, \ldots, X_n)$ in the set of data. No other commonly used measure of central tendency possesses this characteristic.

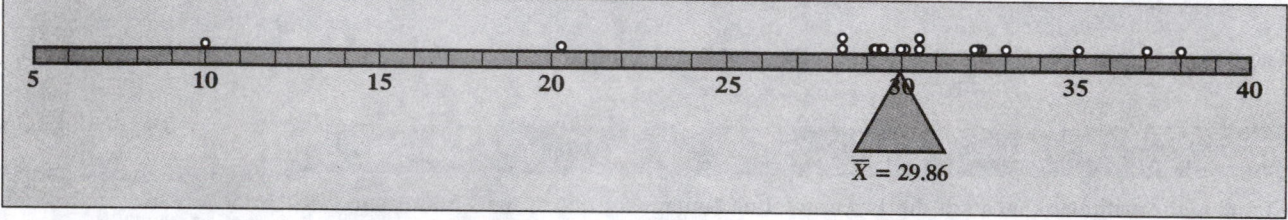

FIGURE 3.1   Dot scale representing the 1-year total percentage returns achieved by 17 domestic general stock funds with marketing fees paid from fund assets

---

COMMENT: *When to Use the Arithmetic Mean*

Because its computation is based on every observation, the arithmetic mean is greatly affected by any extreme value or values. In such instances, the arithmetic mean presents a distorted representation of what the data are conveying; hence, the mean is not the best average to use for describing or summarizing a set of data that has extreme values. This is what happened in Example 3.1.

To demonstrate the effect that extreme values can have when summarizing and describing the property of central tendency, suppose that the outlier Mentor Merger is removed. How the arithmetic mean of the 1-year total percentage returns changes is explored in Example 3.2.

Example 3.2   *Computing the Arithmetic Mean*

Compute the arithmetic mean of the 1-year total percentage returns after removing the outlier Mentor Merger.

SOLUTION

The arithmetic mean for this sample of 16 funds is calculated as follows.

$$\overline{X} = \frac{\sum\limits_{i=1}^{n} X_i}{n} = \frac{32.2 + 29.5 + 29.9 + \cdots + 28.6}{16} = 31.11$$

By the removal of Mentor Merger, the arithmetic mean increased from 29.86 to 31.11.

The dot scale is displayed in Figure 3.2.

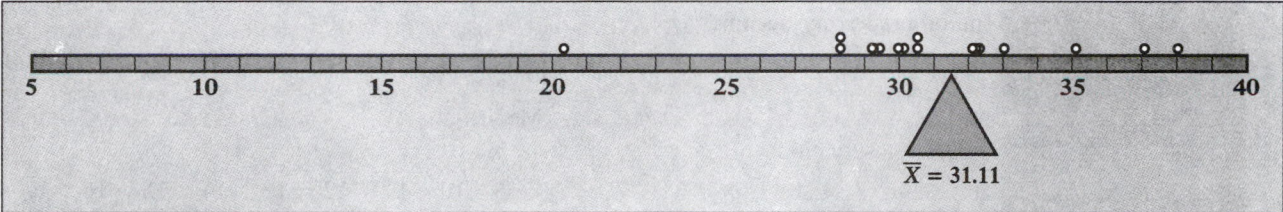

FIGURE 3.2 Dot scale representing the 1-year total percentage returns achieved by 16 domestic general stock funds with marketing fees paid from fund assets

◆ *The Median*   The **median** is the middle value in an ordered array of data. If there are no ties, half the observations will be smaller and half will be larger. The median is unaffected by any extreme observations in a set of data. Thus, whenever an extreme observation is present, it is appropriate to use the median rather than the mean to describe a set of data.

To calculate the median from a set of data, we must first organize the data into an ordered array. Then the median can be obtained as:

Median

The median is the value such that 50% of the observations are smaller and 50% of the observations are larger.

$$\text{Median} = \frac{n+1}{2} \text{ ranked observation} \qquad (3.2)$$

Equation (3.2) is used to find the place in the ordered array that corresponds to the median value by following one of two rules:

RULE 1 If the size of the sample is an *odd* number, then the median is represented by the numerical value corresponding to the positioning point—the $(n + 1)/2$ ordered observation.
RULE 2 If the size of the sample is an *even* number, then the positioning point lies between the two middle observations in the ordered array. The median is the average of the numerical values corresponding to these two middle observations.

Example 3.3 illustrates the computation of the median.

Example 3.3 *Computing the Median from an Odd-Sized Sample*

From our sample of 1-year total percentage returns achieved by domestic general stock funds whose marketing fees are paid from fund assets, the raw data are presented as follows.

32.2 29.5 29.9 32.4 30.5 30.1 32.1 35.2 10.0 20.6 28.6 30.5 38.0 33.0 29.4 37.1 28.6

Compute the median.

SOLUTION

The ordered array becomes

10.0 20.6 28.6 28.6 29.4 29.5 29.9 30.1 **30.5** 30.5 32.1 32.2 32.4 33.0 35.2 37.1 38.0
↑
Median

Ordered observation   ↑

1   2   3   4   5   6   7   8   **9**   10 11 12 13 14 15 16 17

Median = 30.5

For these data, the positioning point is the ninth ordered observation [i.e., $(n + 1)/2 = (17 + 1)/2 = 9$]. Therefore, the median is 30.5.

As can be seen from the ordered array in Example 3.3, the median is unaffected by extreme observations. Regardless of whether the smallest 1-year total percentage return is 1.0, 10.0, or 20.0, the median is still 30.5.

In addition, when computing the median, we ignore the fact that tied values may be present in the data. In Example 3.3, two of the funds (Robertson Stephens ValGrow and Wright Yacktman) achieved the same 1-year total percentage returns of 28.6. However, this amount had no impact on the actual median value. On the other hand, two other funds (Dodge & Cox DominiSo and SSgA S&P500Idx) attained the same 1-year total percentage returns of 30.5, which happens to be equal to the median performance value. Thus, for this odd-sized sample, the median positioning point is the $(n + 1)/2 = 9$th ordered observation and the median is 30.5, the middle value in the ordered sequence, even though the 10th ordered observation is also 30.5. Example 3.4 illustrates the computation of the median for an even-sized sample.

Example 3.4 *Computing the Median from an Even-Sized Sample*

Suppose that our sample consists of the net asset values of 14 domestic general stock funds that are classified as small capitalization blend funds. The raw data, displaying the net asset values (in dollars) for these funds, are as follows.

$$X_1 = \ \ 7.35 \text{ at Baron Funds BanRosSC}$$

$$X_2 = 17.30 \text{ at Citizens Trust CloverEqV}$$

$$X_3 = 11.62 \text{ at DFA US9-10Sm}$$

$$X_4 = 26.10 \text{ at FPA Fasciano}$$

$$X_5 = 21.69 \text{ at GT Global Equity AmerGroA}$$

$X_6 = 21.17$ at GP Global Equity AmerGroB

$X_7 = 14.07$ at Galaxy Retail SmCapVal

$X_8 = 14.09$ at Galaxy Trust SmCapVal

$X_9 = 24.01$ at Heritage SmCapStkA

$X_{10} = 20.34$ at HighMark Fid HomePAGr

$X_{11} = 18.26$ at T Rowe Price OTCSec

$X_{12} = 37.61$ at Princor EmgGro A

$X_{13} = 18.60$ at SSgA SmallCap

$X_{14} = 16.95$ at Wasatch Growth

Compute the median.

SOLUTION

The ordered array becomes

7.35 11.62 14.07 14.09 16.95 17.30 18.26 18.60 20.34 21.17 21.69 24.01 26.10 37.61

Ordered observation

| 1 | 2 | 3 | 4 | 5 | 6 | 7 | 8 | 9 | 10 | 11 | 12 | 13 | 14 |

Median = 18.43

For these data, the positioning point is $(n + 1)/2 = (14 + 1)/2 = 7.5$. Therefore, the median is obtained by averaging the seventh and eighth ordered observations:

$$\frac{18.26 + 18.60}{2} = \$18.43$$

To summarize, the calculation of the median value is affected by the number of observations, not by the magnitude of any extreme(s). Ignoring the possibility of tied data values, which are usually attributable to imprecise measurements, any observation selected at random is just as likely to exceed the median as it is to be exceeded by it.

◆ **The Mode** The **mode** is the value in a set of data that appears most frequently. Unlike the arithmetic mean, the mode is not affected by the occurrence of any extreme values. However, the mode is used only for descriptive purposes because it is more variable from sample to sample than other measures of central tendency. Example 3.5 illustrates the computation of the mode.

Example 3.5 *Obtaining the Mode*

Using the ordered array for the 1-year total percentage returns achieved by the domestic general stock funds whose marketing fees are paid from fund assets (see Example 3.3 on page 158), obtain the mode.

SOLUTION

The ordered array for these data is

10.0 20.6 28.6 28.6 29.4 29.5 29.9 30.1 30.5 30.5 32.1 32.2 32.4 33.0 35.2 37.1 38.0

We see that there are two "most typical" values, or two modes—28.6 and 30.5. Such data are described as *bimodal*.

A set of data may have no mode if none of the values is "most typical." This is illustrated in Example 3.6.

Example 3.6   *Data with No Mode*

Using the ordered array of net asset values (in dollars) of 14 domestic general stock funds that are classified as small capitalization blend funds (see Example 3.4 on page 158), obtain the mode.

SOLUTION

The ordered array for these data is

7.35 11.62 14.07 14.09 16.95 17.30 18.26 18.60 20.34 21.17 21.69 24.01 26.10 37.61

Here there is no mode. None of the net asset values is "most typical."

◆ *The Midrange*   The **midrange** is the average of the *smallest* and *largest* observations in a set of data. This can be written as follows.

Midrange

The midrange is obtained by adding the smallest and largest values and dividing by 2.

$$\text{Midrange} = \frac{X_{\text{smallest}} + X_{\text{largest}}}{2} \tag{3.3}$$

Example 3.7 illustrates the computation of the midrange.

Example 3.7   *Computing the Midrange*

Using the ordered array for the 1-year total percentage returns achieved by the domestic general stock funds whose marketing fees are paid from fund assets (see Example 3.3 on page 158), compute the midrange.

SOLUTION

The ordered array for these data is

10.0 20.6 28.6 28.6 29.4 29.5 29.9 30.1 30.5 30.5 32.1 32.2 32.4 33.0 35.2 37.1 38.0

The midrange is computed from equation (3.3) as

$$\text{Midrange} = \frac{X_{\text{smallest}} + X_{\text{largest}}}{2}$$

$$= \frac{10.0 + 38.0}{2} = 24.0$$

The midrange is often used as a summary measure both by financial analysts and by weather reporters because it can provide an adequate, quick, and simple measure to characterize the *entire* data set. Nevertheless, despite these advantages, the midrange must be used cautiously.

COMMENT: *When to Use the Midrange*

In our dealing with data such as daily closing stock prices or hourly temperature readings, an extreme value is not likely to occur. Nevertheless, in most applications, despite its simplicity, the midrange must be used cautiously. Because it involves only the smallest and largest observations in a data set, the midrange becomes distorted as a summary measure of central tendency if an outlier is present (as in Example 3.7). In such situations, the midrange is inappropriate.

Aside from the measures of central tendency, **quartiles** are the most widely used measures of "noncentral" location (also called **quantiles**) and are employed particularly when summarizing or describing the properties of large sets of numerical data. Whereas the median is a value that splits the ordered array in half (50.0% of the observations are smaller and 50.0% of the observations are larger), the quartiles are descriptive measures that split the ordered data into four quarters. Other often used quantiles are deciles, which split the ordered data into 10ths, and percentiles, which split the data into 100ths.

The quartiles can be defined as in equations (3.4) and (3.5).

First Quartile, Q_1

The **first quartile**, Q_1, is a value such that 25.0% of the observations are smaller and 75.0% of the observations are larger.

$$Q_1 = \frac{(n+1)}{4} \text{ ordered observation} \tag{3.4}$$

Third Quartile, Q_3

The **third quartile**, Q_3, is a value such that 75.0% of the observations are smaller and 25.0% of the observations are larger.

$$Q_3 = \frac{3(n+1)}{4} \text{ ordered observation} \tag{3.5}$$

Three rules are used for obtaining the quartile values.

RULE 1 If the resulting positioning point is an integer, the particular numerical observation corresponding to that positioning point is chosen for the quartile.

RULE 2 If the resulting positioning point is halfway between two integers, the average of their corresponding values is selected.

RULE 3 If the resulting positioning point is neither an integer nor a value halfway between two integers, a simple rule used to approximate the particular quartile is to *round up or down* to the nearest integer positioning point and select the numerical value of the corresponding observation.

The computation of the quartiles is illustrated in Example 3.8.

Example 3.8 *Obtaining the Quartiles*

Using the ordered array for the 1-year total percentage returns achieved by the domestic general stock funds whose marketing fees are paid from fund assets (see Example 3.3 on page 158), compute the quartiles.

SOLUTION

The ordered array is

10.0 20.6 28.6 28.6 29.4 29.5 29.9 30.1 30.5 30.5 32.1 32.2 32.4 33.0 35.2 37.1 38.0

For these data we have

$$Q_1 = \frac{n+1}{4} \text{ ordered observation}$$

$$= \frac{17+1}{4} = 4.5\text{th ordered observation}$$

Therefore, using rule 2, Q_1 can be approximated as the average of the fourth and fifth ordered observations.

$$Q_1 = \frac{28.6 + 29.4}{2} = 29.0$$

In addition,

$$Q_3 = \frac{3(n+1)}{4} \text{ ordered observation}$$

$$= \frac{3(17+1)}{4} = 13.5\text{th ordered observation}$$

Therefore, using rule 2, Q_3 can be approximated as the average of the 13th and 14th ordered observations.

$$Q_3 = \frac{32.4 + 33.0}{2} = 32.7$$

◆ *The Midhinge* The **midhinge** is a summary measure used to overcome potential problems introduced by extreme values in the data. The midhinge is computed as the average of the *first* and *third quartiles* in a set of data.

Midhinge

The midhinge is obtained by adding the first and third quartiles and dividing by 2.

$$\text{Midhinge} = \frac{Q_1 + Q_3}{2} \tag{3.6}$$

where

$$Q_1 = \text{first quartile}$$
$$Q_3 = \text{third quartile}$$

To compute the midhinge, we first need to compute Q_1 and Q_3. This is demonstrated in Example 3.9.

Example 3.9 *Computing the Midhinge*

Using the ordered array for the 1-year total percentage returns achieved by the domestic general stock funds whose marketing fees are paid from fund assets (see Example 3.3 on page 158), compute the midhinge.

SOLUTION

From Example 3.8 we determined that $Q_1 = 29.0$ and $Q_3 = 32.7$. Returning to equation (3.6), we may now compute the midhinge as

$$\text{Midhinge} = \frac{Q_1 + Q_3}{2}$$
$$= \frac{29.0 + 32.7}{2} = 30.85$$

It is important to note that the midhinge, the average of Q_1 and Q_3, two measures of noncentral location, cannot be affected by potential outliers because no observation smaller than Q_1 or larger than Q_3 is considered. Summary measures such as the midhinge and the median, which cannot be affected by outliers, are called **resistant measures.**

Measures of Variation

A second important property that describes a set of numerical data is variation. **Variation** is the amount of **dispersion,** or "spread," in the data. Two sets of data may differ in both central tendency and variation; or, as shown in the polygons of Figure 3.3 on page 164, two sets of data may have the same measures of variation but differ in central tendencies or, as depicted in the polygons of Figure 3.4 on page 164, two sets of data may have the same measures of central tendency but differ greatly in terms of variation. The data set depicted in polygon C of Figure 3.4 is much less variable than that depicted in polygon A.

Five measures of variation include the *range,* the *interquartile range,* the *variance,* the *standard deviation,* and the *coefficient of variation.*

FIGURE 3.3
Two symmetrical bell-shaped distributions differing only in central tendency

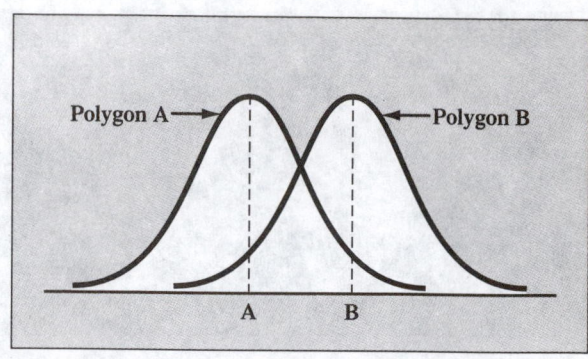

FIGURE 3.4
Two symmetrical bell-shaped distributions differing only in variation

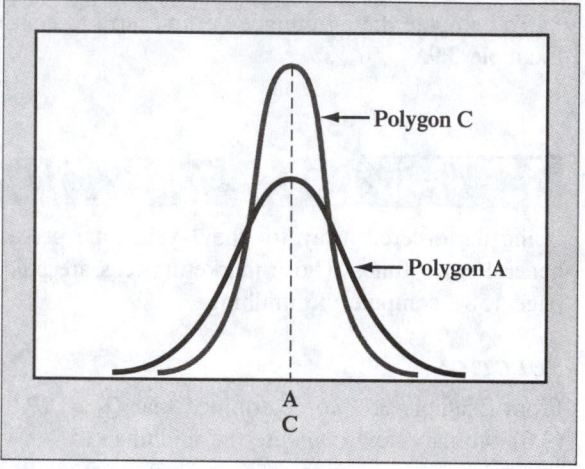

◆ *The Range* The **range** is the difference between the *largest* and *smallest* observations in a set of data.

Range

The range is equal to the largest value minus the smallest value.

$$\text{Range} = X_{\text{largest}} - X_{\text{smallest}} \qquad (3.7)$$

We apply equation (3.7) in Example 3.10.

Example 3.10 *Computing the Range*

Using the ordered array for the 1-year total percentage returns achieved by the domestic general stock funds whose marketing fees are paid from fund assets (see Example 3.3 on page 158), compute the range.

SOLUTION

The ordered array is

10.0 20.6 28.6 28.6 29.4 29.5 29.9 30.1 30.5 30.5 32.1 32.2 32.4 33.0 35.2 37.1 38.0

For these data, the range is $38.0 - 10.0 = 28.0$.

The range measures the *total spread* in the set of data. Although the range is a simple measure of total variation in the data, its distinct weakness is that it does not take into account *how* the data are actually distributed between the smallest and largest values. This can be observed from Figure 3.5. As evidenced in scale C, it would be improper to use the range as a measure of variation when at least one of its components is an extreme observation.

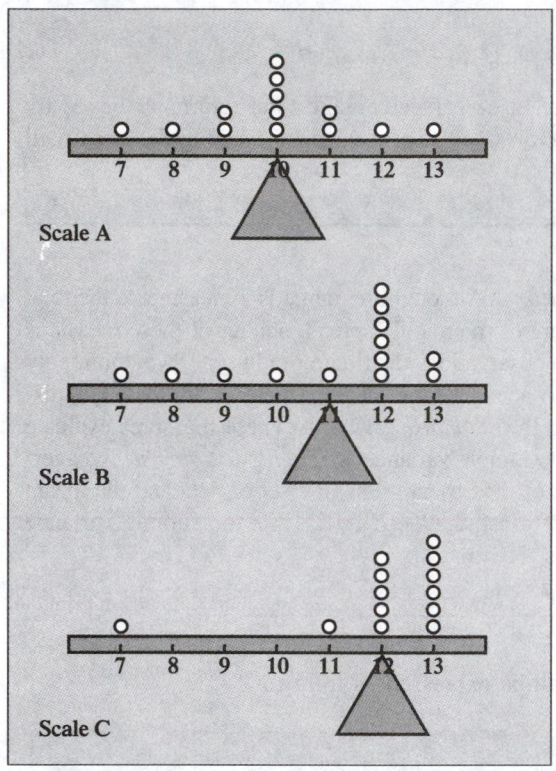

FIGURE 3.5
Comparing three data sets with the same range

◆ *The Interquartile Range* The **interquartile range** (also called **midspread**) is the difference between the *third* and *first quartiles* in a set of data.

Interquartile Range

The interquartile range is obtained by subtracting the first quartile from the third quartile.

$$\text{Interquartile range} = Q_3 - Q_1 \qquad (3.8)$$

This measure considers the spread in the middle 50% of the data; therefore, it is not influenced by extreme values. Example 3.11 illustrates the computation of the interquartile range.

Example 3.11 *Computing the Interquartile Range*

Using the ordered array for the 1-year total percentage returns achieved by the domestic general stock funds whose marketing fees are paid from fund assets (see Example 3.3 on page 158), compute the interquartile range.

SOLUTION

The ordered array is

10.0 20.6 28.6 28.6 29.4 29.5 29.9 30.1 30.5 30.5 32.1 32.2 32.4 33.0 35.2 37.1 38.0

For these data, we have already determined from Example 3.8 on page 162 that $Q_1 = 29.0$ and $Q_3 = 32.7$. Returning to equation (3.8),

$$\text{Interquartile range} = 32.7 - 29.0 = 3.7$$

This is the midspread or interquartile range in 1-year percentage total returns achieved by the *middle group* of the 17 domestic general stock funds whose marketing fees are paid from fund assets.

◆ **The Variance and the Standard Deviation**   Although the range is a measure of the total spread and the interquartile range is a measure of the middle spread, neither of these measures of variation takes into consideration *how* the observations distribute or cluster. Two commonly used measures of variation that do take into account how all the values in the data are distributed are the *variance* and its square root, the *standard deviation*. These measures evaluate how the values fluctuate about the mean. The sample variance is *roughly* (or *almost*) the average of the squared differences between each of the observations in a set of data and the mean.

Thus, for a sample containing n observations, $X_1, X_2, X_3, \ldots, X_n$, the sample variance (given by the symbol S^2) can be written as

$$S^2 = \frac{(X_1 - \overline{X})^2 + (X_2 - \overline{X})^2 + (X_3 - \overline{X})^2 + \cdots + (X_n - \overline{X})^2}{n - 1}$$

Using summation notation, this formula can be expressed as follows:

Sample Variance

The **sample variance** is the sum of the squared differences around the arithmetic mean divided by the sample size minus 1.

$$S^2 = \frac{\sum_{i=1}^{n}(X_i - \overline{X})^2}{n - 1} \tag{3.9}$$

where

$\overline{X}$ = sample arithmetic mean
n = sample size
X_i = ith observation of the random variable X

$\sum_{i=1}^{n}(X_i - \overline{X})^2$ = summation of all the squared differences between the X_i values and $\overline{X}$

Had the denominator been n instead of $n - 1$, the average of the squared differences around the mean would have been obtained. However, $n - 1$ is used here because of certain desirable mathematical properties possessed by the statistic S^2 that make it appropriate for statistical inference (which will be discussed in chapter 6). As the sample size increases, the difference between dividing by n or $n - 1$ becomes smaller and smaller.

Now let us turn our attention to the more practical of the two measures, the **sample standard deviation.** This measure, given by the symbol S, is the square root of the sample variance. It is expressed as follows.

Sample Standard Deviation

The standard deviation is the square root of the sum of the squared differences around the arithmetic mean divided by the sample size minus 1.

$$S = \sqrt{\frac{\sum_{i=1}^{n} (X_i - \overline{X})^2}{n - 1}} \qquad (3.10)$$

The steps for computing both the sample variance and the sample standard deviation are presented in Exhibit 3.1.

Exhibit 3.1 Computing S^2 and S

To compute S^2, the sample variance, complete the following:

✓ **1.** Obtain the difference between each observation and the mean.

✓ **2.** Square each difference.

✓ **3.** Add the squared differences.

✓ **4.** Divide this total by $n - 1$.

To compute S, the sample standard deviation, we take the square root of the variance.

Let us now apply these steps to our domestic general stock funds data in Example 3.12.

Example 3.12 *Computing the Sample Variance and Sample Standard Deviation*

For our sample containing the 17 domestic general stock funds with marketing fees paid from fund assets, the raw data pertaining to achieved 1-year total percentage returns are as follows.

32.2 29.5 29.9 32.4 30.5 30.1 32.1 35.2 10.0 20.6 28.6 30.5 38.0 33.0 29.4 37.1 28.6

The arithmetic mean for this sample is calculated as $\overline{X} = 29.86$. Compute the sample variance S^2 and sample standard deviation S.

SOLUTION

Using the four-step procedure, the sample variance S^2 is computed on the basis of the following tabular layout:

1-year total percentage returns

DOMESTIC GENERAL STOCK FUND	X_i	$\bar{X}$	$(X_i - \bar{X})$	$(X_i - \bar{X})^2$
Amcore Vintage Equity	$X_1 = 32.2$	29.86	+2.34	5.4756
Baron Funds Asset	$X_2 = 29.5$	29.86	−0.36	0.1296
Berger SmCoGrow	$X_3 = 29.9$	29.86	+0.04	0.0016
Chicago Trust GrowInc	$X_4 = 32.4$	29.86	+2.54	6.4516
Dodge & Cox DominiSo	$X_5 = 30.5$	29.86	+0.64	0.4096
Federated Institut MaxCapSvc	$X_6 = 30.1$	29.86	+0.24	0.0576
First Funds GroInc III	$X_7 = 32.1$	29.86	+2.24	5.0176
Harris Insight Inst Haven	$X_8 = 35.2$	29.86	+5.34	28.5156
Mentor Merger	$X_9 = 10.0$	29.86	−19.86	394.4196
Rainler Reich Tang	$X_{10} = 20.6$	29.86	−9.26	85.7476
Robertson Stephens ValGrow	$X_{11} = 28.6$	29.86	−1.26	1.5876
SSgA S&P500Idx	$X_{12} = 30.5$	29.86	+0.64	0.4096
SSgA SmallCap	$X_{13} = 38.0$	29.86	+8.14	66.2596
1784 GrowInc	$X_{14} = 33.0$	29.86	+3.14	9.8596
Stagecoach CorpStk	$X_{15} = 29.4$	29.86	−0.46	0.2116
Westwood Eq R	$X_{16} = 37.1$	29.86	+7.24	52.4176
Wright Yacktman	$X_{17} = 28.6$	29.86	−1.26	1.5876
Total			0^a	658.5592
			$\displaystyle\sum_{i=1}^{n}(X_i - \bar{X})$	$\displaystyle\sum_{i=1}^{n}(X_i - \bar{X})^2$

[a]Result differs from 0 due to rounding.

From equation (3.9), the sample variance is

$$S^2 = \frac{\displaystyle\sum_{i=1}^{n}(X_i - \bar{X})^2}{n - 1}$$

$$= \frac{(32.2 - 29.86)^2 + (29.5 - 29.86)^2 + (29.9 - 29.86)^2 + \cdots + (28.6 - 29.86)^2}{17 - 1}$$

$$= \frac{658.5592}{16}$$

$$= 41.15995$$

From equation (3.10), the sample standard deviation S is computed as

$$S = \sqrt{S^2} = \sqrt{\frac{\displaystyle\sum_{i=1}^{n}(X_i - \bar{X})^2}{n - 1}} = \sqrt{41.15995} = 6.42$$

In making our computations in Example 3.12 we are squaring the differences between each of the observations and the mean; therefore, *neither the variance nor the standard deviation can ever be negative*. The only time S^2 and S can be zero is when there is no variation at all in the data—when each observation in the sample is exactly the same. In such an unusual case, the range and interquartile range would also be zero.

But numerical data are inherently variable—not constant. Any random phenomenon of interest that we can think of usually takes on a variety of values. For example, different domestic general stock funds achieve different 1-year rates of return, attain different 3-year annualized rates of return, have different net asset values, and have different expense ratios. It is because numerical data inherently vary that it becomes so important to study not only measures of central tendency that summarize the data but also measures of variation that reflect how the numerical data are dispersed.

COMMENT: *Interpreting the Variance and the Standard Deviation*

The variance and the standard deviation measure the "average" scatter around the mean; how larger observations fluctuate above it and how smaller observations distribute below it. The variance possesses certain useful mathematical properties. However, its computation results in squared units—squared percents, squared dollars, squared inches, and so on. Thus, our primary measure of variation will be the standard deviation, whose value is in the original units of the data—percent of return, dollars, or inches.

The standard deviation helps tell us how a set of data clusters or distributes around its mean. For almost all sets of data, the majority of the observed values lie within an interval of plus and minus 1 standard deviation above and below the arithmetic mean. This means that the interval between $\overline{X} \pm 1S$ usually captures at least a majority of the data values. Therefore, a knowledge of the arithmetic mean and the standard deviation usually helps define where at least the majority of the data values are clustering.

COMMENT: *What the Standard Deviation Indicates*

In the sample containing the 17 domestic general stock funds whose marketing fees are paid through fund assets, the standard deviation in achieved 1-year total percentage returns is 6.42. This tells us that the 1-year total returns from the majority of the stock funds in this sample are clustering within 6.42 around the mean of 29.86 (i.e., between $\overline{X} - 1S = 23.44$ and $\overline{X} + 1S = 36.28$). In fact, we note that the 1-year total percentage returns of 76.5% of the funds (13 out of the 17) lie within this interval.

We should note that the formulas for variance and standard deviation [equations (3.9) and (3.10)] could not use

$$\sum_{i=1}^{n} (X_i - \overline{X})$$

as a numerator, because, as you may recall, the mean acts as a *balancing point* for observations larger and smaller than it. Therefore, the sum of the deviations about the mean is

always zero[2]; that is,

$$\sum_{i=1}^{n} (X_i - \overline{X}) = 0$$

To demonstrate this, let us again refer to Example 3.12 on page 167. We see from the tabular layout in the fourth column that the summation of the differences between each value and the mean [i.e., $\sum_{i=1}^{n} (X_i - \overline{X})$] is, except for rounding error, equal to zero. The sum of the squared deviations allows us to study the variation in the data. Hence we use

$$\sum_{i=1}^{n} (X_i - \overline{X})^2$$

when computing the variance and standard deviation. In the squaring process, observations that are farther from the mean get more weight than observations closer to the mean.

The respective squared deviations for the 17 domestic general stock funds are displayed in the last column of the tabular layout in Example 3.12 on page 168. We note that the 9th observation, $X_9 = 10.0$ attained by Mentor Merger, is 19.86 lower than the mean performance measure of 29.86 and the 10th observation, $X_{10} = 20.6$ attained by Rainler Reich Tang, is 9.26 lower. In the squaring process, both these values, along with the 13th and 16th observations (SSgA SmallCap and Westwood Eq R), contribute substantially more to the calculation of S^2 and S than do the other observations in the sample, which are much closer to the mean. Therefore, we can generalize as in Exhibit 3.2.

Exhibit 3.2 Understanding Variation in Data

✓ **1.** The more spread out, or dispersed, the data are, the larger will be the range, the interquartile range, the variance, and the standard deviation.

✓ **2.** The more concentrated, or *homogeneous*, the data are, the smaller will be the range, the interquartile range, the variance, and the standard deviation.

✓ **3.** If the observations are all the same (so that there is no variation in the data), the range, interquartile range, variance, and standard deviation will all be zero.

✓ **4.** *None* of the measures of variation (the range, interquartile range, standard deviation, variance) can *ever* be negative.

◆ *The Coefficient of Variation* Unlike the previous measures we have studied, the **coefficient of variation** is a *relative measure* of variation. It is always expressed as a percentage rather than in terms of the units of the particular data.

The coefficient of variation, denoted by the symbol *CV*, measures the scatter in the data relative to the mean. It may be computed as follows:

Coefficient of Variation

The coefficient of variation is equal to the standard deviation divided by the arithmetic mean, multiplied by 100%.

$$CV = \left(\frac{S}{\overline{X}}\right)100\% \tag{3.11}$$

where

S = standard deviation in a set of numerical data

$\overline{X}$ = arithmetic mean in a set of numerical data

Example 3.13 illustrates the computation of the coefficient of variation.

Example 3.13 *Computing the Coefficient of Variation*

For our sample containing the 17 domestic general stock funds with marketing fees paid from fund assets, the raw data pertaining to achieved 1-year total returns (in percents) are

32.2 29.5 29.9 32.4 30.5 30.1 32.1 35.2 10.0 20.6 28.6 30.5 38.0 33.0 29.4 37.1 28.6

Compute the coefficient of variation.

SOLUTION

From these data, the mean 1-year total percentage return $\overline{X}$ is 29.86 and the standard deviation S is 6.42. Using equation (3.11), the coefficient of variation is

$$CV = \left(\frac{S}{\overline{X}}\right)100\% = \left(\frac{6.42}{29.86}\right)100\% = 21.5\%$$

For this sample, the relative size of the "average spread around the mean" to the mean is 21.5%.

As a relative measure, the coefficient of variation is particularly useful when comparing the variability of two or more data sets that are expressed in different units of measurement. This is demonstrated in Example 3.14.

Example 3.14 *Comparing Two Coefficients of Variation*

Suppose that the operations manager of a package delivery service is contemplating the purchase of a new fleet of trucks. When packages are efficiently stored in the trucks in preparation for delivery, two major constraints have to be considered—the weight (in pounds) and the volume (in cubic feet) for each item.

Now suppose that in a sample of 200 packages the average weight is 26.0 pounds with a standard deviation of 3.9 pounds. In addition, suppose that the average volume for each of these packages is 8.8 cubic feet with a standard deviation of 2.2 cubic feet. How can we compare the variation of the weight and the volume?

SOLUTION

Because the units of measurement differ for the weight and volume constraints, to compare the fluctuations in these measurements it would be appropriate for the operations manager to consider the relative variability in the two types of measurements. For weight, the coefficient of variation is $CV_W = (3.9/26.0)100\% = 15.0\%$; for volume, the coefficient of variation is $CV_V = (2.2/8.8)100\% = 25.0\%$. Thus, relative to the mean, the volume of a package is much more variable than the weight of the package.

The coefficient of variation is also very useful when comparing two or more sets of data that are measured in the same units but differ to such an extent that a direct comparison of the respective standard deviations is not very helpful. This is illustrated in Example 3.15.

Example 3.15 *Comparing Two Coefficients of Variation*

Suppose that a potential investor is considering purchasing shares of stock in one of two companies, A or B, that are listed on the New York Stock Exchange. If neither company offers dividends to its stockholders and if both companies are rated equally high (by various investment services) in terms of potential growth, the potential investor might want to consider the volatility (*variability*) of the two stocks to aid in the investment decision.

Now suppose that each share of stock in company A has averaged $50 over the past few months with a standard deviation of $10. In addition, suppose that in this same time period, the price per share for company B stock averaged $12 with a standard deviation of $4. How can the investor determine which stock is more variable?

SOLUTION

In terms of the actual standard deviations, the price of company A shares seems to be more volatile than that of company B shares. However, because the average prices per share for the two stocks are so different, it would be more appropriate for the potential investor to consider the variability in price relative to the average price in order to examine the volatility/stability of the two stocks.

For company A, the coefficient of variation is $CV_A = (\$10/\$50)100\% = 20.0\%$; for company B, the coefficient of variation is $CV_B = (\$4/\$12)100\% = 33.3\%$. Thus, relative to the mean, the price of stock B is much more variable than the price of stock A.

Shape

A third important property of a set of data is its **shape**—the manner in which the data are distributed. Either the distribution of the data is **symmetrical** or it is not. If the distribution of data is not symmetrical, it is called asymmetrical, or skewed.

To describe the shape, we need only compare the mean and the median. If these two measures are equal, we may generally consider the data to be symmetrical (or zero-skewed). On the other hand, if the mean exceeds the median, the data may generally be described as *positive,* or **right-skewed.** If the mean is exceeded by the median, those data can generally be called *negative,* or **left-skewed.** That is, mean > median: positive, or right-skewness

mean = median: symmetry, or zero-skewness

mean < median: negative, or left-skewness

Positive skewness arises when the mean is increased by some unusually high values; negative skewness occurs when the mean is reduced by some extremely low values. Data are symmetrical when there are no really extreme values in a particular direction so that low and high values balance each other out.

Figure 3.6 depicts the shapes of three data sets. The data in panel A are negative, or left-skewed. In this panel we observe a long tail and distortion to the left that is caused by extremely small values. These extremely small values pull the mean downward so that the mean is less than the median.

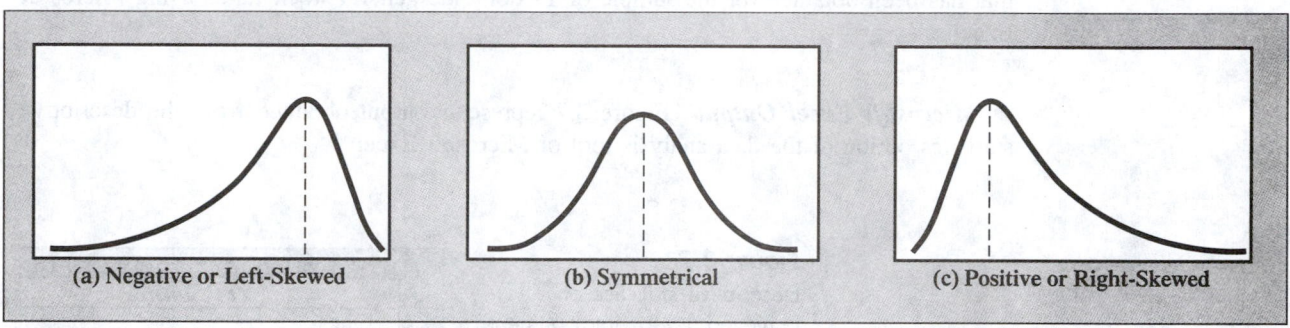

| (a) Negative or Left-Skewed | (b) Symmetrical | (c) Positive or Right-Skewed |

FIGURE 3.6   A comparison of three data sets differing in shape

The data in panel B are symmetrical; each half of the curve is a mirror image of the other half of the curve. The low and high values on the scale balance, and the mean equals the median. The data in panel C are positive or right-skewed. In this panel we observe a long tail on the right of the distribution and a distortion to the right that is caused by extremely large values. These extremely large values pull the mean upward so that the mean is greater than the median. How a data set's shape is determined is discussed in our next example.

Example 3.16 illustrates how to determine the shape of a data set.

Example 3.16   *Determining the Shape of the Data Set*

For our sample containing the 17 domestic general stock funds with marketing fees paid from fund assets, the raw data pertaining to achieved 1-year total percentage returns are

32.2 29.5 29.9 32.4 30.5 30.1 32.1 35.2 10.0 20.6 28.6 30.5 38.0 33.0 29.4 37.1 28.6

and the data are displayed along the dot scale in Figure 3.1 on page 156. What can be said about the shape of this data set?

SOLUTION

Note that there is one outlier in this data set and that the 17 observations do not cluster symmetrically around their arithmetic mean. The 1-year total percentage return of 10.0 attained by Mentor Merger is greatly exceeded by that achieved by all other domestic general stock funds in this sample. In Example 3.1 on page 155 the mean was computed to be 29.86. In Example 3.3 on page 158 the median was computed to be 30.5. Therefore, because the mean is exceeded by the median this data set may be described as negative, or left-skewed.

Interpreting Microsoft Excel Descriptive Statistics Output

Now that we have discussed the characteristics of central tendency, variation, and shape, we can examine the descriptive statistics output of achieved 1-year total percentage returns that has been obtained for the sample of 17 domestic general stock funds using Microsoft Excel.

◆ *Microsoft Excel Output* Figure 3.7 represents output obtained from the descriptive statistics option of the data analysis tool of Microsoft Excel

FIGURE 3.7
Descriptive statistics of achieved 1-year total percentage returns from a sample of 17 domestic general stock funds using Microsoft Excel

	A	B
1	*1-Yr Return*	
2		
3	Mean	29.86470588
4	Standard Error	1.556011615
5	Median	30.5
6	Mode	30.5
7	Standard Deviation	6.415600242
8	Sample Variance	41.15992647
9	Kurtosis	5.546604844
10	Skewness	-2.011159015
11	Range	28
12	Minimum	10
13	Maximum	38
14	Sum	507.7
15	Count	17
16	Largest(1)	38
17	Smallest(1)	10

We note that Excel has provided the (arithmetic) mean, median, mode, standard deviation, variance, range, minimum, maximum, and count (sample size), all of which have been discussed in this section. In addition, Excel has computed the standard error, along with statistics for kurtosis and skewness. The *standard error* is the standard deviation divided by the square root of the sample size and will be discussed in chapter 6. *Skewness* is a measure of lack of symmetry in the data and is based on a statistic that is a function of the *cubed* differences around the arithmetic mean. *Kurtosis* is a measure of the relative concentration of values in the center of the distribution as compared with tails and is based on the differences around the arithmetic mean raised to the fourth power. This measure is not discussed in this text (see reference 4).

3.2E ◆ GENERATING MEASURES OF CENTRAL TENDENCY, VARIATION, AND SHAPE USING MICROSOFT EXCEL

Overview

◆ *For Quick Results Users*   Use the Data Analysis Descriptive Statistics tool to generate measures of central tendency, variation, and shape.

◆ *For Developers*   The Quick Results procedure can be used for one-time results. To examine the effects of changing data values, implement a worksheet that contains formulas that use various Excel statistical functions to obtain a set of descriptive statistics.

The 3-2E.XLS workbook file contains worksheets that display various sample statistics for the 1-year total percentage return data for 17 domestic general stock funds presented in Table 3.1 on page 153.

Quick Results Details

◆ *Generating Measures of Central Tendency, Variation, and Shape Using the Data Analysis Descriptive Statistics Tool*   Use the Data Analysis Descriptive Statistics tool to generate many of the descriptive measures discussed earlier in this chapter. As an example, consider the Starting Point for section 3.2E workbook (STARTING POINT 3-2E.XLS) that contains the 1-year total percentage return data for 17 domestic general stock funds presented in Table 3.1. These data have been entered with a label into the cell range Data!B1:B18, next to the names of the funds. Using this workbook, we can use the Descriptive Statistics tool as follows:

 3-2E.XLS

❶ Open the Starting Point 3-2E workbook and click the Data sheet tab. Verify that the 1-year total percentage return values have been entered in column B through row 18.

❷ Select Tools | Data Analysis. Select Descriptive Statistics from the Analysis Tools list box in the Data Analysis dialog box. Click the OK button.

❸ In the Descriptive Statistics dialog box (see Figure 3E.1):
  a. Enter B1:B18 in the Input Range: edit box.
  b. Select the Grouped By: Columns option button.
  c. Select the Labels in First Row check box.
  d. Select the New Worksheet Ply: option button and enter Descriptive Statistics as the name of the new sheet.
  e. Select the Summary statistics check box.
  f. Select the Kth Largest and the Kth Smallest check boxes, leaving the default values (1) in their edit boxes unchanged. (This obtains the minimum and maximum values.)
  g. Deselect (uncheck) the Confidence Level for Mean: check box. (Confidence levels are discussed in chapter 7.)
  h. Click the OK button.

Microsoft Excel generates a worksheet of sample statistics similar to the one shown in Figure 3.7. The worksheet includes two statistics not discussed in chapter 3: the standard error and kurtosis. The standard error is the standard deviation divided by the square root of the

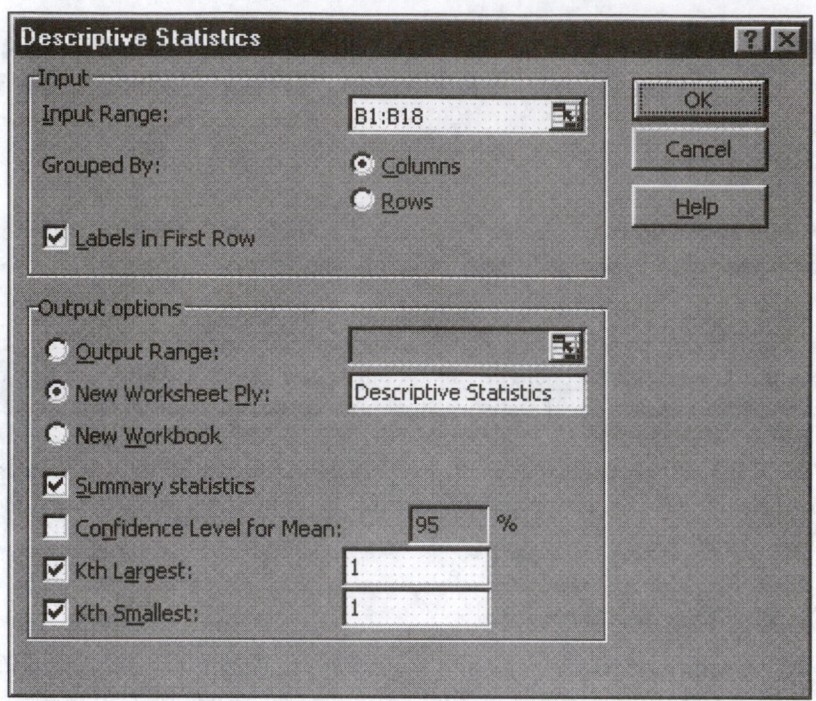

sample size and is discussed in chapter 6. Kurtosis is a measure of the relative concentration of values in the center of the distribution as compared to the tails and is based on the differences around the arithmetic mean raised to the fourth power and is not discussed in this text (see reference 4).

Developer Details

◆ *Implementing a Worksheet to Generate a Set of Descriptive Statistics*   We can create formulas that use a variety of Microsoft Excel functions to implement a worksheet that generates measures of location, variation, and shape. We can use these functions to create formulas that obtain measures of location:

AVERAGE(*cell range*)	Computes the arithmetic mean of the values in the cell range
MEDIAN(*cell range*)	Computes the median of the values in the cell range
MODE(*cell range*)	Computes the mode of the values in the cell range
MIN(*cell range*)	Computes the smallest value of the values in the cell range
MAX(*cell range*)	Computes the largest value of the values in the cell range

We can use these functions to create formulas that obtain measures of variation and shape:

STDEV(*cell range*)	Computes the sample standard deviation of the values in the cell range
VAR(*cell range*)	Computes the sample variance of the values in the cell range
SKEW(*cell range*)	Computes the skewness of the values in the cell range

We can use these functions to implement the rules for obtaining the quartile values as formulas:

IF(*comparison, formula to use if comparison holds, formula to use if comparison fails*) chooses a formula to use based on the result of the algebraic comparison.

COUNT(*cell range*)	Returns the number of cells in the cell range that contain a numeric value
SMALL(*cell range, k*)	Returns the *k*th smallest value of the values in the cell range
INT(*cell*)	Returns the integer portion of the value in cell (always rounds down)
ROUND(*cell*, 0)	Rounds value in cell to nearest integer
FLOOR(*cell, multiple*)	Rounds value in cell down to nearest multiple
CEILING(*cell, multiple*)	Rounds value in cell up to nearest multiple

(The Excel QUARTILE function could also be used for obtaining quartile values. However, as this function can return incorrect or inconsistent results, it is avoided in the following discussion.)

Tables 3E.1 and 3E.2 present a Calculations sheet design that uses these functions as a basis for developing a table of descriptive statistics. This design assumes that the 1-year total percentage return data have been placed in the range Data!B2:B18 as they have been in the Starting Point for section 3.2E workbook (STARTING POINT 3-2E.XLS). In this design, the formulas for cells E8 and E12 use IF functions to implement the rules for calculating quartiles

Table 3E.1 Calculations sheet design for the domestic general stock funds sample

	A	B	C	D	E	F
1	Descriptive Statistics for 1-Year Total Percentage Return Data					
2						
3	Measures of Location					
4	Mean	=AVERAGE(Data!B2:B18)				
5	Median	=MEDIAN(Data!B2:B18)				
6	Mode	=MODE(Data!B2:B18)				
7	Minimum	=MIN(Data!B2:B18)				
8	Maximum	=MAX(Data!B2:B18)				
9	Midrange	=(B7+B8)/2		Quartile calculations		
10	First Quartile	=IF(D9="Rule 2 applies",(E10+F10)/2,E10)		area (see Table 3E.2)		
11	Third Quartile	=IF(D13="Rule 2 applies",(E14+F14)/2,E14)				
12	Midhinge	=(B10+B11)/2				
13						
14	Measures of Variation					
15	Range	=B8-B7				
16	Interquartile Range	=B11-B10				
17	Standard Deviation	=STDEV(Data!B2:B18)				
18	Variance	=VAR(Data!B2:B18)				
19	Coefficient of Variation	=B17/B4				
20						
21	Measure of Shape					
22	Skewness	=SKEW(Data!B2:B18)				

Note: In this table and several others that appear later in the text, open and closed-double-quotation marks have been used for typographic clarity. Both of these quotation marks should be entered using the (straight) double quotation mark key.

Table 3E.2 Quartile Calculations area of the Calculations sheet for the domestic general stock funds sample

	A	B	C	D	E	F
1						
2						
3						
4				Quartile Calculations		
5				Initial first quartile rank	=(COUNT(Data!B2:B18)+1)/4	
6				Initial third quartile rank	=(3*(COUNT(Data!B2:B18)+1))/4	
7						
8				For first quartile:	=IF(E5=INT(E5),"Rule 1 applies", IF(E5=CEILING(E5,0.5), "Rule 2 applies", "Rule 3 applies"))	
9				=IF(E8="Rule 2 applies", "average these ranks:", "use rank:")	=IF(E8="Rule 2 applies", FLOOR(E5,1), ROUND(E5,0))	=IF(E8="Rule 2 applies", CEILING(E5,1), "")
10				=IF(E8="Rule 2 applies", "average these values:", "value of rank:")	=SMALL(Data!B2:B18,E9)	=IF(E8="Rule 2 applies", SMALL(Data!B2:B18,F9), "")
11						
12				For third quartile:	=IF(E6=INT(E6),"Rule 1 applies", IF(E6=CEILING(E6,0.5), "Rule 2 applies", "Rule 3 applies"))	
13				=IF(E12="Rule 2 applies", "average these ranks:", "use rank:")	=IF(E12="Rule 2 applies", FLOOR(E6,1), ROUND(E6,0))	=IF(E12="Rule 2 applies", CEILING(E6,1), "")
14				=IF(E12="Rule 2 applies", "average these values:", "value of rank:")	=SMALL(Data!B2:B18,E13)	=IF(E12="Rule 2 applies", SMALL(Data!B2:B18,F13),"")

discussed in section 3.2. Specifically, these formulas contain nested IF functions that first check to see if rule 1 applies. If this rule does not apply, the second, nested IF determines whether rule 2 or rule 3 applies. Other formulas in the quartile calculations area check the result of the E8 or E12 formula to determine which labels or values should be used or calculated.

To implement this design, do the following:

❶ Open the Starting Point for section 3.2E workbook (STARTING POINT 3-2E.XLS) and click the Data sheet tab. Verify that the 1-year total percentage return values have been entered in column B through row 18.

❷ Select Insert | Worksheet and rename the new sheet Calculations.

❸ Enter the title, the subheadings, and column A value labels.

❹ Enter the formulas for column B.

❺ Enter the column D value labels.

❻ Enter columns D, E, and F formulas shown in Table 3E.2 that are used to calculate quartiles, noting the following:
   a. Enter all formulas as one continuous line (some formulas in the table have been typeset as two lines).
   b. Phrases such as "Rule 2 applies" and "average these ranks:" that appear inside formulas should be entered with a pair of double quotation marks.
   c. The formulas for cells E6, E8, E9, E12, and E13 contain consecutive right parentheses.
   d. The formulas for cells F9, F10, F13, and F14 contain consecutive double quotation marks.

A Calculations sheet, similar to the one illustrated in Figure 3E.2, will be produced.

	A	B	C	D	E	F
1	Descriptive Statistics for One-Year Total Percentage Return Data					
2						
3	Measures of Location					
4	Mean	29.86471		Quartile Calculations		
5	Median	30.5		Initial first quartile rank	4.5	
6	Mode	30.5		Initial third quartile rank	13.5	
7	Minimum	10				
8	Maximum	38		For first quartile:	Rule 2 applies	
9	Midrange	24		average these ranks:	4	5
10	First Quartile	29		average these values:	28.6	29.4
11	Third Quartile	32.7				
12	Midhinge	30.85		For third quartile:	Rule 2 applies	
13				average these ranks:	13	14
14	Measures of Variation			average these values:	32.4	33
15	Range	28				
16	Interquartile Range	3.7				
17	Standard Deviation	6.4156				
18	Variance	41.15993				
19	Coefficient of Variation	21.48%				
20						
21	Measure of Shape					
22	Skewness	-2.01116				

▲ WHAT IF EXAMPLE

We can study the effect of extreme values on the measures of location by pretending that the 1-year total percentage return for the Rainler Reich Tang fund is 74.2 instead of 20.6. In your workbook for this section (or in workbook 3-2E.XLS), change the value of this fund's 1-year return from 20.6 to 74.2 (change cell B11 of the Data sheet if you are using the 3-2E.XLS workbook). Then click the Calculations sheet tab to see the dynamic change set of descriptive statistics illustrated in Figure 3E.3.

	A	B	C	D	E	F
1	Descriptive Statistics for One-Year Total Percentage Return Data					
2						
3	Measures of Location					
4	Mean	33.01765		Quartile Calculations		
5	Median	30.5		Initial first quartile rank	4.5	
6	Mode	30.5		Initial third quartile rank	13.5	
7	Minimum	10				
8	Maximum	74.2		For first quartile:	Rule 2 applies	
9	Midrange	42.1		average these ranks:	4	5
10	First Quartile	29.45		average these values:	29.4	29.5
11	Third Quartile	34.1				
12	Midhinge	31.775		For third quartile:	Rule 2 applies	
13				average these ranks:	13	14
14	Measures of Variation			average these values:	33	35.2
15	Range	64.2				
16	Interquartile Range	4.65				
17	Standard Deviation	12.16898				
18	Variance	148.084				
19	Coefficient of Variation	36.86%				
20						
21	Measure of Shape					
22	Skewness	2.260176				

Compare these results with the statistics originally generated (shown in Figure 3E.2). The arithmetic mean and the midrange are greatly affected by this extreme value of 74.2; the midhinge is much less affected. The arithmetic mean changed to 33.0 and the midrange to 42.1, and the midhinge became 31.8.

We can also observe the effect of this extreme value on the measures of variation and shape. Note that the range and the standard deviation were greatly affected by the extreme value, and the interquartile range was much less affected. The range increased to 64.2 from 28 and the standard deviation increased to 12.17 from 6.42. The interquartile range changed from 3.7 to 4.65.

As first discussed at the end of section 2.2E, we could use the Scenario Manager to store and use sets of alternative data values in order to easily see the effects of many different changes.

Problems for Section 3.2

Learning the Basics

3.1 Given the following set of data from a sample of size $n = 5$:

7   4   9   8   2

(a) Compute the mean, median, mode, midrange, and midhinge.
(b) Compute the range, interquartile range, variance, standard deviation, and coefficient of variation.
(c) Describe the shape.

• **3.2** Given the following set of data from a sample of size $n = 6$:

7   4   9   7   3   12

(a) Compute the mean, median, mode, midrange, and midhinge.
(b) Compute the range, interquartile range, variance, standard deviation, and coefficient of variation.
(c) Describe the shape.

• **3.3** Given the following set of data from a sample of size $n = 7$:

12   7   4   9   0   7   3

(a) Compute the mean, median, mode, midrange, and midhinge.
(b) Compute the range, interquartile range, variance, standard deviation, and coefficient of variation.
(c) Describe the shape.

3.4 Given the following set of data from a sample of size $n = 5$:

7   −5   −8   7   9

(a) Compute the mean, median, mode, midrange, and midhinge.
(b) Compute the range, interquartile range, variance, standard deviation, and coefficient of variation.
(c) Describe the shape.

3.5 Given the following set of data from a sample of size $n = 7$:

3   3   3   3   3   3   3

(a) Compute the mean, median, mode, midrange, and midhinge.

(b) Compute the range, interquartile range, variance, standard deviation, and coefficient of variation.

(c) What is unusual about this set of data?

• **3.6** Given the following two sets of data—each with samples of size $n = 7$:

Set 1:	10	2	3	2	4	2	5
Set 2:	20	12	13	12	14	12	15

(a) For each set, compute the mean, median, mode, midrange, and midhinge.

(b) Compare your results and summarize your findings.

(c) Compare the first sampled item in each set, compare the second sampled item in each set, and so on. Briefly describe your findings here in light of your summary in part (b).

(d) For each set, compute the range, interquartile range, variance, standard deviation, and coefficient of variation.

(e) Describe the shape.

(f) Compare your results in (d) and (e) and discuss your findings.

(g) On the basis of your answers to (a)–(f) above, what can you generalize about the properties of central tendency, variation, and shape?

Applying the Concepts

3.7 The operations manager of a plant that manufactures tires wishes to compare the actual inner diameter of two grades of tires, each of which is expected to be 575 millimeters. A sample of five tires of each grade was selected, and the results representing the inner diameters of the tires, ordered from smallest to largest, were as follows:

Grade X	Grade Y
568 570 575 578 584	573 574 575 577 578

(a) For each of the two grades of tires, compute the
 (1) arithmetic mean (2) median (3) standard deviation

(b) Which grade of tire is providing better quality? Explain.

(c) What would be the effect on your answers in (a) and (b) if the last value for grade Y was 588 instead of 578? Explain.

3.8 Suppose that owing to an error, a data set containing the price-to-earnings (PE) ratios from nine companies traded on the American Stock Exchange was recorded as 13, 15, 14, 17, 13, 16, 15, 16, and 61, where the last value should have been 16 instead of 61.

(a) Show how much the mean, median, and midrange are affected by the error (i.e., compute these statistics for the "bad" and "good" data sets, and compare the results of using different estimators of central tendency).

(b) Compute the range, interquartile range, variance, standard deviation, and coefficient of variation for the data set with the error (61) and then recompute these statistics after the PE ratio is corrected to 16.

 **WHAT IF?**

(c) Discuss the differences in your findings in (b) for each measure of variation.

(d) Which measure in (b) seems to be affected most by the error?

(e) How would you describe the shape of the data set with and without the error?

3.9 A manufacturer of flashlight batteries took a sample of 13 batteries from a day's production and used them continuously until they were drained. The numbers of hours they were used until failure were

342 426 317 545 264 451 1049 631 512 266 492 562 298

 **DATA FILE**
BATTERIES.XLS

(a) Compute the mean, median, mode, midrange, and midhinge. Looking at the distribution of times to failure, which measures of location do you think are best and which worst? (And why?)

(b) In what ways would this information be useful to the manufacturer? Discuss.

(c) Calculate the range, variance, and standard deviation.

(d) For many sets of data the range is about six times the standard deviation. Is this true here? (If not, why do you think it is not?)

(e) Using the information above, what would you advise if the manufacturer wanted to be able to say in advertisements that these batteries "should last 400 hours"? (*Note:* There is no right answer to this question; the point is to consider how to make such a statement precise.)

WHAT IF?

(f) Suppose that the first value was 1,342 instead of 342. Repeat (a), using this value. Comment on the difference in the results.

WHAT IF?

(g) Do (c)–(e) with the first value equal to 1,342 instead of 342. Comment on the difference in the results.

(h) How would you describe the shape of the data set if the first value were 342?

(i) How would you describe the shape of the data set if the first value were 1,342?

3.10 The following data are the monthly rental prices for a sample of 10 unfurnished studio apartments in the center of a large city and a sample of 10 unfurnished studio apartments in an outlying part of the city:

DATA FILE
STUDIO.XLS

Center City

$955 $1,000 $985 $980 $940 $975 $965 $999 $1,247 $1,119

Outlying Area

$750 $775 $725 $705 $694 $725 $690 $745 $575 $800

(a) For each set of data, compute the mean, median, midhinge, range, interquartile range, standard deviation, and coefficient of variation.

(b) What can be said about unfurnished studio apartments renting in the center city versus those renting in an outlying area? Compare and contrast the rents in these two areas.

• **3.11** A bank branch located in a commercial district of a city has developed an improved process for serving customers during the 12:00 P.M. to 1 P.M. peak lunch period. The waiting time in minutes (operationally defined as the time the customer enters the line until he or she is served) of all customers during this hour is recorded over a period of 1 week. A random sample of 15 customers is selected and the results are as follows:

DATA FILE
BANK1.XLS

4.21 5.55 3.02 5.13 4.77 2.34 3.54 3.20 4.50 6.10 0.38 5.12 6.46 6.19 3.79

(a) Compute the

(1) arithmetic mean.
(2) median.
(3) midrange.
(4) first quartile.
(5) third quartile.
(6) midhinge.

(7) range.
(8) interquartile range.
(9) variance.
(10) standard deviation.
(11) coefficient of variation.

(b) Are the data skewed? If so, how?

(c) As a customer walks into the branch office during the lunch hour, she asks the branch manager how long she can expect to wait. The branch manager replies, "Almost certainly not longer than five minutes." On the basis of the results of (a), evaluate this statement.

(d) Suppose that the branch manager would like to guarantee a certain level of service during the peak lunch hour period. Failure to obtain service within a specified time would result in a small monetary payment or gift to the customer. What waiting time do you think should be used as a cutoff above which this small payment or gift would be provided? Explain your answer.

3.12 Suppose that another branch located in a residential area is most concerned with the Friday evening hours from 5 P.M. to 7 P.M. The waiting time in minutes (operationally defined as the time the customer enters the line until he or she is served) of all customers during these hours is recorded over a period of 1 week. A random sample of 15 customers is selected and the results are as follows:

DATA FILE
BANK2.XLS

9.66 5.90 8.02 5.79 8.73 3.82 8.01 8.35 10.49 6.68 5.64 4.08 6.17 9.91 5.47

(a) Compute the
    (1) arithmetic mean.
    (2) median.
    (3) midrange.
    (4) first quartile.
    (5) third quartile.
    (6) midhinge.
    (7) range.
    (8) interquartile range.
    (9) variance.
    (10) standard deviation.
    (11) coefficient of variation.

(b) Are the data skewed? If so, how?

(c) As a customer walks into the branch office during the Friday evening hours, he asks the branch manager how long he can expect to wait. The branch manager replies, "Almost certainly not longer than 5 minutes." On the basis of the results of (a), evaluate this statement.

(d) Suppose that the branch manager would like to guarantee a certain level of service during the Friday evening hours. Failure to obtain service within a specified time would result in a small monetary payment or gift to the customer. What waiting time do you think should be used as a cutoff above which this small payment or gift would be provided? Explain your answer.

(e) What arguments can be raised that would make it inappropriate to compare the waiting times in Problem 3.11 with those in this problem?

3.13 For the last 10 days in June, the "Shore Special" train arrived late at its destination by the times (in minutes) listed here. (A negative number means that the train was early by that number of minutes.)

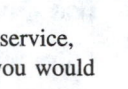

DATA FILE
TRAIN.XLS

$$-3 \quad 6 \quad 4 \quad 10 \quad -4 \quad 124 \quad 2 \quad -1 \quad 4 \quad 1$$

(a) If you were hired by the railroad to show that the railroad is providing good service, what are some of the summary measures pertaining to central tendency that you would use to accomplish this?

(b) If you were hired by a TV station that was producing a documentary to show that the railroad is providing bad service, what summary measures pertaining to central tendency would you use?

(c) If you were trying to be objective and unbiased in assessing the railroad's performance, which summary measures pertaining to central tendency would you use? (This is the hardest part because you cannot answer without making additional assumptions about the relative costs of being late by various amounts of time.)

(d) Compute the range, interquartile range, variance, standard deviation, and coefficient of variation for "lateness" (in minutes).

(e) Discuss the property of variation for these data.

(f) What would be the effect on your conclusions in (a)–(e) if the value of 124 had been incorrectly recorded and should have been 12?

WHAT IF?

(g) Describe the shape of the data shown above.

(h) Describe the shape of the data if the value 124 is replaced by 12.

3.14 In order to estimate how much water will be needed to supply the community of Falling Rock in the next decade, the town council asked the city manager to find out how much water a sample of families currently uses. The sample of 15 families used the following number of gallons (in thousands) in the past year:

DATA FILE
WATER.XLS

11.2 21.5 16.4 19.7 14.6 16.9 32.2 18.2 13.1 23.8 18.3 15.5 18.8 22.7 14.0

(a) What is the mean amount of water used per family? The median? The midrange? The midhinge?

(b) Suppose the town council expects that 10 years from now there will be 4,500 families living in Falling Rock. How many gallons of water will be needed annually if the rate of consumption per family stays the same?

(c) In what ways would the information provided in (a) and (b) be useful to the town council? Discuss.

(d) Why might the town council have used the data from a survey rather than just measuring the total consumption in the town? (Think about what types of users are not yet included in the estimation process.)

(e) Compute the range, interquartile range, variance, standard deviation, and coefficient of variation in water consumption.

(f) Discuss the property of variation for these data.

(g) Describe the shape.

 **3.3** EXPLORATORY DATA ANALYSIS

Now that we have studied the three major properties of numerical data (central tendency, variation, and shape), it is important that we identify and describe the major features of the data in a summarized format. One approach to such an "exploratory data analysis" is to develop a *five-number summary* and to construct a *box-and-whisker plot* (references 5 and 6).

The Five-Number Summary

A **five-number summary** consists of

$$X_{\text{smallest}} \quad Q_1 \quad \text{Median} \quad Q_3 \quad X_{\text{largest}}$$

From the five-number summary we can obtain three measures of central tendency (the median, midhinge, and midrange) and two measures of variation (the interquartile range and range) to provide us with a better idea as to the shape of the distribution.

If the data are perfectly symmetrical, the relationship among the various measures of location can be expressed as in Exhibit 3.3.

Exhibit 3.3 Using the Five-Number Summary to Recognize Symmetry in Data

✓ **1.** The distance from Q_1 to the median equals the distance from the median to Q_3.

✓ **2.** The distance from X_{smallest} to Q_1 equals the distance from Q_3 to X_{largest}.

✓ **3.** The median, the midhinge, and the midrange are all equal. (These measures also equal the mean in the data.)

On the other hand, for nonsymmetrical distributions the relationship among the various measures of location can be expressed as in Exhibit 3.4.

Exhibit 3.4 Using the Five-Number Summary to Recognize Nonsymmetry in Data

✓ **1.** In right-skewed distributions, the distance from Q_3 to $X_{largest}$ greatly exceeds the distance from $X_{smallest}$ to Q_1.

✓ **2.** In right-skewed distributions, the median and midhinge are less than the midrange.

✓ **3.** In left-skewed distributions, the distance from $X_{smallest}$ to Q_1 greatly exceeds the distance from Q_3 to $X_{largest}$.

✓ **4.** In left-skewed distributions, the midrange is less than the median and midhinge.

We determine the five-number summary in Example 3.17.

Example 3.17 *Determining the Five-Number Summary*

For our sample representing the 1-year total percentage returns achieved by the 17 domestic general stock funds whose marketing fees are paid from fund assets, the ordered array is

10.0 20.6 28.6 28.6 29.4 29.5 29.9 30.1 30.5 30.5 32.1 32.2 32.4 33.0 35.2 37.1 38.0

State the five-number summary for these data.

SOLUTION

In Example 3.3 on page 158 the median was computed to be 30.5. In Example 3.8 on page 162 the first quartile was computed to be 29.0 and the third quartile 32.7. Therefore, the five-number summary is

10.0   29.0   30.5   32.7   38.0

We may now use the five-number summary to study the shape of this distribution. From the guidelines presented in Exhibits 3.3 and 3.4, it is clear that the 1-year total returns data for our sample are left-skewed because the distance from $X_{smallest}$ to Q_1 (that is, 19.0) greatly exceeds the distance from Q_3 to $X_{largest}$ (that is, 5.3). Also, if we compare the median (30.5), the midhinge (30.85), and the midrange (24.0), we observe that the midrange is distorted by the outlier 10.0 and is by far the smallest of these summary measures. The midhinge and the median, which are resistant to outliers, are very close in value.

Box-and-Whisker Plot

In its simplest form, a **box-and-whisker plot** provides a graphical representation of the data through its five-number summary. The box-and-whisker plot is depicted in Figure 3.8 for the 1-year total returns achieved by the sample of 17 domestic general stock funds whose marketing fees are paid from fund assets.

The vertical line drawn within the box represents the location of the median value in the data. Note further that the vertical line at the left side of the box represents the location of Q_1 and the vertical line at the right side of the box represents the location of Q_3. Therefore,

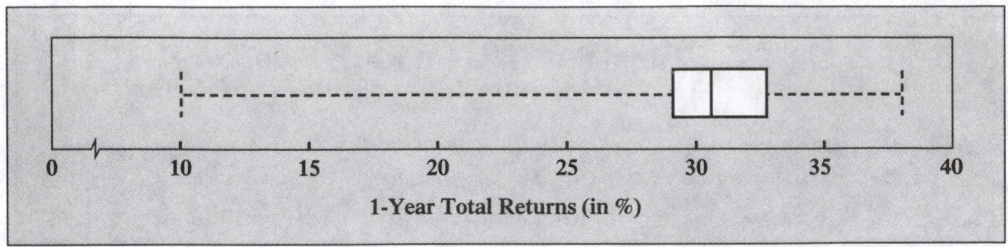

FIGURE 3.8 Box-and-whisker plot representing the 1-year total percentage returns achieved by 17 domestic general stock funds whose marketing fees are paid from fund assets

we see that the box contains the middle 50% of the observations in the distribution. The lower 25% of the data are represented by a dashed line (i.e., a *whisker*) connecting the left side of the box to the location of the smallest value, $X_{smallest}$. Similarly, the upper 25% of the data are represented by a dashed line connecting the right side of the box to $X_{largest}$.

COMMENT: *Interpreting the Box-and-Whisker Plot*

The visual representation of the 1-year total percentage returns depicted in Figure 3.8 indicates that the shape of this data set is left-skewed. Although we observe that the vertical median line is unexpectedly closer to the left side of the box, we also see that the left-side whisker length is clearly much larger than the right-side whisker length. This has occurred here because the 1-year total percentage return of 10.0 attained by Mentor Merger is an outlier.

To summarize what we have learned about graphical representation of our data, Figure 3.9 demonstrates the relationship between exploratory data analysis methods such as the box-and-whisker plot and graphical displays such as polygons. Four different types of distributions are depicted with their box-and-whisker plots and corresponding polygons.

When a data set is perfectly symmetrical, as is the case in Figure 3.9(a) and (d), the mean, median, midrange, and midhinge will be the same. In addition, the length of the left whisker will equal the length of the right whisker, and the median line will divide the box in half. In practice, it is unlikely that we will observe a data set that is perfectly symmetrical. However, we should be able to state that our data set is approximately symmetrical if the lengths of the two whiskers are almost equal and the median line almost divides the box in half.

On the other hand, when our data set is left-skewed as in Figure 3.9(b), the few small observations distort the midrange and mean toward the left tail. In such cases, it would be expected that we have the following sequence among the five measures of central tendency:

$$\text{midrange} < \text{mean} < \text{midhinge} < \text{median} < \text{mode}$$

For this hypothetical left-skewed distribution, we observe from Figure 3.9(b) that the skewed (i.e., distorted) nature of the data set indicates that there is a heavy clustering of observations at the high end of the scale (i.e., the right side); 75% of all data values are found between the left edge of the box (Q_1) and the end of the right whisker ($X_{largest}$). Therefore, the long left whisker contains the distribution of only the smallest 25% of the observations, demonstrating the distortion from symmetry in this data set.

170

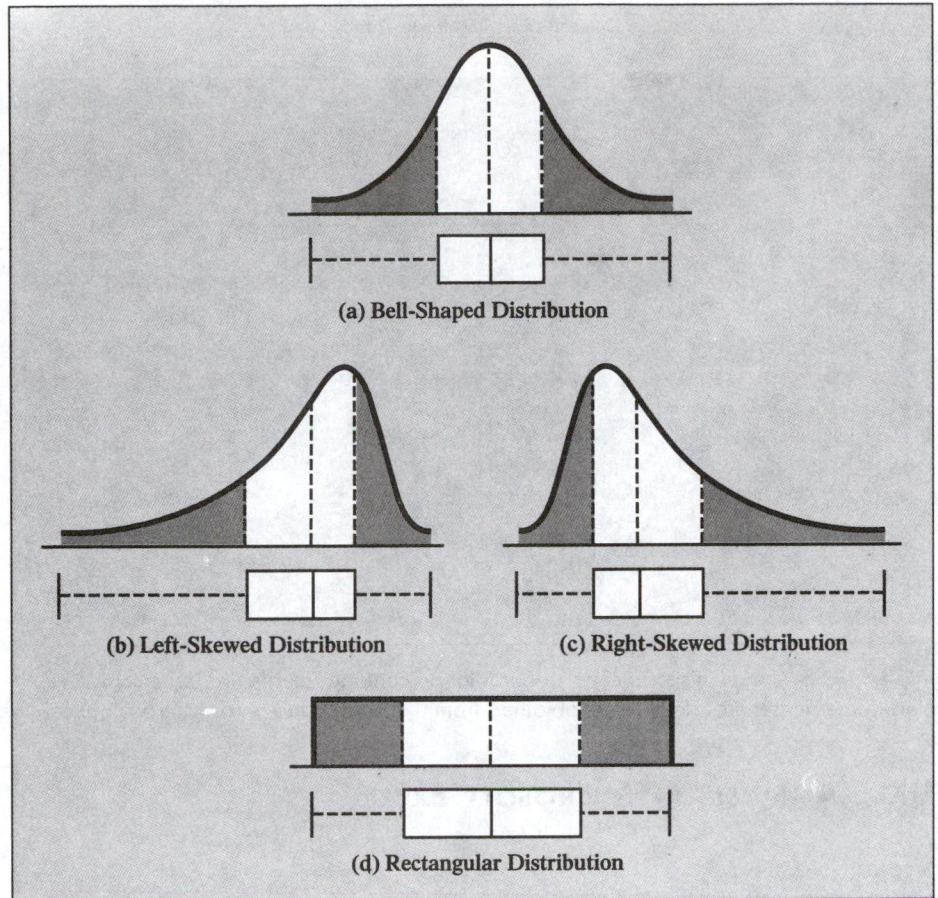

FIGURE 3.9 Four hypothetical distributions examined through their box-and-whisker plots and their corresponding polygons. *Note: Area under each polygon is split into quartiles corresponding to the five-number summary for the respective box-and-whisker plot*

(a) Bell-Shaped Distribution

(b) Left-Skewed Distribution

(c) Right-Skewed Distribution

(d) Rectangular Distribution

If the data set is right-skewed as in Figure 3.9(c), the few large observations distort the midrange and mean toward the right tail. In such cases, it would be expected that we have the following sequence among the five measures of central tendency:

$$\text{mode} < \text{median} < \text{midhinge} < \text{mean} < \text{midrange}$$

For the right-skewed data set in Figure 3.9(c) the concentration of data points will be on the low end of the scale (i.e., the left side of the box-and-whisker plot). Here, 75% of all data values are found between the beginning of the left whisker (X_{smallest}) and the right edge of the box (Q_3), and the remaining 25% of the observations are dispersed along the long right whisker at the upper end of the scale.

Instead of depicting the box-and-whisker plot horizontally from left (low) to right (high), as in Figures 3.8 and 3.9, output from the PHStat add-in for Microsoft Excel displays the box-and-whisker plot vertically from bottom (low) to top (high). Figure 3.10 demonstrates the box-and-whisker plot for the 1-year total percentage returns achieved by our 17 domestic general stock funds using the PHStat add-in for Microsoft Excel.

We observe that the PHStat add-in for Microsoft Excel has plotted a vertical box-and-whisker plot, rather than the horizontal plot of Figure 3.8. Figure 3.10 provides a line that connects the minimum and maximum values shown with a short dash.

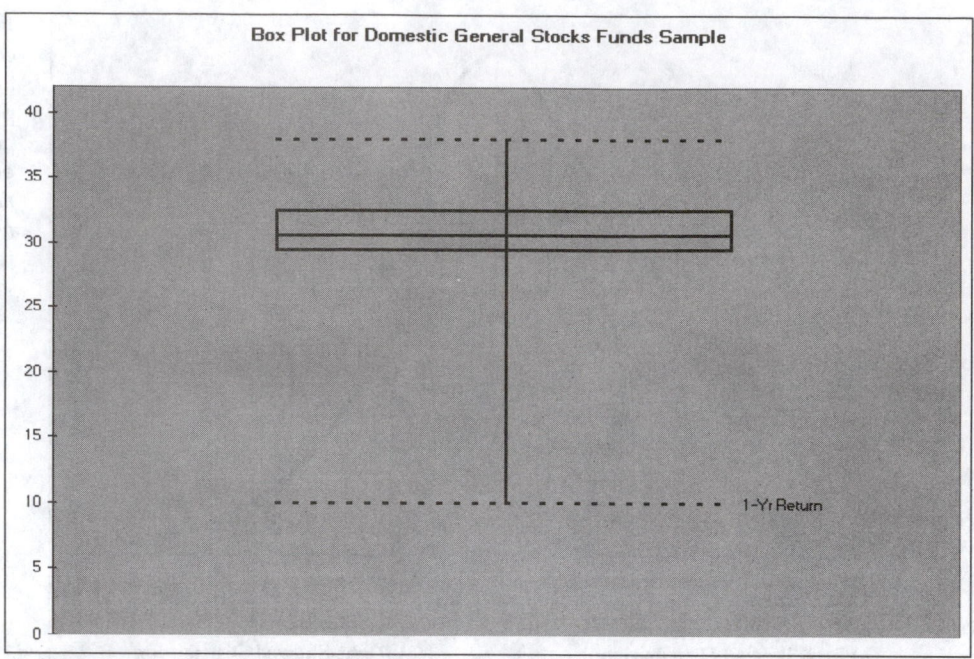

FIGURE 3.10 Box-and-whisker plot of 1-year total percentage returns (in percents) for 17 sampled domestic general stock funds obtained from the PHStat add-in for Microsoft Excel

3.3E EXPLORATORY DATA ANALYSIS IN MICROSOFT EXCEL

Overview

◆ *For Quick Results Users and Developers* Both quick results users and developers should use the PHStat add-in to generate box-and-whisker plots and five-number summaries as there are no basic features of Microsoft Excel that could be used easily to produce these analyses.

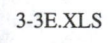 3-3E.XLS

The 3-3E.XLS workbook file contains a box-and-whisker plot and a five-number summary for the 1-year total percentage return data for 17 domestic general stock funds presented in Table 3.1 on page 153.

Details for All Users

◆ *Generating a Box-and-Whisker Plot for the Domestic General Stock Funds Sample Data* Use the Box-and-Whisker Plot choice of the PHStat add-in to generate a box-and-whisker plot and a five-number summary. To generate these displays for the 1-year total percentage return data of Table 3.1, do the following:

❶ If the PHStat add-in has not been previously loaded, load the add-in using the instructions of section S4.2.

❷ Open the Starting Point for section 3.2E workbook (STARTING POINT 3-2E.XLS) and click the Data sheet tab. Verify that the 1-year total percentage return values have been entered in column B through row 18.

172

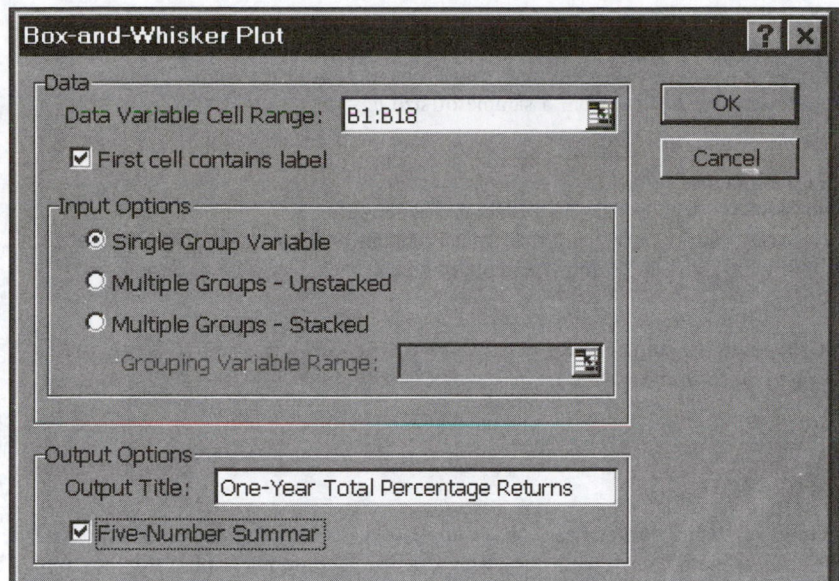

FIGURE 3E.4
Box-and-Whisker Plot dialog box

❸ Select PHStat | Box-and-Whisker Plot.

❹ In the Box-and-Whisker Plot dialog box (see Figure 3E.4):
   a. Select the Single Group Variable option button and enter B1:B18 in the Data Variable Cell Range: edit box.
   b. Enter One-Year Total Percentage Returns in the Output Title: edit box.
   c. Select the Five-Number Summary check box.
   d. Click the OK button.

The add-in produces a box-and-whisker plot similar to the one shown in Figure 3.10 and a five-number summary. The box-and-whisker plot is *not* dynamically changeable, so changes made to the underlying data would require repeating the procedure in order to produce a new chart.

Problems for Section 3.3

Learning the Basics

3.15 Given the following set of data from a sample of size $n = 5$:

   7   4   9   8   2

   (a) List the five-number summary.
   (b) Form the box-and-whisker plot and describe the shape.
   (c) Compare your answer in (b) with that from Problem 3.1(c) on page 180. Discuss.

● **3.16** Given the following set of data from a sample of size $n = 6$:

   7   4   9   7   3   12

   (a) List the five-number summary.
   (b) Form the box-and-whisker plot and describe the shape.
   (c) Compare your answer in (b) with that from Problem 3.2(c) on page 180. Discuss.

● **3.17** Given the following set of data from a sample of size $n = 7$:

   12   7   4   9   0   7   3

(a) List the five-number summary.

(b) Form the box-and-whisker plot and describe the shape.

(c) Compare your answer in (b) with that from Problem 3.3(c) on page 180. Discuss.

3.18 Given the following set of data from a sample of size $n = 5$:

$$7 \quad -5 \quad -8 \quad 7 \quad 9$$

(a) List the five-number summary.

(b) Form the box-and-whisker plot and describe the shape.

(c) Compare your answer in (b) with that from Problem 3.4(c) on page 180. Discuss.

3.19 Given the following set of data from a sample of size $n = 7$:

$$3 \quad 3 \quad 3 \quad 3 \quad 3 \quad 3 \quad 3$$

(a) List the five-number summary.

(b) Why can't you form the box-and-whisker plot here?

Applying the Concepts

• 3.20 A manufacturer of flashlight batteries took a sample of 13 batteries from a day's production and used them continuously until they were drained. The numbers of hours they were used until failure were

342 426 317 545 264 451 1049 631 512 266 492 562 298

(a) List the five-number summary.

(b) Form the box-and-whisker plot and describe the shape.

(c) Compare your answer in (b) with that from Problem 3.9(h) on page 182. Discuss.

• 3.21 The following data are the annual percentage yields on money market accounts from a sample of 15 commercial banks in the New York metropolitan area as of February 12, 1997, the day before the Dow Jones Industrial Average passed 7,000 for the first time:

BANK NAME	MM ACCT. YIELD	BANK NAME	MM ACCT. YIELD
Banco Popular	3.10	Fleet Bank	2.28
Bank of N.Y.	2.63	Key Bank of N.Y.	3.01
Bank of Tokyo-Mitsubishi	3.05	Marine Midland	2.73
Chase Manhattan	2.79	North Fork Bank	2.53
Citibank	3.25	PNC Bank (N.J.)	2.00
CoreStates NJ National Bank	1.90	Republic National	3.05
EAB	2.79	Summit Bank	2.02
First Union	2.90		

(a) List the five-number summary.

(b) Form the box-and-whisker plot and describe the shape.

(c) If someone said to you, "Money market rates don't vary much from bank to bank," on the basis of these data, what would you say?

3.22 The following data are the monthly rental prices for a sample of 10 unfurnished studio apartments in the center of a large city and a sample of 10 unfurnished studio apartments in an outlying part of the city:

Center City

$955 $1,000 $985 $980 $940 $975 $965 $999 $1,247 $1,119

Outlying Area

$750 $775 $725 $705 $694 $725 $690 $745 $575 $800

For each of the two areas,
(a) List the five-number summary.
(b) Form the box-and-whisker plot and describe the shape.
(c) Does the distribution of the rents in the two areas appear to be similar? Explain.

3.23 For the last 10 days in June, the "Shore Special" train arrived late at its destination by the times (in minutes) listed here. (A negative number means that the train was early by that number of minutes.)

DATA FILE
TRAIN.XLS

$$-3 \quad 6 \quad 4 \quad 10 \quad -4 \quad 124 \quad 2 \quad -1 \quad 4 \quad 1$$

(a) List the five-number summary.
(b) Form the box-and-whisker plot and describe the shape.
(c) Compare your answer in (b) with that from Problem 3.13(g) on page 183. Discuss.

3.24 In order to estimate how much water will be needed to supply the community of Falling Rock in the next decade, the town council asked the city manager to find out how much water a sample of families currently uses. The sample of 15 families used the following number of gallons (in thousands) in the past year:

DATA FILE
WATER.XLS

11.2 21.5 16.4 19.7 14.6 16.9 32.2 18.2 13.1 23.8 18.3 15.5 18.8 22.7 14.0

(a) List the five-number summary.
(b) Form the box-and-whisker plot and describe the shape.
(c) Compare your answer in (b) with that from Problem 3.14(g) on page 184. Discuss.

3.4 ◆ OBTAINING DESCRIPTIVE SUMMARY MEASURES FROM A POPULATION

In section 3.2 we examined various *statistics* that summarize or describe numerical information from a *sample*. In particular, we used these statistics to describe the properties of central tendency, variation, and shape.

Suppose, however, that the data set we have access to is not a sample but rather a collection of numerical measurements from an entire *population*. For example, suppose that a wholesale plumbing supply company had a population of 50 sales vouchers from a particular day. The amount (in $) of these vouchers is illustrated in Table 3.2.

Table 3.2 *Amounts for a population of 50 sales vouchers*

127.43	372.68	349.03	213.45	326.55	148.93	213.54	409.61	211.01	290.87
219.76	429.05	328.44	215.62	462.45	389.04	234.65	543.67	176.43	435.32
430.32	278.93	436.72	327.80	354.11	265.76	216.87	654.32	345.45	213.65
399.05	324.55	451.23	287.60	219.06	214.54	278.96	378.90	368.02	319.06
267.90	265.78	345.11	379.01	417.89	267.91	210.32	277.62	321.81	334.22

DATA FILE
VOUCHER.XLS

When dealing with a set of data making up an entire population, we compute population *parameters* for the arithmetic mean, the variance, and the standard deviation.

The Population Mean

The **population mean** is given by the symbol μ, the Greek lowercase letter *mu*. It is obtained as follows.

Population Mean

The population mean is equal to the sum of all the values in the population divided by the population size.

$$\mu = \frac{\sum\limits_{i=1}^{N} X_i}{N} \tag{3.12}$$

where

$$N = \text{population size}$$
$$X_i = i\text{th value of the random variable } X$$

$$\sum_{i=1}^{N} X_i = \text{summation of all } X_i \text{ values in the population}$$

Example 3.18 illustrates how we compute the population mean.

Example 3.18 *Computing the Population Mean*

Compute the population mean of the amount of sales of plumbing supplies for the data of Table 3.2.

SOLUTION

The population mean is computed from equation (3.12) as follows.

$$\mu = \frac{\sum\limits_{i=1}^{N} X_i}{N} = \frac{127.43 + 372.68 + 349.03 + \cdots + 334.22}{50} = \frac{15,950}{50} = \$319.00$$

Thus, the average sales voucher amount in this population is $319.

The Population Variance and Standard Deviation

The **population variance** is given by the symbol σ^2, the Greek lowercase letter *sigma* squared, and the **population standard deviation** is given by the symbol σ. These measures are obtained as follows.

Population Variance

The population variance is equal to the sum of the squared differences around the population mean, divided by the population size.

$$\sigma^2 = \frac{\sum_{i=1}^{N}(X_i - \mu)^2}{N} \tag{3.13}$$

where

N = population size

X_i = ith value of the random variable X

$\sum_{i=1}^{n}(X_i - \mu)^2$ = summation of all the squared differences between the X_i values and μ

and the population standard deviation is the square root of the population variance

Population Standard Deviation

$$\sigma = \sqrt{\frac{\sum_{i=1}^{N}(X_i - \mu)^2}{N}} \tag{3.14}$$

We note that the formulas for the population variance and standard deviation differ from those for the sample variance and standard deviation in that $(n - 1)$ in the denominator of S^2 and S [see equations (3.9) and (3.10)] is replaced by N in the denominator of σ^2 and σ. Example 3.19 demonstrates how to compute the population variance and standard deviation.

Example 3.19 *Computing the Population Variance and Standard Deviation*

Compute the population variance and standard deviation of the amount of sales of plumbing supplies for the data of Table 3.2.

SOLUTION

Using equation (3.13), the population variance is computed as follows.

$$\sigma^2 = \frac{\sum_{i=1}^{N}(X_i - \mu)^2}{N}$$

$$= \frac{(127.43 - 319)^2 + (372.68 - 319)^2 + (349.03 - 319)^2 + \cdots + (334.22 - 319)^2}{50}$$

$$= \frac{518,008.2}{50} = 10,360.16 \text{ (in squared dollars)}$$

The population standard deviation is just the square root of the population variance. Using equation (3.14), we have

$$\sigma = \sqrt{\frac{\sum_{i=1}^{N}(X_i - \mu)^2}{N}} = \sqrt{10,360.16} = \$101.78$$

Using the Standard Deviation: The Empirical Rule

In most data sets, a large portion of the observations tend to cluster somewhat near the median. In right-skewed data sets, this clustering occurs to the left of (i.e., *below*) the median, and in left-skewed data sets, the observations tend to cluster to the right of (i.e., *above*) the median. In symmetrical data sets, where the median and mean are the same, the observations tend to distribute equally around these measures of central tendency. When extreme skewness is not present and such clustering is observed in a data set, we can use the so-called **empirical rule** to examine the property of data variability and get a better sense of what the standard deviation is measuring.

Empirical Rule

The empirical rule states that for most data sets, we will find that roughly two out of every three observations (i.e., 67%) are contained within a distance of 1 standard deviation around the mean and roughly 90% to 95% of the observations are contained within a distance of 2 standard deviations around the mean.

Hence, the standard deviation, as a measure of average variation around the mean, helps us understand how the observations distribute above and below the mean and helps us focus on and flag unusual observations (i.e., *outliers*) when analyzing a set of numerical data.

From the data of Table 3.2 on page 191, for the population of 50 sales vouchers, the mean amount μ is \$319.00 and the standard deviation σ is \$101.78. From Table 3.2 we observe that 31 sales vouchers (62%) are between $\mu - 1\sigma$ and $\mu + 1\sigma$ (i.e., between \$217.22 and \$420.78). We also see that 48 sales vouchers (96%) are between $\mu - 2\sigma$ and $\mu + 2\sigma$ (i.e., between \$115.44 and \$522.56). In addition, we note that all but one of the sales vouchers (98%) are between $\mu - 3\sigma$ and $\mu + 3\sigma$ (i.e., between \$13.66 and \$624.34).

 **GENERATING DESCRIPTIVE SUMMARY MEASURES IN A POPULATION USING MICROSOFT EXCEL**

Overview

◆ *For Quick Results Users* Use the Data Analysis Descriptive Statistics tool as first explained in section 3.2E, but ignore the measures of variation which will be incorrect (the tool produces sample statistics only). Use the Box-and-Whisker Plot choice of the PHStat add-in as also first explained in section 3.3E to generate a box-and-whisker plot and a five-number summary.

◆ *For Developers* The Quick Results procedure can be used for one-time results but be aware that the measures of variation will be incorrectly calculated. To see the correct population measures of variation, or to examine the effects of changing data values, implement the worksheet discussed in section 3.2E, using the STDEVP function instead of the STDEV function in cell B17 and the VARP function instead of VAR in cell B18. As in section 3.3E, there is no alternative to using the Box-and-Whisker Plot choice of the PHStat add-in as the Excel Chart Wizard is not capable of generating box plots.

Problems for Section 3.4

Learning the Basics

3.25 Given the following set of data for a population of size $N = 10$:

7 5 11 8 3 6 2 1 9 8

(a) Compute the mean, median, mode, midrange, and midhinge.
(b) Compute the range, interquartile range, variance, standard deviation, and coefficient of variation.
(c) Are these data skewed? If so, how?

3.26 Given the following set of data for a population of size $N = 10$:

7 5 6 6 6 4 8 6 9 3

(a) Compute the mean, median, mode, midrange, and midhinge.
(b) Compute the range, interquartile range, variance, standard deviation, and coefficient of variation.
(c) Are these data skewed? If so, how?
(d) Compare the measures of central tendency to those of Problem 3.25(a). Discuss.
(e) Compare the measures of variation to those of Problem 3.25(b). Discuss.

Applying the Concepts

•**3.27** The following data represent the quarterly sales tax receipts (in $000) submitted to the comptroller of Gmoserville Township for the period ending March 1998 by all 50 business establishments in that locale:

DATA FILE
TAX.XLS

10.3	11.1	9.6	9.0	14.5
13.0	6.7	11.0	8.4	10.3
13.0	11.2	7.3	5.3	12.5
8.0	11.8	8.7	10.6	9.5
11.1	10.2	11.1	9.9	9.8
11.6	15.1	12.5	6.5	7.5
10.0	12.9	9.2	10.0	12.8
12.5	9.3	10.4	12.7	10.5
9.3	11.5	10.7	11.6	7.8
10.5	7.6	10.1	8.9	8.6

(a) Organize the data into an ordered array or stem-and-leaf display.
(b) Compute the arithmetic mean for this population.
(c) Compute the variance and standard deviation for this population.
(d) What proportion of these businesses have quarterly sales tax receipts

          (1) within ±1 standard deviation of the mean?
          (2) within ±2 standard deviations of the mean?
          (3) within ±3 standard deviations of the mean?
     (e) Are you surprised at the results in (d)? *(Hint:* Compare and contrast your findings versus what would be expected on the basis of the empirical rule.)

3.28 Suppose that the population of 1,024 domestic general stock funds was obtained, and it was determined that μ, the mean 1-year total percentage return achieved by all the funds, is 28.20, and that σ, the standard deviation, is 6.75. In addition, suppose it was determined that the range in the 1-year total returns is from 0.3 to 60.3, and that the quartiles are, respectively, 23.9 (Q_1) and 32.3 (Q_3). According to the empirical rule, what proportion of these funds are expected to be
     (a) within ±1 standard deviation of the mean?
     (b) within ±2 standard deviations of the mean?
     (c) within ±3 standard deviations of the mean?

3.29 The following data are intended to show the gap between families with the highest income and families with the lowest income in each of the 50 states and the District of Columbia as measured by the average of the bottom fifth and the top fifth of families with children during 1994–1996. The results classified by states were as follows:

STATE	BOTTOM FIFTH ($000)	TOP FIFTH ($000)	STATE	BOTTOM FIFTH ($000)	TOP FIFTH ($000)
New York	6.787	132.390	Kansas	10.790	110.341
Louisiana	6.430	102.339	Oregon	9.627	97.589
New Mexico	6.408	91.741	New Jersey	14.211	143.010
Arizona	7.273	103.392	Indiana	11.115	110.876
Connecticut	10.415	147.594	Montana	9.051	89.902
California	9.033	127.719	South Dakota	9.474	93.822
Florida	7.705	107.811	Idaho	10.721	104.725
Kentucky	7.364	99.210	Delaware	12.041	116.965
Alabama	7.531	99.062	Arkansas	8.995	83.434
West Virginia	6.439	84.479	Colorado	14.326	131.368
Tennessee	8.156	106.966	Hawaii	12.735	116.060
Texas	8.642	113.149	Missouri	11.090	100.837
Mississippi	6.257	80.980	Alaska	14.868	129.065
Michigan	9.257	117.107	Wyoming	11.174	94.845
Oklahoma	7.483	94.380	Minnesota	14.655	120.344
Massachusetts	10.694	132.962	Nebraska	12.546	102.992
Georgia	9.978	123.837	Maine	11.275	92.457
Illinois	10.002	123.233	New Hampshire	14.299	116.018
Ohio	9.346	111.894	Nevada	12.276	98.693
South Carolina	8.146	96.712	Iowa	13.148	104.253
Pennsylvania	10.512	124.537	Wisconsin	13.398	103.551
North Carolina	9.363	107.490	Vermont	13.107	97.898
Rhode Island	9.914	111.015	North Dakota	12.424	91.041
Washington	10.116	112.501	Utah	15.709	110.938
Maryland	13.346	147.971	District of Columbia	5.293	149.508
Virginia	10.816	116.202			

DATA FILE
STATEINC.XLS

Source: United States Census Bureau.

For each of these numerical variables
(a) organize the data into an ordered array or stem-and-leaf display.
(b) compute the arithmetic mean for the population.
(c) compute the variance and standard deviation for the population.
(d) What proportion of these states have average incomes
    (1) within ± 1 standard deviation of the mean?
    (2) within ± 2 standard deviations of the mean?
    (3) within ± 3 standard deviations of the mean?
(e) Are you surprised at the results in (d)? (*Hint:* Compare and contrast your findings versus what would be expected based on the empirical rule.)
(f) Remove the District of Columbia from consideration. Do parts (a)–(e) with the District of Columbia removed. How have the results changed?

 **WHAT IF?**

 ## 3.5 RECOGNIZING AND PRACTICING PROPER DESCRIPTIVE SUMMARIZATION AND EXPLORING ETHICAL ISSUES

In this chapter we have studied how a set of numerical data can be characterized by various statistics that measure the properties of central tendency, variation, and shape. The next step is data analysis and interpretation; the former is *objective,* the latter is *subjective.* We must avoid errors that may arise either in the objectivity of what is being analyzed or in the subjectivity of what is being interpreted (references 1 and 3).

Avoiding Errors in Analysis and Interpretation

You may recall that in section 3.1 we examined and described a set of numerical data pertaining to 1-year total percentage returns achieved by 17 domestic general stock funds whose marketing fees are paid from fund assets. Without a knowledge of the contents of this chapter, we attempted to analyze and interpret what the data were trying to convey.

Our analysis was *objective;* we should all have agreed with our limited visual findings: The *modal* values or most typical 1-year total percentage returns were 28.6 and 30.5; the *spread* in the 1-year total return performance indicator ranged from 10.0% to 38.0%; and there was one *outlier* present in the data, the 10.0% return attained by Mentor Merger, and one *potential outlier,* the 20.6% return achieved by Rainler Reich Tang.

Having now read the chapter and thus gained a knowledge about various descriptive summary measures and their strengths and weaknesses, how could we improve on our previous objective analysis? Because the data distribute in a slightly nonsymmetrical manner, shouldn't we report the median or midhinge rather than the mean? Doesn't the standard deviation provide more information about the property of variation than the range? Shouldn't we describe the data set as negative, or left-skewed, in shape? Objectivity in data analysis means reporting the most appropriate summary measures for a given data set—those that best meet the assumptions about the given data set.

On the other hand, our data interpretation was *subjective;* we could have formed different conclusions when interpreting our analytical findings. We all see the world from different perspectives. Some of us will look at the ordered array of 1-year total returns in percents (10.0, 20.6, 28.6, 28.6, 29.4, 29.5, 29.9, 30.1, 30.5, 30.5, 32.1, 32.2, 32.4, 33.0, 35.2, 37.1, and 38.0) and be satisfied with the performances achieved by the domestic general stock funds whose marketing fees are paid from fund assets; others, particularly those who have invested in either Mentor Merger (with its 10.0% return) or Rainler Reich Tang (with its 20.6% return), or those individuals who simply expect greater performance to compensate for investment risks taken, will look at the same data set and conclude that the performance has been too low. Thus, because data interpretation is subjective, it must be done in a fair, neutral, and clear manner.

Ethical Issues

Ethical issues are vitally important to all data analysis. As daily consumers of information, we owe it to ourselves to question what we read in newspapers and magazines and what we hear on the radio or television. Over time, much skepticism has been expressed about the purpose, the focus, and the objectivity of published studies. Perhaps no comment on this topic was ever more telling than a quip often attributed to the famous nineteenth-century British statesman Benjamin Disraeli: "There are three kinds of lies: lies, damned lies, and statistics."

Again, as was mentioned in section 1.9, ethical considerations arise when we are deciding the results to present in a report and those not to present. It is vitally important to document both good and bad results. In addition, when making oral presentations and presenting written reports, it is essential that the results be given in a fair, objective, and neutral manner. Thus, we must try to distinguish between poor presentation of results and unethical presentation. Once more, as in our prior discussions on ethical considerations, the key is intent. When pertinent information is omitted, often it is simply done out of ignorance. However, unethical behavior occurs when one willfully chooses an inappropriate summary measure (e.g., the mean or midrange for a very skewed set of data) to distort the facts in order to support a particular position. In addition, unethical behavior occurs when one selectively fails to report pertinent findings because it would be detrimental to the support of a particular position.

Problems for Section 3.5

Applying the Concepts

3.30 You receive a telephone call from a friend who is also studying statistics this semester. Your friend has just used Microsoft Excel to obtain descriptive summary measures for several numerical variables pertaining to a survey concerning student life on campus. He says, "I've been asked to write a report and prepare a 5-minute classroom presentation on student life on campus. I'm looking at my computer printout—I've got all these descriptive summary measures for each of my seven numerical variables. There's so much information here, I just can't get started. Do you have any suggestions?" You think for a moment, and then reply . . .

3.31 An arbitrator is asked to examine a dispute over salaries paid to professional baseball players. The owner of a particular team claims that the average salary per annum is too high. The agent for the players argues that the average salary for the players on this team is too low. How should the arbitrator evaluate these two conflicting statements? (*Hint:* To which *average* do you think the agent would be referring and to which average do you think the owner would be referring?)

 SUMMARY

As you can see from the following summary chart, this chapter was about data summarization and description. In this and the previous two chapters, we have studied the subject matter of descriptive statistics: how data are collected; presented in tabular or chart format; and then summarized, described, analyzed, and interpreted. In the next chapter, we will study the basic principles of probability in order to bridge the gap between the subject of descriptive statistics and the subject of inferential statistics.

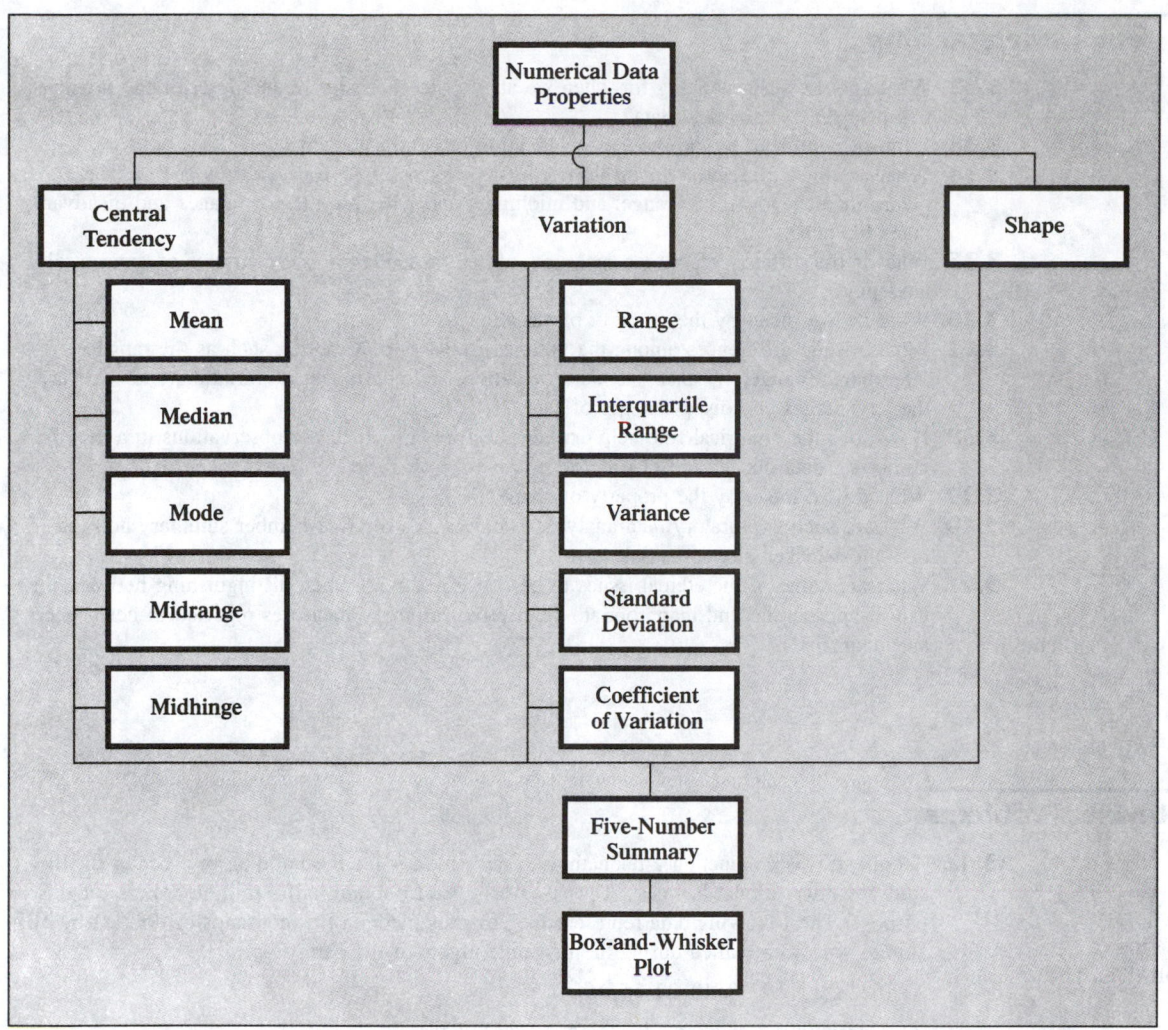

Chapter 3 summary chart

Key Terms

Checking Your Understanding

3.32 What should we be looking for when we attempt to characterize and describe the properties of a set of numerical data?

3.33 What do we mean by the property of location or central tendency?

3.34 What are the differences among the various measures of central tendency such as the mean, median, mode, midrange, and midhinge, and what are the advantages and disadvantages to each?

3.35 What is the difference between measures of central tendency and measures of noncentral tendency?

3.36 What do we mean by the property of variation?

3.37 What are the differences among the various measures of variation such as the range, interquartile range, variance, standard deviation, and coefficient of variation, and what are the advantages and disadvantages of each?

3.38 How does the empirical rule help explain the ways in which the observations in a set of numerical data cluster and distribute?

3.39 What do we mean by the property of shape?

3.40 Why are such exploratory data analysis techniques as the five-number summary and the box-and-whisker plot so useful?

3.41 What are some of the ethical issues to be concerned with when distinguishing between the use of appropriate and inappropriate descriptive summary measures reported in newspapers and magazines?

Chapter Review Problems

3.42 A college was conducting a phonathon to raise money for the building of a center for the study of international business. The provost hoped to obtain half a million dollars for this purpose. The following data represent the amounts pledged (in thousands of dollars) by all alumni who were called during the first nine nights of the campaign.

DATA FILE
FUNDRAISE.XLS

16 18 11 17 13 10 22 15 16

(a) Compute the mean, median, and standard deviation.
(b) Describe the shape of this set of data.
(c) Estimate the total amount that will be pledged (in thousands of dollars) by all alumni if the campaign is to last 30 nights. (*Hint:* Total $= N\overline{X}$.)
(d) Do you think that the phonathon will raise the half million dollars that the provost hoped to obtain? Explain.

• **3.43** The following data represent the monthly long-distance phone rates charged to residential customers across the United States from a sample of 34 different plans (based on 36 calls per month for a total of 318 minutes spread out over the day):

PLAN	RATE	PLAN	RATE
AT&T Dial-1 Standard	74.68	Matrix SBN Flat Rate	54.03
Frontier Dial-1	76.41	Matrix Smartworld Flat	49.46
LCI Basic	71.50	Sprint Sense	55.50
Matrix Dial-1	59.58	Worldcom Home Advantage	49.10
MCI Dial-1 Standard	74.29	AT&T True Reach Savings	56.01
			(continued)

PLAN	RATE	PLAN	RATE
Sprint Dial-1 Standard	74.68	AT&T True Savings	55.21
Worldcom MTS	63.92	Matrix Smartworld Basic	55.76
AT&T One Rate	58.35	MCI Friends and Family	60.98
AT&T One Rate Plus	47.90	MCI Friends and Family Free	58.09
LCI Single Rate	53.97	Sprint Sense With Most Option	53.70
Matrix Flat Rate 1	51.85	Sprint the Most II	60.31
MCI One	47.57	Matrix Smartworld Basic With Discount	52.78
Sprint Sense Day	56.05	Sprint Sense With Cash Back	49.95
AT&T Simple Rate	56.55	Sprint Sense With Most and Cash Back	48.33
Frontier Homesaver	49.36	AT&T One Rate With True Rewards	57.18
LCI All-American	52.71	AT&T True Reach With True Rewards	54.89
LCI Two-Rate	53.10	AT&T True Savings With True Rewards	54.11

DATA FILE
PHONRATE.XLS

Sources: Reprinted by permission of The New York Times. Telecommunications Research and Action Center; The New York Times, March 2, 1997, E5.

(a) Compute all appropriate measures of location.
(b) Compute all appropriate measures of variation.
(c) Construct a box-and-whisker plot.
(d) On the basis of the results of (a) and (c), how would you describe the shape of the distribution? Explain.
(e) On the basis of the results of (a) and (c), if you were summarizing the results in a written report, which measures of location would you provide? Explain.

3.44 The data at the top of page 202 display the price (in dollars), the actual number of cups that can be made, the price to replace a carafe (in dollars), and the type (basic function versus programmable) for a random sample of 19 brands of coffeemakers.

(a) Using the entire sample of 19 coffeemakers, develop stem-and-leaf displays for each of the three numerical variables.
(b) Compute the mean, median, mode, midrange, and midhinge for each of the three numerical variables.
(c) Repeat part (a) for each type of coffeemaker—the 12 basic function machines versus the 7 programmable models.
(d) Repeat part (b) for each type of coffeemaker—the 12 basic function machines versus the 7 programmable models.
(e) Compute the range, interquartile range, variance, standard deviation, and coefficient of variation for each of the three numerical variables.
(f) Repeat part (e) for each type of coffeemaker—the 12 basic function machines versus the 7 programmable models.
(g) Describe the shape for each of the three numerical variables.
(h) Repeat part (g) for each type of coffeemaker—the 12 basic function machines versus the 7 programmable models.
(i) Form the box-and-whisker plot and describe the shape for each of the numerical variables.
(j) Repeat part (i) for each type of coffeemaker—the 12 basic function machines versus the 7 programmable models.
(k) What differences are there in location, variation, and shape between basic and programmable coffeemakers for each of the three variables?

BRAND	PRICE	ACTUAL CUPS	CARAFE	TYPE
Mr. Coffee Accel PR15	22	11.0	9	Basic
Braun KF157	52	10.5	16	Basic
Mr. Coffee AD10	20	9.0	9	Basic
Proctor-Silex 42301	20	11.0	14	Basic
Melitta IBS-10C	50	8.5	11	Basic
Krups CompacTherm 206	85	8.5	45	Basic
Bunn GR	60	8.5	9	Basic
Oster 3272	50	10.5	17	Basic
Black & Decker DCM902	20	9.5	10	Basic
West Bend 56660	30	9.0	7	Basic
Regal K7617	20	9.0	8	Basic
Betty Crocker BC-1733	25	11.5	12	Basic
Braun FlavorSelect KF185	90	10.5	16	Progm
Mr. Coffee Accel PRX20	37	11.0	9	Progm
Krups Crystal Aroma Time 458	90	9.0	20	Progm
Black & Decker ODC300	65	9.5	11	Progm
Proctor-Silex 42461	30	11.0	14	Progm
Black & Decker DCM903	40	9.5	10	Progm
Hamilton Beach 47261	40	11.0	12	Progm

DATA FILE
COFMKR.XLS

Source: "Coffee Makers," Copyright © 1996 by Consumers Union of U.S., Inc. Adapted from CONSUMER REPORTS, *November 1996, 40–41, by permission of Consumers Union of U.S., Inc., Yonkers, NY 10703-1057. Although these data sets originally appeared in* CONSUMER REPORTS, *the selective adaptation and resulting conclusions presented are those of the authors and are not sanctioned or endorsed in any way by Consumers Union, the publisher of* CONSUMER REPORTS.

• **3.45** The following data represent the price, an overall performance score, the battery life, and the battery cost per hour of usage (in cents) for a sample of 22 brands of portable CDs:

BRAND	PRICE (DOLLARS)	SCORE	BATTLIFE	BCOST/HR (CENTS)
RCA RP-7913	86	92	8.50	18
Panasonic SL-S290	134	91	9.50	16
Panasonic SL-S160	92	91	8.50	18
RCA RP-7926A w/car kit	125	87	7.00	21
Panasonic SL-S490	195	85	7.50	20
Sony D-141	89	79	8.00	19
Sony D-335	283	79	8.50	18
Sony D-143	110	72	8.00	19
Craig JC6111	68	72	4.50	33
JVC XL-P41	136	72	5.75	52
Fisher PCD-60 w/car kit	183	72	3.25	46
Sony D-421SP	258	71	6.50	23

(continued)

BRAND	PRICE (DOLLARS)	SCORE	BATTLIFE	BCOST/HR (CENTS)
Aiwa XP-559 w/car kit	152	71	13.50	22
JVC XL-P61CR w/car kit	143	70	6.50	46
Aiwa XP-33	94	69	11.50	26
Onkyo DX-F71 w/car kit	179	68	3.50	43
Kenwood DPC-151	94	63	6.50	46
Optimus CD-3450	130	61	8.00	38
Kenwood DPC-951 w/car kit	263	61	4.75	32
Magnavox AZ 6827C w/car kit	149	61	6.50	46
Kenwood DPC-751 w/car kit	213	60	4.75	32
Emerson HD6825 w/car kit	85	57	7.50	40

DATA FILE
PORTABLECD.XLS

Source: "Portable CDs," Copyright © 1995 by Consumers Union of U.S., Inc. Adapted from CONSUMER REPORTS, *December 1995, 782–783, by permission of Consumers Union of U.S., Inc., Yonkers, NY 10703-1057. Although these data sets originally appeared in* CONSUMER REPORTS, *the selective adaptation and resulting conclusions presented are those of the authors and are not sanctioned or endorsed in any way by Consumers Union, the publisher of* CONSUMER REPORTS.

(a) Develop all the appropriate displays, tables, and charts and thoroughly analyze each of the numerical variables in the data set.

(b) On the basis of your findings in part (a), what conclusions can you reach about the central tendency, variation, and shape of the price, performance score, battery life, and battery cost per hour of these portable CDs?

• **3.46** A problem with a telephone line that prevents a customer from receiving or making calls is disconcerting to both the customer and the telephone company. These problems can be of two types: those that are located inside a central office and those located on lines between the central office and the customer's equipment. The following data represent samples of 20 problems reported to two different offices of a telephone company and the time to clear these problems (in minutes) from the customers' lines:

Central Office I Time to Clear Problems (minutes)

1.48	1.75	0.78	2.85	0.52	1.60	4.15	3.97	1.48	3.10
1.02	0.53	0.93	1.60	0.80	1.05	6.32	3.93	5.45	0.97

Central Office II Time to Clear Problems (minutes)

7.55	3.75	0.10	1.10	0.60	0.52	3.30	2.10	0.58	4.02
3.75	0.65	1.92	0.60	1.53	4.23	0.08	1.48	1.65	0.72

DATA FILE
PHONE.XLS

For each of the two central office locations,
(a) Compute the
  (1) arithmetic mean
  (2) median
  (3) midrange
  (4) first quartile
  (5) third quartile
  (6) midhinge
  (7) range
  (8) interquartile range
  (9) variance
  (10) standard deviation
  (11) coefficient of variation
(b) Construct a box-and-whisker plot.
(c) Are the data skewed? If so, how?
(d) On the basis of the results of (a)–(c), are there any differences between the two central offices? Explain.
(e) What would be the effect on your results and your conclusions if the first value for Central Office II was incorrectly recorded as 27.55 instead of 7.55?

WHAT IF?

3.47 In many manufacturing processes there is a term called work in process (often abbreviated WIP). In a book manufacturing plant this represents the time it takes for sheets from a press to be folded, gathered, sewn, tipped on endsheets, and bound. The following data represent samples of 20 books at each of two production plants and the processing time (operationally defined as the time in days from when the books came off the press to when they were packed in cartons) for these jobs.

Plant A

5.62	5.29	16.25	10.92	11.46	21.62	8.45	8.58	5.41	11.42
11.62	7.29	7.50	7.96	4.42	10.50	7.58	9.29	7.54	8.92

Plant B

9.54	11.46	16.62	12.62	25.75	15.41	14.29	13.13	13.71	10.04
5.75	12.46	9.17	13.21	6.00	2.33	14.25	5.37	6.25	9.71

DATA FILE
WIP.XLS

For each of the two plants,
(a) Compute the

(1) arithmetic mean
(2) median
(3) midrange
(4) first quartile
(5) third quartile
(6) midhinge

(7) range
(8) interquartile range
(9) variance
(10) standard deviation
(11) coefficient of variation

(b) Construct a box-and-whisker plot.
(c) Are the data skewed? If so, how?
(d) On the basis of the results of (a)–(c), are there any differences between the two plants? Explain.

3.48 In New York State, savings banks are permitted to sell a form of life insurance called Savings Bank Life Insurance (SBLI). The approval process consists of underwriting, which includes a review of the application, a medical information bureau check, possible requests for additional medical information and medical exams, and a policy compilation stage where the policy pages are generated and sent to the bank for delivery. The ability to deliver approved policies to customers in a timely manner is critical to the profitability of this service to the bank. During a period of 1 month, a random sample of 27 approved policies was selected and the total processing time in days was recorded with the following results:

```
73 19 16 64 28 28 31 90 60 56 31 56 22 18
45 48 17 17 17 91 92 63 50 51 69 16 17
```

(a) Compute the

(1) arithmetic mean
(2) median
(3) midrange
(4) first quartile
(5) third quartile
(6) midhinge

(7) range
(8) interquartile range
(9) variance
(10) standard deviation
(11) coefficient of variation

(b) Construct a box-and-whisker plot.
(c) Are the data skewed? If so, how?
(d) If a customer enters the bank to purchase this type of insurance policy and asks how long the approval process takes, what would you tell him?

3.49 One of the major measures of the quality of service provided by any organization is the speed with which it responds to customer complaints. A large family-held department store selling furniture and flooring, including carpeting, had undergone a major expansion in the past several years. In particular, the flooring department had expanded from 2 installation crews to an installation supervisor, a measurer, and 15 installation crews. A sample of 50 complaints concerning carpeting installation was selected during a recent year. The following data represent the number of days between the receipt of the complaint and the resolution of the complaint.

```
54   5 35 137 31 27 152   2 123 81 74 27 11 19 126 110 110 29 61 35 94 31 26   5 12
     4 165 32  29 28 29  26 25   1 14 13 13 10  5  27   4  52 30 22 36 26 20 23 33 68
```

(a) Compute the
  (1) arithmetic mean
  (2) median
  (3) midrange
  (4) first quartile
  (5) third quartile
  (6) midhinge
  (7) range
  (8) interquartile range
  (9) variance
  (10) standard deviation
  (11) coefficient of variation

DATA FILE
FURNITURE.XLS

(b) Construct a box-and-whisker plot.
(c) Are the data skewed? If so, how?
(d) On the basis of the results of (a)–(c), if you had to tell the president of the company how long a customer should expect to wait to have a complaint resolved, what would you say? Explain.

3.50 As an illustration of the misuse of statistics, an article by Glenn Kramon ("Coaxing the Stanford Elephant to Dance," *The New York Times* Sunday Business Section, November 11, 1990) implied that costs at Stanford Medical Center had been driven up higher than at competing institutions because the former was more likely to treat indigent, Medicare, Medicaid, sicker, and more complex patients. To illustrate this, a chart was provided that depicted a comparison of average 1989–1990 hospital charges for three medical procedures (coronary bypass, simple birth, and hip replacement) at three competing institutions (El Camino, Sequoia, and Stanford).

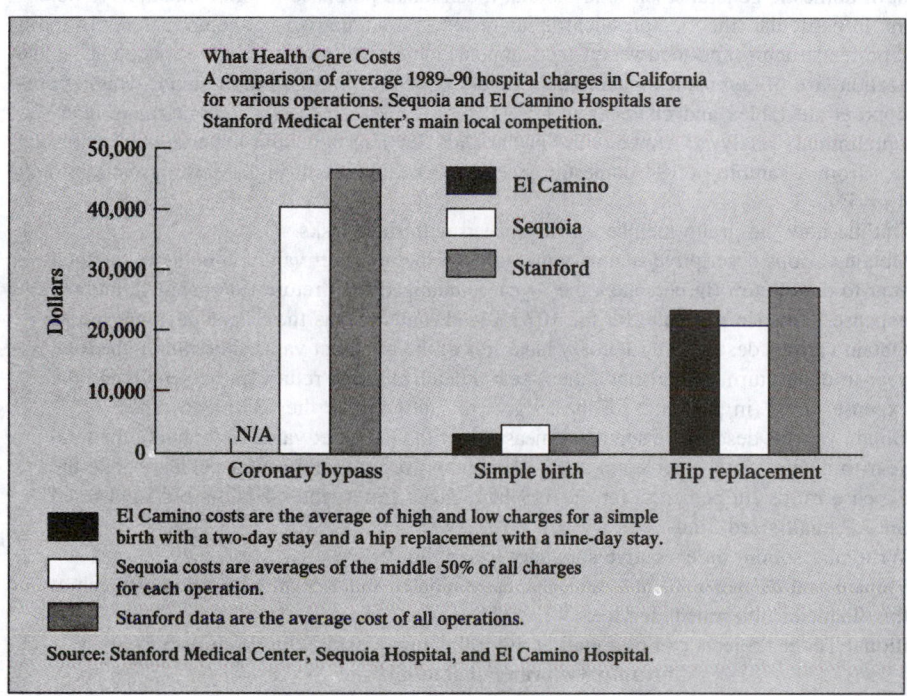

Reprinted by permission of The New York Times.

Suppose you were working in a medical center. Your CEO knows you are currently taking a course in statistics and calls you in to discuss this. She tells you that the article was presented in a discussion group setting as part of a meeting of regional area medical center CEOs last night and that one of them mentioned that this chart was totally meaningless and asked her opinion. She now requests that you prepare her response. You smile, take a deep breath, and reply . . .

3.51 You are planning to study for your statistics examination with a group of classmates, one of whom you particularly want to impress. This individual has volunteered to use a software package to get the needed summary information, tables, and charts for a data set containing several numerical and categorical variables assigned by the instructor for study purposes. This person comes over to you with the printout and exclaims, "I've got it all—the means, the medians, the standard deviations, the stem-and-leafs, the box-and-whisker plots, the pie charts—for all our variables. The problem is, some of the output looks weird—like the stem-and-leafs and the box-and-whisker plots for gender and for major and the pie charts for grade point index and for height. Also, I can't understand why Professor Burke said we can't get the descriptive stats for some of our variables—I got it for everything! See, the mean for height is 68.23, the mean for grade point index is 2.76, the mean for gender is 1.50, the mean for major is 4.33." You look at your would-be friend, take a deep breath, and reply . . .

TEAM PROJECT

DATA FILE
MUTUAL.XLS

TP3.1 Refer to TP2.1 on page 144. Your group, the _____ Corporation, has been hired by the vice president for research at a financial investment service to study the financial characteristics of currently traded domestic general stock funds. The investment service is interested in evaluating the list of domestic general stock funds so that it can make purchase recommendations to potential investors. In particular, the vice president is interested in a comparison of some features of these funds based on fee structure (no-load versus fee payment), objective (growth fund versus blend fund), and capitalization size of companies making up a fund's portfolio (large, mid, or small). Having prepared the appropriate tables and charts (see TP2.1), the _____ Corporation is ready to enhance its preliminary analysis. Armed with Special Data Set 1 of appendix D pertaining to various characteristics from a sample of 194 domestic general stock funds with high Morningstar Inc. dual ratings of 4 or 5:

(a) Outline how the group members will proceed with their tasks.

(b) Obtain various descriptive summary measures of the net asset value (in dollars), the total year-to-date return (in percents), the 3-year annualized total return (in percents), and the expense ratios (in percents) for the 107 no-load funds versus the 87 fee payment funds.

(c) Obtain various descriptive summary measures of the net asset value (in dollars), the total year-to-date return (in percents), the 3-year annualized total return (in percents), and the expense ratios (in percents) for the 59 growth funds versus the 135 blend funds.

(d) Obtain various descriptive summary measures of the net asset value (in dollars), the total year-to-date return (in percents), the 3-year annualized total return (in percents), and the expense ratios (in percents) for the 119 large-sized funds, the 44 mid-sized funds, and the 31 small-sized funds.

(e) Write and submit an executive summary describing the results.

(f) Prepare and deliver a 10-minute oral presentation to the vice president for research at this financial investment service.

Note: Additional Team Projects can be found at the following World Wide Web address:

http://www.prenhall.com/levine

These Team Projects deal with the characteristics of 80 universities and colleges (see the UNIV&COL.XLS file) and the features in 89 automobile models (see the AUTO96.XLS file).

Case Study — STATE ALCOHOLIC BEVERAGES OVERSIGHT BOARD STUDY ON BEERS

Dan Oates, director of research for the State Alcoholic Beverages Oversight Board, held a meeting with Manus Rabb, the newly appointed manager of the Ale and Beer Division, and Dr. Arnold Matlin, professor of nutrition at State University and a leading research consultant on beverage products, whom Mr. Rabb has just hired. "We now have access to the data from the June 1996 *Consumer Reports* beer taste testing study," commented Mr. Oates as he distributed copies of the article and the data. "As I recently said to Mr. Rabb, it is important that the Oversight Board evaluate the data and publish information that would be of interest and use to consumers. The Board has not done such a study during my tenure—the last such evaluation was made in 1990. According to this *Consumer Reports* article (June 1996, 10), one billion dollars worth of beer is sold every week, and in the past decade there has been a ninefold increase in the number of breweries in operation in the United States. As popularity with variants of this beverage has increased, it is important for the consumer to be made aware of the products' features." "I agree with you," Mr. Rabb stated. "I myself was appointed by the Board only 3 weeks ago and since then I have been reading the logs of the various studies directed by my predecessor, David Valinski, who managed this division for almost 40 years. Mr. Oates then asked Dr. Matlin to conduct the study and to report his findings to Mr.

Rabb in 2 weeks. Dr. Matlin requested that a student assistant be provided.

Dr. Matlin has hired you to assist him in this study and has provided you with the *Consumer Reports* data on 69 different beers, broken down by type of product and origin of product. The data represent the price of a six-pack of 12-ounce bottles, the calories per 12 fluid ounces, the percent alcoholic content per 12 fluid ounces, the type of beer (craft lagers, craft ales, imported lagers, regular and ice beers, and light and nonalcoholic beers), and the country of origin (USA versus imported) for each of the 69 beers that were sampled.

Your task is to write a report based on a complete descriptive evaluation of each of the numerical variables—price, calories, and alcoholic content—regardless of type of product or origin. Then perform a similar evaluation comparing and contrasting each of these numerical variables based on type of product—craft lagers, craft ales, imported lagers, regular and ice beers, and light or nonalcoholic beers. In addition, perform a similar evaluation comparing and contrasting each of these numerical variables based on the origins of the beers—those brewed in the United States versus those that were imported. Appended to your report should be all appropriate tables, charts, and descriptive statistical information obtained from the survey results.

DATA FILE
BEER.XLS

PRODUCT	PRICE	CALORIES	PCTALC	TYPE	ORIG
Brooklyn Brand	6.24	159	5.2	Craft Lgrs	USA
Leinenkugel's Red	4.79	160	5.0	Craft Lgrs	USA
Samuel Adams Boston	5.96	160	4.9	Craft Lgrs	USA
George Killian's Irish Red	4.70	162	4.9	Craft Lgrs	USA

Source: From "Beers," Copyright © 1996 by Consumers Union of U.S., Inc. Adapted from CONSUMER REPORTS, June 1996, 10–17, by permission of Consumers Union of U.S., Inc., Yonkers, NY 10703-1057. Although these data sets originally appeared in CONSUMER REPORTS, the selective adaptation and resulting conclusions presented are those of the authors and are not sanctioned or endorsed in any way by Consumers Union, the publisher of CONSUMER REPORTS.

THE SPRINGVILLE HERALD CASE

Walter Fairfax, the head of the computer systems department, realized that in addition to the numerous tables and charts that had been prepared on the basis of Table SH2.4 (see page 148) concerning the number of data cartridges accessed for jobs, various descriptive summary measures relating to location, variation, and skewness will be needed to make any report provided for management more useful.

Exercise

DATA FILE
SH2.XLS

3.1 (a) Compute all descriptive summary measures, stem-and-leaf displays, and box-and-whisker plots relating to the number of data cartridges accessed for jobs you believe would be useful in preparing a report to management.

(b) Write a report to management that summarizes the results obtained from the descriptive summary measures, stem-and-leaf displays, and box-and-whisker plots developed in (a).

References

1. Huff, D., *How to Lie with Statistics* (New York: Norton, 1954).
2. Kendall, M. G., and A. Stuart, *The Advanced Theory of Statistics,* vol. 1 (London: Charles W. Griffin, 1958).
3. Kimble, G. A., *How to Use (and Misuse) Statistics* (Englewood Cliffs, NJ: Prentice Hall, 1978).
4. *Microsoft Excel 97* (Redmond, WA: Microsoft Corporation, 1997).
5. Tukey, J., *Exploratory Data Analysis* (Reading, MA: Addison-Wesley, 1977).
6. Velleman, P. F., and D. C. Hoaglin, *Applications, Basics, and Computing of Exploratory Data Analysis* (Boston, MA: Duxbury Press, 1981).

Chapter 3

Student Solutions Manual

•3.2 (a) Mean = 7   Midrange = (3 + 12)/2 = 7.5
           Median = 7   Midhinge = (4 + 9)/2 = 6.5
           Mode = 7
      (b) Range = 9   Variance = 10.8
           Interquartile range = 5   Standard deviation = 3.286
           Coefficient of variation = (3.286/7)•100% = 46.94%
      (c) Since the mean equals the median, the distribution is symmetrical.

3.4   (a) Mean = 2   Midrange = (– 8 + 9)/2 = 0.5
           Median = 7   Midhinge = (– 6.5 + 8)/2 = 0.75
           Mode = 7
      (b) Range = 17   Variance = 62
           Interquartile range = 14.5   Standard deviation = 7.874
           Coefficient of variation = (7.874/2)•100% = 393.7%
      (c) Since the mean is less than the median, the distribution is left-skewed.

•3.6 (a)

	Set 1	Set 2
Mean	4	14
Median	3	13
Mode	2	12
Midrange	6	16
Midhinge	3.5	13.5

      (b)-(c) The data values in Set 2 are each 10 more than the corresponding values in Set 1.
              The measures of central tendency for Set 2 are all 10 more than the comparable statistics for
              Set 1.
      (d)

	Set 1	Set 2
Range	8	8
Interquartile range	3	3
Variance	8.33*	8.33*
Standard deviation	2.89*	2.89*
Coefficient of variation	72.17%	20.62%

           *Note: Slight differences are due to rounding.
      (e) Since the mean is greater than the median for each data set, the distributions are both right-
           skewed.
      (f) Because the data values in Set 2 are each 10 more than the corresponding values in Set 1, the
           measures of spread among the data values remain the same across the two sets, with the
           exception of the coefficient of variation. The coefficients of variation are different because the
           sample standard deviation is divided by the set's mean; in the case of Set 2, the mean is 10
           more than the mean for Set 1, resulting in a larger denominator and a smaller coefficient. Set 2
           is a reflection of Set 1 simply shifted up the scale 10 units, so the distributions are also
           reflections of each other.

(g)   Generally stated, when a second data set is an additive shift from an original set:
- the measures of central tendency for the second set are equal to the comparable measures for the original set plus the value, or distance, of the shift;
- the measures of spread for the second set are equal to the corresponding measures for the original set, with the exception of the coefficient of variation;
- the shape of the second distribution will be a reflection of the shape of the original distribution.

3.8   (a)

	Bad Data	Good Data
Mean	20	15
Median	15	15
Midrange	37	15

The midrange is drastically affected by the error. But the mean is only moderately altered and the median remains unchanged.

(b)

	Bad Data	Good Data
Range	48	4
Interquartile range	3	2.5
Variance	238.25	2.00
Standard deviation	15.435	1.414
Coefficient of variation	77.18%	9.43%

(c)-(d) The interquartile range is the only measure of variation that is not significantly altered by the error in one of the price-to-earnings ratios reported on the American Stock Exchange. In comparison to what they should be, the range is 12 times bigger, the variance is better than 119 times bigger, the standard deviation almost 11 (or the square root of 119) times bigger, and the coefficient of variation over 8 times bigger.

(e)   The distribution of the bad data set is right-skewed since the mean is greater than the median. But the distribution of the good data set is symmetrical since the mean and the median are equal.

3.10   (a)

	A	B	C
1	Rent ($)		
2		Center City	Outlying Area
3	Mean	1016.5	718.4
4	Standard Error	29.916272	19.37937047
5	Median	982.5	725
6	Mode	#N/A	725
7	Standard Deviation	94.6035588	61.28295032
8	Sample Variance	8949.83333	3755.6
9	Kurtosis	3.82059257	3.088218812
10	Skewness	2.02510267	-1.335483061
11	Range	307	225
12	Minimum	940	575
13	Maximum	1247	800
14	Sum	10165	7184
15	Count	10	10

(b)   Rental prices for unfurnished studio apartments in the center of a large city are much higher than in an outlying part of the city. The lowest rent found in the sample of central city rents ($940) was higher than the highest rent found in sample of rents for the outlying area ($800). All measures of central tendency are higher for the sample taken in the center of the city than the sample taken from the outlying area.

3.12   (a)   (1) Mean = 7.11   (7) Range = 6.67
              (2) Median = 6.68   (8) Interquartile range = 3.09
              (3) Midrange = 7.155   (9) Variance = 4.336
              (4) Q_1 = 5.64   (10) Standard deviation = 2.082
              (5) Q_3 = 8.73   (11) Coefficient of variation = 29.27%
              (6) Midhinge = 7.185

        (b)   Since the mean is greater than the median, the distribution is right-skewed.
        (c)   The mean and median are both well over 5 minutes and the distribution is right-skewed, meaning that there are more unusually high observations than low. Further, 13 of the 15 bank customers sampled (or 86.7%) had wait times in excess of 5 minutes. So, the customer is more likely to experience a wait time in excess of 5 minutes. The manager overstated the bank's service record in responding that the customer would "almost certainly" not wait longer than 5 minutes for service.

3.12
cont.   (e)   The sample reported in Problem 3.11 above was taken from a bank located in a commercial district of a city and was taken between 12 and 1 p.m. The sample reported in this problem was taken from a bank located in a residential area and was taken on Friday evening between 5 and 7 p.m. Customer service needs are likely to be quite different between the two locales, which may influence the length of time required to serve each customer. Differences in time of day and day of week may also influence the mix of customer service needs. Additional information on what week and what month the samples were taken would be useful, as would the customer base and the number of employees for each branch.

3.14   (a)   Mean = 18.46   Midrange = 21.70
               Median = 18.20   Midhinge = 18.05
        (b)   (4,500)(18.46) = 83,070
        (d)   The town council was interested in residential water use, not commercial.
        (e)   Range = 21   Standard deviation = 5.20
               Interquartile range = 6.90   Coefficient of variation = 28.2%
               Variance = 27.00
        (f)   The majority of the data fall within 5.20 thousand gallons on either side of the mean.
        (g)   Since the mean is only slightly larger than the median, the distribution is approximately symmetrical.

•3.16   (a)   Five-number summary: 3 4 7 9 12
        (b)
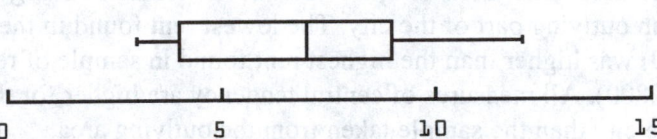
               The distribution is almost symmetrical.
        (c)   The data set is almost symmetrical since the median line almost divides the box in half.

3.18   (a)   Five-number summary: − 8 − 6.5 7 8 9
        (b)
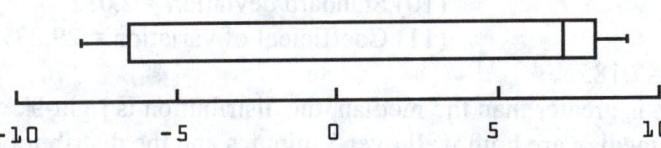
               The distribution is left-skewed.
        (c)   The box-and-whisker plot shows a longer left box from Q_1 to Q_2 than from Q_2 to Q_3, visually confirming our conclusion that the data are left-skewed.

•3.20   (a)   Five-number summary: 264 307.5 451 553.5 1,049
        (b)
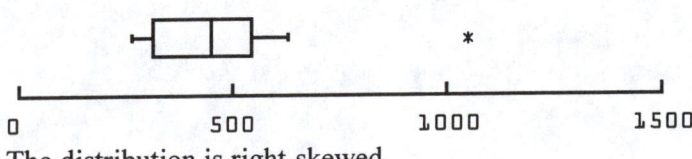
               The distribution is right-skewed.

•3.20

cont.　(c)　Because the data set is small, one very large value (1,049) skews the distribution to the right.

3.22　(a)

	A	B	C
1	Box-and-whisker Plot		
2			
3	Five-number Summary		
4		Center City	Outlying Area
5	Minimum	940	575
6	First Quartile	967.5	696.75
7	Median	982.5	725
8	Third Quartile	999.75	748.75
9	Maximum	1247	800

(b)

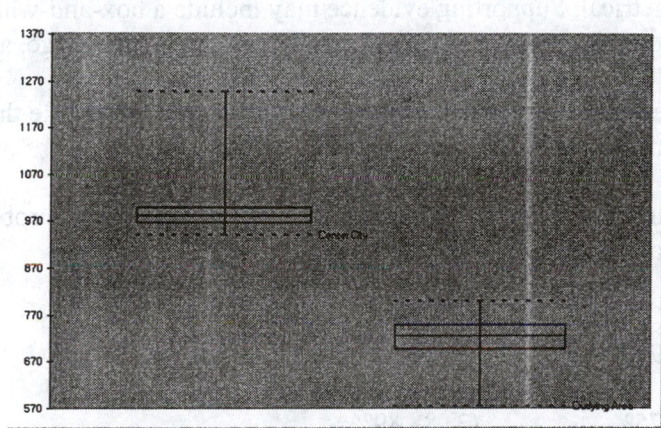

Monthly rental prices in the center of a large city are right-skewed and, because the data set is small, are affected by two extreme outlying values.
Outlying Area:

Monthly rental prices in the area outlying from a large city are left-skewed, affected by one unusually small value.

3.24　(a)　Five-number summary: 11.2　14.6　18.2　21.5　32.2

(b)

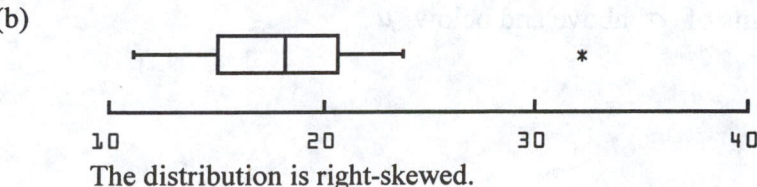

The distribution is right-skewed.

3.24
cont.   (c)   The answer above differs from the answer given in Problem 3.13(g), where we
concluded that the distribution was approximately symmetrical. The box-and-whisker
plot identifies 32.2 as a very large value well beyond the other values in the set. Notice
that the box and whisker below the median are longer than the box and whisker above
the median, which mean that the values in the second quartile are relatively further away
from the median than the values in the third quartile. Normally that would indicate a
distribution is left-skewed. In this instance, however, the existence of one very large
value (32.2) draws the mean (18.46) to the right of the median (18.20), resulting in a
right-skewed distribution.

3.26   (a)   Mean = 6   Midrange = (3 + 9)/2 = 6
               Median = 6   Midhinge = (5 + 7)/2 = 6
               Mode = 6
        (b)   Range = 6   Standard deviation, σ = 1.67
               Interquartile range = 2   Coefficient of variation = 27.89%
               Variance, σ^2 = 2.8
        (c)

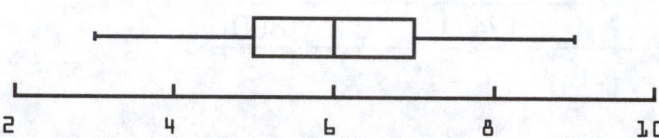

               This distribution is symmetrical. Supporting evidence may include a box-and-whisker
               plot, as indicated above, or a statement that the mean, median, mode, midrange, and
               midhinge are all equal to the same value, 6.
        (d)   Measures of central tendency for Problem 3.25(a) are all close to 6, but, unlike those in
               this problem, do vary from 5.5 (midhinge) to 8 (mode).
        (e)   Measures of variation for Problem 3.25(b) are all larger than comparable
               measures for this problem, a reflection of the fact that the data values in this problem are
               more similar than the data values in Problem 3.25.

3.28   (a)   (1) 67%   (2) 90%-95%   (3) 100%

        (b)   (1) Not calculable   (2) 75%   (3) 88.89%

        (c)   Solving for k,

               $1 - \dfrac{1}{k^2}$ = 0.9375

               $\dfrac{-1}{} = \dfrac{-1}{}$

               $-\dfrac{1}{k^2}$ = -0.0625

               k^2 = 16

               k = ± 4 units of σ above and below μ

200

3.28 (c)
Cont.

Since the population standard deviation is given as 6.75% and the mean is 28.20%, at least 93.75% of the one year returns will fall between the following bounds:

Lower bound: 28.20% − 4(6.75%) =   1.20%
Upper bound: 28.20% + 4(6.75%) =   55.20%

3.42 (a)   Mean = 15.33   Median = 16   Standard deviation = 3.674
(b)   These data are approximately normally distributed, given the mean and median are roughly equivalent.
(c) $460,000.

3.44 (a)

	A	Price($)	Cup Fills	Carafe Price($)
1		Price($)	Cup Fills	Carafe Price($)
2				
3	Mean	44.52632	9.894736842	13.63157895
4	Standard Error	5.50458	0.237490989	1.915175662
5	Median	40	9.5	11
6	Mode	20	11	9
7	Standard Deviation	23.99391	1.035199221	8.348057168
8	Sample Variance	575.7076	1.071637427	69.69005848
9	Kurtosis	-0.43167	-1.647823665	12.01241645
10	Skewness	0.821537	0.020349441	3.221077289
11	Range	70	3	38
12	Minimum	20	8.5	7
13	Maximum	90	11.5	45
14	Sum	846	188	259
15	Count	19	19	19

3.44
cont.

(b)

	Price	Cups	Carafe
Mean	44.53	9.89	13.63
Median	40	9.5	11
Mode	20	11	9
Midrange	55	10	26
Midhinge	41	10	12.5

(c)

Basic Function		Programmable	
Stem-and-leaf for Price		Stem-and-leaf for Price	
2	000025	3	07
3	0	4	00
4		5	
5	002	6	5
6	0	7	
7		8	
8	5	9	00
	$n = 12$		$n = 7$

Stem-and-leaf for Cups		Stem-and-leaf for Cups	
Ones	Tenths	Ones	Tenths
8	555	8	
9	0005	9	055
10	55	10	5
11	005	11	000
	$n = 12$		$n = 7$

Stem-and-leaf for Carafe		Stem-and-leaf for Carafe	
0	78999	0	9
1	012467	1	01246
2		2	0
3			$n = 7$
4	5		
	$n = 19$		

(d)

	Basic Function			Programmable		
	Price	Cups	Carafe	Price	Cups	Carafe
Mean	37.83	9.71	13.92	56	10.21	13.14
Median	27.5	9.25	10.5	40	10.5	12
Mode	20	8.5, 9	9	40, 90	11	none
Midrange	52.5	10	26	60	10	14.5
Midhinge	36	9.75	12.5	63.5	10.25	13

(e)

	Price	Cups	Carafe
Range	70	3	38
Interquartile range	38	2	7
Variance	575.71	1.07	69.69
Standard deviation	23.99	1.04	8.35
Coefficient of variation	53.89%	10.46%	61.24%

3.44 (f)

| | Basic Function | | |
	Price	Cups	Carafe
Range	65	3	38
Interquartile range	32	2.5	7
Variance	451.06	1.25	105.72
Standard deviation	21.24	1.12	10.28
Coefficient of variation	56.14%	11.51%	73.88%

| | Programmable | | |
	Price	Cups	Carafe
Range	60	2	11
Interquartile range	53	1.5	6
Variance	657	0.74	14.81
Standard deviation	25.63	0.86	3.85
Coefficient of variation	45.77%	8.41%	29.28%

(g) **Price:** Since the mean is greater than the median price for coffee makers, the distribution of prices is right-skewed.

Cup capacity: Although the mean cup capacity (9.89) is just slightly higher than the median capacity (9.5), the distance from the first quartile to the median is one third of the distance from the median to the third quartile. The distribution is right-skewed.

Carafe replacement cost: The distribution of carafe replacement costs is right-skewed.

(i) Box-and-whisker plot for **Prices**

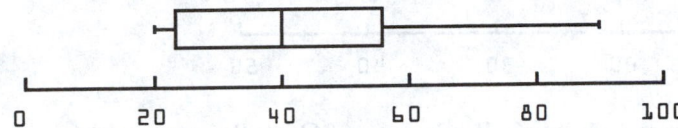

Box-and-whisker plot for **Cup Capacity**

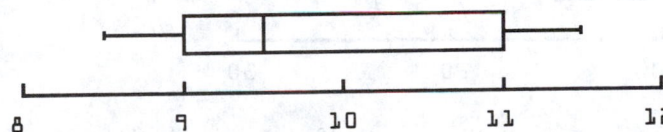

Box-and-whisker plot for **Carafe Replacement Cost**

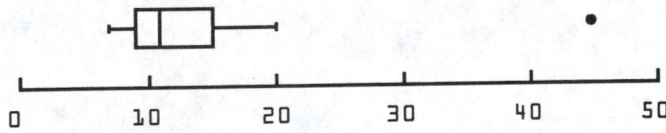

(j) Box-and-whisker plot for **Prices, Basic Function**

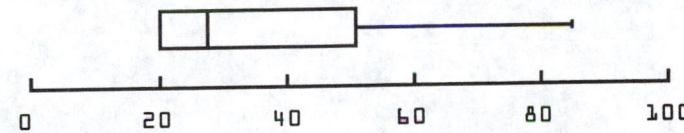

Box-and-whisker plot for **Prices, Programmable**

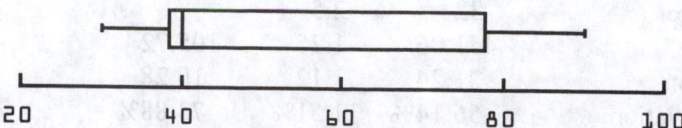

Box-and-whisker plot for **Cup Capacity, Basic Function**

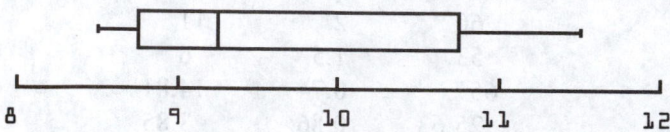

Box-and-whisker plot for **Cup Capacity, Programmable**

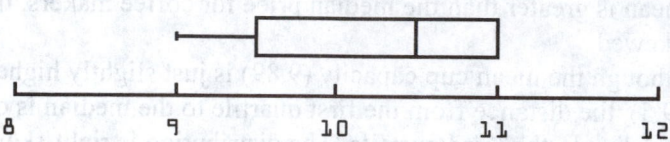

Box-and-whisker plot for **Carafe Replacement Cost, Basic Function**

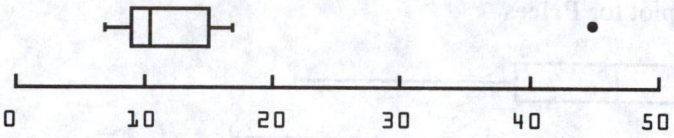

Box-and-whisker plot for **Carafe Replacement Cost, Programmable**

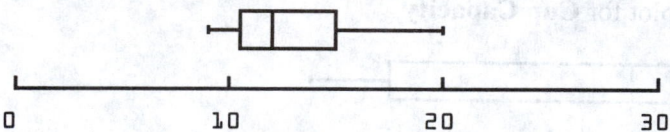

(k) For programmable coffee makers, the upper half of the distribution of prices is more dispersed and the cup capacity tends to be higher on average. The extreme value ($45) greatly affects the mean carafe replacement cost for basic coffee makers.

•3.46   (a)

	A	B	C
1	*Time*		
2		Office I	Office II
3	Mean	2.214	2.0115
4	Standard Error	0.384165	0.422998
5	Median	1.54	1.505
6	Mode	1.48	3.75
7	Standard Deviation	1.718039	1.891706
8	Sample Variance	2.951657	3.57855
9	Kurtosis	0.285677	2.405845
10	Skewness	1.126671	1.466424
11	Range	5.8	7.47
12	Minimum	0.52	0.08
13	Maximum	6.32	7.55
14	Sum	44.28	40.23
15	Count	20	20

(b)

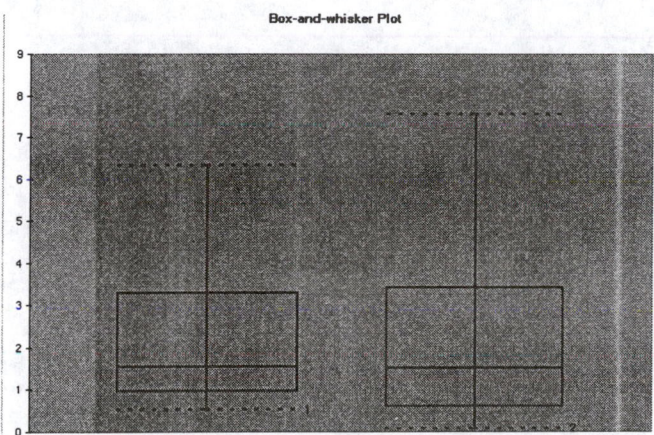

Box-and-whisker Plot

(c)   Times to clear problems at both central offices are right-skewed.

(d)   Times to clear problems for Office I are less dispersed about the mean than times to clear problems for Office II, even though the average for Office I times is higher (2.214) than that for Office II (2.012).

(e)   If the value 7.55 were incorrectly recorded as 27.55, the mean would be one minute higher (from 2.012 to 3.012) and the standard deviation would be over 3 times as large (from 1.892 to 5.936).

3.48 (a)   (1) Mean   43.889
              (2) Median   45
              (3) Midrange   54
              (4) Q_1 18
              (5) Q_3 63
              (6) Midhinge   40.5
              (7) Range   76
              (8) Interquartile range   45
              (9) Variance   639.256
              (10) Standard deviation   25.284
              (11) Coefficient of variation   57.61%

     (b)   Box-and-whisker plot for Bank Processing Times

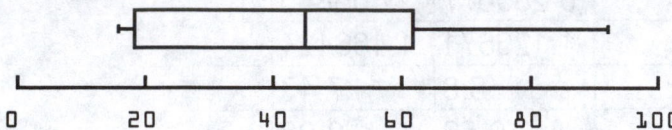

     (c)   Although the whisker covering the upper-most quartile of data is significantly longer
              than the whisker covering the lowest quartile of data, the distribution is left-skewed.
              Supporting evidence includes that the distance from the first quartile to the median is
              much longer (27 units) than the distance from the median to the third quartile (18 units),
              and the mean is to the right of the median.

     (d)   The customer should be told that the typical approval process takes between 18 and 63
              days, with the average around 45 days. No policy has required longer than 92 days for
              approval.

Chapter 4
Basic Probability and Discrete Probability Distributions

207

CHAPTER OBJECTIVES

✓ *To develop an understanding of the basic probability concepts*
✓ *To introduce conditional probability*
✓ *To use Bayes' theorem to revise probabilities in the light of new information*
✓ *To provide an understanding of the basic concepts of discrete probability distributions and their characteristics*
✓ *To develop the concept of mathematical expectation for a discrete random variable*
✓ *To present applications of the binomial distribution in business*
✓ *To present applications of the Poisson distribution in business*
✓ *To present applications of the hypergeometric distribution in business*
✓ *To introduce the covariance and illustrate its application in finance*

Introduction

In chapters 1–3 we studied data collection, tables and charts, and descriptive summary measures. In this chapter we turn our attention to the subject of probability, which serves as the link between describing and presenting information obtained from samples and being able to make inferences to larger populations. We discuss three different approaches to determining the probability of occurrence of different phenomena: *a priori* classical probability, empirical classical probability, and subjective probability. We then learn how to compute a variety of types of probabilities and to revise probabilities in the light of new information. We use this to develop the discrete probability distribution, the concept of mathematical expectation, and the binomial, Poisson, and hypergeometric distributions and illustrate applications in business. Let us begin with a look at how a company uses probability to help plan its activities and processes.

◆ USING STATISTICS: *The Consumer Electronics Company*

Numerous intensive studies have been conducted to analyze consumer planning for the purchase of durable goods such as television sets, refrigerators, washing machines, stoves, and automobiles. Suppose that a marketing director for a consumer electronics company was interested in studying the intention of consumers to purchase new large television sets (defined as 27 inches or larger) in the next 12 months and, as a follow up, whether they in fact actually purchased the television. On the basis of a survey of consumers, some of the questions the marketing director would like to answer include the following:

- What is the probability that the consumer is planning to purchase a large television in the next year?
- What is the probability that the consumer will actually purchase the large television?
- What is the probability that the consumer is planning to purchase the television and actually purchases the television?
- What is the probability that the consumer is planning to purchase the television or actually purchased the television?

- Given that the consumer is planning to purchase the television, what is the probability that the purchase is made?

- Does knowledge of whether the consumer plans to purchase the large television change the likelihood of predicting whether the consumer will purchase the large television?

Answers to these questions and others can help management develop future sales and marketing strategies.

 ## 4.1 BASIC PROBABILITY CONCEPTS

What do we mean by the word probability? **Probability** is the likelihood or chance that a particular event will occur. It could refer to the chance of picking a black card from a deck of cards, the chance that an individual prefers one product over another, or the chance that a new consumer product on the market will be successful. In each of these examples, the probability is a proportion or fraction whose values range between 0 and 1, inclusively. We note that an event that has no chance of occurring (i.e., the **null event**) has a probability of 0, whereas an event that is sure to occur (i.e., the **certain event**) has a probability of 1.

Each of the above examples refers to one of three approaches to the subject of probability. The first is often called the *a priori* **classical probability** approach. Here the probability of success is based on prior knowledge of the process involved. In the simplest case, where each outcome is equally likely, this chance of occurrence of the event is defined as follows:

Probability of Occurrence

$$\text{Probability of occurrence} = \frac{X}{T} \tag{4.1}$$

where

X = number of outcomes in which the event we are looking for occurs

T = total number of possible outcomes

What does this probability tell us? In a standard deck of cards that has 26 red cards and 26 black cards, if we replace each card after it is drawn, does it mean that one out of the next two cards selected will be black? No, because we cannot say for certain what will happen on the next several selections. However, we can say that in the long run, if this selection process is continually repeated, the proportion of black cards selected will approach .50.

In this example, the number of successes and the number of outcomes are known from the composition of the deck of cards. However, in the second approach to probability, called the **empirical classical probability** approach, although the probability is still defined as the ratio of the number of favorable outcomes to the total number of outcomes, these outcomes are based on observed data, not on prior knowledge of a process. This type of probability could refer to the proportion of individuals in a survey who actually purchase a television, who prefer a certain political candidate, or who have a part-time job while attending school.

The third approach to probability is called the **subjective probability** approach. Whereas the probability of a favorable event in the previous two approaches was computed objectively, either from prior knowledge or from actual data, subjective probability refers to the chance of occurrence assigned to an event by a particular individual. This chance may be quite different from the subjective probability assigned by another individual. For example, the inventor of a new toy may assign quite a different probability to the chance of success for the toy than the president of the company that is considering marketing the toy. The assignment of subjective probabilities to various events is usually based on a combination of an individual's past experience, personal opinion, and analysis of a particular situation. Subjective probability is especially useful in making decisions in situations in which the probability of various events cannot be determined empirically.

Sample Spaces and Events

The basic elements of probability theory are the outcomes of the process or phenomenon under study. Each possible type of occurrence is referred to as an event.

> A **simple event** can be described by a single characteristic. The collection of all the possible events is called the **sample space**.

We can achieve a better understanding of these terms by referring to the Consumer Electronics Company discussed in the Using Statistics example on page 210. Suppose that a sample of 1,000 households was initially selected and the respondents were asked whether they planned to purchase a large television. Twelve months later the same respondents were asked whether they actually purchased the television. The results are summarized in Table 4.1.

Table 4.1 *Purchase behavior for large televisions*

	ACTUALLY PURCHASED		
PLANNED TO PURCHASE	YES	NO	TOTAL
Yes	200	50	250
No	100	650	750
Total	300	700	1,000

The sample space consists of the entire set of 1,000 respondents. The events within the sample space depend on how we wish to classify the different outcomes. For example, if we are interested in purchase plans, the events are "plan to purchase" and "do not plan to purchase." If we are interested in actual purchases, the events are "purchase" and "did not purchase." Thus, the manner in which the sample space is subdivided depends on the types of probabilities that are to be determined. With this in mind, it is of interest to define both the complement of an event and a joint event as follows:

> The **complement** of event A includes all events that are not part of event A. It is given by the symbol A'.

The complement of the event "plans to purchase" is "does not plan to purchase."

> A **joint event** is an event that has two or more characteristics.

The event "planned to purchase" and "actually purchases" is a joint event because the respondent must plan to purchase the television *and* actually purchase it.

Contingency Tables

There are several ways in which a particular sample space can be viewed. The method used in this text involves assigning the appropriate events to a **table of cross-classifications** such as that displayed in Table 4.1. Such a table is also called a **contingency table** (see section 2.4). The values in the cells of the table were obtained by subdividing the sample space of 1,000 respondents according to whether someone planned to purchase and actually purchased the large television. Thus, for example, 200 of the respondents planned to purchase a large television and subsequently did purchase the large television.

Simple (Marginal) Probability

Thus far, we have focused on the meaning of probability and on defining and illustrating sample spaces. We will now begin to answer some of the questions posed in the consumer electronics example given at the beginning of the chapter by developing rules for obtaining different types of probability.

The most obvious rule for probabilities is that they range in value from 0 to 1. An impossible event has a probability of occurrence of 0, and an event that is certain to occur has a probability of 1. **Simple probability** refers to the probability of occurrence of a simple event $P(A)$, such as the probability of planning to purchase a large television.

How would we find the probability of selecting a respondent who planned to purchase a large television? Using equation (4.1) we have the following:

$$P(\text{planned to purchase}) = \frac{\text{number who planned to purchase}}{\text{total number of respondents}}$$

$$= \frac{250}{1,000} = .25$$

Thus, there is a .25 (or 25%) chance that a respondent planned to purchase a large television. Let us compute the probability of purchasing a large television set as in Example 4.1.

Example 4.1 *Computing the Probability of Purchasing a Large Television*

Suppose instead that we were interested in those respondents who actually purchased a large television. Find the probability of selecting a respondent who actually purchased a large television.

SOLUTION

$$P(\text{actually purchased}) = \frac{\text{number who actually purchased}}{\text{total number of respondents}}$$

$$= \frac{300}{1,000} = .30$$

There is a .30 (30%) chance of selecting a respondent who actually purchased the television.

Simple probability is also called **marginal probability,** because the total number of successes (those who plan to purchase) can be obtained from the appropriate margin of the contingency table (see Table 4.1 on page 212). Example 4.2 illustrates simple probability.

Example 4.2 *Computing the Probability That a Domestic General Stock Fund Will Be a No-Load Fund*

Suppose we cross-classified the type of fund and whether the fund had a no-load fee structure as in the following table. Find the probability that if a fund is randomly selected, it will have a no-load fee structure.

	FEE SCHEDULE		
FUND OBJECTIVE	NO LOAD	OTHER	TOTAL
Growth	32	27	59
Blend	75	60	135
Total	107	87	194

SOLUTION

$$P(\text{no-load fee structure}) = \frac{\text{number of no-load funds}}{\text{total number of funds}}$$

$$= \frac{107}{194} = .552$$

There is a 55.2% chance that a randomly selected fund will have a no-load fee structure.

Joint Probability

Whereas marginal probability refers to the occurrence of simple events, **joint probability** refers to phenomena containing two or more events, such as the probability of planning to purchase *and* actually purchasing a large television.

Recall that a joint event A and B means that both event A *and* event B must occur simultaneously. With reference to Table 4.1 on page 212, those individuals who planned to purchase and actually purchased a large television consist only of the outcomes in the single cell "yes—planned to purchase *and* yes—actually purchased." Because this consists of 200 respondents, the probability of picking a respondent who planned to purchase *and* actually purchased a large television is

$P(\text{planned to purchase } and \text{ actually purchased})$

$$= \frac{\text{number who planned to purchase } and \text{ actually purchased}}{\text{total number of respondents}}$$

$$= \frac{200}{1,000} = .20$$

Example 4.3 demonstrates how to determine joint probability.

Example 4.3 *Determining the Joint Probability for the Domestic General Stock Fund Example*

In Example 4.2, we cross-classified the type of fund and whether the fund had a no-load fee structure. Find the probability that a randomly selected domestic general stock fund will have a growth objective and a no-load fee structure.

SOLUTION

P(growth objective *and* no-load fee structure)

$$= \frac{\text{number that have growth objective } and \text{ no-load fee structure}}{\text{total number of funds}}$$

$$= \frac{32}{194} = .165$$

There is a 16.5% chance that a randomly selected domestic general stock fund will have a growth objective and a no-load fee structure.

Now that we have discussed the concept of joint probability, the marginal probability of a particular event can be viewed in an alternative manner. In fact, the marginal probability of an event consists of a set of joint probabilities. For example, if B consists of two events, B_1 and B_2, then we can observe that $P(A)$, the probability of event A, consists of the joint probability of event A occurring with event B_1 and the joint probability of event A occurring with event B_2. Thus, in general,

Computing Marginal Probability

$$P(A) = P(A \text{ and } B_1) + P(A \text{ and } B_2) + \cdots + P(A \text{ and } Bk) \qquad (4.2)$$

where $B_1, B_2, \ldots, Bk$ are the k mutually exclusive and collectively exhaustive events.

Two events are **mutually exclusive** if both the events cannot occur.
A set of events are **collectively exhaustive** if one of the events must occur.

For example, being male and being female are mutually exclusive and collectively exhaustive events. No one is both (they are mutually exclusive), and everyone is one or the other (they are collectively exhaustive).

Suppose that we want to use equation (4.2) to compute the marginal probability of planning to purchase a large television. We would have

$$P(\text{planned to purchase}) = P(\text{planned to purchase } and \text{ purchased})$$
$$+ P(\text{planned to purchase } and \text{ did not purchase})$$

$$= \frac{200}{1,000} + \frac{50}{1,000}$$

$$= \frac{250}{1,000} = .25$$

213

This is the same result that we obtain if we add the number of outcomes that make up the simple event "planned to purchase."

General Addition Rule

Having developed a means of finding the probability of event A and the probability of event "A and B," we should like to examine a rule that is used for finding the probability of event "A or B." This rule considers the occurrence of either event A or event B or both A and B.

How would we find the probability that a respondent planned to purchase *or* actually purchased a large television? The event "planned to purchase *or* actually purchased" would include all respondents who had planned to purchase and all respondents who had actually purchased the large television. Each cell of the contingency table (Table 4.1 on page 212) can be examined to determine whether it is part of the event in question. If we want to study the event "planned to purchase *or* actually purchased" from Table 4.1, the cell "planned to purchase *and* did not actually purchase" is part of the event, because it includes respondents who planned to purchase. The cell "did not plan to purchase *and* actually purchased" is included because it contains respondents who actually purchased. Finally, the cell "planned to purchase *and* actually purchased" has both characteristics of interest. Therefore, the probability of planned to purchase *or* actually purchased can be obtained as follows:

P(planned to purchase *or* actually purchased) = P(planned to purchase *and* did not actually purchase)

$+ P$(did not plan to purchase *and* actually purchased)

$+ P$(planned to purchase *and* actually purchased)

$$= \frac{50}{1,000} + \frac{100}{1,000} + \frac{200}{1,000} = \frac{350}{1,000}$$

The computation of $P(A\ or\ B)$, the probability of the event $A\ or\ B$, can be expressed in the following **general addition rule**:

General Addition Rule

The probability of $A\ or\ B$ is equal to the probability of A plus the probability of B minus the probability of $A\ and\ B$.

$$P(A\ or\ B) = P(A) + P(B) - P(A\ and\ B) \tag{4.3}$$

Applying this addition rule to our previous example, we obtain the following result:

P(planned to purchase *or* actually purchased) = P(planned to purchase)

$+ P$(actually purchased)

$- P$(planned to purchase *and* actually purchased)

$$= \frac{250}{1,000} + \frac{300}{1,000} - \frac{200}{1,000}$$

$$= \frac{350}{1,000} = .35$$

The **addition rule** consists of taking the probability of A and adding it to the probability of B; the intersection of $A\ and\ B$ must then be subtracted from this total because it has already been included twice in computing the probability of A and the probability of B. This can be demonstrated by referring to Table 4.1 on page 212. If the outcomes of the event "planned to purchase" are added to those of the event "actually purchased," then the joint event planned to purchase *and* actually purchased has been included in each of these simple events. Therefore, because this has been "double-counted," it must be subtracted to provide the correct result. Example 4.4 illustrates.

Example 4.4 *Using the Addition Rule for the Domestic General Stock Fund Example*

In Example 4.2 on page 214, we cross-classified the type of fund and whether the fund had a no-load fee structure. Find the probability that a randomly selected domestic general stock fund will have a growth objective or a no-load fee structure.

SOLUTION

$$P\left(\begin{array}{c}\text{growth objective } or\\ \text{no-load fee structure}\end{array}\right) = P(\text{growth objective})$$
$$+ P(\text{no-load fee structure})$$
$$- P(\text{growth objective } and \text{ no-load fee structure})$$
$$= \frac{59}{194} + \frac{107}{194} - \frac{32}{194}$$
$$= \frac{134}{194} = .691$$

There is a 69.1% chance that a randomly selected domestic general stock fund will have a growth objective *or* a no-load fee structure.

Addition Rule for Mutually Exclusive Events

In certain circumstances, the joint probability need not be subtracted because it is equal to zero as is discussed in Example 4.5.

Example 4.5 *Using the Addition Rule for Mutually Exclusive Events*

Suppose that we wanted to know the probability of picking either a heart *or* a spade if we were selecting only one card from a standard deck of 52 playing cards. Find this probability using the addition rule.

SOLUTION

$$P(\text{heart } or \text{ spade}) = P(\text{heart}) + P(\text{spade}) - P(\text{heart } and \text{ spade})$$
$$= \frac{13}{52} + \frac{13}{52} - \frac{0}{52} = \frac{26}{52} = .50$$

There is a 50% chance that a randomly selected card is a heart or a spade.

We realize that the probability that a card will be both a heart *and* a spade simultaneously is zero because in a standard deck each card belongs to only one particular suit. The joint occurrence in this case is nonexistent (called the **null set**) because it contains no outcomes—a card cannot be a heart and a spade simultaneously. As mentioned previously, whenever the joint event cannot occur, the events involved are considered to be *mutually exclusive*. This refers to the fact that the occurrence of one event (a heart) means that the other event (a spade) cannot occur. Thus, the addition rule for mutually exclusive events reduces to

Addition Rule for *Mutually Exclusive* Events

The probability of *A or B* is equal to the probability of *A* plus the probability of *B*.

$$P(A \text{ or } B) = P(A) + P(B) \tag{4.4}$$

Addition Rule for Collectively Exhaustive Events

Now consider what the probability would be of selecting a card that was red *or* black. Because red *and* black are mutually exclusive events, using equation (4.4) we have

$$P(\text{red } or \text{ black}) = P(\text{red}) + P(\text{black})$$

$$= \frac{26}{52} + \frac{26}{52} = \frac{52}{52} = 1.0$$

The probability of red or black adds up to 1.0. This means that the card selected must be red or black because these are the only colors in a standard deck. Because one of these events must occur, they are considered to be collectively exhaustive events. Finding such a probability is depicted in Example 4.6.

Example 4.6 *Using the Addition Rule for Collectively Exhaustive Events in the Domestic General Stock Fund Example*

In Example 4.2 on page 214, we cross-classified the type of fund and whether the fund had a no-load fee structure. Find the probability that a randomly selected domestic general stock fund will have a no-load or other fee structure.

SOLUTION

$$P(\text{no-load } or \text{ other}) = P(\text{no-load}) + P(\text{other})$$

$$= \frac{107}{194} + \frac{87}{194}$$

$$= \frac{194}{194} = 1.0$$

The probability that a randomly selected domestic general stock fund will have a no-load *or* other fee structure is 100% because these are the only two categories listed for fund objective.

 **CALCULATING PROBABILITIES USING MICROSOFT EXCEL**

Overview

◆ *For Quick Results Users*   Use the PHStat add-in to generate a worksheet that calculates simple and joint probabilities from data entered into a 2 × 2 contingency table of outcomes.

◆ *For Developers*   Implement a worksheet that uses arithmetic formulas to calculate simple and joint probabilities from a table of outcomes.

The 4-1E.XLS workbook file contains the calculated probabilities for the purchase behavior for large television outcomes data of Table 4.1 on page 212.

 4-1E.XLS

Quick Results Details

To calculate the simple and joint probabilities, use the Data Preparation | Probabilities choice of the PHStat add-in. As an example, consider the large television purchase behavior outcomes of Table 4.1. To calculate probabilities based on these data, do the following:

❶ If the PHStat add-in has not been previously loaded, load the add-in using the instructions of section S4.2.

❷ Select File | New to open a new workbook (or open the existing workbook into which the probabilities worksheet is to be inserted).

❸ Select PHStat | Data Preparation | Probabilities.

The add-in inserts a worksheet containing calculations for simple and joint probabilities.

❹ In the newly inserted worksheet, enter these replacement labels for the Sample Space events:
   a. Enter "planned" (event A) in cell B5.
   b. Enter "did not plan" (event A') in cell B6.
   c. Enter "purchased" (event B) in cell C4.
   d. Enter "did not purchase" (event B') in cell D4.

❺ In the same worksheet:
   a. Enter 200 (the event A *and* B value) in cell C5.
   b. Enter 100 (the event A' *and* B value) in cell C6.
   c. Enter 50 (the event A *and* B' value) in cell D5.
   d. Enter 650 (the event A' *and* B' value) in cell D6.

The inserted worksheet calculates and presents the probabilities in a table similar to the one shown in Figure 4E.1.

	A	B	C	D	E
	Name Box				
1	**Probabilities Calculations**				
2					
3	**Sample Space**			Event B	
4			purchased	did not purchase	Totals
5	Event A	planned	200	50	250
6		did not plan	100	650	750
7		Totals	300	700	1000
8					
9	**Simple Probabilities**				
10	P(planned)	0.25			
11	P(did not plan)	0.75			
12	P(purchased)	0.30			
13	P(did not purchase)	0.70			
14					
15	**Joint Probabilities**				
16	P(planned and purchased)	0.20			
17	P(planned and did not purchase)	0.05			
18	P(did not plan and purchased)	0.10			
19	P(did not plan and did not purchase)	0.65			
20					
21	**Addition Rule**				
22	P(planned or purchased)	0.35			
23	P(planned or did not purchase)	0.90			
24	P(did not plan or purchased)	0.95			
25	P(did not plan or did not purchase)	0.80			

Developer Details

We can use arithmetic formulas to implement a worksheet that calculates probabilities. Table 4E.1 presents a Probabilities sheet design that calculates the simple and joint probabilities for the large television purchase outcome data of Table 4.1. This design features labels for the joint probabilities that change automatically as labels in the sample space table are changed. To implement this design, do the following:

❶ Select File | New to open a new workbook (or open the existing workbook into which the probabilities worksheet is to be inserted).

❷ Select an unused worksheet (or select Insert | Worksheet if there are none) and rename the sheet Probabilities.

❸ Enter the title in row 1, and the headings, labels, and formulas for rows 2 through 8 as shown in Table 4E.1. At this point the totals in the Sample Space table should be equal to zero.

❹ Enter the outcomes from Table 4.1. Enter 200 in cell C5, 100 in cell C6, 50 in cell D5, and 650 in cell D6.

❺ Enter the heading and formulas for rows 9 through 25. Note that the formulas for the cell ranges A16:A19 and A22:A25 each contain three pairs of double quotation marks and four ampersands that must be entered as shown.

At this point, the joint probability labels will be displayed as $P(A$ and $B)$, $P(A$ and $B')$, $P(A'$ and $B)$ and $P(A'$ and $B')$. Less generic labels can be automatically displayed by the

Table 4E.1 Probabilities sheet design for calculating probabilities from the large television purchase outcome data of Table 4.1

	A	B	C	D	E
1	Probabilities Calculations				
2					
3	Sample Space		Event B		
4			B	B′	Totals
5	Event A	A	xx	xx	=C5+D5
6		A′	xx	xx	=C6+D6
7		Totals:	=C5+C6	=D5+D6	=C7+D7
8					
9	Simple Probabilities				
10	="P("&B5&")"	=E5/E7			
11	="P("&B6&")"	=E6/E7			
12	="P("&C4&")"	=C7/E7			
13	="P("&D4&")"	=D7/E7			
14					
15	Joint Probabilities				
16	="P("&B5&" and "&C4&")"	=C5/E7			
17	="P("&B5&" and "&D4&")"	=D5/E7			
18	="P("&B6&" and "&C4&")"	=C6/E7			
19	="P("&B6&" and "&D4&")"	=D6/E7			
20					
21	Addition Rule				
22	="P("&B5&" or "&C4&")"	=B10+B12-B16			
23	="P("&B5&" or "&D4&")"	=B10+B13-B17			
24	="P("&B6&" or "&C4&")"	=B11+B12-B18			
25	="P("&B6&" or "&D4&")"	=B11+B13-B19			

worksheet by changing the event label values in cells B5, B6, C4, and D4 to other values. For example, to mimic the presentation of probabilities in the text for the problem given in section 4.1, do the following:

❻ Enter "planned" in cell B5, "did not plan" in cell B6, "purchased" in cell C4, and "did not purchase" in cell D4.

❼ Select columns A through D, then select Format | Column | AutoFit Selection.

The widths of columns A through D are changed to accommodate the length of the new labels. The resulting worksheet will be similar to the one shown in Figure 4E.1.

Problems for Section 4.1

Learning the Basics

● **4.1** Suppose that two coins are tossed.
  (a) Give an example of a simple event.
  (b) Give an example of a joint event.
  (c) What is the complement of a head on the first toss?

4.2   Suppose that an urn contains 12 red balls and 8 white balls.
  (a) Give an example of a simple event.
  (b) What is the complement of a red ball?

4.3   Suppose that the following contingency table was set up:

	B	*B'*
A	10	20
A'	20	40

What is the probability of
  (a) event *A*?
  (b) event *B*?
  (c) event *A'*?
  (d) event *A and B*?
  (e) event *A and B'*?
  (f) event *A' and B'*?
  (g) event *A or B*?
  (h) event *A or B'*?
  (i) event *A' or B'*?

● **4.4**   Suppose that the following contingency table was set up:

	B	*B'*
A	10	30
A'	25	35

What is the probability of
  (a) event *A*?
  (b) event *B*?
  (c) event *A'*?
  (d) event *A and B*?
  (e) event *A and B'*?
  (f) event *A' and B'*?
  (g) event *A or B*?
  (h) event *A or B'*?
  (i) event *A' or B'*?

Applying the Concepts

4.5   For each of the following, indicate whether the type of probability involved is an example of *a priori* classical probability, empirical classical probability, or subjective probability.
  (a) That the next toss of a fair coin will land on heads
  (b) That Italy will win soccer's World Cup the next time the competition is held
  (c) That the sum of the faces of two dice will be 7
  (d) That the train taking a commuter to work will be more than 10 minutes late
  (e) That a Republican will win the next presidential election in the United States

4.6   For each of the following, state whether the events created are mutually exclusive and/or collectively exhaustive. If they are not, either reword the categories to make them mutually exclusive and/or collectively exhaustive or explain why this would not be useful.
  (a) Registered voters were asked whether they registered as Republicans or Democrats.
  (b) Respondents were classified on car ownership into the categories American, European, Japanese, or none.
  (c) People were asked, "Do you currently live in (i) an apartment, (ii) a house?"
  (d) A product was classified as defective or not defective.
  (e) People were asked, "Do you intend to purchase a new car in the next 6 months?" (i) yes, (ii) no.

4.7   The probability of each of the following events is zero. For each, state why. Tell what common characteristic of these events makes their probability zero.
  (a) A person who is registered as a Republican and a Democrat
  (b) A product that is defective and not defective
  (c) A house that is a ranch style and split-level style

• 4.8 In the past several years, credit card companies have made an aggressive effort to solicit new accounts from college students. Suppose that a sample of 200 students at your college indicated the following information as to whether the student possessed a bank credit card and/or a travel and entertainment credit card:

| BANK CREDIT CARD | TRAVEL AND ENTERTAINMENT CREDIT CARD | |
	YES	NO
Yes	60	60
No	15	65

(a) Give an example of a simple event.
(b) Give an example of a joint event.
(c) What is the complement of having a bank credit card?
(d) Why is "having a bank credit card and having a travel and entertainment credit card" a joint event?

If a student is selected at random, what is the probability that

(e) the student has a bank credit card?
(f) the student has a travel and entertainment credit card?
(g) the student has a bank credit card *and* a travel and entertainment card?
(h) the student has neither a bank credit card *nor* a travel and entertainment card?
(i) the student has a bank credit card *or* has a travel and entertainment card?
(j) the student does not have a bank credit card *or* has a travel and entertainment card?

4.9 The director of a large employment agency wishes to study various characteristics of its job applicants. A sample of 150 applicants has been selected, and the following information is provided as to whether the applicants have had their current jobs for at least 5 years and whether or not the applicants are college graduates:

| HELD CURRENT JOB AT LEAST 5 YEARS | COLLEGE GRADUATE | | |
	YES	NO	TOTAL
Yes	25	45	70
No	55	25	80
Total	80	70	150

(a) Give an example of a simple event.
(b) Give an example of a joint event.
(c) What is the complement of "had current job for at least 5 years"?

If an applicant is selected at random, what is the probability that he or she

(d) is a college graduate?
(e) has held the current job less than 5 years?
(f) is a college graduate *and* has held the current job less than 5 years?
(g) is not a college graduate *and* has held the current job less than 5 years?
(h) is a college graduate *or* has held the current job less than 5 years?
(i) is not a college graduate *or* has held the current job less than 5 years?

• 4.10 A sample of 500 respondents was selected in a large metropolitan area to determine various information concerning consumer behavior. Among the questions asked was "Do you enjoy shopping for clothing?" Of 240 males, 136 answered yes. Of 260 females, 224 answered yes.

(a) Set up a 2 × 2 table to evaluate the probabilities.
(b) Give an example of a simple event.
(c) Give an example of a joint event.

(d) What is the complement of "enjoy shopping for clothing"?

What is the probability that a respondent chosen at random

(e) is a male?

(f) enjoys shopping for clothing?

(g) is a female *and* enjoys shopping for clothing?

(h) is a male *and* does not enjoy shopping for clothing?

(i) is a female *or* enjoys shopping for clothing?

(j) is a male *or* does not enjoy shopping for clothing?

(k) is a male *or* a female?

4.11 A company has made available to its employees (without charge) extensive health club facilities that may be used before work, during the lunch hour, after work, and on weekends. Records for the last year indicate that of 250 employees, 110 used the facilities at some time. Of 170 males employed by the company, 65 used the facilities.

(a) Set up a 2 × 2 table to evaluate the probabilities of using the facilities.

(b) Give an example of a simple event.

(c) Give an example of a joint event.

(d) What is the complement of "used the health club facilities"?

What is the probability that an employee chosen at random

(e) is a male?

(f) has used the health club facilities?

(g) is a female *and* has used the health club facilities?

(h) is a female *and* has not used the health club facilities?

(i) is a female *or* has used the health club facilities?

(j) is a male *or* has not used the health club facilities?

(k) has used the health club facilities *or* has not used the health club facilities?

4.12 Each year, ratings are compiled concerning the performance of new cars during the first 90 days of use. Suppose that the cars have been categorized according to two attributes, whether or not the car needs warranty-related repair (yes or no) and the country in which the company manufacturing the car is based (United States, not United States). Based on the data collected, the probability that the new car needs a warranty repair is .04, the probability that the car is manufactured by an American-based company is .60, and the probability that the new car needs a warranty repair *and* was manufactured by an American-based company is .025.

(a) Set up a 2 × 2 table to evaluate the probabilities of a warranty-related repair.

(b) Give an example of a simple event.

(c) Give an example of a joint event.

(d) What is the complement of "manufactured by an American-based company"?

What is the probability that a new car selected at random

(e) needs a warranty-related repair?

(f) is not manufactured by an American-based company?

(g) needs a warranty repair *and* is manufactured by a company based in the United States?

(h) does not need a warranty repair *and* is not manufactured by a company based in the United States?

(i) needs a warranty repair *or* was manufactured by an American-based company?

(j) needs a warranty repair *or* was not manufactured by an American-based company?

(k) needs a warranty repair *or* does not need a warranty repair?

4.2 ◆ CONDITIONAL PROBABILITY

Computing Conditional Probabilities

Each situation we have examined thus far in this chapter has involved the probability of a particular event when sampling from the entire sample space. However, how would we find various probabilities if certain information about the events involved is already known?

When we are computing the probability of a particular event A, given information about the occurrence of another event B, this probability is referred to as **conditional probability**, $P(A \mid B)$. The conditional probability $P(A \mid B)$ can be defined as follows:

Conditional Probability

The probability of A given B is equal to the probability of A *and* B divided by the probability of B.

$$P(A \mid B) = \frac{P(A \text{ } and \text{ } B)}{P(B)} \tag{4.5a}$$

or

$$P(B \mid A) = \frac{P(A \text{ } and \text{ } B)}{P(A)} \tag{4.5b}$$

where

$$P(A \text{ } and \text{ } B) = \text{joint probability of } A \text{ and } B$$
$$P(A) = \text{marginal probability of } A$$
$$P(B) = \text{marginal probability of } B$$

Before using equation (4.5a) or (4.5b) to find a conditional probability, we could use the contingency table. In the consumer electronics example we have been discussing, suppose we were told that a respondent planned to purchase the large television. What would be the probability that the respondent actually purchased the television? In this example, we wish to find $P(\text{actual purchase} \mid \text{planned to purchase})$. Here the information is given that the respondent planned to purchase the large television. Therefore, the sample space does not consist of all 1,000 respondents in the survey; it consists only of those respondents who planned to purchase the large television. Of 250 such respondents, 200 actually purchased the large television. Therefore, the probability that a respondent actually purchased the large television given that he or she planned to purchase using equation (4.5b) is

$$P(\text{actually purchased} \mid \text{planned to purchase}) = \frac{\text{planned to purchase } and \text{ actually purchased}}{\text{planned to purchase}}$$

$$= \frac{200}{250}$$

This result (200/250) can also be obtained by using equation (4.5b):

If

$$P(B \mid A) = \frac{P(A \text{ } and \text{ } B)}{P(A)}$$

where

$$\text{event } A = \text{planned to purchase}$$
$$\text{event } B = \text{actually purchased}$$

then

$$P(\text{actually purchased} \mid \text{planned to purchase}) = \frac{200/1{,}000}{250/1{,}000}$$

$$= \frac{200}{250} = 0.80$$

Example 4.7 further illustrates this process.

Example 4.7 *Finding a Conditional Probability in the Domestic General Stock Fund Example*

In Example 4.2 on page 214, we cross-classified the type of fund and whether the fund had a no-load fee structure. Suppose we know that a particular fund is a no-load fund. What then is the probability that it has a growth fund objective?

SOLUTION

Because we know that the domestic general stock fund is a no-load fund, the sample space has been reduced to 107 funds. Of these 107 funds, we observe that 32 have growth objectives. Therefore, the probability that a fund has a growth objective given that it is a no-load fund is obtained as follows:

$P(\text{growth objective} \mid \text{no-load fee structure})$

$$= \frac{\text{number of funds with growth objective } and \text{ no-load fee structure}}{\text{number of funds with no-load fee structure}}$$

$$= \frac{32}{107} = .299$$

If we were to use equation (4.5a) and define

$$A = \text{growth fund} \qquad B = \text{no-load fee structure}$$
$$A' = \text{blend fund} \qquad B' = \text{other than no-load fee structure}$$

then, using equation (4.5a)

$$P(\text{growth objective} \mid \text{no-load fee structure}) = \frac{P(\text{growth objective } and \text{ no-load fee structure})}{P(\text{no-load fee structure})}$$

$$= \frac{32/194}{107/194} = \frac{32}{107} = .299$$

Therefore, given that a fund has a no-load fee structure, there is a 29.9% chance that it will have a growth objective.

Decision Trees

In Table 4.1 on page 212 respondents were classified according to whether they planned to purchase and whether they actually purchased a large television. An alternative way to view the breakdown of the possibilities into four cells is through the use of a **decision tree.** Figure 4.1 represents the decision tree for this example.

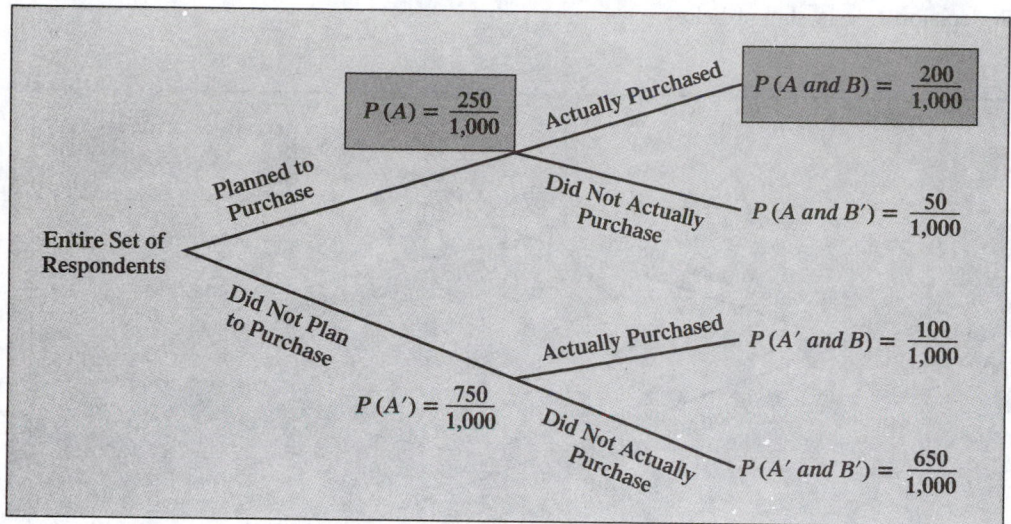

FIGURE 4.1

Decision tree for the consumer electronics example

In Figure 4.1, beginning at the left with the entire set of respondents, there are two "branches" according to whether or not the respondent planned to purchase a large television. Each of these branches has two subbranches, corresponding to whether the respondent actually purchased or did not actually purchase the large television. The probabilities at the end of the initial branches represent the marginal probabilities of A and A'. The probabilities at the end of each of the four subbranches represent the joint probability for each combination of events A and B. The conditional probability can be obtained by dividing the joint probability by the appropriate marginal probability.

For example, to obtain the probability the respondent actually purchased given that he planned to purchase the large television, we would take P(planned to purchase *and* actually purchased) and divide by P(planned to purchase). From Figure 4.1 we would have

$$P(\text{actually purchased} \mid \text{planned to purchase}) = \frac{200/1,000}{250/1,000}$$

$$= \frac{200}{250}$$

Example 4.8 illustrates how to form a decision tree.

Example 4.8 *Forming the Decision Tree for the Domestic General Stock Fund Example*

In Example 4.2 on page 214, we cross-classified the type of fund and whether the fund had a no-load fee structure. Form the decision tree and use the decision tree to find the probability that a fund has a growth fund objective given that the fund is a no-load fund.

SOLUTION

Using the following definitions:

A = growth fund B = no-load fee structure

A' = blend fund B' = other than no-load fee structure the decision tree is shown below.

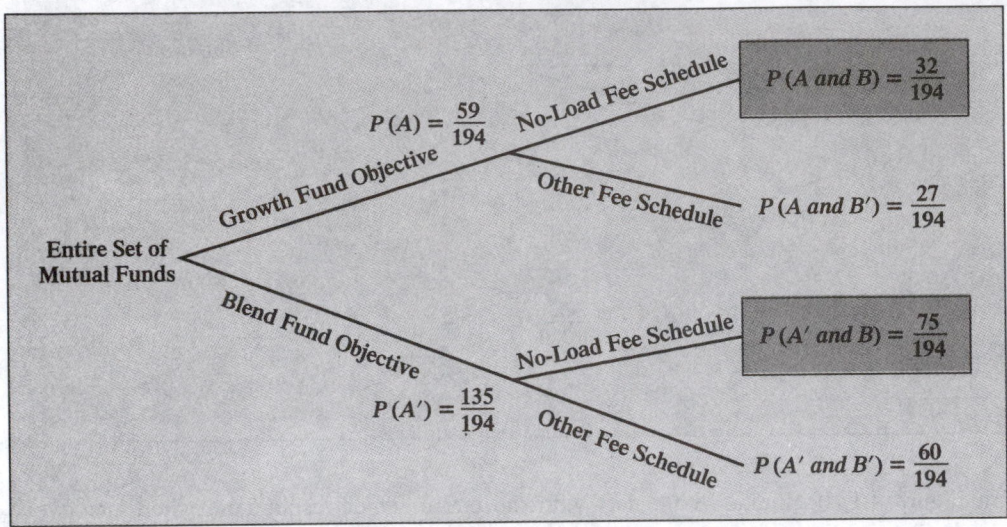

Using equation (4.5a)

$$P(\text{growth objective} \mid \text{no-load fee structure}) = \frac{P(\text{growth objective } and \text{ no-load fee structure})}{P(\text{no-load fee structure})}$$

$$= \frac{32/194}{107/194} = \frac{32}{107} = .299$$

Statistical Independence

In our example concerning the purchase of large televisions, we observed the probability that the respondent selected actually purchased the large television, given that the respondent planned to purchase, is 200/250 = .80. We may remember that the probability of selecting a respondent who actually purchases is 300/1,000, which reduces to .30. This result reveals some important information. The prior knowledge that the respondent planned to purchase affected the probability that the respondent actually purchased the television. The outcome is conditional on prior information. Unlike this example, when the outcome of one event does not affect the probability of occurrence of another event, the events are said to be statistically independent. **Statistical independence** can be defined as follows:

Statistical Independence

$$P(A \mid B) = P(A) \tag{4.6}$$

where

$$P(A \mid B) = \text{conditional probability of } A \text{ given } B$$
$$P(A) = \text{marginal probability of } A$$

Thus, we may note that two events A and B are statistically independent if and only if $P(A \mid B) = P(A)$.

In a 2 × 2 contingency table, once this holds for one combination of A and B, it will be true for all others.[1] Here, "planned to purchase" and "actually purchasing a large television" are not statistically independent events because knowledge of one event affects the probability of the second event. How to determine whether statistical independence exists is presented in Example 4.9.

[1] *In a contingency table with R rows and C columns, the rule would have to be examined for (R − 1)(C − 1) separate combinations of A and B.*

Example 4.9 *Determining Statistical Independence*

Suppose the results of the survey were those summarized in the following table instead of those shown in Table 4.1 on page 212.

Purchase behavior for large televisions

| | ACTUALLY PURCHASED | | |
PLANNED TO PURCHASE	YES	NO	TOTAL
Yes	75	175	250
No	225	525	750
Total	300	700	1,000

Determine whether planning to purchase and actually purchasing a new television are statistically independent.

SOLUTION

For these data,

$$P(\text{actually purchased} \mid \text{planned to purchase}) = \frac{75/1,000}{250/1,000}$$

$$= \frac{75}{250} = .30$$

which is equal to $P(\text{actually purchased}) = 300/1,000 = .30$. Thus, in this case, planning to purchase and actually purchasing a large television are statistically independent. Knowledge of one event in no way affects the probability of the second event.

Multiplication Rule

The formula for conditional probability can be manipulated algebraically so that the joint probability $P(A \text{ and } B)$ can be determined from the conditional probability of an event. Using equation (4.5a),

$$P(A \mid B) = \frac{P(A \text{ and } B)}{P(B)}$$

and solving for the joint probability $P(A \text{ and } B)$, we have the **general multiplication rule**.

General Multiplication Rule

The probability of *A and B* is equal to the probability of *A* given *B* times the probability of *B*.

$$P(A \text{ and } B) = P(A \mid B)P(B) \qquad (4.7)$$

To demonstrate the use of this multiplication rule, we turn to Examples 4.10 and 4.11.

Example 4.10   *Using the Multiplication Rule*

Suppose that 20 marking pens are displayed in a stationery store. Of these, 6 are red and 14 are blue. We are to select 2 markers randomly from the set of 20. Find the probability that both markers selected are red.

SOLUTION

Here the multiplication rule can be used in the following way:

$$P(A \text{ and } B) = P(A \mid B)P(B)$$

Therefore if

$$A_R = \text{second marker selected is red}$$
$$B_R = \text{first marker selected is red}$$

we have

$$P(AR \text{ and } BR) = P(AR \mid BR)P(BR)$$

The probability that the first marker is red is 6/20 because 6 of the 20 markers are red. However, the probability that the second marker is also red depends on the result of the first selection. If the first marker is not returned to the display after its color is determined (sampling *without* replacement), then the number of markers remaining will be 19. If the first marker is red, the probability that the second is also red is 5/19 because 5 red markers remain in the display. Therefore, using equation (4.7), we have the following:

$$P(A_R \text{ and } B_R) = \left(\frac{5}{19}\right)\left(\frac{6}{20}\right)$$
$$= \frac{30}{380} = .079$$

There is a 7.9% chance that both markers will be red.

Example 4.11   *Using the Multiplication Rule When Sampling with Replacement*

What if the first marker selected is returned to the display after its color is determined? Find the probability of selecting red markers on both selections.

228

SOLUTION

In this example, the probability of picking a red marker on the second selection is the same as on the first selection (sampling *with* replacement), because there are 6 red markers out of 20 in the display. Therefore, we have the following:

$$P(A_R \text{ and } B_R) = P(A_R \mid B_R)P(B_R)$$

$$= \left(\frac{6}{20}\right)\left(\frac{6}{20}\right)$$

$$= \frac{36}{400} = .09$$

There is a 9% chance that both markers will be red.

This example of sampling *with* replacement illustrates that the second selection is independent of the first because the second probability was not influenced by the first selection. Therefore, the **multiplication rule for independent events** can be expressed as follows [by substituting $P(A)$ for $P(A \mid B)$].

> If A and B are statistically independent, the probability of A and B is equal to the probability of A times the probability of B.
>
> $$P(A \text{ and } B) = P(A)P(B) \qquad (4.8)$$

If this rule holds for two events, A and B, then A and B are statistically independent. Therefore, there are two ways to determine statistical independence.

1. Events A and B are statistically independent if and only if $P(A \mid B) = P(A)$.

2. Events A and B are statistically independent if and only if $P(A \text{ and } B) = P(A)P(B)$.

It should be noted that for a 2×2 contingency table, if this is true for one joint event, it will be true for all joint events.[2]

Now that we have discussed the multiplication rule, we can write the formula for marginal probability [equation (4.2)] as follows.

If

$$P(A) = P(A \text{ and } B_1) + P(A \text{ and } B_2) + \cdots + P(A \text{ and } Bk)$$

then, using the multiplication rule, we have

Marginal Probability

$$P(A) = P(A \mid B_1)P(B_1) + P(A \mid B_2)P(B_2) + \cdots + P(A \mid Bk)P(Bk) \qquad (4.9)$$

where $B_1, B_2, \ldots, Bk$ are the k mutually exclusive and collectively exhaustive events.

This formula may be illustrated by referring to Table 4.1 on page 212. Using equation (4.9), we compute the probability of planning to purchase as follows:

$$P(A) = P(A \mid B_1)P(B_1) + P(A \mid B_2)P(B_2)$$

[2] *See Footnote 1 on page 229.*

where

$P(A)$ = probability of planned to purchase

$P(B_1)$ = probability of actually purchased

$P(B_2)$ = probability of did not actually purchase

$$P(A) = \left(\frac{200}{300}\right)\left(\frac{300}{1,000}\right) + \left(\frac{50}{700}\right)\left(\frac{700}{1,000}\right)$$

$$= \frac{200}{1,000} + \frac{50}{1,000} = \frac{250}{1,000}$$

Problems for Section 4.2

Learning the Basics

4.13 Suppose that the following contingency table was set up:

	B	B'
A	10	20
A'	20	40

What is the probability of
(a) $A \mid B$?
(b) $A \mid B'$?
(c) $A' \mid B'$?
(d) Are events A and B statistically independent?

•**4.14** Suppose that the following contingency table was set up:

	B	B'
A	10	30
A'	25	35

What is the probability of
(a) $A \mid B$?
(b) $A \mid B'$?
(c) $A' \mid B'$?
(d) Are A and B statistically independent?

•**4.15** If $P(A \text{ and } B) = .4$ and $P(B) = .8$, find $P(A \mid B)$.

4.16 If $P(A) = .7$ and $P(B) = .6$, and A and B are statistically independent, find $P(A \text{ and } B)$?

4.17 If $P(A) = .3$ and $P(B) = .4$, and $P(A \text{ and } B) = .20$, are A and B statistically independent?

Applying the Concepts

•**4.18** In the past several years, credit card companies have made an aggressive effort to solicit new accounts from college students. Suppose that a sample of 200 students at your college indicated the following information as to whether the student possessed a bank credit card and/or a travel and entertainment credit card:

	TRAVEL AND ENTERTAINMENT CREDIT CARD	
BANK CREDIT CARD	YES	NO
Yes	60	60
No	15	65

(a) Assume we know that the student has a bank credit card. What is the probability, then, that he or she has a travel and entertainment card?

(b) Assume that we know that the student does not have a travel and entertainment card. What, then, is the probability that he or she has a bank credit card?

(c) Are the two events, having a bank credit card and having a travel and entertainment card, statistically independent? Explain.

4.19 The director of a large employment agency wishes to study various characteristics of its job applicants. A sample of 150 applicants has been selected and the following information is provided as to whether the applicants have had their current jobs for at least 5 years and whether or not the applicants are college graduates:

HELD CURRENT JOB AT LEAST 5 YEARS	COLLEGE GRADUATE		
	YES	NO	TOTAL
Yes	25	45	70
No	55	25	80
Total	80	70	150

(a) Given that the applicant is a college graduate, what is the probability that he or she has held a current job less than 5 years?

(b) If the applicant has held a current job less than 5 years, what is the probability that he or she is a college graduate?

(c) Explain the difference in the results in (a) and (b).

(d) Are being a college graduate and holding the current job at least 5 years statistically independent? Explain.

•**4.20** A sample of 500 respondents was selected in a large metropolitan area to determine various information concerning consumer behavior. The following contingency table was obtained:

ENJOYS SHOPPING FOR CLOTHING	GENDER		
	MALE	FEMALE	TOTAL
Yes	136	224	360
No	104	36	140
Total	240	260	500

(a) Suppose the respondent chosen is a female. What, then, is the probability that she does not enjoy shopping for clothing?

(b) Suppose the respondent chosen enjoys shopping for clothing. What, then, is the probability that the individual is a male?

(c) Are enjoying shopping for clothing and the gender of the individual statistically independent? Explain.

4.21 A company has made available to its employees (without charge) extensive health club facilities that may be used before work, during the lunch hour, after work, and on weekends. Records for the last year indicate that of 250 employees, 110 used the facilities at some time. Of 170 males employed by the company, 65 used the facilities.

(a) Suppose that we select a female employee of the company. What, then, is the probability that she has used the health club facilities?

(b) Suppose that we select a male employee of the company. What, then, is the probability that he has not used the health club facilities?

(c) Are the gender of the individual and the use of the health club facilities statistically independent? Explain.

4.22 Each year, ratings are compiled concerning the performance of new cars during the first 90 days of use. Suppose that the cars have been categorized according to two attributes, whether or not the car needs warranty-related repair (yes or no) and the country in which the company manufacturing the car is based (United States, not United States). Based on the data collected, the probability that the new car needs a warranty repair is .04, the probability that the car is manufactured by an American-based company is .60, and the probability that the new car needs a warranty repair *and* was manufactured by an American-based company is .025.

(a) Suppose we know that the car was manufactured by a company based in the United States. What, then, is the probability that the car needs a warranty repair?

(b) Suppose we know that the car was not manufactured by a company based in the United States. What, then, is the probability that the car needs a warranty repair?

(c) Are need for a warranty repair and location of the company manufacturing the car statistically independent?

4.23 Suppose you believe the probability that you will get an A in Statistics is .6 and the probability that you will get an A in Organizational Behavior is .8. If these events are independent, what is the probability that you will get an A in both Statistics *and* Organizational Behavior? Give some plausible reasons why these events may not be independent, even though the teachers of these two subjects may not communicate about your work.

4.24 A standard deck of cards is being used to play a game. There are four suits (hearts, diamonds, clubs, and spades), each having 13 faces (ace, 2, 3, 4, 5, 6, 7, 8, 9, 10, jack, queen, and king), making a total of 52 cards. This complete deck is thoroughly mixed, and you will receive the first two cards from the deck without replacement.

(a) What is the probability that both cards are queens?

(b) What is the probability that the first card is a 10 *and* the second card is a 5 or 6?

(c) If we were sampling *with* replacement, what would be the answer in (a)?

(d) In the game of Blackjack, the picture cards (jack, queen, king) count as 10 points and the ace counts as either 1 or 11 points. All other cards are counted at their face value. Blackjack is achieved if your two cards total 21 points. What is the probability of getting blackjack in this problem?

• **4.25** A box of nine golf gloves contains two left-handed gloves and seven right-handed gloves.

(a) If two gloves are randomly selected from the box *without* replacement, what is the probability that
(1) both gloves selected will be right-handed?
(2) there will be one right-handed glove *and* one left-handed glove selected?

(b) If three gloves are selected, what is the probability that all three will be left-handed?

(c) If we were sampling *with* replacement, what would be the answers to (a)(1) and (b)?

4.3 BAYES' THEOREM

Conditional probability takes into account information about the occurrence of one event to find the probability of another event. This concept can be extended to revise probabilities based on new information and to determine the probability that a particular effect was due to a specific cause. The procedure for revising these probabilities is known as **Bayes' theorem,** having been originally developed by the Rev. Thomas Bayes in the eighteenth century (see references 1 and 3).

Bayes' theorem can be applied in the following situation that relates to toy marketing. The marketing manager of a toy manufacturing company is considering the marketing of a new toy. In the past, 40% of the toys introduced by the company have been successful and 60% have been unsuccessful. Before the toy is marketed, market research is conducted and a report, either favorable or unfavorable, is compiled. In the past, 80% of the successful toys received a favorable market research report and 30% of the unsuccessful toys received a favorable market research report. The marketing manager wants to know the probability the toy will be successful if it receives a favorable report.

Bayes' theorem can be developed from the definition of conditional probability [see equations (4.5a and b) on page 225 and (4.9) on page 231]. To find $P(B \mid A)$, we use equation (4.5b) to obtain

$$P(B \mid A) = \frac{P(A \mid B)P(B)}{P(A)}$$

We substitute equation (4.9) for $P(A)$ and obtain Bayes' theorem in equation (4.10).

Bayes' Theorem

$$P(B_i \mid A) = \frac{P(A \mid B_i)P(B_i)}{P(A \mid B_1)P(B_1) + P(A \mid B_2)P(B_2) + \cdots + P(A \mid B_k)P(B_k)} \qquad (4.10)$$

where B_i is the ith event out of k mutually exclusive events.

To use equation (4.10), let

$$\text{event } S = \text{successful toy} \qquad \text{event } F = \text{favorable report}$$
$$\text{event } S' = \text{unsuccessful toy} \qquad \text{event } F' = \text{unfavorable report}$$

and

$$P(S) = .40 \qquad P(F \mid S) = .80$$
$$P(S') = .60 \qquad P(F \mid S') = .30$$

Then, using equation (4.10),

$$P(S \mid F) = \frac{P(F \mid S)P(S)}{P(F \mid S)P(S) + P(F \mid S')P(S)}$$

$$= \frac{(.80)(.40)}{(.80)(.40) + (.30)(.60)}$$

$$= \frac{.32}{.32 + .18} = \frac{.32}{.50}$$

$$= .64$$

The probability of a successful toy, given that a favorable report was received, is .64. Thus, the probability of an unsuccessful toy, given that a favorable report was received, is $1 - .64 = .36$. The computation of the probabilities is summarized in tabular form in Table 4.2 and displayed in the form of a decision tree in Figure 4.2.

Table 4.2 *Bayes' theorem calculations for the toy marketing problem*

Event S_i	PRIOR PROBABILITY $P(S_i)$	CONDITIONAL PROBABILITY $P(F \mid S_i)$	JOINT PROBABILITY $P(F \mid S_i)P(S_i)$	REVISED PROBABILITY $P(S_i \mid F)$
S = successful toy	.40	.80	.32	$.32/.50 = .64 = P(S \mid F)$
S' = unsuccessful toy	.60	.30	.18	$.18/.50 = .36 = P(S' \mid F)$
			.50	

FIGURE 4.2

Decision tree for the
toy marketing problem

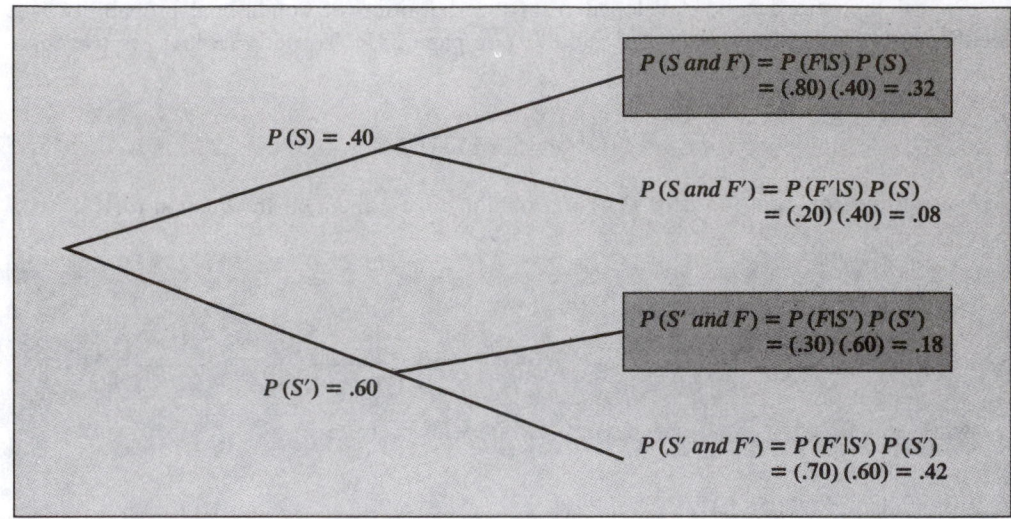

Bayes' theorem is applied to a medical diagnosis problem in Example 4.12.

Example 4.12 *Using Bayes' Theorem in a Medical Diagnosis Problem*

Suppose that the probability a person has a certain disease is .03. Medical diagnostic tests are available to determine whether the person actually has the disease. If the disease is actually present, the probability that the medical diagnostic test will give a positive result (indicating the disease is present) is .90. If the disease is not actually present, the probability of a positive test result (indicating that the disease is present) is .02. Suppose the medical diagnostic test has given a positive result (indicating the disease is present). What is the probability that the disease is actually present?

SOLUTION

Let   event D = has disease   event T = test is positive

            event D' = does not have disease   event T' = test is negative

and $P(D) = .03$ $P(T \mid D) = .90$

            $P(D') = .97$ $P(T \mid D') = .02$

Using equation (4.10), we have

$$P(D \mid T) = \frac{P(T \mid D)P(D)}{P(T \mid D)P(D) + P(T \mid D')P(D')}$$

$$= \frac{(.90)(.03)}{(.90)(.03) + (.02)(.97)}$$

$$= \frac{.0270}{.0270 + .0194} = \frac{.0270}{.0464}$$

$$= .582$$

The probability that the disease is actually present given a positive result has occurred (indicating the disease is present) is .582. The computation of the probabilities is summarized in the following table and also displayed in the form of a decision tree.

Bayes' theorem calculations for the medical diagnosis problem

Event D_i	PRIOR PROBABILITY $P(D_i)$	CONDITIONAL PROBABILITY $P(T \mid D_i)$	JOINT PROBABILITY $P(T \mid D_i)P(D_i)$	REVISED PROBABILITY $P(D_i \mid T)$
D = has disease	.03	.90	.0270	.0270/.0464 = .582 = $P(D \mid T)$
D' = does not have disease	.97	.02	.0194	.0194/.0464 = .418 = $P(D' \mid T)$
			.0464	

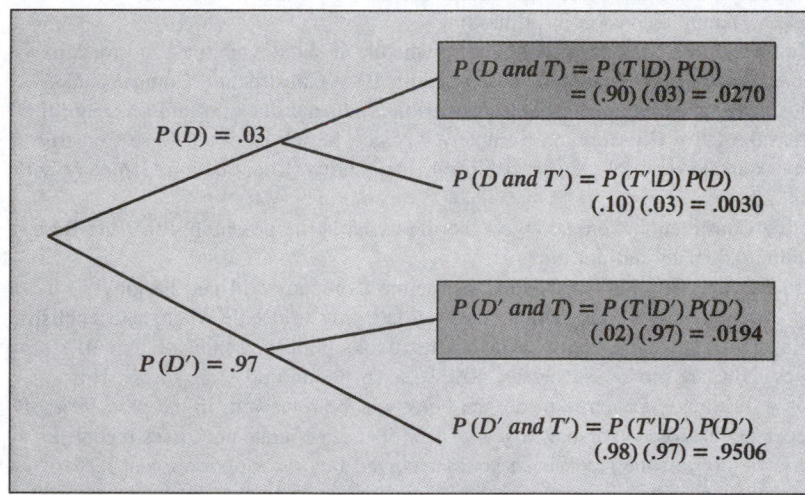

Example 4.13 Finding the Proportion of Medical Diagnostic Tests That Are Positive

Using the data from Example 4.12, determine the proportion of medical diagnostic tests that are positive.

SOLUTION

The denominator in Bayes' theorem represents $P(T)$, the probability of a positive test result, which in this case is .0464. Therefore, the probability of a positive test result is 4.64%.

Problems for Section 4.3

Learning the Basics

• **4.26** If $P(B) = .05$, $P(A \mid B) = .80$, $P(B') = .95$, and $P(A \mid B') = .40$, find $P(B \mid A)$.

4.27 If $P(B) = .30$, $P(A \mid B) = .60$, $P(B') = .70$, and $P(A \mid B') = .50$, find $P(B \mid A)$.

Applying the Concepts

4.28 In Example 4.12, suppose the probability that a medical diagnostic test will give a positive result if the disease is not present is reduced from .02 to .01. Given this information, we would like to know the following:

(a) If the medical diagnostic test has given a positive result (indicating the disease is present), what is the probability that the disease is actually present?

(b) If the medical diagnostic test has given a negative result (indicating the disease is not present), what is the probability that the disease is not present?

• **4.29** An advertising executive is studying television viewing habits of married men and women during prime time hours. On the basis of past viewing records, the executive has determined that during prime time, husbands are watching television 60% of the time. It has also been determined that when the husband is watching television, 40% of the time the wife is also watching. When the husband is not watching television, 30% of the time the wife is watching television. Find the probability that

(a) if the wife is watching television, the husband is also watching television.

(b) the wife is watching television in prime time.

4.30 The Olive Construction Company is determining whether it should submit a bid for a new shopping center. In the past, Olive's main competitor, Base Construction Company, has submitted bids 70% of the time. If Base Construction Company does not bid on a job, the probability that the Olive Construction Company will get the job is .50. If Base Construction Company does bid on a job, the probability that the Olive Construction Company will get the job is .25.

(a) If the Olive Construction Company gets the job, what is the probability that the Base Construction Company did not bid?

(b) What is the probability that the Olive Construction Company will get the job?

4.31 The editor of a major textbook publishing company is trying to decide whether to publish a proposed business statistics textbook. Previous textbooks published indicate that 10% are huge successes, 20% are modest successes, 40% break even, and 30% are losers. However, before a publishing decision is made, the book will be reviewed. In the past, 99% of the huge successes received favorable reviews, 70% of the moderate successes received favorable reviews, 40% of the break-even books received favorable reviews, and 20% of the losers received favorable reviews.

(a) If the proposed text receives a favorable review, how should the editor revise the probabilities of the various outcomes to take this information into account?

(b) What proportion of textbooks receive favorable reviews?

• **4.32** A municipal bond service has three rating categories (A, B, and C). Suppose that in the past year, of the municipal bonds issued throughout the United States, 70% were rated A, 20% were rated B, and 10% were rated C. Of the municipal bonds rated A, 50% were issued by cities, 40% by suburbs, and 10% by rural areas. Of the municipal bonds rated B, 60% were issued by cities, 20% by suburbs, and 20% by rural areas. Of the municipal bonds rated C, 90% were issued by cities, 5% by suburbs, and 5% by rural areas.

(a) If a new municipal bond is to be issued by a city, what is the probability it will receive an A rating?

(b) What proportion of municipal bonds are issued by cities?

(c) What proportion of municipal bonds are issued by suburbs?

◆ 4.4 ◆ THE PROBABILITY DISTRIBUTION FOR A DISCRETE RANDOM VARIABLE

As discussed in section 1.7, a *numerical random variable* is some phenomenon of interest whose responses or outcomes are expressed numerically. We classified numerical random variables as *discrete* or *continuous*—the former arising from a counting process and the latter from a measuring process. The remainder of this chapter deals with some probability distributions that represent discrete random variables.

We define the probability distribution for a discrete random variable as follows:

A **probability distribution for a discrete random variable** is a mutually exclusive listing of all possible numerical outcomes for that random variable such that a particular probability of occurrence is associated with each outcome.

For example, suppose the distribution of the number of mortgages approved per week at the local branch office of a bank is as illustrated in Table 4.3. Because all possible outcomes are included in Table 4.3, the listing is collectively exhaustive, and thus the probabilities must sum to 1. To summarize a discrete probability distribution, we shall compute its major characteristics—the mean and the standard deviation.

A graphical representation of Table 4.3 is illustrated in Figure 4.3.

Table 4.3 *Probability distribution of home mortgages approved per week*

Home Mortgages Approved per Week	Probability
0	.10
1	.10
2	.20
3	.30
4	.15
5	.10
6	.05

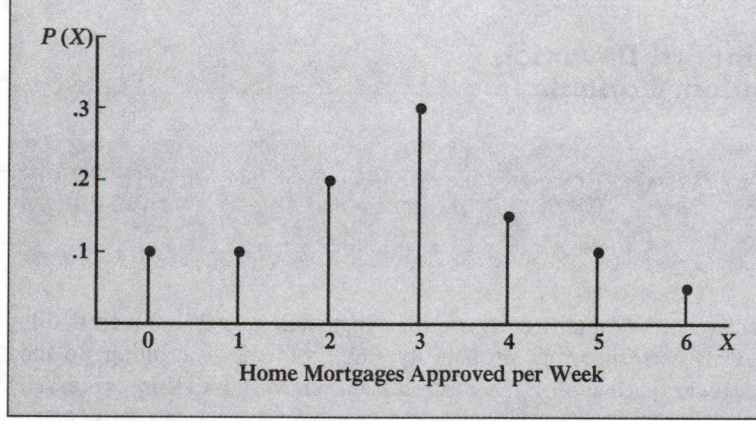

FIGURE 4.3
Probability distribution of the number of home mortgages approved per week

Expected Value of a Discrete Random Variable

The mean (μ) of a probability distribution is the *expected value* of its random variable.

The **expected value of a discrete random variable** is a *weighted* average over all possible outcomes—the weights being the probability associated with each of the outcomes.

This summary measure is obtained by multiplying each possible outcome X by its corresponding probability $P(X)$ and then summing the resulting products. Thus, the expected value of the discrete random variable X, symbolized as $E(X)$, is expressed as follows:

Expected Value of a Discrete Random Variable

$$\mu = E(X) = \sum_{i=1}^{N} X_i P(X_i) \qquad (4.11)$$

where

$\qquad X_i = i$th outcome of X, the discrete random variable of interest

$\qquad P(X_i) =$ probability of occurrence of the ith outcome of X

For the probability distribution of the results of the number of home mortgages approved per week (Table 4.3), the expected value is computed as

$$\mu = E(X) = \sum_{i=1}^{N} X_i P(X_i)$$

$$= (0)(.1) + (1)(.1) + (2)(.2) + (3)(.3) + (4)(.15) + (5)(.1) + (6)(.05)$$

$$= 0 + .1 + .4 + .9 + .6 + .5 + .3$$

$$= 2.8$$

Notice that the expected value of the number of mortgages approved, 2.8, is not "literally meaningful" because the number of mortgages approved per week can never be 2.8. It must be an integer value. The expected value is telling us the average value that we can expect to obtain over many weeks.

Variance and Standard Deviation of a Discrete Random Variable

The **variance (σ^2) of a discrete random variable** is defined as the *weighted* average of the squared differences between each possible outcome and its mean—the weights being the probabilities of each of the respective outcomes.

This summary measure can be obtained by multiplying each possible squared difference $(X_i - E(X))^2$ by its corresponding probability $P(X_i)$ and then summing up the resulting products. Hence, the variance of the discrete random variable X is expressed as follows:

Variance of a Discrete Random Variable

$$\sigma^2 = \sum_{i=1}^{N} [X_i - E(X)]^2 P(X_i) \qquad (4.12)$$

where

$\qquad X_i = i$th outcome of X

$\qquad P(X_i) =$ probability of occurrence of the ith outcome of X

The **standard deviation** σ **of a discrete random variable** is given by

The computation of the variance and the standard deviation of the number of home mortgages approved per week is illustrated in Example 4.14.

Example 4.14 *Computing the Variance and Standard Deviation*

Compute the variance and the standard deviation for the distribution of the number of home mortgages per week in Table 4.3.

SOLUTION

$$\sigma^2 = \sum_{i=1}^{N}[X_i - E(X)]^2 P(X_i)$$

$$= (0 - 2.8)^2(.10) + (1 - 2.8)^2(.10) + (2 - 2.8)^2(.20) + (3 - 2.8)^2(.30)$$
$$+ (4 - 2.8)^2(.15) + (5 - 2.8)^2(.10) + (6 - 2.8)^2(.05)$$

$$= .784 + .324 + .128 + .012 + .216 + .484 + .512$$

$$= 2.46$$

and

$$\sigma = 1.57$$

Thus, for our distribution of the number of mortgages approved per week, the mean is 2.8 and the standard deviation is 1.57.

In the remainder of this chapter we will be concerned mainly with the listing obtained from a mathematical model representing some phenomenon of interest.

A **model** is considered to be a miniature representation of some underlying phenomenon. In particular, a **mathematical model** is a mathematical expression representing some underlying phenomenon. For discrete random variables, this mathematical expression is known as a **probability distribution function.**

When such mathematical expressions are available, the exact probability of occurrence of any particular outcome of the random variable can be computed. In such cases, the entire probability distribution can be obtained and listed. For example, the probability distribution function of the outcome of the selection of a digit from the random number table (see appendix E.1) is one in which the discrete random variable of interest is said to follow the **uniform probability distribution.** The essential characteristic of the uniform distribution is that all outcomes of the random variable are equally likely to occur. Thus, the probability that a random digit of 1 turns up is the same as that for any other result—1/10—because there are 10 possible outcomes.

In addition, other types of mathematical models have been developed to represent various discrete phenomena that occur in the social and natural sciences, in medical research, and in business. One of the most useful of these represents data characterized by the binomial probability distribution. This important distribution is developed in section 4.5.

Problems for Section 4.4

Learning the Basics

•**4.33** Given the following probability distributions:

DISTRIBUTION A		DISTRIBUTION B	
X	P(X)	X	P(X)
0	.50	0	.05
1	.20	1	.10
2	.15	2	.15
3	.10	3	.20
4	.05	4	.50

(a) Compute the expected value for each distribution.
(b) Compute the standard deviation for each distribution.
(c) Compare and contrast the results of distributions A and B. Discuss what you have learned.

4.34 Given the following probability distributions:

DISTRIBUTION C		DISTRIBUTION D	
X	P(X)	X	P(X)
0	.20	0	.10
1	.20	1	.20
2	.20	2	.40
3	.20	3	.20
4	.20	4	.10

(a) Compute the expected value for each distribution.
(b) Compute the standard deviation for each distribution.
(c) Compare and contrast the results of distributions C and D. Discuss what you have learned.

Applying the Concepts

4.35 Using the company records for the past 500 working days, the manager of Torrisi Motors, a suburban automobile dealership, has summarized the number of cars sold per day into the following table:

NUMBER OF CARS SOLD PER DAY	FREQUENCY OF OCCURRENCE
0	40
1	100
2	142
3	66
4	36
5	30
6	26
7	20
8	16
9	14
10	8
11	2
Total	500

(a) Form the empirical probability distribution (i.e., relative frequency distribution) for the discrete random variable X, the number of cars sold per day.
(b) Compute the mean or expected number of cars sold per day.
(c) Compute the standard deviation.
(d) What is the probability that on any given day
    (1) fewer than four cars will be sold?   (4) exactly four cars will be sold?
    (2) at most four cars will be sold?   (5) more than four cars will be sold?
    (3) at least four cars will be sold?

• **4.36** The following table contains the probability distribution of the number of traffic accidents daily in a small city.

NUMBER OF ACCIDENTS DAILY (X)	$P(X)$
0	.10
1	.20
2	.45
3	.15
4	.05
5	.05

(a) Compute the mean or expected number of accidents per day.
(b) Compute the standard deviation.

4.37 The manager of a large computer network has developed a probability distribution of the number of interruptions per day.

INTERRUPTIONS (X)	$P(X)$
0	.32
1	.35
2	.18
3	.08
4	.04
5	.02
6	.01

(a) Compute the mean or expected number of interruptions per day.

(b) Compute the standard deviation.

4.38 In the carnival game Under-or-over-Seven, a pair of fair dice are rolled once, and the resulting sum determines whether or not the player wins or loses his or her bet. For example, the player can bet $1.00 that the sum will be under 7—that is, 2, 3, 4, 5, or 6. For such a bet the player will lose $1.00 if the outcome equals or exceeds 7 or will win $1.00 if the result is under 7. Similarly, the player can bet $1.00 that the sum will be over 7—that is, 8, 9, 10, 11, or 12. Here the player wins $1.00 if the result is over 7 but loses $1.00 if the result is 7 or under. A third method of play is to bet $1.00 on the outcome 7. For this bet the player will win $4.00 if the result of the roll is 7 and lose $1.00 otherwise.

(a) Form the probability distribution function representing the different outcomes that are possible for a $1.00 bet on being under 7.

(b) Form the probability distribution function representing the different outcomes that are possible for a $1.00 bet on being over 7.

(c) Form the probability distribution function representing the different outcomes that are possible for a $1.00 bet on 7.

(d) Show that the expected long-run profit (or loss) to the player is the same—no matter which method of play is used.

◆ 4.5 ◆ BINOMIAL DISTRIBUTION

◆ USING STATISTICS: *The Customer Services Department of a Natural Gas Utility Company*

Customer surveys for natural gas companies have indicated that customer satisfaction is strongly related to a repair response time of no more than 2 hours from the initial call requesting service. The arrival of the repair person within 2 hours is considered an *acceptable* waiting period. Recent data collected by the company indicate that the likelihood is .60 that a repair person will reach the customer's home within a 2-hour period. The company is interested in determining the likelihood of obtaining a certain number of service calls in the acceptable waiting period for a given sample of service calls. How can such probabilities be determined? ◆

The **binomial distribution** is a discrete probability distribution function with many everyday applications. The binomial distribution possesses four essential properties:

Exhibit 4.1 Properties of the Binomial Distribution

✓ **1.** The possible observations may be obtained by two different sampling methods. Each observation may be considered as having been selected either from an *infinite population without replacement* or from a *finite population with replacement.*

✓ **2.** Each observation may be classified into one of two mutually exclusive and collectively exhaustive categories, usually called *success* and *failure.*

✓ **3.** The probability of an observation's being classified as success, p, is constant from observation to observation. Thus, the probability of an observation's being classified as failure, $1 - p$, is constant over all observations.

✓ **4.** The outcome (i.e., success or failure) of any observation is independent of the outcome of any other observation.

The discrete random variable or phenomenon of interest that follows the binomial distribution is the number of successes obtained in a sample of n observations. Returning to our Using Statistics example concerning the natural gas utility company, *success* is the arrival of the repair person within the acceptable 2-hour period and *failure* is any other outcome. In the example, we stated that we were interested in the likelihood that the repair person would arrive within the acceptable period in a certain number of service calls that have been monitored.

What results can occur? If, for example, we consider a sample of four service calls, the service could be provided in an acceptable time period at none of the houses, at one house, at two houses, at three houses, or at all four houses. Can the binomial random variable, the number of arrivals within an acceptable time period, take on any other value? That would be impossible because the number of successful arrivals cannot exceed the sample size n, and it cannot be lower than zero. Hence, the range of a binomial random variable is from 0 to n.

Suppose that the following result is observed in a sample of four monitored service calls:

FIRST SERVICE CALL	SECOND SERVICE CALL	THIRD SERVICE CALL	FOURTH SERVICE CALL
Acceptable	Acceptable	Not Acceptable	Acceptable

What is the probability of obtaining three successes (service calls within an acceptable time) in a sample of four monitored service calls in this particular sequence? Because it may be assumed that making a service call is a stable process with a historical probability of .60 occurring within the acceptable time period, the probability that each service call occurs as noted is

FIRST SERVICE CALL	SECOND SERVICE CALL	THIRD SERVICE CALL	FOURTH SERVICE CALL
$p = .60$	$p = .60$	$1 - p = .40$	$p = .60$

Because each outcome is independent of the others, the probability of obtaining this particular sequence is

$$pp(1 - p)p = p^3(1 - p) = p^3(1 - p)^1 = (.60)^3(.40)^1 = .0864$$

However, this tells us only the probability of obtaining three acceptable service calls within 2 hours (successes) out of a sample of four monitored service calls in a *specific order*. If we now want to find the number of ways of selecting X objects out of n objects *irrespective of order*, we must use the **rule of combinations**.

Combinations

The *number of combinations* of selecting X objects out of n objects is given by

$$\frac{n!}{X!(n - X)!} \tag{4.14}$$

where

$$n! = n(n - 1) \ldots (1) \text{ is called } n \text{ factorial and } 0! = 1.$$

This expression may be denoted by the symbol $\binom{n}{X}$. Therefore, with $n = 4$ and $X = 3$, we have

$$\binom{n}{X} = \frac{n!}{X!(n - X)!} = \frac{4!}{3!(4 - 3)!} = \frac{4 \times 3 \times 2 \times 1}{(3 \times 2 \times 1)(1)} = 4$$

such sequences. These four possible sequences are

> **Sequence 1** = *acceptable, acceptable, acceptable, not acceptable* with probability
> $ppp(1 - p) = p^3(1 - p)^1 = .0864$
>
> **Sequence 2** = *acceptable, acceptable, not acceptable, acceptable* with probability
> $pp(1 - p)p = p^3(1 - p)^1 = .0864$
>
> **Sequence 3** = *acceptable, not acceptable, acceptable, acceptable* with probability
> $p(1 - p)pp = p^3(1 - p)^1 = .0864$
>
> **Sequence 4** = *not acceptable, acceptable, acceptable, acceptable* with probability
> $(1 - p)ppp = p^3(1 - p)^1 = .0864$

Note that ours is the second of these four possible sequences.

Therefore, the probability of obtaining exactly three service calls within the acceptable time period is equal to

$$(\text{Number of possible sequences}) \times (\text{probability of a particular sequence})$$
$$= (4) \times (.0864) = .3456$$

A similar, intuitive derivation can be obtained for the other four possible outcomes of the random variable—zero, one, two, and four service calls within the acceptable time period. However, as *n*, the number of observations, gets large, this type of intuitive approach becomes quite laborious and a mathematical model is more appropriate. In general, the following mathematical model represents the binomial probability distribution for obtaining a number of successes (*X*), given a knowledge of this distribution's parameters *n* and *p*.

Binomial Distribution

$$P(X) = \frac{n!}{X!(n - X)!} p^X (1 - p)^{n-X} \tag{4.15}$$

where

$P(X)$ = the probability of *X* successes given a knowledge of *n* and *p*
n = sample size
p = probability of success
$1 - p$ = probability of failure
X = number of successes in the sample ($X = 0, 1, 2, \ldots, n$)

We note, however, that the generalized form shown in equation (4.15) is merely a restatement of what we had intuitively derived. The binomial random variable *X* can have any integer value *X* from 0 through *n*. In equation (4.15) the product

$$p^X(1 - p)^{n-X}$$

tells us the probability of obtaining exactly *X* successes out of *n* observations in a *particular sequence*; the term

$$\frac{n!}{X!(n - X)!}$$

tells us *how many combinations* of the *X* successes out of *n* observations are possible. Hence, given the number of observations *n* and the probability of success *p*, we determine the probability of *X* successes:

$$P(X) = \text{(number of possible sequences)} \times \text{(probability of a particular sequence)}$$

$$= \frac{n!}{X!(n-X)!} p^X (1-p)^{n-X}$$

by substituting the desired values for n, p, and X and computing the result. In Example 4.15 we illustrate the use of equation (4.15).

Example 4.15 Determining $P(X = 3)$, given $n = 4$ and $p = .6$

If the likelihood of an acceptable service call is .6, what is the probability that three acceptable service calls are made out of the sample of four that are monitored?

SOLUTION

As previously shown, using equation (4.15), the probability of obtaining exactly three service calls within the acceptable time period from a sample of four monitored service calls is

$$P(X = 3) = \frac{4!}{3!(4-3)!}(.6)^3(1-.6)^1$$

$$= \frac{4!}{3!1!}(.6)^3(.4)^1$$

$$= 4(.6)(.6)(.6)(.4) = .3456$$

Other computations concerning different values of X can also be obtained as in Examples 4.16 and 4.17.

Example 4.16 Determining $P(X \geq 3)$, given $n = 4$ and $p = .6$

If the likelihood of an acceptable service call is .6, what is the probability that three or more (i.e., at least three) acceptable service calls are made out of the sample of four that are monitored?

SOLUTION

As previously shown, using equation (4.15), the probability of obtaining exactly three service calls within the acceptable time period from a sample of four monitored service calls is .3456. To obtain the probability of at least three acceptable service calls, we need to add the probability of three acceptable service calls to the probability of four acceptable service calls. The probability of four acceptable service calls is

$$P(X = 4) = \frac{4!}{4!(4-4)!}(.6)^4(1-.6)^0$$

$$= \frac{4!}{4!0!}(.6)^4(.4)^0$$

$$= 1(.6)(.6)(.6)(.6) = .1296$$

Thus the probability of at least three acceptable service calls is

$$P(X \geq 3) = P(X = 3) + P(X = 4)$$
$$= .3456 + .1296$$
$$= .4752$$

There is a .4752 chance that there will be at least three acceptable service calls in a sample of four calls.

Example 4.17 *Determining $P(X < 3)$, given $n = 4$ and $p = .6$*

If the likelihood of an acceptable service call is .6, what is the probability that fewer than three acceptable service calls are made out of the sample of four that are monitored?

SOLUTION

The probability that fewer than three acceptable service calls are made is

$$P(X < 3) = P(X = 0) + P(X = 1) + P(X = 2)$$

Using equation (4.15) to obtain each of these probabilities, we have

$$P(X = 0) = \frac{4!}{0!(4 - 0)!}(.6)^0(1 - .6)^4 = .0256$$
$$P(X = 1) = \frac{4!}{1!(4 - 1)!}(.6)^1(1 - .6)^3 = .1536$$
$$P(X = 2) = \frac{4!}{2!(4 - 2)!}(.6)^2(1 - .6)^2 = .3456$$

Therefore, we have

$$P(X < 3) = .0256 + .1536 + .3456 = .5248$$

Rather than using equation (4.15) to compute $P(X < 3)$, because we have already computed $P(X \geq 3)$, we could obtain $P(X < 3)$ as its complement as follows:

$$P(X < 3) = P(X = 0) + P(X = 1) + P(X = 2)$$
$$= 1 - P(X \geq 3)$$
$$= 1 - .4752 = .5248$$

Such computations can become quite tedious, especially as n gets large. However, we can obtain the probabilities by using Microsoft Excel (see section 4.5E) and thereby avoid any computational drudgery. Figure 4.4 represents output from Microsoft Excel for the binomial distribution example with parameters $n = 4$ and $p = .6$.

	A	B	C	D	E	F	G
1	Binomial Probabilities for Service Calls						
2							
3	Sample size	4					
4	Probability of success	0.6					
5	Mean	2.4					
6	Variance	0.96					
7	Standard deviation	0.979796					
8							
9	Binomial Probabilities Table						
10		X	P(x)	P(<=X)	P(<X)	P(>X)	P(>=X)
11		0	0.0256	0.0256	0	0.9744	1
12		1	0.1536	0.1792	0.0256	0.8208	0.9744
13		2	0.3456	0.5248	0.1792	0.4752	0.8208
14		3	0.3456	0.8704	0.5248	0.1296	0.4752
15		4	0.1296	1	0.8704	0	0.1296

FIGURE 4.4

Binomial distribution calculations for $n = 4$ and $p = .6$ obtained from the PHStat add-in for Microsoft Excel

Characteristics of the Binomial Distribution

Each time a set of parameters—n and p—is specified, a particular binomial probability distribution can be generated.

◆ *Shape* We note that a binomial distribution may be symmetrical or skewed. Whenever $p = .5$, the binomial distribution will be symmetrical regardless of how large or small the value of n. However, when $p \neq .5$, the distribution will be skewed. The closer p is to .5 and the larger the number of observations n, the less skewed the distribution will be. Thus, the distribution of the number of service calls within the acceptable time period is skewed to the left because $p = .60$. This can be observed in Figure 4.5, which is a plot of the distribution for $n = 4$ and $p = .60$.

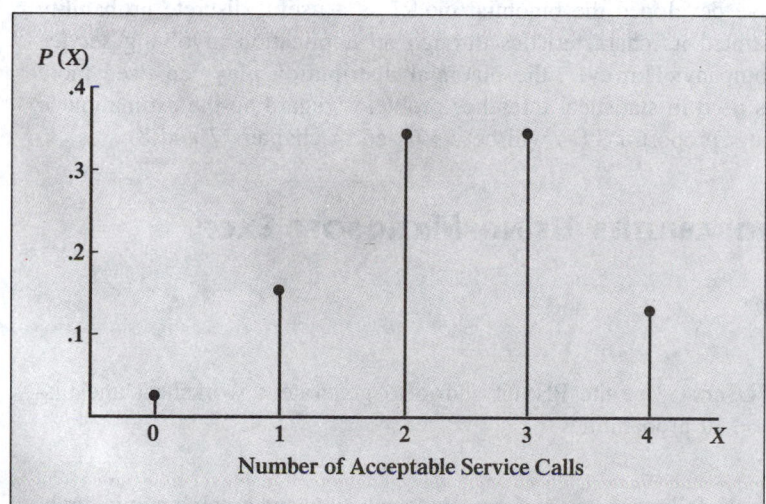

FIGURE 4.5

Binomial distribution for $n = 4$ and $p = .6$

◆ *The Mean* The mean of the binomial distribution can be obtained as the product of its two parameters n and p. Instead of using equation (4.11), which holds for all discrete probability distributions, we use the following to compute the mean for data that are binomially distributed:

The Mean of the Binomial Distribution

The mean μ of the binomial distribution is equal to the sample size n multiplied by the probability of success p.

$$\mu = E(X) = np \tag{4.16}$$

Intuitively, this makes sense. On the average, over the long run, we would theoretically expect $\mu = E(X) = np = (4)(.6) = 2.4$ service calls within the acceptable time period out of a sample of four monitored service calls.

◆ **The Standard Deviation** The standard deviation of the binomial distribution is calculated using the following:

The Standard Deviation of the Binomial Distribution

$$\sigma = \sqrt{\sigma^2} = \sqrt{Var(X)} = \sqrt{np(1-p)} \tag{4.17}$$

Referring to our service calls example, we compute

$$\sigma = \sqrt{(4)(.6)(.4)} = \sqrt{.96} = .98$$

This is the same result that would be obtained from the more general expression shown in equation (4.13).

In this section we have developed the binomial model as a useful discrete probability distribution and demonstrated its characteristics through an application involving service calls by a gas utility company. However, the binomial distribution plays an even more important role when it is used in statistical inference problems regarding the estimation or testing of hypotheses about proportions (as will be discussed in chapters 7 and 8).

4.5E CALCULATING BINOMIAL PROBABILITIES USING MICROSOFT EXCEL

Overview

◆ **For Quick Results Users** Use the PHStat add-in to generate a worksheet and histogram that displays binomial probabilities.

◆ **For Developers** Use the following two-step process to calculate and graph binomial probabilities:

A. Implement a worksheet that uses the BINOMDIST worksheet function to calculate binomial probabilities.

B. Use the Chart Wizard to generate a histogram based on this worksheet.

The 4-5E.XLS workbook file contains the binomial probabilities and a histogram for the natural gas company service call problem discussed in section 4.5.

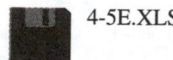 4-5E.XLS

Quick Results Details

To calculate binomial probabilities, use the Probability Distributions | Binomial choice of the PHStat add-in. As an example, consider the natural gas company service call problem discussed in section 4.5. To calculate the binomial probabilities for this problem, do the following:

❶ If the PHStat add-in has not been previously loaded, load the add-in using the instructions of section S4.2.

❷ Select File | New to open a new workbook (or open the existing workbook into which the binomial probabilities worksheet is to be inserted).

❸ Select PHStat | Probability Distributions | Binomial.

❹ In the Binomial Probability Distribution dialog box (see Figure 4E.2):
   a. Enter 4 in the Sample Size: edit box.
   b. Enter 0.6 in the Probability of Success: edit box.
   c. Enter 0 (zero) in the Outcomes From: edit box and enter 4 in the Outcomes To: edit box.
   d. Enter Binomial Probabilities for Service Calls in the Output Title: edit box.
   e. Select the Cumulative Probabilities and Histogram check boxes.
   f. Click the OK button.

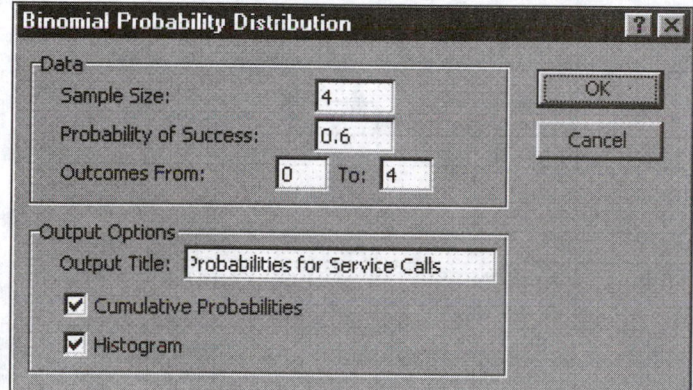

FIGURE 4E.2
PHStat Binomial Probability Distribution dialog box

On separate sheets, the add-in produces a table of binomial probabilities and a histogram similar to those shown in Figures 4.4 and 4.5 on page 249, respectively.

Developer Details

◆ *A. Implementing a Worksheet to Calculate Binomial Probabilities* We can use the BINOMDIST function as the basis for computing binomial probabilities in an Excel worksheet. The format of this function is:

BINOMDIST(*X*, *n*, *p*, *cumulative*)

where

X = the number of successes

n = the sample size

p = the probability of success

cumulative = a True or False value that determines whether the function computes the probability of X of fewer successes (True) or computes the probability of exactly X successes (False)

For example, BINOMDIST(3, 4, .6, False) would calculate the probability of obtaining exactly three service calls within the acceptable period from a sample of four service calls for the natural gas company problem of section 4.5. Changing the False value to True would calculate the probability of 3 or fewer service calls.

Table 4E.2 presents a Binomial sheet design that uses the BINOMDIST function to calculate the binomial probabilities for the natural gas company service call problem discussed in section 4.5. The worksheet also calculates the expected value, variance, and standard deviation for the problem using arithmetic formulas and the square root (SQRT) worksheet function. Because the sample size value determines the length of the binomial probabilities table, this value cannot be interactively changed by the user and therefore is shown as 4—and not as the generalized value .*xx*—in the design. (The probability of success shown as a generalized value could be interactively changed.). To implement this design, do the following:

Table 4E.2 Binomial sheet design for the natural gas service call problem of section 4.2

	A	B	C	D	E	F	G
1	Binomial Probabilities for Service Calls						
2							
3	Sample size	4					
4	Probability of success	.xx					
5	Mean	=B3*B4					
6	Variance	=B5*(1-B4)					
7	Standard deviation	=SQRT(B6)					
8							
9	Binomial Probabilities Table						
10			X	P(X)	P(<=X)	P(<X)	P(>X)	P(>=X)
11			0	=BINOMDIST (B11,$B$3, $B$4,FALSE)	=BINOMDIST (B11,$B$3, $B$4,TRUE)	=D11-C11	=1-D11	=1-E11
12			1	=BINOMDIST (B12,$B$3, $B$4,FALSE)	=BINOMDIST (B12,$B$3, $B$4,TRUE)	=D12-C12	=1-D12	=1-E12
13			2	=BINOMDIST (B13,$B$3, $B$4,FALSE)	=BINOMDIST (B13,$B$3, $B$4,TRUE)	=D13-C13	=1-D13	=1-E13
14			3	=BINOMDIST (B14,$B$3, $B$4,FALSE)	=BINOMDIST (B14,$B$3, $B$4,TRUE)	=D14-C14	=1-D14	=1-E14
15			4	=BINOMDIST (B15,$B$3, $B$4,FALSE)	=BINOMDIST (B15,$B$3, $B$4,TRUE)	=D15-C15	=1-D15	=1-E15

➊ Select File | New to open a new workbook (or open the existing workbook into which the binomial probabilities worksheet is to be inserted).

➋ Select an unused worksheet (or select Insert | Worksheet if there are none) and rename the worksheet Binomial.

➌ Enter the title and headings for column A and row 10 as shown in Table 4E.2.

➍ Enter 4 as the sample size in cell B3 and enter .6 as the probability of success in cell B4. Enter the formulas for the mean, variance, and standard deviation in cells B5, B6, and B7, respectively.

➎ Enter the X values 0 through 4 in the cell range B11:B15.

➏ Enter the columns C through G formulas for row 11. Enter the formulas for cells C11 and D11 as one continuous line using the mix of relative and absolute addresses as shown in the design table. (These formulas in the table have been typeset as three lines.)

➐ Copy the row 11 formulas down through row 15. Verify that the formulas copied to rows 12 through 15 are as shown in the design table.

The completed binomial probabilities worksheet will be similar to the one shown in Figure 4.4.

◆ B. Generating a Histogram of Binomial Probabilities   Once implemented, the binomial probabilities worksheet can serve as the source data for a histogram. To generate this chart for the natural gas service call problem of section 4.5, do the following:

➊ Click the Binomial sheet tab.

➋ Select Insert | Chart.

➌ In the Step 1 dialog box:
   a. Select the Standard Types tab and then select Column from the Chart type: list box. Select the first choice in the first row of choices under the Chart sub-type: heading. When this choice is selected, the phrase "Clustered Column." is displayed in the description box below the sub-types.
   b. Click the Next button.

➍ In the Step 2 dialog box:
   a. Select the Data Range tab. Enter C11:C15 in the Data Range: edit box and select the Columns option button in the Series in: group.
   b. Select the Series tab. Enter =Binomial!B11:B15 in the Category (X) axis labels: edit box. Note that this entry must include the leading equals sign (=).
   c. Click the Next button.

➎ In the Step 3 dialog box:
   a. Select the Titles tab. Enter Binomial Probabilities for Service Calls in the Chart title: edit box, enter Number of Successes (X) in Category (X) axis: edit box, and enter P(X) in the Value (Y) axis: edit box.

b. Select, in turn, the Axes, Gridlines, Legend, Data Labels, and Data Table tabs and verify that their settings match those given in Table 2E.3 on page 88.

c. Click the Next button.

6 In the Step 4 dialog box:

a. Select the As new sheet: option button and enter Binomial Histogram in the edit box to the right of the option button.

b. Click the Finish button to create the histogram. (The floating Chart toolbar that may appear after the chart is created can be closed or dragged to the side of the application window.)

Because this distribution is for a discrete random variable, the bars on the histograms should be drawn as spikes. This can be approximated in Excel by narrowing the bars. To make this change, do the following:

7 Right-click on one of the histogram bars. (The mouse pointer is over a bar when a tool tip that includes the word "Series'" is displayed.)

8 Select Format Data Series from the shortcut menu.

9 In the Format Data Series dialog box, click the Options tab and change the value in the Gap width: edit box to 500. Click OK.

The bars of the histogram are narrowed and the histogram becomes similar to the one shown in Figure 4.5 on page 249.

Problems for Section 4.5

Learning the Basics

• **4.39** Determine the following:
  (a) If $n = 4$ and $p = .12$, then what is $P(X = 0)$?
  (b) If $n = 10$ and $p = .40$, then what is $P(X = 9)$?
  (c) If $n = 10$ and $p = .50$, then what is $P(X = 8)$?
  (d) If $n = 6$ and $p = .83$, then what is $P(X = 5)$?
  (e) If $n = 10$ and $p = .90$, then what is $P(X = 9)$?

4.40 Determine the mean and standard deviation of the random variable X in each of the following binomial distributions:
  (a) If $n = 4$ and $p = .10$
  (b) If $n = 4$ and $p = .40$
  (c) If $n = 5$ and $p = .80$
  (d) If $n = 3$ and $p = .50$

Applying the Concepts

4.41 Suppose that the increase or decrease in the price of a stock between the beginning and the end of a trading day is considered to be an equally likely random event. What is the probability that a stock will show an increase in its closing price on five consecutive days?

4.42 Suppose that warranty records show the probability that a new car needs a warranty repair in the first 90 days is .05. If a sample of three new cars is selected,

(a) what is the probability that
(1) none needs a warranty repair?
(2) at least one needs a warranty repair?
(3) more than one needs a warranty repair?
(b) What assumptions are necessary in (a)?
(c) What are the mean and the standard deviation of the probability distribution in (a)?
(d) What would be your answers to (a)–(c) if the probability of needing a warranty repair was .10?

 **What If?**

•**4.43** Suppose that the likelihood that someone who logs onto a particular site in a "shopping mall" on the World Wide Web will purchase an item is .20. If the site has 10 people accessing it in the next minute, what is the probability that

(a) none of the individuals will purchase an item?
(b) exactly 2 individuals will purchase an item?
(c) at least 2 individuals will purchase an item?
(d) at most 2 individuals will purchase an item?
(e) If 20 people accessed the site in the next minute, what would be your answers to (a)–(d)?
(f) If the probability of purchasing an item was only .10, what would be your answers to (a)–(d)?

 **What If?**

•**4.44** An important part of the customer service responsibilities of a telephone company relate to the speed with which troubles in residential service can be repaired. Suppose past data indicate that the likelihood is .70 that troubles in residential service can be repaired on the same day.

(a) For the first five troubles reported on a given day, what is the probability that
(1) all five will be repaired on the same day?
(2) at least three will be repaired on the same day?
(3) fewer than two will be repaired on the same day?
(b) What assumptions are necessary in (a)?
(c) What are the mean and the standard deviation of the probability distribution in (a)?

 **What If?**

(d) What would be your answers in (a) and (c) if the probability is .80 that troubles in residential service can be repaired on the same day?
(e) Compare the results of (a) and (d).

4.45 Suppose that a student is taking a multiple-choice exam in which each question has four choices. Assuming that she has no knowledge of the correct answers to any of the questions, she has decided on a strategy in which she will place four balls (marked A, B, C, and D) into a box. She randomly selects one ball for each question and replaces the ball in the box. The marking on the ball will determine her answer to the question.

(a) If there are five multiple-choice questions on the exam, what is the probability that she will get
(1) five questions correct?
(2) at least four questions correct?
(3) no questions correct?
(4) no more than two questions correct?
(b) What assumptions are necessary in (a)?
(c) What are the average and the standard deviation of the number of questions that she will get correct in (a)?
(d) Suppose that the exam has 50 multiple-choice questions and 30 or more correct answers is a passing score. What is the probability that she will pass the exam by following her strategy? (Use Microsoft Excel to compute this probability.)

4.6 ▸ POISSON DISTRIBUTION

The **Poisson distribution** is another discrete probability distribution with many important practical applications. Numerous discrete phenomena are represented by a Poisson process as described in Exhibit 4.2.

> **Exhibit 4.2 A Poisson Process**
>
> A **Poisson process** is said to exist if we can observe discrete events in an *area of opportunity*—a continuous interval (of time, length, surface area, etc.)—in such a manner that if we shorten the area of opportunity or interval sufficiently,
>
> ✓ **1.** the probability of observing exactly one success in the interval is stable.
>
> ✓ **2.** the probability of observing more than one success in the interval is 0.
>
> ✓ **3.** the occurrence of a success in any one interval is statistically independent of that in any other interval.

To better understand the Poisson process, suppose we examine the number of customers arriving during the 12 noon to 1 P.M. lunch hour at a bank located in the central business district in a large city. Any arrival of a customer is a *discrete* event at a particular point in time over the *continuous* 1-hour interval. Over such an interval of time, there might be an average of 180 arrivals. Now if we were to break up the 1-hour interval into 3,600 consecutive 1-second intervals,

- the expected (or average) number of customers arriving in any 1-second interval would be .05.

- the probability of having more than one customer arriving in any 1-second interval approaches 0.

- the arrival of one customer in any 1-second interval has no effect on (i.e., is statistically independent of) the arrival of any other customer in any other 1-second interval.

The Poisson distribution has one parameter, called λ (the Greek lowercase letter *lambda*), which is the average or expected number of successes per unit. Interestingly, the variance of a Poisson distribution is also equal to λ and the standard deviation is equal to $\sqrt{\lambda}$. Moreover, the number of successes X of the Poisson random variable ranges from 0 to ∞.

The mathematical expression for the Poisson distribution for obtaining X successes, given that λ successes are expected, is

Poisson Distribution

$$P(X) = \frac{e^{-\lambda}\lambda^X}{X!} \qquad (4.18)$$

where

$P(X)$ = the probability of X successes given a knowledge of λ

λ = expected number of successes

e = mathematical constant approximated by 2.71828

X = number of successes per unit

To demonstrate Poisson model applications, let us return to the example of the bank lunch-hour customer arrivals:

Example 4.18 *Determining Poisson Probabilities*

If, on average, three customers arrive per minute at the bank during the lunch hour, what is the probability that in a given minute exactly two customers will arrive? Also, what is the chance that more than two customers will arrive in a given minute?

SOLUTION

Using equation (4.18), we have, for the first question

$$P(X = 2) = \frac{e^{-3.0}(3.0)^2}{2!} = \frac{9}{(2.71828)^3(2)} = .2240$$

To answer the second question—the probability that in any given minute more than two customers will arrive—we have

$$P(X > 2) = P(X = 3) + P(X = 4) + \cdots + P(X = \infty)$$

Because all the probabilities in a probability distribution must sum to 1, the terms on the right side of the equation $P(X > 2)$ also represent the complement of the probability that X is less than or equal to 2 [i.e., $1 - P(X \le 2)$]. Thus,

$$P(X > 2) = 1 - P(X \le 2) = 1 - [P(X = 0) + P(X = 1) + P(X = 2)]$$

Now, using equation (4.18), we have

$$P(X > 2) = 1 - \left[\frac{e^{-3.0}(3.0)^0}{0!} + \frac{e^{-3.0}(3.0)^1}{1!} + \frac{e^{-3.0}(3.0)^2}{2!} \right]$$

$$= 1 - [.0498 + .1494 + .2240]$$

$$= 1 - .4232 = .5768$$

Thus, we see that there is roughly a 42.3% chance that two or fewer customers will arrive at the bank per minute. Therefore, a 57.7% chance exists that three or more customers will arrive.

Such computations can become quite tedious, especially as λ gets large. However, we can obtain the probabilities by using Microsoft Excel (see section 4.6E) and thereby avoid any computational drudgery. Figure 4.6 represents output from Microsoft Excel for the Poisson distribution example with $\lambda = 3$.

FIGURE 4.6

Poisson distribution calculations for $\lambda = 3$ obtained from the PHStat add-in for Microsoft Excel

	X	P(X)	P(<=X)	P(<X)	P(>X)	P(>=X)
Poisson Probabilities for Customer Arrivals						
Average/Expected number of successes			3			
Poisson Probabilities Table						
	0	0.049787	0.049787	0.000000	0.950213	1.000000
	1	0.149361	0.199148	0.049787	0.800852	0.950213
	2	0.224042	0.423190	0.199148	0.576810	0.800852
	3	0.224042	0.647232	0.423190	0.352768	0.576810
	4	0.168031	0.815263	0.647232	0.184737	0.352768
	5	0.100819	0.916082	0.815263	0.083918	0.184737
	6	0.050409	0.966491	0.916082	0.033509	0.083918
	7	0.021604	0.988095	0.966491	0.011905	0.033509
	8	0.008102	0.996197	0.988095	0.003803	0.011905
	9	0.002701	0.998898	0.996197	0.001102	0.003803
	10	0.000810	0.999708	0.998898	0.000292	0.001102
	11	0.000221	0.999929	0.999708	0.000071	0.000292
	12	0.000055	0.999984	0.999929	0.000016	0.000071
	13	0.000013	0.999997	0.999984	0.000003	0.000016
	14	0.000003	0.999999	0.999997	0.000001	0.000003
	15	0.000001	1.000000	0.999999	0.000000	0.000001

4.6E CALCULATING POISSON PROBABILITIES USING MICROSOFT EXCEL

Overview

◆ *For Quick Results Users* Use the PHStat add-in to generate a worksheet and histogram that displays Poisson probabilities.

◆ *For Developers* Use the following two-step process to calculate and graph Poisson probabilities:

A. Implement a worksheet that uses the POISSON worksheet function to calculate Poisson probabilities.

B. Use the Chart Wizard to generate a histogram based on this worksheet.

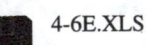 4-6E.XLS

The 4-6E.XLS workbook file contains the Poisson probabilities and a histogram for the bank customer arrivals during the lunch-hour problem discussed in section 4.6.

Quick Results Details

To calculate Poisson probabilities, use the Probability Distributions | Poisson choice of the PHStat add-in. As an example, consider the bank customer arrivals during the lunch-hour problem discussed in section 4.6. To calculate the Poisson probabilities for this problem, do the following:

❶ If the PHStat add-in has not been previously loaded, load the add-in using the instructions of section S4.2.

❷ Select File | New to open a new workbook (or open the existing workbook into which the Poisson probabilities worksheet is to be inserted).

❸ Select PHStat | Probability Distributions | Poisson.

❹ In the Poisson Probability Distribution dialog box (see Figure 4E.3):
   a. Enter 3 in the Average/Expected No. of Successes: edit box.
   b. Enter Poisson Probabilities for Customer Arrivals in the Output Title: edit box.
   c. Select the Cumulative Probabilities and Histogram check boxes.
   d. Click the OK button.

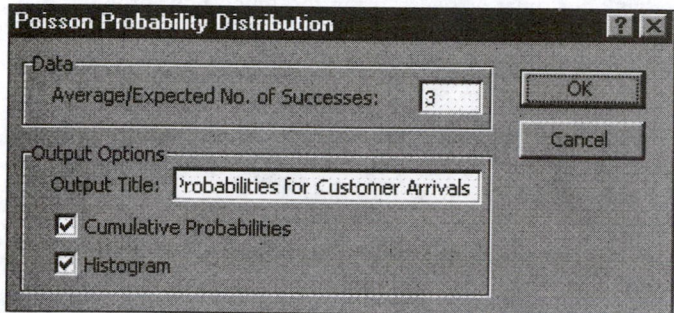

FIGURE 4E.3
PHStat Poisson Probability Distribution dialog box

On separate sheets, the add-in produces a table of Poisson probabilities similar to the one shown in Figure 4.6, and a histogram similar to the one shown in Figure 4E.4.

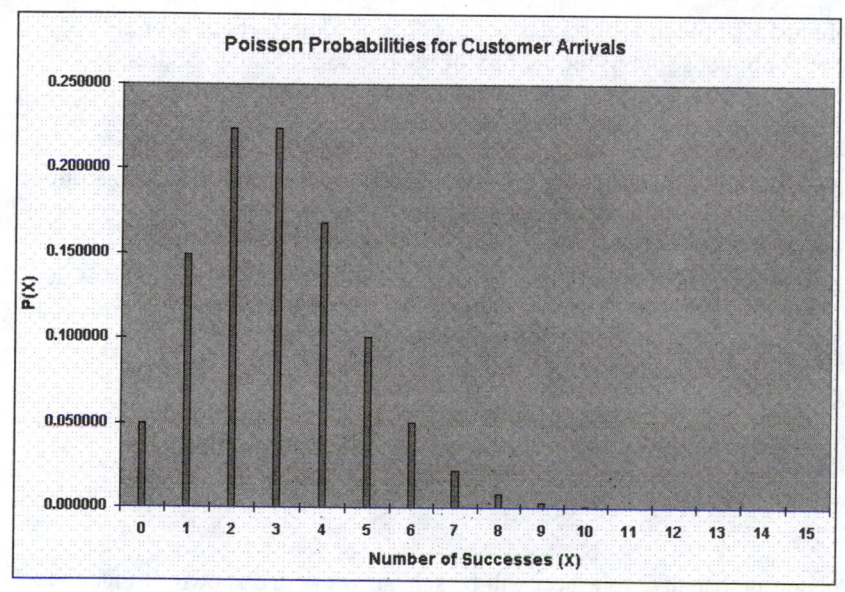

FIGURE 4E.4
Poisson probabilities histogram obtained from the PHStat add-in for Microsoft Excel

Developer Details

◆ *A. Implementing a Worksheet to Calculate Poisson Probabilities*   We can use the POISSON function as the basis for computing Poisson probabilities in an Excel worksheet. The format of this function is:

$$\text{POISSON}(X, \textit{lambda}, \textit{cumulative})$$

where

X = the number of successes

lambda = the average or expected number of successes

cumulative = a True or False value in which True causes the function to compute the probability of X or fewer successes and False, the probability of exactly X successes

Table 4E.3 presents a Poisson sheet design that uses the POISSON function to calculate the Poisson probabilities for the bank customer arrivals during the lunch-hour problem discussed in section 4.6. In this design the number of successes was arbitrarily limited to 15, a reasonable value when the average or expected number of successes is 3. To implement this design, do the following:

❶ Select File | New to open a new workbook (or open the existing workbook into which the Poisson probabilities worksheet is to be inserted).

❷ Select an unused worksheet (or select Insert | Worksheet if there are none) and rename the worksheet Poisson.

❸ Enter the title and headings for column A and row 6 as shown in Table 4E.3.

❹ Enter 3 as the average/expected number of successes value in cell E3.

❺ Enter the X values 0 through 15 in the cell range B7:B22.

❻ Enter the columns C through G formulas for row 7. Enter the formulas for cells C7 and D7 as one continuous line using the mix of relative and absolute addresses as shown in the design table. (These formulas in the table have been typeset as two lines.)

❼ Copy the row 7 formulas down through row 22. Verify that the formulas copied to rows 8 through 22 are as shown in the design table.

The table of numbers produced is hard to read because the values variously display as decimal fractions, numbers in scientific notation, or the integers 0 and 1. To standardize the formatting to enhance the readability of the table, do the following:

❽ Select the cell range C7:G22.

❾ Click the Decrease Decimal button on the Formatting toolbar and then immediately click the Increase Decimal button. Adjust the widths of columns C through G as necessary.

The formatted Poisson probability worksheet will be similar to the one shown in Figure 4.6.

Table 4E.3 Poisson sheet design for the customer arrival problem of section 4.6

	A	B	C	D	E	F	G
1	Poisson Probabilities for Customer Arrivals						
2							
3			Average expected number of successes:		xx		
4							
5	Poisson Probabilities Table						
6		X	P(X)	P(< = X)	P(<X)	P(>X)	P(> = X)
7		0	=POISSON (B7,$E$3,FALSE)	=POISSON (B7,$E$3,TRUE)	=D7-C7	=1-D7	=1-E7
8		1	=POISSON (B8,$E$3,FALSE)	=POISSON (B8,$E$3,TRUE)	=D8-C8	=1-D8	=1-E8
9		2	=POISSON (B9,$E$3,FALSE)	=POISSON (B9,$E$3,TRUE)	=D9-C9	=1-D9	=1-E9
10		3	=POISSON (B10,$E$3,FALSE)	=POISSON (B10,$E$3,TRUE)	=D10-C10	=1-D10	=1-E10
11		4	=POISSON (B11,$E$3,FALSE)	=POISSON (B11,$E$3,TRUE)	=D11-C11	=1-D11	=1-E11
12		5	=POISSON (B12,$E$3,FALSE)	=POISSON (B12,$E$3,TRUE)	=D12-C12	=1-D12	=1-E12
13		6	=POISSON (B13,$E$3,FALSE)	=POISSON (B13,$E$3,TRUE)	=D13-C13	=1-D13	=1-E13
14		7	=POISSON (B14,$E$3,FALSE)	=POISSON (B14,$E$3,TRUE)	=D14-C14	=1-D14	=1-E14
15		8	=POISSON (B15,$E$3,FALSE)	=POISSON (B15,$E$3,TRUE)	=D15-C15	=1-D15	=1-E15
16		9	=POISSON (B16,$E$3,FALSE)	=POISSON (B16,$E$3,TRUE)	=D16-C16	=1-D16	=1-E16
17		10	=POISSON (B17,$E$3,FALSE)	=POISSON (B17,$E$3,TRUE)	=D17-C17	=1-D17	=1-E17
18		11	=POISSON (B18,$E$3,FALSE)	=POISSON (B18,$E$3,TRUE)	=D18-C18	=1-D18	=1-E18
19		12	=POISSON (B19,$E$3,FALSE)	=POISSON (B19,$E$3,TRUE)	=D19-C19	=1-D19	=1-E19
20		13	=POISSON (B20,$E$3,FALSE)	=POISSON (B20,$E$3,TRUE)	=D20-C20	=1-D20	=1-E20
21		14	=POISSON (B21,$E$3,FALSE)	=POISSON (B21,$E$3,TRUE)	=D21-C21	=1-D21	=1-E21
22		15	=POISSON (B22,$E$3,FALSE)	=POISSON (B22,$E$3,TRUE)	=D22-C22	=1-D22	=1-E22

◆ *B. Generating a Histogram of Poisson Probabilities*   Once implemented, the Poisson probabilities worksheet can serve as the source data for a histogram. To generate this chart for the bank customer arrival problem of section 4.6, do the following:

❶ Click the Poisson sheet tab.

❷ Select Insert | Chart.

③ In the Step 1 dialog box:
  a. Select the Standard Types tab and then select Column from the Chart type: list box. Select the first choice in the first row of choices under the Chart sub-type: heading. When this choice is selected, the phrase "Clustered Column." is displayed in the description box below the sub-types.
  b. Click the Next button.

④ In the Step 2 dialog box:
  a. Select the Data Range tab. Enter C7:C22 in the Data range: edit box and select the Columns option button in the Series in: group.
  b. Select the Series tab. Enter =Poisson!B7:B22 in the Category (X) axis labels: edit box. Note that this entry must include the leading equals sign (=).
  c. Click the Next button.

⑤ In the Step 3 dialog box:
  a. Select the Titles tab. Enter Poisson Probabilities for Bank Customer Arrivals in the Chart title: edit box, enter Number of Successes (X) in Category (X) axis: edit box, and enter P(X) in the Value (Y) axis: edit box.
  b. Select, in turn, the Axes, Gridlines, Legend, Data Labels, and Data Table tabs and verify that their settings match those given in Table 2E.3 on page 88.
  c. Click the Next button.

⑥ In the Step 4 dialog box:
  a. Select the As new sheet: option button and enter Poisson Histogram in the edit box to the right of the option button.
  b. Click the Finish button to create the histogram. (The floating Chart toolbar that may appear after the chart is created can be closed or dragged to the side of the application window.)

Because this distribution is for a discrete random variable, the bars on the histograms should be drawn as spikes. This can be approximated in Excel by narrowing the bars. To make this change, do the following:

⑦ Right-click on one of the histogram bars. (The mouse pointer is over a bar when a tool tip that includes the word "Series'" is displayed.)

⑧ Select Format Data Series from the shortcut menu.

⑨ In the Format Data Series dialog box, click the Options tab and change the value in the Gap width: edit box to 500. Click OK.

The bars of the histogram are narrowed and the histogram becomes similar to the one shown in Figure 4E.4 on page 259.

Problems for Section 4.6

Learning the Basics

• **4.46** Determine the following:
  (a) If $\lambda = 2.5$, then what is $P(X = 2)$?
  (b) If $\lambda = 8.0$, then what is $P(X = 8)$?
  (c) If $\lambda = 0.5$, then what is $P(X = 1)$?
  (d) If $\lambda = 3.7$, then what is $P(X = 0)$?
  (e) If $\lambda = 4.4$, then what is $P(X = 7)$?

4.47 Determine the following:

(a) If $\lambda = 2.0$, then what is $P(X \geq 2)$?

(b) If $\lambda = 8.0$, then what is $P(X \geq 3)$?

(c) If $\lambda = 0.5$, then what is $P(X \leq 1)$?

(d) If $\lambda = 4.0$, then what is $P(X \geq 1)$?

(e) If $\lambda = 5.0$, then what is $P(X \leq 3)$?

Applying the Concepts

● **4.48** The average number of claims per hour made to the C-G-N Insurance Company for damages or losses incurred in moving is 3.1. What is the probability that in any given hour

(a) fewer than three claims will be made?

(b) exactly three claims will be made?

(c) three or more claims will be made?

(d) more than three claims will be made?

4.49 Based on past records, the average number of two-car accidents in a New York City police precinct is 3.4 per day. What is the probability that there will be

(a) at least six such accidents in this precinct on any given day?

(b) not more than two such accidents in this precinct on any given day?

(c) fewer than two such accidents in this precinct on any given day?

(d) at least two but no more than six such accidents in this precinct on any given day?

(e) What would be your answers to (a)–(d) if the average is five such accidents per day?

 What If?

● **4.50** The quality control manager of Marilyn's Cookies is inspecting a batch of chocolate-chip cookies that has just been baked. If the production process is in control, the average number of chip parts per cookie is 6.0. What is the probability that in any particular cookie being inspected

(a) fewer than five chip parts will be found?

(b) exactly five chip parts will be found?

(c) five or more chip parts will be found?

(d) four or five chip parts will be found?

(e) What would be your answers to (a)–(d) if the average number of chip parts per cookie is 5.0?

 What If?

4.51 Refer to Problem 4.50. How many cookies in a batch of 100 being sampled should the manager expect to discard if company policy requires that all chocolate-chip cookies sold must have at least four chocolate-chip parts?

4.52 Suppose that the number of claims for missing baggage for a well-known airline in a small city averages nine per day. What is the probability that, on a given day, there will be

(a) seven claims?

(b) seven or eight or nine claims?

(c) fewer than five claims?

4.53 Based on past experience, it is assumed that the number of flaws per foot in rolls of grade 2 paper follows a Poisson distribution with an average of 1 flaw per 5 feet of paper (.2 flaw per foot). What is the probability that in a

(a) 1-foot roll there will be at least 2 flaws?

(b) 12-foot roll there will be at least 1 flaw?

(c) 50-foot roll there will be between 5 and 15 (inclusive) flaws?

4.7 HYPERGEOMETRIC DISTRIBUTION (*OPTIONAL TOPIC*)

Both the binomial distribution and the **hypergeometric distribution** are concerned with the same thing—the number of successes in a sample containing *n* observations. What differentiates these two discrete probability distributions is the manner in which the data are obtained. For the binomial model, the sample data are drawn *with* replacement from a *finite* population or *without* replacement from an *infinite* population. On the other hand, for the hypergeometric model, the sample data are drawn *without* replacement from a *finite* population. Hence, although the probability of success *p* is constant over all observations of a binomial experiment, and the outcome of any particular observation is independent of any other, the same cannot be said for a hypergeometric experiment; here the outcome of one observation is affected by the outcomes of the previous observations.

In general, a mathematical expression of the hypergeometric distribution for obtaining X successes, given a knowledge of the parameters n, N, and A, is

Hypergeometric Distribution

$$P(X) = \frac{\binom{A}{X}\binom{N-A}{n-X}}{\binom{N}{n}}$$ (4.19)

where

$P(X)$ = the probability of X successes given a knowledge of n, N, and A

n = sample size

N = population size

A = number of successes in the population

$N - A$ = number of failures in the population

X = number of successes in the sample

The number of successes in the sample X cannot exceed the number of successes in the population A or the sample size n. Thus, the range of the hypergeometric random variable is limited to the sample size (as was the range for the binomial random variable) or to the number of successes in the population—whichever is smaller.

◆ *The Mean*   Like the binomial distribution, the mean of the hypergeometric distribution can be computed from

The Mean of the Hypergeometric Distribution

$$\mu = E(X) = \frac{nA}{N}$$ (4.20)

◆ *The Standard Deviation*   The standard deviation of the hypergeometric distribution is obtained from equation (4.21).

The Standard Deviation of the Hypergeometric Distribution

$$\sigma = \sqrt{\frac{nA(N-A)}{N^2}} \times \sqrt{\frac{N-n}{N-1}}$$ (4.21)

The expression $\sqrt{\frac{N-n}{N-1}}$ is a **finite population correction factor,** which arises because of the process of sampling without replacement from finite populations. (This correction factor will be discussed in greater detail in section 6.6.) To illustrate the hypergeometric distribution, let us consider Example 4.19.

Example 4.19 *Determining Probabilities from the Hypergeometric Distribution*

Suppose an investment company employs 52 researchers with Ph.D. degrees—13 of whom have their doctorates in finance. A delegation of five researchers is to be randomly chosen by a lottery drawing to attend an international investments conference, all expenses paid. What is the probability that the delegation will contain exactly two researchers with doctorates in finance?

SOLUTION

Here the population of $N = 52$ researchers is finite. In addition, $A = 13$ of these researchers hold doctorates in finance. The delegation to the international conference is to contain $n = 5$ members. Using equation (4.19), we have

$$P(X = 2) = \frac{\binom{13}{2}\binom{39}{3}}{\binom{52}{5}}$$

$$= \frac{\frac{13!}{2!11!} \times \frac{39!}{3!36!}}{\frac{52!}{5!47!}}$$

$$= .2743$$

Thus, the probability that the delegation contains exactly two doctorates in finance is .2743.

Such computations can become quite tedious, especially as N gets large. However, we can obtain the probabilities by using Microsoft Excel (see section 4.7E) and thereby avoid any computational drudgery. Figure 4.7 represents output from Microsoft Excel for the hypergeometric distribution of Example 4.19.

	A	B	C
1	Hypergeometric Probabilities		
2			
3	Sample size	5	
4	No. of successes in population	13	
5	Population size	52	
6			
7	Hypergeometric Probabilities Table		
8		X	P(x)
9		0	0.221534
10		1	0.41142
11		2	0.27428
12		3	0.081543
13		4	0.010729
14		5	0.000495

FIGURE 4.7

Hypergeometric distribution calculations for Example 4.19 obtained from the PHStat add-in for Microsoft Excel

Overview

◆ *For Quick Results Users*   Use the PHStat add-in to generate a worksheet and histogram that displays hypergeometric probabilities.

◆ *For Developers*   Use the following two-step process to tabulate and graph categorical data:

A. Implement a worksheet that uses the HYPGEOMDIST worksheet function to calculate hypergeometric probabilities.

B. Use the Chart Wizard to generate a histogram based on this worksheet.

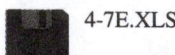 4-7E.XLS   The 4-7E.XLS workbook file contains the hypergeometric probabilities and a histogram for the researcher selection problem discussed in section 4.7.

Quick Results Details

To calculate hypergeometric probabilities, use the Probability Distributions | Hypergeometric choice of the PHStat add-in. As an example, consider the researcher selection problem discussed in section 4.7. To calculate the hypergeometric probabilities for this problem, do the following:

❶ If the PHStat add-in has not been previously loaded, load the add-in using the instructions of section S4.2.

❷ Select File | New to open a new workbook (or open the existing workbook into which the hypergeometric probabilities worksheet is to be inserted).

❸ Select PHStat | Probability Distributions | Hypergeometric.

❹ In the Hypergeometric Probability Distribution dialog box (see Figure 4E.5):
a. Enter 5 in the Sample Size: edit box.
b. Enter 13 in the No. of Successes in Population: edit box.
c. Enter 52 in the Population Size: edit box.
d. Enter Hypergeometric Probabilities for Researcher Selection in the Output Title: edit box.
e. Select the Histogram check box.

FIGURE 4E.5
PHStat Hypergeometric Probability Distribution dialog box

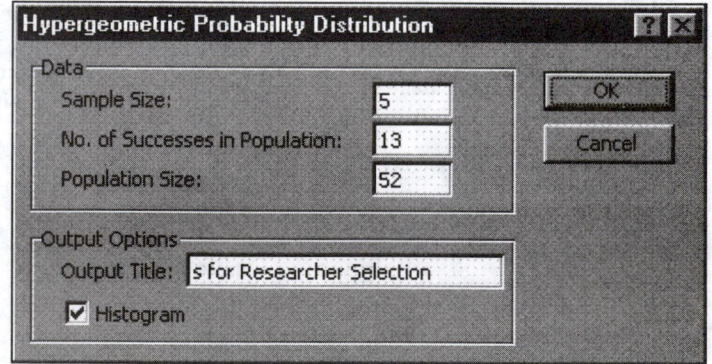

f. Click the OK button.

On separate sheets, the add-in produces a table of hypergeometric probabilities and a histogram similar to those shown in Figures 4.7 and 4E.6, respectively.

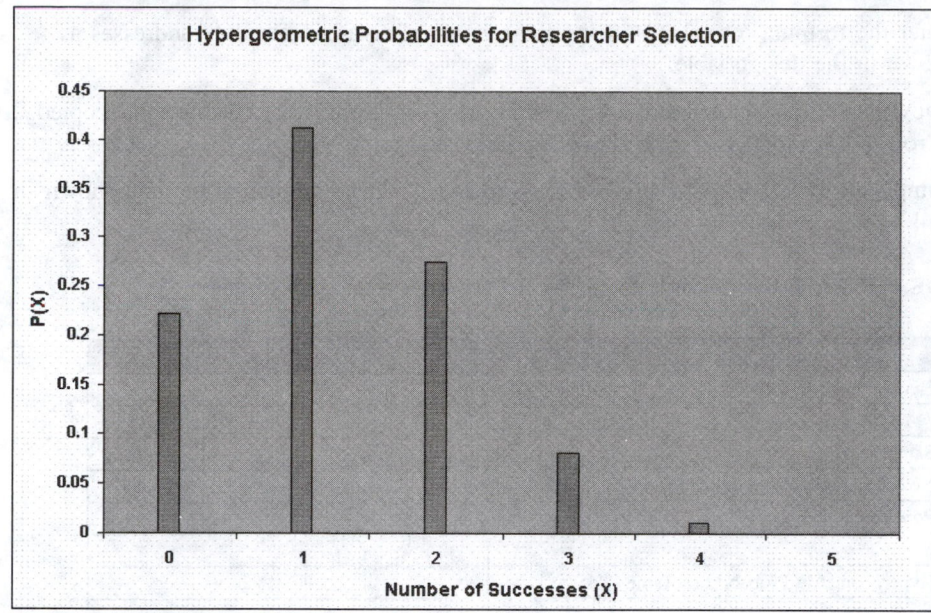

FIGURE 4E.6

Hypergeometric probabilities histogram obtained from the PHStat add-in for Microsoft Excel

Developer Details

◆ **A. Implementing a Worksheet to Calculate Hypergeometric Probabilities** We can use the HYPGEOMDIST function as the basis for computing hypergeometric probabilities in an Excel worksheet. The format of this function is:

$$HYPGEOMDIST(X, n, A, N)$$

where

X = the number of successes

n = the sample size

A = the number of successes in the population

N = the population size

Table 4E.4 presents a Hypergeometric sheet design that uses the HYPGEOMDIST function to calculate the hypergeometric probabilities for the researcher selection problem discussed in section 4.7. Because the sample size value determines the length of the hypergeometric probabilities table, this value cannot be interactively changed by the user and therefore is shown as 5—and not as the generalized value xx—in the design. To implement this design, do the following:

❶ Select File | New to open a new workbook (or open the existing workbook into which the hypergeometric probabilities worksheet is to be inserted).

❷ Select an unused worksheet (or select Insert | Worksheet if there are none) and rename the worksheet Hypergeometric.

❸ Enter the title and headings for column A and row 8 as shown in Table 4E.4.

4 Enter 5 as the sample size value in cell B3, enter 13 as the number of successes in population value in cell B4, and enter 52 as the population size in cell B5.

5 Enter the *X* values 0 through 5 in the cell range B9:B14.

6 Enter the formula for cell C9, using the mix of relative and absolute addresses as shown in the design table.

7 Copy the formula in cell C9 down through row 14. Verify that the formulas copied to rows 10 through 14 are as shown in the design table.

The completed hypergeometric probability worksheet will be similar to the one shown in Figure 4.7 on page 265.

Table 4E.4 Hypergeometric sheet design for the researcher selection problem of section 4.7

	A	B	C
1	Hypergeometric Probabilities for Researcher Selection		
2			
3	Sample size	5	
4	No. of successes in population	xx	
5	Population size	xx	
6			
7	Hypergeometric Probabilities Table		
8		X	P(X)
9		0	=HYPGEOMDIST(B9,$B$3,$B$4,$B$5)
10		1	=HYPGEOMDIST(B10,$B$3,$B$4,$B$5)
11		2	=HYPGEOMDIST(B11,$B$3,$B$4,$B$5)
12		3	=HYPGEOMDIST(B12,$B$3,$B$4,$B$5)
13		4	=HYPGEOMDIST(B13,$B$3,$B$4,$B$5)
14		5	=HYPGEOMDIST(B14,$B$3,$B$4,$B$5)

◆ *B. Generating a Histogram of Hypergeometric Probabilities*   Once implemented, the hypergeometric probabilities worksheet can serve as the source data for a histogram. To generate this chart for the researcher selection problem discussed in section 4.7, do the following:

1 Click the Hypergeometric sheet tab.

2 Select Insert | Chart.

3 In the Step 1 dialog box:
   a. Select the Standard Types tab and then select Column from the Chart type: list box. Select the first choice in the first row of choices under the Chart sub-type: heading. When this choice is selected, the phrase "Clustered Column." is displayed in the description box below the sub-types.
   b. Click the Next button.

4 In the Step 2 dialog box:
   a. Select the Data Range tab. Enter C9:C14 in the Data range: edit box and select the Columns option button in the Series in: group.

b. Select the Series tab. Enter =Hypergeometric!B9:B14 in the Category (X) axis labels: edit box. Note that this entry must include the leading equals sign (=).

c. Click the Next button.

⑤ In the Step 3 dialog box:

a. Select the Titles tab. Enter Hypergeometric Probabilities for Researcher Selection in the Chart title: edit box, enter Number of Successes (X) in Category (X) axis: edit box, and enter P(X) in the Value (Y) axis: edit box.

b. Select, in turn, the Axes, Gridlines, Legend, Data Labels, and Data Table tabs and verify that their settings match those given in Table 2E.3 on page 88.

c. Click the Next button.

⑥ In the Step 4 dialog box:

a. Select the As new sheet: option button and enter Hypergeometric Histogram in the edit box to the right of the option button.

b. Click the Finish button to create the histogram. (The floating Chart toolbar that may appear after the chart is created can be closed or dragged to the side of the application window.)

Because this distribution is for a discrete random variable, the bars on the histograms should be drawn as spikes. This can be approximated in Excel by narrowing the bars. To make this change, do the following:

⑦ Right-click on one of the histogram bars. (The mouse pointer is over a bar when a tool tip that includes the word "Series'" is displayed.)

⑧ Select Format Data Series from the shortcut menu.

⑨ In the Format Data Series dialog box, click the Options tab and change the value in the Gap width: edit box to 500. Click OK.

The bars of the histogram are narrowed and the histogram becomes similar to the one shown in Figure 4E.6 on page 267.

Problems for Section 4.7

Learning the Basics

• **4.54** Determine the following:
   (a) If $n = 4$, $N = 10$, and $A = 5$, then find $P(X = 3)$.
   (b) If $n = 4$, $N = 6$, and $A = 3$, then find $P(X = 1)$.
   (c) If $n = 5$, $N = 12$, and $A = 3$, then find $P(X = 0)$.
   (d) If $n = 3$, $A = 3$, and $N = 10$, then find $P(X = 3)$.

4.55 Refer to Problem 4.54:
   (a) Compute the mean and standard deviation for the hypergeometric distribution described in (a).
   (b) Compute the mean and standard deviation for the hypergeometric distribution described in (b).
   (c) Compute the mean and standard deviation for the hypergeometric distribution described in (c).
   (d) Compute the mean and standard deviation for the hypergeometric distribution described in (d).

Applying the Concepts

•4.56 An auditor for the Internal Revenue Service is selecting a sample of six tax returns from persons in a particular profession for possible audit. If two or more of these indicate "improper" deductions, the entire group (population) of 100 tax returns will be audited.

 **What If?**

   (a) What is the probability that the entire group will be audited if the true number of improper returns in the population is
      (1) 25?   (2) 30?   (3) 5?   (4) 10?
   (b) Discuss the differences in your results depending on the true number of improper returns in the population.

4.57 The dean of a business school wishes to form an executive committee of five from among the 40 tenured faculty members at the school. The selection is to be random, and at the school there are eight tenured faculty members in accounting.

 **What If?**

   (a) What is the probability that the committee will contain
      (1) none of them?
      (2) at least one of them?
      (3) not more than one of them?
   (b) What would be your answers to (a) if the committee consisted of seven members?

4.58 From an inventory of 48 cars being shipped to local automobile dealers, suppose 12 have had defective radios installed.

 **What If?**

   (a) What is the probability that one particular dealership receiving eight cars
      (1) obtains all with defective radios?
      (2) obtains none with defective radios?
      (3) obtains at least one with a defective radio?
   (b) What would be your answers to (a) if six cars have had defective radios installed?

•4.59 A state lottery is conducted in which six winning numbers are selected from a total of 54 numbers. What is the probability that if six numbers are randomly selected,
   (a) all six numbers will be winning numbers?
   (b) five numbers will be winning numbers?
   (c) four numbers will be winning numbers?
   (d) three numbers will be winning numbers?
   (e) none of the numbers will be winning numbers?

 **What If?**

   (f) What would be your answers to (a)–(e) if the six winning numbers were selected from a total of 40 numbers?

4.60 In a shipment of 15 hard disks, five are defective. If four of the disks are inspected,
   (a) what is the probability that
      (1) exactly one is defective?
      (2) at least one is defective?
      (3) no more than two are defective?
   (b) What is the average number of defective hard disks that you would expect to find in the sample of four hard disk drives?

◆ 4.8   COVARIANCE AND ITS APPLICATION IN FINANCE (OPTIONAL TOPIC)

In section 4.4 we studied the expected value, variance, and standard deviation of a discrete random variable in a probability distribution. In this section we introduce the concept of the covariance between two variables and illustrate how it is applied in portfolio management in finance.

The Covariance

The **covariance** σ_{XY} between two discrete random variables X and Y may be defined as:

Covariance

$$\sigma_{XY} = \sum_{i=1}^{N} [X_i - E(X)][Y_i - E(Y)]P(X_iY_i) \qquad (4.22)$$

where

$$X = \text{discrete random variable } X$$
$$X_i = i\text{th outcome of } X$$
$$P(X_iY_i) = \text{probability of occurrence of the } i\text{th outcome of } X \text{ and the } i\text{th outcome of } Y$$
$$Y = \text{discrete random variable } Y$$
$$Y_i = i\text{th outcome of } Y$$
$$i = 1, 2, \ldots, N$$

We illustrate the covariance in Example 4.20.

Example 4.20 *Computing the Covariance*

Suppose that you are deciding between two alternative investments for the coming year. The first investment is a mutual fund whose portfolio consists of a combination of stocks that make up the Dow Jones Industrial Average. The second investment consists of shares of a growth stock. Suppose you estimate the following returns (per $1,000 investment) under three economic conditions, each with a given probability of occurrence.

		INVESTMENT	
$P(X_iY_i)$	ECONOMIC CONDITION	DOW JONES FUND	GROWTH STOCK
.2	Recession	−$100	−$200
.5	Stable economy	+ 100	+ 50
.3	Expanding economy	+ 250	+ 350

Compute the expected value and standard deviation for each investment and the covariance of the two investments.

SOLUTION

If X = Dow Jones fund and Y = growth stock

$$E(X) = \mu_X = (-100)(.2) + (100)(.5) + (250)(.3) = \$105$$
$$E(Y) = \mu_Y = (-200)(.2) + (50)(.5) + (350)(.3) = \$90$$

$$Var(X) = \sigma_X^2 = (.2)(-100 - 105)^2 + (.5)(100 - 105)^2 + (.3)(250 - 105)^2$$

$$\sigma_X^2 = 14{,}725$$

$$\sigma_X = 121.35$$

$$Var(Y) = \sigma_Y^2 = (.2)(-200 - 90)^2 + (.5)(50 - 90)^2 + (.3)(350 - 90)^2$$

$$\sigma_Y^2 = 37{,}900$$

$$\sigma_Y = 194.68$$

$$\sigma_{XY} = (.2)(-100 - 105)(-200 - 90) + (.5)(100 - 105)(50 - 90)$$
$$+ (.3)(250 - 105)(350 - 90)$$
$$= 11{,}890 + 100 + 11{,}310$$
$$= 23{,}300$$

Thus, the Dow Jones fund has a higher expected value or return than the growth fund and also has a lower standard deviation. The covariance of 23,300 between the two investments indicates a large positive relationship in which the two investments are covarying together in the same direction. When the return on one investment is increasing, the return on the other is also increasing.

The Expected Value, Variance, and Standard Deviation of the Sum of Two Random Variables

Having computed the covariance between two variables X and Y, we can now find the **expected value** and **standard deviation of the sum of two random variables.**

Expected Value of the Sum of Two Random Variables

The expected value of the sum of two random variables is equal to the sum of the expected values.

$$E(X + Y) = E(X) + E(Y) \tag{4.23}$$

Variance of the Sum of Two Random Variables

The variance of the sum of two random variables is equal to the sum of the variances plus twice the covariance.

$$Var(X + Y) = \sigma_{X+Y}^2 = \sigma_X^2 + \sigma_Y^2 + 2\sigma_{XY} \tag{4.24}$$

The standard deviation is just the square root of the variance.

Standard Deviation of the Sum of Two Random Variables

$$\sigma_{X+Y} = \sqrt{\sigma_{X+Y}^2} \tag{4.25}$$

The expected value, variance, and standard deviation of the sum of two random variables are illustrated in Example 4.21.

Example 4.21 *The Expected Value, Variance, and Standard Deviation of the Sum of Two Random Variables*

In Example 4.20 on page 271, we studied the expected value, standard deviation, and covariance of two different investments. Suppose that we want to determine the expected value, variance, and standard deviation of the sum of these two investments.

SOLUTION

If X = Dow Jones fund and Y = growth stock, using equations (4.23), (4.24), and (4.25),

$$E(X + Y) = E(X) + E(Y) = 105 + 90 = \$195$$
$$\sigma^2_{X+Y} = \sigma^2_X + \sigma^2_Y + 2\sigma_{XY} = 14{,}725 + 37{,}900 + (2)(23{,}300)$$
$$\sigma^2_{X+Y} = 99{,}225$$
$$\sigma_{X+Y} = \$315$$

The expected return of the sum of the Dow Jones fund and the growth stock is $195 with a standard deviation of $315.

Portfolio Expected Return and Portfolio Risk

Now that we have defined the covariance and the expected return and standard deviation of the sum of two random variables, we are ready to apply these concepts to the study of a group of assets referred to as a **portfolio.** By diversifying their investments, investors combine securities into portfolios to reduce their risk (see references 1 and 2). Often the objective is to maximize the return while minimizing the risk. For such portfolios, rather than studying the sum of two random variables, we weight each investment by the proportion of assets assigned to that investment. This enables us to compute the **portfolio expected return** and the **portfolio risk** as defined in equations (4.26) and (4.27).

Portfolio Expected Return

The portfolio expected return for a two-asset investment is equal to the weight assigned to asset X multiplied by the expected return of asset X plus the weight assigned to asset Y multiplied by the expected return of asset Y.

$$E(P) = wE(X) + (1 - w)E(Y) \tag{4.26}$$

where

$$E(P) = \text{portfolio expected return}$$
$$w = \text{the proportion of portfolio value assigned to asset } X$$
$$(1 - w) = \text{the proportion of portfolio value assigned to asset } Y$$
$$E(X) = \text{expected return of asset } X$$
$$E(Y) = \text{expected return of asset } Y$$

Portfolio Risk

$$\sigma_p = \sqrt{[w^2\sigma_X^2 + (1 - w)^2\sigma_Y^2 + 2w(1 - w)\sigma_{XY}]} \qquad (4.27)$$

The application of equations (4.26) and (4.27) to the evaluation of a portfolio is presented in Example 4.22.

Example 4.22 *Determining Portfolio Expected Return and Portfolio Risk*

In Example 4.20 we evaluated the expected return of two different investments, a Dow Jones indexed fund and a growth stock, and also computed the standard deviation of the return of each investment and the covariance of the two investments. Now suppose that we wish to form a portfolio of these two investments that consists of an equal investment in each of these two assets. Compute the portfolio expected return and the portfolio risk.

SOLUTION

Using equations (4.26) and (4.27), with $w = .50$, $E(X) = \$105$, $E(Y) = \$90$, $\sigma_X^2 = 14{,}725$, $\sigma_Y^2 = 37{,}900$, and $\sigma_{XY} = 23{,}300$

$$E(P) = (.5)(105) + (1 - .5)(90) = \$97.50$$
$$\sigma_p = \sqrt{[(.5)^2(14{,}725) + (1 - .5)^2(37{,}900) + 2(.5)(1 - .5)(23{,}300)]}$$
$$\sigma_p = \sqrt{24{,}806.25} = \$157.50$$

Thus, the portfolio has an expected return of \$97.50 for each \$1,000 invested (a return of 9.75%) but has a portfolio risk of \$157.50. Note here that the portfolio risk is higher than the expected return.

4.8E CALCULATING THE COVARIANCE USING MICROSOFT EXCEL

Overview

◆ *For Quick Results Users*   Use the PHStat add-in to generate a worksheet that calculates the covariance, the portfolio expected return, and risk of two investments.

◆ *For Developers*   Implement a worksheet that uses formulas to calculate the covariance between two variables, the portfolio expected return, and risk of two investments.

4-8E.XLS

The 4-8E.XLS workbook file contains the covariance and portfolio analysis statistics for the investment decision problem of section 4.8.

Quick Results Details

To calculate covariance and portfolio analysis statistics, use the Probability Distributions | Covariance and Portfolio Analysis choice of the PHStat add-in. As an example, consider the investment decision problem discussed in section 4.8. To calculate covariance and portfolio analysis statistics for this problem, do the following:

❶ If the PHStat add-in has not been previously loaded, load the add-in using the instructions of section S4.2.

❷ Select File | New to open a new workbook (or open the existing workbook into which the covariance and portfolio analysis worksheet is to be inserted).

❸ Select PHStat | Probability Distributions | Covariance and Portfolio Management.

❹ In the Covariance and Portfolio Management dialog box (see Figure 4E.7):
   a. Enter 3 in the Number of Outcomes: edit box.
   b. Enter Investment Decision Analysis in the Output Title: edit box.
   c. Select the Portfolio Management Analysis check box.
   d. Click the OK button.

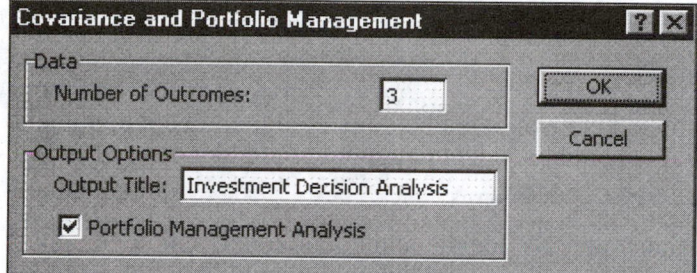

FIGURE 4E.7

PHStat Covariance and Portfolio Management dialog box

The add-in inserts a worksheet containing calculations for covariance and portfolio analysis.

❺ In the newly inserted worksheet:
   a. Enter the outcome probabilities 0.2, 0.5, and 0.3 in cells B4, B5, and B6, respectively.
   b. Enter the estimated returns for the Dow Jones fund investment (-100, 100, 250) in cells C4, C5, and C6, respectively.
   c. Enter the estimated returns for the growth stock fund investment (-200, 50, 350) in cells D4, D5, and D6, respectively.
   d. Enter 0.5 as the weight assigned to the Dow Jones fund investment in cell B20.

The completed worksheet will be similar to the one shown in Figure 4E.8. If desired, descriptive labels for each outcome, such as Recession, Stable economy, and Expanding economy, can be entered in the cell range A4:A6.

	A	B	C	D	E	F	G	H	I
1	Investment Decision Analysis								
2									
3	Probabilities & Outcomes:	P	X	Y					
4		0.2	-100	-200					
5		0.5	100	50					
6		0.3	250	350					
7									
8	Statistics								
9	E(X)	105					Calculations Area		
10	E(Y)	90			For variance and standard deviation:				For covariance:
11	Variance(X)	14725			X-mu	Y-mu	(X-mu)^2	(Y-mu)^2	(X-mu)(Y-mu)
12	Standard Deviation(X)	121.3466			-205	-290	42025	84100	59450
13	Variance(Y)	37900			-5	-40	25	1600	200
14	Standard Deviation(Y)	194.6792			145	260	21025	67600	37700
15	Covariance(XY)	23300							
16	Variance(X+Y)	99225							
17	Standard Deviation(X+Y)	315							
18									
19	Portfolio Management								
20	Weight Assigned to X	0.5							
21	Weight Assigned to Y	0.5							
22	Portfolio Expected Return	97.5							
23	Portfolio Risk	157.5							

FIGURE 4E.8   Investment decision analysis worksheet obtained from the PHStat add-in for Microsoft Excel

Developer Details

We can use arithmetic formulas to implement a worksheet to calculate the covariance between two variables, the portfolio expected return, and risk of two investments. To simplify some of the formulas, we can use the SUMPRODUCT function to sum the products of pairs of numbers. The format of this function is:

SUMPRODUCT(*multiplier cell range, multiplicand cell range*)

where

> *multiplier cell range* = the cell range containing the multipliers
>
> *multiplicand cell range* = the cell range containing the multiplicands

Tables 4E.5 and 4E.6 present a Calculations sheet design for producing this information for the investment decision problem discussed in section 4.8. To implement this design, do the following:

❶ Select File | New to open a new workbook (or open the existing workbook into which the covariance and portfolio analysis worksheet is to be inserted).

❷ Select an unused worksheet (or select Insert | Worksheet if there are none) and rename the worksheet Covariance.

❸ Enter the title, headings, labels, and formulas for columns A through D as shown in Table 4E.5. (Enter the formula for cell B23, which has been typeset in the table as two lines, as one continuous line.)

❹ Enter the headings and formulas for columns E through I as shown in Table 4E.6. (Formulas for row 12 can be entered first and then copied down two rows.)

⑤ Enter the probabilities and outcomes in the cell range B4:D6. Enter 0.2 in cell B4, 0.5 in cell B5, 0.3 in cell B6; −100, 100, and 250 in the cell range C4:C6; and −200, 50, and, 350 in the cell range D4:D6.

⑥ Enter 0.5 as the weight assigned to the Dow Jones fund investment in cell B20.

The completed worksheet will be similar to the one shown in Figure 4E.8. If desired, descriptive labels for each outcome, such as Recession, Stable economy, and Expanding economy, can be entered in the cell range A4:A6.

Table 4E.5 Calculations sheet design (columns A through D) for investment decision problem of section 4.8

	A	B	C	D
1	Investment Decision Analysis			
2				
3	Probabilities & Outcomes:	P	X	Y
4		.xx	xx	xx
5		.xx	xx	xx
6		.xx	xx	xx
7				
8	Statistics			
9	E(X)	=SUMPRODUCT(B4:B6,C4:C6)		
10	E(Y)	=SUMPRODUCT(B4:B6,D4:D6)		
11	Variance(X)	=SUMPRODUCT(B4:B6,G12:G14)		
12	Standard Deviation(X)	=SQRT(B11)		
13	Variance(Y)	=SUMPRODUCT(B4:B6,H12:H14)		
14	Standard Deviation(Y)	=SQRT(B13)		
15	Covariance(XY)	=SUMPRODUCT(B4:B6,I12:I14)		
16	Variance(X+Y)	=B11+B13+2*B15		
17	Standard Deviation($X+Y$)	=SQRT(B16)		
18				
19	Portfolio Management			
20	Weight Assigned to X	.xx		
21	Weight Assigned to Y	=1-B20		
22	Portfolio Expected Return	=B20*B9+B21*B10		
23	Portfolio Risk	=SQRT(B20^2*B11+B21^2*B13 +2*B20*B21*B15)		

Table 4E.6 Calculations sheet design (columns E through I) for investment decision problem of section 4.8

	E	F	G	H	I
1					
2					
3					
4					
5					

(continued)

Table 4E.6 Calculations sheet design (columns E through I) for investment decision problem of section 4.8 (continued)

	E	F	G	H	I
6					
7					
8					
9			Calculations Area		
10	For variance and standard deviation:				For covariance:
11	X-mu	Y-mu	(X-mu)^2	(Y-mu)^2	(X-mu)(Y-mu)
12	=C4-$B$9	=D4-$B$10	=E12^2	=F12^2	=E12*F12
13	=C5-$B$9	=D5-$B$10	=E13^2	=F13^2	=E13*F13
14	=C6-$B$9	=D6-$B$10	=E14^2	=F14^2	=E14*F14

Problems for Section 4.8

Learning the Basics

•**4.61** Given the following probability distributions for variables X and Y

$P(X_i Y_i)$	X	Y
.4	100	200
.6	200	100

Compute

(a) $E(X)$

(b) $E(Y)$

(c) σX

(d) σY

(e) σXY

(f) $E(X + Y)$

(g) $\sigma X + Y$

4.62 Given the following probability distributions for variables X and Y

$P(X_i Y_i)$	X	Y
.2	−100	50
.4	50	30
.3	200	20
.1	300	20

Compute

(a) $E(X)$

(b) $E(Y)$

(c) σX

(d) σY

(e) σXY

(f) $E(X + Y)$

(g) $\sigma X + Y$

4.63 Suppose that two investments X and Y have the following characteristics:
$E(X) = \$50$, $E(Y) = \$100$, $\sigma_X^2 = 9{,}000$, $\sigma_Y^2 = 15{,}000$, and $\sigma XY = 7{,}500$. If the weight assigned to investment X of portfolio assets is .4, compute the

(a) portfolio expected return.

(b) portfolio risk.

Applying the Concepts

4.64 A vendor at a local baseball stadium must determine whether to sell ice cream or soft drinks at today's game. The vendor estimates the following profits that will be made under cool weather and warm weather.

P(EVENT)	EVENT	SELL SOFT DRINKS	SELL ICE CREAM
.4	Cool weather	$50	$30
.6	Warm weather	60	90

Compute the
(a) expected return for selling soft drinks.
(b) expected return for selling ice cream.
(c) standard deviation of selling soft drinks.
(d) standard deviation of selling ice cream.
(e) covariance of selling soft drinks and selling ice cream.
(f) Do you think the vendor should sell soft drinks or ice cream? Explain.
(g) How would you describe the relationship between selling soft drinks and selling ice cream?

4.65 In Example 4.22 on page 274, we assumed that half the portfolio assets were invested in the Dow Jones index fund and half in a growth stock. Recalculate the portfolio expected return and the portfolio risk if

 **What If?**

(a) 30% are invested in the Dow Jones index fund and 70% in the growth stock.
(b) 70% are invested in the Dow Jones index fund and 30% in the growth stock.
(c) Which of the three investment strategies (30%, 50%, or 70% in the Dow Jones index stock) would you recommend? Why?

• 4.66 You are trying to develop a strategy for investing in two different stocks. The anticipated annual return for a $1,000 investment in each stock has the following probability distribution:

	RETURNS	
PROBABILITY	STOCK X	STOCK Y
0.1	−$100	$50
0.3	0	100
0.3	80	−20
0.3	150	100

Compute the
(a) expected return for stock X.
(b) expected return for stock Y.
(c) standard deviation for stock X.
(d) standard deviation for stock Y.
(e) covariance of stock X and stock Y.
(f) Do you think you will invest in stock X or stock Y? Explain.
(g) Suppose you wanted to create a portfolio that consists of stock X and stock Y. Compute the portfolio expected return and portfolio risk for each of the following proportions invested in stock X.

 **What If?**

(1) .10   (4) .70
(2) .30   (5) .90
(3) .50

(h) On the basis of the results of (g), which portfolio would you recommend? Explain.

277

4.67 You are trying to develop a strategy for investing in two different stocks. The anticipated annual return for a $1,000 investment in each stock has the following probability distribution:

| | RETURNS | |
PROBABILITY	STOCK X	STOCK Y
0.1	−$50	−$100
0.3	20	50
0.4	100	130
0.2	150	200

Compute the
(a) expected return for stock X.
(b) expected return for stock Y.
(c) standard deviation for stock X.
(d) standard deviation for stock Y.
(e) covariance of stock X and stock Y.
(f) Do you think you should invest in stock X or stock Y? Explain.
(g) Suppose you wanted to create a portfolio that consists of stock X and stock Y. Compute the portfolio expected return and portfolio risk for each of the following proportions invested in stock X:

 **What If?**

(1) .10   (4) .70
(2) .30   (5) .90
(3) .50

(h) On the basis of the results of (g), which portfolio would you recommend? Explain.

•4.68 You are trying to set up a portfolio that consists of a corporate bond fund and a common stock fund. The following information about the annual return (per $1,000) of each of these investments under different economic conditions is available along with the probability that each of these economic conditions will occur.

PROBABILITY	STATE OF THE ECONOMY	CORPORATE BONDS	COMMON STOCKS
.10	Recession	−$30	−$150
.15	Stagnation	50	−20
.35	Slow growth	90	120
.30	Moderate growth	100	160
.10	High growth	110	250

Compute the
(a) expected return for corporate bonds.
(b) expected return for common stocks.
(c) standard deviation for corporate bonds.
(d) standard deviation for common stocks.
(e) covariance of corporate bonds and common stocks.
(f) Do you think you should invest in corporate bonds or common stocks? Explain.
(g) Suppose you wanted to create a portfolio that consists of corporate bonds and common stocks. Compute the portfolio expected return and portfolio risk for each of the following proportions invested in corporate bonds.

 **What If?**

(1) .10   (4) .70
(2) .30   (5) .90
(3) .50

(h) On the basis of the results of (g), which portfolio would you recommend? Explain.

4.9 ◆ ETHICAL ISSUES AND PROBABILITY

Ethical issues may arise when any statements relating to probability are being presented for public consumption, particularly when these statements are part of an advertising campaign for a product or service. Unfortunately, a substantial portion of the population is not very comfortable with any type of numerical concept (see reference 6) and misinterprets the meaning of the probability. In some instances, the misinterpretation is not intended, but in other cases, advertisements may unethically try to mislead potential customers.

One example of a potentially unethical application of probability relates to the sales of tickets to a state lottery in which the customer typically selects a set of numbers (such as 6) from a larger list of numbers (such as 54). Although virtually all participants know that they are unlikely to win the lottery, they also have very little idea of how unlikely it is for them to select all 6 winning numbers out of the list of 54 numbers. In addition, they have even less of an idea of how likely it is that they can win a consolation prize by selecting either 4 winning numbers or 5 winning numbers.

Given this background, it seems to us that a recent advertising campaign in which a commercial for a state lottery said, "We won't stop until we have made everyone a millionaire," is at the very least deceptive and at the very worst unethical. Actually, given the fact that the lottery brings millions of dollars in revenue into the state treasury, the state is never going to stop running it, although in anyone's lifetime no one can be sure of becoming a millionaire by winning the lottery.

Another example of a potentially unethical application of probability relates to an investment newsletter promising a 20% annual return on investment with a 90% probability. In such a situation, it seems imperative that the investment service needs to (a) explain the basis on which this probability estimate rests, (b) provide the probability statement in another format such as 9 chances in 10, and (c) explain what happens to the investment in the 10% of the cases in which a 20% return is not achieved. Is the entire investment lost?

Problems for Section 4.9

Applying the Concepts

4.69 Write an advertisement for the state lottery that describes in an ethical fashion the probability for winning.

4.70 Write an advertisement for the investment newsletter that states in an ethical fashion the probability for a 20% annual return.

 ## SUMMARY

As shown in the chapter summary chart, this chapter was about basic probability, conditional probability, Bayes' theorem, mathematical expectation, and covariance and the development and application of some important discrete probability distributions—the binomial, Poisson, and hypergeometric distributions. In the following chapter we will concern ouselves with using probability distributions to make decisions among alternative courses of action.

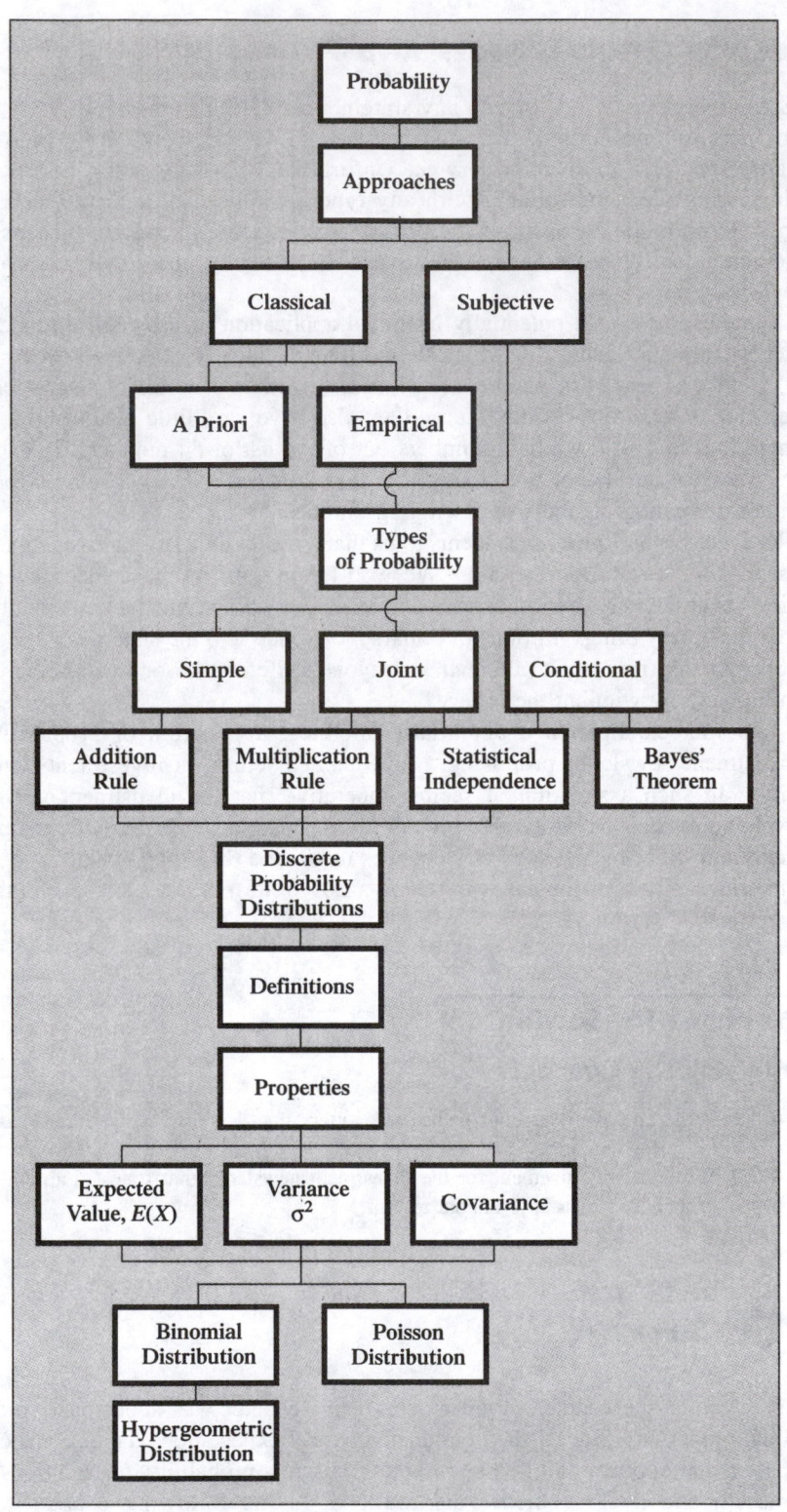

Chapter 4 summary chart

Key Terms

Checking Your Understanding

4.71 What are the differences among *a priori* classical probability, empirical classical probability, and subjective probability?

4.72 What is the difference between a simple event and a joint event?

4.73 How can the addition rule be used to find the probability of occurrence of event *A or B*?

4.74 What is the difference between mutually exclusive events and collectively exhaustive events?

4.75 How does conditional probability relate to the concept of statistical independence?

4.76 How does the multiplication rule differ for events that are and are not independent?

4.77 How can Bayes' theorem be used to revise probabilities in the light of new information?

4.78 What is the meaning of the expected value of a probability distribution?

4.79 What are the assumptions of the binomial distribution?

4.80 What are the assumptions of the Poisson distribution, and how do they differ from those of the binomial distribution?

4.81 Under what circumstances should the hypergeometric distribution be used instead of the binomial distribution?

4.82 Why is it important to know the covariance when determining the portfolio risk?

Chapter Review Problems

4.83 When rolling a die once, what is the probability that
(a) the face of the die is odd?
(b) the face is even *or* odd?
(c) the face is even *or* a 1?
(d) the face is odd *or* a 1?
(e) the face is both even *and* a 1?
(f) given the face is odd, it is a 1?

•4.84 A soft-drink bottling company maintains records concerning the number of unacceptable bottles of soft drink obtained from the filling and capping machines. Based on past data, the probability that a bottle came from machine I and was nonconforming is .01 and the probability that a bottle came from machine II and was nonconforming is .025. Half the bottles are filled on machine I and the other half are filled on machine II.

(a) Give an example of a simple event.

(b) Give an example of a joint event.

(c) If a filled bottle of soft drink is selected at random, what is the probability that
  (1) it is a nonconforming bottle?
  (2) it was filled on machine II?
  (3) it was filled on machine I *and* is a conforming bottle?
  (4) it was filled on machine II *and* is a conforming bottle?
  (5) it was filled on machine I *or* is a conforming bottle?

(d) Suppose we know that the bottle was produced on machine I. What is the probability that it is nonconforming?

(e) Suppose we know that the bottle is nonconforming. What is the probability that it was produced on machine II?

(f) Explain the difference in the answers to (d) and (e).
  (*Hint:* Set up a 2 × 2 table to evaluate the probabilities.)

4.85 Suppose that a survey has been undertaken to determine if there is a relationship between place of residence and ownership of a foreign-made automobile. A random sample of 200 car owners from large cities, 150 from suburbs, and 150 from rural areas was selected with the following results.

	TYPE OF AREA			
CAR OWNERSHIP	LARGE CITY	SUBURB	RURAL	TOTAL
Own foreign car	90	60	25	175
Do not own foreign car	110	90	125	325
Total	200	150	150	500

(a) If a car owner is selected at random, what is the probability that he or she
  (1) owns a foreign car?
  (2) lives in a suburb?
  (3) owns a foreign car *or* lives in a large city?
  (4) lives in a large city *or* a suburb?
  (5) lives in a large city *and* owns a foreign car?
  (6) lives in a rural area *or* does not own a foreign car?

(b) Assume we know that the person selected lives in a suburb. What is the probability that he or she owns a foreign car?

(c) Is area of residence statistically independent of whether the person owns a foreign car? Explain.

4.86 The finance society at a college of business at a large state university would like to determine whether there is a relationship between a student's interest in finance and his or her ability in mathematics. A random sample of 200 students is selected and they are asked whether their interest in finance and ability in mathematics are low, average, or high. The results are as follows:

	ABILITY IN MATHEMATICS			
INTEREST IN FINANCE	LOW	AVERAGE	HIGH	TOTAL
Low	60	15	15	90
Average	15	45	10	70
High	5	10	25	40
Total	80	70	50	200

(a) Give an example of a simple event.
(b) Give an example of a joint event.
(c) Why are high interest in finance *and* high ability in mathematics a joint event?
(d) If a student is selected at random, what is the probability that he or she
  (1) has a high ability in mathematics?
  (2) has an average interest in finance?
  (3) has a low ability in mathematics?
  (4) has a high interest in finance?
  (5) has a low interest in finance *and* a low ability in mathematics?
  (6) has a high interest in finance *and* a high ability in mathematics?
  (7) has a low interest in finance *or* a low ability in mathematics?
  (8) has a high interest in finance *or* a high ability in mathematics?
  (9) has a low ability in mathematics *or* an average ability in mathematics *or* a high ability in mathematics? Are these events mutually exclusive? Are they collectively exhaustive? Explain.
(e) Assume we know that the person selected has a high ability in mathematics. What is the probability that the person has a high interest in finance?
(f) Assume we know that the person selected has a high interest in finance. What is the probability that the person has a high ability in mathematics?
(g) Explain the difference in your answers in (e) and (f).
(h) Are interest in finance and ability in mathematics statistically independent? Explain.

• **4.87** The owner of a restaurant serving continental-style entrées was interested in studying ordering patterns of patrons for the Friday to Sunday weekend time period. Records were maintained that indicated the demand for dessert during the same time period. The owner decided that two other variables were to be studied along with whether a dessert was ordered: the gender of the individual and whether a beef entrée was ordered. The results are as follows:

DESSERT ORDERED	GENDER MALE	FEMALE	TOTAL
Yes	96	40	136
No	224	240	464
Total	320	280	600

DESSERT ORDERED	BEEF ENTRÉE YES	NO	TOTAL
Yes	71	65	136
No	116	348	464
Total	187	413	600

(a) A waiter approaches a table to take an order. What is the probability that the first customer to order at the table
  (1) orders a dessert?
  (2) does not order a beef entrée?
  (3) orders a dessert *or* a beef entrée?
  (4) is a female *and* does not order a dessert?
  (5) orders a dessert *and* a beef entrée?
  (6) is a female *or* does not order a dessert?
(b) Suppose the first person that the waiter takes the dessert order from is a female. What is the probability that she does not order dessert?

(c) Suppose the first person that the waiter takes the dessert order from ordered a beef entrée. What is the probability that this person orders dessert?

(d) Is gender statistically independent of whether the person orders dessert?

(e) Is ordering a beef entrée statistically independent of whether the person orders dessert?

4.88 A natural gas exploration company averages four strikes (that is, natural gas is found) per 100 holes drilled. If 20 holes are to be drilled, what is the probability that

(a) exactly one strike will be made?

(b) at least two strikes will be made?

4.89 On the basis of past experience, 2% of the telephone bills mailed to suburban households are incorrect. If a sample of 20 bills is selected, find the probability that

(a) at least one bill will be incorrect.

(b) at most one bill will be incorrect.

4.90 Records provided by the vice president for human resources at a large urban hospital indicate that on any given workday 10% of the nonclinical workforce (i.e., kitchen, house-keeping and janitorial, electrical and plumbing, security, mailroom, laundry, clerical, and administrative) are absent from work. What is the probability that in a random sample of 15 nonclinical workers,

(a) exactly one will be absent today?

(b) at least two will be absent?

4.91 On the basis of past experience, 15% of the bills of a large mail-order book company are incorrect. A random sample of three current bills is selected.

(a) What is the probability that

(1) exactly two bills are incorrect?

(2) no more than two bills are incorrect?

(3) at least two bills are incorrect?

 What If?

(b) What assumptions about the probability distribution are necessary to solve this problem?

(c) What would be your answers to (a) if the percentage of incorrect bills was 10%?

4.92 On the basis of past experience, printers in a university computer lab are available 90% of the time. If a random sample of 10 time periods is selected

(a) what is the probability that printers are available

(1) exactly 9 times?

(2) at least 9 times?

(3) at most 9 times?

(4) more than 9 times?

(5) fewer than 9 times?

 **What If?**

(b) How many times can the printers be expected to be available?

(c) What would be your answers to (a) and (b) if the printers were available 95% of the time?

• **4.93** Suppose that on a very long mathematics test, the probability is that Lauren would get 70% of the items right.

(a) For a 10-item quiz, calculate the probability that Lauren will get

(1) at least 7 items right.

(2) fewer than 6 items right (and therefore fail the quiz).

(3) 9 or 10 items right (and get an A on the quiz).

(b) What is the expected number of items that Lauren will get right? What proportion of the time will she get that number right?

(c) What is the standard deviation of the number of items that Lauren will get right?

 **What If?**

(d) What would be your answers to (a)–(c) if she typically got 80% correct?

4.94 The manufacturer of the disk drives used in one of the well-known brands of microcomputers expects 2% of the disk drives to malfunction during the microcomputer's warranty period. In a sample of 10 disk drives, what is the probability that

(a) none will malfunction during the warranty period?

(b) exactly one will malfunction during the warranty period?

(c) at least two will malfunction during the warranty period?

 What If?

(d) What would be your answers to (a)–(c) if 1% of the disk drives were expected to malfunction?

THE SPRINGVILLE HERALD CASE

The marketing department of the newspaper has attempted to apply a strategic initiative to increase home-delivery sales through an aggressive direct-marketing campaign that has included mailings, discount coupons, and telephone solicitations. One issue that has emerged on the basis of telephone solicitation of new customers relates to the time at which the newspaper is delivered in the morning. Prospective as well as existing customers are very concerned with obtaining an early delivery of the newspaper, especially during weekdays, primarily for two reasons. First, many customers would like to bring the newspaper with them on their commute to work. Second, and more critically, many customers do not want a newspaper left on their driveways when they are not at home during the day because they believe that an unattended newspaper is a signal to a potential burglar that no one is home.

After several brainstorming sessions, a team consisting of managers and sales associates in the marketing department decided that a policy had to be developed to guarantee delivery by a specific time. After collecting data from a focus group of customers, the team decided that delivery had to be guaranteed by 7 A.M. and that the customer would not be charged for the newspaper on any day that it was not delivered by that time.

Before instituting such a policy, the team needed to determine what percentage of home-delivery customers would be entitled to a free newspaper if this policy was applied to the current delivery process. Al Leslie, the research director, suggested that data maintained on a regular basis could be used to estimate the current delivery times to customers. At a subsequent meeting, he reported to the group that on the basis of data collected from all home-delivery customers in the past week, 7% of the customers did not receive delivery by 7 A.M.

Exercises

4.1 Suppose the group would like to further study the process before instituting any changes. If a sample of 50 customers is selected on a given day, what is the probability that

(a) fewer than 3 customers would receive a free newspaper?

(b) between 2 and 4 customers (inclusive) would receive a free newspaper?

(c) more than 5 customers would receive a free newspaper?

4.2 If the process of delivering newspapers could be improved so that only 5% of the customers did not receive delivery by 7 A.M., what would be the probability that in a sample of 50 customers

(a) fewer than 3 customers would receive a free newspaper?

(b) between 2 and 4 customers (inclusive) would receive a free newspaper?

(c) more than 5 customers would receive a free newspaper?

References

1. Bernstein, P. L., *Against the Gods: The Remarkable Story of Risk* (New York: Wiley, 1996).
2. Emery, D. R., and J. D. Finnerty, *Corporate Financial Management* (Upper Saddle River, NJ: Prentice Hall, 1997).
3. Kirk, R. L., ed., *Statistical Issues: A Reader for the Behavioral Sciences* (Belmont, CA: Wadsworth, 1972).
4. Mendenhall, W., and T. Sincich, *Statistics for Engineering and the Sciences*, 4th ed. (Upper Saddle River, NJ: Prentice Hall, 1995).
5. *Microsoft Excel 97* (Redmond, WA: Microsoft Corp., 1997).
6. Paulos, J. A., *Innumeracy* (New York: Hill and Wang, 1988).
7. Render, B., and R. M. Stair, *Quantitative Analysis for Management*, 6th ed. (Upper Saddle River, NJ: Prentice Hall, 1997).

Chapter 4

Student Solutions Manual

4.2   (a)   Simple events include selecting a red ball or selecting a white ball.

         (b)   Selecting a white ball

•4.4   (a) $40/100 = 2/5 = 0.4$

         (b) $35/100 = 7/20 = 0.35$

         (c) $60/100 = 3/5 = 0.6$

         (d) $10/100 = 1/10 = 0.1$

         (e) $30/100 = 3/10 = 0.3$

         (f) $35/100 = 7/20 = 0.35$

         (g) $\dfrac{40}{100} + \dfrac{35}{100} - \dfrac{10}{100} = \dfrac{65}{100} = \dfrac{13}{20} = 0.65$

         (h) $\dfrac{40}{100} + \dfrac{65}{100} - \dfrac{30}{100} = \dfrac{75}{100} = \dfrac{3}{4} = 0.75$

         (i) $\dfrac{60}{100} + \dfrac{65}{100} - \dfrac{35}{100} = \dfrac{90}{100} = \dfrac{9}{10} = 0.9$

4.6   (a)   Mutually exclusive, not collectively exhaustive

         (b)   Not mutually exclusive, not collectively exhaustive

         (c)   Mutually exclusive, not collectively exhaustive

         (d)   Mutually exclusive, collectively exhaustive

         (e)   Mutually exclusive, collectively exhaustive

•4.8   (a)   Since simple events have only one criterion specified, an example could be any one of the following:

             (1) Having a bank credit card,

             (2) Not having a bank credit card,

             (3) Having a travel/entertainment credit card,

             (4) Not having a travel/entertainment credit card.

         (b)   Since joint events specify two criteria simultaneously, an example could be any one of the following:

             (1) Having a bank credit card and not having a travel/entertainment credit card,

             (2) Not having a bank credit card and not having a travel/entertainment credit card,

             (3) Having a bank credit card and having a travel/entertainment credit card,

             (4) Not having a bank credit card and having a travel/entertainment credit card.

         (c)   "Not having a bank credit card" is the complement of having a bank credit card, since it involves all events other than having a bank credit card.

         (d)   Having a bank credit card and having a travel/entertainment credit card is a joint event because two criteria are specified simultaneously.

         (e) P(has a bank credit card) $= 120/200 = 3/5 = 0.6$

         (f) P(has a travel/entertainment credit card) $= 75/200 = 3/8 = 0.375$

         (g) P(has a bank credit card *and* a travel/entertainment credit card)
             $= 60/200 = 3/10 = 0.3$

         (h) P(does not have a bank credit card *and* does not have a travel/entertainment credit card)
             $= 65/200 = 13/40 = 0.325$

         (i) P(has a bank credit card *or* has a travel/entertainment credit card)
         $= \dfrac{120}{200} + \dfrac{75}{200} - \dfrac{60}{200} = \dfrac{135}{200} = \dfrac{27}{40} = 0.675$

4.8
cont.
(j) P(does not have a bank credit card *or* has a travel/entertainment credit card)
$$= \frac{80}{200} + \frac{75}{200} - \frac{15}{200} = \frac{140}{200} = \frac{7}{10} = 0.7$$

•4.10 (a)

Enjoy Clothes Shopping	Male	Female	Total
Yes	136	224	360
No	104	36	140
Total	240	260	500

(b) Since simple events have only one criterion specified, an example could be any one of the following:
   (1) Being a male,
   (2) Being a female,
   (3) Enjoying clothes shopping,
   (4) Not enjoying clothes shopping.
(c) Since joint events specify two criteria simultaneously, an example could be any one of the following:
   (1) Being a male and enjoying clothes shopping,
   (2) Being a male and not enjoying clothes shopping,
   (3) Being a female and enjoying clothes shopping,
   (4) Being a female and not enjoying clothes shopping.
(d) "Not enjoying clothes shopping" is the complement of "enjoying shopping for clothes," since it involves all events other than enjoying clothes shopping.
(e) P(male) = 240/500 = 12/25 = 0.48
(f) P(enjoys clothes shopping) = 360/500 = 18/25 = 0.72
(g) P(female *and* enjoys clothes shopping) = 224/500 = 56/125 = 0.448
(h) P(male *and* does not enjoy clothes shopping) = 104/500 = 26/125 = 0.208
(i) P(female *or* enjoys clothes shopping) = 396/500 = 99/125 = 0.792
(j) P(male *or* does not enjoy clothes shopping) = 276/500 = 69/125 = 0.552
(k) P(male *or* female) = 500/500 = 1.00

4.12 (a)

Needs Warranty-Related Repair	U.S.	Non-U.S.	Total
Yes	0.025	0.015	0.04
No	0.575	0.385	0.96
Total	0.600	0.400	1.00

(b) Since simple events have only one criterion specified, an example could be any one of the following:
   (1) Needing warranty-related repair,
   (2) Not needing warranty-related repair,
   (3) Manufacturer based in U.S.,
   (4) Manufacturer not based in U.S.
(c) Since joint events specify two criteria simultaneously, an example could be any one of the following:
   (1) Needing warranty-related repair and manufacturer based in U.S.,
   (2) Needing warranty-related repair and manufacturer not based in U.S.,
   (3) Not needing warranty-related repair and manufacturer based in U.S.,
   (4) Not needing warranty-related repair and manufacturer not based in U.S.

4.12   (d)   "Not manufactured by an American-based company" is the complement of
cont.   "manufactured by an American-based company," since it involves all events other than
               the occurrence of vehicles manufactured by an American-based company.
        (e) P(needs warranty repair) = 0.04
        (f) P(manufacturer not based in U.S.) = 0.40
        (g) P(needs warranty repair and manufacturer based in U.S.) = 0.025
        (h) P(does not need warranty repair and manufacturer not based in U.S.) = 0.385
        (i) P(needs warranty repair or manufacturer based in U.S.) = 0.615
        (j) P(needs warranty repair or manufacturer not based in U.S.) = 0.425
        (k) P(needs or does not need warranty repair) = 1.00

• 4.14   (a) $P(A \mid B) = 10/35 = 2/7 = 0.2857$
         (b) $P(A \mid B') = 30/65 = 6/13 = 0.4615$
         (c) $P(A' \mid B') = 35/65 = 7/13 = 0.5385$
         (d)   Since $P(A \mid B) = 0.2857$ and $P(A) = 0.40$, events A and B are not statistically independent.

4.16 $P(A \text{ and } B) = P(A) P(B) = 0.3 \cdot 0.4 = 0.12$

•4.18   (a) P(has travel/entertainment credit card | has bank credit card) = 60/120 = 1/2 = 0.5
         (b) P(has bank credit card | does not have travel/entertainment credit card)
                = 60/125 = 12/25 = 0.48
         (c)   Since P(has travel/entertainment credit card | has bank credit card) = 60/120 or 0.5 and
         P(has travel/entertainment credit card) = 75/200 or 0.375, the two events are not statistically
         independent.

•4.20   (a) P(does not enjoy clothes shopping | female) = 36/260 = 9/65 = 0.1385
         (b) P(male | enjoys clothes shopping) = 136/360 = 17/45 = 0.378
         (c)   Since P(male | enjoys clothes shopping) = 0.378 and P(male) = 240/500 or 0.48, the two
         events are not statistically independent.

4.22   (a) P(needs warranty repair | manufacturer based in U.S.) = 0.025/0.6 = 0.0417
        (b) P(needs warranty repair | manufacturer not based in U.S.) = 0.015/0.4
               = 0.0375
        (c)   Since P(needs warranty repair | manufacturer based in U.S.) = 0.0417 and P(needs
        warranty repair) = 0.04, the two events are not statistically independent.

4.24   (a) $P(\text{both queens}) = \dfrac{4}{52} \cdot \dfrac{3}{51} = \dfrac{12}{2,652} = \dfrac{1}{221} = 0.0045$

        (b) $P(\text{10 followed by 5 or 6}) = \dfrac{4}{52} \cdot \dfrac{8}{51} = \dfrac{32}{2,652} = \dfrac{8}{663} = 0.012$

        (c) $P(\text{both queens}) = \dfrac{4}{52} \cdot \dfrac{4}{52} = \dfrac{16}{2,704} = \dfrac{1}{169} = 0.0059$

        (d) $P(\text{blackjack}) = \dfrac{16}{52} \cdot \dfrac{4}{51} + \dfrac{4}{52} \cdot \dfrac{16}{51} = \dfrac{128}{2,652} = \dfrac{32}{663} = 0.0483$

•4.26

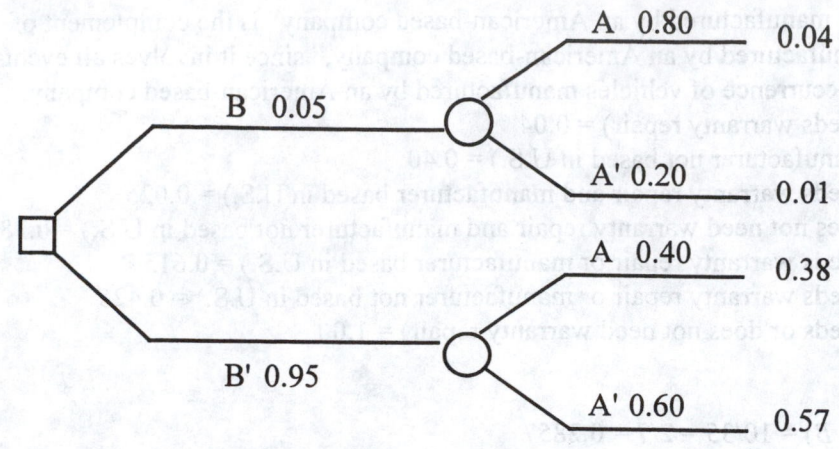

$$P(B \mid A) = \frac{P(A \mid B) \cdot P(B)}{P(A \mid B) \cdot P(B) + P(A \mid B') \cdot P(B')}$$

$$= \frac{0.8 \cdot 0.05}{0.8 \cdot 0.05 + 0.4 \cdot 0.95} = \frac{0.04}{0.42} = \mathbf{0.095}$$

4.28   (a) D = has disease and T = test positive

$$P(D \mid T) = \frac{P(T \mid D) \cdot P(D)}{P(T \mid D) \cdot P(D) + P(T \mid D') \cdot P(D')}$$

$$= \frac{0.9 \cdot 0.03}{0.9 \cdot 0.03 + 0.01 \cdot 0.97} = \frac{0.027}{0.0367} = \mathbf{0.736}$$

(b)

$$P(D' \mid T') = \frac{P(T' \mid D') \cdot P(D')}{P(T' \mid D') \cdot P(D') + P(T' \mid D) \cdot P(D)}$$

$$= \frac{0.99 \cdot 0.97}{0.99 \cdot 0.97 + 0.01 \cdot 0.03} = \frac{0.9603}{0.9633} = \mathbf{0.997}$$

4.30   (a) B = Base Construction Co. enters a bid
         O = Olive Construction Co. wins the contract

$$P(B \mid O) = \frac{P(O \mid B) \cdot P(B)}{P(O \mid B') \cdot P(B') + P(O \mid B) \cdot P(B)}$$

$$= \frac{0.5 \cdot 0.3}{0.5 \cdot 0.3 + 0.25 \cdot 0.7} = \frac{0.15}{0.325} = \mathbf{0.4615}$$

(b) $P(O) = 0.175 + 0.15 = 0.325$

•4.32   (a) $P(A \text{ rating} \mid \text{issued by city}) = 0.35/0.56 = 0.625$
      (b) $P(\text{issued by city}) = 0.5(0.7) + 0.6(0.2) + 0.9(0.1) = 0.56$
      (c) $P(\text{issued by suburb}) = 0.4(0.7) + 0.2(0.2) + 0.05(0.1) = 0.325$

4.34 (a)
Distribution C   Distribution D

X	P(X)	X*P(X)	X	P(X)	X*P(X)
0	0.20	0.00	0	0.10	0.00
1	0.20	0.20	1	0.20	0.20
2	0.20	0.40	2	0.40	0.80
3	0.20	0.60	3	0.20	0.60
4	0.20	0.80	4	0.10	0.40
	1.00	2.00 $\mu = 2.00$		1.00	2.00 $\mu = 2.00$

(b)   Distribution C

X	$(X-\mu)^2$	P(X)	$(X-\mu)^2 * P(X)$
0	$(-1)^2$	0.20	0.80
1	$(0)^2$	0.20	0.20
2	$(1)^2$	0.20	0.00
3	$(2)^2$	0.20	0.20
4	$(3)^2$	0.20	0.80
		$\sigma^2=$	2.00

$$\sigma = \sqrt{\sum (X-\mu)^2 \cdot P(X)} = \sqrt{2.00} = 1.414$$

Distribution D

X	$(X-\mu)^2$	P(X)	$(X-\mu)^2 * P(X)$
0	$(-3)^2$	0.10	0.40
1	$(-2)^2$	0.20	0.20
2	$(-1)^2$	0.40	0.00
3	$(0)^2$	0.20	0.20
4	$(1)^2$	0.10	0.40
		$\sigma^2=$	1.20

$$\sigma = \sqrt{\sum (X-\mu)^2 \cdot P(X)} = \sqrt{1.20} = 1.095$$

(c)   Distribution C is uniform and symmetric; D is unimodal and symmetric.

•4.36 (a)-(b)

X	P(x)	X*P(X)	$(X-\mu_X)^2$	$(X-\mu_X)^2 * P(X)$
0	0.10	0.00	4	0.40
1	0.20	0.20	1	0.20
2	0.45	0.90	0	0.00
3	0.15	0.45	1	0.15
4	0.05	0.20	4	0.20
5	0.05	0.25	9	0.45
	(a) Mean =	2.00	variance =	1.40
			(b) Stdev =	1.18321596
6	0.01	0.06	15.5236	0.155236
	(a) Mean =	2.06	variance =	1.558836
			(b) Stdev =	1.248534

4.38   (a)

X	P(X)
$-1	21/36
$+1	15/36

(b)

X	P(X)
$-1	21/36
$+1	15/36

(c)

X	P(X)
$-1	30/36
$+4	6/36

  (d) -0.167 for each method of play

4.40

	Mean	Standard Deviation
(a)	0.40	0.60
(b)	1.60	0.980
(c)	4.00	0.894
(d)	1.50	0.866

4.42   (a)

	A	B	C	D	E	F	G
1	Warranty Repair						
2							
3	Sample size	3					
4	Probability of success	0.05					
5	Mean	0.15					
6	Variance	0.1425					
7	Standard deviation	0.377492					
8							
9	Binomial Probabilities Table						
10		X	P(X)	P(<=X)	P(<X)	P(>X)	P(>=X)
11		0	0.857375	0.857375	0	0.142625	1
12		1	0.135375	0.99275	0.857375	0.00725	0.142625
13		2	0.007125	0.999875	0.99275	0.000125	0.00725
14		3	0.000125	1	0.999875	0	0.000125

  (b)   Two assumptions: (1) Independence of the cars, (2) Only two outcomes - car needs repair or car does not need repair.

  (c)   Mean: $\mu = 0.15$ Standard deviation: $\sigma = 0.377$

  (d)   If $p = 0.10$ and $n = 3$,

     (1) $P(X = 0) = 0.7290$

     (2) $P(X \geq 1) = 1 - P(X = 0) = 1 - 0.7290 = 0.2710$

     (3) $P(X > 1) = P(X = 2) + P(X = 3) = 0.0270 + 0.0010 = 0.0280$

         Mean: $\mu = 0.30$ Standard deviation: $\sigma = 0.520$

•4.44   (a)

	A	B	C	D	E	F	G
1	Troubles in Residential Service						
2							
3	Sample size	5					
4	Probability of success	0.7					
5	Mean	3.5					
6	Variance	1.05					
7	Standard deviation	1.024695					
8							
9	Binomial Probabilities Table						
10		X	P(X)	P(<=X)	P(<X)	P(>X)	P(>=X)
11		0	0.00243	0.00243	0	0.99757	1
12		1	0.02835	0.03078	0.00243	0.96922	0.99757
13		2	0.1323	0.16308	0.03078	0.83692	0.96922
14		3	0.3087	0.47178	0.16308	0.52822	0.83692
15		4	0.36015	0.83193	0.47178	0.16807	0.52822
16		5	0.16807	1	0.83193	0	0.16807

    (b)   Two assumptions: (1) Independence of the repairs, (2) Only two outcomes - repair accomplished same day or repair not accomplished same day.

    (c)   Mean: $\mu = 3.5$ Standard deviation: $\sigma = 1.0247$

    (d)   If $p = 0.8$ and $n = 5$,

        (1) $P(X = 5) = 0.3277$

        (2) $P(X \geq 3)$ $= P(X = 3) + P(X = 4) + P(X = 5)$
               $= 0.2048 + 0.4096 + 0.3277 = 0.9421$

        (3) $P(X < 2) = P(X = 0) + P(X = 1) = 0.0003 + 0.0064 = 0.0067$
               Mean: $\mu = 4$ Standard deviation: $\sigma = 0.894$

    (e)   The larger p is, the more likely it is that troubles reported on a given day will be repaired on the same day, and the less likely it is that troubles reported on a given day will not be repaired on the same day.

•4.46   (a)   Using the equation, if $\lambda = 2.5$, $P(X = 2) = \dfrac{e^{-2.5} \cdot (2.5)^2}{2!} = 0.2565$

    (b)   If $\lambda = 8.0$, $P(X = 8) = 0.1396$

    (c)   If $\lambda = 0.5$, $P(X = 1) = 0.3033$

    (d)   If $\lambda = 3.7$, $P(X = 0) = 0.0247$

    (e)   If $\lambda = 4.4$, $P(X = 7) = 0.0778$

•4.48

	A	B	C	D	E	F	G
1	Insurance Claims						
2							
3	Average/Expected number of successes:				3.1		
4							
5	Poisson Probabilities Table						
6		X	P(X)	P(<=X)	P(<X)	P(>X)	P(>=X)
7		0	0.045049	0.045049	0.000000	0.954951	1.000000
8		1	0.139653	0.184702	0.045049	0.815298	0.954951
9		2	0.216461	0.401163	0.184702	0.598837	0.815298
10		3	0.223677	0.624840	0.401163	0.375160	0.598837
11		4	0.173350	0.798189	0.624840	0.201811	0.375160
12		5	0.107477	0.905666	0.798189	0.094334	0.201811
13		6	0.055530	0.961196	0.905666	0.038804	0.094334
14		7	0.024592	0.985787	0.961196	0.014213	0.038804
15		8	0.009529	0.995317	0.985787	0.004683	0.014213
16		9	0.003282	0.998599	0.995317	0.001401	0.004683
17		10	0.001018	0.999617	0.998599	0.000383	0.001401
18		11	0.000287	0.999903	0.999617	0.000097	0.000383
19		12	0.000074	0.999977	0.999903	0.000023	0.000097

•4.50 (a) – (d)

	A	B	C	D	E	F	G
1	Chocolate Chip Cookies						
2							
3	Average/Expected number of successes:				6		
4							
5	Poisson Probabilities Table						
6		X	P(X)	P(<=X)	P(<X)	P(>X)	P(>=X)
7		0	0.002479	0.002479	0.000000	0.997521	1.000000
8		1	0.014873	0.017351	0.002479	0.982649	0.997521
9		2	0.044618	0.061969	0.017351	0.938031	0.982649
10		3	0.089235	0.151204	0.061969	0.848796	0.938031
11		4	0.133853	0.285057	0.151204	0.714943	0.848796
12		5	0.160623	0.445680	0.285057	0.554320	0.714943
13		6	0.160623	0.606303	0.445680	0.393697	0.554320
14		7	0.137677	0.743980	0.606303	0.256020	0.393697
15		8	0.103258	0.847237	0.743980	0.152763	0.256020
16		9	0.068838	0.916076	0.847237	0.083924	0.152763
17		10	0.041303	0.957379	0.916076	0.042621	0.083924
18		11	0.022529	0.979908	0.957379	0.020092	0.042621
19		12	0.011264	0.991173	0.979908	0.008827	0.020092
20		13	0.005199	0.996372	0.991173	0.003628	0.008827
21		14	0.002228	0.998600	0.996372	0.001400	0.003628
22		15	0.000891	0.999491	0.998600	0.000509	0.001400
23		16	0.000334	0.999825	0.999491	0.000175	0.000509
24		17	0.000118	0.999943	0.999825	0.000057	0.000175

(e) If $\lambda = 5.0$,
- (a) $P(X < 5) = 0.2650$
- (b) $P(X = 5) = 0.1755$
- (c) $P(X \geq 5) = 0.7350$
- (d) $P(X = 4) + P(X = 5) = 0.3510$

4.52   (a)   If $\lambda = 9.0$, $P(X = 7) = 0.1171$

          (b) $P(7 \leq X \leq 9) = 0.1171 + 0.1318 + 0.1318 = 0.3807$

          (c) $P(X < 5) = 0.0001 + 0.0011 + 0.0050 + 0.0150 + 0.0337 = 0.0549$

•4.54   (a) $P(X = 3) = \dfrac{\binom{5}{3}\binom{10-5}{4-3}}{\binom{10}{4}} = \dfrac{\frac{5 \cdot 4 \cdot 3!}{3!2 \cdot 1} \cdot \frac{5 \cdot 4!}{4!1!}}{\frac{10 \cdot 9 \cdot 8 \cdot 7 \cdot 6!}{6!4 \cdot 3 \cdot 2 \cdot 1}} = \dfrac{5}{3 \cdot 7} = 0.2381$

          (b) $P(X = 1) = \dfrac{\binom{3}{1}\binom{6-3}{4-1}}{\binom{6}{4}} = \dfrac{\frac{3 \cdot 2!}{2! \cdot 1} \cdot \frac{3!}{3! \cdot 0!}}{\frac{6 \cdot 5 \cdot 4!}{4! \cdot 2 \cdot 1}} = \dfrac{1}{5} = 0.2$

          (c) $P(X = 0) = \dfrac{\binom{3}{0}\binom{12-3}{5-0}}{\binom{12}{5}} = \dfrac{\frac{3!}{3! \cdot 0!} \cdot \frac{9 \cdot 8 \cdot 7 \cdot 6 \cdot 5!}{5! 4 \cdot 3 \cdot 2 \cdot 1}}{\frac{12 \cdot 11 \cdot 10 \cdot 9 \cdot 8 \cdot 7!}{7! \cdot 5 \cdot 4 \cdot 3 \cdot 2 \cdot 1}} = \dfrac{7}{44} = 0.1591$

          (d) $P(X = 3) = \dfrac{\binom{3}{3}\binom{7-0}{3-3}}{\binom{10}{3}} = \dfrac{\frac{3!}{3! \cdot 0!} \cdot \frac{7!}{7! \cdot 0!}}{\frac{10 \cdot 9 \cdot 8 \cdot 7!}{7! 3 \cdot 2 \cdot 1}} = \dfrac{1}{120} = 0.0083$

•4.56   (a)   (1)   If $n = 6$, $A = 25$, and $N = 100$, $P(X \geq 2) = 1 - [P(X = 0) + P(X = 1)]$

                       $= 1 - [0.1689 + 0.3620] = 0.4691$

             (2)   If $n = 6$, $A = 30$, and $N = 100$, $P(X \geq 2) = 1 - [P(X = 0) + P(X = 1)]$

                       $= 1 - [0.1100 + 0.3046] = 0.5854$

             (3)   If $n = 6$, $A = 5$, and $N = 100$, $P(X \geq 2) = 1 - [P(X = 0) + P(X = 1)]$

                       $= 1 - [0.7291 + 0.2430] = 0.0279$

             (4)   If $n = 6$, $A = 10$, and $N = 100$, $P(X \geq 2) = 1 - [P(X = 0) + P(X = 1)]$

                       $= 1 - [0.5223 + 0.3687] = 0.1090$

          (b)   The probability that the entire group will be audited is very sensitive to the true number of improper returns in the population. If the true number is very low ($A = 5$), the probability is very low (0.0279). When the true number is increased by a factor of six ($A = 30$), the probability the group will be audited increases by a factor of almost 21 (0.5854).

4.58 (a)

	A	B	C
1	Defective Radios		
2			
3	Sample size	8	
4	No. of successes in population	12	
5	Population size	48	
6			
7	Hypergeometric Probabilities Table		
8		X	P(x)
9		0	0.080192
10		1	0.265463
11		2	0.340677
12		3	0.219792
13		4	0.077271
14		5	0.014986
15		6	0.001543
16		7	7.56E-05
17		8	1.31E-06

(b) (1)   If $n = 6$, $A = 12$, and $N = 48$, $P(X = 8) = 7.5296 \times 10^{-5} \cong .000075$

(2)   If $n = 6$, $A = 12$, and $N = 48$, $P(X = 0) = 0.1587$

(3)   If $n = 6$, $A = 12$, and $N = 48$, $P(X \geq 1) = 1 - 0.1587 = 0.8413$

4.60 (a)

	A	B	C
1	Defective Disks		
2			
3	Sample size	4	
4	No. of successes in population	5	
5	Population size	15	
6			
7	Hypergeometric Probabilities Table		
8		X	P(x)
9		0	0.153846
10		1	0.43956
11		2	0.32967
12		3	0.07326
13		4	0.003663

(b) $\mu = n \cdot p = 4 \cdot (0.333) = 1.33$

4.62 (a) $E(X) = (0.2)(\$ - 100) + (0.4)(\$50) + (0.3)(\$ 200) + (0.1)(\$300) = \$90$

(b) $E(Y) = (0.2)(\$50) + (0.4)(\$30) + (0.3)(\$ 20) + (0.1)(\$20) = \$30$

(c) $\sigma_X = \sqrt{(0.2)(-100 - 90)^2 + (0.4)(50 - 90)^2 + (0.3)(200 - 90)^2 + (0.1)(300 - 90)^2}$

$= \sqrt{15900} = 126.10$

(d) $\sigma_Y = \sqrt{(0.2)(50 - 30)^2 + (0.4)(30 - 30)^2 + (0.3)(20 - 30)^2 + (0.1)(20 - 30)^2}$

$= \sqrt{120} = 10.95$

4.62
cont.   (e) σ_{XY} $= (0.2)(-100 - 90)(50 - 30) + (0.4)(50 - 90)(30 - 30)$
$$+ (0.3)(200 - 90)(20 - 30) + (0.1)(300 - 90)(20 - 30) = -1300$$

(f) $E(X + Y) = E(X) + E(Y) = \$90 + \$30 = \$120$

(g) $\sigma_{X+Y} = \sqrt{15900 + 120 + 2(-1300)} = \sqrt{13420} = 115.85$

4.64   (a) –(e)

	A	B	C	D
1	Covariance Analysis			
2				
3	Probabilities & Outcomes:	P	X	Y
4	Cool Weather	0.4	50	30
5	Warm Weather	0.6	60	90
6				
7	Statistics			
8	E(X)	56		
9	E(Y)	66		
10	Variance(X)	24		
11	Standard Deviation(X)	4.898979		
12	Variance(Y)	864		
13	Standard Deviation(Y)	29.39388		
14	Covariance(XY)	144		

(f)   On the analysis of the expected return alone, you should sell ice cream. But there is more variability in the profits, so there is higher risk with selling ice cream. Both products generate more profits in warmer weather than in cool, but ice cream profits increase more in warm weather.

•4.66   (a) – (e)

	A	B	C	D
1	Covariance Analysis			
2			Stock	
3	Probabilities & Outcomes:	P	X	Y
4		0.1	-100	50
5		0.3	0	100
6		0.3	80	-20
7		0.3	150	100
8				
9	Statistics			
10	E(X)	59		
11	E(Y)	59		
12	Variance(X)	6189		
13	Standard Deviation(X)	78.6702		
14	Variance(Y)	2889		
15	Standard Deviation(Y)	53.74942		
16	Covariance(XY)	39		
17	Variance(X+Y)	9156		
18	Standard Deviation(X+Y)	95.68699		
19				
20	Portfolio Management			
21	Weight Assigned to X	0.1		
22	Weight Assigned to Y	0.9		
23	Portfolio Expected Return	59		
24	Portfolio Risk	49.08156		

(f)   Stock Y gives the investor a lower standard deviation while yielding the same expected return as investing in stock X, so the investor should select stock Y.

•4.66 (g) (1) $E(P) = \$59$ $\sigma_P = 49.01$
cont.   (2) $E(P) = \$59$ $\sigma_P = 44.41$
      (3) $E(P) = \$59$ $\sigma_P = 47.64$
      (4) $E(P) = \$59$ $\sigma_P = 57.38$
      (5) $E(P) = \$59$ $\sigma_P = 71.01$

(h) Based on the results of (g), you should recommend a portfolio with 30% stock X and 70% stock Y because it has the same expected return as other portfolios ($59) but has the smallest portfolio risk ($44.41).

•4.68 (a) $E(X) = \$77$ (b) $E(Y) = \$97$
    (c) $\sigma_X = 39.76$ (d) $\sigma_Y = 108.95$
    (e) $\sigma_{XY} = 4161$
    (f) Stock Y gives the investor a higher expected return than stock X, but also has a standard deviation better than 2.5 times higher than that for stock X. An investor should carefully weigh the increased risk.
    (g) (1) $E(P) = \$95$ $\sigma_P = 101.88$
       (2) $E(P) = \$91$ $\sigma_P = 87.79$
       (3) $E(P) = \$87$ $\sigma_P = 73.78$
       (4) $E(P) = \$83$ $\sigma_P = 59.92$
       (5) $E(P) = \$79$ $\sigma_P = 46.35$

(h) Based on the results of (g), an investor should recognize that as the expected return increases, so does the portfolio risk.

•4.84 (a) Since simple events have only one criterion specified, an example could be any one of the following:
(1) Being filled by machine I,
(2) Being filled by machine II,
(3) Being a conforming bottle,
(4) Being a nonconforming bottle.
(b) Since joint events specify two criteria simultaneously, an example could be any one of the following:
(1) Being filled by machine I and being a conforming bottle,
(2) Being filled by machine I and being a nonconforming bottle,
(3) Being filled by machine II and being a conforming bottle,
(4) Being filled by machine II and being a nonconforming bottle.
(c) (1) $P(\text{nonconforming bottle}) = 0.01 + 0.025 = 0.035$
     (2) $P(\text{machine II}) = 0.5$
     (3) $P(\text{machine I } and \text{ conforming bottle}) = 0.49$
     (4) $P(\text{machine II } and \text{ conforming bottle}) = 0.475$
     (5) $P(\text{machine I } or \text{ conforming bottle}) = 0.5 + 0.475 = 0.975$
(d) $P(\text{nonconforming} \mid \text{machine II}) = 0.02$
(e) $P(\text{machine II} \mid \text{nonconforming}) = 0.01/0.035 = 0.2857$
(f) The conditions are switched. Part (d) answers $P(A|B)$ and part (e) answers $P(B|A)$.

4.86 (a) Since simple events have only one criterion specified, an example could be high interest in finance.
(b) Since joint events specify two criteria simultaneously, an example could be average interest in finance and high ability in mathematics.

4.86   (c)   High interest in finance and high mathematics ability is a joint event because two criteria
cont.   are specified simultaneously.
        (d)   (1) P(high ability, mathematics) = 50/200 = 1/4 = 0.25
               (2) P(average interest, finance) = 70/200 = 0.14
               (3) P(low ability, mathematics) = 80/200 = 2/5 = 0.4
               (4) P(high interest, finance) = 40/200 = 1/5 = 0.2
               (5) P(low interest, finance *and* low ability, mathematics) = 60/200 = 0.3
               (6) P(high interest, finance *and* high ability, mathematics) = 25/200= 0.125
               (7) P(low interest, finance *or* low ability, mathematics) = 110/200= 0.55
               (8) P(high interest, finance *or* low ability, mathematics) = 65/200= 0.325
               (9) P(low, average, *or* high ability, mathematics) = 1.0. These events are mutually
                      exclusive because each individual has only one ability level in mathematics.
                      These events are collectively exhaustive because one of the three must occur;
                      that is, every individual has some ability level in mathematics.
        (e) P(high interest, finance | high ability, mathematics) = 25/50 = 1/2 = 0.5
        (f) P(high ability, mathematics | high interest, finance) = 25/40 = 5/8 = 0.625
        (g)   The conditions are switched. Part (e) answers $P(F|M)$ and part (f) answers $P(M|F)$.
        (h)   Since P(high ability, mathematics | high interest, finance) = 0.625 and P(high ability,
               mathematics) = 0.25, the two events are not statistically independent.

4.88   (a)   If $p = 0.04$ and $n = 20$, $P(X = 1) = 0.3683$
        (b) $P(X \geq 2) = 1 - [P(X = 0) + P(X = 1)] = 1 - [0.4420 + 0.3683] = 0.1897$

4.90   (a)   If $p = 0.10$ and $n = 15$, $P(X = 1) = 0.3432$
        (b) $P(X \geq 2) = 1 - [P(X = 0) + P(X = 1)] = 0.4510$

4.92   (a)   (1)   If $p = 0.90$ and $n = 10$, $P(X = 9) = 0.3874$
               (2) $P(X \geq 9) = P(X = 9) + P(X = 10) = 0.3874 + 0.3487 = 0.7361$
               (3) $P(X \leq 9) = 1 - P(X = 10) = 1 - 0.3487 = 0.6513$
               (4) $P(X > 9) = P(X = 10) = 0.3487$
               (5) $P(X < 9) = 1 - P(X \geq 9) = 1 - 0.7361 = 0.2639$
        (b)   Mean: $\mu = 10 \cdot (0.90) = 9.0$
        (c)   (1)   If $p = 0.95$ and $n = 10$, $P(X = 9) = 0.3151$
               (2) $P(X \geq 9) = P(X = 9) + P(X = 10) = 0.3151 + 0.5987 = 0.9138$
               (3) $P(X \leq 9) = 1 - P(X = 10) = 1 - 0.5987 = 0.4013$
               (4) $P(X > 9) = P(X = 10) = 0.5987$
               (5) $P(X < 9) = 1 - P(X \geq 9) = 1 - 0.9139 = 0.0862$

4.94   (a) $P(X = 0) = (0.98)^{10} = 0.8171$
        (b) $P(X = 1) = 10(0.98)^9 (0.02)^1 = 0.1667$
        (c) $P(X \geq 2) = 1 - P(X = 0) - P(X = 1) = 0.0162$
        (d)   (a) $P(X = 0) = (0.99)^{10} = 0.9044$
               (b) $P(X = 1) = 10(0.99)^9 (0.01)^1 = 0.0914$
               (c) $P(X \geq 2) = 1 - P(X = 0) - P(X = 1) = 0.0043$

Chapter 5
The Normal Distribution and Sampling Distributions

CHAPTER OBJECTIVES

✓ *To show how the normal probability distribution can be used to represent certain types of continuous phenomena*
✓ *To show how the normal probability plot can be used to assess whether a distribution is normally distributed*
✓ *To introduce the exponential distribution*
✓ *To develop the concept of a sampling distribution for both numerical and categorical variables*
✓ *To examine the central limit theorem for cases in which a population is either normally or not normally distributed*
✓ *To provide the foundation for statistical inference*

Introduction

In chapter 4 we developed the concept of a probability distribution for a discrete random variable. In this chapter we turn our attention to the most important probability distribution in statistics, the *Gaussian,* or **normal distribution,** which involves a continuous random variable. The normal distribution is one of many such distributions called **continuous probability density functions**—those that arise because of some measuring process on various phenomena of interest.

When a mathematical expression is available to represent some underlying continuous phenomenon, the probability that various values of the random variable occur within certain ranges or intervals can be calculated. However, the *exact* probability of a *particular value* from a continuous distribution is zero. This is what distinguishes continuous phenomena, which are measured, from discrete phenomena, which are counted. As an example, time (in seconds) is measured, not counted. Thus, we can compute the probability that a task can be completed in between 70 and 80 seconds. Narrowing this interval, we can compute the probability that a task can be accomplished in between 74 and 76 seconds. With more refined or precise measuring instruments, we can also compute the probability that a task can be completed in between 74.99 and 75.01 seconds. However, the probability that a task can be completed in *exactly* 75 seconds is zero.

Continuous models have important applications in engineering and the physical sciences as well as in business and the social sciences. Some examples of continuous random phenomena are height; weight; daily changes in closing prices of stocks; time between arrivals of planes landing on a runway, of telephone calls into a switchboard, of customers at a bank, and so forth; and customer servicing times.

However, obtaining probabilities or computing expected values and standard deviations for continuous phenomena involves mathematical expressions that require a knowledge of integral calculus and are beyond the scope of this book. Nevertheless, the normal distribution is considered so important for applications that special probability tables [such as Tables E.2(a) and E.2(b) of appendix E] were devised to eliminate the need for what otherwise would require laborious mathematical computations. Many continuous random phenomena are either normally distributed or can be approximated by a normal distribution. Thus, we begin this chapter by discussing the properties of the normal distribution. We also study a simple graphical tool, the normal probability plot, that can be used to evaluate whether a set of data appears to be normally distributed. After a brief discussion of the exponential distribution, we develop the concept of the sampling distribution and discuss the central limit theorem.

 **USING STATISTICS:** *Applying the Normal Distribution to Study a Production Process*

Suppose that the operations manager in an automobile assembly plant is interested in studying the process of assembling a particular part of the automobile, with a goal of reducing the amount of time that it takes for assembly. A team of individuals involved in the process is formed. One of the first steps needed for process improvement involves the determination of how long it takes to assemble the part using the current process in which workers are trained with an individual learning approach. After studying the process and collecting data, the team determines that the assembly time approximately follows a normal distribution with an arithmetic mean (μ) of 75 seconds and a standard deviation (σ) of 6 seconds. How can the team use this information to answer questions about the current process, such as what proportion of the parts will be assembled in less than 62 seconds and how many seconds it will take to assemble 10% of the parts?

6.1 THE NORMAL DISTRIBUTION

The normal distribution is vitally important in statistics for three main reasons:

1. Numerous continuous phenomena seem to follow it or can be approximated by it.

2. We can use it to approximate various discrete probability distributions.

3. It provides the basis for *classical statistical inference* because of its relationship to the *central limit theorem* (which we discuss in section 6.5).

The normal distribution has several important theoretical properties as illustrated in Exhibit 6.1.

Exhibit 6.1 Properties of the Normal Distribution

There are four key properties associated with the normal distribution.

✓ **1.** It is bell-shaped (and thus symmetrical) in its appearance.

✓ **2.** Its measures of central tendency (mean, median, mode, midrange, and midhinge) are all identical.

✓ **3.** Its "middle spread" is equal to 1.33 standard deviations. This means that the interquartile range is contained within an interval of two-thirds of a standard deviation below the mean to two-thirds of a standard deviation above the mean.

✓ **4.** Its associated random variable has an infinite range $(-\infty < X < +\infty)$.

In practice, some of the variables we observe may only approximate these theoretical properties. This occurs for two reasons: (1) the underlying population distribution may be only approximately normal, and (2) any actual sample may deviate from the theoretically expected characteristics. When variables are approximately normally distributed, they are only approximately bell-shaped and symmetrical in appearance; their measures of central tendency may differ slightly from each other; their interquartile range may differ slightly from 1.33 standard deviations; and their *practical range* will not be infinite but will generally lie within 3 standard deviations above and below the mean (i.e., range $\cong$ 6 standard deviations).

As a case in point, let us refer to Table 6.1.

Table 6.1 *Thickness of 10,000 brass washers manufactured by Eastern Metal Company*

THICKNESS (INCHES)	RELATIVE FREQUENCY
Under .0180	48/10,000 = .0048
.0180 < .0182	122/10,000 = .0122
.0182 < .0184	325/10,000 = .0325
.0184 < .0186	695/10,000 = .0695
.0186 < .0188	1,198/10,000 = .1198
.0188 < .0190	1,664/10,000 = .1664
.0190 < .0192	1,896/10,000 = .1896
.0192 < .0194	1,664/10,000 = .1664
.0194 < .0196	1,198/10,000 = .1198
.0196 < .0198	695/10,000 = .0695
.0198 < .0200	325/10,000 = .0325
.0200 < .0202	122/10,000 = .0122
.0202 or above	48/10,000 = .0048
Total	1.0000

The data in Table 6.1 represent the thickness (in inches) of 10,000 brass washers manufactured by a large company. The continuous random phenomenon of interest, thickness, is said to follow the Gaussian or **normal probability density function.** The measurements of the thickness of the 10,000 brass washers cluster in the interval (.0190 < .0192) inch and distribute symmetrically around that grouping, forming a "bell-shaped" pattern. As demonstrated in this table, if the nonoverlapping (*mutually exclusive*) listing contains all possible class intervals (is *collectively exhaustive*), the probabilities will again sum to 1. Such a probability distribution may be considered as a relative frequency distribution as described in section 2.2, where, except for the two open-ended classes, the midpoint of every other class interval represents the data in that interval.

Figure 6.1 depicts the relative frequency histogram and polygon for the distribution of the thickness of 10,000 brass washers. For these data, the first three theoretical properties of the normal distribution seem to be satisfied; however, the fourth does not hold. The random variable of interest, thickness, cannot possibly take on values of zero or below, and a washer cannot be so thick that it becomes unusable. From Table 6.1 we note that only 48 out of every 10,000 brass washers manufactured are expected to have a thickness of 0.0202 inch or more, whereas an equal number are expected to have a thickness under 0.0180 inch. Thus, the chance of randomly obtaining a washer so thin or so thick is .0048 + .0048 = .0096—or almost 1 in 100.

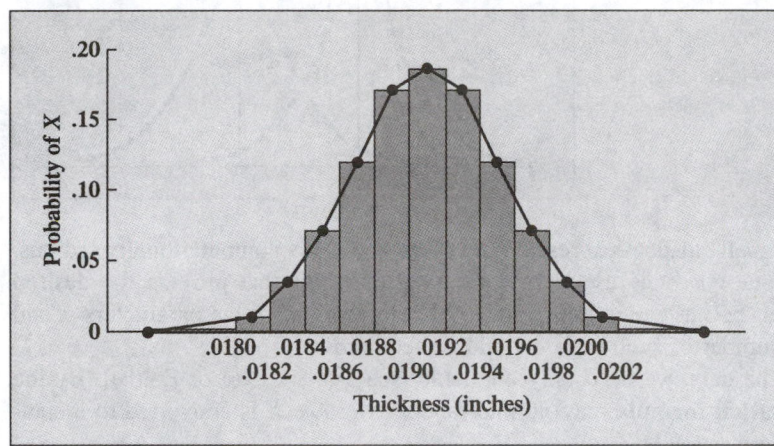

FIGURE 6.1
Relative frequency histogram and polygon of the thickness of 10,000 brass washers
Source: Data are taken from Table 6.1.

The mathematical model or expression representing a probability density function is denoted by the symbol $f(X)$. For the normal distribution, the model used to obtain the desired probabilities is

The Normal Distribution

$$f(X) = \frac{1}{\sqrt{2\pi}\sigma} e^{-(1/2)[(X-\mu)/\sigma]^2} \tag{6.1}$$

where

e = the mathematical constant approximated by 2.71828

π = the mathematical constant approximated by 3.14159

μ = population mean

σ = population standard deviation

X = any value of the continuous random variable, where $-\infty < X < +\infty$

Note that because e and π are mathematical constants, the probabilities of the random variable X are dependent only upon the two parameters of the normal distribution—the population mean μ and the population standard deviation σ. Every time we specify a *particular combination* of μ and σ, a *different* normal probability distribution is generated. We illustrate this in Figure 6.2, which depicts three different normal distributions. Distributions A and B have the same mean (μ) but have different standard deviations. On the other hand, distributions A and C have the same standard deviation (σ) but have different means. Furthermore, distributions B and C depict two normal probability density functions that differ with respect to both μ and σ.

307

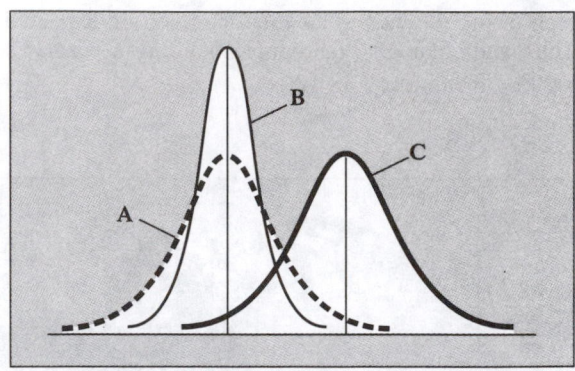

FIGURE 6.2
Three normal distributions having differing parameters μ and σ

Unfortunately, the mathematical expression in equation (6.1) is computationally tedious. To avoid such computations, it is useful to have a set of tables that provide the desired probabilities. However, because an infinite number of combinations of the parameters μ and σ exists, an infinite number of such tables would be required.

By *standardizing* the data, we need only one table [see Table E.2(a) or E.2(b)]. By the use of the **transformation formula,** any normal random variable X is converted to a standardized normal random variable Z.

The Transformation Formula

The Z value is equal to the difference between X and the population mean μ, divided by the standard deviation σ.

$$Z = \frac{X - \mu}{\sigma} \tag{6.2}$$

Whereas the original data for the random variable X had mean μ and standard deviation σ, the standardized random variable Z will always have mean $\mu = 0$ and standard deviation $\sigma = 1$.

A **standardized normal distribution** is one whose random variable Z always has a mean $\mu = 0$ and a standard deviation $\sigma = 1$.

Substituting in equation (6.1), we see that the probability density function of a standard normal variable Z is

The Standardized Normal Distribution

$$f(Z) = \frac{1}{\sqrt{2\pi}} e^{-(1/2)Z^2} \tag{6.3}$$

Thus we can always convert any set of normally distributed data to its standardized form and then determine any desired probabilities either from a table of the standardized normal distribution, Table E.2(a), or the cumulative standardized normal distribution, Table E.2(b).

To see how the transformation formula is applied and the results used to find probabilities from a table of the standardized normal distribution, Table E.2(a) or E.2(b), let us return to the Using Statistics example on page 331 in which the operations manager of an automobile assembly plant has determined that the assembly time for using the current process in which workers receive individual training is normally distributed with a mean μ of 75 seconds and a standard deviation σ of 6 seconds.

We see from Figure 6.3 that every measurement X has a corresponding standardized measurement Z obtained from the transformation formula [equation (6.2)]. Hence, from Figure 6.3, it is clear that the 81 seconds required for a factory worker to complete the task is equivalent to 1 standardized unit (i.e., 1 *standard deviation*) above the mean because

$$Z = \frac{81 - 75}{6} = +1$$

and that the 57 seconds required for a worker to assemble the part is equivalent to 3 standardized units (i.e., 3 *standard deviations*) below the mean because

$$Z = \frac{57 - 75}{6} = -3$$

Thus, the standard deviation has become the unit of measurement. In other words, a time of 81 seconds is 6 seconds (i.e., 1 standard deviation) higher, or *slower,* than the average time of 75 seconds, and a time of 57 seconds is 18 seconds (i.e., 3 standard deviations) lower, or *faster,* than the average time.

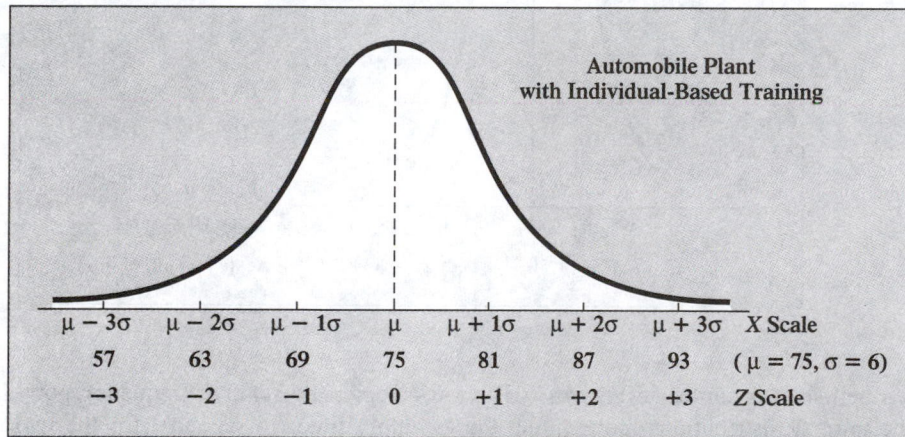

FIGURE 6.3
Transformation of scales

Suppose now that the team decides that it would like to experiment with a small group of workers who will be provided with additional team-based learning. Suppose that after they receive this training, the time to assemble the part for this group of workers is normally distributed with mean μ of 60 seconds and standard deviation σ of 3 seconds. The data are depicted in Figure 6.4. Comparing these results with those of the workers who have individual-based training, we note, for example, that for the workers whose training is team-based, an assembly time of 57 seconds is only 1 standard deviation below the mean for the group because

$$Z = \frac{57 - 60}{3} = -1$$

We may also note that a time of 63 seconds is 1 standard deviation above the mean assembly time because

$$Z = \frac{63 - 60}{3} = +1$$

and a time of 51 seconds is 3 standard deviations below the group mean because

$$Z = \frac{51 - 60}{3} = -3$$

FIGURE 6.4

A different transformation of scales

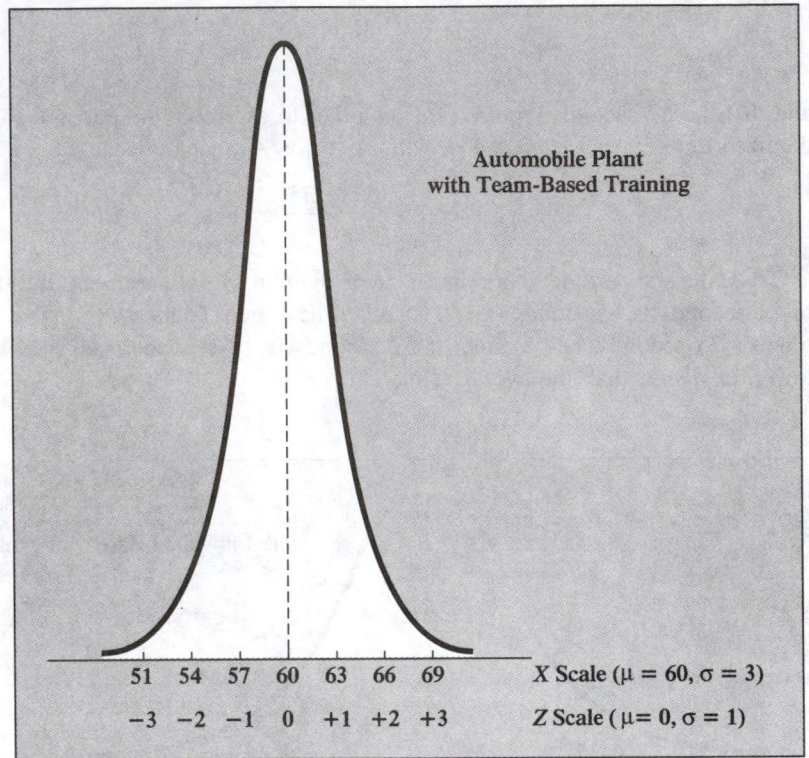

The two bell-shaped curves in Figures 6.3 and 6.4 depict the relative frequency polygons of the normal distributions representing the assembly time (in seconds) for the two groups of workers, one in which workers received individual-based training and the other in which workers received team-based training. Because the times to assemble the part are known for every factory worker in each group, the data represent the entire population and, therefore, the *probabilities* or proportion of area under the entire curve must add up to 1. Thus, the area under the curve between any two reported time values represents only a portion of the total area possible.

Now suppose the operations manager wishes to determine the probability that a factory worker selected at random from those who underwent individual-based training should require between 75 and 81 seconds to assemble the part. That is, what is the likelihood that the worker's time is between the mean and 1 standard deviation above the mean? This answer is found by using Table E.2(a).

Table E.2(a) represents the probabilities, or areas, under the normal curve calculated from the mean μ to the particular values of interest X. Using equation (6.2), this corresponds to the probabilities or areas under the standardized normal curve from 0 to the transformed values of interest Z. Only positive entries for Z are listed in the table because for such a symmetrical distribution having a mean of zero, the area from the mean to $+Z$ (i.e., Z standard deviations above the mean) must be identical to the area from the mean to $-Z$ (i.e., Z standard deviations below the mean).

To use Table E.2(a), we note that all Z values must first be recorded to two decimal places. Thus, our particular Z value of interest is recorded as $+1.00$. To read the probability or area under the curve from the mean to $Z = +1.00$, we scan down the Z column from Table E.2(a) until we locate the Z value of interest (in 10ths). Hence, we stop in the row $Z = 1.0$. Next we read across this row until we intersect the column that contains the 100ths place of the Z value. Therefore, in the body of the table, the tabulated probability for $Z = 1.00$ corresponds to the intersection of the row $Z = 1.0$ with the column $Z = .00$ as shown in Table 6.2, which is a replica of Table E.2(a). This probability is .3413. As depicted in Figure 6.5, there is a 34.13% chance that a factory worker selected at random who has had individual-based training will require between 75 and 81 seconds to assemble the part.

Table 6.2 *Obtaining an area under the normal curve*

Z	.00	.01	.02	.03	.04	.05	.06	.07	.08	.09
0.0	.0000	.0040	.0080	.0120	.0160	.0199	.0239	.0279	.0319	.0359
0.1	.0398	.0438	.0478	.0517	.0557	.0596	.0636	.0675	.0714	.0753
0.2	.0793	.0832	.0871	.0910	.0948	.0987	.1026	.1064	.1103	.1141
0.3	.1179	.1217	.1255	.1293	.1331	.1368	.1406	.1443	.1480	.1517
0.4	.1554	.1591	.1628	.1664	.1700	.1736	.1772	.1808	.1844	.1879
0.5	.1915	.1950	.1985	.2019	.2054	.2088	.2123	.2157	.2190	.2224
0.6	.2257	.2291	.2324	.2357	.2389	.2422	.2454	.2486	.2518	.2549
0.7	.2580	.2612	.2642	.2673	.2704	.2734	.2764	.2794	.2823	.2852
0.8	.2881	.2910	.2939	.2967	.2995	.3023	.3051	.3078	.3106	.3133
0.9	.3159	.3186	.3212	.3238	.3264	.3289	.3315	.3340	.3365	.3389
1.0	.3413	.3438	.3461	.3485	.3508	.3531	.3554	.3577	.3599	.3621

Source: Extracted from Table E.2(a).

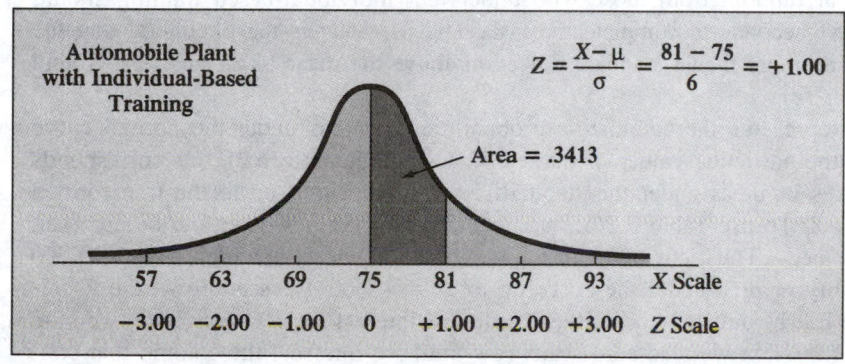

Automobile Plant with Individual-Based Training

$$Z = \frac{X - \mu}{\sigma} = \frac{81 - 75}{6} = +1.00$$

Area = .3413

| X Scale | 57 | 63 | 69 | 75 | 81 | 87 | 93 |
| Z Scale | −3.00 | −2.00 | −1.00 | 0 | +1.00 | +2.00 | +3.00 |

FIGURE 6.5

Determining the area between the mean and Z from a standardized normal distribution

On the other hand, we know from Figure 6.4 that for the workers who received team-based training, a time of 63 seconds is 1 standardized unit above the mean time of 60 seconds. Thus, the likelihood that a randomly selected factory worker who receives team-based training will complete the assemblage in between 60 and 63 seconds is also .3413. These results are illustrated in Figure 6.6, which demonstrates that regardless of the value of the mean μ and standard deviation σ of a particular set of normally distributed data, a transformation to a standardized scale can always be made from equation (6.2) and, by using Table E.2(a), any probability or portion of area under the curve can be obtained. From Figure 6.6 we see that the probability or area under the curve from 60 to 63 seconds for the workers who had team-based training is identical to the probability or area under the curve from 75 to 81 seconds for the workers who had individual-based training.

FIGURE 6.6

Demonstrating a transformation of scales for corresponding portions under two normal curves

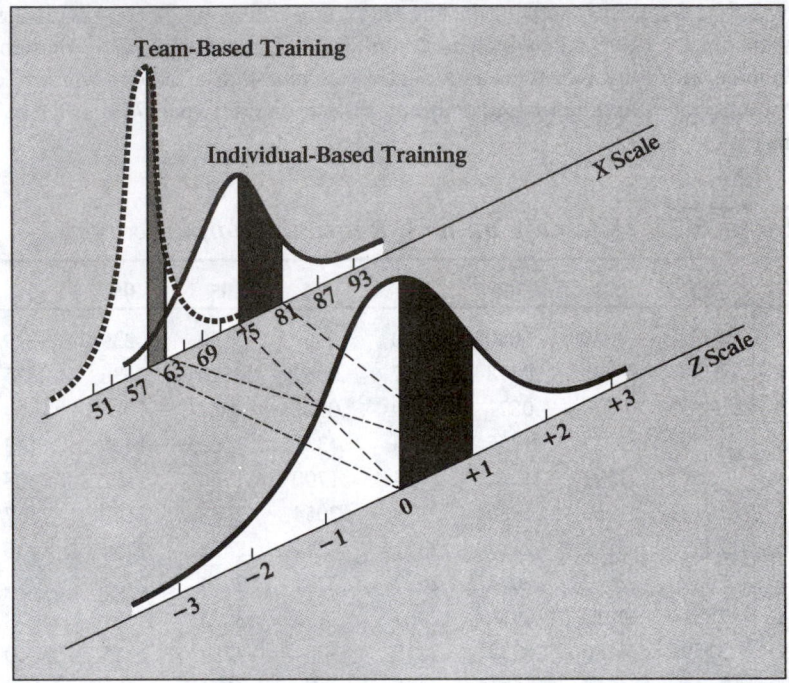

Now suppose the operations manager wishes to determine the probability that a factory worker selected at random from those who underwent individual-based training should require at most 81 seconds to complete the task. That is, what is the likelihood that the worker's time is no more than 1 standard deviation above the mean? This answer is found by using Table E.2(b).

Table E.2(b) represents the cumulative probabilities or areas under the normal curve calculated up to the particular values of interest *X*. Using equation (6.2), this corresponds to the probabilities or areas under the standardized normal curve up to the transformed values of interest *Z*. To use Table E.2(b), we note that all *Z* values must first be recorded to two decimal places. Thus, our particular *Z* value of interest is recorded as +1.00. To read the probability or area under the curve up to *Z* = +1.00, we scan down the *Z* column from Table E.2(b) until we locate the *Z* value of interest (in 10ths). Hence, we stop in the row *Z* = 1.0. Next we read across this row until we intersect the column that contains the 100ths place of the Z value. Therefore, in the body of the table, the tabulated probability for $Z = 1.00$ corresponds to the intersection of the row $Z = 1.0$ with the column $Z = .00$ as shown in Table 6.3, which duplicates Table E.2(b). This probability is .8413. As depicted in Figure 6.7, there is an 84.13% chance that a factory worker selected at random who has had individual-based training will require at most 81 seconds to assemble the part.

Table 6.3 *Obtaining a cumulative area under the normal curve*

Z	.00	.01	.02	.03	.04	.05	.06	.07	.08	.09
0.0	.5000	.5040	.5080	.5120	.5160	.5199	.5239	.5279	.5319	.5359
0.1	.5398	.5438	.5478	.5517	.5557	.5596	.5636	.5675	.5714	.5753
0.2	.5793	.5832	.5871	.5910	.5948	.5987	.6026	.6064	.6103	.6141
0.3	.6179	.6217	.6255	.6293	.6331	.6368	.6406	.6443	.6480	.6517
0.4	.6554	.6591	.6628	.6664	.6700	.6736	.6772	.6808	.6844	.6879
0.5	.6915	.6950	.6985	.7019	.7054	.7088	.7123	.7157	.7190	.7224
0.6	.7257	.7291	.7324	.7357	.7389	.7422	.7454	.7486	.7518	.7549
0.7	.7580	.7612	.7642	.7673	.7704	.7734	.7764	.7794	.7823	.7852
0.8	.7881	.7910	.7939	.7967	.7995	.8023	.8051	.8078	.8106	.8133
0.9	.8159	.8186	.8212	.8238	.8264	.8289	.8315	.8340	.8365	.8389
1.0 →	.8413	.8438	.8461	.8485	.8508	.8531	.8554	.8577	.8599	.8621

Source: Extracted from Table E.2(b).

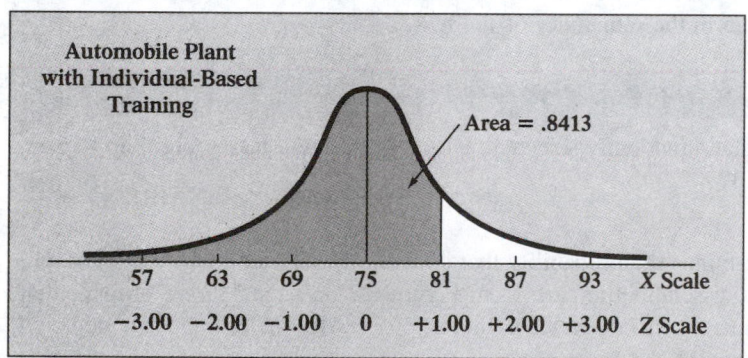

FIGURE 6.7
Determining the area up to Z from a cumulative standardized normal distribution

On the other hand, we know from Figure 6.4 on page 336 that for the workers who received team-based training, a time of 63 seconds is 1 standardized unit above the mean time of 60 seconds. Thus the likelihood that a randomly selected factory worker who received team-based training will assemble the part in at most 63 seconds is also .8413. These results are illustrated in Figure 6.8, which demonstrates that regardless of the value of the mean μ and standard deviation σ of a particular set of normally distributed data, a transformation to a standardized scale can always be made from equation (6.2) and, by using Table E.2(b), any probability or portion of area under the curve can be obtained. From Figure 6.8 we see that the probability or area under the curve up to 63 seconds for the workers who had team-based training is identical to the probability or area under the curve up to 81 seconds for the workers who had individual-based training.

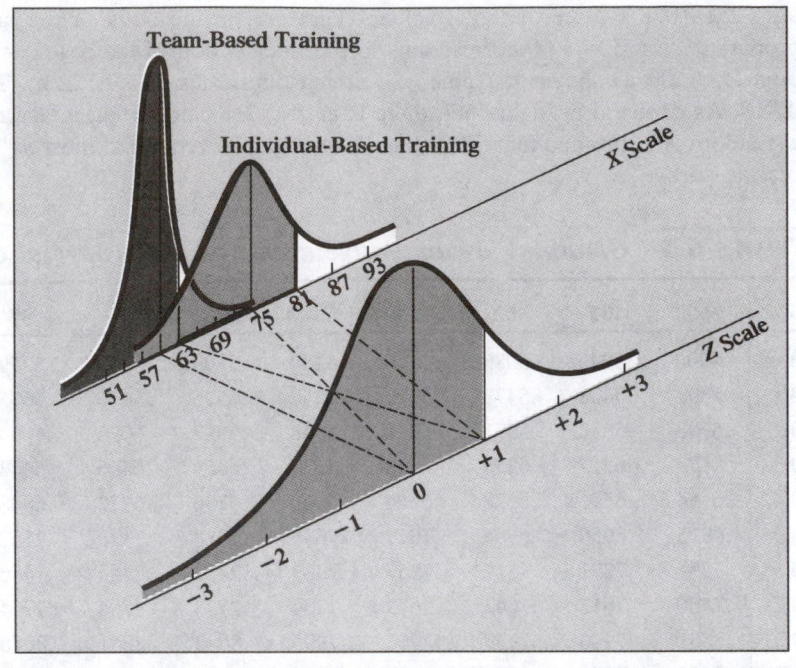

Now that we have learned to use Tables E.2(a) and E.2(b) in conjunction with equation (6.2), many different types of probability questions pertaining to the normal distribution can be resolved. To illustrate this for workers who have individual-based training, we turn to Examples 6.1–6.8. Although either Table E.2(a) or E.2(b) could be used in these examples, we will use Table E.2(b) because we will be referring to this **cumulative standardized normal distribution** table in the remainder of the text.

Example 6.1 *Finding* $P(X > 81)$

What is the probability that a randomly selected factory worker will take more than 81 seconds to assemble the part?

SOLUTION

Because we already determined the probability that a randomly selected factory worker will need up to 81 seconds to assemble the part, from Figure 6.7 on page 339 we observe that our desired probability must be its *complement,* that is, $1 - .8413 = .1587$. This is depicted in the following diagram.

Finding $P(X > 81)$

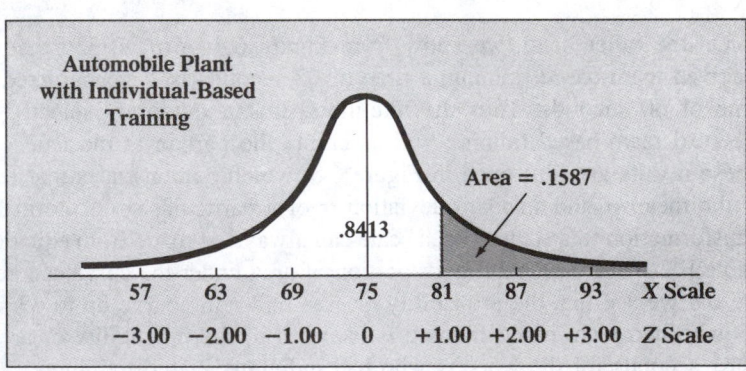

Example 6.2 *Finding P(75 < X < 81)*

What is the probability that a randomly selected factory worker will perform the task in 75 to 81 seconds?

SOLUTION

From Figure 6.7 on page 339 we already determined the probability that a randomly selected factory worker will need up to 81 seconds to assemble the part is .8413. To obtain our desired results we now must determine the probability of assembling the part in under 75 seconds and subtract this from the probability of assembling the part in under 81 seconds. This is depicted in the following diagram.

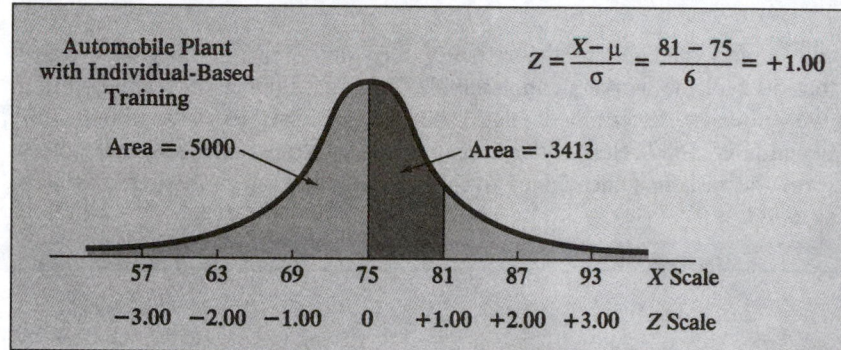

Finding $P(75 < X < 81)$

Automobile Plant with Individual-Based Training

Area = .5000

Area = .3413

$$Z = \frac{X - \mu}{\sigma} = \frac{81 - 75}{6} = +1.00$$

57	63	69	75	81	87	93	X Scale
−3.00	−2.00	−1.00	0	+1.00	+2.00	+3.00	Z Scale

Because the mean and median are theoretically the same for normally distributed data, it follows that 50% of the workers can assemble the part in under 75 seconds. To show this, from equation (6.2) we have

$$Z = \frac{X - \mu}{\sigma} = \frac{75 - 75}{6} = 0.00$$

Using Table E.2(b), we see that the area under the normal curve up to the mean of $Z = 0.00$ is .5000. Hence, the area under the curve between $Z = 0.00$ and $Z = 1.00$ must be $.8413 - .5000 = .3413$. This is the same result we obtained when using the standardized normal distribution table, Table E.2(a), depicted in Figure 6.5 on page 337.

Example 6.3 *Finding P(X < 75 or X > 81)*

What is the probability that a randomly selected factory worker will perform the task in under 75 seconds or over 81 seconds?

SOLUTION

Because we already determined that the probability is .3413 that a randomly selected factory worker will need between 75 and 81 seconds to assemble the part, from Figure 6.5 on page 337 we observe that our desired probability must be its *complement*, that is, $1 - .3413 = .6587$.

Another way to view this problem, however, is to separately obtain both the probability of assembling the part in under 75 seconds and the probability of assembling the part in over 81 seconds and to then add these two probabilities together to obtain the desired result. This is depicted in the following diagram.

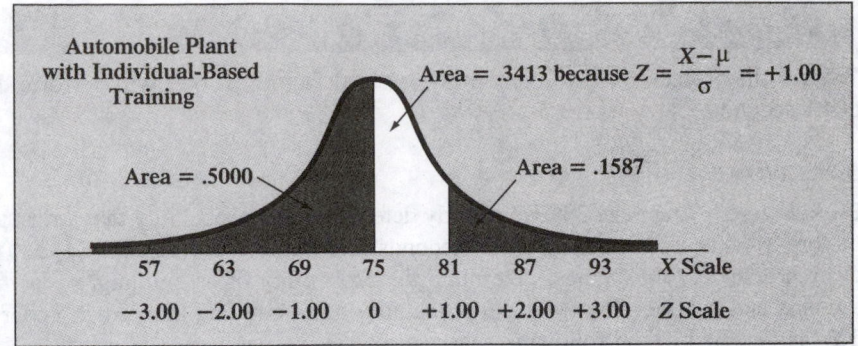

Because the mean and median are theoretically the same for normally distributed data, it follows that 50% of the workers can assemble the part in under 75 seconds. From Example 6.1 we already determined that the probability of assembling the part in over 81 seconds is .1587. Hence, the probability that a randomly selected factory worker will perform the task in under 75 or over 81 seconds, $P(X < 75$ or $X > 81)$, is .5000 + .1587 = .6587.

Example 6.4 *Finding $P(69 < X < 81)$*

What is the probability that a randomly selected factory worker can complete the assembly between 69 and 81 seconds, that is, $P(69 < X < 81)$?

SOLUTION

We note from our diagram at the top of page 343 that one of the values of interest is above the mean assembly time of 75 seconds and the other value is below it. Because our transformation formula equation (6.2) permits us to find probabilities only up to a particular value of interest, we can obtain our desired probability in three steps:

1. Determine the probability up to 81 seconds.
2. Determine the probability up to 69 seconds.
3. Subtract the smaller result from the larger.

For this example, we already completed step 1; the area under the normal curve up to 81 seconds is .8413. To find the area under the normal curve up to 69 seconds (step 2), we have

$$Z = \frac{X - \mu}{\sigma} = \frac{69 - 75}{6} = -1.00$$

Using Table E.2(b), we look up the value $Z = -1.00$ and find the probability to be .1587. Hence, from step 3, the probability that the part can be assembled in between 69 and 81 seconds is .8413 − .1587 = .6826. This is displayed in the following diagram.

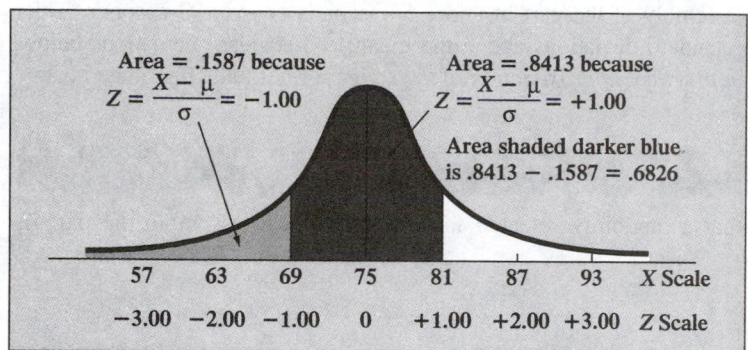

Finding $P(69 \leqslant X \leqslant 81)$

The result of Example 6.4 is rather important and allows us to generalize our findings. We can see that for any normal distribution there is a .6826 chance that a randomly selected item will fall within ±1 standard deviation above or below the mean. We know that slightly more than two out of every three (68.26%) of the factory workers who receive individual-based training can be expected to complete the task within ±1 standard deviation from the mean. Moreover, in Figure 6.9, we note that slightly more than 19 out of every 20 factory workers (95.44%) can be expected to complete the assembly within ±2 standard deviations from the mean (i.e., between 63 and 87 seconds), and, in Figure 6.10 we see that practically all factory workers (99.73%) can be expected to assemble the part within ±3 standard deviations from the mean (i.e., between 57 and 93 seconds).

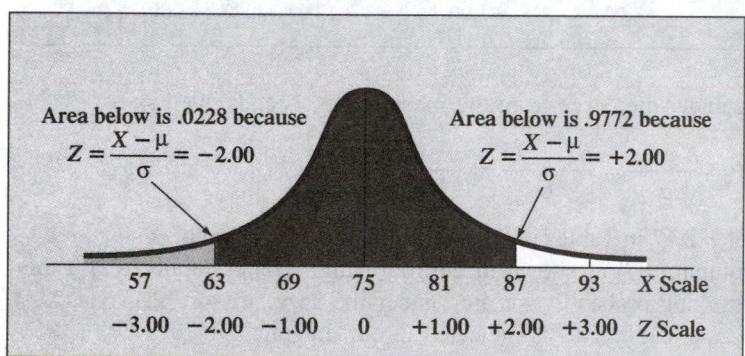

FIGURE 6.9
Finding $P(63 \leqslant X \leqslant 87)$

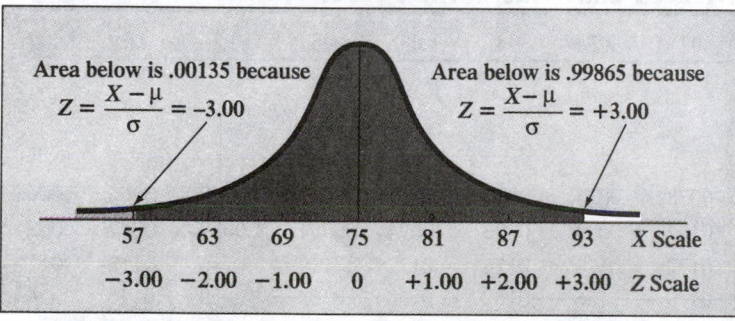

FIGURE 6.10
Finding $P(57 \leqslant X \leqslant 93)$

In Figure 6.10 we note that it is indeed quite unlikely (.0027, or only 27 factory workers in 10,000) a randomly selected factory worker will be so fast or so slow that he or she could be expected to complete the assembly of the part in under 57 seconds or over 93 seconds. Thus it is clear why 6σ (i.e., 3 standard deviations above the mean to 3 standard deviations below the mean) is often used as a *practical approximation of the range* for normally distributed data.

Example 6.5 *Finding P(X < 62)*

What is the probability that a randomly selected factory worker can assemble the part in under 62 seconds?

SOLUTION

We should examine the shaded lower left-tailed region of the accompanying diagram. The transformation formula, equation (6.2), permits us to find areas under the standardized normal distribution up to Z.

Finding $P(X < 62)$

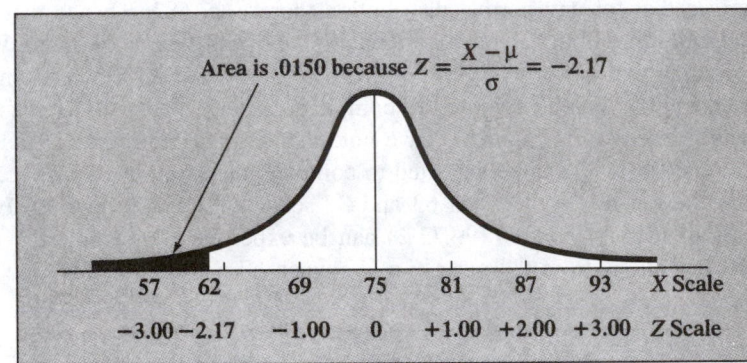

To determine the area under the curve from the mean to 62 seconds, we have

$$Z = \frac{X - \mu}{\sigma} = \frac{62 - 75}{6} = \frac{-13}{6} = -2.17$$

We look up the Z value of -2.17 in Table E.2(b) by matching the appropriate Z row (-2.1) with the appropriate Z column ($.07$) as shown in the following table [a duplicate of Table E.2(b)]. Therefore, the resulting probability or area under the curve up to -2.17 standard deviations below the mean is .0150. This is indicated in our diagram.

Obtaining a cumulative area under the normal curve

Z	.00	.01	.02	.03	.04	.05	.06	.07	.08	.09
.	.	.	.	.	.	.	.	.	.	.
.	.	.	.	.	.	.	.	.	.	.
.	.	.	.	.	.	.	.	.	.	.
−2.4	.0082	.0080	.0078	.0075	.0073	.0071	.0069	.0068	.0066	.0064
−2.3	.0107	.0104	.0102	.0099	.0096	.0094	.0091	.0089	.0087	.0084
−2.2	.0139	.0136	.0132	.0129	.0125	.0122	.0119	.0116	.0113	.0110
−2.1	.0179	.0174	.0170	.0166	.0162	.0158	.0154	.0150	.0146	.0143

Source: Extracted from Table E.2(b).

In Examples 6.1–6.5 we sought to determine the probabilities associated with various measured values. Now we wish to determine particular numerical values of the variable of interest that correspond to known probabilities. To illustrate this, we turn to Examples 6.6–6.8.

Example 6.6 *Finding the Median*

How much time (in seconds) will elapse before 50% of the factory workers assemble the part?

SOLUTION

Because this time value corresponds to the median, and the mean and median are equal in all symmetrical distributions, the median must be 75 seconds.

Finding X

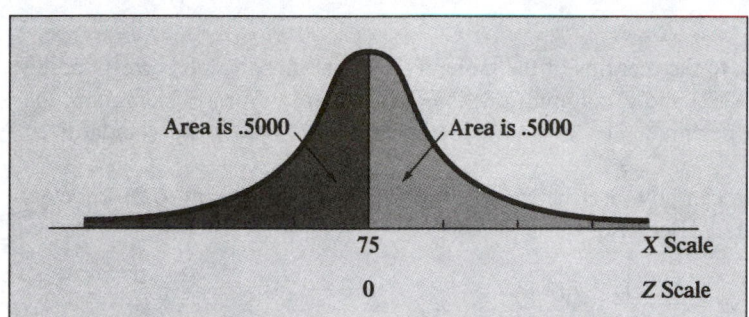

Example 6.7 *Finding the X Value for a Cumulative Probability of .10*

How much time (in seconds) will elapse before 10% of the factory workers assemble the part?

SOLUTION

Finding Z to determine X

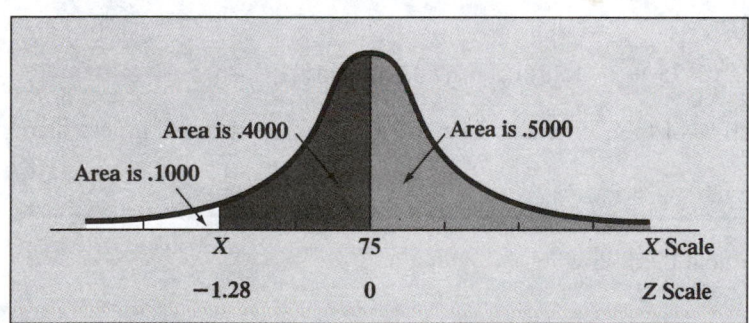

Because 10% of the factory workers are expected to complete the task in under X seconds, the area under the normal curve up to this Z-value must be .1000. Using the body of Table E.2(b), we search for the area or probability .1000. The closest result is .1003, as shown in the accompanying table [a replica of Table E.2(b)].

Obtaining a Z value corresponding to a particular cumulative area (.10) under the normal curve

Z	.00	.01	.02	.03	.04	.05	.06	.07	.08	.09
.	.	.	.	.	.	.	.	.	.	.
.	.	.	.	.	.	.	.	.	.	.
.	.	.	.	.	.	.	.	.	.	.
−1.5	.0668	.0655	.0643	.0630	.0618	.0606	.0594	.0582	.0571	.0559
−1.4	.0808	.0793	.0778	.0764	.0749	.0735	.0721	.0708	.0694	.0681
−1.3	.0968	.0951	.0934	.0918	.0901	.0885	.0869	.0853	.0838	.0823
−1.2	.1151	.1131	.1112	.1093	.1075	.1056	.1038	.1020	.1003	.0985

Source: Extracted from Table E.2(b).

Working from this area to the margins of the table, we see that the Z value corresponding to the particular Z row (−1.2) and Z column (.08) is −1.28. That is, from our diagram, the Z value is recorded as a negative (i.e., $Z = -1.28$) because it is below the standardized mean of 0.

Once Z is obtained, we can now use the transformation formula equation (6.2) to determine the value of interest, X. Because

$$Z = \frac{X - \mu}{\sigma}$$

then

$$Z\sigma = X - \mu$$

and

$$\mu + Z\sigma = X$$

or

$$X = \mu + Z\sigma$$

Substituting, we compute

$$X = 75 + (-1.28)(6) = 67.32 \text{ seconds}$$

Thus we can expect that 10% of the workers will be able to complete the task in less than 67.32 seconds.

From Example 6.7 we note the following:

The X value is equal to the population mean μ plus the product of the Z value and the standard deviation σ.

$$X = \mu + Z\sigma \tag{6.4}$$

To find a *particular* value associated with a known probability, take the steps as displayed in Exhibit 6.2.

Exhibit 6.2 Finding a Particular Value Associated with a Known Probability

To find a particular value associated with a known probability, follow these steps:

✓ **1.** Sketch the normal curve and then place the values for the means on the respective X and Z scales.

✓ **2.** Find the cumulative area less than the desired X.

✓ **3.** Shade the area of interest.

✓ **4.** Using Table E.2(b), determine the appropriate Z value corresponding to the area under the normal curve up to the desired X.

✓ **5.** Using equation (6.4), solve for X; that is,

$$X = \mu + Z\sigma$$

Example 6.8 *Obtaining the Interquartile Range*

What is the interquartile range?

SOLUTION

To obtain the interquartile range, we first find the value for Q_1 and the value for Q_3. Then we subtract the former from the latter.

To find the first quartile value, we determine the time (in seconds) for which only 25% of the factory workers can be expected to assemble the part in less time. This is depicted in the diagram.

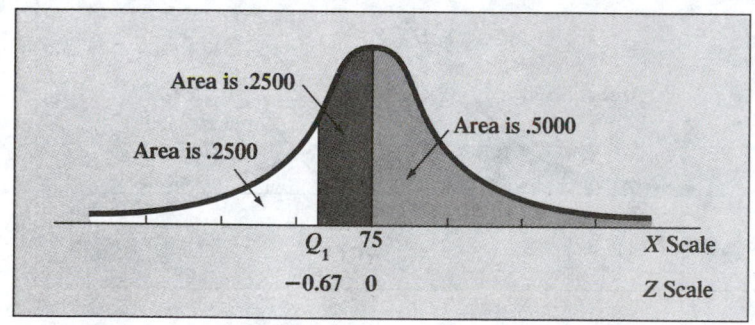

Finding Q_1

Although we do not know Q_1, we can obtain the corresponding standardized value Z because the area under the normal curve up to this Z must be .2500. Using the body of Table E.2(b), we search for the area or probability .2500. The closest result is .2514, as shown in the accompanying table, which is a replica of Table E.2(b).

Obtaining a Z value corresponding to a cumulative area of .25 under the normal curve

Z	.00	.01	.02	.03	.04	.05	.06	.07	.08	.09
.	.	.	.	.	.	.	.	.	.	.
.	.	.	.	.	.	.	.	.	.	.
.	.	.	.	.	.	.	.	.	.	.
−0.9	.1841	.1814	.1788	.1762	.1736	.1711	.1685	.1660	.1635	.1611
−0.8	.2119	.2090	.2061	.2033	.2005	.1977	.1949	.1922	.1894	.1867
−0.7	.2420	.2388	.2358	.2327	.2296	.2266	.2236	.2006	.2177	.2148
−0.6	.2743	.2709	.2676	.2643	.2611	.2578	.2546	→ .2514	.2482	.2451

Source: Extracted from Table E.2(b).

Working from this area to the margins of the table, we see that the Z value corresponding to the particular Z row (-0.6) and Z column ($.07$) is -0.67. From our diagram the Z value is recorded as a negative (i.e., $Z = -0.67$) because it lies to the left of the standardized mean of 0.

Once Z is obtained, the final step is to use equation (6.4). Hence,

$$Q_1 = X = \mu + Z\sigma$$
$$= 75 + (-0.67)(6)$$
$$= 75 - 4$$
$$= 71 \text{ seconds}$$

To find the third quartile, we determine the time (in seconds) for which 75% of the factory workers can be expected to assemble the part in less time (and 25% could complete the task in more time). This is displayed in the following diagram.

Finding Q_3

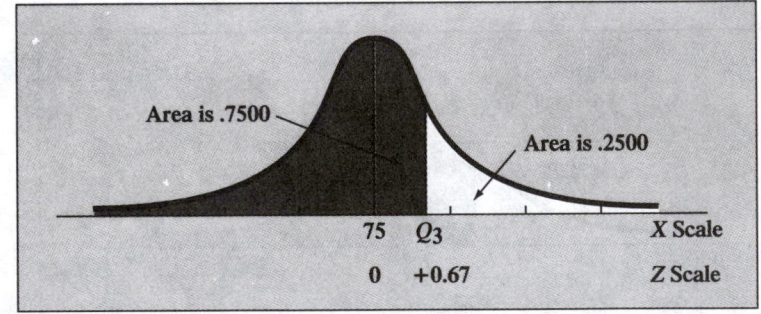

From the symmetry of the normal distribution, our desired Z value must be $+0.67$ (because Z lies to the right of the standardized mean of 0). However, this also can be seen in the following replica of Table E.2(b), where we note that .7486 of the area under the normal curve (i.e., the closest value to .75) is less than the standardized Z value of $+0.67$.

Obtaining a Z value corresponding to a cumulative area of .75 under the normal curve

Z	.00	.01	.02	.03	.04	.05	.06	.07	.08	.09
.	.	.	.	.	.	.	.	.	.	.
.	.	.	.	.	.	.	.	.	.	.
.	.	.	.	.	.	.	.	.	.	.
0.3	.6179	.6217	.6255	.6293	.6331	.6368	.6406	.6443	.6480	.6517
0.4	.6554	.6591	.6628	.6664	.6700	.6736	.6772	.6808	.6844	.6879
0.5	.6915	.6950	.6985	.7019	.7054	.7088	.7123	.7157	.7190	.7224
0.6	.7257	.7291	.7324	.7357	.7389	.7422	.7454 →	.7486	.7518	.7549

Source: *Extracted from Table E.2(b).*

Therefore, using equation (6.4), we compute

$$Q_3 = X = \mu + Z\sigma$$
$$= 75 + (+0.67)(6)$$
$$= 75 + 4$$
$$= 79 \text{ seconds}$$

The interquartile range or middle spread of the distribution is

$$\text{Interquartile range} = Q_3 - Q_1$$
$$= 79 - 71$$
$$= 8 \text{ seconds}$$

Now that we have used Table E.2(a) and E.2(b), we will illustrate how Microsoft Excel can be used to obtain normal probabilities. Panel A of Figure 6.11 illustrates output obtained from the PHStat add-in for Microsoft Excel for Examples 6.4, 6.5, and 6.7, and panel B on page 350 illustrates output for Examples 6.3, 6.4, and 6.7.

	A	B	C	D
1	Normal Probabilities			
2				
3	Mean	75		
4	Standard Deviation	6		
5				
6				
7	Probability for X<=	62		
8	Z Value	-2.166666667		
9	P(X<=62)	0.015130086		
10				
11				
12	Probability for range	69	<= X <=	81
13	Z Value for 69	-1		
14	Z Value for 81	1		
15	P(X<=69)	0.15865526		
16	P(X<=81)	0.84134474		
17	P(69<=X<=81)	0.68268948		
18				
19	Find X and Z			
20	Cumulative Percentage:	10%		
21	Z Value	-1.281550794		
22	X Value	67.31069523		

FIGURE 6.11

Obtaining normal probabilities from the PHStat add-in for Microsoft Excel

PANEL A

	A	B	C	D
3	Mean	75		
4	Standard Deviation	6		
5				
6	Probability for X<=	75		
7	Z Value	0		
8	P(X<=75)	0.5		
9				
10	Probability for X>	81		
11	Z Value	1		
12	P(X>81)	0.15865526		
13				
14	Probability for X<75 or X >81	0.65865526		
15				
16	Probability for range	69	<= X <=	81
17	Z Value for 69	-1		
18	Z Value for 81	1		
19	P(X<=69)	0.15865526		
20	P(X<=81)	0.84134474		
21	P(69<=X<=81)	0.68268948		
22				
23	Find X and Z			
24	Cumulative Percentage:	10%		
25	Z Value	-1.281550794		
26	X Value	67.31069523		

PANEL B

6.1E CALCULATING NORMAL PROBABILITIES USING MICROSOFT EXCEL

Overview

◆ *For Quick Results Users*   Use the PHStat add-in to answer probability questions pertaining to the normal distribution.

◆ *For Developers*   Implement a worksheet that uses several Excel statistical functions to answer probability questions pertaining to the normal distribution.

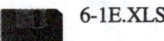

6-1E.XLS

The 6-1E.XLS workbook file contains the answers to three questions about the automobile assembly process asked in section 6.1.

Quick Results Details

To answer probability questions pertaining to the normal distribution, use the Probability Distributions | Normal choice of the PHStat add-in. As an example, consider the following questions posed in section 6.1 for the automobile assembly process that has a mean of 75 seconds and a standard deviation of 6 seconds:

1. What is the probability that a randomly selected worker can complete assembly in under 62 seconds? (See Example 6.5.)

2. What is the probability that a randomly selected worker can complete the assembly between 69 and 81 seconds? (See Example 6.4.)

3. How many seconds elapse before 10% of the workers complete the assembly? (See Example 6.7.)

324

To answer these questions, do the following:

❶ If the PHStat add-in has not been previously loaded, load the add-in using the instructions of section S4.2.

❷ Select File | New to open a new workbook (or open the existing workbook into which the Normal Probabilities worksheet is to be inserted).

❸ Select PHStat | Probability Distributions | Normal.

❹ In the Normal Probability Distribution dialog box (see Figure 6E.1):
   a. Enter 75 in the Mean: edit box.
   b. Enter 6 in the Standard Deviation: edit box.
   c. Select the Probability for: X <= check box and enter 62 in its edit box (for question 1).
   d. Select the Probability for range: check box and enter 69 and 81, respectively, in the two edit boxes for this selection (for question 2).
   e. Select the X for Cumulative Percentage: check box and enter 10 in its edit box (for question 3).
   f. Enter Normal Probabilities in the Output Title: edit box.
   g. Click the OK button.

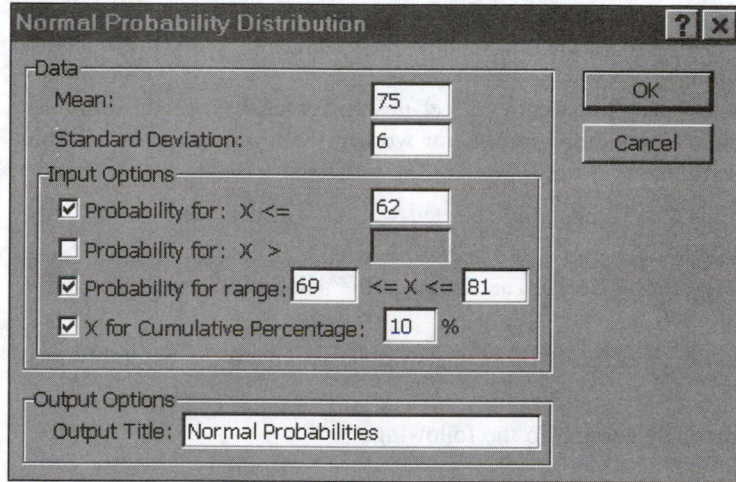

FIGURE 6E.1
PHStat Normal Probability
Distribution dialog box

The add-in inserts a worksheet that reports the probabilities, and the *X* and *Z* values, for these questions as shown in Figure 6.11 panel A, on page 349.

▲ WHAT IF EXAMPLE

We can change the values for the mean and standard deviation in the worksheet inserted by the add-in to see the effects of these values on all probabilities. For example, if we change the standard deviation from 6 to 8, the $P(X<=62)$ increases. Likewise, we can also change the *X* values in cells B7, B12, and/or D12 and the cumulative percentage value in cell B20 and observe their effects on the results.

Developer Details

We can use the STANDARDIZE, NORMDIST, NORMSINV, and NORMINV functions to implement a worksheet that uses formulas to answer probability questions pertaining to the normal distribution. The format of these functions are:

$$\text{STANDARDIZE}(X, \textit{mean}, \textit{standard deviation})$$

where

$$X = X \text{ value of interest}$$

$$\textit{mean} \text{ and } \textit{standard deviation} = \text{mean and standard deviation for a normal}$$
$$\text{probability problem}$$

$$\text{NORMDIST}(X, \textit{mean}, \textit{standard deviation}, \text{True})$$

where X, *mean*, and *standard deviation* are the same as for the STANDARDIZE function

$$\text{NORMSINV}(P < X)$$

where $P < X$ is the area under the curve that is less than X

$$\text{NORMINV}(P < X, \textit{mean}, \textit{standard deviation})$$

where $P < X$, *mean*, and *standard deviation* are as defined above

The STANDARDIZE function returns the Z value for a particular X value, mean, and standard deviation. The NORMDIST returns the area or probability of less than a given X value. The NORMSINV function returns the Z value corresponding to the probability of less than a given X. The NORMINV function returns the X value for a given probability, mean, and standard deviation.

Table 6E.1 presents a Normal Probabilities sheet design for answering the following questions about the automotive assembly process for workers with individual-based training discussed in section 6.1:

1. What is the probability that a randomly selected worker will complete the assembly task in under 75 seconds or over 81 seconds? (See Example 6.3.)

2. What is the probability that a worker can complete the assembly between 69 and 81 seconds? (See Example 6.4.)

3. How many seconds elapse before 10% of the workers complete the assembly task? (See Example 6.7.)

To implement the Table 6E.1 design, do the following:

❶ Select File | New to open a new workbook (or open the existing workbook into which the Normal worksheet is to be inserted).

❷ Select an unused worksheet (or select Insert | Worksheet if there are none) and rename the sheet Normal.

❸ Enter the title, headings, and labels for column A as shown in Table 6E.1, noting the following:
   a. The formulas for cells A8, A12, A14, A19, and A20 contain two pairs of double quotation marks.
   b. The formula for cell A21 contains three pairs of double quotation marks.

❹ Enter the formulas for column B and the label in cell C16.

❺ Enter the values for the mean and standard deviation, 75 and 6, in cells B3 and B4, respectively.

6 Enter the *X*-values and cumulative percentage values in cells B6, B10, B16, D16, and B24. Enter 75 in cell B6, 81 in cell B10, 69 in cell B16, 81 in cell D16, and 0.1 in cell B24.

The completed worksheet will be similar to the one shown in Figure 6.11, panel B.

Table 6E.1 Normal probabilities sheet design for answering probability questions pertaining to the normal distribution

	A	B	C	D
1	Normal Probabilities			
2				
3	Mean		xx	
4	Standard Deviation		xx	
5				
6	Probability for *X*<=		xx	
7	Z Value	=STANDARDIZE(B6,B3,B4)		
8	="P(X<="&B6&")"	=NORMDIST(B6,B3,B4,TRUE)		
9				
10	Probability for *X*>		xx	
11	Z Value	=STANDARDIZE(B10,B3,B4)		
12	="P(X>"&B10&")"	=1-NORMDIST(B10,B3,B4,TRUE)		
13				
14	="Probability for X<"&B6&" or X>"&B10	=B8+B12		
15				
16	Probability for range		xx <= X <=	xx
17	="Z Value for "&B16	=STANDARDIZE(B16,B3,B4)		
18	="Z Value for "&D16	=STANDARDIZE(D16,B3,B4)		
19	="P(X<="&B16&")"	=NORMDIST(B16,B3,B4,TRUE)		
20	="P(X<="&D16&")"	=NORMDIST(D16,B3,B4,TRUE)		
21	="P("&B16&"<=X<="&D16&")"	=ABS(B20-B19)		
22				
23	Find X and Z			
24	Cumulative Percentage:		xx	
25	Z Value	=NORMSINV(B24)		
26	X Value	=NORMINV(B24,B3,B4)		

Problems for Section 6.1

Learning the Basics

6.1 Given a standardized normal distribution [(with a mean of 0 and a standard deviation of 1 as in Table E.2(b)], answer the following:

(a) What is the probability that

(1) *Z* is less than 1.57?

(2) *Z* exceeds 1.84?

(3) *Z* is between 1.57 and 1.84?

(4) *Z* is less than 1.57 or greater than 1.84?

(5) *Z* is between -1.57 and 1.84?

(6) *Z* is less than -1.57 or greater than 1.84?

(b) What is the value of *Z* if 50.0% of all possible *Z* values are larger?

(c) What is the value of *Z* if only 2.5% of all possible *Z* values are larger?

(d) Between what two values of *Z* (symmetrically distributed around the mean) will 68.26% of all possible *Z*-values be contained?

6.2 Given a standardized normal distribution (with a mean of 0 and a standard deviation of 1 as in Table E.2(b), determine the following probabilities.

(a) $P(Z > +1.34)$ (d) $P(Z < -1.17)$

(b) $P(Z < +1.17)$ (e) $P(-1.17 < Z < +1.34)$

(c) $P(0 < Z < +1.17)$ (f) $P(-1.17 < Z < -0.50)$

•**6.3** Given a standardized normal distribution [(with a mean of 0 and a standard deviation of 1 as in Table E.2(b)],

(a) what is the probability that

  (1) Z is between the mean and $+1.08$?   (5) Z is at most $+1.08$?

  (2) Z is less than the mean or greater than $+1.08$?   (6) Z is at least -0.21?

      (7) Z is between -0.21 and $+1.08$?

  (3) Z is between -0.21 and the mean?   (8) Z is less than -0.21 or greater than $+1.08$?

  (4) Z is less than -0.21 or greater than the mean?

(b) determine the following probabilities:

  (1) $P(Z > +1.08)$ (4) $P(-1.96 < Z < +1.08)$

  (2) $P(Z < -0.21)$ (5) $P(+1.08 < Z < +1.96)$

  (3) $P(-1.96 < Z < -0.21)$

(c) what is the value of Z if 50.0% of all possible Z values are smaller?

(d) what is the value of Z if only 15.87% of all possible Z values are smaller?

(e) what is the value of Z if only 15.87% of all possible Z values are larger?

6.4 Verify the following:

(a) The area under the normal curve between the mean and 2 standard deviations above and below it is .9544.

(b) The area under the normal curve between the mean and 3 standard deviations above and below it is .9973.

•**6.5** Given a normal distribution with $\mu = 100$ and $\sigma = 10$,

(a) what is the probability that

  (1) $X > 75$?   (4) $X > 112$?

  (2) $X < 70$?   (5) $X < 80$ or $X > 110$?

  (3) $75 < X < 85$?

(b) 10% of the values are less than what X value?

(c) 80% of the values are between what two X values (symmetrically distributed around the mean)?

(d) 70% of the values will be above what X value?

6.6 Given a normal distribution with $\mu = 50$ and $\sigma = 4$,

(a) what is the probability that

  (1) $X > 43$?   (4) $X > 57.5$?

  (2) $X < 42$?   (5) $X < 40$ or > 55?

  (3) $42 < X < 48$?

(b) 5% of the values are less than what X value?

(c) 60% of the values are between what two X values (symmetrically distributed around the mean)?

(d) 85% of the values will be above what X value?

Applying the Concepts

6.7 Monthly food expenditures for families of four in a large city average $420 with a standard deviation of $80. Assuming the monthly food expenditures are normally distributed,

(a) what percentage of these expenditures are less than $350?

(b) what percentage of these expenditures are between $250 and $350?

(c) what percentage of these expenditures are between $250 and $450?

(d) what percentage of these expenditures are less than $250 or greater than $450?

(e) determine Q_1 and Q_3 from the normal curve.

(f) What will your answers be to (a)–(e) if the standard deviation is $100?

 **WHAT IF?**

• **6.8** Toby's Trucking Company determined that on an annual basis, the distance traveled per truck is normally distributed with a mean of 50.0 thousand miles and a standard deviation of 12.0 thousand miles.

(a) What proportion of trucks can be expected to travel between 34.0 and 50.0 thousand miles in the year?

(b) What is the probability that a randomly selected truck travels between 34.0 and 38.0 thousand miles in the year?

(c) What percentage of trucks can be expected to travel either below 30.0 or above 60.0 thousand miles in the year?

(d) How many of the 1,000 trucks in the fleet are expected to travel between 30.0 and 60.0 thousand miles in the year?

(e) How many miles will be traveled by at least 80% of the trucks?

(f) What will your answers be to (a)–(e) if the standard deviation is 10.0 thousand miles?

 **WHAT IF?**

6.9 Plastic bags used for packaging produce are manufactured so that the breaking strength of the bag is normally distributed with a mean of 5 pounds per square inch and a standard deviation of 1.5 pounds per square inch.

(a) What proportion of the bags produced have a breaking strength of

(1) between 5 and 5.5 pounds per square inch?

(2) between 3.2 and 4.2 pounds per square inch?

(3) at least 3.6 pounds per square inch?

(4) less than 3.17 pounds per square inch?

(b) Between what two values symmetrically distributed around the mean will 95% of the breaking strengths fall?

WHAT IF?

(c) What will your answers be to (a) and (b) if the standard deviation is 1.0 pound per square inch?

6.10 A set of final examination grades in an introductory statistics course was found to be normally distributed with a mean of 73 and a standard deviation of 8.

(a) What is the probability of getting a grade no higher than 91 on this exam?

(b) What percentage of students scored between 65 and 89?

(c) What percentage of students scored between 81 and 89?

(d) Only 5% of the students taking the test scored higher than what grade?

(e) If the professor "curves" (gives A's to the top 10% of the class regardless of the score), are you better off with a grade of 81 on this exam or a grade of 68 on a different exam where the mean is 62 and the standard deviation is 3? Show statistically and explain.

6.11 A statistical analysis of 1,000 long-distance telephone calls made from the headquarters of Johnson & Shurgot Corporation indicates that the length of these calls is normally distributed with $\mu = 240$ seconds and $\sigma = 40$ seconds.

(a) What percentage of these calls lasted less than 180 seconds?

(b) What is the probability that a particular call lasted between 180 and 300 seconds?

(c) How many calls lasted less than 180 seconds or more than 300 seconds?

(d) What percentage of the calls lasted between 110 and 180 seconds?

(e) What is the length of a particular call if only 1% of all calls are shorter?

• **6.12** A building contractor claims he can renovate a 200-square-foot kitchen and dining room in 40 work hours, plus or minus 5 hours (i.e., the mean and standard deviation, respectively). The work includes plumbing, electrical installation, cabinets, flooring, painting, and the installation of new appliances. Assuming, from past experience, that times to complete similar projects are normally distributed with mean and standard deviation as estimated above,

(a) what is the likelihood the project will be completed in less than 35 hours?

(b) what is the likelihood the project will be completed in between 28 hours and 32 hours?

(c) what is the likelihood the project will be completed in between 35 hours and 48 hours?

(d) 10% of such projects require more than how many hours?

(e) determine the midhinge for completion time.

(f) determine the interquartile range for completion time.

(g) What will your answers be to (a)–(f) if the standard deviation is 10 hours?

6.13 Wages for workers in a particular industry average $11.90 per hour and the standard deviation is $0.40. If the wages are assumed to be normally distributed,

(a) what percentage of workers receive wages between $10.90 and $11.90?

(b) what percentage of workers receive wages between $10.80 and $12.40?

(c) what percentage of workers receive wages between $12.20 and $13.10?

(d) what percentage of workers receive wages less than $11.00?

(e) what percentage of workers receive wages more than $12.95?

(f) what percentage of workers receive wages less than $11.00 or more than $12.95?

(g) what must the wage be if only 10% of all workers in this industry earn more?

(h) what must the wage be if 25% of all workers in this industry earn less?

(i) determine the midhinge and interquartile range of the wages in this industry.

6.2 ASSESSING THE NORMALITY ASSUMPTION

Now that we have discussed the importance of the normal distribution and described its properties as well as demonstrated how it may be applied, a very practical question must be considered. That is, we must be able to assess the likelihood that a particular data set can be assumed to be coming from an underlying normal distribution or, at least, can be adequately approximated by it.

The reader must be cautioned—*not all continuous random variables are normally distributed!* Often the continuous random phenomenon that we are interested in studying will neither follow the normal distribution nor be adequately approximated by it. Although some methods for studying such continuous phenomena are outside the scope of this text (see reference 4), *nonparametric* techniques (see reference 3) that do not depend on the particular form of the underlying random variable will be discussed in chapters 9 and 10.

Hence, for a descriptive analysis of any particular set of data, the practical question remains: How can we decide whether our data set seems to follow or at least approximate the normal distribution sufficiently to permit it to be examined using the methodology of this chapter? Two descriptive *exploratory* approaches will be taken here to evaluate the *goodness-of-fit:*

1. A comparison of the data set's characteristics with the properties of an underlying normal distribution

2. The construction of a normal probability plot

More formal *confirmatory* approaches to the goodness-of-fit of a normal distribution can be found in references 3 and 7.

Evaluating the Properties

In section 6.1 we noted that the normal distribution has several theoretical properties. We recall that it is bell-shaped and symmetrical in appearance; its measures of central tendency are all identical; its interquartile range is equal to 1.33 standard deviations; and its random variable is continuous and has an infinite range.

We also noted that, in actual practice, some of the continuous random phenomena we observe may only approximate these theoretical properties, either because the underlying population distribution may be only approximately normal or because any obtained sample data set may deviate from the theoretically expected characteristics. In such circumstances, the data may not be perfectly bell-shaped and symmetrical in appearance. The measures of central tendency will differ slightly, and the interquartile range will not be exactly equal to 1.33 standard deviations. In addition, in practice, the range of the data will not be infinite—it will be approximately equal to 6 standard deviations.

However, many continuous phenomena are neither normally distributed nor approximately normally distributed. For such phenomena, the descriptive characteristics of the respective data sets do not match well with these four properties of a normal distribution.

What then should we do to investigate the assumption of normality in our data? One approach is to check for normality by comparing and contrasting the actual data characteristics against the corresponding properties from an underlying normal distribution as illustrated in Exhibit 6.3.

Exhibit 6.3 Checking for Normality

✓ **1.** Make some tallies and plots and observe their appearances.

    A. For small- or moderate-sized data sets, construct a stem-and-leaf display and box-and-whisker plot.

    B. For large data sets, construct the frequency distribution and plot the histogram or polygon.

✓ **2.** Compute descriptive summary measures and compare the characteristics of the data with the theoretical and practical properties of the normal distribution.

    A. Obtain the mean, median, mode, midrange, and midhinge and note the similarities or differences among these five measures of central tendency.

    B. Obtain the interquartile range and standard deviation. Note how well the interquartile range can be approximated by 1.33 times the standard deviation.

    C. Obtain the range and note how well it can be approximated by 6 times the standard deviation.

✓ **3.** Make some tallies to evaluate how the observations in the data set distribute themselves.

    A. Determine whether approximately two-thirds of the observations lie between the mean ±1 standard deviation.

    B. Determine whether approximately four-fifths of the observations lie between the mean ±1.28 standard deviations.

    C. Determine whether approximately 19 out of every 20 observations lie between the mean ±2 standard deviations.

A second approach to evaluating the assumption of normality in our data is through the construction of a *normal probability plot*.

Constructing the Normal Probability Plot

You may recall that **quantiles** are defined as measures of "noncentral" location that are usually computed for summarizing large sets of numerical data. In section 3.2 we stressed the median (which splits the ordered observations in half) and the quartiles (which split the ordered observations in fourths) and mentioned other quantiles such as the deciles (which split the ordered observations in 10ths) and the percentiles (which split the ordered observations in 100ths). With this in mind, we define a normal probability plot:

> A **normal probability plot** is a two-dimensional plot of the observed data values on the *vertical* axis with their corresponding quantile values from a standardized normal distribution on the *horizontal* axis (see references 2 and 7).

If the plotted points lie either on or close to an imaginary straight line rising from the lower left corner of the graph to the upper right corner, then the data set is (at least approximately) normally distributed. On the other hand, if the plotted points deviate from this imaginary straight line in some patterned fashion, then the data set is not normally distributed and the methodology presented in this chapter may not be appropriate.

To construct and use a normal probability plot, follow the steps presented in Exhibit 6.4:

Exhibit 6.4 Steps Used in Constructing a Normal Probability Plot

To construct a normal probability plot, follow these steps:

✓ **1.** Place the values in the data set into an ordered array.

✓ **2.** Find the corresponding standard normal quantile values.

✓ **3.** Plot the corresponding pairs of points using the observed data values on the vertical axis and the associated standard normal quantile values on the horizontal axis.

✓ **4.** Assess the likelihood that the random variable of interest is (at least approximately) normally distributed by inspecting the plot for evidence of linearity (i.e., a straight line).

These steps will be described in detail.

◆ *Obtaining the Ordered Array*   Because the original data set is likely to be obtained in raw form, the observations must be rearranged from smallest to largest so that the corresponding standard normal quantile values can be obtained. Thus the original data are placed into an ordered array.

◆ *Finding the Standard Normal Quantile Values*   We know that a standard normal distribution is characterized by a mean of 0 and standard deviation of 1. Owing to its symmetry, the median from a standard normal distribution must also be 0. Therefore, in dealing with a standard normal distribution, the quantile values below the median will be negative and the quantile values above the median will be positive. However, the question we must still answer is: How can we obtain the quantile values from this distribution? The process by which we can accomplish this is known as an **inverse normal scores transformation** (see reference 3).

The following is noted: Given a data set containing n observations from a standardized normal distribution, let the symbol O_1 represent the first (and smallest) ordered or quantile value, let the symbol O_2 represent its second smallest ordered value, let the symbol O_i represent the ith smallest ordered value, and let the symbol O_n represent the largest ordered value. Because of symmetry, the standard normal quantiles O_1 and O_n will have the same numerical value—except for sign: O_1 will be negative and O_n will be positive.

- The **first standard normal quantile**, O_1, is the Z value on a standard normal distribution below which the proportion $1/(n + 1)$ of the area under the curve is contained.
- The **second standard normal quantile**, O_2, is the Z value on a standard normal distribution below which the proportion $2/(n + 1)$ of the area under the curve is contained.
- The **ith standard normal quantile**, O_i, is the Z value on a standard normal distribution below which the proportion $i/(n + 1)$ of the area under the curve is contained.
- The **nth (and largest) standard normal quantile**, O_n, is the Z value on a standard normal distribution below which the proportion $n/(n + 1)$ of the area under the curve is contained.

◆ *Making the Inverse Normal Scores Transformation* As in section 6.1, once we know the probability or area under the curve, we may use the body of Table E.2(b) to locate the appropriate area and then its corresponding standard normal ordered value in the margins of this table. Thus, in general, to find the ith standard normal ordered value from a data set containing n observations, we sketch the standard normal distribution and locate the value O_i such that the proportion $i/(n + 1)$ of the area under the curve is contained below that value. We then find this area in the body of Table E.2(b), a table of cumulative normal probabilities, and working to the margins of that table we locate the corresponding standard normal ordered value.

Example 6.9 illustrates how to obtain standard normal ordered values.

Example 6.9 *Obtaining Standard Normal Ordered Values*

Suppose we wish to obtain the 1st, 2nd, and 10th standard normal ordered values corresponding to a sample of 19 observations.

SOLUTION

The first standard normal ordered value, O_1, is that value below which the proportion $\frac{1}{n + 1} = \frac{1}{19 + 1} = \frac{1}{20} = .05$ of the area under the normal curve is contained. From our diagram we see that the area up to O_1 is .05 so that, from the body of our accompanying table, O_1 would fall halfway between -1.65 and -1.64. Because the standard normal ordered values are usually reported with two decimal places, the value -1.65 is chosen here.

Finding the first standard
normal ordered value from a
data set with 19 observations

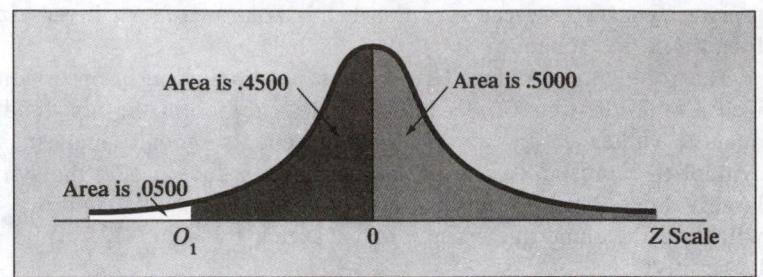

Obtaining a standard normal ordered value corresponding to a particular cumulative area (.05) under the normal curve

Z	.00	.01	.02	.03	.04	.05	.06	.07	.08	.09
.	.	.	.	.	.	.	.	.	.	.
.	.	.	.	.	.	.	.	.	.	.
.	.	.	.	.	.	.	.	.	.	.
−1.9	.0287	.0281	.0274	.0268	.0262	.0256	.0250	.0244	.0239	.0233
−1.8	.0359	.0351	.0344	.0336	.0329	.0322	.0314	.0307	.0301	.0294
−1.7	.0446	.0436	.0427	.0418	.0409	.0401	.0392	.0384	.0375	.0367
−1.6	.0548	.0537	.0526	.0516	→ .0505	.0495	.0485	.0475	.0465	.0455

Source: Extracted from Table E.2(b).

The second standard normal ordered value, O_2, is that value below which the proportion $\frac{2}{n+1} = \frac{2}{19+1} = \frac{2}{20} = .10$ of the area under the normal curve is obtained. From the following diagram and table, O_2 would fall between -1.29 and -1.28 but closer to the latter. Hence, -1.28 is selected here.

Finding the second standard
normal ordered value from a
data set with 19 observations

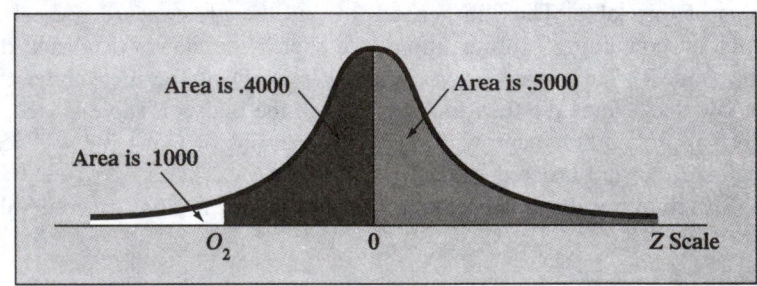

Obtaining a standard normal ordered value corresponding to a particular cumulative area (.10) under the normal curve

Z	.00	.01	.02	.03	.04	.05	.06	.07	.08	.09
.	.	.	.	.	.	.	.	.	.	.
.	.	.	.	.	.	.	.	.	.	.
.	.	.	.	.	.	.	.	.	.	.
−1.5	.0668	.0655	.0643	.0630	.0618	.0606	.0594	.0582	.0571	.0559
−1.4	.0808	.0793	.0778	.0764	.0749	.0735	.0721	.0708	.0694	.0681
−1.3	.0968	.0951	.0934	.0918	.0901	.0885	.0869	.0853	.0838	.0823
−1.2	.1151	.1131	.1112	.1093	.1075	.1056	.1038	.1020 →	.1003	.0985

Source: Extracted from Table E.2(b).

Continuing in a similar manner, for example, the 10th standard normal ordered value, O_{10}, is that value below which the proportion $\dfrac{10}{n+1} = \dfrac{10}{19+1} = \dfrac{10}{20} = .50$ of the area under the normal curve is contained. Because we have located the median, this standard normal ordered value must be 0.00.

Table 6.4 presents ordered arrays of midterm test scores from 19 students in each of four sections (sections I–IV) of a course in introductory finance. Also shown in Table 6.4 are the corresponding standard normal ordered values obtained from the previously described inverse normal scores transformation. If we construct normal probability plots for these four distinct data sets, what would they show us and how would we interpret the plots?

Table 6.4 *Ordered arrays of midterm test scores obtained from 19 students in each of four sections (I–IV) of a course in introductory finance and corresponding standard normal ordered values*

(I) BELL-SHAPED NORMAL DISTRIBUTION	(II) LEFT-SKEWED DISTRIBUTION	(III) RIGHT-SKEWED DISTRIBUTION	(IV) RECTANGULAR-SHAPED DISTRIBUTION	O_i
48	47	47	38	−1.65
52	54	48	41	−1.28
55	58	50	44	−1.04
57	61	51	47	−0.84
58	64	52	50	−0.67
60	66	53	53	−0.52
61	68	53	56	−0.39
62	71	54	59	−0.25
64	73	55	62	−0.13
65	74	56	65	0.00
66	75	57	68	0.13
68	76	59	71	0.25
69	77	62	74	0.39
70	77	64	77	0.52
72	78	66	80	0.67
73	79	69	83	0.84
75	80	72	86	1.04
78	82	76	89	1.28
82	83	83	92	1.65

The normal probability plots for the four class sections are depicted in parts (a)–(d) of Figure 6.12. From (a) we observe that the points appear to deviate from a straight line in a random manner, so we conclude that the data set from class section I is approximately normally distributed. [Note the corresponding polygon and box-and-whisker plot from Figure 3.9(a) on page 187.]

On the other hand, in Figure 6.12(b) we observe a nonlinear pattern to the plot. The points seem to rise somewhat more steeply at first and then increase at a decreasing rate. This pattern is an example of a left-skewed data set. The steepness of the left side of the plot is indicative of the elongated left tail of the distribution of test scores from class section II. [Note the corresponding polygon and box-and-whisker plot from Figure 3.10(b).]

In Figure 6.12(c) we observe the opposite nonlinear pattern. The points here seem to rise more slowly at first and then seem to increase at an increasing rate. This pattern is an example of a right-skewed data set. The steepness of the right side of the plot is indicative of the elongated right tail of the distribution of test scores from class section III. [Note the corresponding polygon and box-and-whisker plot from Figure 3.10(c).]

From Figure 6.12(d) we observe a symmetrical plot with a pattern—that is, the pattern is linear over a large middle portion of the plot. However, on each side of the plot the curve seems to flatten out. This flattening out shows the opposite effect to what was observed in parts (b) and (c) as a result of skewness. Here there are no elongated tails. In fact, there are really no tails—the test scores in class section IV are rectangularly distributed. [Note the respective corresponding polygon and box-and-whisker plot from Figure 3.10(d).]

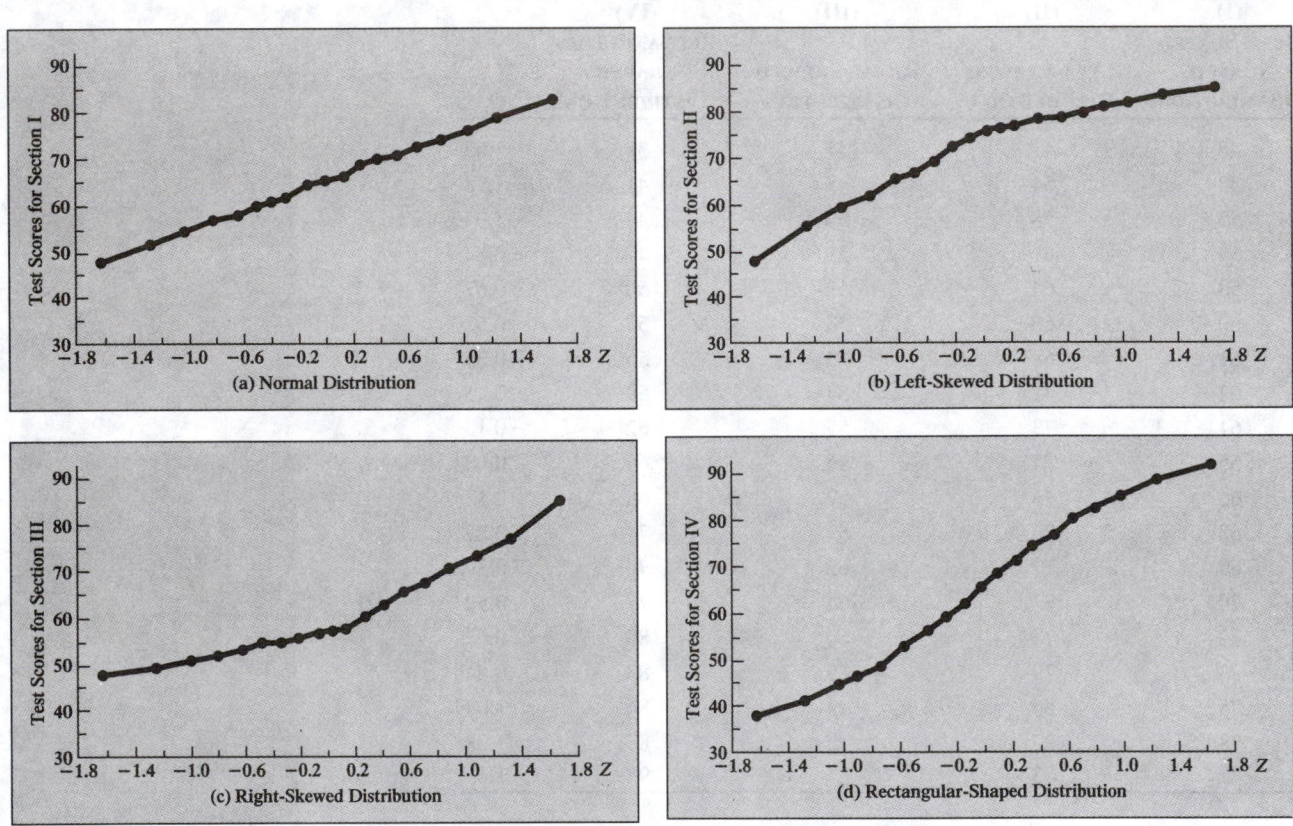

FIGURE 6.12   Normal probability plots for four data sets

Now that we have seen how to assess the normality assumption in a set of exam scores, we can demonstrate the normal probability plot with the set of stock mutual funds discussed in chapter 3. Figure 6.13 depicts the descriptive summary statistics (panel A) and the box-and-whisker plot (panel B) obtained from Microsoft Excel and the PHStat add-in for Microsoft Excel of the net asset values of 194 domestic general stock funds (see Special Data Set 1 of appendix D). Figure 6.14 on page 364 depicts the normal probability plot obtained from the PHStat add-in for Microsoft Excel.

	A	B
1	NAV	
2		
3	Mean	23.02886598
4	Standard Error	0.863628607
5	Median	19.735
6	Mode	23.24
7	Standard Deviation	12.02895457
8	Sample Variance	144.6957479
9	Kurtosis	5.799890073
10	Skewness	2.223081096
11	Range	69.53
12	Minimum	2.99
13	Maximum	72.52
14	Sum	4467.6
15	Count	194
16	Largest(1)	72.52
17	Smallest(1)	2.99

PANEL A

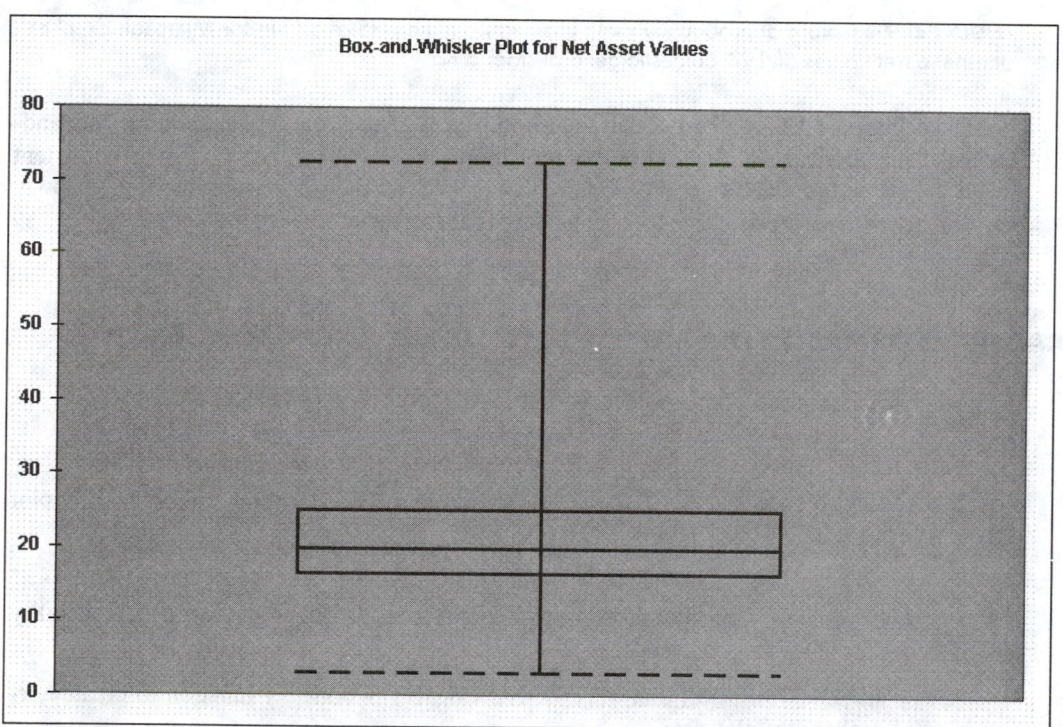

PANEL B

FIGURE 6.13   Descriptive summary measures (panel A) and box-and-whisker plot (panel B) obtained from Microsoft Excel and the PHStat add-in for Microsoft Excel

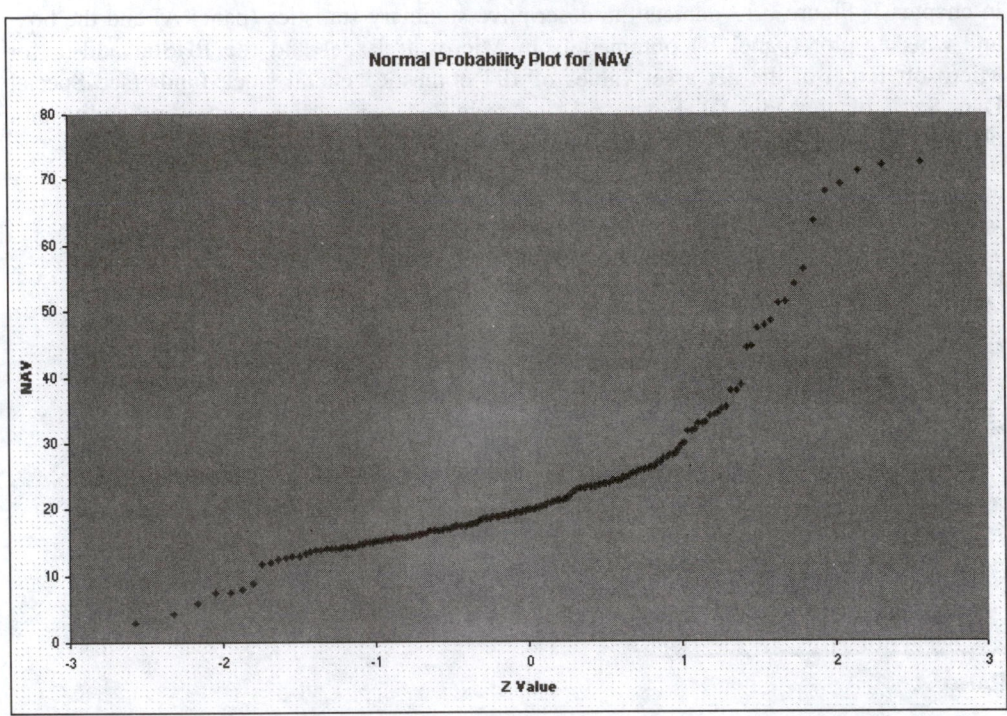

FIGURE 6.14 Normal probability plot obtained from the PHStat add-in for Microsoft Excel of the net asset values of 194 domestic general stock funds

From Figure 6.13 we observe that the mean is greater than the median and the box-and-whisker plot indicates a long tail in the upper portion of the distribution. This is consistent with the normal probability plot of Figure 6.14 that shows the net asset values rising slowly at first and then increasing quickly, indicating that the variable is right-skewed.

6.2E GENERATING NORMAL PROBABILITY PLOTS USING MICROSOFT EXCEL

Overview

◆ *For Quick Results Users* Use the PHStat add-in to generate a normal probability plot from a set of data.

◆ *For Developers* Use the following two-step process to generate a normal probability plot:

A. Implement a worksheet that uses the NORMSINV worksheet function to create normal probability plot data.

B. Use the Chart Wizard to generate a normal probability plot from this data.

6-2E.XLS

The 6-2E.XLS workbook file contains the normal probability plot for the net asset values from the sample of 194 domestic general stock funds first used in section 2.1.

Quick Results Details

To generate a normal probability plot from a set of data, use the Probability Distributions |
Normal Probability Plot choice of the PHStat add-in. As an example, consider the sample
of 194 domestic general stock funds first used in section 2.1. To generate a normal probability plot from the net asset values in this example, do the following:

❶ If the PHStat add-in has not been previously loaded, load the add-in using the
instructions of section S4.2.

❷ Open the Starting Point for Section 6.2E workbook (STARTING POINT 6-2E.XLS)
and click the Data sheet tab. Verify that the net asset values (NAV) appear in column C, sorted in ascending order.

❸ Select PHStat | Probability Distributions | Normal Probability Plot.

❹ In the Normal Probability Plot dialog box (see Figure 6E.2):
a. Enter C1:C195 in the Variable Cell Range: edit box.
b. Select the First cell contains label check box.
c. Enter Normal Probability Plot for NAV in the Output Title: edit box.
d. Click the OK button.

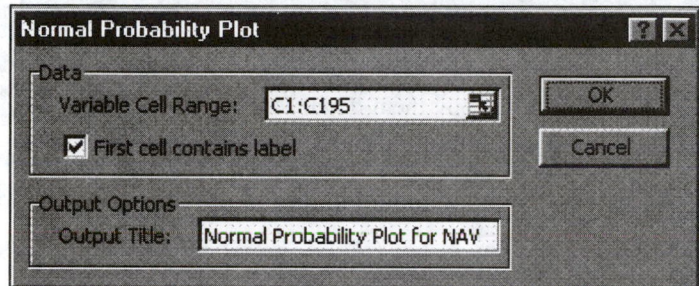

FIGURE 6E.2
PHStat Normal Probability
Plot dialog box

On separate sheets, the add-in inserts a table of ranks, cumulative proportions, Z-values,
and net asset values and a chart containing the normal probability plot for the net asset
value data (see Figure 6.14 on page 364).

Developer Details

◆ *A. Implementing a Worksheet to Generate Normal Probability Plot Data*   We can
use the NORMSINV function introduced in section 6.1E to create data for a normal probability plot. Table 6E.2 presents a partial Plot sheet design to create the normal probability
plot data for the net asset value data of the sample of 194 domestic general stock funds
first used in section 2.5. Note that the column B formulas are the proportions that correspond to each ordered value. (The denominator in these formulas is 195, equal to the sample size, 194, plus 1.) We reserve column D for values to be copied from another sheet (in
step 7). To implement this design, do the following:

❶ Open the Starting Point for Section 6.2E workbook (STARTING POINT 6-2E.XLS)
and click the Data sheet tab. Verify that the net asset values (NAV) appear in column C, sorted in ascending order.

❷ Select Insert | Worksheet and rename the new sheet Plot.

❸ Enter the row 1 headings as shown in Table 6E.2.

❹ Enter the rank values 1 through 194 in cell range A2:A195. To simplify this task, do the following:

a. Enter 1 in cell A2.

b. Select cell A2 and then select Edit | Fill | Series.

c. In the Series dialog box (see Figure 6E.3), select the Columns and Linear option buttons, enter 194 into the Stop value: edit box, and click the OK button.

FIGURE 6E.3
Series dialog box

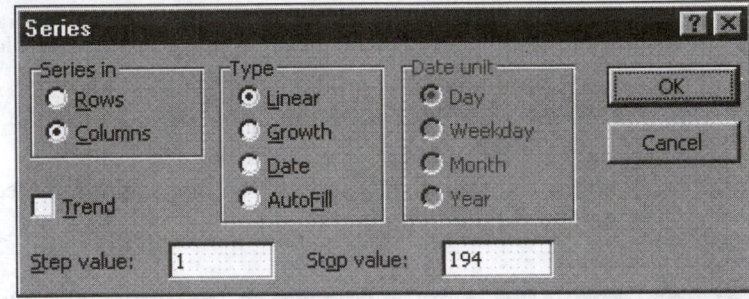

❺ Enter the formulas for cell B2 and C2 and copy them down through row 195.

❻ Click the Data sheet tab and select the range C1:C195. Select Edit | Copy.

❼ Click the Plot sheet tab and select cell D1. Select Edit | Paste.

The completed worksheet can be used as the source data for the normal probability plot as explained in the next procedure.

Table 6E.2 (Partial) Plot sheet design for generating normal probability plot data for the net asset value data from the sample of 194 domestic general stock funds

	A	B	C	D
1	Rank	Proportions	Z Value	
2	1	=A2/195	=NORMSINV(B2)	
3	2	=A3/195	=NORMSINV(B3)	**Reserved**
4	3	=A4/195	=NORMSINV(B4)	
5	4	=A5/195	=NORMSINV(B5)	**for**
6	5	=A6/195	=NORMSINV(B6)	
⋮	⋮	⋮	⋮	⋮
191	190	=A191/195	=NORMSINV(B191)	
192	191	=A192/195	=NORMSINV(B192)	**NAV**
193	192	=A193/195	=NORMSINV(B193)	
194	193	=A194/195	=NORMSINV(B194)	**data**
195	194	=A195/195	=NORMSINV(B195)	

♦ *B. Generating a Normal Probability Plot* Once implemented, the Plot worksheet can serve as the source data for a normal probability plot. To generate this chart for the net asset value data, do the following:

❶ Click the Plot sheet tab.

❷ Select Insert | Chart.

❸ In the Step 1 dialog box:
   a. Select the Standard Types tab and then select XY (Scatter) from the Chart type: list box. Select the first Chart sub-type: choice. When this choice is selected, the phrase "Scatter. Compares pairs of values" is displayed in the description box below the sub-types.
   b. Click the Next button.

❹ In the Step 2 dialog box, select the Data Range tab. Enter C1:D195 in the Data range: edit box and select the Columns option button in the Series in: group.

❺ In the Step 3 dialog box:
   a. Select the Titles tab. Enter Normal Probability Plot for NAV in the Chart title: edit box, enter Z Value in the Category (X) axis: edit box, and enter NAV in the Value (Y) axis: edit box.
   b. Select, in turn, the Axes, Gridlines, Legend, and Data Table tabs and verify that their settings match those given in Table 2E.3 on page 88.
   c. Click the Next button.

❻ In the Step 4 dialog box:
   a. Select the As new sheet: option button and enter Normal Plot in the edit box to the right of the option button.
   b. Click the Finish button to create the chart. (The floating Chart toolbar that may appear after the chart is created can be closed or dragged to the side of the application window.)

The plot generated by the Chart Wizard improperly places the Y axis in the center of the chart. To move this axis to the left side of the chart, do the following:

❼ Right-click the X axis. (The mouse pointer is over the X axis when the tool tip "Value (X) axis" is displayed.)

❽ Select Format Axis from the shortcut menu.

❾ In the Format Axis dialog box, click the Scale tab and enter -3 in the Value (Y) axis Crosses at: edit box. Click OK.

The correct normal probability plot produced will be similar to the one shown in Figure 6.14 on page 364.

Problems for Section 6.2

Learning the Basics

6.14 Show that for a sample of 19 observations, the 18th-smallest (i.e., 2nd-largest) standard normal ordered value obtained from the inverse normal scores transformation is $+1.28$ and the 19th (i.e., largest) standard normal ordered value is $+1.65$.

6.15 Show that for a sample of 39 observations, the smallest and largest standard normal ordered values obtained from the inverse normal scores transformation are, respectively, -1.96 and $+1.96$ and the middle (i.e., 20th) standard normal ordered value is 0.00.

● **6.16** Using the inverse normal scores transformation on a sample of 6 observations, list the 6 expected proportions or areas under the standardized normal curve along with their corresponding standard normal ordered values.

Applying the Concepts

● **6.17** The following ordered array (from left to right) depicts the amount of money (in dollars) withdrawn from a cash machine by 25 customers at a local bank:

40	50	50	70	70	80	80	90	100	100
100	100	100	100	110	110	120	120	130	140
140	150	160	160	200					

DATA FILE
ATM1.XLS

Decide whether or not the data appear to be approximately normally distributed by
(a) evaluating the actual versus theoretical properties.
(b) constructing a normal probability plot.

6.18 The following data indicate the amount spent (in dollars) by a random sample of 28 customers in a local supermarket:

44.24	35.56	45.93	49.92	38.94	41.16	44.84
27.28	50.66	50.97	45.93	46.58	28.73	25.93
24.21	23.84	54.58	52.62	47.36	30.84	48.62
31.15	38.58	34.96	45.32	53.81	40.22	37.19

DATA FILE
GROCERY.XLS

Decide whether or not the data appear to be approximately normally distributed by
(a) evaluating the actual versus theoretical properties.
(b) constructing a normal probability plot.

● **6.19** The following data indicate the amount of gasoline (in gallons) purchased at a highway gasoline station for a random sample of 24 automobile owners:

12.78	8.89	10.09	10.64	15.98	13.95	9.48	10.84
10.88	9.93	7.74	5.80	11.84	10.29	10.89	6.68
12.09	8.28	8.83	7.95	7.33	12.56	8.86	9.15

DATA FILE
GAS1.XLS

Decide whether or not the data appear to be approximately normally distributed by
(a) evaluating the actual versus theoretical properties.
(b) constructing a normal probability plot.

● **6.20** A problem with a telephone line that prevents a customer from receiving or making calls is disconcerting to both the customer and the telephone company. These problems can be of two types: those located inside a central office and those located on lines between the central office and the customer's equipment. The following data represent samples of 20 problems reported to two different offices of a telephone company and the time to clear these problems (in minutes) from the customers' lines:

Central Office I Time to Clear Problems (minutes)

1.48 1.75 0.78 2.85 0.52 1.60 4.15 3.97 1.48 3.10
1.02 0.53 0.93 1.60 0.80 1.05 6.32 3.93 5.45 0.97

Central Office II Time to Clear Problems (minutes)

DATA FILE
PHONE.XLS

7.55 3.75 0.10 1.10 0.60 0.52 3.30 2.10 0.58 4.02
3.75 0.65 1.92 0.60 1.53 4.23 0.08 1.48 1.65 0.72

For each of the two central office locations, decide whether or not the data appear to be approximately normally distributed by
(a) evaluating the actual versus theoretical properties.
(b) constructing a normal probability plot.

6.21 In many manufacturing processes there is a term called *work in process* (often abbreviated as WIP). In a book manufacturing plant, this represents the time it takes for sheets from a press to be folded, gathered, sewn, tipped on endsheets, and bound. The following data represent samples of 20 books at each of two production plants and the processing time (operationally defined as the time in days from when the books came off the press to when they were packed in cartons) for these jobs.

Plant A

| 5.62 | 5.29 | 16.25 | 10.92 | 11.46 | 21.62 | 8.45 | 8.58 | 5.41 | 11.42 |
| 11.62 | 7.29 | 7.50 | 7.96 | 4.42 | 10.50 | 7.58 | 9.29 | 7.54 | 8.92 |

Plant B

| 9.54 | 11.46 | 16.62 | 12.62 | 25.75 | 15.41 | 14.29 | 13.13 | 13.71 | 10.04 |
| 5.75 | 12.46 | 9.17 | 13.21 | 6.00 | 2.33 | 14.25 | 5.37 | 6.25 | 9.71 |

 **DATA FILE**
WIP.XLS

For each of the two plants, decide whether or not the data appear to be approximately normally distributed by
(a) evaluating the actual versus theoretical properties.
(b) constructing a normal probability plot.

6.3 ▶ THE EXPONENTIAL DISTRIBUTION

In this section we introduce another continuous probability distribution, the exponential distribution, that is useful in a variety of circumstances in business, particularly in evaluating manufacturing and service processes. As examples, the exponential distribution has been widely used in waiting line or queuing theory to model the length of time between arrivals in a process such as automobiles at a toll bridge crossing, customers at a bank's automatic teller machine (ATM), clients in a restaurant, or patients in a hospital emergency room.

The **exponential distribution** is defined by a single parameter, its mean λ, the average number of arrivals per unit of time. The probability that the length of time before the next arrival is less than X is given by the following:

The Exponential Distribution

The probability of an arrival in *less than or equal to X* amount of time is equal to 1 minus the mathematical constant e raised to a power equal to minus 1 times the product of the average number of arrivals λ and the value of X.

$$P(\text{arrival time} \leq X) = 1 - e^{-\lambda X} \tag{6.5}$$

where

e = the mathematical constant approximated by 2.71828

λ = the population mean number of arrivals

X = any value of the continuous random variable, where $0 < X < +\infty$

We illustrate the exponential distribution with the following example. Suppose customers arrive at a bank's ATM at the rate of 20 per hour. If a customer has just arrived, what is the probability that the next customer arrives within 6 minutes (that is, 0.1 hour)?

For this example we have $\lambda = 20$ and $X = 0.1$. Using equation (6.5), we have

$$P(\text{arrival time} \leq 0.1) = 1 - e^{-20(.1)}$$

$$P(\text{arrival time} \leq 0.1) = 1 - e^{-2}$$

$$P(\text{arrival time} \leq 0.1) = 1 - .1353 = .8647$$

Thus, the probability that a customer will arrive within 6 minutes is .8647, or 86.47%.

Now that we have used equation (6.5) to obtain exponential probabilities, we will illustrate how Microsoft Excel can be used. Figure 6.15 illustrates output obtained from the PHStat add-in for Microsoft Excel.

FIGURE 6.15

Obtaining exponential probabilities from the PHStat add-in for Microsoft Excel

	A	B	C	D	E
1	Exponential Probability for Customer Arrivals				
2					
3	Mean	20			
4	X Value	0.1			
5	P(<=X)	0.864665			

6.3E GENERATING EXPONENTIAL PROBABILITIES USING MICROSOFT EXCEL

Overview

◆ *For Quick Results Users*   Use the PHStat add-in to answer probability questions pertaining to the exponential distribution.

◆ *For Developers*   Implement a worksheet that uses the EXPONDIST worksheet function to answer probability questions pertaining to the exponential distribution.

 6-3E.XLS   The 6-3E.XLS workbook file contains the answer to the question about bank customer arrival times asked in section 6.3.

Quick Results Details

To answer probability questions pertaining to the exponential distribution, use the Probability Distributions | Exponential choice of the PHStat add-in. As an example, consider the bank customer arrival times example in section 6.3 that has a population mean of 20 per hour. To answer the question, "What is the probability that the next customer arrives within 6 minutes (0.1 hour)?" do the following:

❶ If the PHStat add-in has not been previously loaded, load the add-in using the instructions of section S4.2.

❷ Select File | New to open a new workbook (or open the existing workbook into which the probabilities worksheet is to be inserted).

❸ Select PHStat | Probability Distributions | Exponential.

❹ In the Exponential Probability Distribution dialog box (see Figure 6E.4):
a. Enter 20 in the Mean: edit box.
b. Enter 0.1 in the X Value: edit box.

344

c. Enter Exponential Probabilities for Customer Arrivals in the Output Title: edit box.

d. Click the OK button.

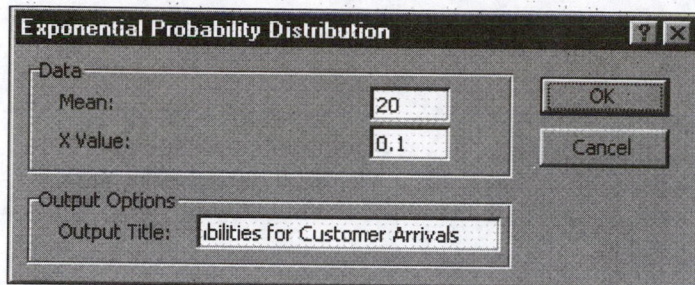

FIGURE 6E.4
PHStat Exponential Probability Distribution dialog box

The add-in inserts a worksheet containing the exponential probability similar to the one shown in Figure 6.15 on page 370.

Developer Details

We can use the EXPONDIST function as the basis for computing exponential probabilities in an Excel worksheet. The format of this function is:

$$\text{EXPONDIST}(X, \textit{mean}, \text{True})$$

where

X = the X-value of interest

$\textit{mean}$ = the population mean λ

This function returns the probability of obtaining a value less than or equal to X. Table 6E.3 presents an Exponential probabilities sheet design that uses the EXPONDIST function to calculate the answer to the question, "What is the probability that the next customer arrives within 6 minutes (0.1 hour)?" for the bank customer arrivals example discussed in section 6.3. To implement this design, do the following:

❶ Select File | New to open a new workbook (or open the existing workbook into which the probabilities worksheet is to be inserted).

❷ Select an unused worksheet (or select Insert | Worksheet if there are none) and rename the sheet Exponential.

❸ Enter the title, headings, and formula as shown in Table 6E.3.

❹ Enter 20 (the mean value) in cell B3 and enter 0.1 in cell B4.

The completed worksheet will be similar to the one shown in Figure 6.15.

Table 6E.3 Exponential probabilities sheet design for the bank customer arrival example of section 6.3

	A	B
1	Exponential Probability for Customer Arrivals	
2		
3	Mean	xx
4	X Value	.xx
5	P(<=X)	=EXPONDIST(B4,B3,TRUE)

Problems for Section 6.3

Learning the Basics

• **6.22** Given an exponential distribution with a mean of $\lambda = 10$, what is the probability that the
(a) arrival time is less than $X = .1$?
(b) arrival time is greater than $X = .1$?
(c) arrival time is between .1 and .2?
(d) arrival time is less than $X = .1$ or greater than $X = .2$?

6.23 Given an exponential distribution with a mean of $\lambda = 30$, what is the probability that the
(a) arrival time is less than $X = .1$?
(b) arrival time is greater than $X = .1$?
(c) arrival time is between .1 and .2?
(d) arrival time is less than $X = .1$ or greater than $X = .2$?

6.24 Given an exponential distribution with a mean of $\lambda = 20$, what is the probability that the
(a) arrival time is less than $X = .4$?
(b) arrival time is greater than $X = .4$?
(c) arrival time is between .4 and .5?
(d) arrival time is less than $X = .4$ or greater than $X = .5$?

Applying the Concepts

6.25 Suppose autos arrive at a toll booth located at the entrance to a bridge at the rate of 50 per minute during the 5–6 P.M. hour. If an auto has just arrived,
(a) what is the probability that the next auto arrives within 3 seconds (0.05 minute)?
(b) what is the probability that the next auto arrives within 1 second (0.0167 minute)?

 WHAT IF?

(c) What would be your answers to (a) and (b) if the rate of arrival of autos was 60 per minute?
(d) What would be your answers to (a) and (b) if the rate of arrival of autos was 30 per minute?

• **6.26** Customers arrive at the drive-up window of a fast-food restaurant at the rate of 2 per minute during the lunch hour.
(a) What is the probability that the next customer arrives within 1 minute?
(b) What is the probability that the next customer arrives within 5 minutes?

 WHAT IF?

(c) During the dinner time period, the arrival rate is 1 per minute. What would be your answers to (a) and (b) for this period?

• **6.27** Telephone calls arrive at the information desk of a large computer software company at the rate of 15 per hour.
(a) What is the probability that the next call arrives within 3 minutes (0.05 hour)?
(b) What is the probability that the next call arrives within 15 minutes (0.25 hour)?

 WHAT IF?

(c) Suppose the company has just introduced an updated version of one of its software programs and telephone calls are now arriving at the rate of 25 per hour. Given this information, what would be your answers to (a) and (b)?

6.28 An on-the-job injury occurs once every 10 days on average at an automobile plant. What is the probability that the next on-the-job injury occurs within
(a) 10 days?   (c) 1 day?
(b) 5 days?

6.29 Suppose golfers arrive at the starter's booth of a public golf course at the rate of 8 per hour during the Monday-to-Friday midweek period. If a golfer has just arrived,
(a) what is the probability that the next golfer arrives within 15 minutes (0.25 hour)?
(b) what is the probability that the next golfer arrives within 3 minutes (0.05 hour)?

 WHAT IF?

(c) Suppose the actual arrival rate on Fridays is 15 per hour. What would be your answers to (a) and (b) on Fridays?

6.4 INTRODUCTION TO SAMPLING DISTRIBUTIONS

A major goal of data analysis is to use statistics such as the sample mean and the sample proportion to estimate the corresponding parameters in the respective populations. We should realize that when making a statistical inference, we are concerned with drawing conclusions about a population, not about a sample. For example, a political pollster is interested in the sample results only as a way of estimating the actual proportion of the votes that each candidate will receive from the population of voters. Likewise, an auditor, in selecting a sample of vouchers, is interested only in using the sample mean for estimating the population average amount.

In practice, a single sample of a predetermined size is selected at random from the population. The items that are to be included in the sample are determined through the use of a random number generator, such as a table of random numbers (see section 1.8 and Table E.1), or by using Microsoft Excel (see section 6.5E2).

Hypothetically, to use the sample statistic to estimate the population parameter, we should examine *every* possible sample that could occur. If this selection of all possible samples were actually done, the distribution of the results would be referred to as a **sampling distribution.**

USING STATISTICS: *Cereal-Fill Packaging Process*

In a food processing plant, each day thousands of boxes of cereal are filled. If the machinery is not working properly, there will be either boxes that are underfilled or boxes that are overfilled. Because it would be too time-consuming, costly, and inefficient to monitor and weigh every single box, the operations manager must plan on taking a sample of boxes and making a decision regarding the likelihood that the cereal-filling process is in control and working properly. Each time a sample of boxes is selected and the individual boxes are weighed, a decision needs to be made as to the likelihood that such a sample with mean $\bar{X}$ could have been randomly drawn from a population whose true mean μ is 368 grams. On the basis of this assessment, a decision will be made as to whether to keep the production process going or shut down the equipment and call in a mechanic.

6.5 SAMPLING DISTRIBUTION OF THE MEAN

In chapter 3, we discussed several measures of central tendency. Undoubtedly, the most widely used measure of central tendency is the arithmetic mean. It is also the best measure if the population can be assumed to be normally distributed.

Properties of the Arithmetic Mean

When computed from a set of normally distributed data, the arithmetic mean contains several important mathematical properties.

Exhibit 6.5 Properties of the Arithmetic Mean

✓ **1.** It is unbiased.

✓ **2.** It is efficient.

✓ **3.** It is consistent.

The first property, **unbiasedness,** involves the fact that the average of all the possible sample means (of a given sample size n) will be equal to the population mean μ.

This property can be demonstrated empirically by the following example: Suppose that each of the four typists making up a population of secretarial support service in a company is asked to type the same page of a manuscript. The number of errors made by each typist is presented in Table 6.5.

Table 6.5 *A listing of number of errors made by each of four typists*

TYPIST	NUMBER OF ERRORS
Ann	$X_1 = 3$
Bob	$X_2 = 2$
Carla	$X_3 = 1$
Dave	$X_4 = 4$

This population distribution is shown in Figure 6.16.

FIGURE 6.16
Number of errors made by a population of four typists

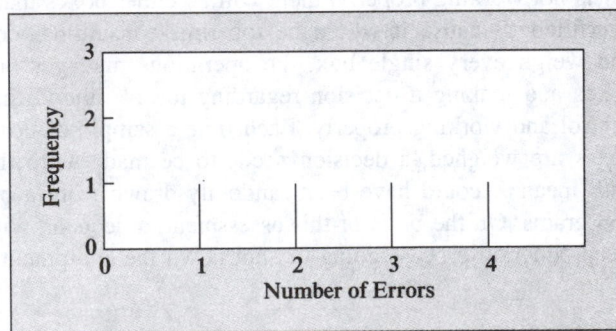

You may recall from section 3.4 that when the data from a population are available, the mean is computed as illustrated in equation (6.6):

Population Mean

The population mean μ is equal to the sum of the X values in the population divided by the population size N.

$$\mu = \frac{\sum_{i=1}^{N} X_i}{N}$$

(6.6)

The standard deviation σ is computed as in equation (6.7):

Population Standard Deviation

$$\sigma = \sqrt{\frac{\sum_{i=1}^{N}(X_i - \mu)^2}{N}}$$

(6.7)

Thus, for the data of Table 6.5,

$$\mu = \frac{3 + 2 + 1 + 4}{4} = 2.5 \text{ errors}$$

and

$$\sigma = \sqrt{\frac{(3 - 2.5)^2 + (2 - 2.5)^2 + (1 - 2.5)^2 + (4 - 2.5)^2}{4}} = 1.12 \text{ errors}$$

If samples of two typists are selected *with* replacement from this population, there are 16 possible samples that are selected ($N^n = 4^2 = 16$). These possible sample outcomes are shown in Table 6.6. If all these 16 sample means are averaged, the mean of these values ($\mu_{\bar{x}}$) is equal to 2.5, which is the mean of the population μ.

Table 6.6 *All 16 samples of n = 2 typists from a population of N = 4 typists when sampling* **with** *replacement*

SAMPLE	TYPISTS	SAMPLE OUTCOMES	SAMPLE MEAN $\bar{X}_i$
1	Ann, Ann	3, 3	$\bar{X}_1 = 3$
2	Ann, Bob	3, 2	$\bar{X}_2 = 2.5$
3	Ann, Carla	3, 1	$\bar{X}_3 = 2$
4	Ann, Dave	3, 4	$\bar{X}_4 = 3.5$
5	Bob, Ann	2, 3	$\bar{X}_5 = 2.5$
6	Bob, Bob	2, 2	$\bar{X}_6 = 2$
7	Bob, Carla	2, 1	$\bar{X}_7 = 1.5$
8	Bob, Dave	2, 4	$\bar{X}_8 = 3$
9	Carla, Ann	1, 3	$\bar{X}_9 = 2$
10	Carla, Bob	1, 2	$\bar{X}_{10} = 1.5$
11	Carla, Carla	1, 1	$\bar{X}_{11} = 1$
12	Carla, Dave	1, 4	$\bar{X}_{12} = 2.5$
13	Dave, Ann	4, 3	$\bar{X}_{13} = 3.5$
14	Dave, Bob	4, 2	$\bar{X}_{14} = 3$
15	Dave, Carla	4, 1	$\bar{X}_{15} = 2.5$
16	Dave, Dave	4, 4	$\bar{X}_{16} = 4$
			$\mu_{\bar{x}} = 2.5$

On the other hand, if sampling was being done *without* replacement, using the *rule of combinations* from section 4.5, we note that there would be six possible samples of two typists:

$$\frac{N!}{n!(N-n)!} = \frac{4!}{2!2!} = 6$$

These six possible samples are listed in Table 6.7.

Table 6.7 *All six possible samples of $n = 2$ typists from population of $N = 4$ typists when sampling* **without** *replacement*

SAMPLE	TYPISTS	SAMPLE OUTCOMES	SAMPLE MEAN $\overline{X}_i$
1	Ann, Bob	3, 2	$\overline{X}_1 = 2.5$
2	Ann, Carla	3, 1	$\overline{X}_2 = 2$
3	Ann, Dave	3, 4	$\overline{X}_3 = 3.5$
4	Bob, Carla	2, 1	$\overline{X}_4 = 1.5$
5	Bob, Dave	2, 4	$\overline{X}_5 = 3$
6	Carla, Dave	1, 4	$\overline{X}_6 = 2.5$
			$\mu_{\overline{x}} = 2.5$

In this case, also, the average of all sample means ($\mu_{\overline{x}}$) is equal to the population mean, 2.5. Therefore, the sample arithmetic mean is an unbiased estimator of the population mean. This tells us that, although we do not know how close the average of any particular sample selected comes to the population mean, we are at least assured that the average of all the possible sample means that could have been selected will be equal to the population mean.

The second property possessed by the mean, **efficiency,** refers to the precision of the sample statistic as an estimator of the population parameter. For distributions such as the normal, the arithmetic mean is considered more stable from sample to sample than are other measures of central tendency. For a sample of size *n,* the sample mean will, on average, come closer to the population mean than any other unbiased estimator, so that the sample mean is a better estimator of the population mean.

The third property, **consistency,** refers to the effect of the sample size on the usefulness of an estimator. As the sample size increases, the variation of the sample mean from the population mean becomes smaller so that the sample arithmetic mean becomes a better estimator of the population mean.

Standard Error of the Mean

The fluctuation in the average number of typing errors obtained from all 16 possible samples when sampling *with* replacement is illustrated in Figure 6.17. In this small example, although we can observe a good deal of fluctuation in the sample mean—depending on which typists were selected—there is not nearly as much fluctuation as in the actual population itself. The fact that the sample means are less variable than the population data follows directly from the **law of large numbers.** A particular sample mean averages together all the values in the sample. A population may consist of individual outcomes that can take on a wide range of values from extremely small to extremely large. However, if an extreme value falls into the sample, although it will have an effect on the mean, the effect will be reduced because it is being averaged in with all the other values in the sample. As the sample size increases, the effect of a single extreme value gets even smaller because it is being averaged with more observations.

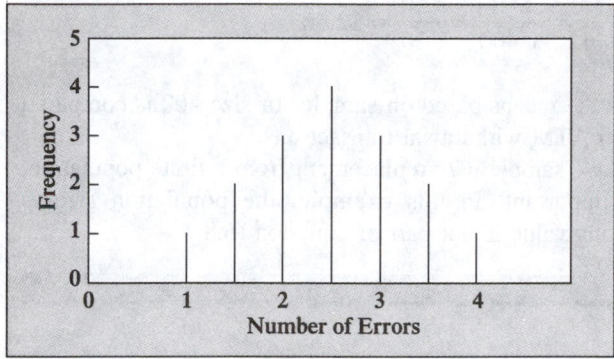

FIGURE 6.17

Sampling distribution of the average number of errors based on all possible samples containing two typists
Source: Data are from Table 6.6

How the arithmetic mean fluctuates from sample to sample is expressed statistically by the value of the standard deviation of all possible sample means. This measure of variability in the mean from sample to sample is referred to as the **standard error of the mean,** $\sigma_{\bar{x}}$. When sampling *with* replacement, we define the standard error of the mean as in equation (6.8):

Standard Error of the Mean

The standard error of the mean $\sigma_{\bar{x}}$ is equal to the standard deviation in the population σ divided by the square root of the sample size n.

$$\sigma_{\bar{x}} = \frac{\sigma}{\sqrt{n}} \qquad (6.8)$$

Therefore, as the sample size increases, the standard error of the mean decreases by a factor equal to the square root of the sample size. This relationship between the standard error of the mean and the sample size is further examined in chapter 7, where we address the issue of sample size determination.

Example 6.10 *Computing the Standard Error of the Mean*

Suppose we consider the various withdrawal transactions made from the ATM (automatic teller machine) of a particular bank over a 1-year period to represent a population and, over the year, 36,500 such transactions are made. If this population's standard deviation of the

351

amount of money withdrawn per transaction is $50, what is the standard error of the mean if samples of size 400 are selected *with* replacement?

SOLUTION

From equation (6.8) we have $\sigma_{\bar{x}} = \dfrac{\sigma}{\sqrt{n}}$. Here, $\sigma = \$50$ and $n = 400$ so that

$$\sigma_{\bar{x}} = \frac{\sigma}{\sqrt{n}} = \frac{\$50}{\sqrt{400}} = \frac{\$50}{20} = \$2.50$$

Note how much smaller the spread in the means based on samples of size 400 is compared with the spread in the original 36,500 ATM withdrawal transactions.

Finally, we should note that when we sample *with* replacement from a finite population, the size of the population is not important. In this example, the population size is $N = 36,500$ transactions. However, this value is not part of equation (6.8).

Sampling from Normally Distributed Populations

Now that we have introduced the idea of a sampling distribution and defined the standard error of the mean, we need to explore the question of what distribution the sample mean $\bar{X}$ will follow if we compute the average from all possible samples, each of size n. It can be shown that if we sample *with* replacement from a population that is normally distributed with mean μ and standard deviation σ, the **sampling distribution of the mean** will also be normally distributed for *any size n* with mean $\mu_{\bar{x}} = \mu$ and have a standard error of the mean $\sigma_{\bar{x}}$.

In the most elementary case, if we draw samples of size $n = 1$, each possible sample mean is a single observation from the population because

$$\bar{X} = \frac{\sum\limits_{i=1}^{n} X_i}{n} = \frac{X_i}{1} = X_i$$

If we know that the population is normally distributed with mean μ and standard deviation σ, then the sampling distribution of $\bar{X}$ for samples of $n = 1$ must also follow the normal distribution with mean $\mu_{\bar{x}} = \mu$ and standard error of the mean $\sigma_{\bar{x}} = \sigma/\sqrt{1} = \sigma$. In addition, we note that, as the sample size increases, the sampling distribution of the mean still follows a normal distribution with mean $\mu_{\bar{x}} = \mu$. However, as the sample size increases the standard error of the mean decreases so that a larger proportion of sample means are closer to the population mean. This can be observed by referring to Figure 6.18 where 500 samples of sizes 1, 2, 4, 8, 16, and 32 were randomly selected from a normally distributed population. We can see clearly from the polygons in Figure 6.18 that, although the sampling distribution of the mean is approximately[1] normal for each sample size, the sample means are distributed more tightly around the population mean as the sample size is increased.

[1] We must remember that "only" 500 samples out of an infinite number of samples have been selected so that the sampling distributions shown are only approximations of the true distributions.

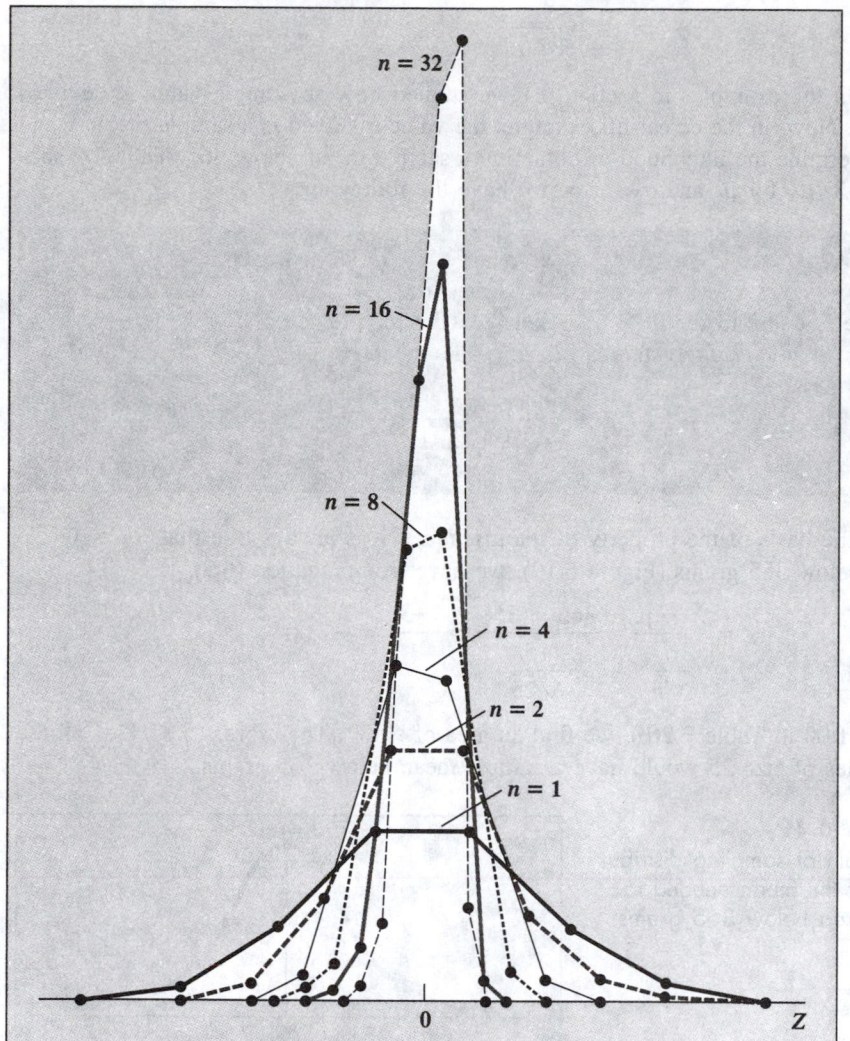

For a deeper insight into the concept of the sampling distribution of the mean let us reexamine the Using Statistics scenario described on page 373. Suppose the packaging equipment in a manufacturing process that is filling 368-gram (i.e., 13-ounce) boxes of cereal is set so that the amount of cereal in a box is normally distributed with a mean of 368 grams. From past experience, the population standard deviation for this filling process is known to be 15 grams.

If a sample of 25 boxes is randomly selected from the many thousands that are filled in a day and the average weight is computed for this sample, what type of result could be expected? For example, do you think that the sample mean would be 368 grams? 200 grams? 365 grams?

The sample acts as a miniature representation of the population so that if the values in the population are normally distributed, the values in the sample should be approximately normally distributed. Thus, if the population mean is 368 grams, the sample mean has a good chance of being close to 368 grams.

To explore this even further, how can we determine the probability that the sample of 25 boxes will have a mean below 365 grams? We know from our study of the normal distribution (section 6.1) that the area below any value X can be found by converting to standardized Z units and finding the appropriate value in the table of the normal distribution,

$$Z = \frac{X - \mu}{\sigma}$$

Table E.2(b). In the examples in section 6.1 we studied how any single value X deviates from the mean. Now, in the cereal-fill example, the value involved is a sample mean $\overline{X}$ and we wish to determine the likelihood of obtaining a sample mean below 365. Thus, by substituting $\overline{X}$ for X, $\mu_{\overline{x}}$ for μ, and $\sigma_{\overline{x}}$ for σ, we have the following:

Finding Z for the Sampling Distribution of the Mean

The Z value is equal to the difference between the sample mean $\overline{X}$ and the population mean μ, divided by the standard error of the mean $\sigma_{\overline{x}}$.

$$Z = \frac{\overline{X} - \mu_{\overline{x}}}{\sigma_{\overline{x}}} = \frac{\overline{X} - \mu}{\dfrac{\sigma}{\sqrt{n}}} \qquad (6.9)$$

Note that, on the basis of the property of *unbiasedness*, it is always true that $\mu_{\overline{x}} = \mu$. To find the area below 365 grams (Figure 6.19), we have from equation (6.9)

$$Z = \frac{\overline{X} - \mu}{\dfrac{\sigma}{\sqrt{n}}} = \frac{365 - 368}{\dfrac{15}{\sqrt{25}}} = \frac{-3}{3} = -1.00$$

Looking up -1.00 in Table E.2(b), we find an area of .1587. Therefore, 15.87% of all the possible samples of size 25 would have a sample mean below 365 grams.

FIGURE 6.19
Diagram of sampling distribution of the mean needed to find area below 365 grams

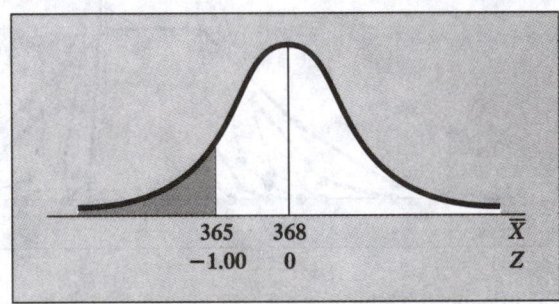

We must realize that this is not the same as saying that a certain percentage of *individual* boxes will have less than 365 grams of cereal. In fact, that percentage can be computed as follows:

$$Z = \frac{X - \mu}{\sigma} = \frac{365 - 368}{15} = \frac{-3}{15} = -0.20$$

The area corresponding to $Z = -0.20$ in Table E.2(b) is .4207. Therefore, 42.07% of the *individual* boxes are expected to contain less than 365 grams. Comparing these results, we may observe that many more *individual boxes* than *sample means* will be below 365 grams. This result can be explained by the fact that each sample consists of 25 different values, some small and some large. The averaging process dilutes the importance of any individual value, particularly when the sample size is large. Thus, the chance that the mean of a sample of 25 will be far away from the population mean is less than the chance that a *single individual* value will be.

Our results are affected by using a different sample size as observed in Examples 6.11 and 6.12.

Example 6.11   *The Effect of Sample Size n on the Computation of $\sigma_{\bar{x}}$*

How is the standard error of the mean affected by increasing the sample size from 25 to 100 boxes?

SOLUTION

If n is 100 boxes, we have the following:

$$\sigma_{\bar{x}} = \frac{\sigma}{\sqrt{n}} = \frac{15}{\sqrt{100}} = \frac{15}{10} = 1.5$$

We observe that a fourfold increase in the sample size from 25 to 100 results in reducing the standard error of the mean by half—from 3 grams to 1.5 grams. This demonstrates that taking a larger sample results in less variability in the possible sample means from sample to sample.

Example 6.12   *The Effect of Sample Size n on the Clustering of Means in the Sampling Distribution*

In the cereal-fill example, if a sample of 100 boxes is selected, what is the likelihood of obtaining a sample mean below 365 grams?

SOLUTION

Using equation (6.9), we have the following:

$$Z = \frac{\bar{X} - \mu}{\dfrac{\sigma}{\sqrt{n}}} = \frac{365 - 368}{\dfrac{15}{\sqrt{100}}} = \frac{-3}{1.5} = -2.00$$

From Table E.2(b), the area less than $Z = -2.00$ is .0228. Therefore, 2.28% of the samples of size 100 would be expected to have means below 365 grams, as compared with 15.87% for samples of size 25.

This important relationship between sample size n and the variability in the sample means from sample to sample in the sampling distribution of the mean is further highlighted in Example 6.13.

Example 6.13   *Computing the Standard Error of the Mean*

In Example 6.10 on page 377, we considered the various withdrawal transactions made from the ATM (automatic teller machine) of a particular bank over a 1-year period to represent a population and, over the year, 36,500 such transactions were made. This population's standard deviation of the amount of money withdrawn per transaction was $50. Using equation (6.8), based on samples of size 400 transactions selected *with* replacement, the standard error of the mean was computed to be

$$\sigma_{\bar{x}} = \frac{\sigma}{\sqrt{n}} = \frac{\$50}{\sqrt{400}} = \frac{\$50}{20} = \$2.50$$

Now suppose the sample size had been smaller. For example, suppose that samples of size $n = 100$ transactions were selected *with* replacement. Here

$$\sigma_{\bar{x}} = \frac{\sigma}{\sqrt{n}} = \frac{\$50}{\sqrt{100}} = \frac{\$50}{10} = \$5.00$$

so that when the sample size is smaller the standard error is larger.

Sometimes we are more interested in finding out the interval within which a fixed proportion of the samples (means) fall. Analogously to section 6.1, we are determining a distance below and above the population mean containing a specific area of the normal curve. From equation (6.9), we have

$$Z_L = \frac{\bar{X}_L - \mu}{\dfrac{\sigma}{\sqrt{n}}}$$

where

$$Z_L = -Z$$

and

$$Z_U = \frac{\bar{X}_U - \mu}{\dfrac{\sigma}{\sqrt{n}}}$$

where

$$Z_U = +Z$$

Therefore, the lower value of $\bar{X}$ is

Finding the Lower Value of $\bar{X}$

$$\bar{X}_L = \mu - Z\frac{\sigma}{\sqrt{n}} \qquad (6.10)$$

and the upper value of $\bar{X}$ is

Finding the Upper Value of $\bar{X}$

$$\bar{X}_U = \mu + Z\frac{\sigma}{\sqrt{n}} \qquad (6.11)$$

This is observed in Example 6.14.

Example 6.14 *Determining the Interval that Includes a Fixed Proportion of the Means*

For our cereal-fill example suppose we want to find an interval around the population mean that will include 95% of the sample means based on samples of 25 boxes.

SOLUTION

The 95% is divided into two equal parts, half below the mean and half above the mean, as shown in the accompanying diagram. Because $\sigma = 15$ and $n = 25$, the value of Z corresponding to an area of .0250 in the lower tail of the normal curve in Table E.2(b) is -1.96, and the value of Z corresponding to a cumulative area of .975 (that is, .025 in the upper tail of the normal curve) in Table E.2(b) is $+1.96$. The lower and upper values of $\overline{X}$ are found using equations (6.10) and (6.11) as follows:

$$\overline{X}_L = 368 - (1.96)\frac{15}{\sqrt{25}} = 368 - 5.88 = 362.12$$

$$\overline{X}_U = 368 + (1.96)\frac{15}{\sqrt{25}} = 368 + 5.88 = 373.88$$

We conclude that 95% of all sample means based on samples of 25 boxes will fall between 362.12 and 373.88 grams.

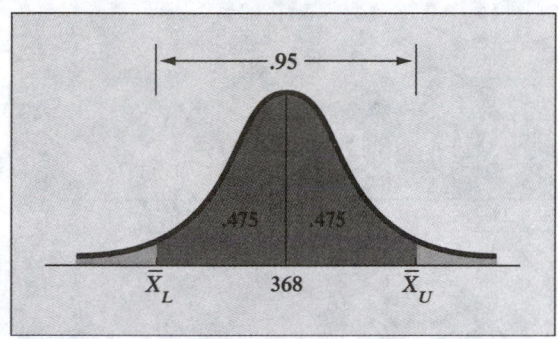

Diagram of sampling distribution of the mean needed to find upper and lower limits to include 95% of sample means

Sampling from Nonnormally Distributed Populations

In this section we have explored the sampling distribution of the mean for the case in which the population itself was normally distributed. However, in many instances, either we will know that the population is not normally distributed or that it is unrealistic to assume a normal distribution. Thus we need to examine the sampling distribution of the mean for populations that are not normally distributed. This issue brings us to an important theorem in statistics, the *central limit theorem*.

> **Central limit theorem:** As the sample size (number of observations in each sample) gets *large enough*, the sampling distribution of the mean can be approximated by the normal distribution. This is true regardless of the shape of the distribution of the individual values in the population.

What sample size is large enough? A great deal of statistical research has gone into this issue. As a general rule, statisticians have found that for many population distributions, once the sample size is at least 30, the sampling distribution of the mean will be approximately normal. However, we may be able to apply the central limit theorem for even smaller sample sizes if some knowledge of the population is available (e.g., if the distribution is symmetrical).[2]

The application of the central limit theorem to different populations can be illustrated by referring to Figures 6.20–6.22 on pages 384–386. Each of the depicted sampling distributions has been obtained by selecting 500 different samples from their respective

[2]*In some cases, sample sizes of 300 may not be enough to ensure normality.*

population distributions. These samples were selected for varying sizes ($n = 2, 4, 8, 16, 32$) from three different continuous distributions (normal, uniform, and exponential).

Figure 6.20 illustrates the sampling distribution of the mean selected from a normal population. In the preceding section we stated that if the population is normally distributed, the sampling distribution of the mean is normally distributed, regardless of the sample size. An examination of the sampling distributions shown in Figure 6.20 gives empirical evidence for this statement. For each sample size studied, the sampling distribution of the mean is *close* to the normal distribution that has been superimposed.

FIGURE 6.20

Normal distribution and sampling distribution of the mean from 500 samples of sizes n = 2, 4, 8, 16, 32

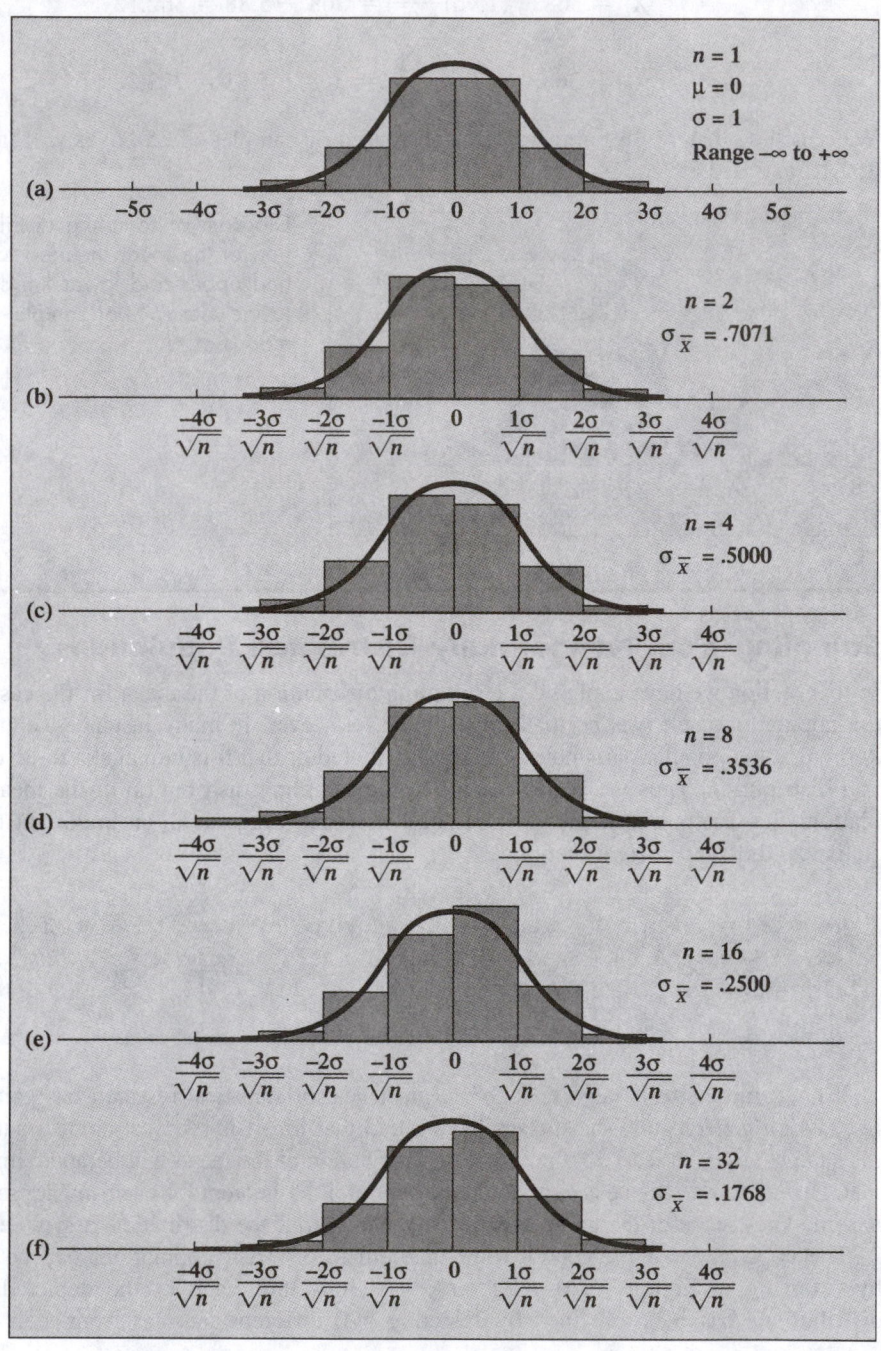

Figure 6.21 presents the sampling distribution of the mean based on a population that follows a continuous uniform (or rectangular) distribution. As depicted in (a), for samples of size $n = 1$, each value in the population is equally likely. However, when samples of only two are selected, there is a peaking or *central limiting* effect already working. In this case, we observe more values close to the mean of the population than far out at the extremes. As the sample size increases, the sampling distribution of the mean rapidly approaches a normal distribution. Once there are samples of at least eight observations, the sample mean approximately follows a normal distribution.

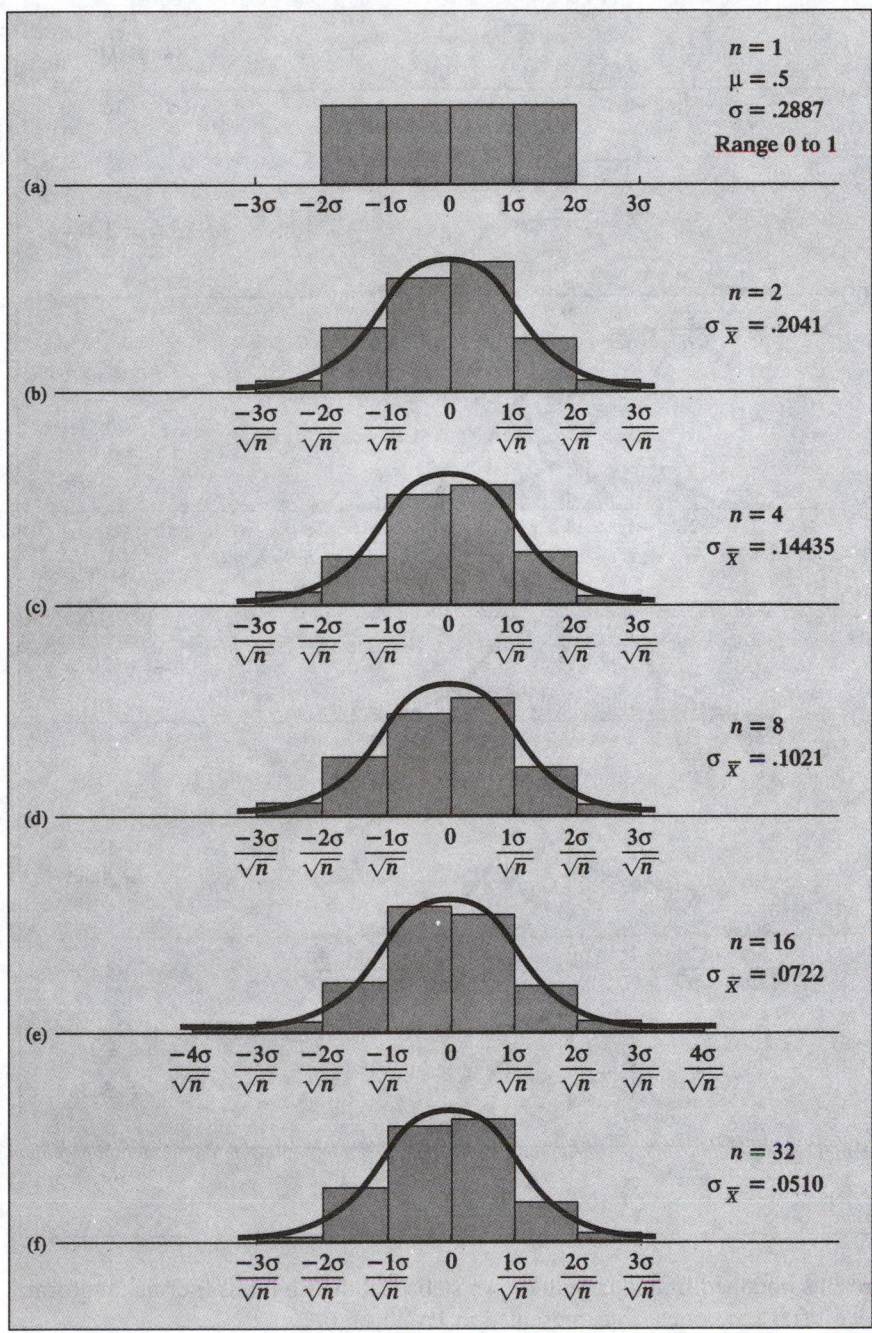

FIGURE 6.21

Continuous uniform (rectangular) distribution and sampling distribution of the mean from 500 samples of sizes $n = 2$, 4, 8, 16, 32

Figure 6.22 depicts the sampling distribution of the mean obtained from an exponential distribution (see section 6.3). From Figure 6.22 we note that as the sample size increases the sampling distribution becomes less skewed. When samples of size 16 are taken, the distribution of the mean is slightly skewed, whereas for samples of size 32, the sampling distribution of the mean appears to be normally distributed.

FIGURE 6.22

Exponential distribution and sampling distribution of the mean from 500 samples of size $n = 2, 4, 8, 16, 32$

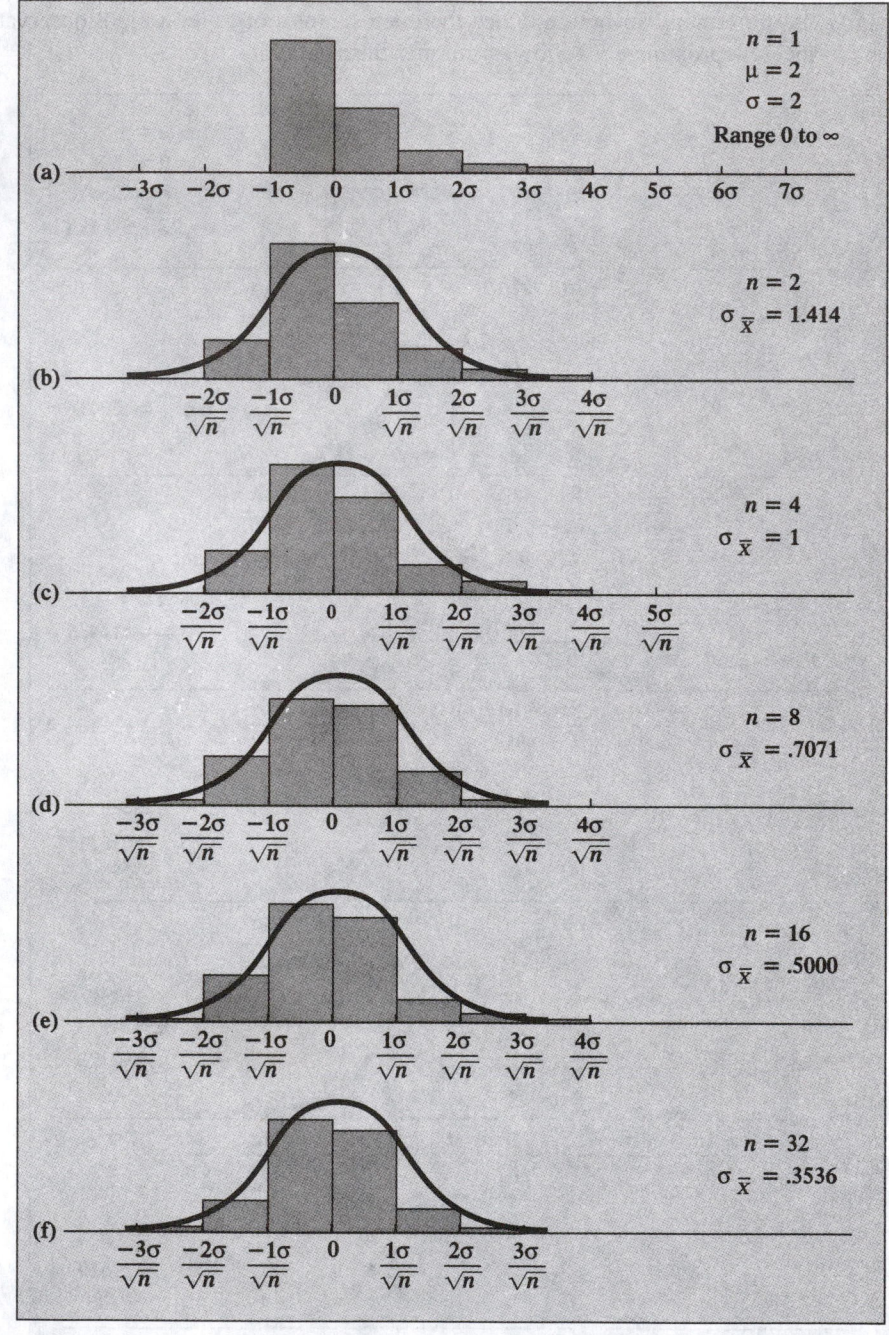

Using the results obtained from our well-known statistical distributions (normal, uniform, exponential), we offer the conclusions presented in Exhibit 6.6.

Exhibit 6.6 Normality and the Sampling Distribution of the Mean

✓ **1.** For most population distributions, regardless of shape, the sampling distribution of the mean is approximately normally distributed if samples of at least 30 observations are selected.

✓ **2.** If the population distribution is fairly symmetrical, the sampling distribution of the mean is approximately normal if samples of at least 15 observations are selected.

✓ **3.** If the population is normally distributed, the sampling distribution of the mean is normally distributed regardless of the sample size.

The central limit theorem is, then, of crucial importance in using statistical inference to draw conclusions about a population. It allows us to make inferences about the population mean without having to know the specific shape of the population distribution.

 **SIMULATING SAMPLING DISTRIBUTIONS USING MICROSOFT EXCEL**

Overview

◆ *For Quick Results Users* Use the PHStat add-in to generate a worksheet containing a simulated sampling distribution of the mean and a chart sheet containing a histogram based on the simulated distribution.

◆ *For Developers* Use either the quick results procedure or this three-step process to generate simulated sampling distributions of the mean:

A. Use the Data Analysis Random Number Generator tool to generate a worksheet containing multiple samples as columns.

B. Add formulas to this worksheet to calculate the sample means and other appropriate measures.

C. Use the Chart Wizard to generate a histogram from the calculated sample means.

The 6-5E.XLS workbook file contains simulated sampling distributions of the mean and histograms for uniformly distributed, standardized normal, and (right-skewed) discrete populations formed from 100 samples of sample size 30.

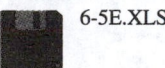 6-5E.XLS

Quick Results Details

To generate simulated sampling distributions from a uniformly distributed, standardized normally distributed, or discrete population (see section 6.5), use the Probability Distributions | Sampling Distributions Simulation choice of the PHStat add-in. For example, to generate a simulated sampling distribution from a uniform or standardized normal population, using 100 samples of sample size 30, do the following:

❶ Select File | New to open a new workbook (or open the existing workbook into which the simulated distribution worksheet is to be inserted).

② Select PHStat | Probability Distributions | Sampling Distributions Simulation.

③ In the Sampling Distributions Simulation dialog box (see Figure 6E.5, panel A):
  a. Enter 100 in the Number of Samples: edit box.
  b. Enter 30 in the Sample Size: edit box.
  c. Select either the Uniform **or** Standardized Normal option button, depending on the type of simulation desired.

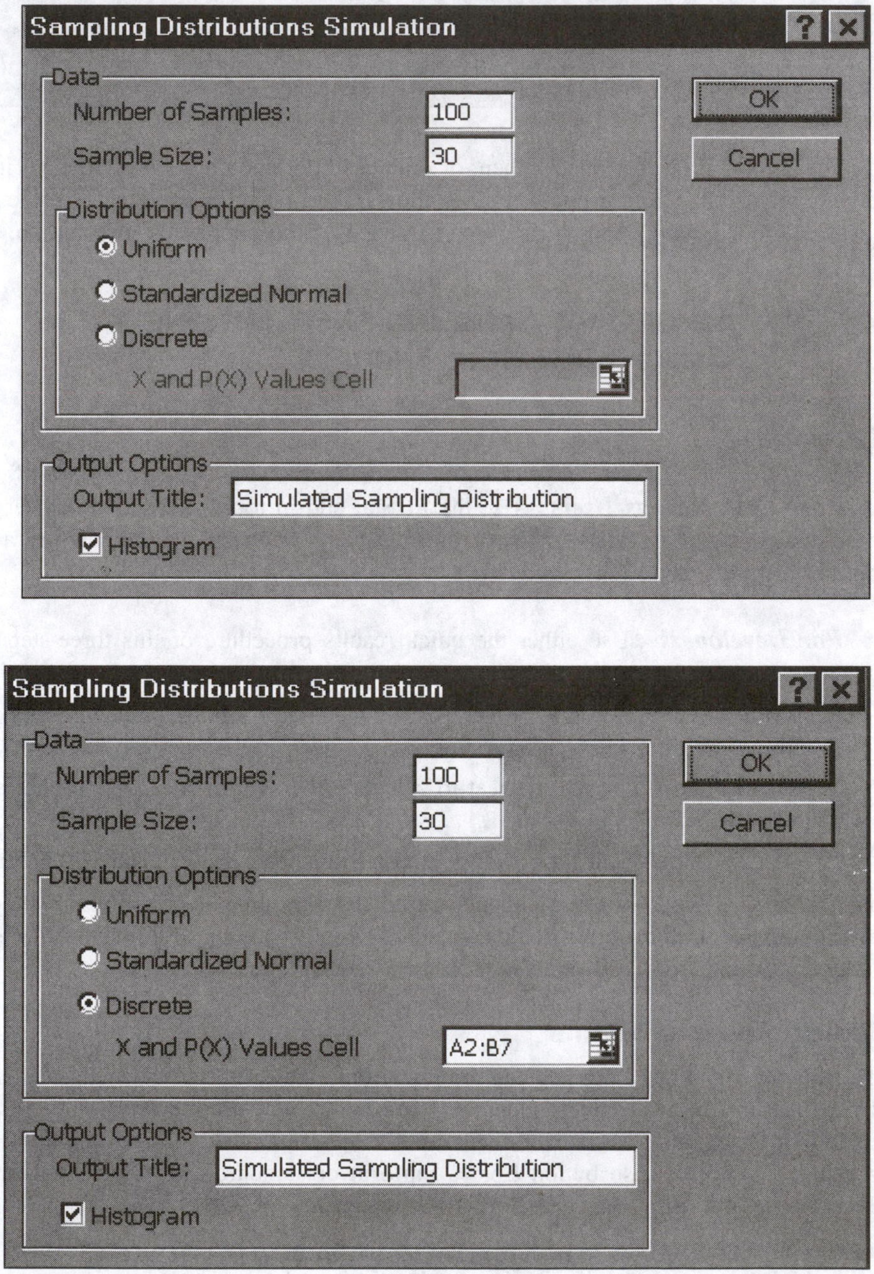

FIGURE 6E.5
Panel A: PHStat Sampling Distributions Simulation dialog box with the Uniform option button selected; panel B: PHStat Sampling Distributions Simulation dialog box with the Discrete option button selected

PANEL A

PANEL B

d. Enter Simulated Sampling Distribution in the Output Title: edit box.

e. Select the Histogram check box.

f. Click the OK button.

The add-in can also generate a simulated sampling distribution from a discrete population. For example, to generate a simulated sampling distribution from a discrete population given in Table 6E.4 using 100 samples of sample size 30, do the following:

❶ Open the Starting Point for section 6.5E workbook (STARTING POINT 6-5E.XLS) and click the Data sheet tab. Verify that the X and P(X) values appear in cell range A2:B7.

❷ Select PHStat | Probability Distributions | Sampling Distributions Simulation.

❸ In the Sampling Distributions Simulation dialog box (see Figure 6E.5, panel B):

a. Enter 100 in the Number of Samples: edit box.

b. Enter 30 in the Sample Size: edit box.

c. Select the Discrete option button.

d. Enter A2:B7 in the X and P(X) Values Cell Range: edit box.

e. Enter Simulated Sampling Distribution in the Output Title: edit box.

f. Select the Histogram check box.

g. Click the OK button.

Table 6E.4 *Table of X-values and P(X)-values to simulate a right-skewed discrete distribution*

X	P(X)
0	0.6
1	0.3
2	0.05
3	0.03
4	0.015
5	0.005

Histograms generated from simulated uniform and standardized normal distributions are shown in Figures 6E.6 and 6E.7, respectively. The histogram generated from the discrete distribution presented in Table 6E.4 and found in the Data sheet of the Starting Point for section 6.5E workbook (STARTING POINT 6-5E.XLS) is shown in Figure 6E.8 on page 390.

Histogram of a simulated
sampling distribution of the
mean for a uniformly distributed population using 100
samples of *n* = 30

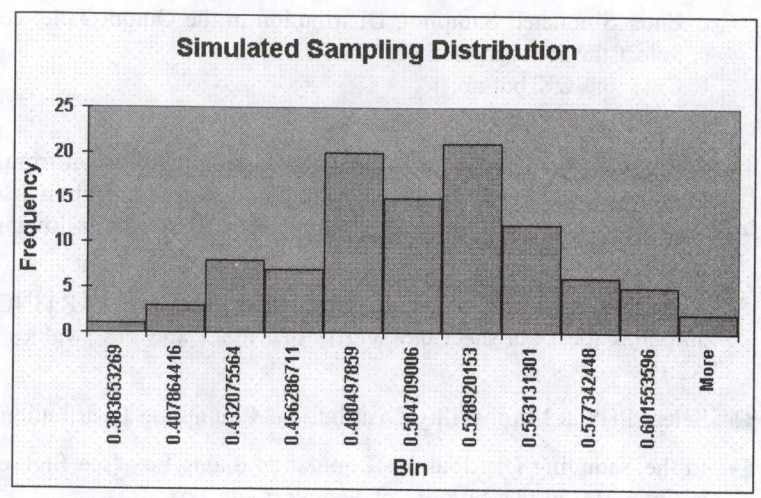

FIGURE 6E.7

Histogram of a simulated
sampling distribution of the
mean for a standardized
normally distributed population using 100 samples of
n = 30

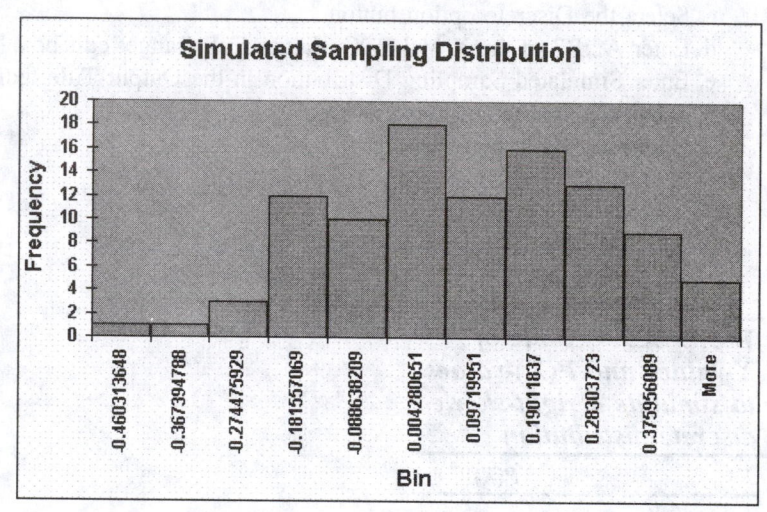

FIGURE 6E.8

Histogram of a simulated
sampling distribution for a
discrete population based on
Table 6E.4, using 100 samples of *n* = 30

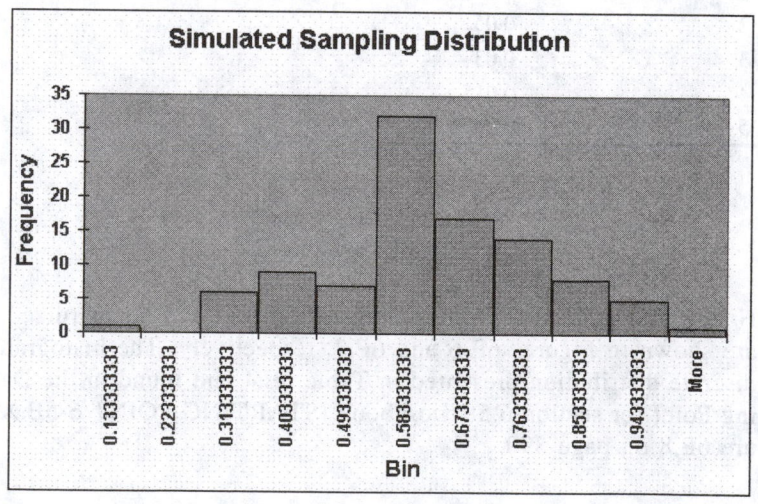

To see the effects of sample size on the shape of the sampling distribution of the mean, repeat the quick results procedure several times, each time varying the value of the sample size. You can also experiment by varying the value for the number of samples and, for a discrete probability, varying the values of $P(X)$.

Developer Details

◆ *A. Generating Multiple Samples Using the Data Analysis Random Number Generation Tool*   We can use the Data Analysis Random Number Generation tool to generate a worksheet containing multiple samples of the same sample size that can be used in step B on page 387 as the basis for simulating a sampling distribution of the mean. Use the appropriate procedure that follows to generate 100 samples of sample size 30 from either a uniform or standardized normal distribution, or for a discrete distribution.

TO GENERATE MULTIPLE SAMPLES FROM A UNIFORMLY DISTRIBUTED POPULATION BETWEEN 0 AND 1:

❶ Select File | New to open a new workbook (or open the existing workbook into which the simulated distribution worksheet is to be inserted).

❷ Select Tools | Data Analysis. Select Random Number Generation from the Analysis Tools list box in the Data Analysis dialog box. Click the OK button.

❸ In the Random Number Generation dialog box (see Figure 6E.9):

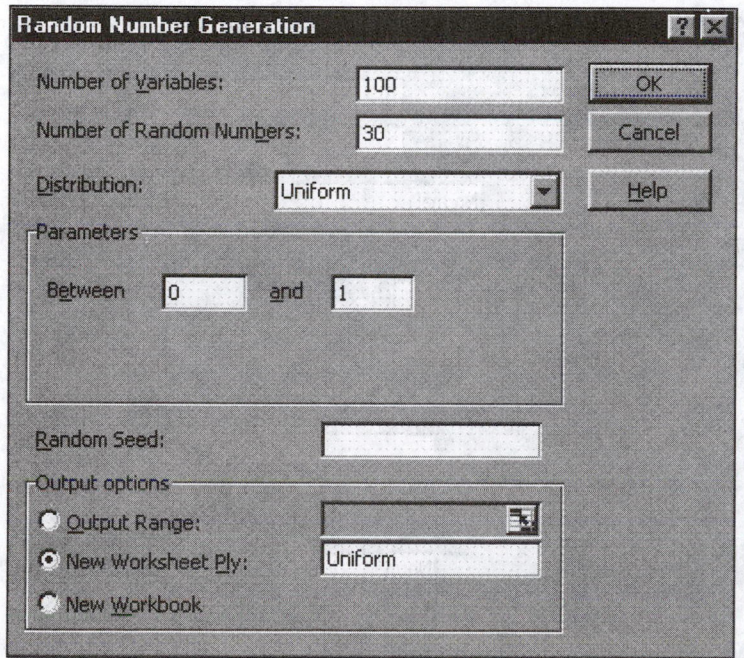

FIGURE 6E.9
Data Analysis tool Random Number Generation dialog box for a uniform distribution

a. Enter 100 in the Number of Variables: edit box.

b. Enter 30 in the Number of Random Numbers: edit box.

c. Select Uniform from the Distribution: drop-down list box.

d. In the Parameters section that appears, enter 0 (zero) and 1 in the Between and "and" edit boxes.

e. Select the New Worksheet Ply: option button and enter Uniform as the name of the new sheet.

f. Click the OK button.

TO GENERATE MULTIPLE SAMPLES FROM A STANDARDIZED NORMALLY DISTRIBUTED POPULATION:

1 Select File | New to open a new workbook (or open the existing workbook into which the simulated distribution worksheet is to be inserted).

2 Select Tools | Data Analysis. Select Random Number Generation from the Analysis Tools list box in the Data Analysis dialog box. Click the OK button.

3 In the Random Number Generation dialog box (see Figure 6E.10):

a. Enter 100 in the Number of Variables: edit box.

b. Enter 30 in the Number of Random Numbers: edit box.

c. Select Normal from the Distribution drop-down list box.

d. In the Parameters section that appears, enter 0 (zero) in the Mean = edit box and enter 1 in the Standard Deviation = edit box.

e. Select the New Worksheet Ply: option button and enter Normal as the name of the new sheet.

f. Click the OK button.

FIGURE 6E.10

Data Analysis tool Random Number Generation dialog box for a normal distribution

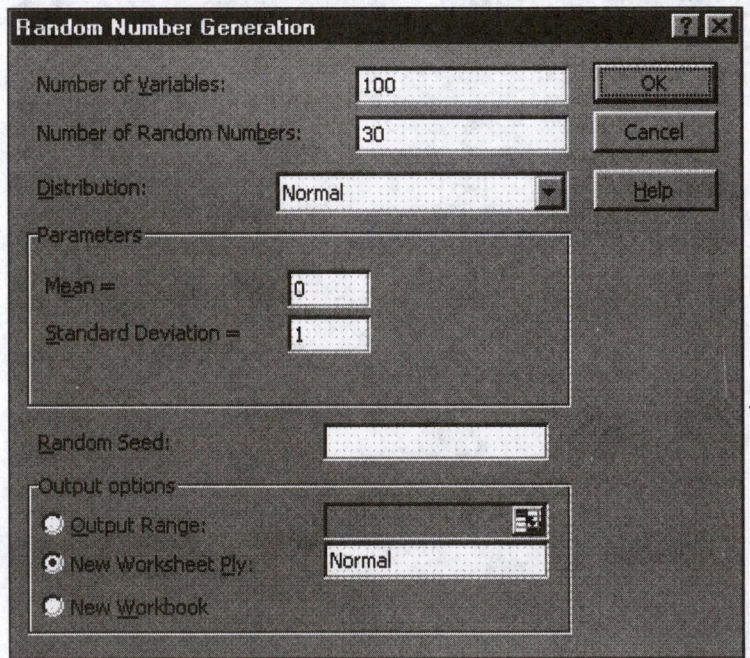

TO GENERATE MULTIPLE SAMPLES FROM A DISCRETE DISTRIBUTION USING THE VALUES OF TABLE 6E.4:

❶ Open the Starting Point for Section 6.5E workbook (STARTING POINT 6-5E.XLS) and click the Data sheet tab. Verify that the *X*- and *P*(*X*)-values from Table 6E.4 have been entered in cell range A2:B7.

❷ Select Tools | Data Analysis. Select Random Number Generation from the Analysis Tools list box in the Data Analysis dialog box. Click the OK button.

❸ In the Random Number Generation dialog box (see Figure 6E.11):
a. Enter 100 in the Number of Variables: edit box.
b. Enter 30 in the Number of Random Numbers: edit box.
c. Select Discrete from the Distribution: drop-down list box.
d. In the Parameters section that appears, enter A2:B7 in the Value and Probability Input Range: edit box.
e. Select the New Worksheet Ply: option button and enter Discrete as the name of the new sheet.
f. Click the OK button.

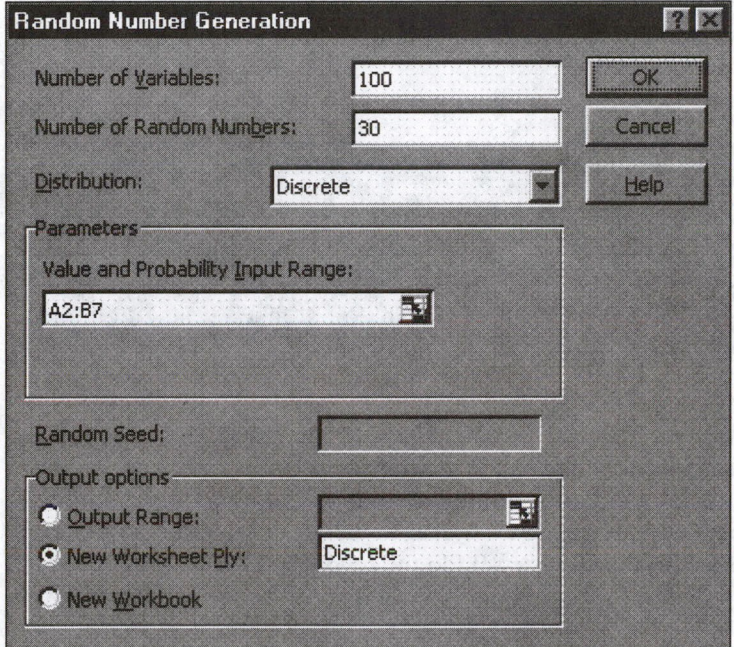

FIGURE 6E.11
Data Analysis tool Random Number Generation dialog box for a discrete distribution

◆ **B. Calculating Sample Means and Other Appropriate Measures from Multiple Samples** Once generated, we can modify a worksheet that contains multiple samples to calculate the sample means and the (overall) average of all sample means. Table 6E.5 presents the modifications for a worksheet that has been generated using one of the step A procedures from page 391. To implement these modifications, do the following:

❶ Click the sheet tab of worksheet generated by a step A procedure.

❷ Enter the labels and formulas for the cell range A31:A34 as shown in Table 6E.5.

❸ Copy the formula in cell A32 across through column CV (the 100th column).

For the worksheets containing multiple samples from either a standardized normal distribution or a discrete distribution, a formula calculating the standard error of the mean can also be inserted. To insert this formula, continue by doing the following:

❹ Enter the label for cell A35 and the formula for cell A36 as shown in Table 6E.5.

Table 6E.5 Modifications for a worksheet containing 100 samples (in columns) of sample size 30

	A	B		CV
31	Sample Means:			
32	=AVERAGE(A1:A30)	=AVERAGE(B1:B30)	· · ·	=AVERAGE(CV1:CV30)
33	Overall Average:			
34	=AVERAGE(A32:CV32)			
35	Standard Error of the Mean:			
36	=STDEV(A32:CV32)			

▲ WHAT IF EXAMPLE

To see the effects of varying the sample size or number of samples on the overall average—and standard error of the mean, if applicable—repeat steps A and B several times with different values for sample size or number of samples

◆ *C. Generating a Histogram from the Sample Means*   After modifying a worksheet in step B, use the Data Analysis Histogram tool to generate a histogram for a sampling distribution of the mean. (We use the Histogram tool here—and not the Chart Wizard—because the Histogram tool does not require class data as the wizard would.) To generate a histogram from a worksheet modified in step B, do the following:

❶ Click the sheet tab of the worksheet modified in step B.

❷ Select Tools | Data Analysis. Select Histogram from the Analysis Tools list box in the Data Analysis dialog box. Click the OK button.

❸ In the Histogram dialog box:
   a. Enter A32:CV32 in the Input Range: edit box.
   b. Leave the Bin Range: edit box blank.
   c. Deselect (uncheck) the Labels check box.
   d. Select the New Worksheet Ply: option button and enter a suitable value in its edit box as the name of the new sheet.

e. Select the Chart Output check box.
f. Deselect (uncheck) the Cumulative Percentage and Pareto check boxes.
g. Click the OK button.

The add-in inserts a new worksheet containing both a frequency distribution table and a histogram. Adjust the bars of the histogram as follows:

❹ Right-click on one of the histogram bars. (A tool tip that includes the words "Series 'Frequency'" appears when the mouse pointer is over a bar.)

❺ Select Format Data Series from the shortcut menu.

❻ In the Format Data Series dialog box, click the Options tab and change the value in the Gap width: edit box to 0. Click OK.

The gaps between the bars are eliminated and the histograms will be similar to the ones shown in Figures 6.E6, 6.E7, and 6.E8 on page 390.

6.5E.2 GENERATING RANDOM SAMPLES USING MICROSOFT EXCEL

Overview

◆ *For Quick Results Users and Developers*   Use the PHStat add-in to generate a random sample without replacement. Use the Data Analysis Sampling tool to generate a random sample with replacement from a population of values. There is no workbook file for this section.

Generating a Random Sample without Replacement

Use the Data Preparation | Random Sample Generator choice of the PHStat add-in to generate a random sample without replacement. We can choose to have the add-in generate a random list of integers from 1 to the population size *or* select values from a population range of values.

To generate a random list of numbers, do the following:

❶ If the PHStat add-in has not been previously loaded, load the add-in using the instructions of section S4.2.

❷ Select File | New to open a new workbook (or open the existing workbook into which the worksheet containing the random list of integers is to be inserted).

❸ Select PHStat | Data Preparation | Random Sample Generator.

❹ In the Random Sample Generator dialog box (see Figure 6E.12):
a. Enter the sample size value in the Sample Size: edit box.
b. Select the Generate list of random numbers option button.
c. Enter the population size value in the Population Size: edit box.
d. Enter a suitable value in the Output Title: edit box.
e. Click the OK button.

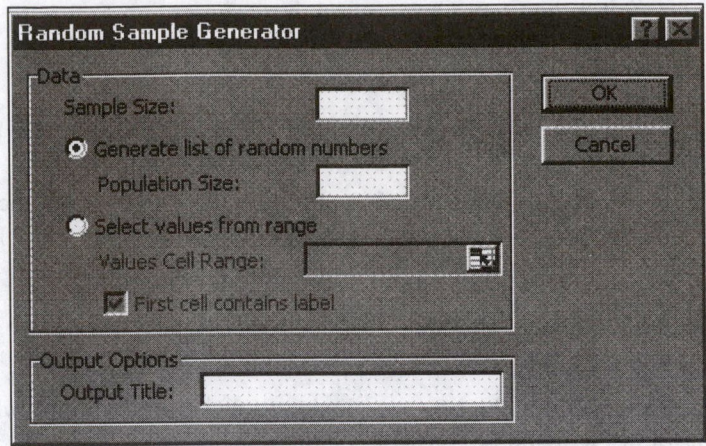

The add-in inserts a worksheet containing the random list of integers.

To randomly select values from a cell range without replacement, do the following:

1 If the PHStat add-in has not been previously loaded, load the add-in using the instructions of section S4.2.

2 Click the sheet tab of the worksheet that contains the population data.

3 Select PHStat | Data Preparation | Random Sample Generator.

4 In the Random Sample Generator dialog box:
  a. Enter the sample size value in the Sample Size: edit box.
  b. Select the Select values from range option button.
  c. Enter the cell range containing the population of values in the Values Cell Range: edit box.
  d. Select the First cell contains label check box if appropriate.
  e. Enter a suitable value in the Output Title: edit box.
  f. Click the OK button.

The add-in inserts a worksheet containing the random sample of values.

Generating a Random Sample with Replacement

Use the Data Analysis Sampling tool to generate a random sample with replacement from a population of values. To use this tool, do the following:

1 Click the sheet tab of the worksheet modified in step B (pages 393–394).

2 Select Tools | Data Analysis. Select Sampling from the Analysis Tools list box in the Data Analysis dialog box. Click the OK button.

3 In the Sampling dialog box (see Figure 6E.13):
  a. Enter the cell range of the population data in the Input Range: edit box.
  b. Select the Labels check box if the first cell of the cell range contains a label.
  c. Select the Random option button and enter the sample size in the Number of Samples: edit box.

d. Select the New Worksheet Ply option button and enter a suitable value in its edit box as the name of the new sheet.

e. Click the OK button.

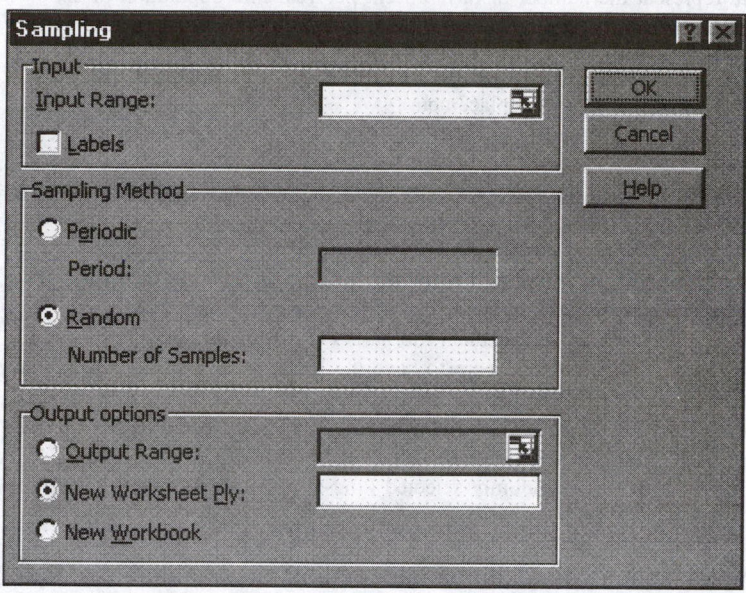

FIGURE 6E.13
Sampling dialog box

The add-in inserts a worksheet containing the random sample of values. To avoid several possible errors, the population cell range entered in step 3a should be either a single row or column of data.

Problems for Section 6.5

Learning the Basics

•**6.30** Given a normal distribution with $\mu = 100$ and $\sigma = 10$, if a sample of $n = 25$ is selected, what is the probability that $\overline{X}$ is
(a) less than 95?
(b) between 95 and 97.5?
(c) above 102.2?
(d) between 99 and 101?
(e) There is a 65% chance that $\overline{X}$ will be above what value?
(f) What would be your answers to (a)–(e) if $n = 16$?

WHAT IF?

6.31 Given a normal distribution with $\mu = 50$ and $\sigma = 5$, if a sample of $n = 100$ is selected, what is the probability that $\overline{X}$ is
(a) less than 47?
(b) between 47 and 49.5?
(c) above 51.1?
(d) between 49 and 51?
(e) There is a 35% chance that $\overline{X}$ will be above what value?
(f) What would be your answers to (a)–(e) if $n = 25$?

WHAT IF?

Applying the Concepts

6.32 For each of the following three populations, indicate what the sampling distribution for samples of 25 would consist of.
(a) Travel expense vouchers for a university in an academic year

(b) Absentee records (days absent per year) in 1997 for employees of a large manufacturing company

(c) Yearly sales (in gallons) of unleaded gasoline at service stations located in a particular county

6.33 The following data represent the number of days absent per year in a population of six employees of a small company:

1, 3, 6, 7, 7, 12

(a) Assuming that you sample *without* replacement
  (1) Select all possible samples of size 2 and set up the sampling distribution of the mean.
  (2) Compute the mean of all the sample means and also compute the population mean. Are they equal? What is this property called?
  (3) Do parts (1) and (2) for all possible samples of size 3.
  (4) Compare the shape of the sampling distribution of the mean obtained in parts (1) and (3). Which sampling distribution seems to have less variability? Why?

(b) Assuming that you sample *with* replacement, do parts (1)–(4) of (a) and compare the results. Which sampling distributions seem to have the least variability, those in (a) or (b)? Why?

● **6.34** The diameter of Ping-Pong balls manufactured at a large factory is expected to be approximately normally distributed with a mean of 1.30 inches and a standard deviation of 0.04 inch. What is the probability that a randomly selected Ping-Pong ball will have a diameter

(a) between 1.28 and 1.30 inches?

(b) between 1.31 and 1.33 inches?

(c) Between what two values (symmetrically distributed around the mean) will 60% of the Ping-Pong balls fall (in terms of the diameter)?

(d) If many random samples of 16 Ping-Pong balls are selected,
  (1) what will the mean and standard error of the mean be expected to be?
  (2) what distribution will the sample means follow?
  (3) what proportion of the sample means will be between 1.28 and 1.30 inches?
  (4) what proportion of the sample means will be between 1.31 and 1.33 inches?
  (5) Between what two values will 60% of the sample means be?

(e) Compare the answers of (a) with (d)(3) and (b) with (d)(4). Discuss.

(f) Explain the difference in the results of (c) and (d)(5).

(g) Which is more likely to occur—an individual ball above 1.34 inches, a sample mean above 1.32 inches in a sample of size 4, or a sample mean above 1.31 inches in a sample of size 16? Explain.

6.35 Time spent using e-mail per session is normally distributed with $\mu = 8$ minutes and $\sigma = 2$ minutes. If random samples of 25 sessions were selected,

(a) compute $\sigma_{\bar{x}}$.

(b) what proportion of the sample means would be between 7.8 and 8.2 minutes?

(c) what proportion of the sample means would be between 7.5 and 8 minutes?

(d) If random samples of 100 sessions were selected, what proportion of the sample means would be between 7.8 and 8.2 minutes?

(e) Explain the difference in the results of (b) and (d).

(f) Which is more likely to occur—a particular session of e-mail for more than 11 minutes, a sample mean above 9 minutes in a sample of 25 sessions, or a sample mean above 8.6 minutes in a sample of 100 sessions? Explain.

6.36 The amount of time a bank teller spends with each customer has a population mean $\mu = 3.10$ minutes and standard deviation $\sigma = 0.40$ minute. If a random sample of 16 customers is selected,

(a) what is the probability that the average time spent per customer will be at least 3 minutes?

(b) there is an 85% chance that the sample mean will be below how many minutes?

(c) What assumption must be made in order to solve (a) and (b)?

(d) If a random sample of 64 customers is selected, there is an 85% chance that the sample mean will be below how many minutes?

(e) What assumption must be made in order to solve (d)?

(f) Which is more likely to occur—an individual service time below 2 minutes, a sample mean above 3.4 minutes in a sample of 16 customers, or a sample mean below 2.9 minutes in a sample of 64 customers? Explain.

• **6.37** Toby's Trucking Company determined that, on an annual basis, the distance traveled per truck is normally distributed with a mean of 50.0 thousand miles and a standard deviation of 12.0 thousand miles. If a sample of 16 trucks is selected,

(a) what is the probability that the average distance traveled is below 45.0 thousand miles in the year?

(b) what is the probability that the average distance traveled is between 44.0 and 48.0 thousand miles in the year?

(c) What assumption must be made in order to solve (a) and (b)?

(d) If a random sample of 64 trucks is selected, there is a 95% chance that the sample mean will be below how many miles?

(e) What assumption must be made in order to solve (d)?

(f) What will your answers be to (a)–(d) if the standard deviation is 10.0 thousand miles?

 WHAT IF?

6.38 Plastic bags used for packaging produce are manufactured so that the breaking strength of the bag is normally distributed with a mean of 5 pounds per square inch and a standard deviation of 1.5 pounds per square inch. If a sample of 25 bags is selected,

(a) what is the probability that the average breaking strength is

(1) between 5 and 5.5 pounds per square inch?

(2) between 4.2 and 4.5 pounds per square inch?

(3) less than 4.6 pounds per square inch?

(b) Between what two values symmetrically distributed around the mean will 95% of the average breaking strengths fall?

(c) What will your answers be to (a) and (b) if the standard deviation is 1.0 pound per square inch?

 WHAT IF?

 ## 6.6 SAMPLING DISTRIBUTION OF THE PROPORTION

When dealing with a categorical variable where each individual or item in the population can be classified as either possessing or not possessing a particular characteristic such as male or female, or prefer brand A or do not prefer brand A, the two possible outcomes could be assigned scores of 1 or 0 to represent the presence or absence of the characteristic. If only a random sample of n individuals is available, the sample mean for such a categorical variable is found by summing all the 1 and 0 scores and then dividing by n. For example if, in a sample of five individuals, three preferred brand A and two did not, there are three 1's and two 0's. Summing the three 1's and two 0's and dividing by the sample size of 5 give us a mean of 0.60, which is also the proportion of individuals in the sample who prefer brand A. Therefore, when dealing with categorical data, the sample mean $\overline{X}$ (of the 1 and 0 scores) is the sample proportion p_s having the characteristic of interest. Thus, the sample proportion p_s is defined as follows.

The Sample Proportion

$$p_s = \frac{X}{n} = \frac{\text{number of successes}}{\text{sample size}} \qquad (6.12)$$

The sample proportion p_s has the special property that it must be between 0 and 1. If all individuals possess the characteristic, each is assigned a score of 1, and p_s is equal to 1. If half the individuals possess the characteristic, half are assigned a score of 1, the other half are assigned a score of 0, and p_s is equal to .5. If none of the individuals possesses the characteristic, each is assigned a score of 0, and p_s is equal to 0.

Whereas the sample mean $\overline{X}$ is an estimator of the population mean μ, the statistic p_s is an estimator of the population proportion p. By analogy to the sampling distribution of the mean, the **standard error of the proportion** σ_{p_s} is

Standard Error of the Proportion

$$\sigma_{p_s} = \sqrt{\frac{p(1-p)}{n}} \qquad (6.13)$$

When sampling *with* replacement from a finite population, the **sampling distribution of the proportion** follows the binomial distribution discussed in section 4.5. However, the normal distribution can be used to approximate the binomial distribution when np and $n(1-p)$ are each at least 5. In most cases in which inferences are made about the proportion, the sample size is substantial enough to meet the conditions for using the normal approximation (see reference 1). Therefore, in many instances, we can use the normal distribution to evaluate the sampling distribution of the proportion. Thus, we have

$$Z = \frac{\overline{X} - \mu_{\overline{x}}}{\sigma_{\overline{x}}} = \frac{\overline{X} - \mu}{\dfrac{\sigma}{\sqrt{n}}}$$

Because we are dealing with sample proportions (not sample means), we have

$$p_s = \text{sample proportion}$$
$$p = \text{population proportion}$$
$$\sigma_{p_s} = \sqrt{\frac{p(1-p)}{n}}$$

Substituting p_s for $\overline{X}$, p for $\mu_{\overline{x}}$, and $\sigma_{p_s} = \sqrt{p(1-p)/n}$ for $\sigma_{\overline{x}}$, we have

Difference between the Sample and Population Proportion in Standardized Normal Units

$$Z \cong \frac{p_s - p}{\sqrt{\dfrac{p(1-p)}{n}}} \qquad (6.14)$$

To illustrate the sampling distribution of the proportion, we will refer to the following example.

Example 6.15 *The Sampling Distribution of the Proportion*

The manager of the local branch of a savings bank determines that 40% of all depositors have multiple accounts at the bank. If a random sample of 200 depositors is selected, what is the probability that the sample proportion of depositors with multiple accounts is no more than .30?

SOLUTION

Because $np = 200(.40) = 80$ and $n(1 - p) = 200(.60) = 120$, the sampling distribution of the proportion is assumed to be approximately normally distributed. Using equation (6.14),

$$Z \cong \frac{p_s - p}{\sqrt{\dfrac{p(1 - p)}{n}}}$$

$$= \frac{.30 - .40}{\sqrt{\dfrac{(.40)(.60)}{200}}} = \frac{-.10}{\sqrt{\dfrac{.24}{200}}} = \frac{-.10}{.0346}$$

$$= -2.89$$

Using Table E.2(b), the area under the normal curve up to $Z = -2.89$ is .0019. Therefore, the probability of obtaining a sample proportion of, at most, .30 is .0019—a highly unlikely event. That is, if the true proportion of successes in the population were .40, then less than one-fifth of 1% of the samples of size 200 are expected to have sample proportions of, at most, .30. This is depicted in the accompanying diagram.

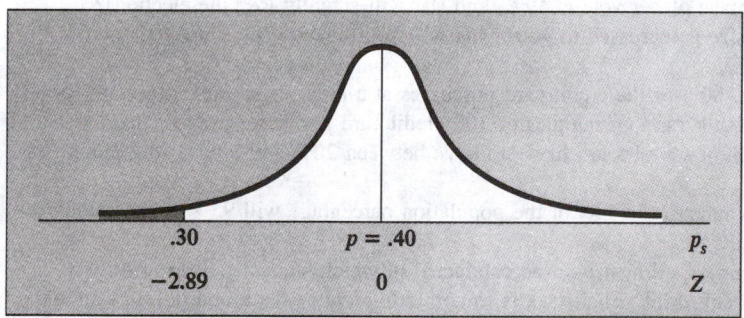

Diagram of the sampling distribution of the proportion to find area up to sample proportion of .30

Problems for Section 6.6

Learning the Basics

• **6.39** In a random sample of 64 people, 48 were classified as "successful." If the population proportion is .70,
  (a) determine the sample proportion p_s of "successful" people.
  (b) determine the standard error of the sample proportion σ_{p_s}.

6.40 A random sample of 50 households was selected for a telephone survey. The key question asked was, "Do you or any member of your household own a cellular telephone?" Of the 50 respondents, 15 said yes and 35 said no. If the population proportion is .40,

(a) determine the sample proportion p_s of households with cellular telephones.

(b) determine the standard error of the sample proportion σ_{p_s}.

6.41 The following raw data represent the responses (Y for yes and N for no) from a sample of 40 college students to the question, "Do you currently own shares in any stocks?"

N N Y N N Y N Y N Y N N Y N Y Y N N N Y

N Y N N N N Y N N Y Y N N N Y N N Y N N

If the population proportion is .30,

(a) determine the sample proportion p_s of college students who own shares of stock.

(b) determine the standard error of the sample proportion σ_{p_s}.

Applying the Concepts

WHAT IF?

• **6.42** Historically, 10% of a large shipment of machine parts are defective. If random samples of 400 parts are selected, what proportion of the samples will have

(a) between 9% and 10% defective parts?

(b) less than 8% defective parts?

(c) If a sample size of only 100 were selected, what would your answers have been in (a) and (b)?

(d) Which is more likely to occur—a percent defective above 13% in a sample of 100 or a percent defective above 10.5% in a sample of 400? Explain.

6.43 A political pollster is conducting an analysis of sample results in order to make predictions on election night. Assuming a two-candidate election, if a specific candidate receives at least 55% of the vote in the sample, then that candidate will be forecast as the winner of the election. If a random sample of 100 voters is selected, what is the probability that a candidate will be forecast as the winner when

(a) the true percentage of her vote is 50.1%?

(b) the true percentage of her vote is 60%?

(c) the true percentage of her vote is 49% (and she will actually lose the election)?

WHAT IF?

(d) If the sample size is increased to 400, what will your answers be to (a), (b), and (c)? Discuss.

6.44 Based on past data, 30% of the credit card purchases at a large department store are for amounts above $100. If random samples of 100 credit card purchases are selected,

(a) what proportion of samples are likely to have between 20% and 30% of the purchases over $100?

(b) within what symmetrical limits of the population percentage will 95% of the sample percentages fall?

• **6.45** Suppose a marketing experiment is to be conducted in which students are to taste two different brands of soft drink. Their task is to correctly identify the brand tasted. Random samples of 200 students are selected and it is assumed that the students have no ability to distinguish between the two brands. (*Hint:* If an individual has no ability to distinguish between the two soft drinks, then each one is equally likely to be selected.)

(a) What proportion of the samples will have between 50% and 60% of the identifications correct?

(b) Within what symmetrical limits of the population percentage will 90% of the sample percentages fall?

(c) What is the probability of obtaining a sample percentage of correct identifications in excess of 65%?

(d) Which is more likely to occur—more than 60% correct identifications in the sample of 200 or more than 55% correct identifications in a sample of 1,000? Explain.

6.46 Historically, 93% of the deliveries of an overnight mail service arrive before 10:30 the following morning. If random samples of 500 deliveries are selected, what proportion of the samples will have

(a) between 93% and 95% of the deliveries arriving before 10:30 the following morning?

(b) more than 95% of the deliveries arriving before 10:30 the following morning?

(c) If samples of size 1,000 are selected, what will your answers be in (a) and (b)?

(d) Which is more likely to occur—more than 95% of the deliveries in a sample of 500 or less than 90% in a sample of 1,000 arriving before 10:30 the following morning? Explain.

WHAT IF?

6.7 ▸ SAMPLING FROM FINITE POPULATIONS

The central limit theorem and the standard errors of the mean and of the proportion are based on the premise that the samples selected are chosen *with* replacement. However, in virtually all survey research, sampling is conducted *without* replacement from populations that are of a finite size N. In these cases, particularly when the sample size n is not *small* in comparison with the population size N (i.e., more than 5% of the population is sampled) so that $n/N > .05$, a **finite population correction factor (fpc)** is used to define both the standard error of the mean and the standard error of the proportion. The finite population correction factor is expressed as

Finite Population Correction Factor (*fpc*)

$$fpc = \sqrt{\frac{N-n}{N-1}} \tag{6.15}$$

where

$$n = \text{sample size}$$
$$N = \text{population size}$$

Therefore, when dealing with means, we have

Standard Error of the Mean for Finite Populations

$$\sigma_{\bar{x}} = \frac{\sigma}{\sqrt{n}}\sqrt{\frac{N-n}{N-1}} \tag{6.16}$$

When we are referring to proportions, we have

Standard Error of the Proportion for Finite Populations

$$\sigma_{p_s} = \sqrt{\frac{p(1-p)}{n}}\sqrt{\frac{N-n}{N-1}} \tag{6.17}$$

Examining the formula for the finite population correction factor [equation (6.15)], we see that the numerator is always smaller than the denominator, so the correction factor is less than 1. Because this finite population correction factor is multiplied by the standard error, the standard error becomes smaller when corrected. This means that we get more precise estimates because we are sampling a larger segment of the population.

We illustrate the application of the finite population correction factor using two examples previously discussed in this chapter.

Example 6.16 *Using the Finite Population Correction Factor with the Mean*

In the cereal-filling example on pages 379–380, a sample of 25 cereal boxes was selected from a filling process. Suppose that 2,000 boxes (i.e., the population) are filled on this particular day. Using the finite population correction factor, determine the probability of obtaining a sample whose mean is below 365 grams.

SOLUTION

Using the finite population correction factor, we have

$$\sigma = 15, n = 25, N = 2,000$$

so that

$$\sigma_{\bar{x}} = \frac{\sigma}{\sqrt{n}}\sqrt{\frac{N - n}{N - 1}}$$

$$= \frac{15}{\sqrt{25}}\sqrt{\frac{2,000 - 25}{2,000 - 1}}$$

$$= 3\sqrt{.988} = 2.982$$

The probability of obtaining a sample whose mean is between 365 and 368 grams is computed as follows.

$$Z = \frac{\bar{X} - \mu_{\bar{x}}}{\sigma_{\bar{x}}} = \frac{-3}{2.982} = -1.01$$

From Table E.2(b), the area below 365 grams is .1562.

It is evident in this example that the use of the finite population correction factor has a very small effect on the standard error of the mean and the subsequent area under the normal curve because the sample size (i.e., $n = 25$ boxes) is only 1.25% of the population size (i.e., $N = 2,000$ boxes).

Example 6.17 *Using the Finite Population Correction Factor with the Proportion*

In Example 6.15 on page 401, concerning multiple banking accounts, suppose there are a total of 1,000 different depositors at the bank. Using the finite population correction factor, determine the probability of obtaining a sample whose proportion is less than .30.

SOLUTION

When we use the finite population correction factor, the previous sample of size 200 out of this finite population results in the following.

$$\sigma_{p_s} = \sqrt{\frac{p(1-p)}{n}}\sqrt{\frac{N-n}{N-1}}$$

$$= \sqrt{\frac{(.40)(.60)}{200}}\sqrt{\frac{1{,}000-200}{1{,}000-1}}$$

$$= \sqrt{\frac{.24}{200}}\sqrt{\frac{800}{999}} = \sqrt{.0012}\sqrt{.801}$$

$$= (.0346)(.895) = .031$$

When we use $\sigma_{p_s} = .031$ as the standard error of the sample proportion, from equation (6.13) we have, $Z = -.10/.031 = -3.23$. From Table E.2(b), the appropriate area below $Ps = .30$ is .00062. In this example, the use of the finite population correction factor has a moderate effect on the standard error of the proportion and on the area under the normal curve because the sample size is 20% (i.e., $n/N = .20$) of the population.

Problems for Section 6.7

Learning the Basics

•**6.47** Given that $N = 80$ and $n = 10$ and the sample is obtained *without* replacement, determine the finite population correction factor.

6.48 Which of the following finite population factors will have a greater effect in reducing the standard error—one based on a sample of size 100 selected *without* replacement from a population of size 400 or one based on a sample of size 200 selected *without* replacement from a population of size 900? Explain.

6.49 Given that $N = 60$ and $n = 20$ and the sample is obtained *with* replacement, should the finite population correction factor be used? Explain.

Applying the Concepts

•**6.50** The diameter of Ping-Pong balls manufactured at a large factory is expected to be approximately normally distributed with a mean of 1.30 inches and a standard deviation of 0.04 inch. If many random samples of 16 Ping-Pong balls are selected from a population of 200 Ping-Pong balls *without* replacement, what proportion of the sample means would be between 1.31 and 1.33 inches?

6.51 The amount of time a bank teller spends with each customer has a population mean $\mu = 3.10$ minutes and standard deviation $\sigma = 0.40$ minute. If a random sample of 16 customers is selected *without* replacement from a population of 500 customers,
  (a) what is the probability that the average time spent per customer will be at least 3 minutes?
  (b) there is an 85% chance that the sample mean will be below how many minutes?

•**6.52** Historically, 10% of a large shipment of machine parts are defective. If random samples of 400 parts are selected *without* replacement from a shipment that included 5,000 machine parts, what proportion of the samples will have
  (a) between 9% and 10% defective parts?
  (b) less than 8% defective parts?

6.53 Historically, 93% of the deliveries of an overnight mail service arrive before 10:30 the following morning. If random samples of 500 deliveries are selected *without* replacement from a population that consisted of 10,000 deliveries, what proportion of the samples will have
  (a) between 93% and 95% of the deliveries arriving before 10:30 the following morning?
  (b) more than 95% of the deliveries arriving before 10:30 the following morning?

◆ SUMMARY

As seen in the summary chart, in this chapter we have studied the normal distribution, the exponential distribution, the sampling distribution of the sample mean, and the sampling distribution of the sample proportion. The importance of the normal distribution in statistics was further emphasized by examining the central limit theorem. Knowledge of a population distribution is not always necessary in drawing conclusions from a sampling distribution of the mean or proportion.

The concepts concerning sampling distributions are central to the development of statistical inference. The main objective of statistical inference is to take information based only on a sample and use this information to draw conclusions and make decisions about various population values. The statistical techniques developed to achieve these objectives are discussed fully in the next five chapters (confidence intervals and tests of hypotheses).

Key Terms

Checking Your Understanding

6.54 Why is it that only one table of the normal distribution such as Table E.2(a) or E.2(b) is needed to find any probability under the normal curve?

6.55 How would you find the area between two values under the normal curve when both values are on the same side of the mean?

6.56 How would you find the X-value that corresponds to a given percentile of the normal distribution?

6.57 Why do the individual observations have to be converted to standard normal ordered values in order to develop a normal probability plot?

6.58 What are some of the distinguishing properties of a normal distribution?

6.59 How does the normal probability plot allow one to evaluate whether a set of data is normally distributed?

6.60 Under what circumstances can the exponential distribution be used?

6.61 Why is the sample arithmetic mean an unbiased estimator of the population arithmetic mean?

6.62 Why does the standard error of the mean decrease as the sample size n increases?

6.63 Why does the sampling distribution of the mean follow a normal distribution for a *large enough* sample size even though the population may not be normally distributed?

6.64 Explain why a statistician would be interested in drawing conclusions about a population rather than merely describing the results of a sample.

6.65 What is the difference between a probability distribution and a sampling distribution?

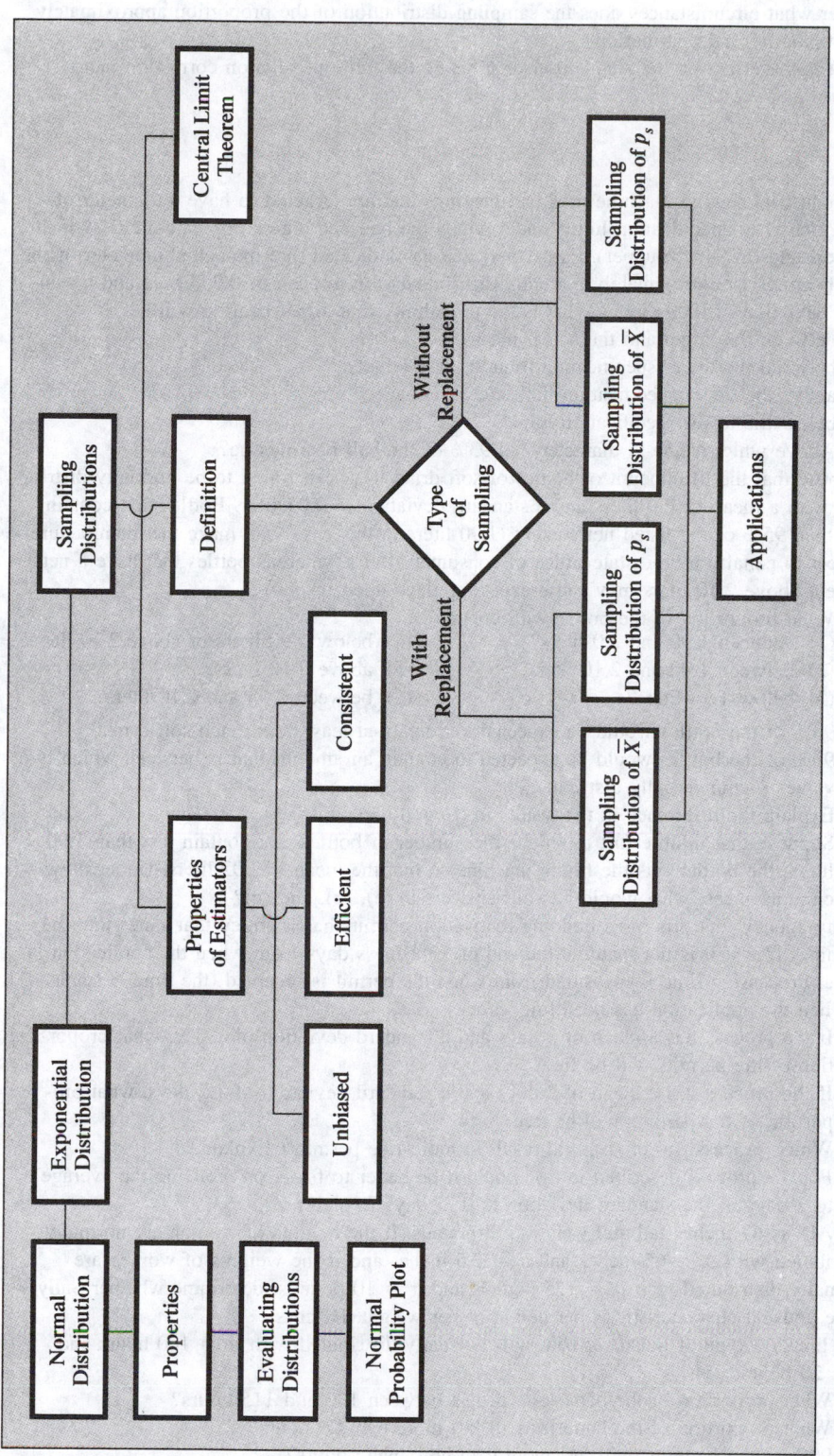

Chapter 6 summary chart

6.66 Under what circumstances does the sampling distribution of the proportion approximately follow the normal distribution?

6.67 What is the effect on the standard error of using the finite population correction factor?

Chapter Review Problems

6.68 An industrial sewing machine uses ball bearings that are targeted to have a diameter of 0.75 inch. The specification limits under which the ball bearing can operate are 0.74 inch (lower) and 0.76 inch (upper). Past experience has indicated that the actual diameter of the ball bearings is approximately normally distributed with a mean of 0.753 inch and a standard deviation of 0.004 inch. What is the probability that a ball bearing will be
  (a) between the target and the actual mean?
  (b) between the lower specification limit and the target?
  (c) above the upper specification limit?
  (d) below the lower specification limit?
  (e) Above which value in diameter will 93% of the ball bearings be?

6.69 Suppose that the fill amount of bottles of soft drink has been found to be normally distributed with a mean of 2.0 liters and a standard deviation of 0.05 liter. Bottles that contain less than 95% of the listed net content (1.90 liters in this case) can make the manufacturer subject to penalty by the state office of consumer affairs, whereas bottles that have a net content above 2.10 liters may cause excess spillage upon opening.
  (a) What proportion of the bottles will contain
      (1) between 1.90 and 2.0 liters?
      (2) between 1.90 and 2.10 liters?
      (3) below 1.90 liters?
      (4) below 1.90 liters or above 2.10 liters?
      (5) above 2.10 liters?
      (6) between 2.05 and 2.10 liters?
  (b) 99% of the bottles would be expected to contain at least how much soft drink?
  (c) 99% of the bottles would be expected to contain an amount that is between which two values (symmetrically distributed)?
  (d) Explain the difference in the results in (b) and (c).

 WHAT IF?

  (e) Suppose that in an effort to reduce the number of bottles that contain less than 1.90 liters, the bottler sets the filling machine so that the mean is 2.02 liters. Under these circumstances, what would be your answers in (a), (b), and (c)?

● **6.70** A city agency that processes building renovation permits has a policy that states that the permit is free if it is not ready at the end of 5 business days from when the application is made. Processing time is measured from when the permit is received (the time is stamped) to when the application has been fully processed.
  (a) If the process has a mean of 3 days and a standard deviation of 1 day, what proportion of the permits will be free?
  (b) If the process has a mean of 2 days and a standard deviation of 1.5 days, what proportion of the permits will be free?
  (c) Which process, (a) or (b), will result in more free permits? Explain.
  (d) For the process described in (a), would it be better to focus on reducing the average to 2 days or the standard deviation to 0.75 day? Explain.

6.71 Sally D. is 67 inches tall and weighs 135 pounds. If the heights of women are normally distributed with $\mu = 65$ inches and $\sigma = 2.5$ inches and if the weights of women are normally distributed with $\mu = 125$ pounds and $\sigma = 10$ pounds, determine whether Sally's more unusual characteristic is her height or her weight. Discuss.

6.72 The life of a type of transistor battery is normally distributed with $\mu = 100$ hours and $\sigma = 20$ hours.
  (a) What proportion of the batteries will last between 100 and 115 hours?
  (b) What proportion of the batteries will last more than 90 hours?
  (c) 90% of the batteries will last more than how many hours?
  (d) What is the interquartile range in battery life?

(e) If random samples of 16 batteries are selected,
    (1) what proportion of the sample means will be between 100 and 115 hours?
    (2) what proportion of the sample means will be more than 90 hours?
    (3) within what limits around the population mean will 90% of sample means fall?
(f) Is the central limit theorem necessary to answer (e)(1), (2), and (3)? Explain.

• 6.73 An orange juice producer buys all his oranges from a large orange orchard. The amount of juice squeezed from each of these oranges is approximately normally distributed with a mean of 4.70 ounces and a standard deviation of 0.40 ounce.
 (a) What is the probability that a randomly selected orange will contain between 4.70 and 5.00 ounces?
 (b) What is the probability that a randomly selected orange will contain between 5.00 and 5.50 ounces?
 (c) 77% of the oranges will contain at least how many ounces of juice?
 (d) Between what two values (in ounces) symmetrically distributed around the population mean will 80% of the oranges fall?
 Suppose that a sample of 25 oranges is selected:
 (e) What is the probability that the sample mean will be at least 4.60 ounces?
 (f) Between what two values symmetrically distributed around the population mean will 70% of the sample means fall?
 (g) 77% of the sample means will be above what value?
 (h) Are the results of (c) and (g) different? Explain why.

6.74 Suppose the governor projects that, on a weekly basis, a state football lottery program she has proposed is expected to average 10.0 million dollars in profits (to be turned over to the state for educational programs) with a standard deviation of 2.5 million dollars. Suppose further that the weekly profits data are assumed to be (approximately) normally distributed. The following questions may be raised (or anticipated at the governor's next press conference):
 (a) What is the probability that in any given week profits will be
    (1) between 10.0 and 12.5 million   (4) at least 7.5 million dollars?
       dollars?   (5) under 7.5 million dollars?
    (2) between 7.5 and 10.0 million dollars?   (6) between 12.5 and 14.3 million
    (3) between 7.5 and 12.5 million dollars?   dollars?
 (b) 50% of the time weekly profits (in millions of dollars) are expected to be above what value?
 (c) 90% of the time weekly profits (in millions of dollars) are expected to be above what value?
 (d) What is the interquartile range in weekly profits expected from the state football lottery program?

6.75 (Class Project) According to Burton G. Malkiel, the daily changes in the closing price of stock follow a *random walk*—that is, these daily events are independent of each other and move upward or downward in a random manner—and can be approximated by a normal distribution.
 To test this theory, each student should use either a newspaper or the Internet to select one company traded on the New York Stock Exchange, one company traded on the American Stock Exchange, and one company traded "over the counter" (NASDAQ national market) and then do the following:
1. Record the daily closing stock price of each of these companies for 6 consecutive weeks (so that you have 30 observations per company).
2. Record the daily changes in the closing stock price of each of these companies for 6 consecutive weeks (so that you have 30 observations per company).
 For each of your six data sets, decide whether or not the data appear to be approximately normally distributed by
 (a) examining the stem-and-leaf, histogram or polygon, and box-and-whisker plot.
 (b) evaluating the actual versus theoretical properties.
 (c) constructing a normal probability plot.

(d) Discuss the results of (a), (b), and (c).

Note: The random-walk theory pertains to the daily changes in the closing stock price, not the daily closing stock price.

(e) Based on your conclusions in (d), what can you now say about your three distributions of each type—daily closing prices and daily changes in closing prices? Which, if any, of the data sets appear to be approximately normally distributed?

6.76 (Class Project) The table of random numbers is an example of a uniform distribution because each digit is equally likely to occur. Starting in the row corresponding to the day of the month in which you were born, use the table of random numbers (Table E.1) to take *one digit* at a time.

Select samples of sizes $n = 2$, $n = 5$, $n = 10$. Compute the sample mean $\overline{X}$ of each sample. For each sample size, each student selects five different samples so that a frequency distribution of the sample means is developed for the results of the entire class based on samples of sizes $n = 2$, $n = 5$, and $n = 10$.

What can be said about the shape of the sampling distribution for each of these sample sizes?

6.77 (Class Project) A coin having one side heads and the other side tails is to be tossed 10 times and the number of heads obtained is to be recorded. If each student performs this experiment five times, a frequency distribution of the number of heads can be developed from the results of the entire class. Does this distribution seem to approximate the normal distribution?

6.78 (Class Project) The number of cars waiting in line at a car wash is distributed as follows:

Length of waiting line

NUMBER OF CARS	PROBABILITY
0	.25
1	.40
2	.20
3	.10
4	.04
5	.01

The table of random numbers (Table E.1) can be used to select samples from this distribution by assigning numbers as follows:

1. Start in the row corresponding to the day of the month in which you were born.
2. *Two-digit* random numbers are to be selected.
3. If a random number between 00 and 24 is selected, record a length of 0; if between 25 and 64, record a length of 1; if between 65 and 84, record a length of 2; if between 85 and 94, record a length of 3; if between 95 and 98, record a length of 4; if it is 99, record a length of 5.

Select samples of sizes $n = 2$, $n = 10$, $n = 25$. Compute the sample mean for each sample. For example, if a sample of size 2 results in random numbers 18 and 46, these would correspond to lengths of 0 and 1, respectively, producing a sample mean of 0.5. If each student selects five different samples for each sample size, a frequency distribution of the sample means (for each sample size) can be developed from the results of the entire class.

What conclusions can you draw about the sampling distribution of the mean as the sample size is increased?

6.79 (Class Project) The table of random numbers can simulate the selection of different colored balls from a bowl as follows:

1. Start in the row corresponding to the day of the month in which you were born.
2. Select *one-digit* numbers.
3. If a random digit between 0 and 6 is selected, consider the ball white; if a random digit is a 7, 8, or 9, consider the ball red.

Select samples of sizes $n = 10$, $n = 25$, and $n = 50$ digits. In each sample, count the number of white balls and compute the proportion of white balls in the sample. If each student in the class selects five different samples for each sample size, a frequency distribution of the proportion of white balls (for each sample size) can be developed from the results of the entire class.

What conclusions can be drawn about the sampling distribution of the proportion as the sample size is increased?

6.80 **(Class Project)** Suppose that step 3 of Problem 6.79 uses the following rule:

If a random digit between 0 and 8 is selected, consider the ball to be white; if a random digit of 9 is selected, consider the ball to be red. Compare and contrast the results obtained in this problem and in Problem 6.79.

TEAM PROJECT

TP6.1 Refer to TP2.1 on page 144. A financial investment service is evaluating the list of domestic general stock funds so that it can make purchase recommendations to potential investors. The vice president for research has hired your group, the _____ Corporation, to study the financial characteristics of currently traded domestic general stock funds and gives you access to Special Data Set 1 of appendix D containing various characteristics from a sample of 194 such funds. In particular, the vice president is interested in a comparison of some features of these funds based on fee structure (no-load versus fee payment). As part of your analysis of the 107 no-load funds versus the 87 fee payment funds, you want to decide whether or not both the net asset value (in dollars) and the total year-to-date return (in percentage rates) appear to be approximately normally distributed. You accomplish this by

DATA FILE
MUTUAL

(a) evaluating the actual versus theoretical properties.
(b) constructing a normal probability plot.
(c) stating your conclusions based on (a) and (b).

Note: Additional team projects can be found at the following World Wide Web address:

http://www.prenhall.com/levine

THE SPRINGVILLE HERALD CASE

The production department of the newspaper has embarked on a quality improvement effort and has chosen as its first project an issue that relates to the blackness of the newspaper print. Each day a determination needs to be made concerning how "black" the newspaper is printed. This is measured on a standard scale in which the target value is 1.0. Data collected over the past year indicates that the blackness is normally distributed with an average of 1.005 and a standard deviation of 0.10.

Exercises

6.1 Each day, one spot on the first newspaper printed is chosen and the blackness of the spot is measured. Suppose the blackness of the newspaper is considered acceptable if the blackness of the spot is between 0.95 and 1.05. Assuming that the distribution has not changed from what it was in the past year, what is the probability that the blackness of the spot is

(a) less than 1.0?

(c) between 1.0 and 1.05?

(b) between .95 and 1.0?

(d) less than .95 or greater than 1.05?

6.2 If the objective of the production team is to reduce the probability that the blackness is below 0.95 or above 1.05, would it be better off focusing on process improvement that lowered the blackness to the target value of 1.0 or on process improvement that reduced the standard deviation to 0.075? Explain.

6.3 Suppose that instead of selecting only one spot on the newspaper on which to measure blackness, a sample of 25 spots is selected. Assuming that the distribution has *not* changed from what it was in the past year, what is the probability that the average blackness of the spots is

(a) less than 1.0?

(c) between 1.0 and 1.05?

(b) between .95 and 1.0?

(d) less than .95 or greater than 1.05?

6.4 Suppose that the average blackness of today's sample of 25 spots is 0.952. What conclusion would you draw about the blackness of the newspaper based on this result? Explain.

References

1. Cochran, W. G., *Sampling Techniques*, 3d ed. (New York: Wiley, 1977).

2. Gunter, B., "Q-Q Plots," *Quality Progress* (February 1994), 81–86.

3. Marascuilo, L. A., and M. McSweeney, *Nonparametric and Distribution-Free Methods for the Social Sciences* (Monterey, CA: Brooks/Cole, 1977).

4. Mendenhall, W., and T. Sincich, *Statistics for Engineering and the Sciences*, 4th ed. (Englewood Cliffs, NJ: Prentice Hall, 1995).

5. *Microsoft Excel 97* (Redmond, WA: Microsoft Corp., 1997).

6. Ramsey, P. P., and P. H. Ramsey, "Simple Tests of Normality in Small Samples," *Journal of Quality Technology* 22 (1990): 299–309.

7. Sievers, G. L., "Probability Plotting." In *Encyclopedia of Statistical Sciences*, vol. 7, edited by S. Kotz and N. L. Johnson (New York: Wiley, 1986), 232–237.

Chapter 5

Student Solutions Manual

6.2   (a) $P(Z > 1.34) = 0.5 - 0.4099 = 0.0901$

      (b) $P(Z < 1.17) = 0.5 + 0.3790 = 0.8790$

      (c) $P(0 < Z < 1.17) = 0.3790$

      (d) $P(Z < -1.17) = 0.5 - 0.3790 = 0.1210$

      (e) $P(-1.17 < Z < 1.34) = 0.3790 + 0.4099 = 0.7889$

      (f) $P(-1.17 < Z < -0.50) = 0.3790 - 0.1915 = 0.1875$

6.4   (a) $P(-2.00 < Z < 2.00) = P(-2.00 < Z < 0) + P(0 < Z < 2.00)$

                          $= 0.4772 + 0.4772 = 0.9544$

      (b) $P(-3.00 < Z < 3.00) = P(-3.00 < Z < 0) + P(0 < Z < 3.00)$

                          $= 0.49865 + 0.49865 = 0.9973$

6.6   (a)   (1) $P(X > 43) = P(Z > -1.75) = 0.4599 + 0.5 = 0.9599$

           (2) $P(X < 42) = P(Z < -2.00) = 0.5 - 0.4772 = 0.0228$

           (3) $P(42 < X < 48) = P(-2.00 < Z < -0.50) = 0.4772 - 0.1915 = 0.2857$

           (4) $P(X > 57.5) = P(Z > 1.88) = 0.5 - 0.4699 = 0.0301$

           (5) $P(X < 40) = P(Z < -2.50) = 0.5 - 0.4938 = 0.0062$

                  $P(X > 55) = P(Z > 1.25) = 0.5 - 0.3944 = 0.1056$

                  $P(X < 40) + P(X > 55) = 0.0062 + 0.1056 = 0.1118$

      (b) $P(X < A) = 0.05$, so $P(A < X < 0) = 0.45$ and $P(-1.645 < Z < 0) = 0.45$

$$Z = -1.645 = \frac{X - 50}{4} \qquad X = 50 - 1.645(4) = 43.42$$

      (c) $P(-A < X < A) = 0.60$

           $P(-0.84 < Z < 0) = 0.30$ and $P(0 < Z < 0.84) = 0.30$

$$Z = -0.84 = \frac{X_{lower} - 50}{4} \qquad Z = +0.84 = \frac{X_{upper} - 50}{4}$$

           $X_{lower} = 50 - 0.84(4) = 46.64$ and $X_{upper} = 50 + 0.84(4) = 53.36$

      (d) $P(X > A) = 0.85$, so $P(A < X < 0) = 0.35$

$$P(-1.04 < Z < 0) = 0.35 \qquad\qquad Z = -1.35 = \frac{X - 50}{4}$$

           $X = 50 - 1.35(4) = 44.60$

•6.8   (a) $P(34 < X < 50) = P(-1.33 < Z < 0) = 0.4082$

      (b) $P(34 < X < 38) = P(-1.33 < Z < -1.00) = 0.4082 - 0.3413 = 0.0669$

      (c) $P(X < 30) + P(X > 60) = P(Z < -1.67) + P(Z > 0.83)$

                  $= (0.5 - 0.4525) + (0.5 - 0.2967) = 0.2508$

      (d) $1000(1 - 0.2508) = 749.2$ trucks

    (e)$P(X > A) = 0.80$ $P(-0.84 < Z < 0) \cong 0.30$ $Z = -0.84 = \dfrac{X - 50}{12}$

         $X = 50 - 0.84(12) = 39.92$ thousand miles or 39,920 miles

•6.8 (f)   The larger standard deviation makes the Z-values smaller.
cont.   (a) $P(34 < X < 50) = P(-1.60 < Z < 0) = 0.4452$
          (b) $P(34 < X < 38) = P(-1.60 < Z < -1.20) = 0.4452 - 0.3849$
                $= 0.0603$
          (c) $P(X < 30) + P(X > 60) = P(Z < -2.00) + P(Z > 1.00)$
                $= (0.5 - 0.4772) + (0.5 - 0.3413) = 0.1815$
          (d) $1000(1 - 0.1815) = 818.5$ trucks
          (e) $X = 50 - 0.84(10) = 41.6$ thousand miles or 41,600 miles

6.10   (a) $P(X < 91) = P(Z < 2.25) = 0.4878 + 0.5 = 0.9878$
       (b) $P(65 < X < 89) = P(-1.00 < Z < 2.00) = 0.3413 + 0.4772 = 0.8185$
       (c) $P(81 < X < 89) = P(1.00 < Z < 2.00) = 0.4772 - 0.3413 = 0.1359$
       (d) $P(X > A) = 0.05\ P(0 < Z < 1.645) = 0.4500$

$$Z = 1.645 = \frac{X - 73}{8} \qquad X = 73 + 1.645(8) = 86.16\%$$

(e) Option 1: $P(X > A) = 0.10 \qquad P(0 < Z < 1.28) \cong 0.4000$

$$Z = \frac{81 - 73}{8} = 1.00$$

Since your score of 81% on this exam represents a Z-score of 1.00, which is below the minimum Z-score of 1.28, you will not earn an "A" grade on the exam under this grading option.

$$\text{Option 2: } Z = \frac{68 - 62}{3} = 2.00$$

Since your score of 68% on this exam represents a Z-score of 2.00, which is well above the minimum Z-score of 1.28, you will earn an "A" grade on the exam under this grading option. You should prefer Option 2.

•6.12   (a) $P(X < 35) = P(Z < -1.00) = 0.5 - 0.3414 = 0.1587$
        (b) $P(28 < X < 32) = P(-2.40 < Z < -1.60) = 0.4918 - 0.4452 = 0.0466$
        (c) $P(35 < X < 48) = P(-1.00 < Z < 1.60) = 0.3413 + 0.4452 = 0.7865$
        (d) $\qquad P(X > A) = 0.10 \qquad\qquad P(0 < Z < 1.28) \cong 0.4000$
              $X = 40 + 1.28(5) = 46.40$ hours
        (e) $Q_2 = 40$
        (f) $Q_1 = 40 - 0.67(5) = 36.65 \qquad Q_3 = 40 + 0.67(5) = 43.35$
              Interquartile range $= Q_3 - Q_1 = 43.35 - 36.65 = 6.7$ hours
        (g)   The larger standard deviation cuts the prior Z-values in half.
              (a) $P(X < 35) = P(Z < -0.50) = 0.5 - 0.1915 = 0.3085$
              (b) $P(28 < X < 32) = P(-1.20 < Z < -0.80) = 0.3849 - 0.2881 = 0.0968$
              (c) $P(35 < X < 48) = P(-0.50 < Z < 0.80) = 0.1915 + 0.2881 = 0.4796$
              (d) $X = 40 + 1.28(10) = 52.80$ hours
              (e) $Q_2 = 40$
              (f) $Q_1 = 40 - 0.67(10) = 33.30 \qquad Q_3 = 40 + 0.67(10) = 46.70$
                    Interquartile range $= Q_3 - Q_1 = 46.70 - 33.30 = 13.4$ hours

6.14   With 19 observations, the 18th largest observation covers an area under the normal curve of .90. The corresponding Z-value is +1.28. The largest observation covers an area under the normal curve of .95. The corresponding Z-value is either + 1.645, + 1.64, or + 1.65 depending on the "rule" used for this selection.

•6.16   Area under normal curve covered: 0.1429 0.2857 0.4286 0.5714 0.7143 0.8571
        Standardized normal quantile value: -1.07 -0.57 -0.18 $+0.18$ $+0.57$ $+1.07$

6.18   (a) $\overline{X} = \$40.71$ $S = \$9.45$

            Five-number summary $23.84 \ \$32.10 \ \$42.70 \ \$48.31 \ \$54.58$

            Because the mean is slightly less than the median, the distribution is slightly left-skewed.

  (b)

Z-value	$	Z-value	$
-1.82	23.84	0.04	44.24
-1.48	24.21	0.13	44.84
-1.26	25.93	0.22	45.32
-1.09	27.28	0.35	45.93
-0.94	28.73	0.35	45.93
-0.82	30.84	0.50	46.58
-0.70	31.15	0.60	47.36
-0.60	34.96	0.70	48.62
-0.50	35.56	0.82	49.92
-0.40	37.19	0.94	50.66
-0.31	38.58	1.09	50.97
-0.22	38.94	1.26	52.62
-0.13	40.22	1.48	53.81
-0.04	41.16	1.82	54.58

The Normal Probability Plot

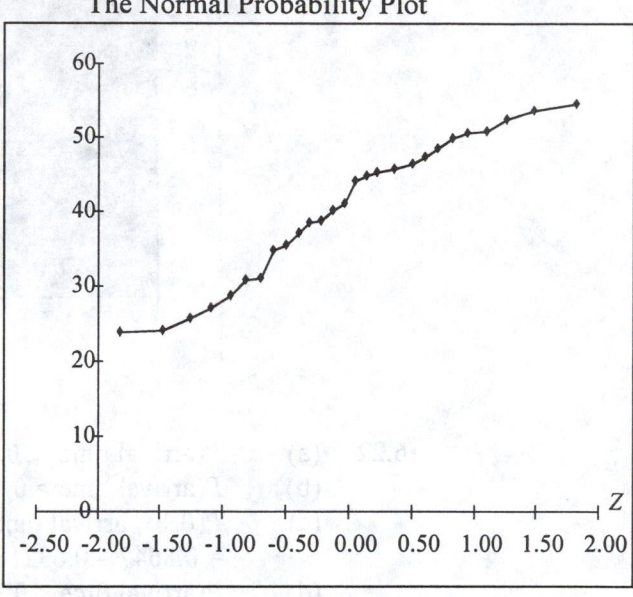

•6.20   (a)   Office I: $\overline{X} = 2.214$ $S = 1.718$

                  Five-number summary   0.52 0.93 1.54 3.93 6.32

                  The distribution is right-skewed.

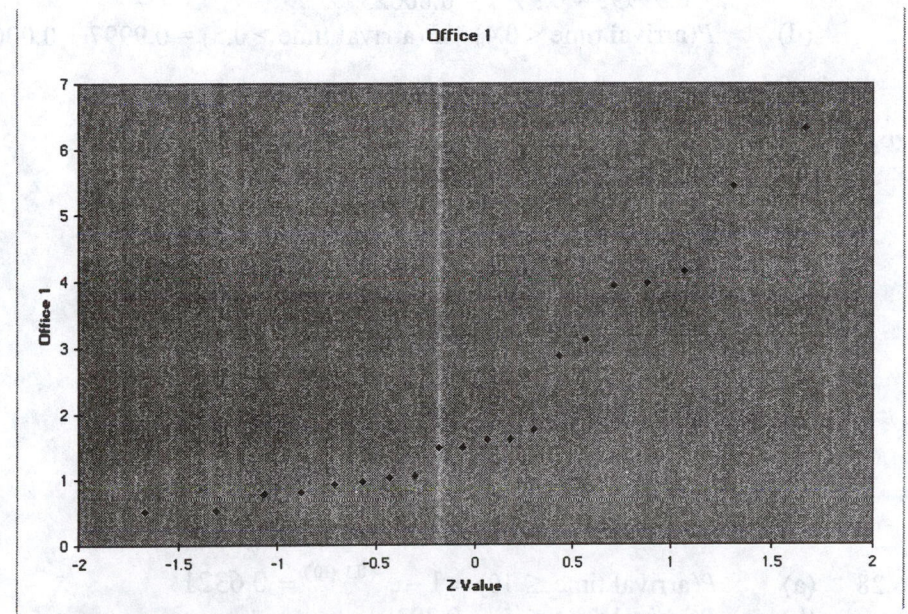

        Office II: $\overline{X} = 2.011$ $S = 1.892$

                  Five-number summary   0.08 0.60 1.505 3.75 7.55

                  The distribution is right-skewed.

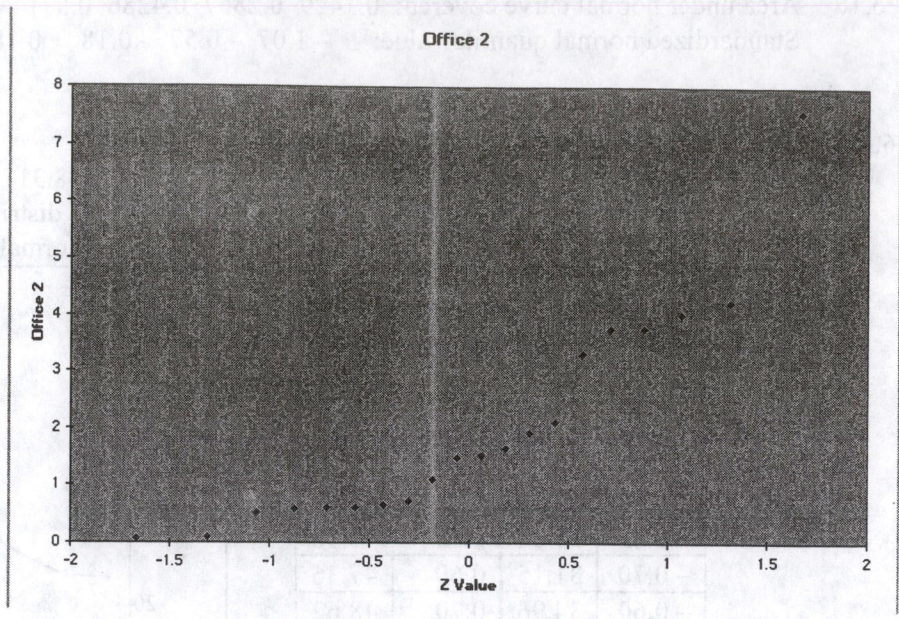

Office 2

•6.22 (a) $P(\text{arrival time} \leq 0.1) = 1 - e^{-\lambda x} = 1 - e^{-(10)(0.1)} = 0.6321$
(b) $P(\text{arrival time} > 0.1) = 1 - P(\text{arrival time} \leq 0.1) = 1 - 0.6321 = 0.3679$
(c) $P(0.1 < \text{arrival time} < 0.2) = P(\text{arrival time} < 0.2) - P(\text{arrival time} < 0.1)$
$= 0.8647 - 0.6321 = 0.2326$
(d) $P(\text{arrival time} < 0.1) + P(\text{arrival time} > 0.2) = 0.6321 + 0.1353 = 0.7674$

6.24 (a) $P(\text{arrival time} \leq 0.4) = 1 - e^{-(20)(0.4)} = 0.9997$
(b) $P(\text{arrival time} > 0.4) = 1 - P(\text{arrival time} \leq 0.4) = 1 - 0.9997 = 0.0003$
(c) $P(0.4 < \text{arrival time} < 0.5) = P(\text{arrival time} < 0.5) - P(\text{arrival time} < 0.4)$
$= 0.99995 - 0.9997 = 0.00025$
(d) $P(\text{arrival time} < 0.4) + P(\text{arrival time} > 0.5) = 0.9997 + 0.000045 = 0.999745$

•6.26 (a)

	A	B
1	Fast Food Arrivals	
2		
3	Mean	2
4	X Value	1
5	P(<=X)	0.864665

(b) $P(\text{arrival time} \leq 5) = 0.99996$
(c) If $\lambda = 1$, $P(\text{arrival time} \leq 1) = 0.6321$,
$P(\text{arrival time} \leq 5) = 0.9933$

6.28 (a) $P(\text{arrival time} \leq 10) = 1 - e^{-(0.1)(10)} = 0.6321$
(b) $P(\text{arrival time} \leq 5) = 0.3935$
(c) $P(\text{arrival time} \leq 1) = 0.0952$

•6.30 (a) $P(\bar{X} < 95) = P(Z < -2.50) = 0.5 - 0.4938 = 0.0062$

(b) $P(95 < \bar{X} < 97.5) = P(-2.50 < Z < -1.25) = 0.4938 - 0.3944 = 0.0994$

(c) $P(\bar{X} > 102.2) = P(Z > 1.10) = 0.5 - 0.3643 = 0.1357$

(d) $P(99 < \bar{X} < 101) = P(-0.50 < Z < 0.50) = 0.1915 + 0.1915 = 0.3830$

(e) $P(\bar{X} > A) = P(Z > -0.39) = 0.65$ $\bar{X} = 100 - 0.39\left(\dfrac{10}{\sqrt{25}}\right) = 99.22$

(f) (a) $P(\bar{X} < 95) = P(Z < -2.00) = 0.5 - 0.4772 = 0.0228$

(b) $P(95 < \bar{X} < 97.5) = P(-2.00 < Z < -1.00)$
$= 0.4772 - 0.3413 = 0.1359$

(c) $P(\bar{X} > 102.2) = P(Z > 0.88) = 0.5 - 0.3106 = 0.1894$

(d) $P(99 < \bar{X} < 101) = P(-0.40 < Z < 0.40)$
$= 0.1554 + 0.1554 = 0.3108$

(e) $P(\bar{X} > A) = P(Z > -0.39) = 0.65$ $\bar{X} = 100 - 0.39\left(\dfrac{10}{\sqrt{16}}\right) = 99.025$

6.32

(a) For samples of 25 travel expense vouchers for a university in an academic year, the sampling distribution of sample means is the distribution of means from all possible samples of 25 vouchers that could occur.

(b) For samples of 25 absentee records in 1997 for employees of a large manufacturing company, the sampling distribution of sample means is the distribution of means from all possible samples of 25 records that could occur.

(c) For samples of 25 sales of unleaded gasoline at service stations located in a particular county, the sampling distribution of sample means is the distribution of means from all possible samples of 25 sales that could occur.

•6.34 (a) $P(1.28 < X < 1.30) = P(-0.50 < Z < 0) = 0.1915$

(b) $P(1.31 < X < 1.33) = P(0.25 < Z < 0.75) = 0.2734 - 0.0987 = 0.1747$

(c) $P(-A < X < A) = P(0.84 < Z < 0.84) = 0.60$
$X = 1.30 - 0.84(0.04) = 1.2664$ $X = 1.30 + 0.84(0.04) = 1.3336$

(d) (1) $\mu_{\bar{X}} = 1.30$ $\sigma_{\bar{X}} = 0.01$

(2) Because the population diameter of Ping-Pong balls is approximately normally distributed, the sampling distribution of samples of 16 will also be approximately normally distributed.

(3) $P(1.28 < \bar{X} < 1.30) = P(-2.00 < Z < 0) = 0.4772$

(4) $P(1.31 < \bar{X} < 1.33) = P(1.00 < Z < 3.00) = 0.49865 - 0.3413 = 0.15735$

(5) $P(-A < \bar{X} < A) = P(0.84 < Z < 0.84) = 0.60$
Lower bound: $\bar{X} = 1.30 - 0.84(0.01) = 1.2916$
Upper bound: $\bar{X} = 1.30 + 0.84(0.01) = 1.3084$

(e)-(f) When samples of size 16 are taken rather than individual values (samples of $n = 1$), more values lie closer to the mean and fewer values lie farther away from the mean with the increased sample size. This occurs because the standard deviation of the sampling distribution, the standard error, is given by:

$$\sigma_{\bar{X}} = \frac{\sigma}{\sqrt{n}}$$

As n increases, the value of the denominator increases, resulting in a smaller value of the overall fraction. The standard error for the distribution of sample means of size 16 is 1/4 of the population standard deviation of individual values and means that the sampling distribution is more concentrated around the population mean.

(g)   They are equally likely to occur (probability = 0.1587) since as n increases, more sample means will be closer to the mean of the distribution.

6.36   (a) $P(\overline{X} > 3) = P(Z > -1.00) = 0.3413 + 0.5 = 0.8413$
       (b) $P(\overline{X} < A) = P(Z < 1.04) = 0.85$ $\overline{X} = 3.10 + 1.04(0.1) = 3.204$
       (c)   To be able to use the standard normal distribution as an approximation for the area under the curve, we must assume that the population is symmetrically distributed such that the central limit theorem will likely hold for samples of $n = 16$.
       (d) $P(\overline{X} < A) = P(Z < 1.04) = 0.85$ $\overline{X} = 3.10 + 1.04(0.05) = 3.152$
       (e)   To be able to use the standard normal distribution as an approximation for the area under the curve, we must assume that the central limit theorem will hold for samples of $n = 64$.
       (f)   For $n = 1$, $P(X < 2) = P(Z < -2.75) = 0.5 - 0.4970 = 0.0030$
              For $n = 16$, $P(\overline{X} > 3.4) = P(Z > 3.00) = 0.5 - 0.49865 = 0.00135$
              For $n = 100$, $P(\overline{X} < 2.9) = P(Z < -4.00) =$ virtually zero
              It is more likely to have an individual service time below 2 minutes.

6.38   (a)(1) $P(5 < \overline{X} < 5.5) = P(0 < Z < 1.67) = 0.4525$
       (2) $P(4.2 < \overline{X} < 4.5) = P(-2.67 < Z < -1.67)$
                  $= 0.4962 - 0.4525 = 0.0437$
       (3) $P(\overline{X} < 4.6) = P(Z < -1.33) = 0.5 - 0.4082 = 0.0918$

       (b) $P(-1.96 < Z < 1.96) = 0.95$
              $\overline{X} = 5 - 1.96(0.3) = 4.412$ $\overline{X} = 5 + 1.96(0.3) = 5.588$

       (c)(a) (1) $P(5 < \overline{X} < 5.5) = P(0 < Z < 2.50) = 0.4938$
              (2) $P(4.2 < \overline{X} < 4.5) = P(-4.00 < Z < -2.50)$
                         $= 0.5 - 0.4938 = 0.0062$
              (3) $P(\overline{X} < 4.6) = P(Z < -2.00) = 0.5 - 0.4772 = 0.0228$
       (b) $P(-1.96 < Z < 1.96) = 0.95$
              $\overline{X} = 5 - 1.96(0.2) = 4.608$ $\overline{X} = 5 + 1.96(0.2) = 5.392$

6.40   (a) $p_s = 15/50 = 0.30$ (b) $\sigma_{p_s} = \sqrt{\dfrac{0.40(0.60)}{50}} = 0.0693$

•6.42 (a) $P(0.09 < p_s < 0.10) = P(-0.67 < Z < 0) = 0.2486$
       (b) $P(p_s < 0.08) = P(-1.33 < Z) = 0.5 - 0.4082 = 0.0918$
       (c)   Decreasing the sample size by a factor of 4 increases the standard error by a factor of 2.
              (a) $P(0.09 < p_s < 0.10) = P(-0.33 < Z < 0) = 0.1293$
              (b) $P(p_s < 0.08) = P(Z < -0.67) = 0.5 - 0.2486 = 0.2514$
       (d)   If $n = 100$, $P(p_s > 0.13) = P(Z > 1.00) = 0.5 - 0.3413 = 0.1587$
              If $n = 400$, $P(p_s > 0.105) = P(Z > 0.33) = 0.5 - 0.1293 = 0.3707$
              It is more likely that a percent defective above 10.5% occur among samples of size 400.

6.44   (a) $P(0.20 < p_s < 0.30) = P(-2.18 < Z < 0) = 0.4854$

        (b) $P(-1.96 < Z < 1.96) = 0.95$

              $p_s = .30 - 1.96(0.0458) = 0.2102$ $p_s = .30 + 1.96(0.0458) = 0.3898$

6.46   (a) $P(0.93 < p_s < 0.95) = P(0 < Z < 1.75) = 0.4599$

        (b) $P(p_s > 0.95) = P(Z > 1.75) = 0.5 - 0.4599 = 0.0401$

        (c)   (a) $P(0.93 < p_s < 0.95) = P(0 < Z < 2.48) = 0.4934$

               (b) $P(p_s > 0.95) = P(Z > 2.48) = 0.5 - 0.4934 = 0.0066$

        (d)   If $n = 500$, $P(p_s > 0.95) = P(Z > 1.75) = 0.5 - 0.4599 = 0.0401$

              If $n = 1000$, $P(p_s < 0.90) = P(Z < -3.72) = 0.5 - 0.4999 = 0.0001$

              More than 95% in a sample of 500 is more likely than less than 90% in a sample of 1000.

6.48   If $N = 400$ and $n = 100$, fpc $= 0.8671$.

      If $N = 900$ and $n = 200$, fpc $= 0.8824$.

      The finite population correction done for a population of 400 sampling 100 has a greater effect in reducing the standard error.

•6.50 $P(1.31 < \overline{X} < 1.33) = P(1.04 < Z < 3.12) = 0.4991 - 0.3508 = 0.1483$

$$\text{where } Z = \frac{1.31 - 1.30}{\frac{0.04}{\sqrt{16}} \cdot \sqrt{\frac{200-16}{200-1}}} = 1.04 \quad \text{and} \quad Z = \frac{1.33 - 1.30}{\frac{0.04}{\sqrt{16}} \cdot \sqrt{\frac{200-16}{200-1}}} = 3.12$$

•6.52   (a) $P(0.09 < p_s < 0.10) = P(-0.69 < Z < 0) = 0.2549$

$$\text{where } Z = \frac{0.09 - 0.10}{\sqrt{\frac{0.1(0.9)}{400}} \cdot \sqrt{\frac{5000-400}{5000-1}}} = -0.69$$

        (b) $P(p_s < 0.08) = P(Z < -1.39) = 0.5 - 0.4177 = 0.0823$

$$\text{where } Z = \frac{0.08 - 0.10}{\sqrt{\frac{0.1(0.9)}{400}} \cdot \sqrt{\frac{5000-400}{5000-1}}} = -1.39$$

6.68   (a) $P(0.75 < X < 0.753) = P(-0.75 < Z < 0) = 0.2734$

        (b) $P(0.74 < X < 0.75) = P(-3.25 < Z < -0.75) = 0.49942 - 0.2734 = 0.2260$

        (c) $P(X > 0.76) = P(Z > 1.75) = 0.5 - 0.4599 = 0.0401$

        (d) $P(X < 0.74) = P(Z < -3.25) = 0.5 - 0.49942 = 0.00058$

        (e) $P(X < A) = P(Z < -1.48) = 0.07 \ X = 0.753 - 1.48(0.004) = 0.7471$

6.70 (a) $P(X > 5) = P(Z > 2.00) = 0.5 - 0.4772 = 0.0228$

(b) $P(X > 5) = P(Z > 2.00) = 0.0228$

(c)   Although the mean processing time is reduced from 3 days to 2, the increased standard deviation offsets any increase, resulting in the same calculated value of Z.

(d)   The city agency would do better reducing the average to 2 days since the proportion of free permits would be 0.00135 instead of 0.0038.

6.72 (a) $P(100 < X < 115) = P(0 < Z < 0.75) = 0.2734$

(b) $P(X > 90) = P(Z > -0.50) = 0.5 + 0.1915 = 0.6915$

(c) $P(X > A) = P(Z > -1.28) = 0.90$ $X = 100 - 1.28(20) = 74.40$ hours

(d) $Q_1 = 100 - 0.67(20) = 86.60$ $Q_3 = 100 + 0.67(20) = 113.40$
Interquartile range $= Q_3 - Q_1 = 113.40 - 86.60 = 26.80$

(e)(1) $P(100 < \bar{X} < 115) = P(0 < Z < 3.00) = 0.49865$

   (2) $P(\bar{X} > 90) = P(Z > -2.00) = 0.5 + 0.4772 = 0.9772$

   (3) $P(-A < \bar{X} < A) = P(-1.645 < Z < 1.645) = 0.90$
       $\bar{X} = 100 - 1.645(5) = 91.775$ $\bar{X} = 100 + 1.645(5) = 108.225$

(f)   The central limit theorem is not necessary because the population is normally distributed. So the sampling distribution of the mean will also be normally distributed regardless of the sample size.

•6.74 (a)   (1) $P(10 < X < 12.5) = P(0 < Z < 1.00) = 0.3413$

       (2) $P(7.5 < X < 10) = P(-1.00 < Z < 0) = 0.3413$

       (3) $P(7.5 < X < 12.5) = 0.3413 + 0.3413 = 0.6826$

       (4) $P(X > 7.5) = P(Z > -1.00) = 0.5 + 0.3413 = 0.8413$

       (5) $P(X < 7.5) = 1 - P(X > 7.5) = 1 - 0.8413 = 0.1587$

       (6) $P(12.5 < X < 14.3) = P(1.00 < Z < 1.72) = 0.4573 + 0.3413 = 0.1160$

(b) $P(X > A) = P(Z > 0) = 0.50$ $X = 10$ million dollars

(c) $P(X > A) = P(Z > -1.28) = 0.90$ $X = 10 - 1.28(2.5) = 6.8$ million dollars

(d) $Q_1 = 10 - 0.67(2.5) = 8.325$ $Q_3 = 10 + 0.67(2.5) = 11.675$
Interquartile range $= Q_3 - Q_1 = 11.675 - 8.325 = 3.35$ million dollars

Chapter **6**
Confidence Interval Estimation

397

CHAPTER OBJECTIVES

✓ *To develop confidence interval estimates for the mean and the proportion*
✓ *To determine the sample size necessary to obtain a desired confidence interval*
✓ *To examine the effect of a finite population on the confidence interval and the sample size*
✓ *To demonstrate the application of confidence intervals in auditing*

Introduction

Statistical inference is the process of using sample results to draw conclusions about the characteristics of a population. In this chapter we examine statistical procedures that will enable us to *estimate* either a population mean or a population proportion.

There are two major types of estimates: point estimates and interval estimates. A **point estimate** consists of a single sample statistic that is used to estimate the true value of a population parameter. For example, the sample mean $\overline{X}$ is a point estimate of the population mean μ and the sample variance S^2 is a point estimate of the population variance σ^2. Recall from section 6.5 that the sample mean $\overline{X}$ possesses the highly desirable properties of *unbiasedness* and *efficiency*. Although in practice only one sample is selected, we know that the average value of all possible sample means is μ, the true population parameter.[1] A sample statistic such as $\overline{X}$ varies from sample to sample because it depends on the items selected in the sample, so we must take this into consideration when providing an estimate of the population characteristic. To accomplish this, we develop an **interval estimate** of the true population mean by taking into account the sampling distribution of the mean. The interval that we construct will have a specified confidence or probability of correctly estimating the true value of the population parameter μ. Similar interval estimates are also developed for the population proportion, *p,* and the population total.

In this chapter, we also discuss how to determine the size of the sample that should be selected in a survey. In addition, we discuss how a finite population affects the width of the confidence interval developed and the sample size selected.

We begin with a look at how a business uses confidence interval estimation to draw conclusions.

[1] *It is for this reason that the denominator of the sample variance is n − 1 instead of n so that S^2 is an unbiased estimator of σ^2, that is, the average of all possible sample variances is σ^2.*

◆ **USING STATISTICS:** *Auditing Sales Invoices at Saxon Plumbing Company*

Saxon Plumbing Company is a large distributor of wholesale plumbing supplies in a suburban area outside a city in the northeastern United States. In an effort to maintain internal controls on sales, the company has prenumbered sales invoices that include a warehouse removal slip that must be used for each sale. Goods are not to be removed from the warehouse without an authorized warehouse removal slip. At the end of each month a sample of the sales invoices is selected to determine the following:

- The average amount listed on the sales invoices for the warehouse in that month.
- The total amount listed on the sales invoices for the warehouse in that month.

- Any differences between the actual amounts on the sales invoice and the amounts entered into the accounting system for the warehouse.

- The frequency of occurrence of various types of errors that violate the internal control policy of the warehouse. These errors may include failure to use the warehouse removal slip, failure to attach a duplicate sales invoice to the item being shipped, failure to include the correct account number of the customer, and shipment of the incorrect plumbing supply item.

 **7.1** ## CONFIDENCE INTERVAL ESTIMATION OF THE MEAN (σ KNOWN)

In section 6.5 we observed that we can use the central limit theorem and/or knowledge of the population distribution to determine the percentage of sample means that fall within certain distances of the population mean. For instance, in the cereal-filling-process example (see page 383) we observed that 95% of all sample means fall between 362.12 and 373.88 grams. This statement is based on *deductive reasoning*. However, it is exactly opposite to the type of reasoning that is needed here: *inductive reasoning*.

Inductive reasoning is needed because, in statistical inference, we must take the results of a single sample and draw conclusions about the population, not vice versa. In practice, the population mean is the unknown quantity that is to be estimated. Suppose that, in the cereal-filling-process example, the true population mean μ is unknown but the true population standard deviation σ is known to be 15 grams. Thus, rather than taking $\mu \pm (1.96)(\sigma/\sqrt{n})$ to find the upper and lower limits around μ as in section 6.5, we determine the consequences of substituting the sample mean $\overline{X}$ for the unknown μ and using $\overline{X} \pm (1.96)(\sigma/\sqrt{n})$ as an interval within which we estimate the unknown μ. Although in practice a single sample of size n is selected and the mean $\overline{X}$ is computed, we need to obtain a hypothetical set of all possible samples, each of size n, in order to understand the full meaning of the interval estimate to be obtained.

Suppose, for example, that our sample of size $n = 25$ boxes has a mean of 362.3 grams. The interval developed to estimate μ is $362.3 \pm (1.96)(15)/(\sqrt{25})$ or 362.3 ± 5.88. That is, the estimate of μ is

$$356.42 \leq \mu \leq 368.18$$

Because the population mean μ (equal to 368) is included within the interval, this sample has led to a correct statement about μ (see Figure 7.1 on page 416).

To continue our hypothetical example, suppose that for a different sample of $n = 25$ boxes, the mean is 369.5. The interval developed from this sample is $369.5 \pm (1.96)(15)/(\sqrt{25})$ or 369.5 ± 5.88. That is, the estimate is

$$363.62 \leq \mu \leq 375.38$$

Because the true population mean μ (equal to 368) is also included within this interval, we conclude that this statement about μ is correct.

Now, before we begin to think that we will *always* make correct statements about μ by developing a confidence interval estimate from the sample $\overline{X}$, suppose we draw a third hypothetical sample of size $n = 25$ boxes in which the sample mean is equal to 360 grams. The interval developed here is $360 \pm (1.96)(15)/(\sqrt{25})$ or 360 ± 5.88. In this case, the estimate of μ is

$$354.12 \leq \mu \leq 365.88$$

FIGURE 7.1

Confidence interval estimates from five different samples of size $n = 25$ taken from population where $\mu = 368$ and $\sigma = 15$

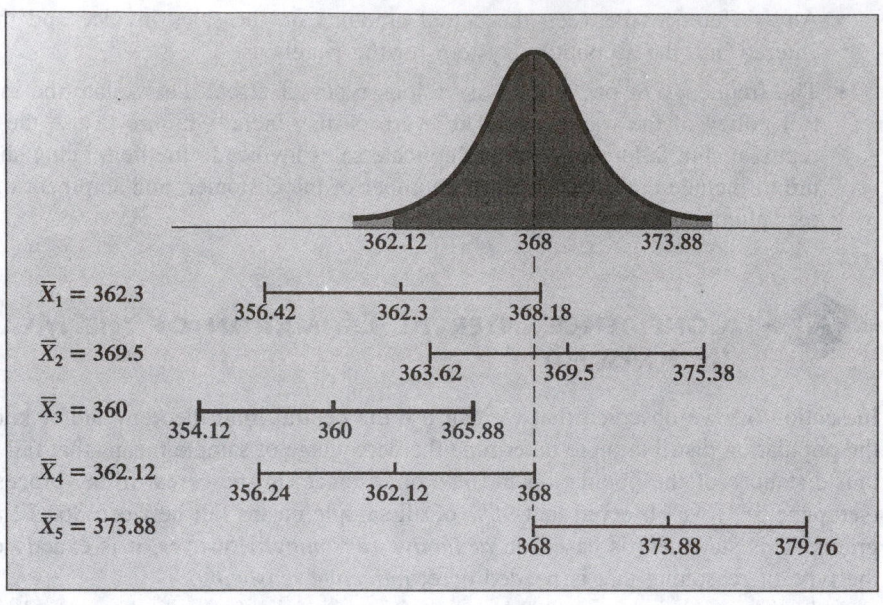

Observe that this estimate is *not* a correct statement, because the population mean μ is not included in the interval developed from this sample (see Figure 7.1). Thus, we are faced with a dilemma. For some samples the interval estimate of μ will be correct, and for others it will be incorrect. In addition, we must realize that in practice we select only one sample, and because we do not know the true population mean, we cannot determine whether our particular statement is correct.

What can we do to resolve this dilemma? We can determine the proportion of samples producing intervals that result in correct statements about the population mean μ. To do this, we need to examine two other hypothetical samples: the case in which $\overline{X} = 362.12$ grams and the case in which $\overline{X} = 373.88$ grams. If $\overline{X} = 362.12$, the interval is $362.12 \pm (1.96)(15)/(\sqrt{25})$ or 362.12 ± 5.88. That is,

$$356.24 \le \mu \le 368.00$$

Because the population mean of 368 is at the upper limit of the interval, the statement is a correct one (see Figure 7.1).

Finally, if $\overline{X} = 373.88$, the interval is $373.88 \pm (1.96)(15)/(\sqrt{25})$ or 373.88 ± 5.88. That is,

$$368.00 \le \mu \le 379.76$$

In this case, because the population mean of 368 is included at the lower limit of the interval, the statement is a correct one.

Thus, from these examples (see Figure 7.1), we can determine that, if the sample mean based on a sample of $n = 25$ boxes falls anywhere between 362.12 and 373.88 grams, the population mean is included *somewhere* within the interval. However, we know from our discussion of the sampling distribution in section 6.5 that 95% of the sample means fall between 362.12 and 373.88 grams. Therefore, 95% of all samples of $n = 25$ boxes have sample means that include the population mean within the interval developed. The interval from 362.12 to 373.88 is referred to as a 95% confidence interval.

In general, a 95% **confidence interval estimate** is interpreted as follows: If all possible samples of the same size n are taken and their sample means are computed, 95% of them include the true population mean somewhere within the interval around their sample means and only 5% of them do not.

Because only one sample is selected in practice and μ is unknown, we never know for sure whether our specific interval obtained includes the population mean. However, we can state that we have 95% confidence that we have selected a sample whose interval does include the population mean.

In some situations, we might desire a higher degree of assurance (such as 99%) of including the population mean within the interval. In other cases, we might be willing to accept less assurance (such as 90%) of correctly estimating the population mean.

In general, the **level of confidence** is symbolized by $(1 - \alpha) \times 100\%$, where α is the proportion in the tails of the distribution that is outside the confidence interval. Therefore, to obtain the $(1 - \alpha) \times 100\%$ confidence interval estimate of the mean with σ known, we have

Confidence Interval for a Mean (σ Known)

$$\bar{X} \pm Z\frac{\sigma}{\sqrt{n}}$$

or

$$\bar{X} - Z\frac{\sigma}{\sqrt{n}} \leq \mu \leq \bar{X} + Z\frac{\sigma}{\sqrt{n}} \tag{7.1}$$

where

$Z =$ the value corresponding to an area of $(1 - \alpha)/2$ from the center of a standardized normal distribution.

To construct a 95% confidence interval estimate of the mean, the Z value corresponding to an area of $.95/2 = .4750$ from the center of the standard normal distribution is 1.96 because there is .025 in the upper tail of the distribution and the cumulative area less than $+Z$ is .975. The value of Z selected for constructing such a confidence interval is called the **critical value** for the distribution.

There is a different critical value for each level of confidence $1 - \alpha$. A level of confidence of 95% leads to a Z value of ± 1.96 (see Figure 7.2 on page 418). If a level of confidence of 99% is desired, the area of .99 is divided in half, leaving .495 between each limit and μ (see Figure 7.3 on page 418). The Z value corresponding to an area of .495 from the center of the normal curve is approximately 2.58 because the upper tail area is .005 and the cumulative area below $+Z$ is .995.

Now that we have considered various levels of confidence, why would we not want to make the confidence level as close to 100% as possible? We would not because any increase in the level of confidence is achieved only by simultaneously widening (and making less precise) the confidence interval obtained. Thus, we would have more confidence that the population mean is within a broader range of values but this is less useful for decision-making purposes. This trade-off between the width of the confidence interval and the level of confidence will be discussed in greater depth when we investigate how the sample size n is determined (see section 7.4).

FIGURE 7.2
Normal curve for determining the Z value needed for 95% confidence

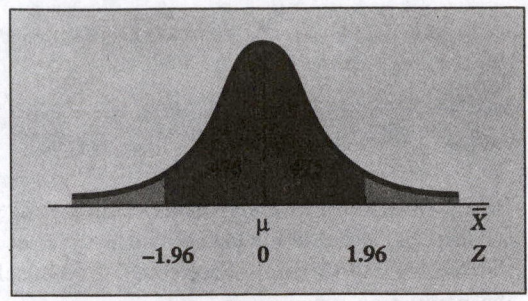

FIGURE 7.3
Normal curve for determining the Z value needed for 99% confidence

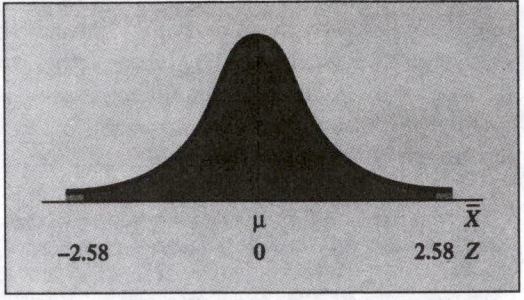

To illustrate the application of the confidence interval estimate, we turn to Example 7.1.

Example 7.1 *Estimating the Mean Paper Length with 95% Confidence*

A manufacturer of computer paper has a production process that operates continuously throughout an entire production shift. The paper is expected to have an average length of 11 inches and the standard deviation is known to be 0.02 inch. At periodic intervals, samples are selected to determine whether the average paper length is still equal to 11 inches or whether something has gone wrong in the production process to change the length of the paper produced. If such a situation has occurred, corrective action is needed. Suppose a random sample of 100 sheets is selected, and the average paper length is found to be 10.998 inches. Set up a 95% confidence interval estimate of the population average paper length.

SOLUTION

Using equation (7.1), with $Z = 1.96$ for 95% confidence, we have

$$\bar{X} \pm Z\frac{\sigma}{\sqrt{n}} = 10.998 \pm (1.96)\frac{0.02}{\sqrt{100}}$$

$$= 10.998 \pm .00392$$

$$10.99408 \leq \mu \leq 11.00192$$

Thus, we estimate, with 95% confidence, that the population mean is between 10.99408 and 11.00192 inches. Because 11, the value that indicates the production process is working properly, is included within the interval, there is no reason to believe that anything is wrong with the production process. There is 95% confidence that the sample selected is one in which the true population mean is included somewhere within the interval developed.

To observe the effect of using a 99% confidence interval, let us look at Example 7.2.

Example 7.2 *Estimating the Mean Paper Length with 99% Confidence*

Suppose that 99% confidence is desired. Set up a 99% confidence interval estimate of the population average paper length.

SOLUTION

Using equation (7.1), with $Z = 2.58$, we have

$$\bar{X} \pm Z \frac{\sigma}{\sqrt{n}} = 10.998 \pm (2.58)\frac{0.02}{\sqrt{100}}$$

$$= 10.998 \pm .00516$$

$$10.99284 \leq \mu \leq 11.00316$$

Once again, because 11 is included within this wider interval, there is no reason to believe that anything is wrong with the production process.

 **7.1E** **CALCULATING THE CONFIDENCE INTERVAL ESTIMATE FOR THE MEAN (σ KNOWN) USING MICROSOFT EXCEL**

Overview

◆ *For Quick Results Users*   Use the PHStat add-in to calculate the confidence interval estimate for the mean when σ is known.

◆ *For Developers*   Implement a worksheet that uses the NORMSINV and CONFIDENCE worksheet functions to calculate the confidence interval estimate for the mean when σ is known.

The 7-1E.XLS workbook file contains the confidence interval estimate for the mean paper length for the paper production problem of Example 7.1.

 7-1E.XLS

Quick Results Details

To calculate the confidence interval estimate for the mean (σ known), use the Confidence Intervals | Estimate for the Mean, sigma known, choice of the PHStat add-in. As an example, consider the paper production problem of Example 7.1. To calculate the confidence interval estimate for the mean paper length for this problem, do the following:

❶ If the PHStat add-in has not been previously loaded, load the add-in using the instructions of section S4.2.

❷ Select File | New to open a new workbook (or open the existing workbook into which the confidence interval estimate worksheet is to be inserted).

❸ Select PHStat | Confidence Intervals | Estimate for the Mean, sigma known.

❹ In the Estimate for the Mean, sigma known, dialog box (see Figure 7E.1):
   a. Enter 0.02 in the Population Standard Deviation: edit box.
   b. Enter 95 in the Confidence Level: edit box.
   c. Select the Sample Statistics Known option button and enter 100 in the Sample Size: edit box and 10.998 in the Sample Mean: edit box.
   d. Enter Confidence Interval Estimate for the Mean Paper Length in the Output Title: edit box.
   e. Click the OK button.

FIGURE 7E.1
PHStat Estimate for the Mean, sigma known, dialog box

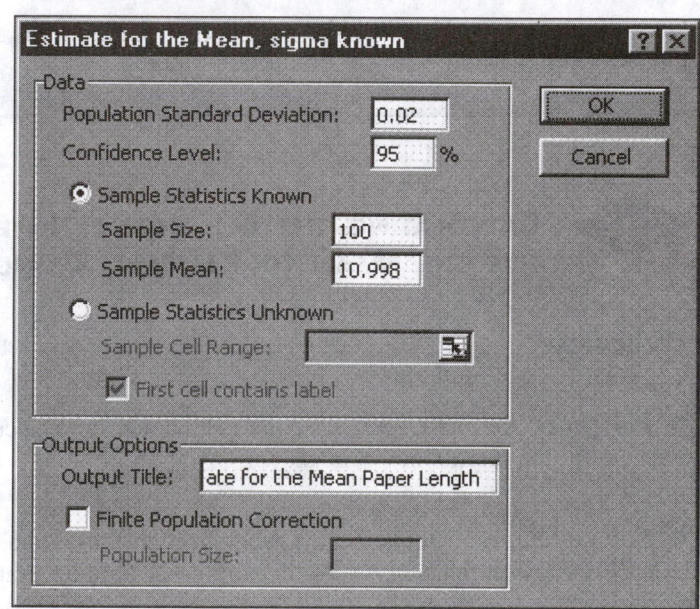

The add-in inserts a worksheet containing calculations for the confidence interval estimate for the mean similar to the one shown in Figure 7E.2. (*Note:* For other problems in which the sample mean is not known, select the Sample Statistics Unknown option button in step 4c and enter the cell range of the sample data in the Sample Cell Range: edit box, shown dimmed in Figure 7E.1).

	A	B	C
1	**Confidence Interval Estimate for the Mean Paper Length**		
2			
3	**Population Standard Deviation**	0.02	
4	**Sample Mean**	10.998	
5	**Sample Size**	100	
6	**Confidence Level**	95%	
7	Standard Error of the Mean	0.002	
8	Z Value	-1.95996108	
9	Interval Half Width	0.003919922	
10	**Interval Lower Limit**	10.99408008	
11	**Interval Upper Limit**	11.00191992	
12			
13			
14	**Finite Populations**		
15	**Population Size**	10000	
16	FPC Factor	0.99503719	
17	Interval Half Width	0.003900468	
18	**Interval Lower Limit**	10.99409953	
19	**Interval Upper Limit**	11.00190047	

FIGURE 7E.2

Confidence interval estimate of the mean paper length for the paper production problem (Example 7.1)

▲ WHAT IF EXAMPLE

We can change the values for the population standard deviation, sample mean, sample size, and confidence level in the worksheet inserted by the add-in to see their effects on the confidence interval estimate. For example, if we change the confidence level in cell B6 from 95% to 90%, we will observe that the interval half width narrows to .0033 and the confidence interval estimate changes to between 10.9947 and 11.0013. To facilitate comparisons among many different alternative values, we can create scenarios using the Scenario Manager as discussed at the end of section 2.2E.

Developer Details

We can use the CONFIDENCE worksheet function as the basis for calculating the confidence interval of the mean when σ is known. The format of this function is:

CONFIDENCE(1 − *confidence level, population standard deviation, sample size*)

Table 7E.1 on page 422 presents a Confidence interval estimate sheet design for calculating the confidence interval estimate for the paper production problem of Example 7.1 on page 418. This design takes the half width of the confidence interval returned by the CONFIDENCE function and subtracts and adds it to the sample mean to calculate the lower and upper limits of the confidence interval. (For reference, the design also includes calculations for the standard error of the mean and the appropriate Z value because we can multiply these values to get the half width returned by the CONFIDENCE function.)

To implement the Table 7E.1 design, do the following:

❶ Select File | New to open a new workbook (or open the existing workbook into which the confidence interval estimate worksheet is to be inserted).

❷ Rename the new worksheet Confidence.

❸ Enter the title and labels for column A as shown in Table 7E.1.

❹ Enter the population standard deviation, sample mean, sample size, and confidence level in the cell range B3:B6. Enter 0.02 in cell B3, 10.998 in cell B4, 100 in cell B5, and 0.95 in cell B6.

❺ Select cell B6 and click the Percent style button on the formatting toolbar (see section S3.11) to format the decimal value 0.95 as 95%.

❻ Enter the formulas for the cell range B7:B11 as shown in Table 7E.1. Note that the formula for cell B8 contains consecutive left parentheses.

The completed worksheet will be similar to the one shown in Figure 7E.2. Had the sample mean been unknown, we could have entered a formula using the AVERAGE function in cell B4 to calculate it.

Table 7E.1 Confidence interval estimate sheet design for the mean paper length for the paper production problem (Example 7.1)

	A	B
1	Confidence Interval Estimate for the Mean Paper Length	
2		
3	Population Standard Deviation	xxx
4	Sample Mean	xxx
5	Sample Size	xxx
6	Confidence Level	.xx
7	Standard Error of the Mean	=B3/SQRT(B5)
8	Z Value	=NORMSINV((1-B6)/2)
9	Interval Half Width	=CONFIDENCE(1-B6,B3,B5)
10	Interval Lower Limit	=B4-B9
11	Interval Upper Limit	=B4+B9

▲ WHAT IF EXAMPLE

As with the worksheet inserted by the add-in using the quick results method of this section, we can change the values for the population standard deviation, sample mean, sample size, and confidence level in this worksheet to see their effects on the confidence interval estimate. (For further details, see the What If Example on page 421.)

Problems for Section 7.1

Learning the Basics

• 7.1 If $\overline{X} = 85$, $\sigma = 8$, and $n = 64$, set up a 95% confidence interval estimate of the population mean μ.

7.2 If $\overline{X} = 125$, $\sigma = 24$, and $n = 36$, set up a 99% confidence interval estimate of the population mean μ.

7.3 A market researcher states that she has 95% confidence that the true average monthly sales of a product are between $170,000 and $200,000. Explain the meaning of this statement.

7.4 Why is it not possible for the production manager in Example 7.1 on page 418 to have 100% confidence? Explain.

7.5 From the results of Example 7.1 on page 418 regarding paper production, is it true that 95% of the sample means will fall between 10.99408 and 11.00192 inches? Explain.

7.6 Is it true in Example 7.1 on page 418 that we do not know for sure whether the true population mean is between 10.99408 and 11.00192 inches? Explain.

Applying the Concepts

• **7.7** Suppose that the manager of a paint supply store wants to estimate the actual amount of paint contained in 1-gallon cans purchased from a nationally known manufacturer. It is known from the manufacturer's specifications that the standard deviation of the amount of paint is equal to 0.02 gallon. A random sample of 50 cans is selected, and the average amount of paint per 1-gallon can is 0.995 gallon.

  (a) Set up a 99% confidence interval estimate of the true population average amount of paint included in a 1-gallon can.

  (b) On the basis of your results, do you think that the store owner has a right to complain to the manufacturer? Why?

  (c) Does the population amount of paint per can have to be normally distributed here? Explain.

  (d) Explain why an observed value of 0.98 gallon for an individual can is not unusual, even though it is outside the confidence interval you calculated.

  (e) Suppose that you used a 95% confidence interval estimate. What would be your answers to (a) and (b)?

 WHAT IF?

7.8 The quality control manager at a light bulb factory needs to estimate the average life of a large shipment of light bulbs. The process standard deviation is known to be 100 hours. A random sample of 64 light bulbs indicated a sample average life of 350 hours.

  (a) Set up a 95% confidence interval estimate of the true average life of light bulbs in this shipment.

  (b) Do you think that the manufacturer has the right to state that the light bulbs last an average of 400 hours? Explain.

  (c) Does the population of light bulb life have to be normally distributed here? Explain.

  (d) Explain why an observed value of 320 hours is not unusual, even though it is outside the confidence interval you calculated.

  (e) Suppose that the process standard deviation changed to 80 hours. What would be your answers in (a) and (b)?

 WHAT IF?

7.9 The inspection division of the Lee County Weights and Measures Department is interested in estimating the actual amount of soft drink that is placed in 2-liter bottles at the local bottling plant of a large nationally known soft-drink company. The bottling plant has informed the inspection division that the standard deviation for 2-liter bottles is 0.05 liter. A random sample of one hundred 2-liter bottles obtained from this bottling plant indicates a sample average of 1.99 liters.

  (a) Set up a 95% confidence interval estimate of the true average amount of soft drink in each bottle.

  (b) Does the population of soft-drink fill have to be normally distributed here? Explain.

  (c) Explain why an observed value of 2.02 liters is not unusual, even though it is outside the confidence interval you calculated.

 WHAT IF?

  (d) Suppose that the sample average had been 1.97 liters. What would be your answer to (a)?

CONFIDENCE INTERVAL ESTIMATION OF THE MEAN (σ UNKNOWN)

Just as the mean of the population μ is usually not known, the actual standard deviation of the population σ is also usually unknown. Therefore, we need to obtain a confidence interval estimate of μ by using only the sample statistics of $\overline{X}$ and S. To achieve this, we turn to the work of William S. Gosset.

Student's *t* Distribution

At the turn of the twentieth century, a statistician named William S. Gosset, an employee of Guinness Breweries in Ireland (see reference 3), was interested in making inferences about the mean when σ was unknown. Because Guinness employees were not permitted to publish research work under their own names, Gosset adopted the pseudonym "Student." The distribution that he developed has come to be known as **Student's *t* distribution.**

If the random variable X is normally distributed, then the statistic

$$\frac{\overline{X} - \mu}{\dfrac{S}{\sqrt{n}}}$$

has a t distribution with $n - 1$ **degrees of freedom.** Notice that this expression has the same form as equation (6.9) on page 380, except that S is used to estimate σ, which is presumed to be unknown in this case.

Properties of the *t* Distribution

In appearance, the t distribution is very similar to the normal distribution. Both distributions are bell-shaped and symmetrical. However, the t distribution has more area in the tails and less in the center than does the normal distribution (see Figure 7.4). This is because σ is unknown and we are using S to estimate it. Because we are uncertain of the value σ, the values of t that we observe will be more variable than for Z.

FIGURE 7.4
Standard normal distribution and t distribution for 5 degrees of freedom

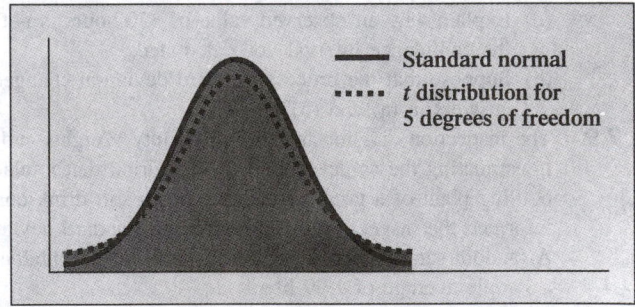

However, as the number of degrees of freedom increases, the t distribution gradually approaches the normal distribution until the two are virtually identical. This happens because S becomes a better estimate of σ as the sample size gets larger. With a sample size of about 120 or more, S estimates σ precisely enough that there is little difference between the t and Z distributions. For this reason, most statisticians use Z instead of t when the sample size is over 120.

COMMENT: *Checking the Assumptions*

Recall that the t distribution assumes that the random variable X being studied is normally distributed. In practice, however, as long as the sample size is large enough and the population is not very skewed, the t distribution can be used to estimate the population mean when σ is unknown. We should be concerned about the validity of the confidence interval primarily when we are dealing with a small sample size and a skewed population distribution. However, we can assess the assumption of normality in the population by evaluating the shape of the sample data using a histogram, stem-and-leaf display, box-and-whisker plot, or normal probability plot (see section 6.2).

The critical values of t for the appropriate degrees of freedom can be obtained from the table of the t distribution (see Table E.3). The top of each column of the table indicates the area in the right tail of the t distribution (because positive entries for t are supplied, the values for t are for the upper tail); each row represents the particular t value for each specific degree of freedom. For example, with 99 degrees of freedom, if 95% confidence is desired, the appropriate value of t is found in the manner shown in Table 7.1. The 95% confidence level means that 2.5% of the values (an area of .025) are in each tail of the distribution. Looking in the column for an upper-tailed area of .025 and in the row corresponding to 99 degrees of freedom results in a critical value for t of 1.9842. Because t is a symmetrical distribution with a mean of 0, if the upper-tailed value is $+1.9842$, the value for the lower-tailed area (lower .025) will be -1.9842. A t value of 1.9842 means that the probability that t exceeds $+1.9842$ is .025 or 2.5% (see Figure 7.5).

Table 7.1   *Determining the critical value from the t table for an area of .025 in each tail with 99 degrees of freedom*

DEGREES OF FREEDOM	UPPER-TAILED AREAS					
	.25	.10	.05	.025	.01	.005
1	1.0000	3.0777	6.3138	12.7062	31.8207	63.6574
2	0.8165	1.8856	2.9200	4.3027	6.9646	9.9248
3	0.7649	1.6377	2.3534	3.1824	4.5407	5.8409
4	0.7407	1.5332	2.1318	2.7764	3.7469	4.6041
5	0.7267	1.4759	2.0150	2.5706	3.3649	4.0322
.	.	.	.	.	.	.
.	.	.	.	.	.	.
.	.	.	.	.	.	.
96	0.6771	1.2904	1.6609	1.9850	2.3658	2.6280
97	0.6770	1.2903	1.6607	1.9847	2.3654	2.6275
98	0.6770	1.2902	1.6606	1.9845	2.3650	2.6269
99	0.6770	1.2902	1.6604	1.9842	2.3646	2.6264
100	0.6770	1.2901	1.6602	1.9840	2.3642	2.6259

Source: Extracted from Table E.3.

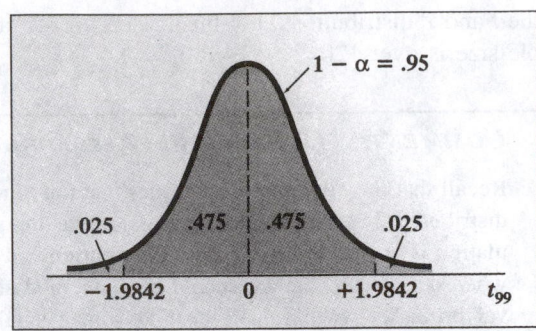

FIGURE 7.5
t distribution with 99 degrees of freedom

The Concept of Degrees of Freedom

You may recall from chapter 3 that the sample variance S^2 requires the computation of

$$\sum_{i=1}^{n}(X_i - \overline{X})^2$$

Thus, in order to compute S^2, we first need to know $\overline{X}$. Therefore, we can say that only $n - 1$ of the sample values are free to vary. That is, there are $n - 1$ degrees of freedom.

This concept is illustrated as follows. Suppose we have a sample of five values that have a mean of 20. How many distinct values do we need to know before we can obtain the remainder? The fact that $n = 5$ and $\overline{X} = 20$ also tells us that

$$\sum_{i=1}^{n} X_i = 100$$

because

$$\sum_{i=1}^{n} X_i / n = \overline{X}$$

Thus, once we know four of the values, the fifth one will *not* be free to vary because the sum must add to 100. For example, if four of the values are 18, 24, 19, and 16, the fifth value can only be 23 so that the sum equals 100.

The Confidence Interval Statement

The $(1 - \alpha) \times 100\%$ confidence interval estimate for the mean with σ unknown is expressed as follows:

Confidence Interval for a Mean (σ Unknown)

$$\overline{X} \pm t_{n-1}\frac{S}{\sqrt{n}}$$

or

$$\overline{X} - t_{n-1}\frac{S}{\sqrt{n}} \leq \mu \leq \overline{X} + t_{n-1}\frac{S}{\sqrt{n}} \qquad (7.2)$$

where
t_{n-1} is the critical value of the t distribution with $n - 1$ degrees of freedom for an area of $\alpha/2$ in the upper tail.

410

To illustrate the application of the confidence interval estimate for the mean when the standard deviation σ is unknown, let us return to the Saxon Plumbing Company example presented earlier. Suppose that we select a sample of 100 sales invoices from the population of sales invoices during the month and the average amount of each sales invoice is $110.27 with a sample standard deviation of $28.95. The auditor in this case wants 95% confidence of estimating the population mean.

Using $\overline{X} = \$110.27$ and $S = \$28.95$, we obtain the critical value from the t distribution as shown in Table 7.1. The critical value of t is 1.9842. Using equation (7.2), we have

$$\overline{X} \pm t_{n-1}\frac{S}{\sqrt{n}}$$

$$= 110.27 \pm (1.9842)\frac{28.95}{\sqrt{100}}$$

$$= 110.27 \pm 5.74$$

$$\$104.53 \le \mu \le \$116.01$$

Thus, we conclude with 95% confidence that the average amount of the sales invoice is between $104.53 and $116.01. The 95% confidence interval states that we are 95% sure that the sample we have selected is one in which the population mean μ is located within the interval. This 95% confidence means that if all possible samples of size 100 were selected (something that would never be done in practice), 95% of the intervals developed would include the true population mean somewhere within the interval. The validity of this confidence interval estimate depends on the assumption of normality for the distribution of the amount of the sales invoices. With a sample of 100, as we noted on pages 424–425, the use of the t distribution is likely to be appropriate. In practice, the results of this interval on the basis of the actual invoices would be compared with the average invoice as entered into the computer system of the company. If there appeared to be large differences, further investigation would be undertaken.

To further illustrate how confidence intervals for a mean are constructed when the population standard deviation is unknown, let us turn to Example 7.3.

Example 7.3   *Estimating the Mean Annual Usage of Home Heating Oil*

A marketing manager for a company that supplies home heating oil wants to estimate the average annual usage (in gallons) by single-family homes in a particular geographic area. A random sample of 35 single-family homes is selected, and the annual usage for these homes is summarized in the following table.

ANNUAL AMOUNT OF HEATING OIL CONSUMED (IN GALLONS) IN A SAMPLE OF 35
SINGLE-FAMILY HOUSES

1150.25	1352.67	983.45	1365.11	942.71	1577.77	330.00
872.37	1126.57	1184.17	1046.35	1110.50	1050.86	851.60
1459.56	1252.01	373.91	1047.40	1064.46	1018.23	996.92
941.96	767.37	1598.57	1598.66	1343.29	1617.73	1300.76
1013.27	1402.59	1069.32	1108.94	1326.19	1074.86	975.86

DATA FILE
OILUSE.XLS

411

Set up a 95% confidence interval estimate of the population average amount of heating oil consumed per year.

SOLUTION

For these data, the accompanying figure, obtained from the PHStat add-in, shows that the sample average is $\bar{X} = 1,122.75$ gallons and the sample standard deviation is $S = 295.72$ gallons. To obtain the confidence interval of 1,021.17 to 1,224.33 provided by Microsoft Excel, we first determine the critical value from the t table for an area of .025 in each tail with 34 degrees of freedom. From Table E.3 we have $t_{34} = 2.0322$.

	A	B	C	D	E	F
1	Confidence Interval Estimate of the Average Amount of Heating Oil Consumed					
2						
3	Sample Standard Deviation	295.7232382				
4	Sample Mean	1122.749714				
5	Sample Size	35				
6	Confidence Level	95%				
7	Standard Error of the Mean	49.98635059				
8	Degrees of Freedom	34				
9	t Value	2.032243174				
10	Interval Half Width	101.5844198				
11	Interval Lower Limit	1021.17				
12	Interval Upper Limit	1224.33				

Thus, using $\bar{X} = 1,122.75$, $S = 295.7$, $n = 35$, and $t_{34} = 2.0322$, we have

$$\bar{X} \pm t_{n-1}\frac{S}{\sqrt{n}}$$

$$= 1,122.75 \pm (2.0322)\frac{295.72}{\sqrt{35}}$$

$$= 1,122.75 \pm 101.58$$

$$1,021.17 \le \mu \le 1,224.33$$

We thus conclude with 95% confidence that the average amount of heating oil consumed per year is between 1,021.17 and 1,224.33 gallons. The 95% confidence interval states that we are 95% sure that the sample we have selected is one in which the population mean μ is located within the interval. This 95% confidence means that if all possible samples of size 35 were selected (something that would never be done in practice), 95% of the intervals developed would include the true population mean. The validity of this confidence interval estimate depends on the assumption of normality of the heating oil usage data. With a sample of 35, the use of the t distribution is likely to be appropriate.

 **7.2E** **CALCULATING THE CONFIDENCE INTERVAL ESTIMATE FOR THE MEAN (σ UNKNOWN) USING MICROSOFT EXCEL**

Overview

◆ *For Quick Results Users*   Use the PHStat add-in to calculate the confidence interval estimate for the mean when σ is unknown.

◆ *For Developers*   Implement a worksheet that uses the TINV worksheet function and arithmetic formulas to calculate the confidence interval estimate for the mean when σ is unknown.

The 7-2E.XLS workbook file contains the confidence interval estimate for the mean sales invoice amount for the Saxon Plumbing Company problem of section 7.2.

 7-2E.XLS

Quick Results Details

To calculate a confidence interval estimate for the mean (σ unknown), use the Confidence Intervals | Estimate for the Mean, sigma unknown, choice of the PHStat add-in. As an example, consider the Saxon Plumbing Company problem of section 7.2. To calculate the confidence interval estimate for the mean sales invoice amount for this problem, do the following:

❶ If the PHStat add-in has not been previously loaded, load the add-in using the instructions of Section S4.2.

❷ Select File | New to open a new workbook (or open the existing workbook into which the confidence interval estimate worksheet is to be inserted).

❸ Select PHStat | Confidence Intervals | Estimate for the Mean, sigma unknown.

❹ In the Estimate for the Mean, sigma unknown, dialog box (see Figure 7E.3):

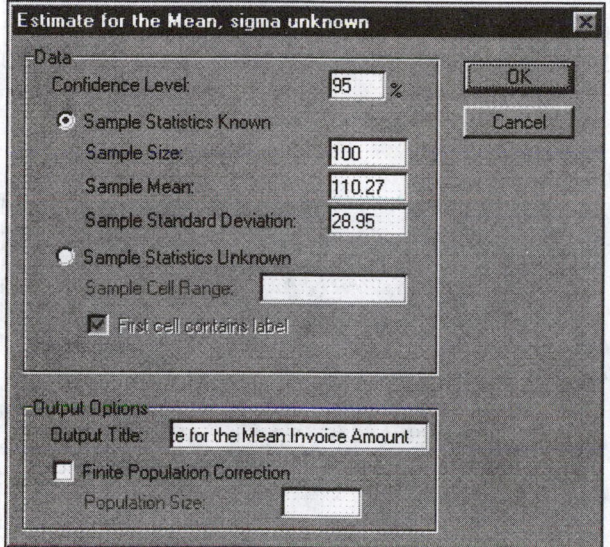

FIGURE 7E.3
PHStat Estimate for the Mean, sigma unknown, dialog box

a. Enter 95 in the Confidence Level: edit box.
b. Select the Sample Statistics Known option button and enter 100 in the Sample Size: edit box, 110.27 in the Sample Mean: edit box, and 28.95 in the Sample Standard Deviation: edit box.
c. Enter Confidence Interval Estimate for the Mean Invoice Amount in the Output Title: edit box.
d. Click the OK button.

The add-in inserts a worksheet containing calculations for a confidence interval estimate of the mean similar to the one shown in Figure 7E.4. (*Note:* For other problems in which the sample mean and sample standard deviation are not known, select the Sample Statistics Unknown option button in step 4b and enter the cell range of the sample data in the Sample Cell Range: edit box, dimmed in Figure 7E.3).

FIGURE 7E.4
Confidence interval estimate for the mean sales invoice amount for the Saxon Plumbing Company problem of section 7.2

	A	B	C	D
1	Confidence Interval Estimate for the Mean Invoice Amount			
2				
3	Sample Standard Deviation	28.95		
4	Sample Mean	110.27		
5	Sample Size	100		
6	Confidence Level	95%		
7	Standard Error of the Mean	2.895		
8	Degrees of Freedom	99		
9	t Value	1.984217306		
10	Interval Half Width	5.744309101		
11	Interval Lower Limit	104.53		
12	Interval Upper Limit	116.01		
13				
14				
15	Finite Populations			
16	Population Size	5000		
17	FPC Factor	0.990048503		
18	Interval Half Width	5.687144629		
19	Interval Lower Limit	104.58		
20	Interval Upper Limit	115.96		

▲ WHAT IF EXAMPLE

We can change the values for the sample standard deviation, sample mean, sample size, and confidence level in the worksheet inserted by the add-in to see their effects on the confidence interval estimate. For example, if we change the confidence level in cell B6 from 95% to 90%, we will observe that the interval half width narrows to 4.80683 and the confidence interval estimate changes to between 105.46 and 115.08. To facilitate comparisons among many different alternative values, we can create scenarios using the Scenario Manager as discussed at the end of section 2.2E.

Developer Details

We can use the TINV worksheet function in calculating the confidence interval of the mean when σ is unknown. The format of this function is:

$$\text{TINV}(1 - \textit{confidence level, degrees of freedom})$$

Table 7E.2 presents a Confidence interval estimate sheet design for calculating the confidence interval estimate for the Saxon Plumbing Company problem of section 7.2. This design takes the t-value returned by the TINV function and multiplies it by the standard error of the mean to produce half the width of the confidence interval. This half width is then subtracted from and added to the sample mean to calculate the lower and upper limits of the confidence interval. To implement the Table 7E.2 design, do the following:

❶ Select File | New to open a new workbook (or open the existing workbook into which the confidence interval estimate worksheet is to be inserted).

❷ Rename the new worksheet Confidence.

❸ Enter the title and labels for column A as shown in Table 7E.2.

❹ Enter the sample standard deviation, sample mean, sample size, and confidence level in the cell range B3:B6. Enter 28.95 in cell B3, 110.27 in cell B4, 100 in cell B5, and 0.95 in cell B6.

❺ Select cell B6 and click the Percent style button on the formatting toolbar (see section S3.11) to format the decimal value 0.95 as 95%.

❻ Enter the formulas for the cell range B7:B12 as shown in Table 7E.2.

❼ Select cells B11 and B12 and click the Decrease decimal button on the formatting toolbar until the values (which represent currency amounts) contain only two decimal places.

The completed worksheet will be similar to the one shown in Figure 7E.4. Had the sample standard deviation and sample mean been unknown, we could have entered formulas using the STDEV and AVERAGE functions in cells B3 and B4 to calculate these statistics.

Table 7E.2 Confidence interval estimate sheet design for the mean sales invoice amount for the Saxon Plumbing Company problem of section 7.2

	A	B
1	Confidence Interval Estimation of the Mean Invoice Amount	
2		
3	Sample Standard Deviation	xxx
4	Sample Mean	xxx
5	Sample Size	xxx
6	Confidence Level	.xx
7	Standard Error of the Mean	=B3/SQRT(B5)
8	Degrees of Freedom	=B5-1
9	t Value	=TINV(1-B6,B8)
10	Interval Half Width	=B9*B7
11	Interval Lower Limit	=B4-B10
12	Interval Upper Limit	=B4+B10

Problems for Section 7.2

Learning the Basics

7.10 Determine the critical value of t in each of the following circumstances:
(a) $1 - \alpha = .95$, $n = 10$.
(b) $1 - \alpha = .99$, $n = 10$.
(c) $1 - \alpha = .95$, $n = 32$.
(d) $1 - \alpha = .95$, $n = 65$.
(e) $1 - \alpha = .90$, $n = 16$.

•**7.11** If $\bar{X} = 75$, $S = 24$, and $n = 36$, assuming the population is normally distributed, set up a 95% confidence interval estimate of the population mean μ.

7.12 If $\bar{X} = 50$, $S = 15$, and $n = 16$, assuming the population is normally distributed, set up a 99% confidence interval estimate of the population mean μ.

7.13 Set up a 95% confidence interval estimate for the population mean, based on each of the following sets of data, assuming the population is normally distributed:

Set 1: 1, 1, 1, 1, 8, 8, 8, 8

Set 2: 1, 2, 3, 4, 5, 6, 7, 8

Explain why they have different confidence intervals even though they have the same mean and range.

7.14 Compute a 95% confidence interval for the population mean, based on the numbers 1, 2, 3, 4, 5, 6, 20. Change the number 20 to 7 and recalculate the confidence interval. Using these results, describe the effect of an outlier (or extreme value) on the confidence interval.

Applying the Concepts

•**7.15** The United States Department of Transportation requires tire manufacturers to provide tire performance information on the sidewall of the tire so that a prospective customer can be better informed when making a purchasing decision. One very important measure of tire performance is the tread wear index, which indicates the tire's resistance to tread wear compared with a tire graded with a base of 100. This means that a tire with a grade of 200 should last twice as long, on average, as a tire graded with a base of 100. Suppose that a consumer organization wishes to estimate the actual tread wear index of a brand name of tires graded 200 that are produced by a certain manufacturer. A random sample of 18 of these tires indicates a sample average tread wear index of 195.3 and a sample standard deviation of 21.4.
(a) Assuming the population of tread wear indices is normally distributed, set up a 95% confidence interval estimate of the population average tread wear index for tires produced by this manufacturer under this brand name.
(b) Do you think that the consumer organization should accuse the manufacturer of producing tires that do not meet the performance information provided on the sidewall of the tire? Explain.
(c) Tell why an observed tread wear index of 210 for a particular tire is not unusual, even though it is outside the confidence interval developed in (a).

7.16 The manager of a branch of a local savings bank wants to estimate the average amount held in passbook savings accounts by depositors at the bank. A random sample of 30 depositors is selected, and the results indicate a sample average of $4,750 and a sample standard deviation of $1,200.

(a) Assuming a normal distribution, set up a 95% confidence interval estimate of the average amount held in all passbook savings accounts.

(b) If an individual has $4,000 in a passbook savings account, is this considered unusual? Explain your answer.

7.17 A stationery store wants to estimate the average retail value of greeting cards that it has in its inventory. A random sample of 20 greeting cards indicates an average value of $1.67 and a standard deviation of $0.32.

(a) Assuming a normal distribution, set up a 95% confidence interval estimate of the average value of all greeting cards in the store's inventory.

(b) How might the results obtained in (a) be useful in assisting the store owner to estimate the total value of her inventory?

7.18 The personnel department of a large corporation wants to estimate the family dental expenses of its employees to determine the feasibility of providing a dental insurance plan. A random sample of 10 employees reveals the following family dental expenses (in dollars) for the preceding year:

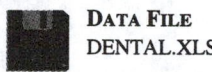

DATA FILE
DENTAL.XLS

| 110 | 362 | 246 | 85 | 510 | 208 | 173 | 425 | 316 | 179 |

(a) Set up a 90% confidence interval estimate of the average family dental expenses for all employees of this corporation.

(b) What assumption about the population distribution must be made in (a)?

(c) Give an example of a family dental expense that is outside the confidence interval but is not unusual for an individual family, and explain why this is not a contradiction.

(d) Suppose you used a 95% confidence interval in (a). What would be your answer to (a)?

WHAT IF?

(e) Suppose the fourth value was $585 instead of $85. What would be your answer to (a)? What effect does this change have on the confidence interval?

●7.19 The customer service department of a local gas utility wants to estimate the average length of time between the entry of the service request and the connection of service. A random sample of 15 houses is selected from the records available during the past year. The results recorded in number of days are as follows:

DATA FILE
GASERVE.XLS

| 114 | 78 | 96 | 137 | 78 | 103 | 117 |
| 126 | 86 | 99 | 114 | 72 | 104 | 73 | 86 |

(a) Set up a 95% confidence interval estimate of the population average waiting time in the past year.

(b) If the service department has been advising customers that the average waiting time is 90 days, can you say that the results obtained in (a) are consistent with such advice? Explain.

(c) What assumption about the population distribution must be made in (a)?

WHAT IF?

(d) Suppose the last value was 286 days instead of 86 days. What would be your answers to (a) and (b)? What effect does this change have on the confidence interval?

7.20 The director of patient services of a large health maintenance organization wants to evaluate patient waiting time at a local facility. A random sample of 25 patients is selected from the appointment book. The waiting time is defined as the time from when the patient signs in to when he or she is seen by the doctor. The data at the top of page 434 represent the waiting times (in minutes).

(a) Set up a 95% confidence interval estimate of the population average waiting time.

(b) The director of patient services at the health maintenance organization wants to tell prospective patients that the average waiting time is 15 minutes. On the basis of the results of (a) can this statement be made? Explain.

19.5	30.5	45.6	39.8	29.6
25.4	21.8	28.6	52.0	25.4
26.1	31.1	43.1	4.9	12.7
10.7	12.1	1.9	45.9	42.5
41.3	13.8	17.4	39.0	36.6

(c) What assumption about the population distribution must be made in (a)?

(d) Suppose that the recorded value of 1.9 minutes was actually 101.9 minutes. What would be your answers to (a) and (b)? What effect does this change have on the confidence interval?

7.3 ▸ CONFIDENCE INTERVAL ESTIMATION FOR THE PROPORTION

In this section we extend the concept of the confidence interval to categorical data to estimate the population proportion p from the sample proportion $p_s = X/n$. Recall from section 6.6 that when both np and $n(1 - p)$ are at least 5, the binomial distribution can be approximated by the normal distribution. Hence, we can set up the following $(1 - \alpha) \times 100\%$ confidence interval estimate for the population proportion p:

Confidence Interval Estimate for the Proportion

$$p_s \pm Z\sqrt{\frac{p_s(1 - p_s)}{n}}$$

or

$$p_s - Z\sqrt{\frac{p_s(1 - p_s)}{n}} \le p \le p_s + Z\sqrt{\frac{p_s(1 - p_s)}{n}} \qquad (7.3)$$

where

p_s = sample proportion

p = population proportion

Z = critical value from the normal distribution

n = sample size

To see how this confidence interval estimate of the proportion is utilized, we return to our Saxon Plumbing Company example. Recall that the auditor wants to determine the frequency of occurrence of various types of errors that violate the internal control policy of the warehouse. Suppose that in the sample of 100 sales invoices, 10 contain an error. The auditor wants to develop a 95% confidence interval estimate of the population proportion of sales invoices that contain errors in violation of company policy. For these data we have $p_s = 10/100 = .10$. With 95% confidence $Z = 1.96$ so that using equation (7.3),

$$p_s \pm Z\sqrt{\frac{p_s(1 - p_s)}{n}} = .10 \pm (1.96)\sqrt{\frac{(.10)(.90)}{100}}$$

$$= .10 \pm (1.96)(.03)$$

$$= .10 \pm .0588$$

$$.0412 \le p \le .1588$$

Therefore, the auditor estimates with 95% confidence that between 4.12% and 15.88% of the sales invoices have errors that violate company policy.

To study another application of a confidence interval estimate for the proportion, consider Example 7.4.

Example 7.4 *Estimating the Proportion of Nonconforming Newspapers Printed*

The operations manager for a large city newspaper wants to determine the proportion of newspapers printed that have a nonconforming attribute, such as excessive ruboff, improper page setup, missing pages, duplicate pages, and so on. The operations manager determines that a random sample of 200 newspapers should be selected for analysis. Suppose that, of this sample of 200, thirty-five contain some type of nonconformance. If the operations manager wants to have 90% confidence in estimating the true population proportion, set up the confidence interval estimate.

SOLUTION

The confidence interval is computed as follows:

$p_s = 35/200 = .175$, and with a 90% level of confidence $Z = 1.645$

Using equation (7.3), we have

$$p_s \pm Z \sqrt{\frac{p_s(1 - p_s)}{n}}$$

$$= .175 \pm (1.645)\sqrt{\frac{(.175)(.825)}{200}}$$

$$= .175 \pm (1.645)(.0269)$$

$$= .175 \pm .0442$$

$$.1308 \leq p \leq .2192$$

Therefore, the operations manager estimates with 90% confidence that between 13.08% and 21.92% of the newspapers printed on that day have some type of nonconformance.

COMMENT: *Checking the Assumptions*

In Example 7.4 the number of successes and failures is sufficiently large that the normal distribution provides an excellent approximation for the binomial distribution. However, if the sample size is not large or the percentage of successes is either very low or very high, then the binomial distribution is used rather than the normal distribution (references 1 and 6). The exact confidence intervals for various sample sizes and proportions of successes have been tabulated by Fisher and Yates (reference 2).

For a given sample size, confidence intervals for proportions often seem to be wider than those for continuous variables. With continuous variables, the measurement on each respondent contributes more information than for a categorical variable. In other words, a categorical variable with only two possible values is a very crude measure compared with a continuous variable, so each observation contributes less information about the parameter we are estimating.

Overview

◆ *For Quick Results Users*   Use the PHStat add-in to calculate the confidence interval estimate for the proportion.

◆ *For Developers*   Implement a worksheet that uses the NORMSINV worksheet function and arithmetic formulas to calculate the confidence interval estimate for the proportion.

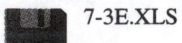 **7-3E.XLS**   The 7-3E.XLS workbook file contains the confidence interval estimate for the proportion of sales invoices that contain errors in violation of company policy for the Saxon Plumbing Company problem of section 7.3.

Quick Results Details

To calculate a confidence interval estimate for the proportion, use the Confidence Intervals | Estimate for the Proportion choice of the PHStat add-in. As an example, consider the Saxon Plumbing Company problem of section 7.3. To calculate the confidence interval estimate for the proportion of sales invoices that contain errors that are in violation of company policy, do the following:

❶   If the PHStat add-in has not been previously loaded, load the add-in using the instructions of section S4.2.

❷   Select File | New to open a new workbook (or open the existing workbook into which the confidence interval estimate worksheet is to be inserted).

❸   Select PHStat | Confidence Intervals | Estimate for the Proportion.

❹   In the Estimate for the Proportion dialog box (see Figure 7E.5):

FIGURE 7E.5
PHStat Estimate for the Proportion dialog box

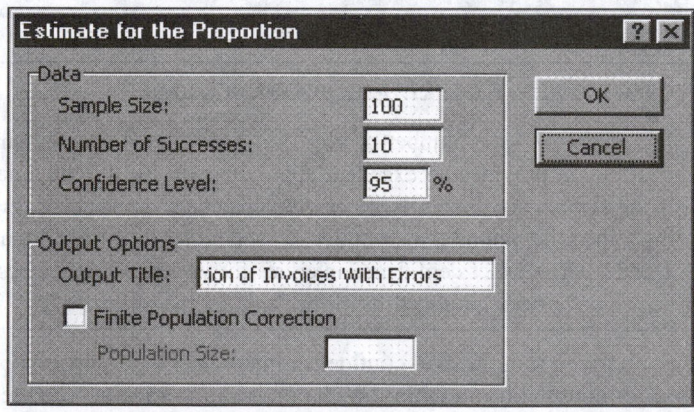

a. Enter 100 in the Sample Size: edit box.
b. Enter 10 in the Number of Successes: edit box.
c. Enter 95 in the Confidence Level: edit box.

d. Enter Confidence Interval Estimation for the Proportion of Invoices With Errors in the Output Title: edit box.

e. Click the OK button.

The add-in inserts a worksheet containing calculations for a confidence interval estimate for the proportion similar to the one shown in Figure 7E.6.

	A	B	C	D	E
1	Confidence Interval Estimation for the Proportion of Invoices With Errors				
2					
3	Sample Size	100			
4	Number of Successes	10			
5	Confidence Level	95%			
6	Sample Proportion	0.1			
7	Z Value	-1.95996108			
8	Standard Error of the Proportion	0.03			
9	Interval Half Width	0.058798832			
10	Interval Lower Limit	0.041201168			
11	Interval Upper Limit	0.158798832			

FIGURE 7E.6
Confidence interval estimate for the proportion of sales invoices that contain errors that are in violation of company policy for the Saxon Plumbing Company problem of section 7.3

▲ WHAT IF EXAMPLE

We can change the values for the sample size, number of successes, and confidence level in the worksheet inserted by the add-in to see their effects on the confidence interval estimate. For example, if we change the confidence level in cell B5 from 95% to 90%, we will observe that the interval half width narrows to .049335 and the confidence interval estimate changes to between .05065 and .14935. To facilitate comparisons among many different alternative values, we can create scenarios using the Scenario Manager as discussed at the end of section 2.2E.

Developer Details

We can use the NORMSINV worksheet function (see section 6.1E) as the basis for calculating the confidence interval of the proportion. Table 7E.3 on page 438 presents a Confidence interval estimate sheet design for calculating the confidence interval estimate for the Saxon Plumbing Company problem. This design takes the Z value returned by the NORMSINV function and multiplies it by the standard error of the proportion to produce half the width of the confidence interval. The half width is then subtracted from and added to the sample proportion to calculate the lower and upper limits of the confidence interval. (The design also uses the ABS function in cell B9 to ensure that the half-width value will be positive.) To implement the Table 7E.3 design, do the following:

❶ Select File | New to open a new workbook (or open the existing workbook into which the probabilities worksheet is to be inserted).

❷ Select an unused worksheet (or select Insert | Worksheet if there are none) and rename the sheet Confidence.

❸ Enter the title and labels for column A as shown in Table 7E.3.

❹ Enter the sample size, number of successes, and confidence level in the cell range B3:B5. Enter 100 in cell B3, 10 in cell B4, and 0.95 in cell B5.

⑤ Select cell B5 and click the Percent style button on the formatting toolbar (see section S3.11) to format the decimal value 0.95 as 95%.

⑥ Enter the formulas for the cell range B6:B11 as shown in Table 7E.3.

The completed worksheet will be similar to the one shown in Figure 7E.6 on page 437.

Table 7E.3 Confidence interval estimate sheet design for the proportion of sales invoices that contain errors in violation of company policy for the Saxon Plumbing Company problem of section 7.3

	A	B
1	Confidence Interval Estimation for the Proportion of Invoices With Errors	
2		
3	Sample Size	xxx
4	Number of Successes	xxx
5	Confidence Level	.xx
6	Sample Proportion	=B4/B3
7	Z Value	=NORMSINV((1-B5)/2)
8	Standard Error of the Proportion	=SQRT(B6*(1-B6)/B3)
9	Interval Half Width	=ABS(B7*B8)
10	Interval Lower Limit	=B6-B9
11	Interval Upper Limit	=B6+B9

▲ WHAT IF EXAMPLE

As with the worksheet inserted by the add-in using the quick results method of this section, we can change the values for the sample size, number of successes, and confidence level in this worksheet to see their effects on the confidence interval estimate. (For further details, see the What If Example on page 437.)

Problems for Section 7.3

Learning the Basics

• **7.21** If $n = 200$ and $X = 50$, set up a 95% confidence interval estimate of the population proportion.

7.22 If $n = 400$ and $X = 25$, set up a 99% confidence interval estimate of the population proportion.

Applying the Concepts

• **7.23** The manager of a bank in a small city wants to determine the proportion of its depositors who have more than one account at the bank. A random sample of 100 depositors is selected and 30 state that they have more than one account at the bank.

(a) Set up a 90% confidence interval estimate of the population proportion of the bank's depositors who have more than one account at the bank.

(b) How can the results in (a) be used in a marketing strategy for a new type of investment that targets current depositors at the bank?

7.24 An auditor for the state insurance department wants to determine the proportion of claims that are paid by a health insurance company within 2 months of receipt of the claim. A random sample of 200 claims is selected, and it is determined that 80 were paid out within 2 months of the receipt of the claim.

(a) Set up a 99% confidence interval estimate of the population proportion of the claims paid within 2 months.

(b) If 90% or more claims are supposed to be paid within 2 months of their receipt, what should the auditor report to the state insurance department about the payment performance of the health insurance company?

●**7.25** An automobile dealer wants to estimate the proportion of customers who still own the cars they purchased 5 years earlier. A random sample of 200 customers selected from the automobile dealer's records indicates that 82 still own cars that were purchased 5 years earlier.

(a) Set up a 95% confidence interval estimate of the population proportion of all customers who still own the cars 5 years after they were purchased.

(b) How can the results in (a) be used by the automobile dealer to study satisfaction with cars purchased at the dealership?

7.26 A stationery supply store receives a shipment of a certain brand of inexpensive ball-point pens from a manufacturer. The owner of the store wishes to estimate the proportion of pens that are defective. A random sample of 300 pens is tested, and 30 are found to be defective.

(a) Set up a 90% confidence interval estimate of the proportion of defective pens in the shipment.

(b) The shipment can be returned if it is more than 5% defective; on the basis of the sample results, can the owner return this shipment?

(c) Suppose that a 99% confidence interval estimate was desired in (a). What would be the effect of this change on your answers to (a) and (b)?

 **WHAT IF?**

7.27 The marketing manager for a fast-food chain wants to estimate the proportion of high school students who have eaten at one of the chain's restaurants in the last month. A random sample of 400 high school students indicates that 90 have eaten at one of the chain's restaurants in the last month.

(a) Set up a 95% confidence interval estimate of the population proportion of high school students who have eaten at one of the chain's restaurants in the last month.

(b) How can the marketing manager use the results in (a) to increase the proportion of high school students who will eat at one of the chain's restaurants?

7.28 The telephone company wants to estimate the proportion of households that would purchase an additional telephone line if it were made available at a substantially reduced installation cost. A random sample of 500 households is selected. The results indicate that 135 of the households would purchase the additional telephone line at a reduced installation cost.

(a) Set up a 99% confidence interval estimate of the population proportion of households that would purchase the additional telephone line.

(b) How would the manager in charge of promotional programs concerning residential customers use the results in (a)?

7.29 The dean of a graduate school of business wishes to estimate the proportion of MBA students enrolled who have access to a personal computer outside the school (either at home or at work). A sample of 150 students reveals that 135 have access to personal computers outside the school (either at home or at work).

(a) Set up a 90% confidence interval estimate of the population proportion of students who have access to personal computers outside the school.

(b) How can the dean use the results in (a) to determine whether additional personal computers should be purchased for the computer lab?

In each of our examples concerning confidence interval estimation, the sample size was arbitrarily determined without regard to the size of the confidence interval. In the business world, the determination of the proper sample size is a complicated procedure subject to the constraints of budget, time, and ease of selection. As a case in point, if, in the Saxon Plumbing Company example, the auditor wants to estimate the average sales invoice or the proportion of sales invoices that contain errors, he would try to determine in advance how good an estimate would be required. This means that he must decide how much error he is willing to allow in estimating each of these variables. The auditor must also determine in advance how sure (confident) he wants to be of correctly estimating the true population parameter.

Sample Size Determination for the Mean

To determine the sample size needed for estimating the mean, we must keep in mind the amount of sampling error we are willing to accept and the level of confidence desired. In addition, we need some information about the standard deviation.

To develop a formula for determining sample size, recall equation (6.9) on page 380:

$$Z = \frac{\overline{X} - \mu}{\dfrac{\sigma}{\sqrt{n}}}$$

where Z is the critical value corresponding to an area of $(1 - \alpha)/2$ from the center of a standardized normal distribution. Multiplying both sides of equation (6.9) by $\sigma/\sqrt{n}$, we have

$$Z \frac{\sigma}{\sqrt{n}} = \overline{X} - \mu$$

Thus, the value of Z is positive or negative depending on whether $\overline{X}$ is larger or smaller than μ. The difference between the sample mean $\overline{X}$ and the population mean μ, denoted by e, is called the **sampling error.** The sampling error e is defined as

$$e = Z \frac{\sigma}{\sqrt{n}}$$

Solving for n gives us the sample size needed to develop the confidence interval estimate for the mean.

Sample Size Determination for the Mean

The sample size n is equal to the product of the Z value squared and the variance σ^2, divided by the sampling error e squared.

$$n = \frac{Z^2 \sigma^2}{e^2} \tag{7.4}$$

To determine the sample size, we must know the following three factors:

1. The desired confidence level, which determines the value of Z, the critical value from the normal distribution[2]
2. The acceptable sampling error e
3. The standard deviation σ

[2]We use Z instead of t because, to determine the critical value of t, we would need to know the sample size, which we don't know yet, and for most studies, the sample size needed will be large enough that the normal distribution is a good approximation of the t distribution.

In practice, it is usually not easy to determine these three quantities. How can we know what level of confidence to use and what sampling error is desired? Typically, these questions are answered only by the subject matter expert, that is, the individual most familiar with the variables to be analyzed. Although 95% is the most common confidence level used (in which case $Z = 1.96$), if a greater confidence is desired, then 99% might be more appropriate; if less confidence is deemed acceptable, then 90% might be used. For the sampling error, we should be thinking not of how much sampling error we would like to have (we really do not want any error) but of how much we can tolerate and still be able to provide adequate conclusions from the data.

Even when the confidence level and the sampling error are specified, an estimate of the standard deviation must be available. Unfortunately, the population standard deviation σ is rarely known. In some instances, we can estimate the standard deviation from past data. In other situations, we can develop an educated guess by taking into account the range and distribution of the variable. For example, if we assume a normal distribution, the range is approximately equal to 6 σ (i.e., $\pm 3\sigma$ around the mean) so that σ is estimated as range/6. If we cannot estimate σ in this manner, a *pilot* study can be conducted and the standard deviation estimated from the resulting data.

To explore how to determine the sample size needed for estimating the population mean, let us again consider the audit to be performed this month at Saxon Plumbing Company. In section 7.2, we saw that a sample of 100 sales invoices was selected and a 95% confidence interval estimate of the population average sales invoice amount was developed. How was this sample size determined? Should we have selected a different sample size?

Suppose that, after consultation with company officials, it is determined that a sampling error of no more than $\pm\$5$ is desired along with 95% confidence. The auditor notes that the standard deviation of the sales amount has been approximately $25 for a substantial period of time. Thus, $e = \$5$, $\sigma = \$25$, and $Z = 1.96$ (for 95% confidence). Using equation (7.4), we have

$$n = \frac{Z^2\sigma^2}{e^2} = \frac{(1.96)^2(25)^2}{(5)^2}$$

$$n = 96.04$$

Therefore, $n = 97$ because the general rule is to round the sample size up to the next whole integer to slightly oversatisfy the criteria. Thus, the sample size of 100 that was taken is close to the one necessary to satisfy the needs of the company based on the estimated standard deviation, desired confidence level, and sampling error. However, we note from the confidence interval obtained on page 427 that because the sample standard deviation is $28.95, the width of the confidence interval was slightly more than was desired.

We provide another application on how to determine the sample size needed to develop a confidence interval estimate for the mean in Example 7.5.

Example 7.5 *Determining the Sample Size for the Mean*

To return to Example 7.3 on page 427, suppose the marketing manager wishes to estimate the population mean annual usage of home heating oil to within ±50 gallons of the true value and he desires to be 95% confident of correctly estimating the true mean. On the basis of a study taken the previous year, he believes that the standard deviation can be estimated as 325 gallons. Find the sample size needed.

SOLUTION

With this information, the sample size is determined in the following manner for $e = 50$, $\sigma = 325$, and 95% confidence ($Z = 1.96$):

$$n = \frac{Z^2 \sigma^2}{e^2} = \frac{(1.96)^2 (325)^2}{(50)^2}$$

$$n = 162.31$$

Therefore, $n = 163$. We choose a sample size of 163 homes because the general rule for determining sample size is to always round up to the *next integer value* in order to slightly oversatisfy the criteria desired.

We may note that if the marketing manager uses these criteria, a sample of 163 homes will be taken—not a sample of 35 as was used in Example 7.3. However, the standard deviation that was used was estimated at 325 based on a previous survey. If the standard deviation obtained in the actual survey is very different from this value, the computed sampling error will be directly affected.

Sample Size Determination for a Proportion

Thus far in this section we have discussed how to determine the sample size needed for estimating the population mean. Now suppose that the auditor of Saxon Plumbing Company also wishes to determine the sample size necessary for estimating the population proportion of sales invoices that have errors violating company policy. The methods of sample size determination that are used in estimating a population proportion are similar to those employed in estimating a mean.

To develop a formula for determining sample size, recall from equation (6.14 on page 400) that

$$Z \cong \frac{p_s - p}{\sqrt{\dfrac{p(1 - p)}{n}}}$$

where Z is the critical value corresponding to an area of $(1 - \alpha)/2$ from the center of a standardized normal distribution. Multiplying both sides of equation (6.14) by $\sqrt{p(1 - p)/n}$, we have

$$Z \sqrt{\frac{p(1 - p)}{n}} = p_s - p$$

The sampling error e is equal to $(p_s - p)$, the difference between the sample proportion p_s and the parameter to be estimated p. This sampling error can be defined as

$$e = Z \sqrt{\frac{p(1-p)}{n}}$$

Solving for *n*, we obtain the sample size necessary to develop a confidence interval estimate for a proportion.

Sample Size Determination for a Proportion

The sample size *n* is equal to the *Z* value squared times the true proportion *p*, times 1 minus the true proportion *p*, divided by the sampling error *e* squared.

$$n = \frac{Z^2 p(1-p)}{e^2} \tag{7.5}$$

To determine the sample size for estimating a proportion, three unknowns must be defined:

1. The desired level of confidence
2. The acceptable sampling error, *e*
3. The true proportion of "success" *p*

In practice, the selection of these quantities requires some planning. Once we determine the desired level of confidence, we can obtain the appropriate *Z* value from the normal distribution. The sampling error *e* indicates the amount of error that we are willing to accept or tolerate in estimating the population proportion. The third quantity—the true proportion of success *p*—is actually the population parameter that we want to find! Thus, how do we state a value for the very thing that we are taking a sample in order to determine?

Here there are two alternatives. First, in many situations, past information or relevant experiences may be available that enable us to provide an educated estimate of *p*. Second, if past information or relevant experiences are not available, we can try to provide a value for *p* that would never *underestimate* the sample size needed. Referring to equation (7.5), we observe that the quantity $p(1-p)$ appears in the numerator. Thus we must determine the value of *p* that will make the quantity $p(1-p)$ as large as possible. It can be shown that when $p = .5$, then the product $p(1-p)$ achieves its maximum result. Several values of *p* along with the accompanying products of $p(1-p)$ are

When $p = .9$, then $p(1-p) = (.9)(.1) = .09$

When $p = .7$, then $p(1-p) = (.7)(.3) = .21$

When $p = .5$, then $p(1-p) = (.5)(.5) = .25$

When $p = .3$, then $p(1-p) = (.3)(.7) = .21$

When $p = .1$, then $p(1-p) = (.1)(.9) = .09$

Therefore, when we have no prior knowledge or estimate of the true proportion *p*, we should use $p = .5$ as the most conservative way of determining the sample size. This produces the largest possible sample size but unfortunately results in the highest possible cost of sampling. That is, the use of $p = .5$ may result in an overestimate of the sample size because the actual sample proportion is used in developing the confidence interval. If the actual sample proportion is very different from .5, the width of the confidence interval may be substantially narrower than originally intended. The increased precision comes at the cost of spending more time and money for an increased sample size.

Returning to the auditor for Saxon Plumbing Company, suppose that he wants to have 95% confidence of estimating the population proportion of sales invoices with errors to within ±.07 of the true population proportion. The results from past months indicate that the largest proportion has been no more than .15. Thus, $e = .07$, $p = .15$, and $Z = 1.96$ (for 95% confidence). Using equation (7.5), we have

$$n = \frac{Z^2 p(1 - p)}{e^2}$$

$$= \frac{(1.96)^2(.15)(.85)}{(.07)^2}$$

$$= 99.96$$

Therefore, $n = 100$ because the general rule is to round the sample size up to the next whole integer to slightly oversatisfy the criteria. Thus, the sample size of 100 that was taken was exactly what was needed to satisfy the requirements of the company based on the estimated proportion, desired confidence level, and sampling error. However, we note from the confidence interval obtained on page 434 that, because the sample proportion was .10, the width of the confidence level is actually narrower than desired.

A second application of determining the sample size for estimating the population proportion is provided in Example 7.6.

Example 7.6 Determining the Sample Size for the Proportion

In Example 7.4 on page 435, suppose the operations manager wants to have 90% confidence of estimating the proportion of nonconforming newspapers to within ±.05 of its true value. In addition, because the publisher of the newspaper has not previously undertaken such a survey, no information is available from past data. Determine the sample size needed.

SOLUTION

Because there is no information available from past data, we will set $p = .50$. With this and other criteria in mind, the sample size needed can be determined in the following manner when $e = .05$, $p = .5$, and for 90% confidence, $Z = 1.645$.

$$n = \frac{(1.645)^2(.5)(.5)}{(.05)^2}$$

$$= 270.6$$

Thus, $n = 271$. Therefore, in order to be 90% confident of estimating the proportion to within ±.05 of its true value, a sample size of 271 newspapers is needed.

 **7.4E.1** **DETERMINING THE SAMPLE SIZE FOR ESTIMATING THE MEAN USING MICROSOFT EXCEL**

Overview

◆ *For Quick Results Users*   Use the PHStat add-in to determine the sample size needed for estimating the mean.

◆ *For Developers*   Implement a worksheet that uses the NORMSINV and ROUNDUP worksheet functions and arithmetic formulas to determine the sample size needed for estimating the mean.

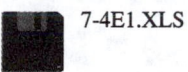 7-4E1.XLS

The 7-4E1.XLS workbook file contains the determination of the sample size needed for estimating the mean sales invoice amount for the Saxon Plumbing Company problem of section 7.4.

Quick Results Details

To determine the sample size needed for estimating the mean, use the Sample Size | Determination for the Mean choice of the PHStat add-in. As an example, consider the Saxon Plumbing Company problem of section 7.4. To determine the sample size needed for estimating the mean sales invoice amount, do the following:

❶ If the PHStat add-in has not been previously loaded, load the add-in using the instructions of section S4.2.

❷ Select File | New to open a new workbook (or open the existing workbook into which the sample size determination worksheet is to be inserted).

❸ Select PHStat | Sample Size | Determination for the Mean.

❹ In the Determination for the Mean dialog box (see Figure 7E.7):

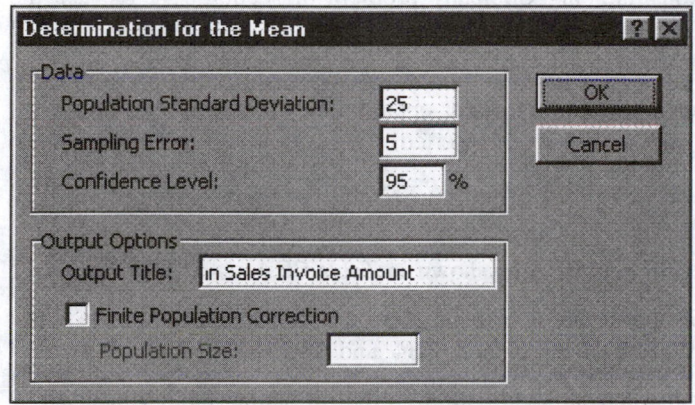

FIGURE 7E.7
PHStat Determination for the Mean dialog box

a. Enter 25 in the Population Standard Deviation: edit box.
b. Enter 5 in the Sampling Error: edit box.
c. Enter 95 in the Confidence Level: edit box.

d. Enter Sample Size for Estimating the Mean Sales Invoice Amount in the Output Title: edit box.

e. Click the OK button.

The add-in inserts a worksheet containing calculations for determining the sample size needed for estimating the mean similar to the one shown in Figure 7E.8.

FIGURE 7E.8

Sample size determination for estimating the mean sales invoice amount for the Saxon Plumbing Company problem of section 7.4

	A	B	C	D
1	**Sample Size for Estimating the Mean Sales Invoice Amount**			
2				
3	**Population Standard Deviation**	25		
4	**Sampling Error**	5		
5	**Confidence Level**	95%		
6	Z Value	-1.95996108		
7	Calculated Sample Size	96.03618611		
8	**Sample Size Needed**	97		

▲ WHAT IF EXAMPLE

We can change the values for the population standard deviation, sampling error, and confidence level in the worksheet inserted by the add-in to see their effects on the sample size needed. For example, if we change the sampling error in cell B4 from 5 to 10, we will observe that the sample size needed changes from 97 to 25. To facilitate comparisons among many different alternative values, we can create scenarios using the Scenario Manager as discussed at the end of section 2.2E.

Developer Details

We can use the NORMSINV worksheet function (see section 6.1E) to return the Z value needed to determine the sample size for estimating the mean. Table 7E.4 presents a Sample Size sheet design for determining the sample size needed to estimate the mean sales invoice amount for the Saxon Plumbing Company problem of section 7.4. We use the ROUNDUP function in cell B8 to round up the result of the sample size calculation of cell B7 to the next integer. To implement the Table 7E.4 design, do the following:

❶ Select File | New to open a new workbook (or open the existing workbook into which the Sample Size worksheet is to be inserted).

❷ Select an unused worksheet (or select Insert | Worksheet if there are none) and rename the sheet Sample Size.

❸ Enter the title and labels for column A as shown in Table 7E.4.

❹ Enter the population standard deviation, sampling error, and confidence level in the cell range B3:B5. Enter 25 in cell B3, 5 in cell B4, and 0.95 in cell B5.

❺ Select cell B5 and click the Percent style button on the formatting toolbar (see section S3.11) to format the decimal value 0.95 as 95%.

❻ Enter the formulas for the cell range B6:B8 as shown in Table 7E.4.

The completed worksheet will be similar to the one shown in Figure 7E.8.

Table 7E.4 Sample Size determination sheet design for estimating the mean sales invoice for the Saxon Plumbing Company problem of section 7.4

	A	B
1	Sample Size for Estimating the Mean Sales Invoice Amount	
2		
3	Population Standard Deviation	xxx
4	Sampling Error	xxx
5	Confidence Level	.xx
6	Z Value	=NORMSINV((1-B5)/2)
7	Calculated Sample Size	=((B6*B3)/B4)^2
8	Sample Size Needed	=ROUNDUP(B7,0)

▲ WHAT IF EXAMPLE

As with the worksheet inserted by the add-in using the quick results method of this section, we can change the values for population standard deviation, sampling error, and confidence level in this worksheet to see their effects on the sample size needed. (For further details, see the What If Example on page 446.)

7.4E.2 DETERMINING THE SAMPLE SIZE FOR ESTIMATING THE PROPORTION USING MICROSOFT EXCEL

Overview

◆ *For Quick Results Users*   Use the PHStat add-in to determine the sample size needed for estimating the proportion.

◆ *For Developers*   Implement a worksheet that uses the NORMSINV and ROUNDUP worksheet functions and arithmetic formulas to determine the sample size needed for estimating the proportion.

The 7-4E2.XLS workbook file contains the determination of the sample size needed for estimating the proportion of sales invoices with errors for the Saxon Plumbing Company problem of section 7.4.

7-4E2.XLS

Quick Results Details

To determine the sample size needed for estimating the proportion, use the Sample Size | Determination for the Proportion choice of the PHStat add-in. As an example, consider the Saxon Plumbing Company problem of section 7.4. To determine the sample size needed for estimating the proportion of sales invoices, do the following:

❶ If the PHStat add-in has not been previously loaded, load the add-in using the instructions of section S4.2.

❷ Select File | New to open a new workbook (or open the existing workbook into which the sample size determination worksheet is to be inserted).

❸ Select PHStat | Sample Size | Determination for the Proportion.

❹ In the Determination for the Proportion dialog box (see Figure 7E.9):

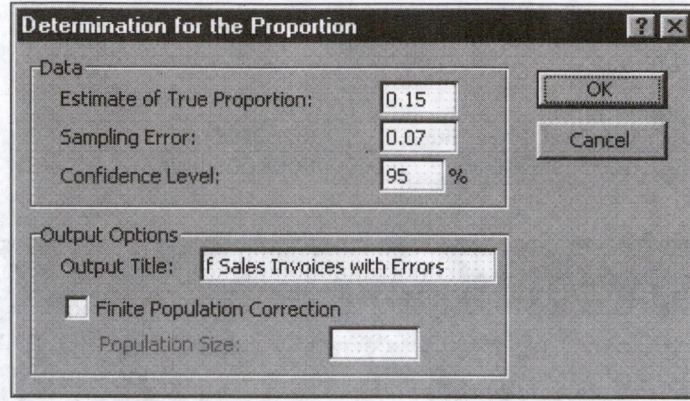

FIGURE 7E.9
PHStat Determination for the Proportion dialog box

a. Enter 0.15 in the Estimate of True Proportion: edit box.
b. Enter 0.07 in the Sampling Error: edit box.
c. Enter 95 in the Confidence Level: edit box.
d. Enter Sample Size for Estimating the Proportion of Sales Invoices with Errors in the Output Title: edit box.
e. Click the OK button.

The add-in inserts a worksheet containing calculations for determining the sample size needed for estimating the proportion similar to the one shown in Figure 7E.10.

FIGURE 7E.10
Sample size determination for estimating the proportion of sales invoices with errors for the Saxon Plumbing Company problem of section 7.4

	A	B	C	D	E
1	Sample Size for Estimating the Proportion of Sales Invoices with Errors				
2					
3	Estimate of True Proportion	0.15			
4	Sampling Error	0.07			
5	Confidence Level	95%			
6	Z Value	-1.95996108			
7	Calculated Sample Size	99.95603044			
8	Sample Size Needed	100			

▲ WHAT IF EXAMPLE

We can change the values for the estimate of the true proportion, sampling error, and confidence level in the worksheet inserted by the add-in to see their effects on the sample size needed. For example, if we change the sampling error in cell B4 from .07 to .05, we will observe that the sample size needed changes from 100 to 196. To facilitate comparisons among many different alternative values, we could create scenarios using the Scenario Manager as discussed at the end of section 2.2E.

Developer Details

We can use the NORMSINV worksheet function (see section 6.1E) to return the Z value needed to determine the sample size for estimating the proportion. Table 7E.5 presents a Sample Size sheet design for determining the sample size needed for estimating the proportion of sales invoices with errors for the Saxon Plumbing Company problem of section 7.4. As was first done in the Table 7E.4 design of section 7.4E1, we use the ROUNDUP function in cell B8 to round up the result of the sample size calculation to the next integer. To implement the Table 7E.5 design, do the following:

❶ Select File | New to open a new workbook (or open the existing workbook into which the Sample Size worksheet is to be inserted).

❷ Select an unused worksheet (or select Insert | Worksheet if there are none) and rename the sheet Sample Size.

❸ Enter the title and labels for column A as shown in Table 7E.5.

❹ Enter the estimate of the true proportion, sampling error, and confidence level in the cell range B3:B5. Enter 0.15 in cell B3, 0.07 in cell B4, and 0.95 in cell B5.

❺ Select cell B5 and click the Percent style button on the formatting toolbar (see section S3.11) to format the decimal value 0.95 as 95%.

❻ Enter the formulas for the cell range B6:B8.

The completed worksheet will be similar to the one shown in Figure 7E.10.

Table 7E.5 Sample Size determination sheet design for estimating the proportion of sales invoices with errors for the Saxon Plumbing Company problem of section 7.4

	A	B
1	Sample Size for Estimating the Proportion of Sales Invoices with Errors	
2		
3	Estimate of True Proportion	.xx
4	Sampling Error	xxx
5	Confidence Level	.xx
6	Z Value	=NORMSINV((1-B5)/2)
7	Calculated Sample Size	=(B6^2*B3*(1-B3))/B4^2
8	Sample Size Needed	=ROUNDUP(B7,0)

▲ WHAT IF EXAMPLE

As with the worksheet inserted by the add-in using the quick results method of this section, we can change the values for the estimate of the true proportion, sampling error, and confidence level in this worksheet to see their effects on the sample size needed. (For further details, see the What If Example on page 448.)

Problems for Section 7.4

Learning the Basics

•7.30 If you want to be 95% confident of estimating the population mean to within a sampling error of ±5 and the standard deviation is assumed to be equal to 15, what sample size is required?

7.31 If you want to be 99% confident of estimating the population mean to within a sampling error of ±20 and the standard deviation is assumed to be equal to 100, what sample size is required?

7.32 If you want to be 99% confident of estimating the population proportion to within an error of ±.04, what sample size is needed?

7.33 If you want to be 95% confident of estimating the population proportion to within an error of ±.02 and there is historical evidence that the population proportion is approximately .40, what sample size is needed?

Applying the Concepts

7.34 A survey is planned to determine the average annual family medical expenses of employees of a large company. The management of the company wishes to be 95% confident that the sample average is correct to within ±$50 of the true average annual family medical expenses. A pilot study indicates that the standard deviation can be estimated as $400.

  WHAT IF?
(a) How large a sample size is necessary?
(b) If management wants to be correct to within ±$25, what sample size is necessary?

7.35 If the manager of a paint supply store wants to estimate the average amount in a 1-gallon can to within ±0.004 gallon with 95% confidence and also assumes that the standard deviation is 0.02 gallon, what sample size is needed?

•7.36 If a quality control manager wants to estimate the average life of light bulbs to within ±20 hours with 95% confidence and also assumes that the process standard deviation is 100 hours, what sample size is needed?

7.37 If the inspection division of a county weights and measures department wants to estimate the average amount of soft-drink fill in 2-liter bottles to within ±0.01 liter with 95% confidence and also assumes that the standard deviation is 0.05 liter, what sample size is needed?

•7.38 A consumer group wishes to estimate the average electric bills for the month of July for single-family homes in a large city. Based on studies conducted in other cities, the standard deviation is assumed to be $25. The group wants to estimate the average bill for July to within ±$5 of the true average with 99% confidence.

  WHAT IF?
(a) What sample size is needed?
(b) If 95% confidence is desired, what sample size is necessary?

7.39 A pharmaceutical company is considering a request to pay for the continuing education of its research scientists. It would like to estimate the average amount spent by these scientists for professional memberships. Based on a pilot study, the standard deviation is estimated to be $35.

  WHAT IF?
(a) What sample size is required to be 90% confident of being correct to within ±$10?
(b) If 95% confidence of being correct to within ±$20 is desired, what sample size is necessary?

7.40 An advertising agency that serves a major radio station wants to estimate the average amount of time that the station's audience spends listening to the radio on a daily basis. From past studies, the standard deviation is estimated as 45 minutes.

  WHAT IF?
(a) What sample size is needed if the agency wants to be 90% confident of being correct to within ±5 minutes?
(b) If 99% confidence is desired, what sample size is necessary?

7.41 Suppose that a gas utility wishes to estimate its average waiting time for installation of service to within ±5 days with 95% confidence. Because it does not have access to previous data, it makes its own independent estimate of the standard deviation, which it believes to be 20 days. What sample size is needed?

7.42 A political pollster wants to estimate the proportion of voters who will vote for the Democratic candidate in a presidential campaign. The pollster wishes to have 90% confidence that her prediction is correct to within ±.04 of the population proportion.

    (a) What sample size is needed?

    (b) If the pollster wants to have 95% confidence, what sample size is needed?

    (c) If she wants to have 95% confidence and a sampling error of ±.03, what sample size is needed?

    (d) On the basis of your answers to (a)–(c), what general conclusions can be reached about the effects of the confidence level desired and the acceptable sampling error on the sample size needed? Discuss.

 **WHAT IF?**

• **7.43** A cable television company wants to estimate the proportion of its customers who would purchase a cable television program guide. The company would like to have 95% confidence that its estimate is correct to within ±.05 of the true proportion. Past experience in other areas indicates that 30% of the customers will purchase the program guide. What sample size is needed?

• **7.44** A bank manager wants to be 90% confident of being correct to within ±.05 of the true population proportion of depositors who have both savings and checking accounts at the bank. How many depositors need to be sampled?

7.45 An audit test to establish the percentage of occurrence of failures to follow a specific internal control procedure is to be undertaken. The auditor decides that the maximum tolerable error rate that is permissible is 5%.

    (a) What size sample is required to achieve a sample precision of ±2% with 99% confidence?

    (b) What would be your answer in (a) if the maximum tolerable error rate is

        (1) 10%?

        (2) 15%?

        (3) 20%?

 WHAT IF?

7.46 A large shipment of air filters is received by Joe's Auto Supply Company. The air filters are to be sampled to estimate the proportion that are unusable. From past experience the proportion of unusable air filters is estimated to be .10.

    (a) How large a random sample should be taken to estimate the true proportion of unusable air filters to within ±.07 with 99% confidence?

    (b) If 95% confidence is desired with a sampling error of ±.06, what sample size should be taken?

 **WHAT IF?**

7.47 Suppose that Matt's Motors wants to conduct a survey to determine the proportion of its customers who still own their cars 5 years after purchasing them. Suppose it wants to be 95% confident of being correct to within ±.025 of the true proportion. What sample size is needed?

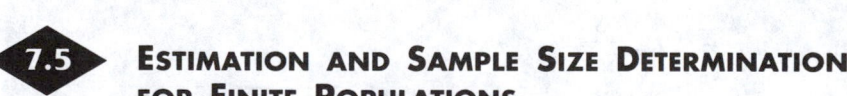

7.5   ESTIMATION AND SAMPLE SIZE DETERMINATION FOR FINITE POPULATIONS

Estimating the Mean

In section 6.7 we saw that in sampling without replacement from finite populations, the **finite population correction (fpc) factor** serves to reduce the standard error by a value equal to $\sqrt{(N - n)/(N - 1)}$. When developing confidence interval estimates for population parameters, the fpc factor is used when samples are selected without replacement. Thus, the $(1 - \alpha) \times 100\%$ confidence interval estimate for the mean is calculated as in equation (7.6).

Confidence Interval Estimate for the Mean (σ Unknown) for a Finite Population

$$\overline{X} \pm t_{n-1}\frac{S}{\sqrt{n}}\sqrt{\frac{N-n}{N-1}}$$

(7.6)

To illustrate the finite population correction factor, we refer to the confidence interval estimate for the mean developed for Saxon Plumbing Company on page 427. Suppose that in this month there are 5,000 sales invoices. Using $\overline{X} = \$110.27$, $S = \$28.95$, $N = 5,000$, $n = 100$, and with 95% confidence, $t_{99} = 1.9842$. From equation (7.6) we have

$$\overline{X} \pm t_{n-1}\frac{S}{\sqrt{n}}\sqrt{\frac{N-n}{N-1}}$$

$$= 110.27 \pm (1.9842)\frac{28.95}{\sqrt{100}}\sqrt{\frac{5,000-100}{5,000-1}}$$

$$= 110.27 \pm 5.74(.99)$$

$$= 110.27 \pm 5.68$$

$$\$104.59 \leq \mu \leq \$115.95$$

In this case, because the sample is a very small fraction of the population, the correction factor has a minimal effect on the width of the confidence interval. To examine the effect of the correction factor when the sample size is more than 5% of the population size, we present Example 7.7.

Example 7.7   *Estimating the Mean Annual Usage for Home Heating Oil*

In Example 7.3 on page 427, a sample of 35 single-family homes was selected. Suppose there is a population of 500 single-family homes served by the company. Set up a 95% confidence interval estimate of the population mean.

SOLUTION

Using the finite population correction factor, we have, with $\overline{X} = 1,122.7$ gallons, $S = 295.72$, $n = 35$, $N = 500$, and $t_{34} = 2.0322$ (for 95% confidence):

$$\overline{X} \pm t_{n-1}\frac{S}{\sqrt{n}}$$

$$= 1,122.7 \pm (2.0322)\frac{295.72}{\sqrt{35}}\sqrt{\frac{500-35}{500-1}}$$

$$= 1,122.7 \pm 101.58(.9653)$$

$$= 1,122.7 \pm 98.05$$

$$1,024.65 \leq \mu \leq 1,220.75$$

Here, because more than 5% of the population is to be sampled, the fpc factor has a moderate effect on the confidence interval estimate.

Estimating the Proportion

In sampling without replacement, the $(1 - \alpha) \times 100\%$ confidence interval estimate of the proportion is found as in equation (7.7).

Confidence Interval Estimate for the Proportion Using the Finite Population Correction Factor

$$p_s \pm Z \sqrt{\frac{p_s(1 - p_s)}{n}} \sqrt{\frac{N - n}{N - 1}} \qquad (7.7)$$

To illustrate the use of the finite population correction factor when developing a confidence interval estimate of the population proportion, consider again the estimate developed for Saxon Plumbing Company on page 434. For these data, we have $N = 5,000$, $n = 100$, $p_s = 10/100 = .10$, and with 95% confidence, $Z = 1.96$. Using equation (7.7),

$$p_s \pm Z \sqrt{\frac{p_s(1 - p_s)}{n}} \sqrt{\frac{N - n}{N - 1}} = .10 \pm (1.96) \sqrt{\frac{(.10)(.90)}{100}} \sqrt{\frac{5,000 - 100}{5,000 - 1}}$$

$$= .10 \pm (1.96)(.03)(.99)$$

$$= .10 \pm .0582$$

$$.0418 \leq p \leq .1582$$

In this case, because the sample is a very small fraction of the population, the fpc factor has virtually no effect on the confidence interval estimate.

Determining the Sample Size

Just as the fpc factor is used to develop confidence interval estimates, it also is used to determine sample size when sampling without replacement. For example, in estimating the mean, the sampling error is

$$e = \frac{Z\sigma}{\sqrt{n}} \sqrt{\frac{N - n}{N - 1}}$$

and in estimating the proportion, the sampling error is

$$e = Z \sqrt{\frac{p(1 - p)}{n}} \sqrt{\frac{N - n}{N - 1}}$$

To determine the sample size in estimating the mean or the proportion we have, from equations (7.4) and (7.5),

$$n_0 = \frac{Z^2\sigma^2}{e^2} \quad \text{and} \quad n_0 = \frac{Z^2 p(1 - p)}{e^2}$$

where n_0 is the sample size without considering the finite population correction factor. Applying the fpc factor to this results in the actual sample size n, computed as in equation (7.8).

In the auditor's determination of sample size for Saxon Plumbing Company, a sample size of 97 was needed for the mean and a sample of 100 was needed for the proportion (see pages 441 and 444). Using the fpc factor in equation (7.8) for the mean, with $N = 5,000$, $e = \$5$, $\sigma = \$25$, and $Z = 1.96$ (for 95% confidence), leads to

$$n = \frac{(96.04)(5,000)}{96.04 + (5,000 - 1)} = 94.25$$

Thus, $n = 95$.

Using the fpc factor in equation (7.8) for the proportion, with $N = 5,000$, $e = .07$, $p = .15$, and $Z = 1.96$ (for 95% confidence),

$$n = \frac{(99.96)(5,000)}{99.96 + (5,000 - 1)} = 98.02$$

Thus, $n = 99$.

To satisfy *both requirements simultaneously with one sample,* we need to take the larger sample size of 99.

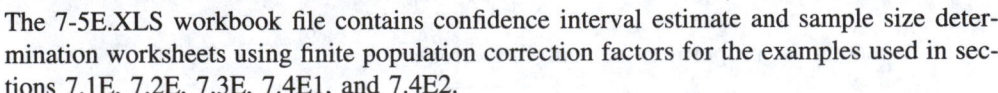

7.5E ◆ USING THE FINITE POPULATION CORRECTION FACTOR IN MICROSOFT EXCEL

Overview

◆ *For Quick Results Users* Use the PHStat add-in to calculate a confidence interval estimate or to determine a sample size needed for estimating the mean or proportion using the finite population correction factor.

◆ *For Developers* Modify a worksheet implemented earlier in this chapter to include the finite population correction calculations.

7-5E.XLS The 7-5E.XLS workbook file contains confidence interval estimate and sample size determination worksheets using finite population correction factors for the examples used in sections 7.1E, 7.2E, 7.3E, 7.4E1, and 7.4E2.

Quick Results Details

To calculate a confidence interval estimate or a sample size using the finite population correction factor, select the appropriate choice of the Confidence Intervals or Sample Size submenus of the PHStat add-in. Use the sections 7.1E, 7.2E, 7.3E, 7.4E1, and 7.4E2 procedures, modifying step 4 of these procedures to include:

> Select the Finite Population Correction check box and enter the population size in the Population Size: edit box.

For example, to use the finite population correction in section 7.2E, select the Finite Population Correction check box and enter 5000 in the Population Size: edit box after step 4c and before clicking the OK button in step 4d.

Worksheets containing calculations using the finite population correction factor will be similar to those shown in Figures 7E.11 and 7E.12.

	A	B	C	D
1	Confidence Interval Estimation for the Mean Invoice Amount			
2				
3	Sample Standard Deviation	28.95		
4	Sample Mean	110.27		
5	Sample Size	100		
6	Confidence Level	95%		
7	Standard Error of the Mean	2.895		
8	Degrees of Freedom	99		
9	t Value	1.984217306		
10	Interval Half Width	5.744309101		
11	Interval Lower Limit	104.53		
12	Interval Upper Limit	116.01		
13				
14				
15	Finite Populations			
16	Population Size	5000		
17	FPC Factor	0.990048503		
18	Interval Half Width	5.687144629		
19	Interval Lower Limit	104.58		
20	Interval Upper Limit	115.96		

FIGURE 7E.11

Confidence interval estimate for the mean sales invoice amount with the finite population correction for the Saxon Plumbing Company problem of section 7.2

	A	B	C	D
1	Sample Size for Estimating the Mean Sales Invoice Amount			
2				
3	Population Standard Deviation	25		
4	Sampling Error	5		
5	Confidence Level	95%		
6	Z Value	-1.95996108		
7	Calculated Sample Size	96.03618611		
8	Sample Size Needed	97		
9				
10				
11	Finite Populations			
12	Population Size	5000		
13	Sample Size with FPC	94.24485185		
14	Sample Size Needed	95		

FIGURE 7E.12

Sample Size for estimating the mean sales invoice amount with the finite population correction for the Saxon Plumbing Company problem of section 7.4

▲ WHAT IF EXAMPLE

We can change the values for the population size (as well as the values noted in earlier sections) in the worksheet inserted by the add-in to see their effects on the confidence interval estimate or sample size needed. For example, if we change the population size in cell B16 from 5,000 to 1,000 for the worksheet for the Saxon Plumbing Company problem of section 7.2, we will observe that the interval half width narrows to 5.4523 and the confidence interval estimate changes to between 104.82 and 115.72. To facilitate comparisons among many different alternative values, we could create scenarios using the Scenario Manager as discussed at the end of section 2.2E.

Developer Details

We can add formulas to the worksheet designs of Tables 7E.1–7E.5 of sections 7.1E, 7.2E, 7.3E, 7.4E1, and 7.4E2 in order to calculate the finite population correction factor in each worksheet. To implement the finite population correction factor in one of the worksheets described earlier in this chapter, do the following:

❶ Implement a worksheet based on the design of Tables 7E.1, 7E.2, 7E.3, 7E.4, or 7E.5, using the procedures of sections 7.1E, 7.2E, 7.3E, 7.4E1, or 7.4E2, respectively.

❷ Add the labels and formulas shown in the appropriate matching table (either Table 7E.6, 7E.7, 7E.8, or 7E.9) to the worksheet implemented in step 1.

❸ Use the Merge and Center button on the formatting toolbar (see section S3.11) to center the title Finite Populations over two columns.

❹ Enter 5000 as the population size in the appropriate cell if implementing Tables 7E.7, 7E.8, or 7E.9. If implementing Table 7.E6, enter 10000 as the population size. (*Note:* The actual population size for this problem is not provided in Example 7.1.)

The completed worksheet will be similar to the ones shown in Figures 7E.11 and 7E.12.

▲ WHAT IF EXAMPLE

As with the worksheet inserted by the add-in using the quick results method of this section, we can change the value for the population size to see its effect on the confidence interval estimate or sample size needed. (For further details, see the What If Example on page 455.)

Table 7E.6 Finite population correction modifications for the confidence interval estimate of the mean (σ known) sheet design of Table 7E.1

	A	B
12		
13		
14	Finite Populations	
15	Population Size	xxx
16	FPC Factor	=SQRT((B15-B5)/(B15-1))
17	Interval Half Width	=B9*B16
18	Interval Lower Limit	=B4-B17
19	Interval Upper Limit	=B4+B17

Table 7E.7 Finite population correction modifications for the confidence interval estimate of the mean (σ unknown) sheet design of Table 7E.2

	A	B
13		
14		
15	Finite Populations	
16	Population Size	xxx
17	FPC Factor	=SQRT((B16-B5)/(B16-1))
18	Interval Half Width	=B10*B17
19	Interval Lower Limit	=B4-B18
20	Interval Upper Limit	=B4+B18

Table 7E.8 Finite population correction modifications for the confidence interval estimate for the proportion sheet designs of Table 7E.3

	A	B
12		
13		
14	Finite Populations	
15	Population Size	xxx
16	FPC Factor	=SQRT((B15-B3)/(B15-1))
17	Interval Half Width	=B9*B16
18	Interval Lower Limit	=B6-B17
19	Interval Upper Limit	=B6+B17

Table 7E.9 Finite population correction modifications for the sample size determination sheet designs of Tables 7E.4 and 7E.5

	A	B
9		
10		
11	Finite Populations	
12	Population Size	xxx
13	Calculated Sample Size	=(B7*B12)/(B7+B12-1)
14	Sample Size Needed	=ROUNDUP(B13,0)

Problems for Section 7.5

Learning the Basics

• **7.48** If $\bar{X} = 75$, $S = 24$, $n = 36$, and $N = 200$, set up a 95% confidence interval estimate of the population mean μ if sampling is done *without* replacement.

7.49 For a population of 1,000, we want 95% confidence with a sampling error of 5, and the standard deviation is assumed equal to 20. What sample size would be required if sampling is done *without* replacement?

Applying the Concepts

• **7.50** The quality control manager at a light bulb factory needs to estimate the average life of a large shipment of light bulbs. The process standard deviation is known to be 100 hours. Assuming the shipment contains a total of 2,000 light bulbs, if sampling is done *without* replacement

   (a) set up a 95% confidence interval estimate of the true average life of light bulbs in this shipment if a random sample of 50 light bulbs selected from the shipment indicates a sample average life of 350 hours.

   (b) determine the sample size needed to estimate the average life to within ±20 hours with 95% confidence.

   (c) What are your answers in (a) and (b) if the shipment contains 1,000 light bulbs?

 **WHAT IF?**

7.51 A survey is planned to determine the average annual family medical expenses of employees of a large company. The management of the company wishes to be 95% confident that the sample average is correct to within ±$50 of the true average annual family medical expenses. A pilot study indicates that the standard deviation is estimated as $400. How large a sample size is necessary if the company has 3,000 employees and if sampling is done *without* replacement?

• **7.52** The manager of a bank that has 1,000 depositors in a small city wants to determine the proportion of its depositors with more than one account at the bank.

   (a) Set up a 90% confidence interval estimate of the population proportion of the bank's depositors who have more than one account at the bank if a random sample of 100 depositors is selected *without* replacement and 30 state that they have more than one account at the bank.

WHAT IF?

(b) A bank manager wants to be 90% confident of being correct to within ±.05 of the true population proportion of depositors who have more than one account at the bank. What sample size is needed if sampling is done *without* replacement?

(c) What are your answers to (a) and (b) if the bank has 2,000 depositors?

•7.53 An automobile dealer wants to estimate the proportion of customers who still own the cars they purchased 5 years earlier. Sales records indicate that the population of owners is 4,000.

(a) Set up a 95% confidence interval estimate of the population proportion of all customers who still own their cars 5 years after they were purchased if a random sample of 200 customers selected *without* replacement from the automobile dealer's records indicate that 82 still own cars that were purchased 5 years earlier.

WHAT IF?

(b) What sample size is necessary to estimate the true proportion to within ±.025 with 95% confidence?

(c) What are your answers to (a) and (b) if the population consists of 6,000 owners?

7.54 The inspection division of the Lee County Weights and Measures Department is interested in estimating the actual amount of soft drink that is placed in 2-liter bottles at the local bottling plant of a large nationally known soft-drink company. The population consists of 2,000 bottles. The bottling plant has informed the inspection division that the standard deviation for 2-liter bottles is 0.05 liter.

(a) Set up a 95% confidence interval estimate of the population average amount of soft drink in each bottle if a random sample of one hundred 2-liter bottles obtained *without* replacement from this bottling plant indicates a sample average of 1.99 liters.

WHAT IF?

(b) Determine the sample size necessary to estimate the population average amount to within ±0.01 liter with 95% confidence.

(c) What are your answers to (a) and (b) if the population consists of 1,000 bottles?

7.55 A stationery store wishes to estimate the average retail value of greeting cards that it has in its inventory, which consists of 300 greeting cards.

(a) Set up a 95% confidence interval estimate of the population average value of all greeting cards that are in its inventory if a random sample of 20 greeting cards selected *without* replacement indicates an average value of $1.67 and a standard deviation of $0.32.

WHAT IF?

(b) What is your answer to (a) if the store has 500 greeting cards in its inventory?

7.6 APPLICATIONS OF CONFIDENCE INTERVAL ESTIMATION IN AUDITING

In our discussion of estimation procedures, we have focused on estimating either the population mean or the population proportion when sampling *with* replacement or *without* replacement from finite populations. One of the areas in business that makes widespread use of statistical sampling for the purposes of estimation is auditing.

> **Auditing** may be defined as the collection and evaluation of evidence about information relating to an economic entity such as a sole business proprietor, a partnership, a corporation, or a government agency in order to determine and report on how well the information obtained corresponds to established criteria.

Exhibit 7.1 lists six advantages of statistical sampling in auditing.

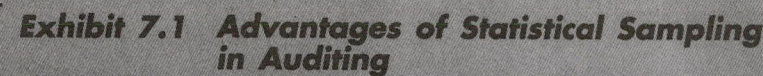

Exhibit 7.1 Advantages of Statistical Sampling in Auditing

✓ **1.** The sample result is objective and defensible. Because the sample size is based on demonstrable statistical principles, the audit is defensible before one's superiors and in a court of law.

✓ **2.** The method provides a way of estimating the sample size in advance on an objective basis.

✓ **3.** The method provides an estimate of the sampling error.

✓ **4.** This approach may turn out to be more accurate in drawing conclusions about the population because of the fact that the examination of large populations may be time consuming and subject to more nonsampling error than a statistical sample.

✓ **5.** Statistical samples may be combined and evaluated even though accomplished by different auditors. This is because there is a scientific basis for the sample so that the samples may be treated as if they have been done by a single auditor.

✓ **6.** Objective evaluation of the results of an audit is possible. The results can be projected with a known sampling error.

Estimating the Population Total Amount

In auditing applications, we are more interested in obtaining estimates of the population **total amount** than the population mean. Equation (7.9) shows how to estimate a population total amount.

Estimating the Population Total Amount

The population total is equal to the population size N times the sample mean $\bar{X}$.

$$Total = N\bar{X} \qquad (7.9)$$

Because the estimated total is $N\bar{X}$, a confidence interval estimate for the population total can be obtained as in equation (7.10).

Confidence Interval Estimate for the Total

$$N\bar{X} \pm N(t_{n-1})\frac{S}{\sqrt{n}}\sqrt{\frac{N-n}{N-1}} \qquad (7.10)$$

To demonstrate the application of the confidence interval estimate for the total, let us return to the Saxon Plumbing Company example. The auditor needs an estimate of the total amount listed on all sales invoices for the warehouse in that month (i.e., the population). Using equations (7.9) and (7.10), with $N = 5,000$, $\bar{X} = \$110.27$, and $S = \$28.95$, for 95% confidence, $t_{99} = 1.9842$, and we have

$$\text{Total} = (5,000)(\$110.27) = \$551,350$$

so that

$$N\overline{X} \pm N(t_{n-1})\frac{S}{\sqrt{n}}\sqrt{\frac{N-n}{N-1}} = 551,350 \pm (5,000)(1.9842)\frac{28.95}{\sqrt{100}}\sqrt{\frac{5,000-100}{5,000-1}}$$

$$= 551,350 \pm 28,721.3(.99)$$

$$= 551,350 \pm 28,434.09$$

$$\$522,915.91 \le \text{population total} \le \$579,784.09$$

Thus we estimate with 95% confidence that the total amount of sales invoices is between $522,915.91 and $579,784.09.

To further illustrate the population total, let us turn to Example 7.8.

Example 7.8 Developing a Confidence Interval Estimate for the Population Total

Suppose that an auditor is faced with a population of 1,000 vouchers and wishes to estimate the total value of the population of vouchers. A sample of 50 vouchers is selected with the following results:

$$\text{Average voucher amount } (\overline{X}) = \$1,076.39$$

$$\text{Standard deviation } (S) = \$273.62$$

Set up a 95% confidence interval estimate of the total amount for the population of vouchers.

SOLUTION

Using equation (7.9), the total is computed as

$$\text{Total} = (1,000)(1,076.39) = \$1,076,390$$

From equation (7.10), a 95% confidence interval estimate of the population total amount is obtained as follows:

$$(1,000)(1,076.39) \pm (1,000)(2.0096)\frac{(273.62)}{\sqrt{50}}\sqrt{\frac{1,000-50}{1,000-1}}$$

$$= 1,076,390 \pm 77,762.9(.975)$$

$$= 1,076,390 \pm 75,818.83$$

$$\$1,000,571.17 \le \text{population total} \le \$1,152,208.83$$

Thus, we estimate with 95% confidence that the total amount of the vouchers is between $1,000,571.17 and $1,152,208.83.

Difference Estimation

Difference estimation is used when an auditor believes that errors exist in a set of items being audited and the auditor wishes to estimate the magnitude of the errors based only on a sample. The following steps are used in difference estimation.

1. Determine the sample size required.
2. Compute the average difference in the sample $(\overline{D})$ by dividing the total difference by the sample size as shown in equation (7.11).

Average Difference

$$\overline{D} = \frac{\sum_{i=1}^{n} D_i}{n}$$

(7.11)

3. Compute the standard deviation of the differences (S_D) as shown in equation (7.12).

Standard Deviation of the Difference

$$S_D = \sqrt{\frac{\sum_{i=1}^{n} (D_i - \overline{D})^2}{n - 1}}$$

(7.12)

Be sure to remember that any item that is not in error has a difference value of 0.

4. Set up a confidence interval estimate of the total difference in the population, as shown in equation (7.13).

Confidence Interval Estimate for the Total Difference

$$N\overline{D} \pm N(t_{n-1}) \frac{S_D}{\sqrt{n}} \sqrt{\frac{N - n}{N - 1}}$$

(7.13)

Recall from the Saxon Plumbing Company example that the auditor wants to obtain a 95% confidence interval estimate of the difference between the actual amounts on the sales invoice and the amounts entered into the accounting system for the warehouse. Suppose that in the sample of 100 sales invoices, there are 12 invoices in which the actual amount on the sales invoice and the amount entered into the accounting system for the warehouse is different. These 12 differences are:

$9.03 $7.47 $17.32 $8.30 $5.21 $10.80 $6.22 $5.63 $4.97 $7.43 $2.99 $4.63

DATA FILE
PLUMBINV.XLS

The other 88 invoices are not in error. The *differences* are each 0. Thus,

$$\overline{D} = \frac{\sum_{i=1}^{n} D_i}{n} = \frac{90}{100} = .90$$

and

$$S_D = \sqrt{\frac{\sum_{i=1}^{n}(D_i - \overline{D})^2}{n-1}}$$

$$S_D = \sqrt{\frac{(9.03 - .9)^2 + (7.47 - .9)^2 + \cdots + (0 - .9)^2}{100 - 1}}$$

$$S_D = 2.752$$

To obtain the confidence interval estimate for the total difference in the population, we use equation (7.13), so that we have

$$(5,000)(.90) \pm 5,000(1.9842)\frac{2.752}{\sqrt{100}}\sqrt{\frac{5,000 - 100}{5,000 - 1}}$$

$$= 4,500 \pm 2,703.09$$

$$\$1,796.91 \le \text{total difference} \le \$7,203.09$$

Thus, the auditor estimates with 95% confidence that the total difference between the sales invoices and the actual amount entered into the accounting system is between \$1,796.91 and \$7,203.09.

In the Saxon Plumbing Company example, all 12 differences are positive because the actual amount on the sales invoice is more than the amount entered into the accounting system. It is possible that the error could have been negative. To illustrate such an occurrence, we present Example 7.9.

Example 7.9 *Difference Estimation*

Returning to Example 7.8 on page 460, suppose that in the sample of 50 vouchers there are 14 vouchers that contain errors. Suppose that the values of the 14 errors are as follows, in which two differences are negative.

DATA FILE
DIFFTEST.XLS

| $ 75.41 | $38.97 | $108.54 | −$37.18 | $62.75 | $118.32 | −$88.84 |
| $127.74 | $55.42 | $ 39.03 | $29.41 | $47.99 | $ 28.73 | $84.05 |

Set up a 95% confidence interval estimate of the total difference in the population of vouchers.

SOLUTION

For these data,

$$\overline{D} = \frac{\sum_{i=1}^{n}D_i}{n} = \frac{690.34}{50} = 13.8068$$

and

$$S_D = \sqrt{\frac{\sum_{i=1}^{n}(D_i - \overline{D})^2}{n-1}}$$

$$S_D = \sqrt{\frac{(75.41 - 13.8068)^2 + (38.97 - 13.8068)^2 + \cdots + (0 - 13.8068)^2}{50 - 1}}$$

$$S_D = 37.427$$

To obtain the confidence interval estimate for the total difference in the population we use equation (7.13) so that we have

$$(1{,}000)(13.8068) \pm 1{,}000(2.0096)\frac{37.427}{\sqrt{50}}\sqrt{\frac{1{,}000-50}{1{,}000-1}}$$

$$= 13{,}806.8 \pm 10{,}372.63$$
$$\$3{,}434.17 \le \text{total difference} \le \$24{,}179.43$$

Thus, we estimate with 95% confidence that the total difference in the population of vouchers is between $3,434.17 and $24,179.43.

CALCULATING THE CONFIDENCE INTERVAL ESTIMATE FOR THE POPULATION TOTAL USING MICROSOFT EXCEL

Overview

◆ *For Quick Results Users*   Use the PHStat add-in to determine the confidence interval estimate for the population total.

◆ *For Developers*   Implement a worksheet that uses the TINV worksheet function and arithmetic formulas to determine the confidence interval estimate for the population total.

The 7-6E1.XLS workbook file contains the confidence interval estimate for the total of all sales invoice amounts for a warehouse for the Saxon Plumbing Company problem of section 7.6.

 7-6E1.XLS

Quick Results Details

To calculate the confidence interval estimate for the population total, use the Confidence Intervals | Estimate for the Population Total choice of the PHStat add-in. As an example, consider the Saxon Plumbing Company problem of section 7.6. To calculate the confidence interval estimate for the total of all sales invoice amounts for a warehouse, do the following:

❶ If the PHStat add-in has not been previously loaded, load the add-in using the instructions of section S4.2.

❷ Select File | New to open a new workbook (or open the existing workbook into which the confidence interval estimate worksheet is to be inserted).

❸ Select PHStat | Confidence Intervals | Estimate for the Population Total.

❹ In the Estimate for the Population Total dialog box (see Figure 7E.13):
   a. Enter 5000 in the Population Size: edit box.
   b. Enter 95 in the Confidence Level: edit box.
   c. Select the Sample Statistics Known option button and enter 100 in the Sample Size: edit box, 110.27 in the Sample Mean: edit box, and 28.95 in the Sample Standard Deviation: edit box.
   d. Enter Confidence Interval Estimate for the Total Amount of All Invoices in the Output Title: edit box.
   e. Click the OK button.

FIGURE 7E.13

PHStat Estimate for the Population Total dialog box

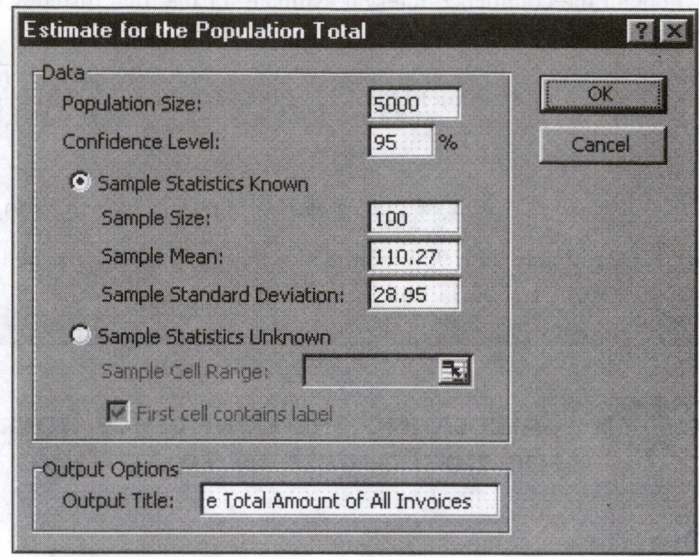

The add-in inserts a worksheet containing calculations for the confidence interval estimate for the total similar to the one shown in Figure 7E.14. (*Note:* For other problems in which the sample mean and sample standard deviation are not known, select the Sample Statistics Unknown option button in step 4c and enter the cell range of the sample data in the Sample Cell Range: edit box, shown dimmed in Figure 7E.13.)

FIGURE 7E.14

Confidence interval estimate for the total of all invoices for the Saxon Plumbing Company problem of section 7.6

	A	B	C	D	E
1	Confidence Interval Estimate for the Total Amount of All Invoices				
2					
3	Population Size	5000			
4	Sample Mean	110.27			
5	Sample Size	100			
6	Sample Standard Deviation	28.95			
7	Confidence Level	95%			
8	Population Total	551350.00			
9	FPC Factor	0.990048503			
10	Standard Error of the Total	14330.95209			
11	Degrees of Freedom	99			
12	t Value	1.984217306			
13	Interval Half Width	28435.72			
14	Interval Lower Limit	522914.28			
15	Interval Upper Limit	579785.72			

Developer Details

We can use the TINV worksheet function (see section 7.2E) as the basis for calculating the confidence interval estimate for the population total. Table 7E.10 presents a Confidence interval estimate sheet design for calculating the total amount of all sales invoices for the Saxon Plumbing Company problem of section 7.6. This design takes the *t*-value returned by the TINV function and multiplies it by the standard error of the total to produce half the width of the confidence interval. This half width is then subtracted from and added to the sample total to calculate the lower and upper limits of the confidence interval. (The design calculates the finite population correction factor as well.) To implement the Table 7E.10 design, do the following:

❶ Select File | New to open a new workbook (or open the existing workbook into which the confidence interval estimate worksheet is to be inserted).

❷ Select an unused worksheet (or select Insert | Worksheet if there are none) and rename the sheet Confidence.

❸ Enter the title and labels for column A as shown in Table 7E.10.

❹ Enter the population size, sample mean, sample size, sample standard deviation, and confidence level in the cell range B3:B7. Enter 5000 in cell B3, 110.27 in cell B4, 100 in cell B5, 28.95 in cell B6, and 0.95 in cell B7.

❺ Select cell B7 and click the Percent style button on the formatting toolbar (see section S3.11) to format the decimal value 0.95 as 95%.

❻ Enter the formulas for the cell range B8:B15 as shown in Table 7E.10.

The completed worksheet will be similar to the one shown in Figure 7E.14. Had the sample mean and sample standard deviation been unknown, we could have entered formulas using the AVERAGE and STDEV functions in cells B4 and B6 to calculate these statistics.

Table 7E.10 Confidence interval estimate sheet design for the total of all invoices for the Saxon Plumbing Company problem of section 7.6

	A	B
1	Confidence Interval Estimate for the Total Amount of All Invoices	
2		
3	Population Size	xxx
4	Sample Mean	xxx
5	Sample Size	xxx
6	Sample Standard Deviation	xxx
7	Confidence Level	.xx
8	Population Total	=B3*B4
9	FPC Factor	=SQRT((B3-B5)/(B3-1))
10	Standard Error of the Total	=(B3*B6*B9)/SQRT(B5)
11	Degrees of Freedom	=B5-1
12	t Value	=TINV(1-B7,B11)
13	Interval Half Width	=B12*B10
14	Interval Lower Limit	=B8-B13
15	Interval Upper Limit	=B8+B13

 **7.6E.2 CALCULATING THE CONFIDENCE INTERVAL ESTIMATE FOR THE TOTAL DIFFERENCE USING MICROSOFT EXCEL**

Overview

◆ *For Quick Results Users* Use the PHStat add-in to determine the confidence interval estimate for the total difference.

◆ *For Developers* Implement a worksheet that uses the TINV worksheet function and arithmetic formulas to determine the confidence interval estimate for the population total difference.

7-6E2.XLS

The 7-6E2.XLS workbook file contains the confidence interval estimate for the total difference of all sales invoice amounts for a warehouse for the Saxon Plumbing Company problem of section 7.6.

Quick Results Details

To calculate the confidence interval estimate for the total difference, use the Confidence Intervals | Estimate for Total Difference choice of the PHStat add-in. As an example, consider the Saxon Plumbing Company problem of section 7.6. To calculate the confidence interval estimate for the total difference between the invoice amounts and the amounts entered in the Saxon accounting system, do the following:

❶ If the PHStat add-in has not been previously loaded, load the add-in using the instructions of section S4.2.

❷ Open the Starting Point for Section 7.6E2 workbook (STARTING POINT 7-6E2.XLS) and click the Data sheet tab. Verify that the differences for the Saxon Plumbing Company problem of section 7.6 appear in column A.

❸ Select PHStat | Confidence Intervals | Estimate for the Total Difference.

❹ In the Estimate for the Total Difference dialog box (see Figure 7E.15):

FIGURE 7E.15
PHStat Estimate for the Total Difference dialog box

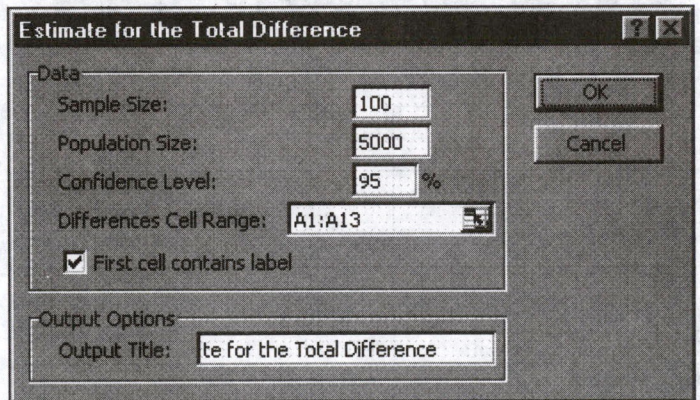

a. Enter 100 in the Sample Size: edit box.
b. Enter 5000 in the Population Size: edit box.
c. Enter 95 in the Confidence Level: edit box.
d. Enter A1:A13 in the Differences Cell Range: edit box and select the First cell contains label check box.
e. Enter Confidence Interval Estimate for the Total Difference in the Output Title: edit box.
f. Click the OK button.

The add-in adds a column of formulas on the Data sheet and inserts a worksheet containing calculations for the confidence interval estimate for the total difference similar to the one shown in Figure 7E.16.

	A	B	C	D	E
1	**Confidence Interval Estimate for the Total Difference**				
2					
3	**Population Size**	**5000**			
4	**Sample Size**	**100**			
5	**Confidence Level**	**95%**			
6	Sum of Differences	90		Calculations Area	
7	Average Difference in Sample	0.9		For standard deviation of differences:	
8	Total Difference	4500		Number of Differences Not = 0	12
9	Standard Deviation of Differences	2.751797		Number of Differences = 0	88
10	FPC Factor	0.990049		SS for Differences Not = 0	678.3864
11	Standard Error of the Mean	1362.206		SS for Differences = 0	71.28
12	Degrees of Freedom	99		Sum of Squares	749.6664
13	*t* Value	1.984217		Variance of Differences	7.572388
14	Interval Half Width	2702.913			
15	**Interval Lower Limit**	**1797.09**			
16	**Interval Upper Limit**	**7202.91**			

FIGURE 7E.16 Confidence interval estimate for the total difference between the invoice amounts and the amounts entered in the accounting system for the Saxon Plumbing Company problem of section 7.6

Developer Details

To simplify intermediate calculations, we can use a two-sheet design for calculating the confidence interval estimate for the total difference. Tables 7E.11 and 7E.12 on page 468 present such a design for calculating the total difference between the invoice amounts and the amounts entered in the accounting system for the Saxon Plumbing Company problem of section 7.6. The design assumes that the nonzero differences are entered in column A of a sheet named Data.

Using this two-sheet design permits us to break up the calculation of the standard deviation of the difference into several steps. First the values for the squares of the differences for those invoices with nonzero errors are calculated on the Data sheet. Then, on the Confidence Interval sheet, the sum of these squares and the sum of squares for those invoices that are not in error (difference = 0) are combined in cell E12 to form the numerator of the expression beneath the square root sign in equation (7.12). This cell E12 value is then divided by the degrees of freedom (cell B12) to get the variance of the difference. Finally, the formula in cell B9 takes the square root of this variance to calculate the standard deviation of the difference.

This two-sheet design also takes the *t*-value returned by the TINV function and multiplies it by the standard error of the difference to produce half the width of the confidence interval. This half width is then subtracted from and added to the sample total difference to calculate the lower and upper limits of the confidence interval. To implement the two-sheet design of Tables 7E.11 and 7E.12 design, do the following:

❶ Open the Starting Point for Section 7.6E2 workbook (STARTING POINT 7-6E2.XLS) and click the Data sheet tab. Verify that the differences for the Saxon Plumbing Company problem of section 7.6 appear in column A as shown in Table 7E.11.

❷ Select an unused worksheet (or select Insert | Worksheet if there are none) and rename the sheet Confidence.

❸ Click the Data sheet tab. Enter the label for column B for this sheet as shown in Table 7E.11.

❹ Enter the formula for cell B2 as shown in Table 7E.11 and copy it down through row 13.

⑤ Click the Confidence sheet tab.

⑥ Enter the title and labels for columns A and D for this sheet as shown in Table 7E.12.

⑦ Enter the population size, sample size, and confidence level in the cell range B3:B5. Enter 5000 in cell B3, 100 in cell B4, and 0.95 in cell B5.

⑧ Select cell B5 and click the Percent style button on the formatting toolbar (see section S3.11) to format the decimal value 0.95 as 95%.

⑨ Enter the formulas for the cell range B6:B16 and E8:E13 as shown in Table 7E.12.

The completed worksheet will be similar to the one shown in Figure 7E.16.

Table 7E.11 Data sheet design necessary for confidence interval estimate design of Table 7E.12

	A	B
1	Differences	(D-DBar)^2
2	9.03	=(A2-Confidence!$B$7)^2
3	7.47	=(A3-Confidence!$B$7)^2
4	17.32	=(A4-Confidence!$B$7)^2
5	8.30	=(A5-Confidence!$B$7)^2
6	5.21	=(A6-Confidence!$B$7)^2
7	10.80	=(A7-Confidence!$B$7)^2
8	6.22	=(A8-Confidence!$B$7)^2
9	5.63	=(A9-Confidence!$B$7)^2
10	4.97	=(A10-Confidence!$B$7)^2
11	7.43	=(A11-Confidence!$B$7)^2
12	2.99	=(A12-Confidence!$B$7)^2
13	4.63	=(A13-Confidence!$B$7)^2

Table 7E.12 Confidence interval estimate sheet design for the total difference between invoice amounts and amounts entered in the accounting system for the Saxon Plumbing Company problem of section 7.6

	A	B	C	D	E
1	Confidence Interval Estimation for the Total Difference				
2					
3	Population Size	xxx			
4	Sample Size	xxx			
5	Confidence Level	.xx			
6	Sum of Differences	=SUM(Data!A:A)		Calculations Area	
7	Average Difference in Sample	=B6/B4		For standard deviation of differences:	
8	Total Difference	=B3*B7		Number of Differences Not = 0	=COUNT(Data!A:A)
9	Standard Deviation of Differences	=SQRT(E13)		Number of Differences = 0	=B4-E8
10	FPC Factor	=SQRT((B3-B4)/(B3-1))		SS for Differences Not = 0	=SUM(Data!B:B)
11	Standard Error of the Total Difference	=(B3*B9*B10)/SQRT(B4)		SS for Differences = 0	=E9*(-B7)^2
12	Degrees of Freedom	=B4-1		Sum of Squares	=E10+E11
13	t Value	=TINV(1-B5,B12)		Variance of Differences	=E12/B12
14	Interval Half Width	=B13*B11			
15	Interval Lower Limit	=B8-B14			
16	Interval Upper Limit	=B8+B14			

Problems for Section 7.6

Learning the Basics

• **7.56** Suppose that a sample of 25 is selected from a population of 500 items. The sample mean is 25.7 and the sample standard deviation is 7.8. Set up a 99% confidence interval estimate of the population total.

7.57 Suppose that a sample of 200 items is selected from a population of 10,000 items. Ten items are found to have errors of the following amounts:

13.76   42.87   34.65   11.09   14.54   22.87   25.52   9.81   10.03   15.49

Set up a 95% confidence interval estimate of the total difference in the population.

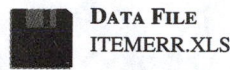

DATA FILE
ITEMERR.XLS

Applying the Concepts

7.58 The manager of a branch of a local savings bank that has 2,000 depositors wants to estimate the total amount held in passbook savings accounts by depositors at the bank. A random sample of 30 depositors is selected, and the results indicate a sample average of $4,750 and a sample standard deviation of $1,200. Set up a 95% confidence interval estimate of the total amount held in passbook savings accounts in the branch.

• **7.59** A stationery store wants to estimate the total retail value of greeting cards that it has in its inventory, which consists of 300 greeting cards. Set up a 95% confidence interval estimate of the population total value of all greeting cards that are in its inventory if a random sample of 20 greeting cards indicates an average value of $1.67 and a standard deviation of $0.32.

7.60 The personnel department of a large corporation employing 3,000 workers wishes to estimate the family dental expenses of its employees to determine the feasibility of providing a dental insurance plan. A random sample of 10 employees reveals the following family dental expenses (in dollars) for the preceding year:

110   362   246   85   510   208   173   425   316   179

Set up a 90% confidence interval estimate of the total family dental expenses for all employees in the preceding year.

DATA FILE
DENTAL.XLS

7.61 A branch of a chain of large electronics stores is conducting an end-of-month inventory of the merchandise in stock. It was determined that there are 1,546 items in inventory at that time. A sample size of 50 items was randomly selected and an audit was conducted with the following results:

Value of Merchandise
$\overline{X} = \$252.28 \quad S = \$93.67$

Set up a 95% confidence interval estimate of the total estimated value of the merchandise in inventory at the end of the month.

• **7.62** A customer in the wholesale garment trade is often entitled to a discount for a cash payment for goods. The amount of discount varies by vendor. A sample of 150 items selected from a population of 4,000 invoices at the end of a period of time revealed that in 13 cases the customer failed to take the discount to which he or she was entitled. The amounts of the 13 discounts that were not taken were as follows:

$6.45 15.32 97.36 230.63 104.18 84.92 132.76 66.12 26.55 129.43 88.32 47.81 89.01

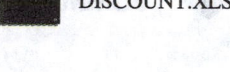

DATA FILE
DISCOUNT.XLS

Set up a 99% confidence interval estimate of the population total amount of discount not taken.

7.63 Econe Dresses is a small company manufacturing women's dresses for sale to specialty stores. There are 1,200 inventory items in which the historical cost is recorded on a first in, first out (FIFO) basis. In the past, approximately 15% of the inventory items were incorrectly priced. However, any misstatements were usually not significant. A sample of 120 items was selected and the historical cost of each item was compared with the audited value. The results indicated that 15 items differed in their historical costs and audited values. These differences were as follows:

SAMPLE NUMBER	HISTORICAL COST ($)	AUDITED VALUE ($)	SAMPLE NUMBER	HISTORICAL COST ($)	AUDITED VALUE ($)
5	261	240	60	21	210
9	87	105	73	140	152
17	201	276	86	129	112
18	121	110	95	340	216
28	315	298	96	341	402
35	411	356	107	135	97
43	249	211	119	228	220
51	216	305			

DATA FILE
FIFO.XLS

Set up a 95% confidence interval estimate of the total population difference in the historical cost and audited value.

7.7 CONFIDENCE INTERVAL ESTIMATION AND ETHICAL ISSUES

Ethical issues relating to the selection of samples and the inferences that accompany them from sample surveys can arise in several ways. The major ethical issue relates to whether or not confidence interval estimates are provided along with the point estimates of the sample statistics obtained from a survey. To indicate a point estimate of a sample statistic without also including the confidence interval limits (typically set at 95%), the sample size used, and an interpretation of the meaning of the confidence interval in terms that a layperson can understand raises ethical issues because of their omission. Failure to include a confidence interval estimate might mislead the user of the survey results into thinking that the point estimate obtained from the sample is all that is needed to predict the population characteristics with certainty. Thus, it is important that the interval estimate be indicated in a prominent place in any written communication along with a simple explanation of the meaning of the confidence interval. In addition, the size of the sample should be highlighted so that the reader clearly understands the magnitude of the survey that has been undertaken.

One of the most common areas where ethical issues concerning estimation from sample surveys occurs is in the publication of the results of political polls. All too often, the results of the polls are highlighted on page 1 of the newspaper and the sampling error involved along with the methodology used is printed on the page where the article is typically continued (often in the middle of the newspaper). To ensure an ethical interpretation of statistical results, the confidence levels, sample size, and confidence limits should be made available for all surveys.

SUMMARY

As observed in the summary chart on page 471, in this chapter we have developed a confidence interval approach for estimating the characteristics of a population. In addition, we have investigated how we can determine the necessary sample size for a survey.

Now that we have made estimates of population characteristics such as the mean, proportion, total, and difference using confidence intervals, in the next four chapters we turn to a hypothesis-testing approach in which we make decisions about population parameters.

454

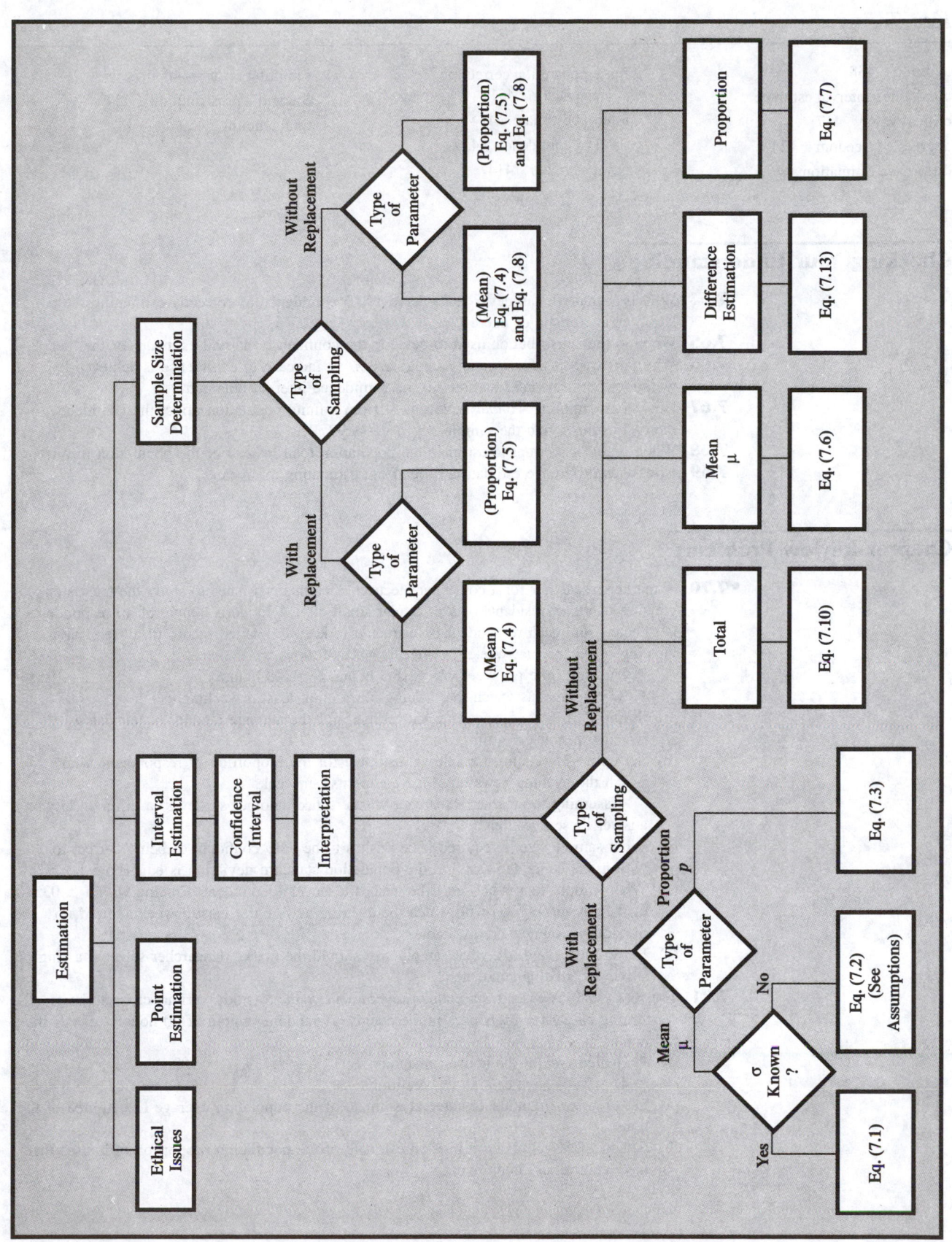

Key Terms

auditing 458
confidence interval estimate 417
critical value 417
degrees of freedom 424
difference estimation 460

finite population correction (fpc) factor 451
interval estimate 414
level of confidence 417
point estimate 414

sampling error 440
Student's t distribution 424
total amount 459

Checking Your Understanding

7.64 Why is it that we can never really have 100% confidence of correctly estimating the population characteristic of interest?

7.65 When is the t distribution used to develop the confidence interval estimate for the mean?

7.66 Why is it true that for a given sample size n, an increase in confidence is achieved by widening (and making less precise) the confidence interval obtained?

7.67 How does sampling *without* replacement from a finite population affect the confidence interval estimate and the sample size necessary?

7.68 When would you want to estimate the population total instead of the population mean?

7.69 How does difference estimation differ from estimating the mean?

Chapter Review Problems

• **7.70** A market researcher for a consumer electronics company wishes to study the television viewing habits of residents of a particular small city. A random sample of 40 respondents is selected, and each respondent is instructed to keep a detailed record of all television viewing in a particular week. The results are as follows:
  • Viewing time per week: $\overline{X} = 15.3$ hours, $S = 3.8$ hours.
  • 27 respondents watch the evening news on at least 3 weeknights.
  (a) Set up a 95% confidence interval estimate for the average amount of television watched per week in this city.
  (b) Set up a 95% confidence interval estimate for the proportion of respondents who watch the evening news on at least 3 nights per week.
      Assuming the market researcher wants to take another survey in a different city, answer these questions:
  (c) What sample size is required if he wishes to be 95% confident of being correct to within ±2 hours and assumes the population standard deviation is equal to 5 hours?
  (d) What sample size is needed if he wishes to be 95% confident of being within ±.035 of the true proportion who watch the evening news on at least 3 weeknights if no previous estimate were available?
  (e) Based on (c) and (d), what sample size should the market researcher select if a single survey was being conducted?

7.71 The real estate assessor for a county government wants to study various characteristics concerning single-family houses in the county. A random sample of 70 houses reveals the following:
  • Heated area of the house (in square feet): $\overline{X} = 1,759$, $S = 380$.
  • 42 houses have central air conditioning.
  (a) Set up a 99% confidence interval estimate of the population average heated area of the house.
  (b) Set up a 95% confidence interval estimate of the population proportion of houses that have central air conditioning.

7.72 The personnel director of a large corporation wishes to study absenteeism among clerical workers at the corporation's central office during the year. A random sample of 25 clerical workers reveals the following:

- Absenteeism: $\overline{X} = 9.7$ days, $S = 4.0$ days.
- 12 clerical workers were absent more than 10 days.

(a) Set up a 95% confidence interval estimate of the average number of absences for clerical workers last year.

(b) Set up a 95% confidence interval estimate of the population proportion of clerical workers absent more than 10 days last year.

Assuming that the personnel director also wishes to take a survey in a branch office, answer these questions:

(c) What sample size is needed if the director wishes to be 95% confident of being correct to within ± 1.5 days and the population standard deviation is assumed to be 4.5 days?

(d) What sample size is needed if the director wishes to be 90% confident of being correct to within $\pm .075$ of the true proportion of workers who are absent more than 10 days if no previous estimate is available?

(e) Based on (c) and (d), what sample size should the personnel director select if a single survey is being conducted?

7.73 The market research director for Dotty's department store wants to study women's spending per year on cosmetics. A survey is to be sent to a sample of the store's credit card holders to determine

- the average yearly amount that women spend on cosmetics.
- the population proportion of women who purchase their cosmetics primarily from Dotty's department store.

(a) If the market researcher wants to have 99% confidence of estimating the true population average to within $\pm \$5$ and the standard deviation is assumed to be \$18 (based on previous surveys), what sample size is needed?

(b) If the market researcher wishes to have 90% confidence of estimating the true proportion to within $\pm .045$, what sample size is needed?

(c) Based on the results in (a) and (b), how many of the store's credit card holders should be sampled? Explain.

• 7.74 The branch manager of an outlet of a large, nationwide bookstore chain wants to study characteristics of customers of her store, which is located near the campus of a large state university. In particular, she decides to focus on two variables: the amount of money spent by customers and whether the customers would consider purchasing educational videotapes relating to specific courses such as statistics, accounting, or calculus or graduate preparation exams such as GMAT, GRE, or LSAT. The results from a sample of 70 customers are as follows:

- Amount spent: $\overline{X} = \$28.52$, $S = \$11.39$.
- 28 customers stated that they would consider purchasing educational videotapes.

(a) Set up a 95% confidence interval estimate of the population average amount spent in the bookstore.

(b) Set up a 90% confidence interval estimate of the population proportion of customers who would consider purchasing educational videotapes.

Assuming the branch manager of another store from a different bookstore chain wishes to conduct a similar survey in his store (which is located near another university), answer these questions:

(c) If he wants to have 95% confidence of estimating the true population average amount spent in his store to within $\pm \$2$ and the standard deviation is assumed to be \$10, what sample size is needed?

(d) If he wants to have 90% confidence of estimating the true proportion of shoppers who would consider the purchase of videotapes to within $\pm .04$, what sample size is needed?

(e) Based on your answers to (c) and (d), what size sample should be taken?

7.75 The branch manager of an outlet (store 1) of a large, nationwide chain of pet supply stores wants to study characteristics of customers of her store. In particular, she decides to focus on two variables: the amount of money spent by customers and whether the customers own only one dog, only one cat, or more than one dog and/or one cat. The results from a sample of 70 customers are as follows:

- Amount of money spent: $\overline{X} = \$21.34$, $S = \$9.22$.
- 37 customers own only a dog.
- 26 customers own only a cat.
- 7 customers own more than one dog and/or one cat.

(a) Set up a 95% confidence interval estimate of the population average amount spent in the pet supply store.

(b) Set up a 90% confidence interval estimate of the proportion of customers who own only a cat.

Assuming the branch manager of another outlet (store 2) wishes to conduct a similar survey in his store (and does not have any access to the information generated by the manager of store 1), answer these questions:

(c) If he wants to have 95% confidence of estimating the true population average amount spent in his store to within $\pm\$1.50$ and the standard deviation is assumed to be $10, what sample size is needed?

(d) If he wants to have 90% confidence of estimating the true proportion of customers who own only a cat to within $\pm.045$, what sample size is needed?

(e) Based on your answers to (c) and (d), what size sample should be taken?

7.76 The owner of a restaurant serving continental food wants to study characteristics of customers of his restaurant. In particular, he decides to focus on two variables: the amount of money spent by customers and whether or not customers order dessert. The results from a sample of 60 customers are as follows:

- Amount spent: $\overline{X} = \$38.54$, $S = \$7.26$.
- 18 customers purchased dessert.

(a) Set up a 95% confidence interval estimate of the population average amount spent per customer in the restaurant.

(b) Set up a 90% confidence interval estimate of the population proportion of customers who purchase dessert.

Assuming the owner of a competing restaurant wishes to conduct a similar survey in her restaurant (and does not have access to the information obtained by the owner of the first restaurant), answer these questions:

(c) If she wants to have 95% confidence of estimating the true population average amount spent in her restaurant to within $\pm\$1.50$ and the standard deviation is assumed to be $8, what sample size is needed?

(d) If she wants to have 90% confidence of estimating the true proportion of customers who purchase dessert to within $\pm.04$, what sample size is needed?

(e) Based on your answers to (c) and (d), what size sample should be taken?

7.77 A representative for a large chain of hardware stores is interested in testing the product claims of a manufacturer of a product called "Ice Melt," which is reported to melt snow and ice at temperatures as low as 0° Fahrenheit. A shipment of 400 five-pound bags is purchased by the chain for distribution. The representative wants to know with 95% confidence, within $\pm.05$, what proportion of bags of Ice Melt perform the job as claimed by the manufacturer.

(a) How many bags does the representative need to test? What assumption should be made concerning the true proportion in the population? (This is called *destructive testing;* that is, the product being tested is destroyed by the test and is then unavailable to be sold.)

The representative tests 50 bags, and 42 do the job as claimed.

(b) Construct a 95% confidence interval estimate for the population proportion that will do the job as claimed.

(c) How can the representative use the results of (b) to determine whether to sell the Ice Melt product?

7.78 An audit test to establish the percentage of occurrence of failure to follow a specific internal control procedure is to be undertaken. Suppose a sample of 50 from a population of 1,000 items is selected and it is determined that in seven instances the internal control procedure was not followed.

(a) Set up a 90% confidence interval estimate of the population proportion of items in which the internal control procedure was not followed.

(b) Suppose that, in an effort to objectively determine the sample size, the auditor decides that she would like to have 95% confidence of correctly estimating the population proportion to within $\pm.04$ (4%) of the true population value. She also assumes from past experience that the maximum error rate is 10%. What sample size needs to be selected?

7.79 In a test of an attribute from a population of 15,000 items, if the maximum rate of occurrence is not expected to exceed 5% and a sample precision of $\pm 2\%$ is deemed satisfactory,

(a) what sample size is required for 95% confidence if sampling *without* replacement?

(b) what sample size is required for 99% confidence if sampling *without* replacement?

(c) Compare the difference in your answers to (a) and (b).

(d) If you desire a precision of $\pm 3\%$, what sample size is required for 95% confidence?

(e) What sample size is required in (a) if the population size is 5,000?

7.80 An auditor for a government agency is assigned the task of evaluating reimbursement for office visits to doctors paid by Medicare. The audit is to be conducted for all Medicare payments in a particular geographic zip code during a certain month. A population of 25,056 visits exists in this area for this month. The auditor primarily wants to estimate the total amount paid by Medicare in this area for the month. He wants a 95% confidence of estimating the true population average to within $\pm\$5$. On the basis of past experience, he estimates the standard deviation to be \$30.

(a) What sample size should be selected?

Using the sample size selected in (a), an audit is conducted. It is discovered that in 12 of the office visits, an incorrect amount of reimbursement was provided.

Amount of Reimbursement
$\overline{X} = \$93.70 \qquad S = \$34.55$

For the 12 office visits in which incorrect reimbursement was provided, the differences between the amount reimbursed and the amount that the auditor determined should have been reimbursed were:

$17 \quad \$25 \quad \$14 \quad -\$10 \quad \$20 \quad \$40 \quad \$35 \quad \$30 \quad \$28 \quad \$22 \quad \$15 \quad \$5

 **DATA FILE**
MEDICARE.XLS

(b) Set up a 90% confidence interval estimate of the population proportion of reimbursements that contain errors.

(c) Set up 95% confidence interval estimates of the
(1) average reimbursement per office visit.
(2) the total amount of reimbursements for this geographic area in this month.

(d) Set up a 95% confidence interval estimate of the total difference between the amount reimbursed and the amount that the auditor determined should have been reimbursed.

7.81 A large computer store is conducting an end-of-month inventory of the computers in stock. It is determined that there were 258 computers in inventory at that time. An auditor for the store wants to estimate the average value of the computers in inventory at that time. She wants to have 99% confidence that her estimate of the average value is correct to within $\pm\$200$. On the basis of past experience, she estimates that the standard deviation of the value of a computer is \$400.

(a) What sample size should be selected?

Using the sample size selected in (a), an audit was conducted with the following results:

Value of Computers
$$\overline{X} = \$3{,}054.13 \quad S = \$384.62$$

(b) Set up a 99% confidence interval estimate of the total estimated value of the computers in inventory at the end of the month.

TEAM PROJECT

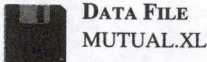
TP7.1 Refer to TP2.1 on page 144. Set up all appropriate estimates of population characteristics of currently traded domestic general stock funds. Include these estimates in any written and oral presentation to be made to the vice president for research at the financial investment service.
Note: Additional team projects can be found on the World Wide Web site for this text at

http://www.prenhall.com/levine

These team projects deal with characteristics of 80 universities and colleges (see the UNIV&COL file) and the features of 89 automobiles (see the AUTO96 file).

THE SPRINGVILLE HERALD CASE

The marketing department team met collectively to study the results of the new subscriptions forecasting model. Paul Kravitz, the market research director, asked the group to consider how new subscriptions could be increased and how those customers who agreed to trial subscription plans could be kept as subscribers after the trial period ended. Lauren Alfonso, a district sales manager, suggested that a survey be undertaken to determine various characteristics of readers of the newspaper who were not home-delivery subscribers. After a great deal of discussion by the group, a questionnaire was developed as follows.

1. Do you or a member of your household ever purchase the *Springville Herald*?
   (1) Yes (2) No
   [If the respondent answers no, the interview is terminated.]
2. Do you receive the *Springville Herald* via home delivery?
   (1) Yes (2) No
   [If no, skip to question 4.]
3. Do you receive the *Springville Herald*:
   (1) Monday–Saturday (2) Sunday only
   (3) Every day
   [Skip to question 9.]

4. How often during the Monday–Saturday period do you purchase the *Springville Herald*?
   (1) Every day (2) Most days
   (3) Occasionally
5. How often do you purchase the *Springville Herald* on Sundays?
   (1) Every Sunday
   (2) 2–3 Sundays a month
   (3) Once a month
6. Where are you most likely to purchase the *Springville Herald*?
   (1) Convenience store/delicatessen
   (2) Stationery/candy store
   (3) Vending machine (4) Supermarket
   (5) Other _____
7. Would you consider subscribing to the *Springville Herald* for a trial period if a discount was offered?
   (1) Yes (2) No
   [If no, skip to question 9.]
8. The *Springville Herald* currently costs 50 cents on Monday–Saturday and $1.50 on Sunday for a total of $4.50 per week. How much would you be willing to pay per week to obtain home delivery for a 90-day trial period? _____
9. Do you read a daily newspaper other than the *Springville Herald*?
   (1) Yes (2) No

10. As an incentive to long-term subscribers, the newspaper is considering the possibility of offering all subscribers who pay for 6 months of home delivery in advance (about $100) a card that would provide discounts at a list of restaurants in Springville. Would you want to obtain such a card under the terms of this offer? (1) Yes   (2) No

A random sample of 500 households located in Springville was selected for a telephone survey that was to be conducted using the *random-digit dialing* method. In this approach, used to reach households that have unlisted telephone numbers, the last four digits of a telephone number are randomly selected to go with the telephone exchange that consists of the first three numbers. Only those exchanges that are contained in Springville are used. (For example, if the exchanges 676, 759, and 923 are available in Springville, the exchange would first be randomly selected among these three, and then the last four digits would be randomly selected to produce the telephone number to be dialed.)

Of the 500 households selected, 94 either refused to participate, could not be contacted after repeated attempts, or represented telephone numbers that were not in service or belonged to nonresidences (remember that random digit dialing was used). The summary results were as follows:

HOUSEHOLDS THAT PURCHASE THE *SPRINGVILLE HERALD*	FREQUENCY
Yes	352
No	54

HOUSEHOLDS THAT RECEIVE THE *SPRINGVILLE HERALD* VIA HOME DELIVERY	FREQUENCY
Yes	136
No	216

TYPE OF SUBSCRIPTION FOR HOME DELIVERY	FREQUENCY
Monday–Saturday	18
Sunday only	25
Every day	93

PURCHASE BEHAVIOR OF NONSUBSCRIBERS— SATURDAY EDITIONS	FREQUENCY
Every day	78
Most days	95
Occasionally	43

PURCHASE BEHAVIOR OF NONSUBSCRIBERS— SUNDAY EDITION	FREQUENCY
Every Sunday	138
2–3 Sundays a month	54
Once a month	24

LOCATION OF PURCHASE	FREQUENCY
Convenience store/delicatessen	74
Stationery/candy store	95
Vending machine	21
Supermarket	13
Other	13

CONSIDER TRIAL PERIOD SUBSCRIPTION WITH DISCOUNT	FREQUENCY
Yes	46
No	170

How much would you be willing to pay per week to obtain home delivery for a 90-day trial period?

$4.15 3.60 4.10 3.60 3.60 3.60 4.40
 3.15 4.00 3.75 4.00 3.25 3.75 3.30
 3.75 3.65 4.00 4.10 3.90 3.50 3.75
 3.00 3.40 4.00 3.80 3.50 4.10 4.25
 3.50 3.90 3.95 4.30 4.20 3.50 3.75
 3.30 3.85 3.20 4.40 3.80 3.40 3.50
 2.85 3.75 3.80 3.90

READ DAILY NEWSPAPER OTHER THAN THE *SPRINGVILLE HERALD*	FREQUENCY
Yes	138
No	214

WOULD PREPAY 6 MONTHS TO RECEIVE RESTAURANT DISCOUNT CARD	FREQUENCY
Yes	66
No	286

DATA FILE
SH7.XLS

Exercises

7.1 Some members of the marketing department team raised questions about the use of the random-digit dialing method used in the survey. Prepare a short report for the team that

(a) indicates the advantages and disadvantages of conducting the survey using a random-digit dialing telephone survey.

(b) suggests an alternative approach for conducting the survey. What would be the advantages and disadvantages of conducting the survey based on this alternative approach?

7.2 Analyze the results of the survey of households in Springville. Write a report for the marketing department team that discusses the marketing implication of the survey results for the *Springville Herald*. In addition to the written report, prepare a summary that can be presented orally in less than 10 minutes.

References

1. Cochran, W. G., *Sampling Techniques*, 3d ed. (New York: Wiley, 1977).
2. Fisher, R. A., and F. Yates, *Statistical Tables for Biological, Agricultural and Medical Research*, 5th ed. (Edinburgh: Oliver & Boyd, 1957).
3. Kirk, R. E., ed., *Statistical Issues: A Reader for the Behavioral Sciences* (Belmont, CA: Wadsworth, 1972).
4. Larsen, R. L., and M. L. Marx, *An Introduction to Mathematical Statistics and Its Applications*, 2d ed. (Englewood Cliffs, NJ: Prentice Hall, 1986).
5. *Microsoft Excel 97* (Redmond, WA: Microsoft Corporation, 1997).
6. Snedecor, G. W., and W. G. Cochran, *Statistical Methods*, 7th ed. (Ames, IA: Iowa State University Press, 1980).

Chapter 6

Student Solutions Manual

7.2 $\bar{X} \pm Z \cdot \dfrac{\sigma}{\sqrt{n}} = 125 \pm 2.58 \cdot \dfrac{24}{\sqrt{36}}$ $114.68 \leq \mu \leq 135.32$

7.4   Since the results of only one sample are used to indicate whether something has gone wrong in the production process, the manufacturer can never know with 100% certainty that the specific interval obtained from the sample includes the true population mean. In order to have every possible interval estimate of the true mean, the entire population (sample size N) would have to be selected.

7.6   Approximately 5% of the intervals will not include the true population mean somewhere in the interval. Since the true population mean is not known, we do not know for certain whether it is in the one interval we have developed, between 10.99408 and 11.00192 inches.

7.8   (a)

	A	B
1	Light Bulbs	
2		
3	Population Standard Deviation	100
4	Sample Mean	350
5	Sample Size	64
6	Confidence Level	95%
7	Standard Error of the Mean	12.5
8	Z Value	-1.95996108
9	Interval Half Width	24.49951353
10	Interval Lower Limit	325.5004865
11	Interval Upper Limit	374.4995135

      (b)   No. The manufacturer cannot support a claim that the bulbs last an average 400 hours. Based on the data from the sample, a mean of 400 hours would represent a distance of 4 standard deviations above the sample mean of 350 hours.

      (c)   No. Since σ is known and $n = 64$, from the central limit theorem, we may assume that the sampling distribution of $\bar{X}$ is approximately normal.

      (d)   An individual value of 320 is only 0.30 standard deviations below the sample mean of 350. The confidence interval represents bounds on the estimate of a sample of 64, not an individual value.

      (e)   The confidence interval is narrower based on a process standard deviation of 80 hours rather than the original assumption of 100 hours.

      (a) $\bar{X} \pm Z \cdot \dfrac{\sigma}{\sqrt{n}} = 350 \pm 1.96 \cdot \dfrac{80}{\sqrt{64}}$ $330.4 \leq \mu \leq 369.6$

      (b)   Based on the smaller standard deviation, a mean of 400 hours would represent a distance of 5 standard deviations above the sample mean of 350 hours. No, the manufacturer cannot support a claim that the bulbs last an average of 400 hours.

7.10 (a) $t_9 = 2.2622$
(b) $t_9 = 3.2498$
(c) $t_{31} = 2.0395$
(d) $t_{64} = 1.9977$
(e) $t_{15} = 1.7531$

7.12 $\bar{X} \pm t \cdot \dfrac{s}{\sqrt{n}} = 50 \pm 2.9467 \cdot \dfrac{15}{\sqrt{16}}$ $38.9499 \leq \mu \leq 61.0501$

7.14 Original data: $5.8571 \pm 2.4469 \cdot \dfrac{6.4660}{\sqrt{7}}$ $-0.1229 \leq \mu \leq 11.8371$

Altered data: $4.00 \pm 2.4469 \cdot \dfrac{2.1602}{\sqrt{7}}$ $2.0022 \leq \mu \leq 5.9978$

The presence of an outlier in the original data increases the value of the sample mean and greatly inflates the sample standard deviation.

7.16 (a) $\bar{X} \pm t \cdot \dfrac{s}{\sqrt{n}} = 4750 \pm 2.0452 \cdot \dfrac{1200}{\sqrt{30}}$ $\$4,301.96 \leq \mu \leq \$5,198.04$

(b) It is not unusual. An individual account of $4,000 is only 0.625 standard deviation below the sample mean of $4,750.

7.18 (a)

Dental Expenses	
Sample Standard Deviation	138.8045789
Sample Mean	261.4
Sample Size	10
Confidence Level	90%
Standard Error of the Mean	43.89386188
Degrees of Freedom	9
t Value	1.833113856
Interval Half Width	80.46244642
Interval Lower Limit	180.94
Interval Upper Limit	341.86

(b) The population of dental expenses must be approximately normally distributed.

(d) $\bar{X} \pm t \cdot \dfrac{s}{\sqrt{n}} = 261.40 \pm 2.2622 \cdot \dfrac{138.80}{\sqrt{10}}$ $\$162.11 \leq \mu \leq \$360.69$

(e) The additional $500 in dental expenses, divided across the sample of 10, raises the mean by $50 and increases the standard deviation by nearly $20. The interval half-width increases over $11 in the process. The new interval is:

$\bar{X} \pm t \cdot \dfrac{s}{\sqrt{n}} = 311.40 \pm 1.8331 \cdot \dfrac{157.056}{\sqrt{10}}$ $\$220.36 \leq \mu \leq \$402.22$

7.20   (a) $\bar{X} \pm t \cdot \dfrac{s}{\sqrt{n}} = 27.892 \pm 2.0639 \cdot \dfrac{13.9697}{\sqrt{25}}$ $22.167 \leq \mu \leq 33.617$

    (b)   The director should not make the statement that average waiting time at the HMO is 15 minutes. Based on these sample data, 15 minutes is 4.65 standard deviations below the sample mean of 27.892 minutes, rendering the statement extremely unlikely.

    (c)   Waiting times are normally distributed.

    (d)   Inflating one data value by 100 minutes raises the average sample waiting time by 4 minutes and increases the standard deviation from 13.9 to 19.4 minutes. The interval half-width increases from 5.725 minutes to just over 8 minutes. The new interval is:

$$\bar{X} \pm t \cdot \frac{s}{\sqrt{n}} = 31.892 \pm 2.0639 \cdot \frac{19.3847}{\sqrt{25}} \qquad 23.890 \leq \mu \leq 39.894$$

7.22 $p_s = \dfrac{X}{n} = \dfrac{25}{400} = 0.0625$ $p_s \pm Z \cdot \sqrt{\dfrac{p_s(1-p_s)}{n}} = 0.25 \pm 2.58\sqrt{\dfrac{0.0625(0.9375)}{400}}$

                                                 $0.0313 \leq p \leq 0.0937$

7.24   (a) $p_s = 0.4$ $p_s \pm Z \cdot \sqrt{\dfrac{p_s(1-p_s)}{n}} = 0.4 \pm 2.58 \cdot \sqrt{\dfrac{0.4(0.6)}{200}}$

                                                 $0.3106 \leq p \leq 0.4894$

    (b)   The auditor should report that actual payment performance is well below the goal of 90% or more of claims that should be paid within two months. In fact, the target of 90% is 14.43 standard deviations above the sample proportion of 40%.

7.26   (a) $p_s = 0.10$ $p_s \pm Z \cdot \sqrt{\dfrac{p_s(1-p_s)}{n}} = 0.10 \pm 1.645 \cdot \sqrt{\dfrac{0.10(0.90)}{300}}$

                                                 $0.0715 \leq p \leq 0.1285$

    (b)   Based on the results of this sample, the owner can return this shipment because it is likely that the shipment contains more than 5% defective pens. It is possible but highly unlikely that the owner got a 10% defective rate from a sample of 300 pens that really had only a 5% defective rate.

    (c)   The center of the confidence interval remains anchored at $p_s = 0.10$, but the increased confidence level increases the width of the half-interval from 0.0285 to 0.0447. The new interval is:

    $p_s = 0.10$ $p_s \pm Z \cdot \sqrt{\dfrac{p_s(1-p_s)}{n}} = 0.10 \pm 2.58 \cdot \sqrt{\dfrac{0.10(0.90)}{300}}$

                                                 $0.0553 \leq p \leq 0.1447$

7.28 (a)

	Purchase Additional Telephone Line	
2		
3	Sample Size	500
4	Number of Successes	135
5	Confidence Level	99%
6	Sample Proportion	0.27
7	Z Value	-2.57583451
8	Standard Error of the Proportion	0.019854471
9	Interval Half Width	0.05114183
10	Interval Lower Limit	0.21885817
11	Interval Upper Limit	0.32114183

•7.30 $\quad n = \dfrac{Z^2\sigma^2}{e^2} = \dfrac{1.96^2 \cdot 15^2}{5^2} = 34.57$ $\qquad\qquad$ Use $n = 35$

7.32 $\quad n = \dfrac{Z^2 p(1-p)}{e^2} = \dfrac{2.58^2(0.5)(0.5)}{(0.04)^2} = 1{,}040.06$ $\qquad$ Use $n = 1{,}041$

7.34 $\quad n = \dfrac{Z^2\sigma^2}{e^2} = \dfrac{1.96^2 \cdot 400^2}{50^2} = 245.86$ $\qquad$ Use $n = 246$

•7.36 $\quad n = \dfrac{Z^2\sigma^2}{e^2} = \dfrac{1.96^2 \cdot (100)^2}{(20)^2} = 96.04$ $\qquad$ Use $n = 97$

•7.38 (a) $n = \dfrac{Z^2\sigma^2}{e^2} = \dfrac{2.58^2 \cdot 25^2}{5^2} = 166.41$ $\qquad$ Use $n = 167$

$\quad$ (b) $n = \dfrac{Z^2\sigma^2}{e^2} = \dfrac{1.96^2 \cdot 25^2}{5^2} = 96.04$ $\qquad$ Use $n = 97$

7.40 (a) $n = \dfrac{Z^2\sigma^2}{e^2} = \dfrac{1.645^2 \cdot 45^2}{5^2} = 219.19$ $\qquad$ Use $n = 220$

$\quad$ (b) $n = \dfrac{Z^2\sigma^2}{e^2} = \dfrac{2.58^2 \cdot 45^2}{5^2} = 539.17$ $\qquad$ Use $n = 540$

7.42 (a) $n = \dfrac{Z^2 p(1-p)}{e^2} = \dfrac{1.645^2(0.5)(0.5)}{(0.04)^2} = 422.82$ $\qquad$ Use $n = 423$

$\quad$ (b) $n = \dfrac{Z^2 p(1-p)}{e^2} = \dfrac{1.96^2(0.5)(0.5)}{(0.04)^2} = 600.25$ $\qquad$ Use $n = 601$

•7.44 $n = \dfrac{Z^2 p(1-p)}{e^2} = \dfrac{1.645^2(0.5)(0.5)}{(0.05)^2} = 270.60$ Use $n = 271$

7.46 (a) $n = \dfrac{Z^2 p(1-p)}{e^2} = \dfrac{2.58^2(0.10)(0.90)}{(0.07)^2} = 122.26$ Use $n = 123$

(b) $n = \dfrac{Z^2 p(1-p)}{e^2} = \dfrac{1.96^2(0.10)(0.90)}{(0.06)^2} = 96.04$ Use $n = 97$

•7.48 $\bar{X} \pm t \cdot \dfrac{s}{\sqrt{n}} \sqrt{\dfrac{N-n}{N-1}} = 75 \pm 2.0301 \cdot \dfrac{24}{\sqrt{36}} \sqrt{\dfrac{200-36}{200-1}}$ $67.6282 \le \mu \le 82.3718$

•7.50 (a) $\bar{X} \pm Z \cdot \dfrac{\sigma}{\sqrt{n}} \sqrt{\dfrac{N-n}{N-1}} = 350 \pm 1.96 \cdot \dfrac{100}{\sqrt{50}} \sqrt{\dfrac{2000-50}{2000-1}}$

$322.62 \le \mu \le 377.38$

Note: Because the process standard deviation is known, use a Z rather than a t to build the confidence interval.

(b) $n_0 = \dfrac{Z^2 \sigma^2}{e^2} = \dfrac{1.96^2 \cdot 100^2}{20^2} = 96.04$

$n = \dfrac{n_0 N}{n_0 + (N-1)} = \dfrac{96.04 \cdot 2000}{96.04 + (2000-1)} = 91.68$ Use $n = 92$

(c)(a) $\bar{X} \pm Z \cdot \dfrac{\sigma}{\sqrt{n}} \sqrt{\dfrac{N-n}{N-1}} = 350 \pm 1.96 \cdot \dfrac{100}{\sqrt{50}} \sqrt{\dfrac{1000-50}{1000-1}}$

$322.97 \le \mu \le 377.03$

(b) $n = \dfrac{n_0 N}{n_0 + (N-1)} = \dfrac{96.04 \cdot 1000}{96.04 + (1000-1)} = 87.70$ Use $n = 88$

•7.52 (a) $p_s \pm Z \sqrt{\dfrac{p_s(1-p_s)}{n}} \sqrt{\dfrac{N-n}{N-1}} = 0.3 \pm 1.645 \cdot \sqrt{\dfrac{0.3(0.7)}{100}} \sqrt{\dfrac{1000-100}{1000-1}}$

$0.2284 \le p \le 0.3716$

(b)
$n_0 = \dfrac{Z^2 p(1-p)}{e^2} = \dfrac{1.645^2(0.5)(0.5)}{(0.05)^2} = 270.6025$

$n = \dfrac{n_0 N}{n_0 + (N-1)} = \dfrac{270.6025 \cdot 1000}{270.6025 + (1000-1)} = 213.14$

Use $n = 214$

•7.52 (b) *Note:* To be the most conservative, assume in part (b) that the population proportion is
cont. not known, using an estimate of 0.5 for p. An estimate of 0.5 maximizes the
 sample size by maximizing $p(1-p)$ to produce the largest n_0 necessary.

(c) (a)

$$0.3 \pm 1.645 \cdot \sqrt{\frac{0.3(0.7)}{100}} \sqrt{\frac{2000-100}{2000-1}} \qquad 0.2265 \le p \le 0.3735$$

(b) $\quad n = \dfrac{n_0 N}{n_0 + (N-1)} = \dfrac{270.6025 \cdot 2000}{270.6025 + (2000-1)} = 238.46 \qquad$ Use $n = 239$

7.54 (a) $\bar{X} \pm Z \cdot \dfrac{\sigma}{\sqrt{n}} \sqrt{\dfrac{N-n}{N-1}} = 1.99 \pm 1.96 \cdot \dfrac{0.05}{\sqrt{100}} \cdot \sqrt{\dfrac{2000-100}{2000-1}}$

$$1.9804 \le \mu \le 1.9996$$

(b) $\qquad n_0 = \dfrac{Z^2 \sigma^2}{e^2} = \dfrac{1.96^2 \cdot (0.05)^2}{(0.01)^2} = 96.04$

$\qquad n = \dfrac{n_0 N}{n_0 + (N-1)} = \dfrac{96.04 \cdot 2000}{96.04 + 1999} = 91.68 \qquad$ Use $n = 92$

(c)(a) $\quad \bar{X} \pm t \cdot \dfrac{s}{\sqrt{n}} \sqrt{\dfrac{N-n}{N-1}} = 1.99 \pm 1.96 \cdot \dfrac{0.05}{\sqrt{100}} \cdot \sqrt{\dfrac{1000-100}{1000-1}}$

$$1.9807 \le \mu \le 1.9993$$

(b) $\qquad n_0 = \dfrac{Z^2 \sigma^2}{e^2} = \dfrac{1.96^2 \cdot (0.05)^2}{(0.01)^2} = 96.04$

$\qquad n = \dfrac{n_0 N}{n_0 + (N-1)} = \dfrac{96.04 \cdot 1000}{96.04 + 999} = 87.70 \qquad$ Use $n = 88$

•7.56 (b) $\quad N \cdot \bar{X} \pm N \cdot t \cdot \dfrac{s}{\sqrt{n}} \sqrt{\dfrac{N-n}{N-1}} = 500 \cdot 25.7 \pm 500 \cdot 2.7969 \cdot \dfrac{7.8}{\sqrt{25}} \cdot \sqrt{\dfrac{500-25}{500-1}}$

$$\$10{,}721.53 \le \text{Population Total} \le \$14{,}978.47$$

7.58

$$N \cdot \bar{X} \pm N \cdot t \cdot \dfrac{s}{\sqrt{n}} \sqrt{\dfrac{N-n}{N-1}}$$

$$= 2000 \cdot \$4{,}750 \pm 2000 \cdot 2.0452 \cdot \dfrac{\$1{,}200}{\sqrt{30}} \cdot \sqrt{\dfrac{2000-30}{2000-1}}$$

$$\$8{,}610{,}362.43 \le \text{Population Total} \le \$10{,}389{,}637.57$$

7.60

$$N \cdot \overline{X} \pm N \cdot t \cdot \frac{s}{\sqrt{n}} \sqrt{\frac{N-n}{N-1}}$$

$$= 3000 \cdot \$261.40 \pm 3000 \cdot 1.8331 \cdot \frac{\$138.8046}{\sqrt{10}} \cdot \sqrt{\frac{3000-10}{3000-1}}$$

$$\$543,176.96 \leq \text{Population Total} \leq \$1,025,223.04$$

•7.62

$$N \cdot \overline{D} \pm N \cdot t \cdot \frac{s_D}{\sqrt{n}} \sqrt{\frac{N-n}{N-1}}$$

$$= 4000 \cdot \$7.45907 \pm 4000 \cdot 2.60923 \cdot \frac{\$29.55234}{\sqrt{150}} \cdot \sqrt{\frac{4000-150}{4000-1}}$$

$$\$5,125.99 \leq \text{Total Difference in the Population} \leq \$54,546.57$$

Note: The *t*-value of 2.60923 for 95% confidence and *df* = 149 was derived on Excel.

•7.70 (a) $\overline{X} \pm t \cdot \frac{s}{\sqrt{n}} = 15.3 \pm 2.0227 \cdot \frac{3.8}{\sqrt{40}}$ $14.085 \leq \mu \leq 16.515$

(b) $p_s \pm Z \cdot \sqrt{\frac{p_s(1-p_s)}{n}} = 0.675 \pm 1.96 \cdot \sqrt{\frac{0.675(0.325)}{40}}$ $0.530 \leq p \leq 0.820$

(c) $n = \frac{Z^2 \cdot \sigma^2}{e^2} = \frac{1.96^2 \cdot 5^2}{2^2} = 24.01$ Use $n = 25$

(d) $n = \frac{Z^2 \cdot p \cdot (1-p)}{e^2} = \frac{1.96^2 \cdot (0.5) \cdot (0.5)}{(0.035)^2} = 784$ Use $n = 784$

(e) If a single sample were to be selected for both purposes, the larger of the two sample sizes ($n = 784$) should be used.

7.72 (a) $\overline{X} \pm t \cdot \frac{s}{\sqrt{n}} = 9.7 \pm 2.0639 \cdot \frac{4}{\sqrt{25}}$ $8.049 \leq \mu \leq 11.351$

(b) $p_s \pm Z \cdot \sqrt{\frac{p_s(1-p_s)}{n}} = 0.48 \pm 1.96 \cdot \sqrt{\frac{0.48(0.52)}{25}}$ $0.284 \leq p \leq 0.676$

(c) $n = \frac{Z^2 \cdot \sigma^2}{e^2} = \frac{1.96^2 \cdot 4.5^2}{1.5^2} = 34.57$ Use $n = 35$

(d) $n = \frac{Z^2 \cdot p \cdot (1-p)}{e^2} = \frac{1.645^2 \cdot (0.5) \cdot (0.5)}{(0.075)^2} = 120.268$ Use $n = 121$

(e) If a single sample were to be selected for both purposes, the larger of the two sample sizes ($n = 121$) should be used.

•7.74 (a) $\overline{X} \pm t \cdot \dfrac{s}{\sqrt{n}} = \$28.52 \pm 1.9949 \cdot \dfrac{\$11.39}{\sqrt{70}}$ $\$25.80 \le \mu \le \$31.24$

(b) $p_s \pm Z \cdot \sqrt{\dfrac{p_s(1-p_s)}{n}} = 0.40 \pm 1.645 \cdot \sqrt{\dfrac{0.40(0.60)}{70}}$ $0.3037 \le p \le 0.4963$

(c) $n = \dfrac{Z^2 \cdot \sigma^2}{e^2} = \dfrac{1.96^2 \cdot 10^2}{2^2} = 96.04$ Use $n = 97$

(d) $n = \dfrac{Z^2 \cdot p \cdot (1-p)}{e^2} = \dfrac{1.645^2 \cdot (0.5) \cdot (0.5)}{(0.04)^2} = 422.82$ Use $n = 423$

(e) If a single sample were to be selected for both purposes, the larger of the two sample sizes ($n = 423$) should be used.

7.76 (a) $\overline{X} \pm t \cdot \dfrac{s}{\sqrt{n}} = \$38.54 \pm 2.0010 \cdot \dfrac{\$7.26}{\sqrt{60}}$ $\$36.66 \le \mu \le \$40.42$

(b) $p_s \pm Z \cdot \sqrt{\dfrac{p_s(1-p_s)}{n}} = 0.30 \pm 1.645 \cdot \sqrt{\dfrac{0.30(0.70)}{60}}$ $0.2027 \le p \le 0.3973$

(c) $n = \dfrac{Z^2 \cdot \sigma^2}{e^2} = \dfrac{1.96^2 \cdot 8^2}{1.5^2} = 109.27$ Use $n = 110$

(d) $n = \dfrac{Z^2 \cdot p \cdot (1-p)}{e^2} = \dfrac{1.645^2 \cdot (0.5) \cdot (0.5)}{(0.04)^2} = 422.82$ Use $n = 423$

(e) If a single sample were to be selected for both purposes, the larger of the two sample sizes ($n = 423$) should be used.

7.78 (a) $p \pm Z \cdot \sqrt{\dfrac{p(1-p)}{n}} \cdot \sqrt{\dfrac{N-n}{N-1}} = 0.14 \pm 1.96 \cdot \sqrt{\dfrac{0.14(0.86)}{50}} \cdot \sqrt{\dfrac{1000-50}{1000-1}}$

$0.0613 \le p \le 0.2187$

(b) $n = \dfrac{Z^2 \cdot p \cdot (1-p)}{e^2} = \dfrac{1.96^2 \cdot (0.10) \cdot (0.90)}{(0.04)^2} = 216.09$ Use $n = 217$

7.80 (a) $n_0 = \dfrac{Z^2 \cdot \sigma^2}{e^2} = \dfrac{1.96^2 \cdot 30^2}{5^2} = 138.2976$

$n = \dfrac{n_0 N}{n_0 + (N-1)} = \dfrac{138.2976 \cdot 25{,}056}{138.2976 + (25{,}056 - 1)} = 137.54$ Use $n = 138$

7.80

cont. (b) $p_s = \dfrac{7}{138} = 0.0507$

$$p \pm Z \cdot \sqrt{\dfrac{p(1-p)}{n}} \cdot \sqrt{\dfrac{N-n}{N-1}}$$

$$= 0.0507 \pm 1.645 \cdot \sqrt{\dfrac{0.0507(0.9493)}{138}} \cdot \sqrt{\dfrac{25,056-138}{25,056-1}}$$

$$0.0201 \le p \le 0.0813$$

(c) (1) Using Excel, we find the t-value for 95% confidence and 137 degrees of freedom is $t = 1.9774$.

$$\bar{X} \pm t \cdot \dfrac{s}{\sqrt{n}} \cdot \sqrt{\dfrac{N-n}{N-1}} = \$93.70 \pm 1.9774 \cdot \dfrac{\$34.55}{\sqrt{138}} \cdot \sqrt{\dfrac{25,056-138}{25,056-1}}$$

$$\$87.90 \le \mu_X \le \$99.50$$

(2)

$$N \cdot \bar{X} \pm N \cdot t \cdot \dfrac{s}{\sqrt{n}} \cdot \sqrt{\dfrac{N-n}{N-1}}$$

$$= 25,056 \cdot \$93.70 \pm 25,056 \cdot 1.9774 \cdot \dfrac{\$34.55}{\sqrt{138}} \cdot \sqrt{\dfrac{25,056-138}{25,056-1}}$$

$$\$2,202,427.61 \le \text{Population Total} \le \$2,493,066.79$$

(d) $\bar{D} = \dfrac{\sum D}{n} = \dfrac{241}{138} = 1.7463768 \qquad s_D = \sqrt{\dfrac{\sum(D-\bar{D})^2}{n-1}} = \sqrt{\dfrac{6432.12}{137}} = 6.85199$

$$N \cdot \bar{D} \pm N \cdot t \cdot \dfrac{s_D}{\sqrt{n}} \cdot \sqrt{\dfrac{N-n}{N-1}}$$

$$= 25,056 \cdot \$1.7463768 \pm 25,056 \cdot 1.9774 \cdot \dfrac{\$6.85199}{\sqrt{138}} \cdot \sqrt{\dfrac{25,056-138}{25,056-1}}$$

$$\$14,937.30 \le \text{Total Difference in the Population} \le \$72,577.14$$

Chapter 7
Fundamentals of Hypothesis Testing: One-Sample Tests

✓ *To develop hypothesis-testing methodology as a technique for making decisions about population parameters based on sample statistics*

✓ *To determine the risks involved in making these decisions based only on sample information*

✓ *To describe various practical tests of hypothesis in dealing with one sample*

✓ *To understand the conceptual differences between the traditional critical value approach and the p-value approach to hypothesis testing*

Introduction

In chapter 6 we began our discussion of statistical inference by developing the concept of a sampling distribution. In chapter 7 we considered studies in which a statistic (such as the sample mean or sample proportion) obtained from a random sample is used to *estimate* its corresponding population parameter.

In this chapter we focus on another phase of statistical inference that is also based on sample information—hypothesis testing. We develop a step-by-step methodology that enables us to make inferences about a population parameter by *analyzing differences* between the results we observe (our sample statistic) and the results we expect to obtain if some underlying hypothesis is actually true. Emphasis here is placed on the fundamental and conceptual underpinnings of **hypothesis-testing methodology.** In the three chapters that follow we present various hypothesis-testing procedures that are frequently employed in the analysis of data obtained from studies and experiments designed under a variety of conditions.

◆ **USING STATISTICS:** *Testing a Manufacturer's Claim Regarding Product Specifications*

To develop hypothesis-testing methodology, we continue our focus in this chapter on some issues based on the cereal box–filling process discussed in chapters 6 and 7. For example, the operations manager is concerned with evaluating whether or not the process is working to ensure that, on average, the proper amount of cereal (i.e., 368 grams) is being filled in each box. He decides to select a random sample of 25 boxes from the filling process and examines their weights to determine how close each of these boxes comes to the company's specification of an average of 368 grams per box. The operations manager hopes to find that the process is working properly. However, he might find that the sampled boxes weigh too little or perhaps too much. As a result, he may then decide to halt the production process until the reason for the failure to adhere to the specified weight of 368 grams is determined.

By analyzing the differences between the weights obtained from the sample and the 368-gram expectation obtained from the company's specifications, he can reach a decision based on this sample information, and one of the following two conclusions can be drawn:

1. The average fill in the entire process is 368 grams. No corrective action is needed.

2. The average fill is not 368 grams; either it is less than 368 grams or it is more than 368 grams. Corrective action is needed.

 HYPOTHESIS-TESTING METHODOLOGY

The Null and Alternative Hypotheses

Hypothesis testing typically begins with some theory, claim, or assertion about a particular parameter of a population. For example, for purposes of statistical analysis, the operations manager at our cereal company chooses as his initial hypothesis that the process is in control; that is, the average fill is 368 grams and no corrective action is needed.

The hypothesis that the population parameter is equal to the company specification is referred to as the **null hypothesis.** A null hypothesis is always one of status quo or no difference. We commonly identify the null hypothesis by the symbol H_0. Our operations manager establishes as his null hypothesis that the filling process is in control and working properly, that the mean fill per box is the 368-gram specification. This can be stated as

$$H_0: \mu = 368$$

Note that even though the operations manager has information only from the sample, the null hypothesis is written in terms of the population parameter. This is because he is interested in the entire filling process, that is, (the population of) all cereal boxes being filled. The sample statistic will be used to make inferences about the entire filling process. One such inference may be that the results observed from the sample data indicate that the null hypothesis is false. If the null hypothesis is considered false, something else must be true. To anticipate this possibility, whenever we specify a null hypothesis, we must also specify an **alternative hypothesis,** or one that must be true if the null hypothesis is found to be false. The alternative hypothesis (H_1) is the opposite of the null hypothesis (H_0). For the operations manager, this is stated as

$$H_1: \mu \neq 368$$

The alternative hypothesis represents the conclusion reached by rejecting the null hypothesis if there is sufficient evidence from the sample information to decide that the null hypothesis is unlikely to be true. In our example, if the weights of the sampled boxes are sufficiently above or below the expected 368-gram average specified by the company, the operations manager rejects the null hypothesis in favor of the alternative hypothesis that the average amount of fill is different from 368 grams. He then stops production and takes whatever action is necessary to correct the problem.

Hypothesis-testing methodology is designed so that our rejection of the null hypothesis is based on evidence from the sample that our alternative hypothesis is far more likely to be true. However, failure to reject the null hypothesis is not proof that it is true. We can never prove that the null hypothesis is correct because our decision is based only on the sample information, not on the entire population. Therefore, if we fail to reject the null hypothesis, we can only conclude that there is insufficient evidence to warrant its rejection. A summary of the null and alternative hypotheses is presented in Exhibit 8.1.

To illustrate the null and alternative hypotheses, consider Examples 8.1 and 8.2:

Example 8.1   *Stating the Null and Alternative Hypotheses*

Suppose Toyota claims that a new model car will average 30 miles per gallon in highway driving. If you are planning an experiment to test this claim, what are the null and alternative hypotheses?

SOLUTION

$$H_0: \mu = 30 \text{ miles per gallon}$$
$$H_1: \mu \neq 30 \text{ miles per gallon}$$

Example 8.2   *Stating the Null and Alternative Hypotheses*

In the past the average age of policyholders of term life insurance at Empire Insurance Company was 48 years. As the company expanded and provided more term policies across the nation, the chief financial officer came to believe the average age may have shifted. If you were planning a survey to test this claim, what would be the null and alternative hypotheses?

$$H_0: \mu = 48 \text{ years}$$
$$H_1: \mu \neq 48 \text{ years}$$

The Critical Value of the Test Statistic

We can develop the logic behind the hypothesis-testing methodology by contemplating how we can use only sample information to determine the plausibility of the null hypothesis.

Our operations manager states as his null hypothesis that the average amount of cereal per box over the entire filling process is 368 grams (i.e., the population parameter specified by the company). He then obtains a sample of boxes from the filling process, weighs each box, and computes the sample mean. We recall that a statistic from a sample is an estimate of the corresponding parameter from the population from which the sample is drawn. Even if the null hypothesis is in fact true, this statistic will likely differ from the actual parameter value because of chance or sampling error. Nevertheless, under such circumstances we expect the sample statistic to be close to the population parameter. In such a situation there would be insufficient evidence to reject the null hypothesis. If, for example, the sample mean is 367.9, our instinct is to conclude that the population mean has not changed (i.e., $\mu = 368$) because a sample mean of 367.9 is very close to the hypothesized value of 368. Intuitively, we might think that it is likely that we could obtain a sample mean of 367.9 from a population whose mean is 368.

On the other hand, if there is a large discrepancy between the value of the statistic and its corresponding hypothesized parameter, our instinct is to conclude that the null hypothesis is unlikely to be true. For example, if the sample average is 320, our instinct is to conclude that the true population average is not 368 (i.e., $\mu \neq 368$) because the sample mean is very far from the hypothesized value of 368. In such a case we reason that it is very unlikely that the sample mean of 320 can be obtained if the population mean is really 368 and, therefore, more logical to conclude that the population mean is not equal to 368. Here we reject the null hypothesis.

In either case our decision is reached because of our belief that randomly selected samples are truly representative of the underlying populations from which they are drawn.

Unfortunately, the decision-making process is not always so clear-cut and cannot be left to an individual's subjective judgment as to the meaning of "very close" or "very different." Determining what is very close and what is very different is very arbitrary without using operational definitions. Hypothesis-testing methodology provides clear definitions for evaluating such differences and enables us to quantify the decision-making process so that the probability of obtaining a given sample result can be found if the null hypothesis is true. This is achieved by first determining the sampling distribution for the sample statistic of interest (i.e., the sample mean) and then computing the particular *test statistic* based on the given sample result. Because the sampling distribution for the test statistic often follows a well-known statistical distribution, such as the normal or *t* distribution, we can use these distributions to determine the likelihood of a null hypothesis being true.

Regions of Rejection and Nonrejection

The sampling distribution of the test statistic is divided into two regions, a **region of rejection** (sometimes called the **critical region**) and a **region of nonrejection** (see Figure 8.1). If the test statistic falls into the region of nonrejection, the null hypothesis cannot be rejected. In our example, the operations manager concludes that the average fill amount is not changed. If the test statistic falls into the rejection region, the null hypothesis is rejected. Here the operations manager concludes that the population mean is not 368.

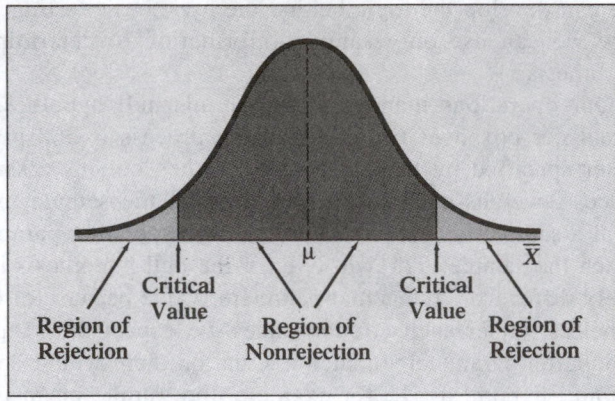

FIGURE 8.1
Regions of rejection and nonrejection in hypothesis testing

The region of rejection may be thought of as consisting of the values of the test statistic that are unlikely to occur if the null hypothesis is true. On the other hand, these values are not so unlikely to occur if the null hypothesis is false. Therefore, if we observe a value of the test statistic that falls into this *rejection region,* we reject the null hypothesis because that value is unlikely if the null hypothesis is true.

To make a decision concerning the null hypothesis, we first determine the **critical value** of the test statistic. The critical value divides the nonrejection region from the rejection region. However, the determination of this critical value depends on the size of the rejection region. As we will see in the following section, the size of the rejection region is directly related to the risks involved in using only sample evidence to make decisions about a population parameter.

Risks in Decision Making Using Hypothesis-Testing Methodology

When using a sample statistic to make decisions about a population parameter, there is a risk that an incorrect conclusion will be reached. Indeed, two different types of errors can occur when applying hypothesis-testing methodology, Type I and Type II.

A **Type I error** occurs if the null hypothesis H_0 is rejected when in fact it is true and should not be rejected. The probability of a Type I error occurring is α.

A **Type II error** occurs if the null hypothesis H_0 is not rejected when in fact it is false and should be rejected. The probability of a Type II error occurring is β.

In our cereal-filling-process example, the Type I error occurs if the operations manager concludes (based on sample information) that the average population amount filled is *not* 368 when in fact it is 368. On the other hand, the Type II error occurs if he concludes (based on sample information) that the average population fill amount is 368 when in fact it is not 368.

◆ *The Level of Significance*   The probability of committing a Type I error, denoted by α (the lowercase Greek letter alpha), is referred to as the **level of significance** of the statistical test. Traditionally, one controls the Type I error rate by deciding the risk level α he or she is willing to tolerate in rejecting the null hypothesis when it is in fact true. Because the level of significance is specified before the hypothesis test is performed, the risk of committing a Type I error, α, is directly under the control of the individual performing the test. Researchers traditionally select α levels of .05 or smaller. The choice of selecting a particular risk level for making a Type I error is dependent on the cost of making a Type I error. Once the value for α is specified, the size of the rejection region is known because α is the probability of rejection under the null hypothesis. From this fact, the critical value or values that divide the rejection and nonrejection regions are determined.

◆ *The Confidence Coefficient*   The complement $(1 - \alpha)$ of the probability of a Type I error is called the confidence coefficient, which, when multiplied by 100%, yields the confidence level that we studied in section 7.1.

The **confidence coefficient**, denoted by $1 - \alpha$, is the probability that the null hypothesis H_0 is not rejected when in fact it is true and should not be rejected.

In terms of hypothesis-testing methodology, this coefficient represents the probability of concluding that the specified value of the parameter being tested under the null hypothesis may be plausible when, in fact, it is true. In our cereal-filling-process example, the confidence coefficient measures the probability of concluding that the average fill per box is 368 grams when in fact it is 368 grams.

◆ *The β Risk*   The probability of committing a Type II error, denoted by β (the lowercase Greek letter beta), is often referred to as the *consumer's risk* level. Unlike the Type I error, which we control by our selection of α, the probability of making a Type II error is dependent on the difference between the hypothesized and actual value of the population parameter. Because large differences are easier to find, if the difference between the sample statistic and the corresponding population parameter is large, β, the probability of committing a Type II error, will likely be small. For example, if the true population average (which is unknown to us) is 320 grams, there is a small chance (β) of concluding that the average has not changed from 368. On the other hand, if the difference between the statistic and the corresponding parameter value is small, the probability of committing a Type II error is large. Thus, if the true population average is really 367 grams, there is a high probability of concluding that the population average fill amount has not changed from the specified 368 grams (and we would be making a Type II error).

◆ **The Power of a Test**   The complement $(1 - \beta)$ of the probability of a Type II error is called the power of a statistical test.

> The **power of a statistical test**, denoted by $1 - \beta$, is the probability of rejecting the null hypothesis when in fact it is false and should be rejected.

In our cereal-filling-process example, the power of the test is the probability of concluding that the average fill amount is not 368 grams when in fact it actually is not 368 grams.

◆ **Risks in Decision Making: A Delicate Balance**   Table 8.1 illustrates the results of the two possible decisions (do not reject H_0 or reject H_0) that can occur in any hypothesis test. Depending on the specific decision, one of two types of errors may occur[1] or one of two types of correct conclusions may be reached.

Table 8.1   *Hypothesis testing and decision making*

	ACTUAL SITUATION	
STATISTICAL DECISION	H_0 TRUE	H_0 FALSE
Do not reject H_0	Correct decision P (confidence) $= 1 - \alpha$	Type II error P (Type II error) $= \beta$
Reject H_0	Type I error P (Type I error) $= \alpha$	Correct decision P (power) $= 1 - \beta$

One way in which we can control and reduce the probability of making a Type II error in a study is to increase the size of the sample. Larger sample sizes generally permit us to detect even very small differences between the sample statistics and the population parameters. For a given level of α, increasing the sample size will decrease β and therefore will increase the power of the test to detect that the null hypothesis H_0 is false. Of course, however, there is always a limit to our resources and this will affect the decision as to how large a sample we can take. Thus, for a given sample size we must consider the trade-offs between the two possible types of errors. Because we can directly control our risk of Type I error, we can reduce our risk by selecting a lower level for α (e.g., .01 instead of .05). However, when α is decreased, β will be increased, so a reduction in risk of Type I error will result in an increased risk of Type II error. If, on the other hand, we wish to reduce β, our risk of Type II error, we could select a larger value for α (e.g., .05 instead of .01).

In our cereal-filling-process example, the risk of a Type I error involves concluding that the average fill per box has changed from the hypothesized 368 grams when in fact it has not changed. The risk of a Type II error involves concluding that the average fill per box has not changed from the hypothesized 368 grams when in truth it has changed. The choice of reasonable values for α and β depends on the costs inherent in each type of error. For example, if it were very costly to change the status quo, then we would want to be very sure that a change would be beneficial, so the risk of a Type I error might be most important and would be kept very low. On the other hand, if we wanted to be very certain of detecting changes from a hypothesized mean, the risk of a Type II error would be most important and we might choose a higher level of α.

Problems for Section 8.1

Learning the Basics

- **8.1** The symbol H_0 is used to denote which hypothesis?
- **8.2** The symbol H_1 is used to denote which hypothesis?
- **8.3** The level of significance or chance of committing a Type I error is denoted by what symbol?
- **8.4** The consumer's risk or chance of committing a Type II error is denoted by what symbol?
- **8.5** What does $1 - \beta$ represent?
- **8.6** What is the relationship of α to the Type I error?
- **8.7** What is the relationship of β to the Type II error?
- **8.8** How is power related to the probability of making a Type II error?

Applying the Concepts

- **8.9** Why is it possible for the null hypothesis to be rejected when in fact it is true?
- **8.10** Why is it possible that the null hypothesis will not always be rejected when it is false?
- **8.11** For a given sample size, if α is reduced from .05 to .01, what will happen to β?
- **8.12** For H_0: $\mu = 100$, H_1: $\mu \neq 100$, and for a sample of size n, β will be larger if the actual value of μ is 90 than if the actual value of μ is 75. Why?
- **8.13** In the American legal system, a defendant is presumed innocent until proved guilty. Consider a null hypothesis H_0 that the defendant is innocent and an alternative hypothesis H_1 that the defendant is guilty. A jury has two possible decisions: convict the defendant (i.e., reject the null hypothesis) or do not convict the defendant (i.e., do not reject the null hypothesis). Explain the meaning of the risks of committing either a Type I or Type II error in this example.
- **8.14** Suppose the defendant in Problem 8.13 is presumed guilty until proved innocent as in some other judicial systems. How do the null and alternative hypotheses differ from those in Problem 8.13? What are the meanings of the risks of committing either a Type I or Type II error here?
- **8.15** The CEO of a national clothing manufacturer claims that average profits per store are anticipated to be \$1 million in the first quarter of the year. As market research director, you must evaluate the CEO's claim. State the null hypothesis H_0 and the alternative hypothesis H_1.
- **8.16** Owing to complaints from both students and faculty about lateness, the registrar at a large university wants to adjust the scheduled class times to allow for adequate travel time between classes and is ready to undertake a study. Up until now, the registrar believed 20 minutes between scheduled classes should be sufficient. State the null hypothesis H_0 and the alternative hypothesis H_1.
- **8.17** The manager of a local branch of a commercial bank believes that over the past few years the bank has been catering to a different clientele and that the average amount withdrawn from its ATMs is no longer \$140. State the null hypothesis H_0 and the alternative hypothesis H_1.

8.2 Z TEST OF HYPOTHESIS FOR THE MEAN (σ KNOWN)

Now that we have described the hypothesis-testing methodology, let us return to the question of interest to the operations manager at the cereal-packaging plant. You may recall that he wants to determine whether or not the cereal-filling process is in control—that the average fill per box throughout the entire packaging process remains at the specified 368 grams and no corrective action is needed. To study this, he plans to take a random sample of 25 boxes, weigh each one, and then evaluate the difference between the sample statistic and the hypothesized population parameter by comparing the mean weight (in grams) from the sample to the expected mean of 368 grams specified by the company. For this filling process, the null and alternative hypotheses are

$$H_0: \mu = 368$$
$$H_1: \mu \neq 368$$

If we assume that the standard deviation σ is known, then based on the central limit theorem, the sampling distribution of the mean follows the normal distribution resulting in the following **Z-test statistic:**

Z Test of Hypothesis for a Population Mean (σ Known)

$$Z = \frac{\overline{X} - \mu}{\dfrac{\sigma}{\sqrt{n}}} \tag{8.1}$$

In this equation the numerator measures how far (in an absolute sense) the observed sample mean $\overline{X}$ is from the hypothesized mean μ. The denominator is the standard error of the mean, so Z represents how many standard errors $\overline{X}$ is from μ.

The Critical Value Approach to Hypothesis Testing

If the operations manager decides to choose a level of significance of .05, the size of the rejection region would be .05 and the critical values of the normal distribution could be determined. These critical values can be expressed as standardized Z values (i.e., in standard-deviation units). Because the rejection region is divided into the two tails of the distribution (this is called a **two-tailed test**), the .05 is divided into two equal parts of .025 each. A rejection region of .025 in each tail of the normal distribution results in a cumulative area of .025 below the lower critical value and a cumulative area of .975 below the upper critical value. Looking up these areas in the normal distribution [Table E.2(b)], we find that the critical values that divide the rejection and nonrejection regions are (in standard-deviation units) -1.96 and $+1.96$. Figure 8.2 illustrates this case; it shows that if the mean is actually 368 grams, as H_0 claims, then the values of the test statistic Z have a standard normal distribution centered at $\mu = 368$ (which corresponds to a standardized Z value of 0). Observed values of Z greater than $+1.96$ or less than -1.96 indicate that $\overline{X}$ is so far from the hypothesized $\mu = 368$ that it is unlikely that such a value would occur if H_0 were true. Therefore, the decision rule is

Reject H_0 if $Z > +1.96$

or if $Z < -1.96$;

otherwise do not reject H_0.

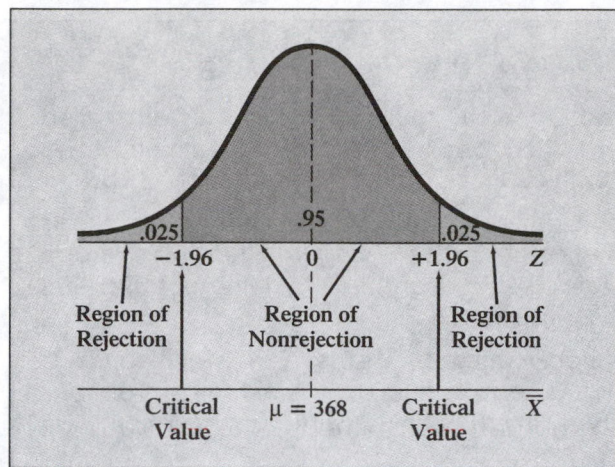

FIGURE 8.2
Testing hypothesis about
mean (σ known) at .05 level
of significance

Suppose that the sample of 25 cereal boxes indicates a sample mean $\overline{X}$ of 372.5 grams and the population standard deviation σ is assumed to remain at 15 grams as specified by the company (see section 6.5). Using equation (8.1), we have

$$Z = \frac{\overline{X} - \mu}{\dfrac{\sigma}{\sqrt{n}}} = \frac{372.5 - 368}{\dfrac{15}{\sqrt{25}}} = +1.50$$

Because $Z = +1.50$, we see that $-1.96 < +1.50 < +1.96$. Thus, as seen in Figure 8.3, our decision is not to reject H_0. We would conclude that the average fill amount is 368 grams. Alternatively, to take into account the possibility of a Type II error, we may phrase the conclusion as "there is no evidence that the average fill is different from 368 grams."

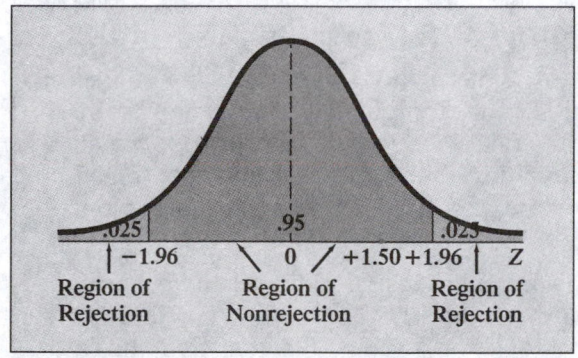

FIGURE 8.3
Testing hypothesis about
mean (σ known) at .05 level
of significance

Now that we have used hypothesis-testing methodology to draw a conclusion about the population mean in situations where the population standard deviation is known, we summarize the steps involved in Exhibit 8.2.

Exhibit 8.2 The Steps in Hypothesis Testing

✓ **1.** State the null hypothesis H_0. The null hypothesis must be stated in statistical terms. In testing whether the average amount filled is 368 grams, the null hypothesis states that μ equals 368.

✓ **2.** State the alternative hypothesis H_1. The alternative hypothesis must be stated in statistical terms. In testing whether the average amount filled is 368 grams, the alternative hypothesis states that μ is not equal to 368 grams.

✓ **3.** Choose the level of significance α. The level of significance is specified according to the relative importance of the risks of committing Type I and Type II errors in a particular situation. We chose $\alpha = .05$.

✓ **4.** Choose the sample size n. The sample size is determined after taking into account the specified risks of committing Type I and Type II errors (i.e., selected levels of α and β) and considering budget constraints in carrying out the study. Here 25 cereal boxes were randomly selected.

✓ **5.** Determine the appropriate statistical technique and corresponding test statistic to use. Because σ is known (i.e., specified by the company to be 15 grams), a Z test was selected.

✓ **6.** Set up the critical values that divide the rejection and nonrejection regions. Once we specify the null and alternative hypotheses and we determine the level of significance and the sample size, the critical values for the appropriate statistical distribution can be found so that the rejection and nonrejection regions can be indicated. Here the values $+1.96$ and -1.96 were used to define these regions because the Z-test statistic refers to the standard normal distribution.

✓ **7.** Collect the data and compute the sample value of the appropriate test statistic. Here, $\overline{X} = 372.5$ grams, so $Z = +1.50$.

✓ **8.** Determine whether the test statistic has fallen into the rejection or the nonrejection region. The computed value of the test statistic is compared with the critical values for the appropriate sampling distribution to determine whether it falls into the rejection or nonrejection region. Here, $Z = +1.50$ was in the region of nonrejection because $-1.96 < Z = +1.50 < +1.96$.

✓ **9.** Make the statistical decision. If the test statistic falls into the nonrejection region, the null hypothesis H_0 cannot be rejected. If the test statistic falls into the rejection region, the null hypothesis is rejected. Here, H_0 was not rejected.

✓ **10.** Express the statistical decision in terms of a particular situation. In our cereal-filling-process example, we concluded that there was no evidence that the average amount of cereal fill was different from 368 grams. No corrective action need be taken.

Problems for Section 8.2

Learning the Basics

8.18 If a .05 level of significance is used in a (two-tailed) hypothesis test, what will you decide if the computed value of the test statistic Z is $+2.21$?

8.19 If a .10 level of significance is used in a (two-tailed) hypothesis test, what will be your decision rule for rejecting a null hypothesis that the population mean is 500 if you use the Z test?

● **8.20** If a .01 level of significance were used in a (two-tailed) hypothesis test, what would be your decision rule for rejecting H_0: $\mu = 12.5$ if you were to use the Z test?

● **8.21** What would be your decision in Problem 8.20 if the computed value of the test statistic Z were -2.61?

Applying the Concepts

8.22 Suppose the director of manufacturing at a clothing factory needs to determine whether a new machine is producing a particular type of cloth according to the manufacturer's specifications, which indicate that the cloth should have a mean breaking strength of 70 pounds and a standard deviation of 3.5 pounds. A sample of 49 pieces reveals a sample mean of 69.1 pounds.
(a) State the null and alternative hypotheses.
(b) Is there evidence that the machine is not meeting the manufacturer's specifications for average breaking strength? (Use a .05 level of significance.)
(c) What will your answer be in (b) if the standard deviation is specified as 1.75 pounds? **WHAT IF?**
(d) What will your answer be in (b) if the sample mean is 69 pounds?

● **8.23** The purchase of a coin-operated laundry is being considered by a potential entrepreneur. **WHAT IF?** The present owner claims that over the past 5 years the average daily revenue has been $675 with a standard deviation of $75. A sample of 30 selected days reveals a daily average revenue of $625.
(a) State the null and alternative hypotheses.
(b) Is there evidence that the claim of the present owner is not valid? (Use a .01 level of significance.)
(c) What will your answer be in (b) if the standard deviation is now $100? **WHAT IF?**
(d) What will your answer be in (b) if the sample mean is $650?

● **8.24** A manufacturer of salad dressings uses machines to dispense liquid ingredients into bottles **WHAT IF?** that move along a filling line. The machine that dispenses dressings is working properly when 8 ounces are dispensed. The standard deviation of the process is 0.15 ounce. A sample of 50 bottles is selected periodically, and the filling line is stopped if there is evidence that the average amount dispensed is different from 8 ounces. Suppose that the average amount dispensed in a particular sample of 50 bottles is 7.983 ounces.
(a) State the null and alternative hypotheses.
(b) Is there evidence that the population average amount is different from 8 ounces? (Use a .05 level of significance.)
(c) What will your answer be in (b) if the standard deviation is specified as 0.05 ounce? **WHAT IF?**
(d) What will your answer be in (b) if the sample mean is 7.952 ounces?

● **8.25** ATMs must be stocked with enough cash to satisfy customers making withdrawals over an **WHAT IF?** entire weekend. On the other hand, if too much cash is unnecessarily kept in the ATMs, the bank is forgoing the opportunity of investing the money and earning interest. Suppose that at a particular branch the expected (i.e., population) average amount of money withdrawn from ATM machines per customer transaction over the weekend is $160 with an expected (i.e., population) standard deviation of $30.
(a) State the null and alternative hypotheses.

WHAT IF?

WHAT IF?

(b) If a random sample of 36 customer transactions is examined and it is observed that the sample mean withdrawal is $172, is there evidence to believe that the true average withdrawal is no longer $160? (Use a .05 level of significance.)

(c) What will your answer be in (b) if the standard deviation is really $24?

(d) What will your answer be in (b) if you use a .01 level of significance?

8.3 THE *p*-VALUE APPROACH TO HYPOTHESIS TESTING

In recent years, with the advent of widely available statistical and spreadsheet software, the concept of the *p*-value is an approach to hypothesis testing that has increasingly gained acceptance.

> The **p-value** is the probability of obtaining a test statistic equal to or more extreme than the result obtained from the sample data, given that the null hypothesis H_0 is really true.

The *p*-value is often referred to as the *observed level of significance,* which is the smallest level at which H_0 can be rejected for a given set of data. The decision rule for rejecting H_0 in the *p*-value approach follows:

- If the *p*-value is greater than or equal to α, the null hypothesis is not rejected.

- If the *p*-value is smaller than α, the null hypothesis is rejected.

To understand the *p*-value approach, let us refer to the cereal-filling-process example of section 8.2. In that section, we tested whether or not the average fill amount was equal to 368 grams (pages 488–489). We obtained a Z value of +1.50 and did not reject the null hypothesis because +1.50 was greater than the lower critical value of −1.96 but less than the upper critical value of +1.96.

Now we use the *p*-value approach. For our *two-tailed test,* we wish to find the probability of obtaining a test statistic Z that is equal to or *more extreme* than 1.50 standard deviation units from the center of a standardized normal distribution. This means that we need to compute the probability of obtaining a Z value greater than +1.50 along with the probability of obtaining a Z value less than −1.50. From Table E.2(b), the probability of obtaining a Z value below −1.50 is .0668. The probability of obtaining a value below +1.50 is .9332. Therefore, the probability of obtaining a value above +1.50 is 1 − .9332 = .0668. Thus, the *p*-value for this two-tailed test is .0668 + .0668 = .1336 (see Figure 8.4). This result may be interpreted to mean that the probability of obtaining a result equal to or more extreme than the one observed is .1336. Because this is greater than $\alpha = .05$, the null hypothesis is not rejected.

FIGURE 8.4

Finding *p*-value for two-tailed test

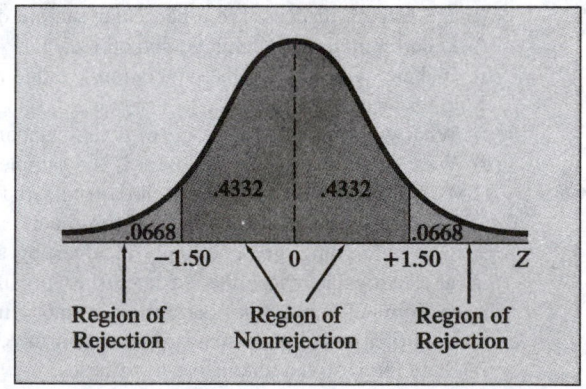

Unless we are dealing with a test statistic that follows the normal distribution, the computation of the p-value can be very difficult. Thus, it is fortunate that software such as Microsoft Excel (see reference 4) routinely presents the p-value as part of the output for hypothesis-testing procedures.

A summary of the p-value approach for hypothesis testing is displayed in Exhibit 8.3.

Exhibit 8.3 Steps in Determining the p-Value

✓ **1.** State the null hypothesis H_0.

✓ **2.** State the alternative hypothesis H_1.

✓ **3.** Choose the level of significance α.

✓ **4.** Choose the sample size n.

✓ **5.** Determine the appropriate statistical technique and corresponding test statistic to use.

✓ **6.** Collect the data and compute the sample value of the appropriate test statistic.

✓ **7.** Calculate the p-value based on the test statistic. This involves

    (a) sketching the distribution under the null hypothesis H_0

    (b) placing the test statistic on the horizontal axis

    (c) shading in the appropriate area under the curve, on the basis of the alternative hypothesis H_1

✓ **8.** Compare the p-value to α.

✓ **9.** Make the statistical decision. If the p-value is greater than or equal to α, the null hypothesis is not rejected. If the p-value is smaller than α, the null hypothesis is rejected.

✓ **10.** Express the statistical decision in terms of the particular situation.

8.3E PERFORMING THE Z TEST OF HYPOTHESIS FOR THE MEAN (σ KNOWN) USING MICROSOFT EXCEL

Overview

◆ *For Quick Results Users*   Use the PHStat add-in to perform a Z test of the hypothesis for the mean when σ is known.

◆ *For Developers*   Implement a worksheet that uses the NORMSINV and NORMSDIST worksheet functions and arithmetic and logical formulas to perform a Z test of the hypothesis for the mean when σ is known.

The 8-3E.XLS workbook file contains the Z test of the hypothesis for the cereal-filling process of section 8.2.

8-3E.XLS

Quick Results Details

To perform a Z test of hypothesis for the mean (σ known), use the One-Sample Tests | Z Test for the Mean, sigma known, choice of the PHStat add-in. As an example, consider the cereal-filling-process problem of section 8.2. To test the hypothesis that the cereal-packaging process is in control and that the packaging process remains at the specified 368 grams, do the following:

❶ If the PHStat add-in has not been previously loaded, load the add-in using the instructions of section S4.2.

❷ Select File | New to open a new workbook (or open the existing workbook into which the hypothesis testing worksheet is to be inserted).

❸ Select PHStat | One-Sample Tests | Z Test for the Mean, sigma known.

❹ In the Z Test for the Mean, sigma known, dialog box (see Figure 8E.1):
  a. Enter 368 in the Null Hypothesis: edit box.
  b. Enter 0.05 in the Level of Significance: edit box.
  c. Enter 15 in the Population Standard Deviation: edit box.
  d. Select the Sample Statistics Known option button and enter 25 in the Sample Size: edit box and 372.5 in the Sample Mean: edit box.
  e. Select the Two-Tailed Test option button.
  f. Enter Z Test of the Hypothesis for the Mean Cereal Packaging Weight in the Output Title: edit box.
  g. Click the OK button.

FIGURE 8E.1
PHStat Z Test for the Mean, sigma known, dialog box

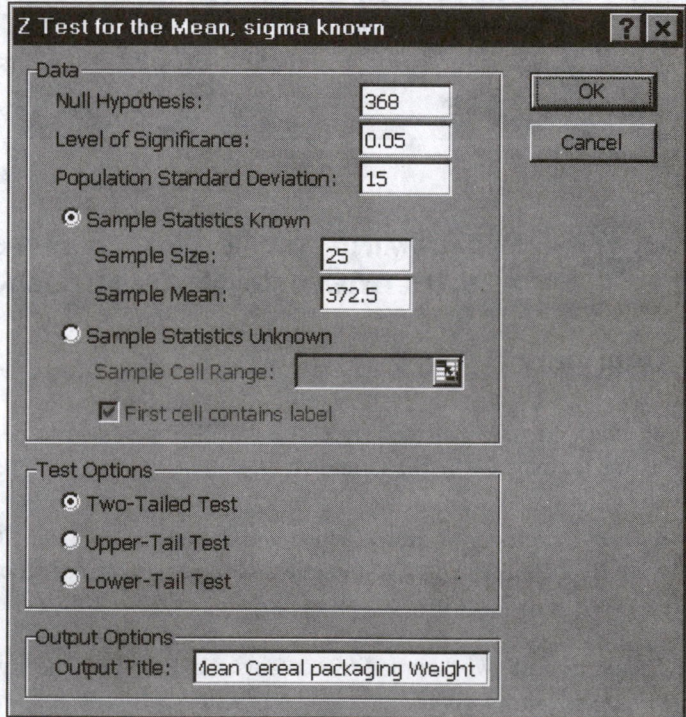

490

The add-in inserts a worksheet containing calculations for the Z test of the hypothesis for the mean similar to the one shown in Figure 8E.2. (*Note:* For other problems in which the sample mean is not known and needs to be calculated, select the Sample Statistics Unknown option button in step 4d and enter the cell range of the sample data in the Sample Cell Range: edit box, dimmed in Figure 8E.1.)

	A	B	C	D
1	Z Test of the Hypothesis for the Mean Cereal Packaging Weight			
2				
3	Null Hypothesis $\mu=$	368		
4	Level of Significance	0.05		
5	Population Standard Deviation	15		
6	Sample Size	25		
7	Sample Mean	372.5		
8	Standard Error of the Mean	3		
9	Z Test Statistic	1.5		
10				
11	Two-Tailed Test			
12	Lower Critical Value	-1.959961082		
13	Upper Critical Value	1.959961082		
14	p-Value	0.133614458		
15	Do not reject the null hypothesis			

FIGURE 8E.2

PHStat add-in for Microsoft Excel output for the Z test of the hypothesis for the mean for the cereal-filling-process problem of section 8.2

▲ WHAT IF EXAMPLE

We can change the values for the level of significance, population standard deviation, sample size, and sample mean in the worksheet inserted by the add-in to see their effects on the test of hypothesis for the mean. For example, if we change the population standard deviation in cell B5 from 15 to 10, we will observe that the two-tailed p-value changes from .1336 to .0244, thereby causing us to reject the null hypothesis. To facilitate comparisons among many different alternative values, we can create scenarios using the Scenario Manager as discussed at the end of section 2.2E.

Developer Details

We can use the NORMSINV (see section 6.1E) and NORMSDIST worksheet functions as the basis for performing the Z test of the hypothesis for the mean when σ is known. Table 8E.1 presents a Hypothesis Test sheet design that performs a Z test for the hypothesis for the mean for the cereal-filling-process problem of section 8.2. In this design, we use the NORMSINV function to return the lower and upper critical values and use the NORMSDIST function, the format of which is:

$$\text{NORMSDIST}(Z\ value)$$

to return the probability of less than the Z value computed in cell B9. Because the design needs to perform a two-tailed test, we can multiply by 2 the result of subtracting the NORMSDIST value from 1—representing the probability of one of the tails—in cell B14 to calculate the p-value.

This design also uses the IF function (see section 3.2E) to compare the p-value of cell B14 to the level of significance entered in cell B4. Based on this comparison, the function displays a message informing the user whether or not to reject the null hypothesis.

To implement the Table 8E.1 design, do the following:

❶ Select File | New to open a new workbook (or open the existing workbook into which the hypothesis testing worksheet is to be inserted).

❷ Select an unused worksheet (or select Insert | Worksheet if there are none) and rename the sheet Hypothesis.

❸ Enter the title and labels for column A as shown in Table 8E.1. (The Symbol font can be used to enter the μ symbol in cell A3.)

❹ Select the cell range A11:B11 and click the Merge and Center button on the formatting toolbar (see section S3.11).

❺ Enter the null hypothesis value, the level of significance, population standard deviation, sample size, and sample mean in the cell range B3:B7. Enter 368 in cell B3, 0.05 in cell B4, 15 in cell B5, 25 in cell B6, and 372.5 in cell B7.

❻ Enter the formulas for the cell ranges B8:B9 and B12:B14 as shown in Table 8E.1.

❼ Enter the formula for cell A15 as a single continuous line (in Table 8E.1, this formula has been typeset as two lines).

❽ Select the cell range A15:B15 and click the Merge and Center button on the formatting toolbar.

The completed worksheet will be similar to the one shown in Figure 8E.2. Had the sample mean been unknown, we could have entered a formula using the AVERAGE function in cell B7 to calculate it.

Table 8E.1 Hypothesis test sheet design for the cereal-filling-process problem of section 8.2

	A	B
1	Z Test of the Hypothesis for the Mean Cereal Packaging Weight	
2		
3	Null Hypothesis $\mu=$	xxx
4	Level of Significance	.xx
5	Population Standard Deviation	xxx
6	Sample Size	xxx
7	Sample Mean	xxxx
8	Standard Error of the Mean	=B5/SQRT(B6)
9	Z Test Statistic	=(B7-B3)/B8
10		
11	Two-Tailed Test	
12	Lower Critical Value	=NORMSINV(B4/2)
13	Upper Critical Value	=NORMSINV(1-B4/2)
14	*p*-Value	=2*(1-NORMSDIST(ABS(B9)))
15	=IF(B14<$B$4,"Reject the null hypothesis", "Do not reject the null hypothesis")	

As with the worksheet inserted by the add-in using the quick results method of this section, we can change the values for the level of significance, population standard deviation, sample size, and sample mean in this worksheet to see their effects on the test of hypothesis for the mean. (For further details, see the What If Example on page 495.)

Problems for Section 8.3

Learning the Basics

• **8.26** Suppose that in a two-tailed hypothesis test you compute the value of the test statistic Z as $+2.00$. What is the p-value?

• **8.27** In Problem 8.26, what is your statistical decision if you test the null hypothesis at the .10 level of significance?

8.28 Suppose that in a two-tailed hypothesis test you compute the value of the test statistic Z as -1.38. What is the p-value?

8.29 In Problem 8.28, what is your statistical decision if you test the null hypothesis at the .01 level of significance?

Applying the Concepts

8.30 The director of manufacturing at a clothing factory needs to determine whether a new machine is producing a particular type of cloth according to the manufacturer's specifications, which indicate that the cloth should have a mean breaking strength of 70 pounds and a standard deviation of 3.5 pounds. A sample of 49 pieces reveals a sample mean of 69.1 pounds.
  (a) Compute the p-value and interpret its meaning.
  (b) What is your statistical decision if you test the null hypothesis at the .05 level of significance?
  (c) Is there evidence that the machine is not meeting the manufacturer's specifications for average breaking strength?
  (d) Compare your conclusions here with those of (b) in Problem 8.22 on page 491.

• **8.31** An entrepreneur is considering the purchase of a coin-operated laundry. The present owner claims that over the past 5 years, the average daily revenue has been $675 with a standard deviation of $75. A sample of 30 selected days reveals a daily average revenue of $625.
  (a) Compute the p-value and interpret its meaning.
  (b) What is your statistical decision if you test the null hypothesis at the .01 level of significance?
  (c) Is there evidence that the claim of the present owner is not valid?
  (d) Compare your conclusions here with those of (b) in Problem 8.23 on page 491.

• **8.32** A manufacturer of salad dressings uses machines to dispense liquid ingredients into bottles that move along a filling line. The machine that dispenses dressings is working properly when 8 ounces are dispensed. The standard deviation of the process is 0.15 ounce. A sample of 50 bottles is selected periodically, and the filling line is stopped if there is evidence that the average amount dispensed is different from 8 ounces. Suppose that the average amount dispensed in a particular sample of 50 bottles is 7.983 ounces.
  (a) Compute the p-value and interpret its meaning.
  (b) What is your statistical decision if you test the null hypothesis at the .05 level of significance?

(c) Is there evidence that the population average amount is different from 8 ounces?

(d) Compare your conclusions here with those of (b) in Problem 8.24 on page 491.

• **8.33** ATMs must be stocked with enough cash to satisfy customers making withdrawals over an entire weekend. On the other hand, if too much cash is unnecessarily kept in the ATMs, the bank is forgoing the opportunity of investing the money and earning interest. Suppose that at a particular branch the expected (i.e., population) average amount of money withdrawn from ATM machines per customer transaction over the weekend is $160 with an expected (i.e., population) standard deviation of $30. Suppose that a random sample of 36 customer transactions is examined and it is observed that the sample mean withdrawal is $172.

(a) Compute the *p*-value and interpret its meaning.

(b) What is your statistical decision if you test the null hypothesis at the .05 level of significance?

(c) Is there evidence to believe that the true average withdrawal is no longer $160?

(d) Compare your conclusions here with those of (b) in Problem 8.25 on page 491.

8.4   A CONNECTION BETWEEN CONFIDENCE INTERVAL ESTIMATION AND HYPOTHESIS TESTING

Both in this chapter and in chapter 7 we examined the two major components of statistical inference—confidence interval estimation and hypothesis testing. Although they are based on the same set of concepts, we used them for different purposes. In chapter 7 we used confidence intervals to estimate parameters and in this chapter we used hypothesis testing for making decisions about specified values of population parameters.

For example, in section 8.2 we first attempted to determine whether the population average fill amount was different from 368 grams by using equation (8.1)

$$Z = \frac{\overline{X} - \mu}{\dfrac{\sigma}{\sqrt{n}}}$$

Instead of testing the null hypothesis that $\mu = 368$ grams, we can also reach the conclusion by obtaining a confidence-interval estimate of μ. If the hypothesized value of $\mu = 368$ falls into the interval, the null hypothesis is not rejected. That is, the value 368 would not be considered unusual for the data observed. On the other hand, if the hypothesized value does not fall into the interval, the null hypothesis is rejected, because 368 grams are then considered an unusual value. Using equation (7.1), the confidence interval estimate is set up from the following data:

$$n = 25, \quad \overline{X} = 372.5 \text{ grams}, \quad \sigma = 15 \text{ grams (specified by the company)}$$

For a confidence level of 95% (corresponding to a .05 level of significance—i.e., $\alpha = .05$), we have

$$\overline{X} \pm Z\frac{\sigma}{\sqrt{n}}$$

$$372.5 \pm (1.96)\,\frac{15}{\sqrt{25}}$$

$$372.5 \pm 5.88$$

so that

$$366.62 \leq \mu \leq 378.38$$

Because the interval includes the hypothesized value of 368 grams, we do not reject the null hypothesis, and we conclude that there is no evidence the mean fill amount over the entire filling process is not 368 grams. This is the same decision we reached by using hypothesis-testing methodology.

Problems for Section 8.4

Applying the Concepts

• **8.34** The manager of a paint supply store wants to estimate the correct amount of paint contained in 1-gallon cans purchased from a nationally known manufacturer. It is known from the manufacturer's specifications that the standard deviation of the amount of paint is equal to .02 gallon. A random sample of 50 cans is selected, and the average amount of paint per 1-gallon can is 0.995 gallon.
  (a) State the null and alternative hypotheses.
  (b) Is there evidence that the average amount is different from 1.0 gallon (use $\alpha = .01$)?
  (c) Compare the conclusions reached in (b) with those from Problem 7.7 on page 423. Are the conclusions the same? Why?

8.35 The quality control manager at a light bulb factory needs to estimate the average life of a large shipment of light bulbs. The process standard deviation is known to be 100 hours. A random sample of 64 light bulbs indicates a sample average life of 350 hours.
  (a) State the null and alternative hypotheses.
  (b) At the .05 level of significance is there evidence that the average life is different from 375 hours?
  (c) Compare the conclusions reached in (b) with those from Problem 7.8 on page 423. Are the conclusions the same? Why?

8.36 The inspection division of the Lee County Weights and Measures Department is interested in estimating the actual amount of soft drink that is placed in 2-liter bottles at the local bottling plant of a large nationally known soft-drink company. The bottling plant has informed the inspection division that the standard deviation for 2-liter bottles is .05 liter. A random sample of one hundred 2-liter bottles obtained from this bottling plant indicates a sample average of 1.99 liters.
  (a) State the null and alternative hypotheses.
  (b) At the .05 level of significance is there evidence that the average amount in the bottles is different from 2.0 liters?
  (c) Compare the conclusions reached in (b) with those from Problem 7.9 on page 423. Are the conclusions the same? Why?

◆ 8.5 ◆ ONE-TAILED TESTS

In section 8.2 we used hypothesis-testing methodology to examine the question of whether or not the average fill amount over the entire filling process (i.e., the population) is 368 grams. The alternative hypothesis ($H_1: \mu \neq 368$) contains two possibilities: Either the average is less than 368 grams or the average is more than 368 grams. For this reason, the rejection region is divided into the two tails of the sampling distribution of the mean. Furthermore, as we have just observed in the previous section, because a confidence interval estimate of the mean contains a lower and upper limit, respectively, corresponding to the left- and right-tail critical values from the sampling distribution of the mean, we are able to use the confidence interval to do a test of the null hypothesis that the average amount of fill over the entire filling process is 368 grams.

In some situations, however, the alternative hypothesis focuses in a *particular direction*. For example, the chief financial officer (CFO) of the food packaging company is mainly concerned with excess because if more than 368 grams of cereal are actually being filled per box but the price charged to the customer is based on the 368 grams labeled on the box, the company is losing money unnecessarily. Therefore, she is interested in whether the average fill for the entire filling process was *above* 368 grams. To her, strictly from a financial point of view with respect to her responsibility as CFO for the company, unless the sample mean is significantly above 368 grams, the process is working properly.

The Critical Value Approach

For the CFO the null and alternative hypotheses are stated as follows:

$$H_0: \mu \leq 368 \text{ (process is working properly)}$$
$$H_1: \mu > 368 \text{ (process is not working properly)}$$

The rejection region here is entirely contained in the upper tail of the sampling distribution of the mean because we want to reject H_0 only when the sample mean is significantly above 368 grams. When such a situation occurs where the entire rejection region is contained in one tail of the sampling distribution of the test statistic, it is called a **one-tailed** or **directional test.** If we again choose a level of significance α of .05, the critical value on the Z distribution can be determined. As seen from Table 8.2 and Figure 8.5, because the entire rejection region is in the upper tail of the standard normal distribution and contains an area of .05, the area below the critical value must be .95; thus, the critical value of the Z-test statistic is $+1.645$, the average of $+1.64$ and $+1.65$. (We should note here that some statisticians *round off* to two decimal places and select $+1.64$ as the critical value, whereas others *round up* to $+1.65$. We prefer to interpolate between the areas .9495 and .9505 so as to select the critical value with upper-tail area as close to .05 as possible. Thus, we take the average of $+1.64$ and $+1.65$.) The decision rule is

Reject H_0 if $Z > +1.645$;

otherwise do not reject H_0.

Table 8.2 *Obtaining the critical value of the Z test statistic from the standard normal distribution for a one-tailed test with* $\alpha = .05$

Z	.00	.01	.02	.03	.04	.05	.06	.07	.08	.09
⋮	⋮	⋮	⋮	⋮	⋮	⋮	⋮	⋮	⋮	⋮
1.3	.9032	.9049	.9066	.9082	.9099	.9115	.9131	.9147	.9162	.9177
1.4	.9192	.9207	.9222	.9236	.9251	.9265	.9279	.9292	.9306	.9319
1.5	.9332	.9345	.9357	.9370	.9382	.9394	.9406	.9418	.9429	.9441
1.6	.9452	.9463	.9474	.9484 →	.9495	.9505	.9515	.9525	.9535	.9545

Source: Extracted from Table E.2(b).

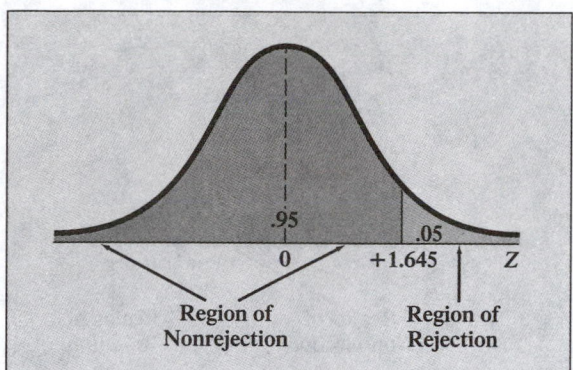

FIGURE 8.5
One-tailed test of hypothesis
for mean (σ known) at the
.05 level of significance

Using the Z test given by equation (8.1) on the information obtained from the sample drawn by the operations manager,

$$n = 25, \quad \overline{X} = 372.5 \text{ grams}, \quad \sigma = 15 \text{ grams (specified by the company)}$$

we have

$$Z = \frac{\overline{X} - \mu}{\dfrac{\sigma}{\sqrt{n}}} = \frac{372.5 - 368}{\dfrac{15}{\sqrt{25}}} = +1.50$$

Because $Z = +1.50 < +1.645$, our decision is not to reject H_0, and we conclude that there is no evidence that the average fill per box over the entire filling process is above 368 grams. That is, even though the sample mean $\overline{X}$ exceeded 368 grams, the result from the sample is deemed due to *chance* or *sampling error;* it is not statistically significant.

The *p*-Value Approach

To understand the *p*-value approach for the one-tailed test, we compute the probability of obtaining a value *either* greater than the computed test statistic *or* less than the computed test statistic, depending on the direction of the alternative hypothesis. To illustrate the computation of the *p*-value for the one-tailed test, we again refer to the cereal-filling-process example. For the chief financial officer (CFO), the null and alternative hypotheses are

$$H_0: \mu \leq 368 \text{ (process is working properly)}$$
$$H_1: \mu > 368 \text{ (process is not working properly)}$$

Because the alternative hypothesis indicates a rejection region entirely in the *upper* tail of the sampling distribution of the Z-test statistic, we need only find the probability of obtaining a Z value above $+1.50$. From Table E.2(b), the probability of obtaining a Z value above $+1.50$ is $1 - .9332 = .0668$ (see Figure 8.6). Because this *p*-value is greater than the selected level of significance ($\alpha = .05$), the null hypothesis is not rejected.

FIGURE 8.6

Determining *p*-value for one-tailed test

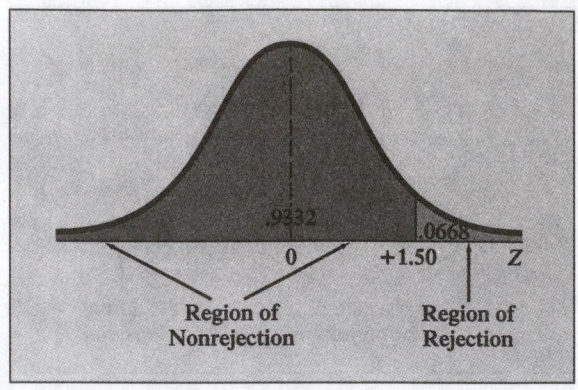

To perform one-tailed tests of hypotheses, we must properly formulate H_0 and H_1. A summary of the null and alternative hypotheses for one-tailed tests is presented in Exhibit 8.4.

Exhibit 8.4   The Null and Alternative Hypotheses in One-Tailed Tests

To summarize some key points about the null and alternative hypotheses in one-tailed tests:

✓ **1.** The null hypothesis H_0 is the hypothesis that is always tested.

✓ **2.** The alternative hypothesis H_1 is set up as the opposite of the null hypothesis and represents the conclusion supported if the null hypothesis is rejected.

✓ **3.** The null hypothesis H_0 always refers to a specified value of the *population parameter* (such as μ), not a *sample statistic* (such as $\overline{X}$).

✓ **4.** The statement of the null hypothesis *always* contains an equal sign regarding the specified value of the parameter (e.g., H_0: $\mu \leq 368$ grams).

✓ **5.** The statement of the alternative hypothesis *never* contains an equal sign regarding the specified value of the parameter (e.g., H_1: $\mu > 368$ grams).

To further illustrate these points, consider Examples 8.3 and 8.4:

Example 8.3   *Stating the Null and Alternative Hypotheses in One-Tailed Tests*

Toyota claims that a new model car will average *at least* 30 miles per gallon in highway driving. If you were planning an experiment to test this claim, what would be the null and alternative hypotheses?

SOLUTION

The words "at least" contain an equal sign so this must be part of the null hypothesis. Therefore,

$$H_0: \mu \geq 30 \text{ miles per gallon}$$
$$H_1: \mu < 30 \text{ miles per gallon}$$

Example 8.4 *Stating the Null and Alternative Hypotheses in One-Tailed Tests*

The average age of policyholders of term life insurance at Empire Insurance Company was 48 for several years. As the company has expanded and provided more term policies across the nation, the chief financial officer would like to find evidence that the average age has *decreased*. If you were planning a survey to test this claim, what would be the null and alternative hypotheses?

SOLUTION

The word "decreased" in what we would like to find evidence of does not have an equal sign so it must be part of the alternative hypothesis. Therefore,

$$H_0: \mu \geq 48 \text{ years}$$
$$H_1: \mu < 48 \text{ years}$$

8.5E PERFORMING A ONE-TAILED Z TEST OF HYPOTHESIS FOR THE MEAN (σ KNOWN) USING MICROSOFT EXCEL

OVERVIEW

◆ *For Quick Results Users*   Use the PHStat add-in to perform a one-tailed Z test of the hypothesis for the mean when σ is known.

◆ *For Developers*   Modify the worksheet implemented in section 8.3E by adding formulas that use the NORMSINV, NORMSDIST, and IF worksheet functions and arithmetic and logical formulas to perform one-tailed Z tests of the hypothesis for the mean when σ is known.

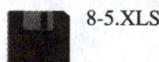
8-5.XLS

The 8-5E.XLS workbook file contains calculations for one-tailed (and two-tailed) Z tests of the hypothesis for the cereal-filling process of sections 8.2 and 8.5.

Quick Results Details

To perform a one-tailed Z test of the hypothesis for the mean, use the quick results procedure of section 8.3E, selecting the Upper-Tail Test or Lower-Tail Test option button, as appropriate, instead of the Two-Tailed Test option button in step 4e. For the cereal-filling-process problem of section 8.5, the add-in would produce a worksheet similar to the one shown in Figure 8E.3.

FIGURE 8E.3

One-tailed *Z* test of the
hypothesis for the mean for
the cereal-filling-process problem of section 8.5

	A	B	C	D
1	Z Test of the Hypothesis for the Mean Cereal Packaging Weight			
2				
3	Null Hypothesis $\mu=$	368		
4	Level of Significance	0.05		
5	Population Standard Deviation	15		
6	Sample Size	25		
7	Sample Mean	372.5		
8	Standard Error of the Mean	3		
9	Z Test Statistic	1.5		
10				
11	Upper-Tail Test			
12	Upper Critical Value	1.644853		
13	p-Value	0.066807229		
14	Do not reject the null hypothesis			

Developer Details

We modify the worksheet design of Table 8E.1 of section 8.3E in order to perform one-
tailed *Z* tests of the hypothesis for the mean when σ is known. Table 8E.2 presents modifications to the earlier design for both a lower-tail test and an upper-tail test. To implement
these additions, do the following:

❶ Implement the worksheet based on the design of Table 8E.1 of section 8.3E.

❷ Enter the labels and formulas for either the lower-tail or upper-tail test (or both) as
shown in Table 8E.2. Note that the formulas for cells A20 and A25, which have
been typeset as two lines, should be entered as single continuous lines.

❸ Select the cell range A17:B17 and click the Merge and Center button on the formatting toolbar. Repeat this step for the cell ranges A20:B20, A22:B22, and A25:B25.

A completed worksheet with both the one-tailed test and the original two-tailed test will be
similar to the one shown in Figure 8E.4.

**Table 8E.2 Modifications for the Hypothesis Testing sheet design
of Table 8E.1**

	A	B
16		
17	Lower-Tail Test	
18	Lower Critical Value	=NORMSINV(B4)
19	*p*-Value	=NORMSDIST(B9)
20	=IF(B19<$B$4,"Reject the null hypothesis", "Do not reject the null hypothesis")	
21		
22	Upper-Tail Test	
23	Upper Critical Value	=NORMSINV(1-B4)
24	*p*-Value	=1−NORMSDIST(B9)
25	=IF(B24<$B$4,"Reject the null hypothesis", "Do not reject the null hypothesis")	

	A	B	C	D
1	Z Test of the Hypothesis for the Mean Cereal Packaging Weight			
2				
3	Null Hypothesis $\mu=$	368		
4	Level of Significance	0.05		
5	Population Standard Deviation	15		
6	Sample Size	25		
7	Sample Mean	372.5		
8	Standard Error of the Mean	3		
9	Z Test Statistic	1.5		
10				
11	Two-Tailed Test			
12	Lower Critical Value	-1.959961082		
13	Upper Critical Value	1.959961082		
14	p-Value	0.133614458		
15	Do not reject the null hypothesis			
16				
17	Lower-Tail Test			
18	Lower Critical Value	-1.644853		
19	p-Value	0.933192771		
20	Do not reject the null hypothesis			
21				
22	Upper-Tail Test			
23	Upper Critical Value	1.644853		
24	p-Value	0.066807229		
25	Do not reject the null hypothesis			

FIGURE 8E.4

Worksheet containing one-tailed and two-tailed Z tests of the hypothesis for the mean

Problems for Section 8.5

Learning the Basics

8.37 What is the *upper-tail* critical value of the Z-test statistic at the .01 level of significance?

8.38 In Problem 8.37 what is the statistical decision if the computed value of the Z-test statistic is 2.39?

•**8.39** What is the *lower-tail* critical value of the Z-test statistic at the .10 level of significance?

•**8.40** In Problem 8.39 what is the statistical decision if the computed value of the Z-test statistic is -1.15?

•**8.41** Suppose that in a one-tailed hypothesis test where you reject H_0 only in the *upper* tail, you compute the value of the test statistic Z to be $+2.00$. What is the p-value?

•**8.42** In Problem 8.41 what would be your statistical decision if you tested the null hypothesis at the .10 level of significance?

8.43 Suppose that in a one-tailed hypothesis test where you reject H_0 only in the *lower* tail, you compute the value of the test statistic Z as -1.38. What would the p-value be?

8.44 In Problem 8.43, what is your statistical decision if you tested the null hypothesis at the .01 level of significance?

Applying the Concepts

8.45 The CFO of a company selling computer software through megastores claims that average monthly profits throughout the country will *exceed* $1 billion. You are asked to test this claim by examining the monthly profits from a sample of megastores. State the null hypothesis H_0 and the alternative hypothesis H_1.

•**8.46** The Glen Valley Steel Company manufactures steel bars. If the production process is working properly, it turns out steel bars with an average length of *at least* 2.8 feet with a standard deviation of 0.20 foot (as determined from engineering specifications on the production equipment involved). Longer steel bars can be used or altered, but shorter bars must be scrapped. A sample of 25 bars is selected from the production line. The sample indicates an average length of 2.73 feet. The company wishes to determine whether the production equipment needs an immediate adjustment.

(a) State the null and alternative hypotheses.
(b) If the company wishes to test the hypothesis at the .05 level of significance, what decision would it make using the critical value approach to hypothesis testing?
(c) If the company wishes to test the hypothesis at the .05 level of significance, what decision would it make using the p-value approach to hypothesis testing?
(d) Interpret the meaning of the p-value in this problem.
(e) Compare your conclusions in (b) and (c).

8.47 The director of manufacturing at a clothing factory needs to determine whether a new machine is producing a particular type of cloth according to the manufacturer's specifications, which indicate that the cloth should have a mean breaking strength of 70 pounds and a standard deviation of 3.5 pounds. The director is concerned that if the mean breaking strength is actually *less* than 70 pounds, the company will face too many lawsuits. A sample of 49 pieces reveals a sample mean of 69.1 pounds.

(a) State the null and alternative hypotheses.
(b) At the .05 level of significance, using the critical value approach to hypothesis testing, is there evidence that the mean breaking strength is less than 70 pounds?
(c) At the .05 level of significance, using the p-value approach to hypothesis testing, is there evidence that the mean breaking strength is less than 70 pounds?
(d) Interpret the meaning of the p-value in this problem.
(e) Compare your conclusions in (b) and (c).

•**8.48** A manufacturer of salad dressings uses machines to dispense liquid ingredients into bottles that move along a filling line. The machine that dispenses dressings is working properly when 8 ounces are dispensed. The standard deviation of the process is 0.15 ounce. A sample of 50 bottles is selected periodically, and the filling line is stopped if there is evidence that the average amount dispensed is actually *less* than 8 ounces. Suppose that the average amount dispensed in a particular sample of 50 bottles is 7.983 ounces.

(a) State the null and alternative hypotheses.
(b) At the .05 level of significance, using the critical value approach to hypothesis testing, is there evidence that the average amount dispensed is less than 8 ounces?
(c) At the .05 level of significance, using the p-value approach to hypothesis testing, is there evidence that the average amount dispensed is less than 8 ounces?
(d) Interpret the meaning of the p-value in this problem.
(e) Compare your conclusions in (b) and (c).

•**8.49** The policy of a particular bank branch is that its ATMs must be stocked with enough cash to satisfy customers making withdrawals over an entire weekend. Customer goodwill depends on such services meeting customer needs. At this branch the expected (i.e., population) average amount of money withdrawn from ATM machines per customer transaction over the weekend is $160 with an expected (i.e., population) standard deviation of $30. Suppose that a random sample of 36 customer transactions is examined and it is observed that the sample mean withdrawal is $172.

(a) State the null and alternative hypotheses.
(b) At the .05 level of significance, using the critical value approach to hypothesis testing, is there evidence to believe that the true average withdrawal is greater than $160?
(c) At the .05 level of significance, using the p-value approach to hypothesis testing, is there evidence to believe that the true average withdrawal is greater than $160?
(d) Interpret the meaning of the p-value in this problem.
(e) Compare your conclusions in (b) and (c).

8.50 A pharmaceutical company claims to have produced a pill that if taken daily for 1 month will *reduce* systolic blood pressure of hypertensive patients by an average of 25 mg/mm. If you were asked to evaluate an experimental trial conducted by this company on a random sample of patients who consent to participate, what would the null hypothesis H_0 and alternative hypothesis H_1 be?

 ## 8.6 *t* TEST OF HYPOTHESIS FOR THE MEAN (σ UNKNOWN)

In most hypothesis-testing situations dealing with numerical data, the standard deviation σ of the population is unknown. However, the actual standard deviation of the population is estimated by computing S, the standard deviation of the sample. If the population is assumed to be normally distributed, you may recall from section 7.2 that the sampling distribution of the mean will follow a t distribution with $n - 1$ degrees of freedom. The test statistic t for determining the difference between the sample mean $\overline{X}$ and the population mean μ when the sample standard deviation S is used is given by

t Test of Hypothesis for a Population Mean (σ Unknown)

$$t = \frac{\overline{X} - \mu}{\dfrac{S}{\sqrt{n}}} \qquad (8.2)$$

where the test statistic t follows a t distribution having $n - 1$ degrees of freedom.

To illustrate the use of this t test, we return to the Saxon Plumbing Company example presented on page 427. In one of the efforts to maintain internal controls on sales, the auditor takes a sample of sales invoices at the end of the month to evaluate the average amount listed on the sales invoices for the warehouse in that month. Over the past five years, the average amount of each sales invoice for customers outside the suburban area in which Saxon Plumbing Company is located is $120. Because shipping costs are affected by delivery distance, it is important that the auditor carefully monitor the average sales amount. The following data display the amounts listed (in dollars) in a random sample of 12 sales invoices that were selected from the population of sales invoices for customers outside the suburban area during the past month.

108.98	152.22	111.45	110.59	127.46	107.26
93.32	91.97	111.56	75.71	128.58	135.11

DATA FILE
INVOICES.XLS

Because the auditor is interested in whether or not there is evidence of a change in the average amount of these sales invoices from the $120 amount that has been the monthly average over the past few years, the test is two-tailed and the following null and alternative hypotheses are established:

$$H_0: \mu = \$120$$
$$H_1: \mu \neq \$120$$

◆ *Critical Value Approach*   For a given sample size n, the test statistic t follows a t distribution with $n - 1$ degrees of freedom. If a level of significance of $\alpha = .05$ is selected, the critical values of the t distribution with $12 - 1 = 11$ degrees of freedom can be obtained from Table E.3, as illustrated in Figure 8.7 and Table 8.3. Because the alternative hypothesis H_1 that $\mu \neq \$120$ is *nondirectional,* the area in the rejection region of the t distribution's left (lower) tail is .025 and the area in the rejection region of the t distribution's right (upper) tail is also .025.

From the t table as given in Table E.3, a replica of which is shown in Table 8.3, the critical values are ±2.2010. The decision rule is

$$\text{Reject } H_0 \text{ if } t < t_{11} = -2.2010$$
$$\text{or if } t > t_{11} = +2.2010;$$
$$\text{otherwise do not reject } H_0.$$

FIGURE 8.7
Testing hypothesis about mean (σ unknown) at .05 level of significance with 11 degrees of freedom

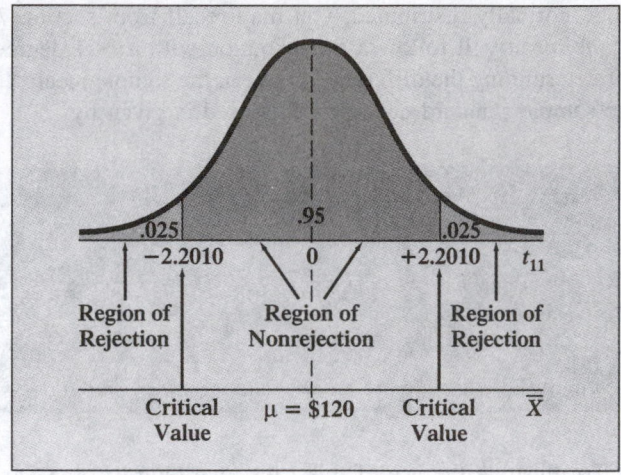

Table 8.3 *Determining the critical value from the t table for an area of .025 in each tail with 11 degrees of freedom*

DEGREES OF FREEDOM	UPPER-TAIL AREAS					
	.25	.10	.05	.025	.01	.005
1	1.0000	3.0777	6.3138	12.7062	31.8207	63.6574
2	0.8165	1.8856	2.9200	4.3027	6.9646	9.9248
3	0.7649	1.6377	2.3534	3.1824	4.5407	5.8409
4	0.7407	1.5332	2.1318	2.7764	3.7469	4.6041
5	0.7267	1.4759	2.0150	2.5706	3.3649	4.0322
6	0.7176	1.4398	1.9432	2.4469	3.1427	3.7074
7	0.7111	1.4149	1.8946	2.3646	2.9980	3.4995
8	0.7064	1.3968	1.8595	2.3060	2.8965	3.3554
9	0.7027	1.3830	1.8331	2.2622	2.8214	3.2498
10	0.6998	1.3722	1.8125	2.2281	2.7638	3.1693
11	0.6974	1.3634	1.7959	→ 2.2010	2.7181	3.1058

Source: Extracted from Table E.3.

For the data set pertaining to the random sample of $n = 12$ sales invoices, using equations (3.1) and (3.10) on pages 155 and 167, respectively, we compute

$$\overline{X} = \frac{\displaystyle\sum_{i=1}^{n} X_i}{n} = \$112.85 \quad \text{and} \quad S = \sqrt{\frac{\displaystyle\sum_{i=1}^{n} (X_i - \overline{X})^2}{n - 1}} = \$20.80$$

From equation (8.2) we then obtain

$$t = \frac{\overline{X} - \mu}{\dfrac{S}{\sqrt{n}}} = \frac{112.85 - 120}{\dfrac{20.80}{\sqrt{12}}} = -1.19$$

Because $t = -1.19$ falls within the nonrejection region between the critical values of $t_{11} = \pm 2.2010$, we cannot reject H_0. There is no evidence to believe that the average monthly sales invoice has changed from the long-term average of \$120—our observed difference is insignificant and due to chance.

◆ **p-Value Approach** Software such as Microsoft Excel (reference 4) provides p-values as part of the output for various hypothesis-testing procedures. As an example, using the PHStat add-in for Microsoft Excel for the sales invoice data, the two-tailed p-value is equal to .26 (see Figure 8.8). Because our p-value or observed *level of significance* is greater than α, our *specified level of significance,* we cannot reject H_0. (That is, .26 > .05, so we don't reject H_0.) If the null hypothesis were true, the probability that a sample of 12 invoices could have a monthly average that differs by \$7.15 or more from the stated \$120 is .26. This leads us to conclude that there is no evidence to believe that the average monthly sales invoice has changed from the long-term average of \$120. The auditor need not make any recommendation to management about altering shipping policies for items purchased outside the suburban community.

	A	B	C
1	*t* Test for the Hypothesis of the Mean Invoice Amount		
2			
3	Null Hypothesis $\mu=$	120	
4	Level of Significance	0.05	
5	Sample Size	12	
6	Sample Mean	112.8508333	
7	Sample Standard Deviation	20.7979918	
8	Standard Error of the Mean	6.003863082	
9	Degrees of Freedom	11	
10	*t* Test Statistic	-1.19076111	
11			
12	Two-Tailed Test		
13	Lower Critical Value	-2.200986273	
14	Upper Critical Value	2.200986273	
15	*p*-Value	0.258809315	
16	Do not reject the null hypothesis		

FIGURE 8.8

PHStat add-in output with p-value for one-sample t test of sales invoices

Example 8.5 applies the 10 steps of hypothesis testing displayed in Exhibits 8.2 and 8.3 to the sales invoices example.

Example 8.5 *t Test of Hypothesis for the Mean (σ Unknown)*

Using the sales invoice data on page 507, test the claim that the average monthly amount for items sold outside the suburban region is \$120.

SOLUTION

Steps 1 and 2:
$$H_0: \mu = \$120$$
$$H_1: \mu \neq \$120$$

Step 3:
$$\alpha = .05$$

Step 4:
$$n = 12$$

Step 5: Because σ is unknown, we choose the one-sample t test with test statistic t given by equation (8.2):

$$t = \frac{\overline{X} - \mu}{\dfrac{S}{\sqrt{n}}}$$

Step 6: We use Table E.3 to develop the following decision rule as illustrated in Figure 8.7 on page 508:

Reject H_0 if $t < t_{11} = -2.2010$
or if $t > t_{11} = +2.2010$;

otherwise do not reject H_0.

Step 7: We collect the data and evaluate the assumptions of the t test (to be described in the following section). We then compute the t-test statistic:

$$t = \frac{\overline{X} - \mu}{\dfrac{S}{\sqrt{n}}} = \frac{112.85 - 120}{\dfrac{20.80}{\sqrt{12}}} = -1.19$$

Steps 8, 9, and 10: The critical values for the t-test statistic are $t_{11} = \pm 2.2010$. Because $-2.2010 < t = -1.19 < +2.2010$, we cannot reject H_0. (Alternatively, the p-value, .26, is greater than α of .05, so we cannot reject H_0.) There is no evidence to believe that the average monthly sales invoice has changed from the long-term average of \$120—our observed difference is insignificant and due to chance. The auditor need not make any recommendation to management about altering shipping policies for items purchased outside the suburban community.

The one-sample t test can be either a two-tailed test or a one-tailed test, depending on whether the alternative hypothesis is *nondirectional* or *directional,* respectively. If the alternative hypothesis is nondirectional, as in our sales invoice example, we are looking to reject the null hypothesis that the value of the parameter is a specified amount such as $\mu = \$120$. In this two-tailed test we reject H_0 if there is evidence from the sample that the value of the parameter being tested is likely to be either significantly more or significantly less than this hypothesized amount. In a one-tailed test, the alternative hypothesis is directional. We reject H_0 only if there is evidence from the sample that the value of the parameter being tested is too small or too large, depending on the direction specified in the alternative hypothesis. The regions of rejection and nonrejection for these one-sample t tests are depicted in Figure 8.9.

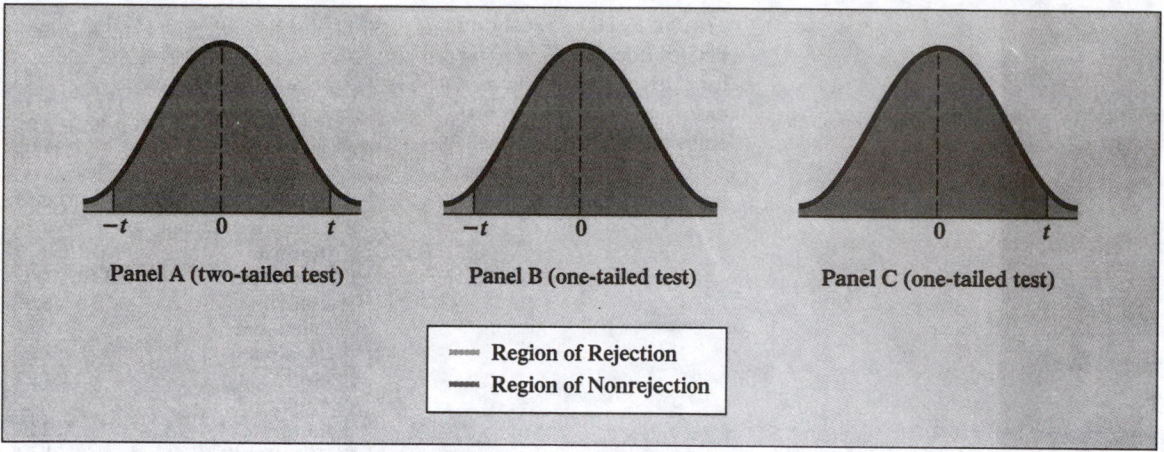

FIGURE 8.9 Regions of rejection and nonrejection for one-sample *t* test .

The one-sample *t* test is considered a *classical parametric* procedure—one that makes a variety of stringent assumptions that must hold if we are to be assured that the results we obtain from employing the test are valid. Assumptions of the **one-sample *t* test** are presented in the accompanying Comment box.

COMMENT: *Assumptions of the One-Sample t Test*

To use the one-sample *t* test, it is assumed that the obtained numerical data are independently drawn and represent a random sample from a population that is normally distributed. In practice, it has been found that as long as the sample size is not very small and the population is not very skewed, the *t* distribution gives a good approximation to the sampling distribution of the mean when σ is unknown.

Software such as Microsoft Excel enables us to evaluate the assumptions necessary for using the *t* test on the sales invoice data. As we learned in section 6.2, the normality assumption can be checked in several ways. A determination of how closely the actual data match the normal distribution's theoretical properties can be made by a descriptive analysis of the obtained statistics along with a graphical analysis to provide a visual interpretation. Thus, by exploring the sample data through a study of its descriptive summary measures along with a graphical analysis (a histogram, a stem-and-leaf display, a box-and-whisker plot, and a normal probability plot), we may draw our own conclusions as to the likelihood that the underlying population is at least approximately normally distributed.

Using the sales invoice data, Figure 8.10 presents Microsoft Excel output depicting the descriptive summary measures and a normal probability plot. From these, because the mean is very close to the median and the points on the normal probability plot appear to be increasing in an approximate straight line, there is no reason to believe that the assumption of underlying population normality of the sales invoice data is violated to any great degree, and we may conclude that the results obtained by the auditor are valid.

FIGURE 8.10 Excel and PHStat add-in for Excel output for studying assumptions necessary to employ *t* test for sales invoice data: panel A—descriptive statistics; panel B—normal probability plot

	A	B
1	**Invoice Amount**	
2		
3	Mean	112.8508333
4	Standard Error	6.003863082
5	Median	111.02
6	Mode	#N/A
7	Standard Deviation	20.7979918
8	Sample Variance	432.5564629
9	Kurtosis	0.172707598
10	Skewness	0.13363802
11	Range	76.51
12	Minimum	75.71
13	Maximum	152.22
14	Sum	1354.21
15	Count	12
16	Largest(1)	152.22
17	Smallest(1)	75.71

PANEL A

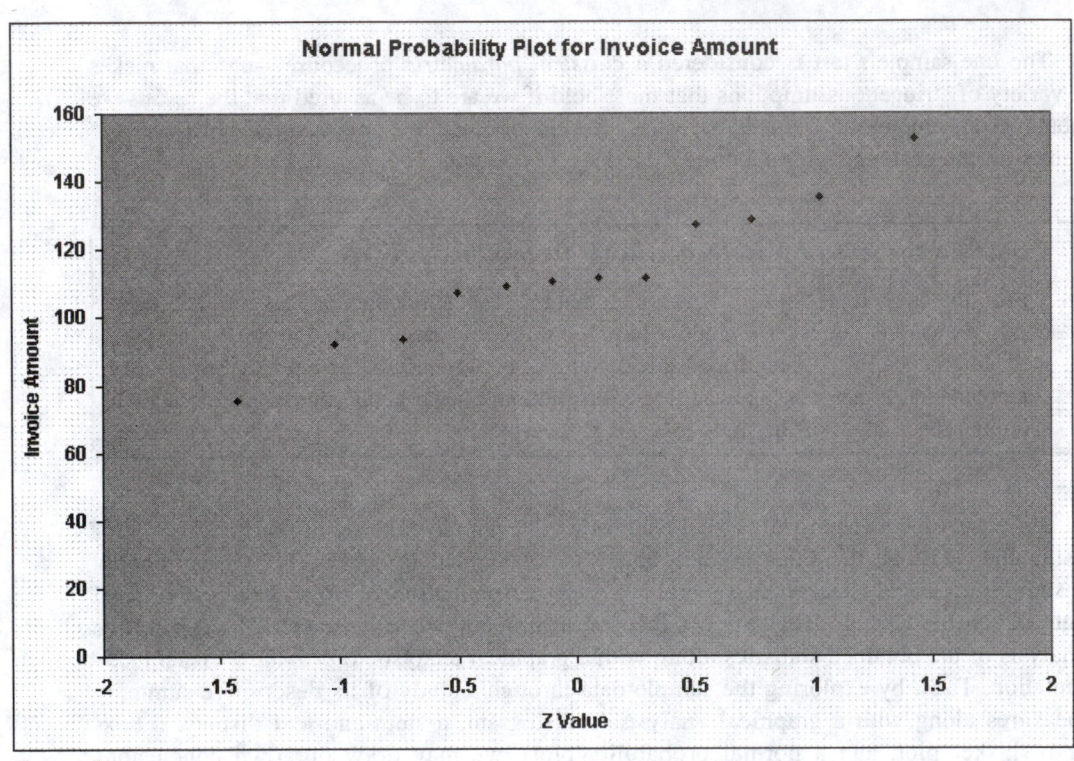

PANEL B

The *t* test is a **robust** test. That is, it does not lose power if the shape of the population from which the sample is drawn departs somewhat from a normal distribution, particularly when the sample size is large enough to enable the test statistic *t* to be influenced by the central limit theorem (see section 6.5). However, erroneous conclusions may be

508

drawn and statistical power can be lost if the *t* test is incorrectly used. If the sample size *n* is small (i.e., less than 30) and we cannot easily make the assumption that the underlying population from which the sample was drawn is at least approximately normally distributed, other, *nonparametric* testing procedures are likely to be more powerful (see references 1 and 2).

8.6E PERFORMING THE *t* TEST OF HYPOTHESIS FOR THE MEAN (σ UNKNOWN) USING MICROSOFT EXCEL

Overview

◆ *For Quick Results Users*   Use the PHStat add-in to perform a *t* test of the hypothesis for the mean when σ is unknown.

◆ *For Developers*   Implement a worksheet that uses the TINV and TDIST worksheet functions and arithmetic and logical formulas to perform a *t* test of the hypothesis for the mean when σ is unknown.

The 8-6E.XLS workbook file contains the *t* test of the hypothesis for the Saxon Plumbing Company problem of section 8.6.

8-6E.XLS

Quick Results Details

To perform a *t* test of the hypothesis for the mean (σ unknown), use the One-Sample Tests | t Test for the Mean, sigma unknown, choice of the PHStat add-in. As an example, consider the Saxon Plumbing Company process problem of section 8.6. To test the hypothesis of whether or not there is evidence of a change in the average amount of the invoices from the past average of $120, do the following:

❶ If the PHStat add-in has not been previously loaded, load the add-in using the instructions of section S4.2.

❷ Open the Starting Point for Section 8.6E workbook (STARTING POINT 8-6E.XLS) and click the Data sheet tab. Verify that the invoice amounts appear in column A.

❸ Select PHStat | One-Sample Tests | t Test for the Mean, sigma unknown.

❹ In the t Test for the Mean, sigma unknown, dialog box (see Figure 8E.5):
  a. Enter 120 in the Null Hypothesis: edit box.
  b. Enter 0.05 in the Level of Significance: edit box.
  c. Select the Sample Statistics Unknown option button and enter A1:A13 in the Sample Cell Range: edit box and select the First cell contains label check box.
  d. Select the Two-Tailed Test option button.

e. Enter t Test of the Hypothesis for the Mean Invoice Amount in the Output Title: edit box.

f. Click the OK button.

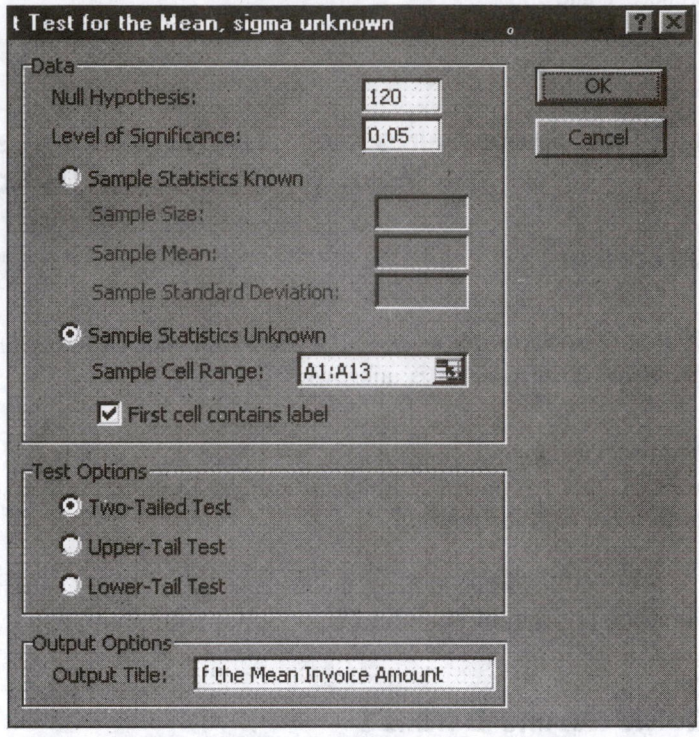

FIGURE 8E.5
PHStat *t* Test for the Mean, sigma unknown, dialog box

The add-in inserts a worksheet containing calculations for the *t* test of the hypothesis for the mean similar to the one shown in Figure 8E.6. (*Note:* For other problems in which the sample size, sample mean, and sample standard deviation are previously known, select the Sample Statistics Known option button in step 4c and enter these values in the appropriate edit boxes, shown dimmed in Figure 8E.5.)

FIGURE 8E.6
PHStat add-in for Microsoft Excel output for the *t* test of the hypothesis for the mean for the Saxon Plumbing Company problem of section 8.6

	A	B	C
1	*t* Test of the Hypothesis for the Mean Invoice Amount		
2			
3	Null Hypothesis $\mu=$	120	
4	Level of Significance	0.05	
5	Sample Size	12	
6	Sample Mean	112.8508333	
7	Sample Standard Deviation	20.7979918	
8	Standard Error of the Mean	6.003863082	
9	Degrees of Freedom	11	
10	T Test Statistic	-1.19076111	
11			
12	Two-Tailed Test		
13	Lower Critical Value	-2.200986273	
14	Upper Critical Value	2.200986273	
15	*p*-Value	0.258809315	
16	Do not reject the null hypothesis		

We can change the value for the level of significance in the worksheet inserted by the add-in or any of the sample data on the Data sheet to see their effects on the t test of hypothesis for the mean. For example, if we change the largest invoice amount in cell B2 of the Data sheet from 152.22 to 52.22, we will observe that the t-test statistic changes from -1.1908 to -2.2867 and the two-tailed p-value changes from .259 to .043, thereby causing us to reject the null hypothesis. To facilitate comparisons among many different alternative values, we can create scenarios using the Scenario Manager as discussed at the end of section 2.2E.

Developer Details

We can use the TINV (see section 7.2E) and TDIST worksheet functions as the basis for performing the t test of the hypothesis for the mean when σ is unknown. Tables 8E.3 and 8E.4 on pages 516–517 present a Hypothesis test sheet design that performs a t test of the hypothesis for the mean for the Saxon Plumbing Company problem of section 8.6. In this design, we use the TINV function to return the t values for the lower and upper critical values and use the TDIST function to return probability values. The format of the TDIST function is:

$$\text{TDIST(ABS}(t), \textit{degrees of freedom, tails)}$$

where

$$\text{ABS}(t) = \text{the absolute value of the } t\text{-test statistic}$$
$$\textit{degrees of freedom} = \text{the degrees of freedom}$$
$$\textit{tails} = \text{either 1, for a one-tailed test, or 2, for a two-tailed test}$$

In the Two-Tailed Test area of Table 8E.3, we use TDIST with *tails* set to 2; in the Calculations Area of Table 8E.4, we use TDIST with *tails* set to 1.

The Tables 8E.3 and 8E.4 design assumes that the invoice sample data are entered in column A of a sheet named Data. For general utility, the design includes both two-tailed and one-tailed tests, though only the two-tailed is necessary for the Saxon Plumbing Company problem of section 8.6. Similar to the design of Table 8E.1, this design also includes a formula that uses the IF function to display a message informing the user whether or not to reject the null hypothesis.

To implement the Tables 8E.3 and 8E.4 design, do the following:

❶ Open the Starting Point for Section 8.6E workbook (STARTING POINT 8-6E.XLS) and click the Data sheet tab. Verify that the invoice amounts appear in column A.

❷ Select an unused worksheet (or select Insert | Worksheet if there are none) and rename the sheet Hypothesis.

❸ Enter the title, labels, and formulas for columns A and B as shown in Table 8E.3, noting the following:
a. The Symbol font can be used to enter the μ symbol in cell A3.
b. The formulas for cells A16, A21, and A26, typeset as two lines, should be entered as single, continuous lines.

❹ Enter the title, labels, and formulas for columns D and E as shown in Table 8E.4. (Do not make any entries in column C.)

❺ Select the cell range A12:B12 and click the Merge and Center button on the formatting toolbar (see section S3.11). Repeat this step for the cell ranges A16:B16, A18:B18, A21:B21, A23:B23, A26:B26, and D17:E17.

❻ Enter the null hypothesis value, 120, and the level of significance, 0.05, in cells B3 and B4, respectively.

The completed worksheet will be similar to the one shown in Figure 8E.7. Had we wanted to use summary statistics instead of the sample data, we could have directly entered those statistics into cells B5, B6, and B7, in place of the formulas shown for those cells in Table 8E.3.

Table 8E.3 Hypothesis test sheet design for columns A and B for the one-tailed and two-tailed t tests of hypothesis for the mean (σ unknown)

	A	B
1	t Test for the Hypothesis of the Mean Invoice Amount	
2		
3	Null Hypothesis $\mu=$	xxx
4	Level of Significance	.xx
5	Sample Size	=COUNT(Data!A:A)
6	Sample Mean	=AVERAGE(Data!A:A)
7	Sample Standard Deviation	=STDEV(Data!A:A)
8	Standard Error of the Mean	=B7/SQRT(B5)
9	Degrees of Freedom	=B5-1
10	t Test Statistic	=(B6-B3)/B8
11		
12	Two-Tailed Test	
13	Lower Critical Value	=-(TINV(B4,B9))
14	Upper Critical Value	=TINV(B4,B9)
15	p-Value	=TDIST(ABS(B10),B9,2)
16	=IF(B15<$B$4,"Reject the null hypothesis", "Do not reject the null hypothesis")	
17		
18	Lower-Tail Test	
19	Lower Critical Value	=-(TINV(2*B4,B9))
20	p-Value	=IF(B10<0,E19,E20)
21	=IF(B20<$B$4,"Reject the null hypothesis", "Do not reject the null hypothesis")	
22		
23	Upper-Tail Test	
24	Upper Critical Value	=(TINV(2*B4,B9))
25	p-Value	=IF(B10<0,E20,E19)
26	=IF(B25<$B$4,"Reject the null hypothesis", "Do not reject the null hypothesis")	

Table 8E.4 Hypothesis test sheet design for columns D and E for the one-tailed and two-tailed *t* tests of hypothesis for the mean (σ unknown)

	D	E
16		
17	Calculations Area	
18	For one-tailed tests:	
19	TDIST value	=TDIST(ABS(B10),B9,1)
20	1-TDIST value	=1-E19

	A	B	C	D	E
1	*t* Test for the Hypothesis of the Mean Invoice Amount				
2					
3	Null Hypothesis μ=	120			
4	Level of Significance	0.05			
5	Sample Size	12			
6	Sample Mean	112.8508333			
7	Sample Standard Deviation	20.7979918			
8	Standard Error of the Mean	6.003863082			
9	Degrees of Freedom	11			
10	*t* Test Statistic	-1.19076111			
11					
12	Two-Tailed Test				
13	Lower Critical Value	-2.200986273			
14	Upper Critical Value	2.200986273			
15	*p*-Value	0.258809315			
16	Do not reject the null hypothesis				
17				Calculations Area	
18	Lower-Tail Test			For one-tailed tests:	
19	Lower Critical Value	-1.795883691		TDIST value	0.129405
20	*p*-Value	0.129404658		1-TDIST value	0.870595
21	Do not reject the null hypothesis				
22					
23	Upper-Tail Test				
24	Upper Critical Value	1.795883691			
25	*p*-Value	0.870595342			
26	Do not reject the null hypothesis				

FIGURE 8E.7

Worksheet containing one-tailed and two-tailed *t* tests of the hypothesis for the mean for the Saxon Plumbing Company problem of section 8.6

▲ WHAT IF EXAMPLE

As with the worksheet inserted by the add-in using the quick results method of this section, we can change the value for the level of significance in this worksheet or any of the sample data on the Data sheet to see their effects on the *t* test of the hypothesis for the mean. (For further details, see the What If Example on page 515.)

Problems for Section 8.6

Learning the Basics

- **8.51** If, in a sample of size $n = 16$ selected from an underlying normal population, the sample mean is $\bar{X} = 56$ and the sample standard deviation is $S = 12$, what is the value of the t-test statistic if we are testing the null hypothesis H_0 that $\mu = 50$?

- **8.52** In Problem 8.51, how many degrees of freedom would there be in the one-sample t test?

  8.53 In Problems 8.51 and 8.52, what are the critical values from the t table if the level of significance α is chosen to be .05 and the alternative hypothesis H_1 is as follows:
  (a) $\mu \neq 50$?
  (b) $\mu > 50$?

  8.54 In Problems 8.51, 8.52, and 8.53, what is your statistical decision if your alternative hypothesis H_1 is as follows:
  (a) $\mu \neq 50$?
  (b) $\mu > 50$?

- **8.55** If, in a sample of size $n = 16$ selected from a left-skewed population, the sample mean is $\bar{X} = 65$ and the sample standard deviation is $S = 21$, would you use the t test to test the null hypothesis H_0 that $\mu = 60$? Discuss.

  8.56 If, in a sample of size $n = 160$ selected from a left-skewed population, the sample mean is $\bar{X} = 65$ and the sample standard deviation is $S = 21$, would you use the t test to test the null hypothesis H_0 that $\mu = 60$? Discuss.

Applying the Concepts

- **8.57** The manager of the credit department for an oil company would like to determine whether the average monthly balance of credit card holders is equal to $75. An auditor selects a random sample of 100 accounts and finds that the average owed is $83.40 with a sample standard deviation of $23.65.
   WHAT IF?
  (a) Using the .05 level of significance, should the auditor conclude that there is evidence the average balance is different from $75?
   WHAT IF?
  (b) What is your answer in (a) if the standard deviation is $37.26?
  (c) What is your answer in (a) if the sample mean is $78.81?

  8.58 A manufacturer of detergent claims that the mean weight of a particular box of detergent is 3.25 pounds. A random sample of 64 boxes reveals a sample average of 3.238 pounds and a sample standard deviation of 0.117 pound.
   WHAT IF?
  (a) Using the .01 level of significance, is there evidence that the average weight of the boxes is different from 3.25 pounds?
   WHAT IF?
  (b) What is your answer in (a) if the standard deviation is 0.05 pound?
  (c) What is your answer in (a) if the sample mean is 3.211 pounds?

  8.59 The director of admissions at a large university advises parents of incoming students about the cost of textbooks during a typical semester. A sample of 100 students enrolled in the university indicates a sample average cost of $315.40 with a sample standard deviation of $43.20.
  (a) Using the .10 level of significance, is there evidence that the population average is above $300?
   WHAT IF?
  (b) What is your answer in (a) if the standard deviation is $75 and the .05 level of significance is used?
   WHAT IF?
  (c) What is your answer in (a) if the sample average is $305.11?

- **8.60** A consumers' advocate group would like to evaluate the average energy efficiency rating (EER) of window-mounted, large-capacity (i.e., in excess of 7,000 Btu) air-conditioning units. A random sample of 36 such air-conditioning units is selected and tested for a fixed period of time. Their EER records are as follows:

8.9	9.1	9.2	9.1	8.4	9.5	9.0	9.6	9.3
9.3	8.9	9.7	8.7	9.4	8.5	8.9	8.4	9.5
9.3	9.3	8.8	9.4	8.9	9.3	9.0	9.2	9.1
9.8	9.6	9.3	9.2	9.1	9.6	9.8	9.5	10.0

DATA FILE
EER.XLS

(a) Using the .05 level of significance, is there evidence that the average EER is different from 9.0?
(b) What assumptions are made to perform this test?
(c) Find the p-value and interpret its meaning.
(d) What will be your answer in (a) if the last data value is 8.0 instead of 10.0?

WHAT IF?

8.61 A manufacturer of plastics wants to evaluate the durability of rectangularly molded plastic blocks that are to be used in furniture. A random sample of 50 such plastic blocks is examined, and the hardness measurements (in Brinell units) are recorded as follows:

283.6	273.3	278.8	238.7	334.9	302.6	239.9	254.6	281.9	270.4
269.1	250.1	301.6	289.2	240.8	267.5	279.3	228.4	265.2	285.9
279.3	252.3	271.7	235.0	313.2	277.8	243.8	295.5	249.3	228.7
255.3	267.2	255.3	281.0	302.1	256.3	233.0	194.4	291.9	263.7
273.6	267.7	283.1	260.9	274.8	277.4	276.9	259.5	262.0	263.5

DATA FILE
PLASTIC.XLS

(a) Using the .05 level of significance, is there evidence that the average hardness of the plastic blocks exceeds 260 (in Brinell units)?
(b) What assumptions are made to perform this test?
(c) Find the p-value and interpret its meaning.
(d) What will be your answer in (a) if the first data value is 233.6 instead of 283.6?

8.62 A machine being used for packaging seedless golden raisins has been set so that, on average, 15 ounces of raisins will be packaged per box. The operations manager wishes to test the machine setting and selects a sample of 30 consecutive raisin packages filled during the production process. Their weights are recorded as follows:

15.2	15.3	15.1	15.7	15.3	15.0	15.1	14.3	14.6	14.5
15.0	15.2	15.4	15.6	15.7	15.4	15.3	14.9	14.8	14.6
14.3	14.4	15.5	15.4	15.2	15.5	15.6	15.1	15.3	15.1

DATA FILE
RAISINS.XLS

(a) Is there evidence that the mean weight per box is different from 15 ounces? (Use $\alpha = .05$.)
(b) To perform the test in (a), we assume that the observed sequence in which the data were collected is random. What other assumptions must be made to perform the test? Discuss.
(c) What is your answer in (a) if the weights for the last two packages are 16.3 and 16.1 instead of 15.3 and 15.1?

WHAT IF?

• **8.63** A manufacturer claims that the average capacity of a certain type of battery the company produces is at least 140 ampere-hours. An independent consumer protection agency wishes to test the credibility of the manufacturer's claim and measures the capacity of a random sample of 20 batteries from a recently produced batch. The results, in ampere-hours, are as follows:

| 137.4 | 140.0 | 138.8 | 139.1 | 144.4 | 139.2 | 141.8 | 137.3 | 133.5 | 138.2 |
| 141.1 | 139.7 | 136.7 | 136.3 | 135.6 | 138.0 | 140.9 | 140.6 | 136.7 | 134.1 |

DATA FILE
AMPHRS.XLS

(a) Using the .05 level of significance, is there evidence that the manufacturer's claim is being overstated?
(b) What assumption must hold in order to perform the test in (a)?
(c) Evaluate this assumption through a graphical approach. Discuss.
(d) What is your answer in (a) if the last two values are 146.7 and 144.1 instead of 136.7 and 134.1?

WHAT IF?

8.64 Suppose that the manager of a 500-car taxi fleet in a large city wishes to reevaluate the maintenance contract on its vehicles. A major part of the analysis considers the "wear and tear" on the vehicles—that is, the daily usage as represented by miles logged per taxi per day. Upon examination of his contract, the manager decides that he wants to renegotiate or change the contract if the taxis are averaging more than 70 miles per day. Because an odometer recording is made by management as each taxi departs and then returns to the terminal, the difference represents the total mileage driven per taxi per day. A sample of 16 taxis is selected at random from the fleet. The following table records their mileage for a particular day:

DATA FILE
TAXI.XLS

107.1	121.0	71.2	76.1	95.7	92.8	74.8	92.1
94.4	42.5	82.3	56.5	74.6	91.7	63.7	62.9

(a) The manager knows you are studying statistics and asks you to completely analyze these data. Using an $\alpha = .05$ level of significance, what would you conclude about average daily mileage? What would you recommend to the manager?

(b) Present your findings in a memo to the manager. Be sure to attach an appendix describing the assumption that must hold in order for you to perform the test in (a). Be sure to provide an evaluation of this assumption through a graphical approach.

8.7 ◆ Z TEST OF HYPOTHESIS FOR THE PROPORTION

In some situations, we want to test a hypothesis pertaining to the population proportion p of values that are in a particular category rather than the population mean value. A random sample can be selected from the population, and the sample proportion, $p_s = X/n$, computed. The value of this statistic is then compared to the hypothesized value of the parameter p so that a decision pertaining to the hypothesis can be made.

If certain assumptions are met, the sampling distribution of a proportion follows a standardized normal distribution (see section 6.6). To perform the hypothesis test in order to evaluate the magnitude of the difference between the sample proportion p_s and the hypothesized population proportion p, the test statistic Z given in equation (8.3) can be used.

One-Sample Z Test for the Proportion

$$Z \cong \frac{p_s - p}{\sqrt{\dfrac{p(1 - p)}{n}}} \qquad (8.3)$$

where

$$p_s = \frac{X}{n} = \frac{\text{number of successes in sample}}{\text{sample size}}$$

$$= \text{observed proportion of successes}$$

$$p = \text{hypothesized proportion of successes}$$

This test statistic Z is approximately normally distributed.

Alternatively, instead of examining the *proportion* of successes in a sample, as in equation (8.3), we can study the *number* of successes in a sample. The test statistic Z for determining the magnitude of the difference between the number of successes in a sample and the hypothesized or expected number of successes in the population are presented in equation (8.4).

One-Sample Z Test for the Proportion

$$Z \cong \frac{X - np}{\sqrt{np(1 - p)}} \qquad \qquad (8.4)$$

Aside from possible rounding errors, the test statistic Z given by equations (8.3) and (8.4) provides exactly the same results. The two alternative forms of the test statistic are equivalent because the numerator of equation (8.4) is n times the numerator of equation (8.3) and the denominator of equation (8.4) is also n times the denominator of equation (8.3). The choice of which of these two formulas to employ is up to the user.

To illustrate the (one-sample) Z test for a hypothesized proportion, let us return to the cereal-filling-process example discussed earlier in this chapter. Suppose that the operations manager is also concerned with the sealing process for filled boxes. Once the package inside the box is filled, it is supposed to be sealed so that it is airtight. On the basis of past experience, however, it is known that 1 out of 10 packages (i.e., 10% or .10) initially do not meet standards for sealing and must be "reworked" in order to pass inspection. To alter this situation, suppose the operations manager implements a newly developed sealing system on a trial basis. After a 1-day "break-in" period, he takes a random sample of 200 boxes that represent daily output at the plant and, through inspection, finds that 11 need rework. The operations manager wants to determine whether there is evidence that the proportion of defective packages has improved under the new sealing system (i.e., has decreased below .10).

In terms of proportions (rather than percentages), the null and alternative hypotheses can be stated as follows:

$$H_0: p \geq .10$$

$$H_1: p < .10$$

◆ *Critical Value Approach* Because the operations manager is interested in whether or not there has been a significant reduction in the proportion of defective packages owing to the new sealing system, the test is one-tailed. If a level of significance α of .05 is selected, the rejection and nonrejection regions are set up as in Figure 8.11, and the decision rule is

Reject H_0 if $Z < -1.645$;

otherwise do not reject H_0.

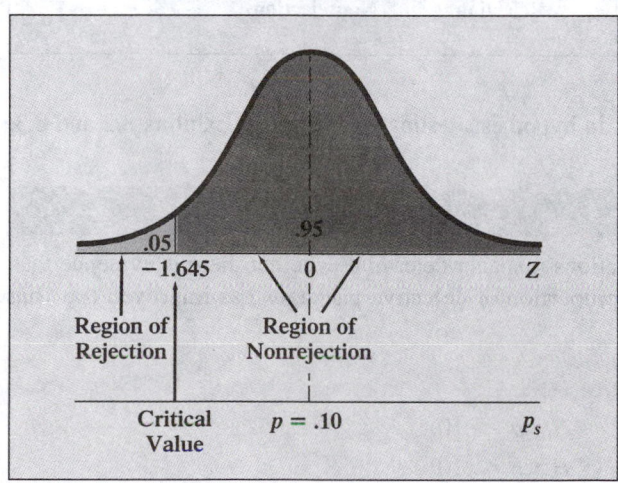

FIGURE 8.11

One-tailed test of hypothesis for proportion at .05 level of significance

517

From our data,

$$p_s = \frac{11}{200} = .055$$

Using equation (8.3), we have

$$Z \cong \frac{p_s - p}{\sqrt{\dfrac{p(1 - p)}{n}}} = \frac{.055 - .10}{\sqrt{\dfrac{(.10)(.90)}{200}}} = \frac{-.045}{\sqrt{.00045}} = \frac{-.045}{.0212} = -2.12$$

or, using equation (8.4), we have

$$Z \cong \frac{X - np}{\sqrt{np(1 - p)}} = \frac{11 - (200)(.10)}{\sqrt{(200)(.10)(.90)}} = \frac{11 - 20}{\sqrt{18}} = \frac{-9}{4.243} = -2.12$$

Because $-2.12 < -1.645$, we reject H_0. Thus, the manager may conclude that there is evidence that the proportion of defectives with the new system is less than .10.

◆ *p-Value Approach* As an alternative approach toward making a hypothesis-testing decision, we may also compute the *p*-value for this situation (see sections 8.3 and 8.5). Because a one-tailed test is involved in which the rejection region is located only in the lower tail (see Figure 8.12), we need to find the area below a Z value of -2.12. From Table E.2(b), this probability will be .0170—our *observed* level of significance. Because this value is less than the *selected* level of significance ($\alpha = .05$), the null hypothesis can be rejected.

FIGURE 8.12
Determining *p*-value for one-tailed test

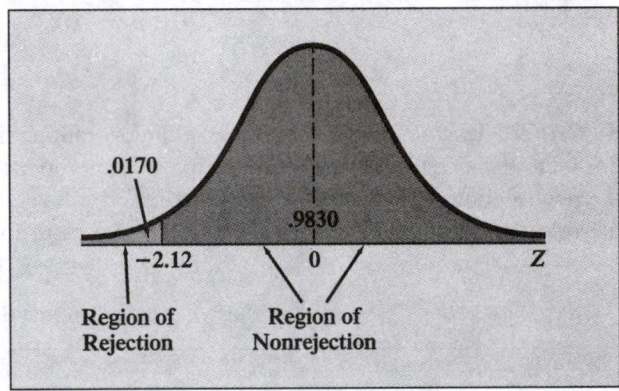

Example 8.6 applies the 10 steps in hypothesis testing developed in Exhibits 8.2 and 8.3.

Example 8.6   *One-Sample Z Test for the Proportion*

Using the data available to the operations manager determine whether there is evidence that, under the new sealing system, the proportion of defective packages has improved (i.e., has decreased below .10).

SOLUTION

Steps 1 and 2: H_0: $p \geq .10$

H_1: p $< .10$

Step 3: $\alpha = .05$

Step 4: $n = 200$

Step 5: We choose the one-sample Z test with test statistic Z given by equation (8.3):

$$Z \cong \frac{p_s - p}{\sqrt{\dfrac{p(1 - p)}{n}}}$$

Step 6: We use Table E.2(b) to develop the following decision rule as illustrated in Figure 8.11 on page 521:

$$\text{Reject } H_0 \text{ if } Z < -1.645;$$

$$\text{otherwise do not reject } H_0.$$

Step 7: We collect the data and compute the Z-test statistic:

$$Z \cong \frac{p_s - p}{\sqrt{\dfrac{p(1 - p)}{n}}} = \frac{.055 - .10}{\sqrt{\dfrac{(.10)(.90)}{200}}} = \frac{-.045}{\sqrt{.00045}} = \frac{-.045}{.0212} = -2.12$$

Steps 8, 9, and 10: Because $-2.12 < -1.645$, we reject H_0. (Alternatively, the p-value, .0170, is less than α of .05, so we reject H_0.) Thus, the manager may conclude there is evidence that the proportion of defectives with the new system is less than .10.

The one-sample Z test for the proportion can be either a two-tailed test or a one-tailed test, depending on whether the alternative hypothesis is *nondirectional* or *directional*, respectively. If the alternative hypothesis is directional, as in our box-sealing example, we reject only H_0 if there is evidence from the sample that the value of the parameter being tested is too small or too large, depending on the direction specified in the alternative hypothesis. In a two-tailed test, the alternative hypothesis is nondirectional. We reject H_0 if there is evidence from the sample that the value of the parameter being tested is likely to be either significantly more or significantly less than this hypothesized amount. The regions of rejection and nonrejection for these one-sample Z tests are depicted in Figure 8.13.

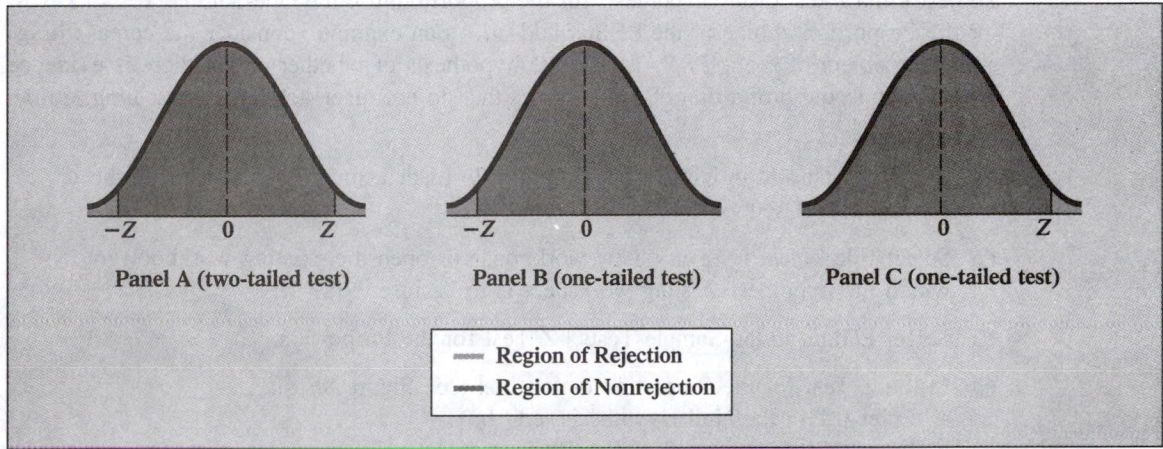

FIGURE 8.13   Regions of rejection and nonrejection for one-sample Z tests for proportion

The one-sample Z test for a proportion, sometimes called the *binomial test,* is considered a nonparametric or distribution-free procedure—one whose test statistic does not depend on the form of the underlying population distribution from which the sample data were drawn and one whose data are categorical. Assumptions for the Z test for a proportion are presented in the Comment box.

**COMMENT: *Checking the Assumptions of the Z Test
for a Proportion***

The test statistic Z given in equations (8.3) and (8.4) is approximately normally distributed. You may recall from section 6.6 that although the random variable X (the number of successes in the sample) follows a binomial distribution, if the sample size is large enough so that both $np \geq 5$ and $n(1 - p) \geq 5$, the normal distribution provides a good approximation of the binomial distribution.

8.7E PERFORMING THE *Z* TEST OF HYPOTHESIS
FOR THE PROPORTION USING MICROSOFT EXCEL

Overview

◆ *For Quick Results Users*   Use the PHStat add-in to perform a Z test of the hypothesis for the proportion.

◆ *For Developers*   Implement a worksheet that uses the NORMSINV and NORMSDIST worksheet functions and arithmetic and logical formulas to perform a Z test of the hypothesis for the proportion.

 8-7E.XLS

The 8-7E.XLS workbook file contains the Z test of the hypothesis for the proportion for the cereal-filling-process problem of section 8.7.

Quick Results Details

To perform a Z test of the hypothesis for the proportion, use the One-Sample Tests | Z Test for the Proportion choice of the PHStat add-in. As an example, consider the cereal-filling-process problem of section 8.7. To test the hypothesis of whether or not there is evidence of a change in the proportion of cereal boxes that do not meet standards for sealing, do the following:

❶ If the PHStat add-in has not been previously loaded, load the add-in using the instructions of section S4.2.

❷ Select File | New to open a new workbook (or open the existing workbook into which the hypothesis testing worksheet is to be inserted).

❸ Select PHStat | One-Sample Tests | Z Test for the Proportion.

❹ In the Z Test for the Proportion dialog box (see Figure 8E.8):
  a. Enter 0.1 in the Null Hypothesis: edit box.
  b. Enter 0.05 in the Level of Significance: edit box.

c. Enter 11 in the Number of Successes: edit box.

d. Enter 200 in the Sample Size: edit box.

e. Select the Lower-Tail Test option button.

f. Enter Z Test of Hypothesis for the Proportion in the Output Title: edit box.

g. Click the OK button.

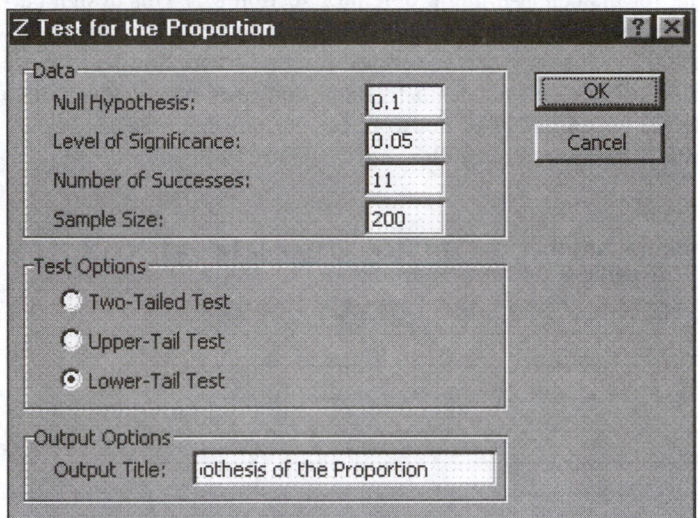

FIGURE 8E.8
PHStat Z test for the Proportion dialog box

The add-in inserts a worksheet containing calculations for the Z test of the hypothesis for the proportion similar to the one shown in Figure 8E.9.

	A	B
1	**Z Test of Hypothesis for the Proportion**	
2		
3	**Null Hypothesis** $p=$	0.1
4	**Level of Significance**	0.05
5	**Number of Successes**	11
6	**Sample Size**	200
7	Sample Proportion	0.055
8	Standard Error	0.021213203
9	Z Test Statistic	-2.121320344
10		
11	**Lower-Tail Test**	
12	**Lower Critical Value**	-1.644853
13	*p*-**Value**	0.016947366
14	**Reject the null hypothesis**	

FIGURE 8E.9
PHStat add-in for Microsoft Excel output for the Z test of the hypothesis for the proportion for the cereal-filling-process problem of section 8.7

▲ WHAT IF EXAMPLE

We can change the values for the level of significance, number of successes, and sample size in the worksheet inserted by the add-in to see their effects on the test of hypothesis for the proportion. For example, if we change the number of successes in cell B5 from 11 to 15, we will observe that the lower tail *p*-value changes from .0169 to .1193, thereby causing us not to reject the null hypothesis. To facilitate comparisons among many different alternative values, we can create scenarios using the Scenario Manager as discussed at the end of section 2.2E.

Developer Details

We can use the NORMSINV and NORMSDIST worksheet functions (see section 8.3E) as the basis for performing the Z test of the hypothesis for the proportion. Table 8E.5 presents a Hypothesis Test sheet design that performs a Z test for the proportion of boxes that do not meet standards for sealing for the cereal-filling problem of section 8.7. This design parallels the designs of Tables 8E.1 and 8E.2, and includes both two-tailed and one-tailed tests, though only the lower-tailed test is necessary for the cereal-filling problem. Similar to those other designs, the Table 8E.5 design also includes a formula that uses the IF function to display a message informing the user whether or not to reject the null hypothesis.

To implement the Table 8E.5 design, do the following:

Table 8E.5 Hypothesis sheet design for columns A and B for the one-tailed and two-tailed Z test of hypothesis for the proportion

	A	B
1	Z Test of Hypothesis for the Proportion	
2		
3	Null Hypothesis $p=$	.xx
4	Level of Significance	.xx
5	Number of Successes	xxx
6	Sample Size	xxx
7	Sample Proportion	=B5/B6
8	Standard Error	=SQRT(B3*(1-B3)/B6)
9	Z Test Statistic	=(B7-B3)/B8
10		
11	Two-Tailed Test	
12	Lower Critical Value	=NORMSINV(B4/2)
13	Upper Critical Value	=NORMSINV(1-B4/2)
14	p-Value	=2*(1-NORMSDIST(ABS(B9)))
15	=IF(B14<$B$4,"Reject the null hypothesis", "Do not reject the null hypothesis")	
16		
17	Lower-Tail Test	
18	Lower Critical Value	=NORMSINV(B4)
19	p-Value	=NORMSDIST(B9)
20	=IF(B19<$B$4,"Reject the null hypothesis", "Do not reject the null hypothesis")	
21		
22	Upper-Tail Test	
23	Upper Critical Value	=NORMSINV(1-B4)
24	p-Value	=1-NORMSDIST(B9)
25	=IF(B24<$B$4,"Reject the null hypothesis", "Do not reject the null hypothesis")	

❶ Select File | New to open a new workbook (or open the existing workbook into which the hypothesis testing worksheet is to be inserted).

❷ Select an unused worksheet (or select Insert | Worksheet if there are none) and rename the sheet Hypothesis.

③ Enter the title, labels, and formulas for column A as shown in Table 8E.5, noting that the formulas for cells A15, A20, and A25, typeset as two lines, should be entered as single, continuous lines.

④ Enter the null hypothesis value, the level of significance, number of successes, and sample size, in the cell range B3:B6. Enter 0.1 in cell B3, 0.05 in cell B4, 11 in cell B5, and 200 in cell B6.

⑤ Enter the formulas for cell ranges B7:B9, B12:B14, B18:B19, and B23:B24 as shown in Table 8E.5.

⑥ Select the cell range A11:B11 and click the Merge and Center button on the formatting toolbar (see section S3.11). Repeat this step for the cell ranges A15:B15, A17:B17, A20:B20, A22:B22, and A25:B25.

The completed worksheet will be similar to the one shown in Figure 8E.10.

	A	B
1	Z Test of Hypothesis for the Proportion	
2		
3	Null Hypothesis p=	0.1
4	Level of Significance	0.05
5	Number of Successes	11
6	Sample Size	200
7	Sample Proportion	0.055
8	Standard Error	0.021213203
9	Z Test Statistic	-2.121320344
10		
11	Two-Tailed Test	
12	Lower Critical Value	-1.959961082
13	Upper Critical value	1.959961082
14	p-Value	0.033894732
15	Reject the null hypothesis	
16		
17	Lower-Tail Test	
18	Lower Critical Value	-1.644853
19	p-Value	0.016947366
20	Reject the null hypothesis	
21		
22	Upper-Tail Test	
23	Upper Critical Value	1.644853
24	p-Value	0.983052634
25	Do not reject the null hypothesis	

FIGURE 8E.10

Worksheet containing one-tailed and two-tailed Z test of the hypothesis for the proportion for the cereal-filling-process problem of section 8.7

▲ WHAT IF EXAMPLE

As with the worksheet inserted by the add-in using the quick results method of this section, we can change the values for the level of significance, number of successes, and sample size in this worksheet to see their effects on the test of hypothesis for the proportion. (For further details, see the What If Example on page 525.)

Problems for Section 8.7

Learning the Basics

• **8.65** If in a random sample of 400 items, 88 are found to be defective, what is the sample proportion of defective items?

• **8.66** In Problem 8.65, if it is hypothesized that 20% of the items in the population are defective, what is the value of the Z-test statistic
(a) computed from equation (8.3)?
(b) computed from equation (8.4)?

8.67 In Problems 8.65 and 8.66, suppose you are testing the null hypothesis H_0: $p = .20$ against the two-tailed alternative hypothesis H_1: $p \neq .20$ and you choose the level of significance α to be .05. What is your statistical decision?

Applying the Concepts

8.68 A television manufacturer claims in its warranty that in the past not more than 10% of its television sets needed any repair during their first 2 years of operation. To test the validity of this claim, a government testing agency selects a sample of 100 sets and finds that 14 sets required some repair within their first 2 years of operation. Using the .01 level of significance,

 WHAT IF?

(a) is the manufacturer's claim valid or is there evidence that the claim is not valid?
(b) compute the p-value and interpret its meaning.
(c) What is your answer in (a) if 18 sets required some repair?

8.69 The Giansante Company, provider of extermination services, claims that no more than 15% of its customers need repeated treatment after a 90-day warranty period. To determine the validity of this claim, a consumer organization selects a sample of 100 customers and finds that 22 needed repeated treatment after the 90-day warranty period.
(a) Is there evidence at the .05 level of significance that the claim is not valid (i.e., that the proportion needing treatment is greater than .15)?

 WHAT IF?

(b) Compute the p-value and interpret its meaning.

 WHAT IF?

(c) What is your answer to (a) if the .01 level of significance is used?
(d) What is your answer to (a) if 18 homes needed repeated treatment?

• **8.70** The personnel director of a large insurance company is interested in reducing the turnover rate of data processing clerks in the first year of employment. Past records indicate that 25% of all new hires in this area are no longer employed at the end of 1 year. Extensive new training approaches are implemented for a sample of 150 new data processing clerks. At the end of a 1-year period, 29 of these 150 individuals are no longer employed.
(a) At the .01 level of significance, is there evidence that the proportion of data processing clerks who have gone through the new training and are no longer employed is less than .25?

 WHAT IF?

(b) Compute the p-value and interpret its meaning.
(c) What is your answer to (a) if 22 of the individuals are no longer employed?

• **8.71** The marketing manager for an automobile manufacturer is interested in determining the proportion of new compact-car owners who would have purchased a passenger-side inflatable air bag if it had been available for an additional cost of $300. The manager believes from previous information that the proportion is .30. Suppose that a survey of 200 new compact-car owners is selected and 79 indicate that they would have purchased the inflatable air bags.
(a) At the .10 level of significance, is there evidence that the population proportion is different from .30?

 WHAT IF?

(b) Compute the p-value and interpret its meaning.
(c) What is your answer to (a) if 70 new owners indicated that they would have purchased inflatable air bags?

8.72 The marketing branch of the Mexican Tourist Bureau would like to increase the proportion of tourists who purchase silver jewelry while vacationing in Mexico from its present estimated value of .40. Toward this end, promotional literature describing both the beauty and value of the jewelry is prepared and distributed to all passengers on airplanes arriving at a certain seaside resort during a 1-week period. A sample of 500 passengers returning at the end of the 1-week period is randomly selected, and 227 of these passengers indicate that they purchased silver jewelry.

    (a) At the .05 level of significance, is there evidence that the proportion has increased above the previous value of .40?

    (b) Compute the *p*-value and interpret its meaning.

 **WHAT IF?**

    (c) What is your answer to (a) if 213 passengers indicate that they purchased silver jewelry?

8.73 On the basis of industry sales of $1.5 billion recorded for the 1-year period ending May 25, 1997, *The New York Times* reported (June 20, 1997, D4) that Crest toothpaste was the market leader with a share of 26.3%.

    (a) Suppose that a recently taken random sample of 250 individuals indicates that 68 are using Crest toothpaste. At the .05 level of significance, is there evidence that the proportion has changed from the previous 1996–1997 market share?

    (b) What is the *p*-value? Interpret its meaning.

    (c) **(Class Project)** Consider your class to be a sample of all students at your school. Determine the proportion of students in your class who use Crest toothpaste. At the .05 level of significance, is there evidence that this proportion is different from the 1996–1997 market share?

8.74 There is an expression "the more things change, the more they stay the same." Over the 1-year period ending October 1992, *The New York Times* reported (November 17, 1992, D4) that Kellogg had been the market leader for ready-to-eat breakfast cereals with a share of 37.8%.

    (a) Suppose that a recently taken random sample of 200 individuals indicates that 78 preferred Kellogg products to those from all other companies producing ready-to-eat cereals. At the .05 level of significance, is there evidence that the proportion has changed from the 1992 market share? On the basis of your findings, comment on the above expression.

    (b) What is the *p*-value? Interpret its meaning.

    (c) **(Class Project)** Consider your class to be a sample of all students at your school. Determine the proportion of students in your class who prefer Kellogg's products to those from all other companies producing ready-to-eat cereals. At the .05 level of significance, is there evidence that this proportion is different from the 1992 market share?

 ## 8.8   POTENTIAL HYPOTHESIS-TESTING PITFALLS AND ETHICAL ISSUES

To this point, we have studied the fundamental concepts of hypothesis-testing methodology. We have learned how to use it for analyzing differences among sample estimates (i.e., statistics) of hypothesized population characteristics (i.e., parameters) in order to make decisions about the underlying characteristics. We have also learned how to evaluate the risks involved in making these decisions.

    When planning to carry out a test of the hypothesis based on some designed experiment or research study under investigation, several questions must be asked to ensure that proper methodology is used. A listing of these questions appears in Exhibit 8.5.

Exhibit 8.5   Questions to Consider in the Planning Stage of Hypothesis Testing

✓ **1.** What is the goal of the experiment or research? Can it be translated into a null and alternative hypothesis?

✓ **2.** Is the hypothesis test going to be two-tailed or one-tailed?

✓ **3.** Can a random sample be drawn from the underlying population of interest?

✓ **4.** What kinds of measurements will be obtained from the sample? Are the sampled outcomes of the random variable going to be numerical or categorical?

✓ **5.** At what significance level, or risk of committing a Type I error, should the hypothesis test be conducted?

✓ **6.** Is the intended sample size large enough to achieve the desired power of the test for the level of significance chosen?

✓ **7.** What statistical test procedure is to be used on the sampled data and why?

✓ **8.** What kinds of conclusions and interpretations can be drawn from the results of the hypothesis test?

Questions like these need to be raised and answered in the planning stage of a survey or designed experiment, so a person with substantial statistical training should be consulted and involved early in the process. All too often such an individual is consulted far too late in the process, after the data have been collected. Typically, all that can be done at such a late stage is to choose the statistical test procedure that would be best for the obtained data. We are forced to assume that certain biases built into the study (because of poor planning) are negligible. But this is a large assumption. Good research involves good planning. To avoid biases, adequate controls must be built in from the beginning.

We need to distinguish between what is poor research methodology and what is unethical behavior. Ethical considerations arise when a researcher is manipulative of the hypothesis-testing process. Some of the ethical issues that arise when dealing with hypothesis-testing methodology include the data collection method, informed consent from human subjects being "treated," the type of test—two-tailed or one-tailed, the choice of level of significance α, data snooping, the cleansing and discarding of data, and reporting of findings.

◆ *Data Collection Method—Randomization*   To eliminate the possibility of potential biases in the results, we must use proper data collection methods. To draw meaningful conclusions, the data we obtain must be the outcomes of a random sample from some underlying population or the outcomes from some experiment in which a **randomization** process was employed. Potential subjects should not be permitted to self-select for a study. In a similar manner, a researcher should not be permitted to purposely select the subjects for the study. Aside from the potential ethical issues that may be raised, such a lack of randomization can result in serious coverage errors or selection biases that destroy the value of any study.

◆ *Informed Consent from Human Subjects Being "Treated"*   Ethical considerations require that any individual who is to be subjected to some "treatment" in an experiment be apprised of the research endeavor and any potential behavioral or physical side effects and provide informed consent with respect to participation. A researcher is not permitted to dupe or manipulate the subjects in a study.

◆ *Type of Test—Two-Tailed or One-Tailed*   If we have prior information that leads us to test the null hypothesis against a specifically directed alternative, then a one-tailed test will be more powerful than a two-tailed test. On the other hand, we should realize that if we are interested only in *differences* from the null hypothesis, not in the *direction* of the difference, the two-tailed test is the appropriate procedure to use. This is an important point. For example, if previous research and statistical testing have already established the difference in a particular direction or if an established scientific theory states that it is possible for results to occur in only one direction, then a one-tailed or directional test may be employed. However, these conditions are not often satisfied in practice, and it is recommended that one-tailed tests be used cautiously.

Using arguments based on ethical principles, Fleiss (see reference 3) and other statisticians have stated that, in the overwhelming majority of research studies, a two-tailed test should be employed, particularly if the intention is to report the results to professional colleagues at meetings or in published journal articles. A major reason for this more conservative approach to testing is to enable us to draw more appropriate conclusions on data that may yield unexpected, counterintuitive results.

◆ *Choice of Level of Significance* α In a well-designed experiment or study, the level of significance α is selected in advance of data collection. One cannot be permitted to alter the level of significance, after the fact, to achieve a specific result. This would be **data snooping.** One answer to this issue of level of significance is to always report the *p*-value, not just the results of the test.

◆ *Data Snooping*   Data snooping is never permissible. It is unethical to perform a hypothesis test on a set of data, look at the results, and then select whether the test should be two-tailed or one-tailed and/or choose the level of significance. These steps must be done first, as part of the planned experiment or study, before the data are collected, for the conclusions drawn to have meaning. In those situations in which a statistician is consulted by a researcher late in the process, with data already available, it is imperative that the null and alternative hypotheses be established and the level of significance chosen prior to carrying out the hypothesis test.

◆ **Cleansing and Discarding of Data** Data cleansing is not data snooping. Data cleansing is an important part of an overall analysis—remember GIGO (garbage in, garbage out). In the data preparation stage of editing, coding, and transcribing, one has an opportunity to review the data for any observation whose measurement seems to be extreme or unusual. After this has been done, the outcomes of the numerical variables in the data set should be organized into stem-and-leaf displays and box-and-whisker plots in preparation for further data presentation and *confirmatory analysis*. This *exploratory data analysis* stage gives us another opportunity to cleanse the data set by flagging outlier observations that need to be checked against the original data. In addition, the exploratory data analysis enables us to examine the data graphically with respect to the assumptions underlying a particular hypothesis test procedure needed for confirmatory analysis.

The process of data cleansing raises a major ethical question. Should an observation be removed from a study? The answer is a qualified yes. If it can be determined that a measurement is incomplete or grossly in error because of some equipment problem or unusual behavioral occurrence unrelated to the study, a decision to discard the observation may be made. Sometimes there is no choice—an individual may decide to quit a particular study he or she has been participating in before a final measurement can be made. In a well-designed experiment or study, the researcher would decide, in advance, on all rules regarding the possible discarding of data.

◆ **Reporting of Findings** In conducting research, it is vitally important to document both good and bad results so that individuals who follow up on such research do not have to "reinvent the wheel." It is inappropriate to report the results of hypothesis tests that show statistical significance but not those for which there was insufficient evidence in the findings.

To summarize, we conclude that in discussing ethical issues concerning hypothesis-testing methodology, the key is *intent*. We must distinguish between poor confirmatory data analysis and unethical practice. Unethical behavior occurs when a researcher willfully causes a selection bias in data collection, manipulates the treatment of human subjects without informed consent, uses data snooping to select the type of test (two-tailed or one-tailed) and/or level of significance to his or her advantage, hides the facts by discarding observations that do not support a stated hypothesis, or fails to report pertinent findings.

 **SUMMARY**

As observed in the summary chart, this chapter presented the fundamental underpinnings of hypothesis-testing methodology. In the three chapters that follow we shall be building on the foundations of hypothesis testing that we have discussed here. We will present a set of procedures that may be employed to verify or confirm statistically the results of studies and experiments designed under a variety of conditions.

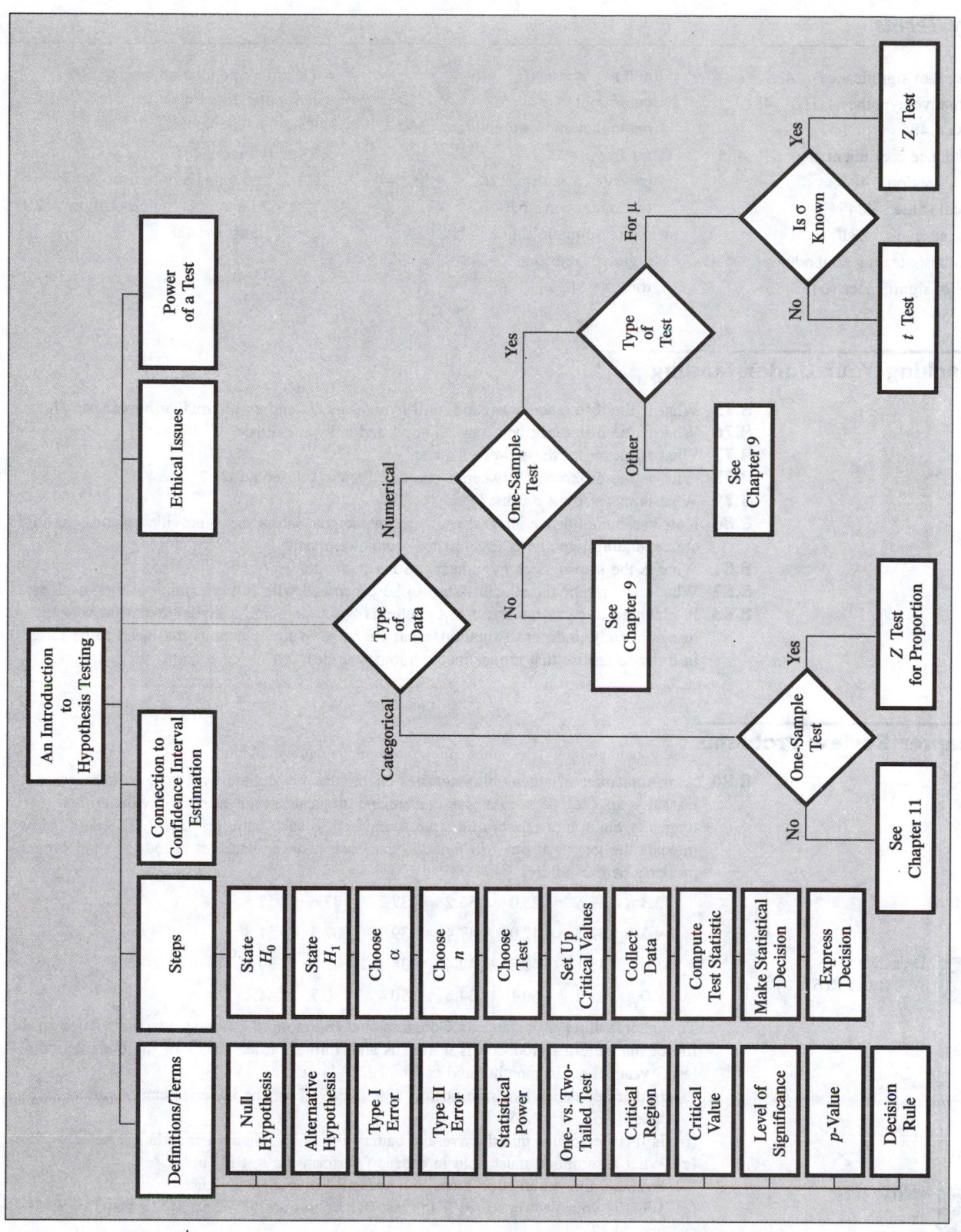

summary chart

529

Key Terms

Checking Your Understanding

8.75 What is the difference between a null hypothesis H_0 and an alternative hypothesis H_1?

8.76 What is the difference between a Type I and a Type II error?

8.77 What is meant by the power of a test?

8.78 What is the difference between a one-tailed and a two-tailed test?

8.79 What is meant by a p-value?

8.80 How can a confidence interval estimate for the population mean provide conclusions to the corresponding hypothesis test for the population mean?

8.81 What is the step-by-step hypothesis-testing methodology?

8.82 What are some of the ethical issues to be concerned with in performing a hypothesis test?

8.83 In planning to carry out a test of hypothesis based on some designed experiment or research study under investigation, what are some of the questions that need to be raised in order to ensure that proper methodology will be used?

Chapter Review Problems

8.84 A manufacturer of automobile batteries claims that his product will last, on average, at least 4 years (i.e., 48 months). A consumers' advocate group wants to evaluate this longevity claim and selects a random sample of 28 such batteries to test. The data below indicate the length of time (in months) that each of these batteries lasted (i.e., performed properly before failure).

42.3	39.6	25.0	56.2	37.2	47.4	57.5
39.3	39.2	47.0	47.4	39.7	57.3	51.8
31.6	45.1	40.8	42.4	38.9	42.9	34.1
49.0	41.5	60.1	34.6	50.4	30.7	44.1

DATA FILE
AUTOBAT.XLS

The manufacturer also stated in Congressional testimony that the standard deviation in the life of the batteries produced is 9 months and, further, at least 90% of the batteries will last 3 years (i.e., 36 months) and can be called "reliable."

(a) Is there evidence that significantly less than 90% of the batteries can be called "reliable"? (Use $\alpha = .05$.)

(b) Is there evidence that the average battery life is less than 48 months? (Use $\alpha = .05$.)

(c) What assumption must hold in order to perform the test in (b)?

WHAT IF?

(d) Evaluate this assumption through a graphical approach. Discuss.

(e) What is your answer to (b) if the last two values are 50.7 and 54.1 instead of 30.7 and 44.1?

8.85 The owner of a gasoline service station wants to study gasoline purchasing habits by motorists at his station. A random sample of 60 motorists during a certain week is selected with the following results:

- Amount purchased: $\overline{X}$ = 11.3 gallons, S = 3.1 gallons.
- 11 motorists purchased superunleaded gasoline.

(a) At the .05 level of significance, is there evidence that the average purchase is different from 10 gallons?
(b) Find the *p*-value in (a).
(c) At the .05 level of significance, is there evidence that less than 20% of the motorists purchase superunleaded gasoline?
(d) What is your answer to (a) if the sample average is 10.3 gallons?
(e) What is your answer to (c) if seven motorists purchased superunleaded gasoline?

 WHAT IF?

 WHAT IF?

• **8.86** An auditor for a government agency is assigned the task of evaluating reimbursement for office visits to doctors paid by Medicare. The audit is to be conducted for all Medicare payments in a particular geographic area during a certain month. Suppose the audit is conducted on a sample of 75 of the reimbursements with the following results:

- In 12 of the office visits, an incorrect amount of reimbursement was provided; the amount of reimbursement: $\overline{X}$ = $93.70, S = $34.55.

(a) At the .05 level of significance, is there evidence that the average reimbursement is less than $100?
(b) At the .05 level of significance, is there evidence that the proportion of incorrect reimbursements in the population is greater than .10?
(c) Discuss the underlying assumptions of the test used in (a).
(d) What is your answer to (a) if the sample average is $90?
(e) What is your answer to (b) if 15 office visits had incorrect reimbursements?

 WHAT IF?

 WHAT IF?

• **8.87** A bank branch located in a commercial district of a city has developed an improved process for serving customers during the 12 noon to 1 P.M. peak lunch period. The waiting time (operationally defined as the time the customer enters the line until he or she is served) of all customers during this hour is recorded over a period of 1 week. A random sample of 15 customers is selected, and the results are as follows:

4.21 5.55 3.02 5.13 4.77 2.34 3.54 3.20 4.50 6.10 0.38 5.12 6.46 6.19 3.79

(a) At the .05 level of significance, is there evidence that the average waiting time is less than 5 minutes?
(b) What assumption must hold in order to perform the test in (a)?
(c) Evaluate this assumption through a graphical approach. Discuss.
(d) As a customer walks into the branch office during the lunch hour, she asks the branch manager how long she can expect to wait. The branch manager replies, "Almost certainly not longer than 5 minutes." On the basis of the results of (a), evaluate this statement.

 **DATA FILE**
BANK1.XLS

8.88 One of the major measures of the quality of service provided by any organization is the speed with which it responds to customer complaints. A large family-held department store selling furniture and flooring including carpeting had undergone a major expansion in the past several years. In particular, the flooring department had expanded from two installation crews to an installation supervisor, a measurer, and 15 installation crews. A sample of 50 complaints in a recent year concerning carpeting installation was selected. The following data represent the number of days between the receipt of the complaint and the resolution of the complaint.

54	5	35	137	31	27	152	2	123	81	74	27	11	19	126	110	110
29	61	35	94	31	26	5	12	4	165	32	29	28	29	26	25	1
14	13	13	10	5	27	4	52	30	22	36	26	20	23	33	68	

 **DATA FILE**
FURNCOMP.XLS

(a) At the .05 level of significance, is there evidence that the average number of days between the receipt of the complaint and the resolution of the complaint is greater than 20?

(b) What assumption must hold in order to perform the test in (a)?

(c) Evaluate this assumption through a graphical approach. Discuss.

(d) Suppose a customer calls the department store with a complaint after the carpet has been installed. She asks the flooring department manager how long she can expect to wait to have her complaint resolved. The flooring department manager replies, "Almost certainly not longer than 20 days." On the basis of the results of (a), evaluate this statement.

References

1. Bradley, J. V., *Distribution-Free Statistical Tests* (Englewood Cliffs, NJ: Prentice Hall, 1968).

2. Daniel, W., *Applied Nonparametric Statistics*, 2d ed. (Boston, MA: Houghton Mifflin, 1990).

3. Fleiss, J. L., *Statistical Methods for Rates and Proportions*, 2d ed. (New York: Wiley, 1981).

4. *Microsoft Excel 97* (Redmond, WA: Microsoft Corporation, 1997).

Chapter 7

Student Solutions Manual

•8.2 H_1 is used to denote the alternative hypothesis.

•8.4 β is used to denote the consumer's risk, or the chance of committing a Type II error.

8.6 α is the probability of making a Type I error – that is, the probability of incorrectly rejecting the null hypothesis when in reality the null hypothesis is true and should not be rejected.

8.8   The power of a test is the complement of the probability β of making a Type II error.

8.10   It is possible to incorrectly fail to reject a false null hypothesis because it is possible for the mean of a single sample to fall in the nonrejection region even though the hypothesized population mean is false.

8.12   Other things being equal, the closer the *hypothesized* mean is to the *actual* mean, the larger is the risk of committing a Type II error.

8.14   Under the French judicial system, unlike ours in the United States, the null hypothesis is that the defendant is assumed to be guilty, the alternative hypothesis is that the defendant is innocent. The meaning of α and β risks would also be switched.

•8.16 H_0: $\mu = 20$ minutes. 20 minutes is adequate travel time between classes.
        H_1: $\mu \neq 20$ minutes. 20 minutes is not adequate travel time between classes.

8.18   Decision rule: Reject H_0 if $Z < -1.96$ or $Z > +1.96$.
        Decision: Since $Z_{calc} = +2.21$ is greater than $Z_{crit} = +1.96$, reject H_0.

•8.20   Decision rule: Reject H_0 if $Z < -2.58$ or $Z > +2.58$.

8.22   (a)H_0: $\mu = 70$ pounds. The cloth has an average breaking strength of 70 pounds.
        H_1: $\mu \neq 70$ pounds. The cloth has an average breaking strength that differs from 70 pounds.

        (b)   Decision rule: Reject H_0 if $Z < -1.96$ or $Z > +1.96$.

$$\text{Test statistic: } Z = \frac{\overline{X} - \mu}{\sigma / \sqrt{n}} = \frac{69.1 - 70}{3.5 / \sqrt{49}} = -1.80$$

        Decision: Since $Z_{calc} = -1.80$ is between the critical bounds of ± 1.96, do not reject H_0. There is not enough evidence to conclude that the cloth has an average breaking strength that differs from 70 pounds.

8.22
cont. **(c)** Decision rule: Reject H_0 if $Z < -1.96$ or $Z > +1.96$.

Test statistic: $Z = \dfrac{\overline{X} - \mu}{\sigma/\sqrt{n}} = \dfrac{69.1 - 70}{1.75/\sqrt{49}} = -3.60$

Decision: Since $Z_{calc} = -3.60$ is less than the lower critical bound of -1.96, reject H_0. There is enough evidence to conclude that the cloth has an average breaking strength that differs from 70 pounds.

(d) Decision rule: Reject H_0 if $Z < -1.96$ or $Z > +1.96$.

Test statistic: $Z = \dfrac{\overline{X} - \mu}{\sigma/\sqrt{n}} = \dfrac{69 - 70}{3.5/\sqrt{49}} = -2.00$

Decision: Since $Z_{calc} = -2.00$ is less than the lower critical bound of -1.96, reject H_0. There is enough evidence to conclude that the cloth has an average breaking strength that differs from 70 pounds.

•8.24 **(a), (b)**

	A	B
1	Salad Dressings	
2		
3	Null Hypothesis μ=	8
4	Level of Significance	0.05
5	Population Standard Deviation	0.15
6	Sample Size	50
7	Sample Mean	7.983
8	Standard Error of the Mean	0.021213203
9	Z Test Statistic	-0.80138769
10		
11	Two-Tailed Test	
12	Lower Critical Value	-1.95996108
13	Upper Critical Value	1.959961082
14	p-Value	0.422907113
15	Do not reject the null hypothesis	

(c) Decision rule: Reject H_0 if $Z < -1.96$ or $Z > +1.96$.

Test statistic: $Z = \dfrac{\overline{X} - \mu}{\sigma/\sqrt{n}} = \dfrac{7.983 - 8}{0.05/\sqrt{50}} = -2.40$

Decision: Since $Z_{calc} = -2.40$ is less than the lower critical bound of -1.96, reject H_0. There is enough evidence to conclude that the machine is filling bottles improperly.

(d) Decision rule: Reject H_0 if $Z < -1.96$ or $Z > +1.96$.

Test statistic: $Z = \dfrac{\overline{X} - \mu}{\sigma/\sqrt{n}} = \dfrac{7.952 - 8}{0.15/\sqrt{50}} = -2.26$

Decision: Since $Z_{calc} = -2.26$ is less than the lower critical bound of -1.96, reject H_0. There is enough evidence to conclude that the machine is filling bottles improperly.

•8.26 p value $= 2(0.5 - 0.4772) = 0.0456$

8.28 p value $= 2(0.5 - 0.4147) = 0.1706$

8.30   (a)   Test statistic: $Z = \dfrac{\overline{X} - \mu}{\sigma / \sqrt{n}} = \dfrac{69.1 - 70}{3.5 / \sqrt{49}} = -1.80$

p value $= 2(0.5 - 0.4641) = 0.0718$

Interpretation: The probability of getting a sample of 49 pieces that yield a mean strength that is farther away from the hypothesized population mean than this sample is 0.0718 or 7.18%.

(b)   Decision: Since p value $= 0.0718$ is greater than $\alpha = 0.05$, do not reject H_0.

(c)   At the 0.05 level of significance, a p value of 0.0718 does not provide sufficient evidence to conclude that the machine is not meeting the manufacturer's specifications in terms of the average breaking strength of the cloth produced.

(d)   Our conclusion has not changed. It is not affected by whether we base our decision on the comparison of Z-values or p values.

•8.32   (a)   Test statistic: $Z = \dfrac{\overline{X} - \mu}{\sigma / \sqrt{n}} = \dfrac{7.983 - 8}{0.15 / \sqrt{50}} = -0.80$

p value $= 2(0.5 - 0.2882) = 0.4238$

Interpretation: The probability of getting a sample of 50 bottles of salad dressing that are more different from the hypothesized population mean of 8 ounces than this sample is 0.4238 or 42.38%.

(b)   Decision: Since p value $= 0.4238$ is greater than $\alpha = 0.05$, do not reject H_0.

(c)   At the 0.05 level of significance, a p value of 0.4238 does not provide enough evidence to support a conclusion that the machine is dispensing an average amount of dressing different from 8 ounces.

(d)   Our conclusion has not changed. It is not affected by whether we base our decision on the comparison of Z-values or p values.

•8.34   (a) H_0: $\mu = 1.00$. The average amount of paint per one-gallon can is one gallon.
H_1: $\mu \neq 1.00$. The average amount of paint per one-gallon can differs from one gallon.

(b)   Decision rule: Reject H_0 if $Z < -2.58$ or $Z > +2.58$.

Test statistic: $Z = \dfrac{\overline{X} - \mu}{\sigma / \sqrt{n}} = \dfrac{0.995 - 1.00}{0.02 / \sqrt{50}} = -1.77$

Decision: Since $Z_{calc} = -1.77$ is between the critical bounds of ± 2.58, do not reject H_0. There is not enough evidence to conclude that the average amount of paint per one-gallon can differs from one gallon.

(c)   Same decision. The confidence interval includes the hypothesized value of 1.00.

8.36   (a) H_0: $\mu = 2.00$ liters. The average amount of soft drink placed in 2-liter bottles at the local bottling plant is equal to 2 liters.

                H_1: $\mu \neq 2.00$ liters. The average amount of soft drink placed in 2-liter bottles at the local bottling plant differs from 2 liters.

    (b)   Decision rule: Reject H_0 if $Z < -1.96$ or $Z > +1.96$.

$$\text{Test statistic: } Z = \frac{\bar{X} - \mu}{\sigma / \sqrt{n}} = \frac{1.99 - 2.00}{0.05 / \sqrt{100}} = -2.00$$

        Decision: Since $Z_{calc} = -2.00$ is below the lower critical bounds of -1.96, reject H_0. There is enough evidence to conclude that the amount of soft drink placed in 2-liter bottles at the local bottling plant differs from 2 liters.

    (c)   The results here are the same as those found in Problem 7.9. The confidence interval formed in Problem 7.9 does not include 2.00.

8.38   Since $Z_{calc} = 2.39$ is greater than $Z_{crit} = 2.33$, reject H_0.

•8.40   Since $Z_{calc} = -1.15$ is greater than $Z_{crit} = -1.28$, do not reject H_0.

•8.42   Since the p value $= 0.0228$ is less than $\alpha = 0.10$, reject H_0.

8.44   Since the p value $= 0.0838$ is greater than $\alpha = 0.01$, reject H_0.

•8.46   (a) H_0: $\mu \geq 2.8$ feet.

                The average length of steel bars produced is at least 2.8 feet and the production equipment does not need immediate adjustment.

                H_1: $\mu < 2.8$ feet.

                The average length of steel bars produced is less than 2.8 feet and the production equipment does need immediate adjustment.

    (b)   Decision rule: If $Z < -1.645$, reject H_0.

$$\text{Test statistic: } Z = \frac{\bar{X} - \mu}{\sigma / \sqrt{n}} = \frac{2.73 - 2.8}{0.2 / \sqrt{25}} = -1.75$$

        Decision: Since $Z_{calc} = -1.75$ is less than $Z_{crit} = -1.645$, reject H_0. There is enough evidence to conclude the production equipment needs adjustment.

    (c)   Decision rule: If p value < 0.05, reject H_0.

$$\text{Test statistic: } Z = \frac{\bar{X} - \mu}{\sigma / \sqrt{n}} = \frac{2.73 - 2.8}{0.2 / \sqrt{25}} = -1.75$$

$$p \text{ value} = 0.5 - 0.4599 = 0.0401$$

        Decision: Since p value $= 0.0401$ is less than $\alpha = 0.05$, reject H_0. There is enough evidence to conclude the production equipment needs adjustment.

    (d)   The probability of obtaining a sample whose mean is 2.73 feet or less when the null hypothesis is true is 0.0401.

    (e)   The conclusions are the same.

•8.48 (a) H_0: $\mu \geq 8$ ounces.

The mean amount of salad dressing dispensed is at least 8 ounces. The machine is working properly. No work stoppage should occur.

H_1: $\mu < 8$ ounces.

The mean amount of salad dressing dispensed is less than 8 ounces. The machine is not working properly. The filling line should be stopped.

(b)   Decision rule: If $Z < -1.645$, reject H_0.

Test statistic: $Z = \dfrac{\overline{X} - \mu}{\sigma/\sqrt{n}} = \dfrac{7.983 - 8}{0.15/\sqrt{50}} = -0.80$

Decision: Since $Z_{calc} = -0.80$ is greater than $Z_{crit} = -1.645$, do not reject H_0. There is not enough evidence to conclude that the mean amount of salad dressing dispensed is less than 8 ounces. There is insufficient evidence to conclude the machine is not working properly. The filling line should not be stopped.

(c)   Decision rule: If p value < 0.05, reject H_0.

Test statistic: $Z = \dfrac{\overline{X} - \mu}{\sigma/\sqrt{n}} = \dfrac{7.983 - 8}{0.15/\sqrt{50}} = -0.80$

p value $= 0.5 - 0.2881 = 0.2119$

Decision: Since p value $= 0.2119$ is greater than $\alpha = 0.05$, do not reject H_0. There is not enough evidence to conclude that the mean amount of salad dressing dispensed is less than 8 ounces. There is insufficient evidence to conclude the machine is not working properly. The filling line should not be stopped.

(d)   The probability of obtaining a sample whose mean is 7.983 ounces or less when the null hypothesis is true is 0.2119.

(e)   The conclusions are the same.

8.50 H_0: $\mu \leq 25$ mgs/mm.
The average margin by which a new pill reduces the systolic blood pressure of hypertensive patients is less than or equal to 25 mgs/mm.

H_1: $\mu > 25$ mgs/mm.
The average margin by which a new pill reduces the systolic blood pressure of hypertensive patients is greater than 25 mgs/mm.

•8.52 $df = n - 1 = 16 - 1 = 15$

8.54   (a)   Since $t_{calc} = 2.00$ is between the critical bounds of $t_{crit} = \pm 2.1315$, do not reject H_0.
       (b)   Since $t_{calc} = 2.00$ is above the critical bound of $t_{crit} = +1.7531$, reject H_0.

8.56   Yes, you may use the t test to test the null hypothesis that $\mu = 60$ on a population is left-skewed if the sample size is sufficiently large($n = 160$). The t test assumes that, if the underlying population is not normally distributed, the sample size is sufficiently large to enable the test statistic t to be influenced by the central limit theorem. With large sample sizes ($n > 30$), the t test may be used because the sampling distribution of the mean does meet the requirements of the central limit theorem.

538

8.58   (a)

Detergent Weights		
	A	B
Null Hypothesis $\mu=$		3.25
Level of Significance		0.01
Sample Size		64
Sample Mean		3.238
Sample Standard Deviation		0.117
Standard Error of the Mean		0.014625
Degrees of Freedom		63
t Test Statistic		-0.82051282
Two-Tailed Test		
Lower Critical Value		-2.6561429
Upper Critical Value		2.656142897
p-Value		0.415017678
Do not reject the null hypothesis		

(b) H_0: $\mu = 3.25$ pounds. The average weight of the boxes is 3.25 pounds.

H_1: $\mu \neq 3.25$ pounds. The average weight of the boxes of detergent is not equal to 3.25 pounds.

Decision rule: $df = 63$. If $t > 2.6561$ or $t < -2.6561$, reject H_0.

Test statistic: $t = \dfrac{\bar{X} - \mu}{S/\sqrt{n}} = \dfrac{3.238 - 3.25}{0.05/\sqrt{64}} = -1.9200$

Decision: Since $t_{calc} = -1.9200$ is between the critical bounds of $t = \pm 2.6561$, do not reject H_0. There is not enough evidence to conclude that the average weight of the boxes of detergent is not equal to 3.25 pounds.

(c) H_0: $\mu = 3.25$ pounds. The average weight of the boxes is 3.25 pounds.

H_1: $\mu \neq 3.25$ pounds. The average weight of the boxes of detergent is not equal to 3.25 pounds.

Decision rule: $df = 63$. If $t > 2.6561$ or $t < -2.6561$, reject H_0.

Test statistic: $t = \dfrac{\bar{X} - \mu}{S/\sqrt{n}} = \dfrac{3.211 - 3.25}{0.117/\sqrt{64}} = -2.6667$

Decision: Since $t_{calc} = -2.6667$ is below the critical bound of $t = -2.6561$, reject H_0. There is enough evidence to conclude that the average weight of the boxes of detergent is not equal to 3.25 pounds.

•8.60 (a) H_0: $\mu = 9.0$ The average energy efficiency rating of window-mounted large capacity air conditioning units is equal to 9.0.
H_1: $\mu \neq 9.0$ The average energy efficiency rating of window-mounted large-capacity air-conditioning units is not equal to 9.0.
Decision rule: $df = 35$. If $t > 2.0301$ or $t < -2.0301$, reject H_0.

Test statistic: $t = \dfrac{\overline{X} - \mu}{S/\sqrt{n}} = \dfrac{9.2111 - 9.0}{0.3838/\sqrt{36}} = 3.3002$

Decision: Since $t_{calc} = 3.3002$ is above the upper critical bound of $t = 2.0301$, reject H_0. There is enough evidence to conclude that the average energy efficiency rating of window-mounted large-capacity air-conditioning units is not equal to 9.0.

(b)   To perform this test, you must assume that (1) the observed sequence in which the data were collected is random, and (2) the sample size is sufficiently large for the central limit theorem to apply, meaning that the sampling distribution of the mean is approximately normally distributed.

(c) p value = 0.0022. The probability of obtaining a sample whose mean is further away from the hypothesized value of 9.0 than 9.2111 is 0.0022.

(d) H_0: $\mu = 9.0$ The average energy efficiency rating of window-mounted large capacity air-conditioning units is equal to 9.0.
H_1: $\mu \neq 9.0$ The average energy efficiency rating of window-mounted large-capacity air-conditioning units is not equal to 9.0.
Decision rule: $df = 35$. If $t > 2.0301$ or $t < -2.0301$, reject H_0.

Test statistic: $t = \dfrac{\overline{X} - \mu}{S/\sqrt{n}} = \dfrac{9.1556 - 9.0}{0.4102/\sqrt{36}} = 2.2754$

Decision: Since $t_{calc} = 2.2754$ is above the critical bound of $t = 2.0301$, reject H_0. There is enough evidence to conclude that the average energy efficiency rating of window-mounted large-capacity air-conditioning units is not equal to 9.0.

8.62 (a) H_0: $\mu = 15$ ounces. The average weight of raisins is 15 ounces per box.
H_1: $\mu \neq 15$ ounces. The average weight of raisins is not equal to 15 ounces per box.
Decision rule: $df = 29$. If $t > 2.0452$ or $t < -2.0452$, reject H_0.

Test statistic: $t = \dfrac{\overline{X} - \mu}{S/\sqrt{n}} = \dfrac{15.1133 - 15.0}{0.4058/\sqrt{30}} = 1.5298$

Decision: Since $t_{calc} = 1.5298$ is between the critical bounds of $t = \pm 2.0452$, do not reject H_0. There is not enough evidence to conclude that the average weight of raisins is not equal to 15 ounces per box.

(b)   In addition to assuming that the observed sequence in which the data were collected is random, to perform this test, you must assume that the sample size is sufficiently large for the central limit theorem to apply, meaning that the sampling distribution of the mean is approximately normally distributed.

8.62   (c)H_0: $\mu = 15$ ounces. The average weight of raisins is 15 ounces per box.
cont. H_1: $\mu \neq 15$ ounces. The average weight of raisins is not equal to 15 ounces per box.
        Decision rule: $df = 29$. If $t > 2.0452$ or $t < -2.0452$, reject H_0.

$$\text{Test statistic: } t = \frac{\overline{X} - \mu}{S/\sqrt{n}} = \frac{15.18 - 15.0}{0.4909/\sqrt{30}} = 2.0084$$

        Decision: Since $t_{calc} = 2.0084$ is between the critical bounds of $t = \pm 2.0452$, do not reject H_0. There is not enough evidence to conclude that the average weight of raisins is not equal to 15 ounces per box.

8.64   (a)H_0: $\mu \leq 70$ miles. The average miles logged per taxi per day is no more than 70 miles. There is no need to alter the contract.
        H_1: $\mu > 70$ miles. The average miles logged per taxi per day is more than 70 miles. There is a need to alter the contract.
        Decision rule: $df = 15$. If $t > 1.7531$, reject H_0.

$$\text{Test statistic: } t = \frac{\overline{X} - \mu}{S/\sqrt{n}} = \frac{81.2125 - 70}{19.9393/\sqrt{16}} = 2.2493$$

        Decision: Since $t_{calc} = 2.2493$ is above the critical bound of $t = 1.7531$, reject H_0. There is enough evidence to conclude that the average miles logged per taxi per day is more than 70 miles. There is a need to alter the contract.

    (b)   Box-and-whisker plot:

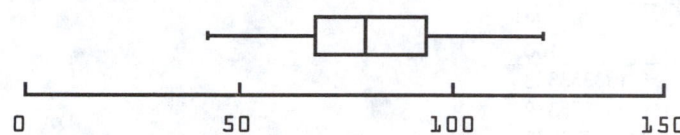

        The assumption of normality appears to hold.

•8.66   (a)
$$Z = \frac{p_s - p}{\sqrt{\dfrac{p(1-p)}{n}}} = \frac{0.22 - 0.20}{\sqrt{\dfrac{0.20(0.80)}{400}}} = 1.00$$

    (b)
$$Z = \frac{X - np}{\sqrt{n \cdot p \cdot (1-p)}} = \frac{88 - 400 \cdot (0.20)}{\sqrt{400 \cdot (0.20) \cdot (0.80)}} = 1.00$$

8.68   (a) H_0: $p \leq 0.10$. No more than 10% of the television sets manufactured required repair during their first 2 years of operation. There is no reason to doubt the manufacturer's claim.
        H_1: $p > 0.10$. More than 10% of the television sets manufactured required repair during their first 2 years of operation. There is enough evidence to doubt the manufacturer's claim.
        Decision rule: If $Z > 2.33$, reject H_0.

$$\text{Test statistic: } Z = \frac{p_s - p}{\sqrt{\dfrac{p(1-p)}{n}}} = \frac{0.14 - 0.10}{\sqrt{\dfrac{0.10(0.90)}{100}}} = 1.33$$

        Decision: Since $Z_{calc} = 1.33$ is below the critical bound of $Z = 2.33$, do not reject H_0. There is not enough evidence to cast doubt on the manufacturer's claim.

8.68
cont.
(b) p value = 0.0918. The probability of getting a sample with a higher rate of television sets requiring repair during their first 2 years of operation when the null hypothesis is true is 0.0918.

(c) $H_0: p \leq 0.10$. No more than 10% of the television sets manufactured required repair during their first 2 years of operation. There is no reason to doubt the manufacturer's claim.
$H_1: p > 0.10$. More than 10% of the television sets manufactured required repair during their first 2 years of operation. There is enough evidence to doubt the manufacturer's claim.
Decision rule: If $Z > 2.33$, reject H_0.
Test statistic: $Z = \dfrac{p_s - p}{\sqrt{\dfrac{p(1-p)}{n}}} = \dfrac{0.18 - 0.10}{\sqrt{\dfrac{0.10(0.90)}{100}}} = 2.67$

Decision: Since $Z_{calc} = 2.67$ is above the critical bound of $Z = 2.33$, reject H_0. There is enough evidence to doubt the manufacturer's claim.

•8.70 (a) (b)

	A	B
1	Training of Clerks	
2		
3	Null Hypothesis $\quad p=$	0.25
4	Level of Significance	0.01
5	Number of Successes	29
6	Sample Size	150
7	Sample Proportion	0.193333333
8	Standard Error	0.035355339
9	Z Test Statistic	-1.60277537
10		
11	Lower-Tail Test	
12	Lower Critical Value	-2.32634193
13	p-Value	0.054492125
14	Do not reject the null hypothesis	

(c) $H_0: p \geq 0.25$. At least 25% of the data processing clerks who have gone through the new training are no longer employed at the company after one year.
$H_1: p < 0.25$. Less than 25% of the data processing clerks who have gone through the new training are no longer employed at the company after one year.
Decision rule: If $Z < -2.33$, reject H_0.
Test statistic: $Z = \dfrac{p_s - p}{\sqrt{\dfrac{p(1-p)}{n}}} = \dfrac{0.1467 - 0.25}{\sqrt{\dfrac{0.25(0.75)}{150}}} = -2.92$

Decision: Since $Z_{calc} = -2.92$ is below the critical bound of $Z = -2.33$, reject H_0. There is enough evidence to show that the new training approaches are effective in reducing the turnover rate.

8.72   (a) $H_0: p \leq 0.40$. The proportion of tourists who purchase silver jewelry while vacationing in Mexico is no more than 40%.

$H_1: p > 0.40$. The proportion of tourists who purchase silver jewelry while vacationing in Mexico is greater than 40%.

Decision rule: If $Z > 1.645$, reject H_0.

Test statistic: $Z = \dfrac{p_s - p}{\sqrt{\dfrac{p(1-p)}{n}}} = \dfrac{0.454 - 0.40}{\sqrt{\dfrac{0.40(0.60)}{500}}} = 2.46$

Decision: Since $Z_{calc} = 2.46$ is above the critical bound of $Z = 1.645$, reject H_0. There is enough evidence to conclude that the proportion of tourists who purchase silver jewelry while vacationing in Mexico is greater than 40%.

(b) p value = 0.0069. The probability of obtaining a sample that has a proportion further away from 0.40 when the null hypothesis is true is 0.0069.

(c) $H_0: p \leq 0.40$. The proportion of tourists who purchase silver jewelry while vacationing in Mexico is no more than 40%.

$H_1: p > 0.40$. The proportion of tourists who purchase silver jewelry while vacationing in Mexico is greater than 40%.

Decision rule: If $Z > 1.645$, reject H_0.

Test statistic: $Z = \dfrac{p_s - p}{\sqrt{\dfrac{p(1-p)}{n}}} = \dfrac{0.426 - 0.40}{\sqrt{\dfrac{0.40(0.60)}{500}}} = 1.19$

Decision: Since $Z_{calc} = 1.19$ is below the critical bound of $Z = 1.645$, do not reject H_0. There is not enough evidence to conclude that the proportion of tourists who purchase silver jewelry while vacationing in Mexico is greater than 40%.

8.74   (a) $H_0: p = 0.378$. The proportion of market share for Kellogg ready-to-eat breakfast cereals is 37.8%.

$H_1: p \neq 0.378$. The proportion of market share for Kellogg ready-to-eat breakfast cereals differs from 37.8%.

Decision rule: If $Z < -1.96$ or $Z > 1.96$, reject H_0.

Test statistic: $Z = \dfrac{p_s - p}{\sqrt{\dfrac{p(1-p)}{n}}} = \dfrac{0.39 - 0.378}{\sqrt{\dfrac{0.378(0.622)}{200}}} = 0.35$

Decision: Since $Z_{calc} = 0.35$ is between the critical bounds of $Z = \pm 1.96$, do not reject H_0. There is not enough evidence to conclude that the proportion of market share for Kellogg ready-to-eat breakfast cereals differs from the previous 1996-1997 market share.

(b) p value = $2(0.5 - 0.1368) = 0.7264$. The probability of obtaining a sample that has a proportion further away from 0.378 than this sample proportion when the null hypothesis is true is 0.7264.

8.84 (a) $H_0: p \geq 0.90$. At least 90% of the batteries last three years and can be called "reliable."
$H_1: p < 0.90$. Less than 90% of the batteries last three years and can be called "reliable."
Decision rule: If $Z < -1.645$, reject H_0.

Test statistic: $Z = \dfrac{p_s - p}{\sqrt{\dfrac{p(1-p)}{n}}} = \dfrac{0.8214 - 0.90}{\sqrt{\dfrac{0.90(0.10)}{28}}} = -1.39$

Decision: Since $Z_{calc} = -1.39$ is above the critical bound of $Z = -1.645$, do not reject H_0. There is not enough evidence to conclude that less than 90% of the batteries last three years and can be called "reliable."

(b) $H_0: \mu \geq 48$ months. The average useful life of a certain type of battery is at least 48 months.
$H_1: \mu < 48$ months. The average useful life of a certain type of battery is less than 48 months.
Decision rule: $df = 27$. If $t < -1.7033$, reject H_0.

Test statistic: $t = \dfrac{\bar{X} - \mu}{S/\sqrt{n}} = \dfrac{43.325 - 48}{8.5439/\sqrt{28}} = -2.8954$

Decision: Since $t_{calc} = -2.8954$ is below the critical bound of $t = -1.7033$, reject H_0. There is enough evidence to conclude that the average useful life of a certain type of battery is less than 48 months.

(c) To perform the t-test, you must assume that the data are approximately normally distributed.

(d) Box-and-whisker plot:

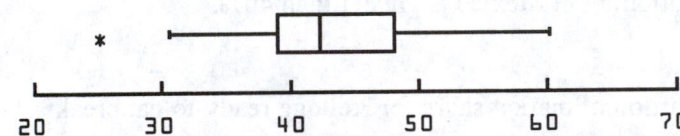

The data seem slightly right-skewed. But the sample size is nearly 30 so the accuracy of the test is not seriously affected.

(e) (b) $H_0: \mu \geq 48$ months. The average useful life of a certain type of battery is at least 48 months.
$H_1: \mu < 48$ months. The average useful life of a certain type of battery is less than 48 months.
Decision rule: $df = 27$. If $t < -1.7033$, reject H_0.

Test statistic: $t = \dfrac{\bar{X} - \mu}{S/\sqrt{n}} = \dfrac{44.3964 - 48}{8.4969/\sqrt{28}} = -2.2441$

Decision: Since $t_{calc} = -2.2441$ is below the critical bound of $t = -1.7033$, reject H_0. There is enough evidence to conclude that the average useful life of a certain type of battery is less than 48 months.

•8.86 (a) H_0: $\mu \geq \$100$. The average reimbursement for office visits to doctors paid by Medicare is at least $100.

H_1: $\mu < \$100$. The average reimbursement for office visits to doctors paid by Medicare is less than $100.

Decision rule: $df = 74$. If $t < -1.6657$, reject H_0.

Test statistic: $t = \dfrac{\overline{X} - \mu}{S/\sqrt{n}} = \dfrac{\$93.70 - \$100}{\$34.55/\sqrt{75}} = -1.5791$

Decision: Since the test statistic of $t_{calc} = -1.5791$ is above the critical bound of $t = -1.6657$, do not reject H_0. There is not enough evidence to conclude that the average reimbursement for office visits to doctors paid by Medicare is less than $100.

(b) H_0: $p \leq 0.10$. At most 10% of all reimbursements for office visits to doctors paid by Medicare are incorrect.

H_1: $p > 0.10$. More than 10% of all reimbursements for office visits to doctors paid by Medicare are incorrect.

Decision rule: If $Z > 1.645$, reject H_0.

Test statistic: $Z = \dfrac{p_s - p}{\sqrt{\dfrac{p(1-p)}{n}}} = \dfrac{0.16 - 0.10}{\sqrt{\dfrac{0.10(0.90)}{75}}} = 1.73$

Decision: Since the test statistic of $Z_{calc} = 1.73$ is above the critical bound of $Z = 1.645$, reject H_0. There is sufficient evidence to conclude that more than 10% of all reimbursements for office visits to doctors paid by Medicare are incorrect.

(c)   To perform the t-test on the population mean, you must assume that the observed sequence in which the data were collected is random and that the data are approximately normally distributed.

(d) H_0: $\mu \geq \$100$. The average reimbursement for office visits to doctors paid by Medicare is at least $100.

H_1: $\mu < \$100$. The average reimbursement for office visits to doctors paid by Medicare is less than $100.

Decision rule: $df = 74$. If $t < -1.6657$, reject H_0.

Test statistic: $t = \dfrac{\overline{X} - \mu}{S/\sqrt{n}} = \dfrac{\$90 - \$100}{\$34.55/\sqrt{75}} = -2.5066$

Decision: Since the test statistic of $t_{calc} = -2.5066$ is below the critical bound of $t = -1.6657$, reject H_0. There is enough evidence to conclude that the average reimbursement for office visits to doctors paid by Medicare is less than $100.

(e) H_0: $p \le 0.10$. At most 10% of all reimbursements for office visits to doctors paid by Medicare are incorrect.

H_1: $p > 0.10$. More than 10% of all reimbursements for office visits to doctors paid by Medicare are incorrect.

Decision rule: If $Z > 1.645$, reject H_0.

Test statistic: $Z = \dfrac{p_s - p}{\sqrt{\dfrac{p(1-p)}{n}}} = \dfrac{0.20 - 0.10}{\sqrt{\dfrac{0.10(0.90)}{75}}} = 2.89$

Decision: Since the test statistic of $Z_{calc} = 2.89$ is above the critical bound of $Z = 1.645$, reject H_0. There is sufficient evidence to conclude that more than 10% of all reimbursements for office visits to doctors paid by Medicare are incorrect.

8.88   (a) H_0: $\mu \le 20$ days. The average length of time to resolve a formal complaint is no more than 20 days.

H_1: $\mu > 20$ days. The average length of time to resolve a formal complaint is more than 20 days.

Decision rule: $df = 49$. If $t > 1.6766$, reject H_0.

Test statistic: $t = \dfrac{\bar{X} - \mu}{S / \sqrt{n}} = \dfrac{43.04 - 20}{41.926 / \sqrt{50}} = 3.8858$

Decision: Since the test statistic of $t_{calc} = 3.8858$ is above the critical value of $t = 1.6766$, reject H_0. There is sufficient evidence on which to conclude that the average length of time to resolve a formal complaint is more than 20 days.

(b)   To perform the t-test on the population mean, you must assume that the observed sequence in which the data were collected is random and that the sample size is sufficiently large for the central limit theorem to apply, meaning the sampling distribution of the mean is approximately normal.

(c)   Box-and-whisker plot:

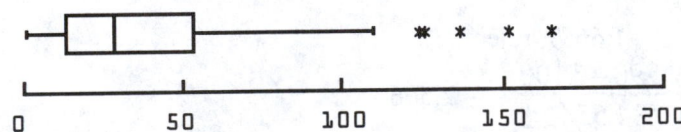

Although there are 5 extreme points in the upper tail of the distribution, the sample size of $n = 50$ is sufficiently large for the central limit theorem to apply.

(d) p value = 0.00015. The probability of obtaining a sample whose mean is at most 20 days when the null hypothesis is true is 0.00015, a very small likelihood. The floor manager is not making a judicious statement.

Note: The p value was found using Excel.

Chapter 8
Simple Linear Regression and Correlation

CHAPTER OBJECTIVES

✓ *To develop the simple linear regression model as a means of using one variable to predict another variable*
✓ *To assess the fit of the simple linear regression model*
✓ *To study the pitfalls involved in using regression models*
✓ *To introduce correlation as a measure of the strength of the association between two variables*

Introduction

In previous chapters we focused primarily on a single numerical response variable, such as the rate of return of mutual funds. We studied various measures of statistical description (see chapter 3) and applied different techniques of statistical inference to make estimates and draw conclusions about our numerical response variable (see chapters 7–10). In this and the following chapter we will concern ourselves with situations involving two or more numerical variables as a means of viewing the relationships that exist between them. Two techniques will be discussed: regression and correlation.

Regression analysis is used primarily for the purpose of prediction. Our goal in regression analysis is the development of a statistical model that can be used to predict the values of a **dependent** or **response variable** based on the values of at least one **explanatory** or **independent variable**. In this chapter we focus on a *simple* linear regression model—one that uses a *single* numerical independent variable X to predict the numerical dependent variable Y. In chapter 14 we develop *multiple* regression models that use *several* explanatory variables $(X_1, X_2, \ldots, X_p)$ to predict a numerical dependent variable Y.[1]

Correlation analysis, in contrast to regression, is used to measure the strength of the association between numerical variables. For example, in section 13.11 we will determine the correlation between the value of the German mark and the Japanese yen over a 10-year period. In this instance the objective is not to use one variable to predict another but rather to measure the strength of the association or covariation that exists between two numerical variables.

[1] *Regression models in which the dependent variable is categorical involve the use of logistic regression (see reference 4).*

◆ USING STATISTICS: *Forecasting Sales for a Clothing Store*

Over the past 25 years a chain of discount women's clothing stores has increased market share by increasing the number of locations in the chain. A systematic approach to site selection was never used. Site selection was primarily based on what was considered to be a great location or a great lease. This year, with a strategic plan for opening several new stores, the director of special projects and planning is being asked to develop an approach to forecasting annual sales for all new stores.

13.1 TYPES OF REGRESSION MODELS

In chapter 2, when information concerning mutual funds was studied, various graphs were used for data presentation. In regression analysis, a **scatter diagram** is used to plot the independent variable on the X axis and the dependent variable on the Y axis. The nature of the relationship between two variables can take many forms, ranging from simple to extremely complicated mathematical functions. The simplest relationship consists of a straight-line or **linear relationship.** An example of this relationship is shown in Figure 13.1.

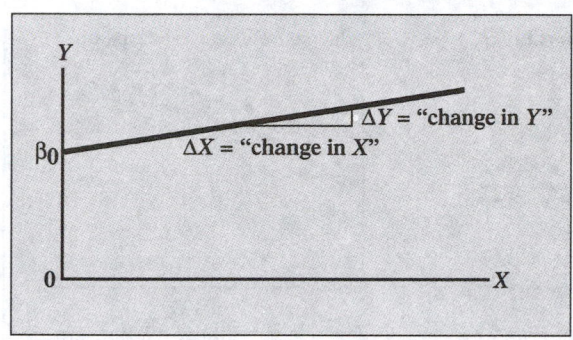

FIGURE 13.1
A positive straight-line relationship

The straight-line (linear) model can be represented as

Simple Linear Regression Model

$$Y_i = \beta_0 + \beta_1 X_i + \epsilon_i \tag{13.1}$$

where

$\beta_0 = Y$ intercept for the population

$\beta_1 =$ slope for the population

$\epsilon_i =$ random error in Y for observation i

In this model, the **slope** of the line β_1 represents the expected change in Y per unit change in X. It represents the average amount that Y changes (either positively or negatively) for a particular unit change in X. The **Y intercept** β_0 represents the average value of Y when X equals 0. The last component of the model, ϵ_i, represents the random error in Y for each observation i that occurs.

The selection of the proper mathematical model is influenced by the distribution of the X and Y values on the scatter diagram. This can be seen readily from an examination of panels A–F in Figure 13.2. In panel A we note that the values of Y are generally increasing linearly as X increases. This panel is similar to Figure 13.3 on page 776,

549

which illustrates the positive relationship between the store size (i.e., square footage available) and the annual sales at branches of a women's clothing store.

FIGURE 13.2

Examples of types of relationships found in scatter diagrams

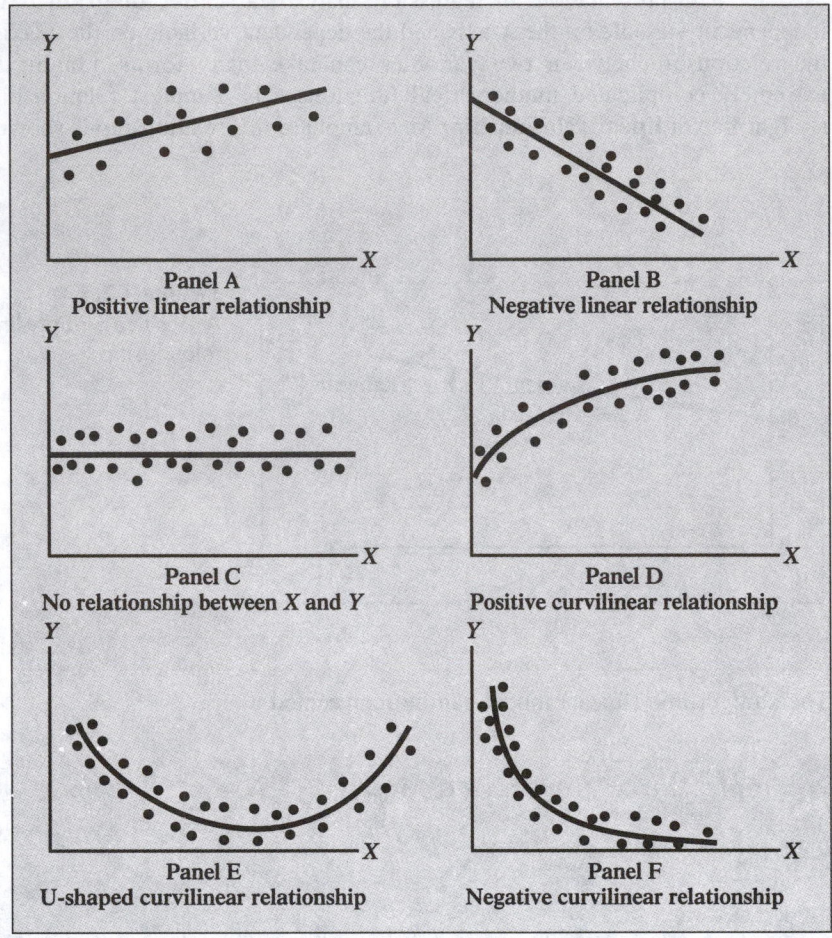

Panel A
Positive linear relationship

Panel B
Negative linear relationship

Panel C
No relationship between X and Y

Panel D
Positive curvilinear relationship

Panel E
U-shaped curvilinear relationship

Panel F
Negative curvilinear relationship

Panel B is an example of a *negative* linear relationship. As X increases, we note that the values of Y are decreasing. An example of this type of relationship might be the price of a particular product and the amount of sales. Panel C shows a set of data in which there is very little or no relationship between X and Y. High and low values of Y appear at each value of X.

The data in panel D show a positive curvilinear relationship between X and Y. The values of Y are increasing as X increases, but this increase tapers off beyond certain values of X. An example of this positive curvilinear relationship might be the age and maintenance cost of a machine. As a machine gets older, the maintenance cost may rise rapidly at first but then level off beyond a certain number of years.

Panel E shows a parabolic or U-shaped relationship between X and Y. As X increases, at first Y decreases; but as X continues to increase, Y not only stops decreasing but actually increases above its minimum value. An example of this type of relationship could be the number of errors per hour at a task and the number of hours worked. The number of errors per hour would decrease as the individual becomes more proficient at the task but then would increase beyond a certain point because of factors such as fatigue and boredom.

Finally, panel F indicates an exponential or negative curvilinear relationship between X and Y. In this case, Y decreases very rapidly as X first increases but then decreases much less rapidly as X increases further. An example of this exponential relationship could be the resale value of a particular type of automobile and its age. In the first year the resale value drops drastically from its original price; however, the resale value then decreases much less rapidly in subsequent years.

In this section we have briefly examined a variety of different models that could be used to represent the relationship between two variables. Although scatter diagrams can be extremely helpful in determining the mathematical form of the relationship, more sophisticated statistical procedures are available to determine the most appropriate model for a set of variables. In subsequent sections of this chapter, we primarily focus on building statistical models for fitting linear relationships between variables.

13.2 ▶ DETERMINING THE SIMPLE LINEAR REGRESSION EQUATION

In the Using Statistics example introduced earlier we stated that the director of special projects wanted to develop a strategy for forecasting annual sales for all new stores. Suppose that he decided to examine the relationship between the size (i.e., square footage) of a store and its annual sales by selecting a sample of 14 stores. The results for these 14 stores are summarized in Table 13.1.

Table 13.1 *Square footage and annual sales ($000) for sample of 14 branches of women's clothing store chain*

STORE	SQUARE FEET	ANNUAL SALES ($000)	STORE	SQUARE FEET	ANNUAL SALES ($000)
1	1,726	3,681	8	1,102	2,694
2	1,642	3,895	9	3,151	5,468
3	2,816	6,653	10	1,516	2,898
4	5,555	9,543	11	5,161	10,674
5	1,292	3,418	12	4,567	7,585
6	2,208	5,563	13	5,841	11,760
7	1,313	3,660	14	3,008	4,085

DATA FILE
SITE.XLS

The scatter diagram for the data in Table 13.1 is shown in Figure 13.3. An examination of Figure 13.3 indicates a clearly increasing relationship between square feet (X) and annual sales (Y). As the size of the store as measured by its square footage increases, annual sales increase approximately as a straight line. On this basis, if we assume that a straight line provides a useful mathematical model of this relationship, the question in regression analysis becomes the determination of the particular straight-line model that is the best fit to these data.

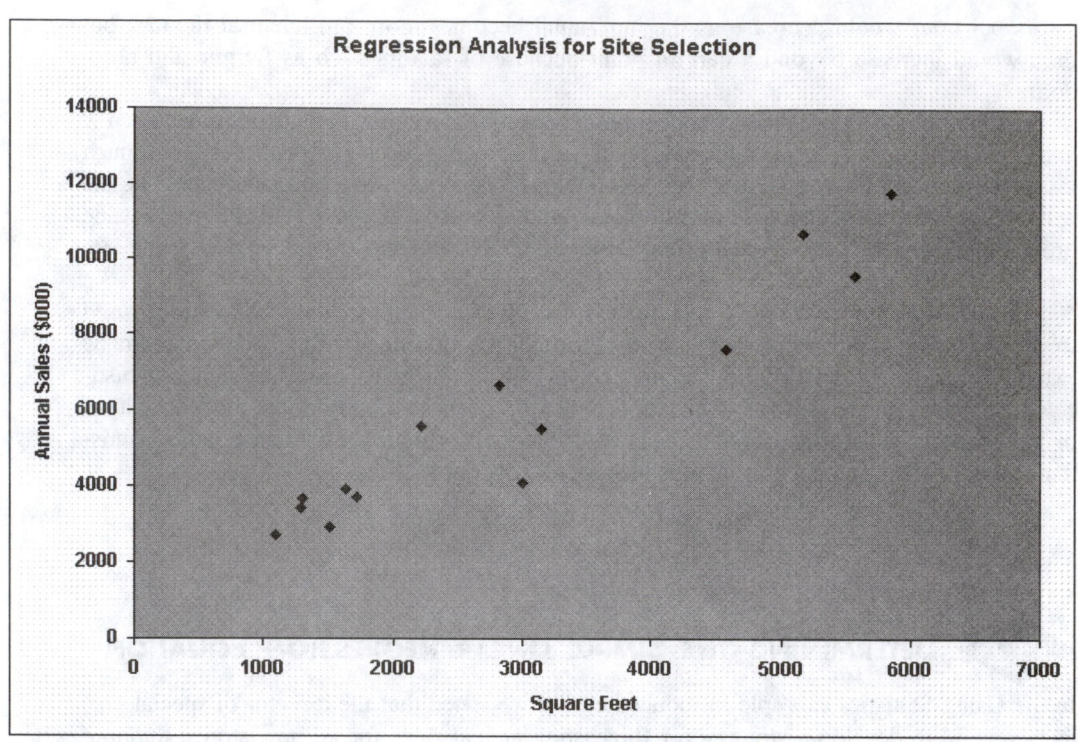

FIGURE 13.3 Scatter diagram for site selection data obtained from Microsoft Excel

The Least-Squares Method

In the preceding section we hypothesized a statistical model to represent the relationship between two variables, square footage and sales, in a chain of women's clothing stores. However, as shown in Table 13.1 on page 775, we have obtained data from only a random sample of the population of stores. If certain assumptions are valid (see section 13.4), the sample Y intercept b_0 and the sample slope b_1 can be used as estimates of the respective population parameters β_0 and β_1. Thus, the sample regression equation representing the straight-line regression model is

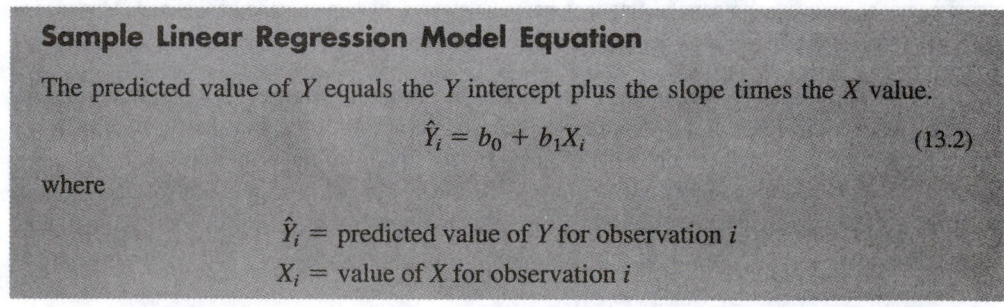

Sample Linear Regression Model Equation

The predicted value of Y equals the Y intercept plus the slope times the X value.

$$\hat{Y}_i = b_0 + b_1 X_i \qquad (13.2)$$

where

$$\hat{Y}_i = \text{predicted value of } Y \text{ for observation } i$$
$$X_i = \text{value of } X \text{ for observation } i$$

This equation requires the determination of two **regression coefficients**—b_0 (the Y intercept) and b_1 (the slope) in order to predict values of Y. Once b_0 and b_1 are obtained, the straight line is known and can be plotted on the scatter diagram. Then we can make a visual comparison of how well our particular statistical model (a straight line) fits the original data by observing whether the original data lie close to the fitted line or deviate greatly from the fitted line.

Simple linear regression analysis is concerned with finding the straight line that fits the data best. The *best* fit could be defined in a variety of ways. Perhaps the simplest way would involve finding the straight line for which the differences between the actual values (Y_i) and the values that would be predicted from the fitted line of regression ($\hat{Y}_i$) are as small as possible. However, because these differences will be positive for some observations and negative for other observations, mathematically we *minimize* the sum of the squared differences

$$\sum_{i=1}^{n}(Y_i - \hat{Y}_i)^2$$

where

$$Y_i = \text{actual value of } Y \text{ for observation } i$$
$$\hat{Y}_i = \text{predicted value of } Y \text{ for observation } i$$

Because $\hat{Y}_i = b_0 + b_1 X_i$, we are minimizing

$$\sum_{i=1}^{n}[Y_i - (b_0 + b_1 X_i)]^2$$

which has two unknowns, b_0 and b_1.

A mathematical technique that determines the values of b_0 and b_1 that minimizes this difference is known as the **least-squares method.** Any values for b_0 and b_1 other than those determined by the least-squares method result in a greater sum of squared differences between the actual value of Y and the predicted value of Y. In using the least-squares method, we obtain the following set of equations:

Equations from the Least-Squares Method

$$\sum_{i=1}^{n}Y_i = nb_0 + b_1\sum_{i=1}^{n}X_i \qquad (13.3a)$$

$$\sum_{i=1}^{n}X_iY_i = b_0\sum_{i=1}^{n}X_i + b_1\sum_{i=1}^{n}X_i^2 \qquad (13.3b)$$

From these two equations we must solve for b_1 and b_0. In this text we take the view that the Excel spreadsheet software will be used to perform the calculations. However, to understand how the results displayed in the output of this software have been computed for the case of simple linear regression, in section 13.10 we illustrate many of the computations involved. Figure 13.4 represents output from Microsoft Excel for the data of Table 13.1.

FIGURE 13.4 Microsoft Excel output for site selection problem

From Figure 13.4 we observe that $b_1 = 1.686$ and $b_0 = 901.247$. Thus, the equation for the best straight line for these data is

$$\hat{Y}_i = 901.247 + 1.686\,X_i$$

The slope b_1 was computed as $+1.686$. This means that for each increase of 1 unit in X, the average value of Y is estimated to increase by 1.686 units. In other words, for each increase of 1 square foot in the size of the store, the fitted model predicts that the expected annual sales are estimated to increase by 1.686 thousands of dollars, or \$1,686. Thus, the slope can be viewed as representing the portion of the annual sales that are estimated to vary according to the size of the store.

The Y intercept b_0 was computed to be $+901.247$ (thousands of dollars). The Y intercept represents the average value of Y when X equals 0. Because the square footage size of the store cannot be 0, this Y intercept can be viewed as representing the portion of the annual sales that varies with factors other than the size of the store.

To illustrate a situation where there is a direct interpretation for the Y intercept b_0, we turn to Example 13.1.

Example 13.1 *Interpreting the Y Intercept b_0 and the Slope b_1*

Suppose that an economist wanted to use the yearly rate of productivity growth in the United States (X) to predict the percentage change in the Standard and Poor's index of 500 stocks. Suppose that a regression model was fit based upon annual data for a period of 50 years with the following results:

$$\hat{Y}_i = -5.0 + 7X_i$$

What is the interpretation of the Y intercept b_0 and the slope b_1?

SOLUTION

The Y intercept $b_0 = -5.0$ tells us that when the rate of productivity growth is zero, the expected change in the Standard and Poor's stock index is -5.0, meaning that the index is predicted to decrease by 5% during the year. The slope $b_1 = 7$ tells us that for each increase in productivity of 1%, we predict that the expected change in the Standard and Poor's stock index is +7.0, meaning that the index is predicted to increase by 7% for each 1% increase in productivity.

The regression model that has been fit to the site selection data can now be used to predict the average annual sales as illustrated in Example 13.2.

Example 13.2 *Predicting Average Annual Sales Based on Square Footage*

Suppose that we would like to use the fitted model to predict the average annual sales for a store with 4,000 square feet.

SOLUTION

We can determine the predicted value by substituting $X = 4,000$ into our regression equation,

$$\hat{Y}_i = 901.247 + 1.686X_i$$
$$\hat{Y}_i = 901.247 + 1.686(4,000) = 7,645.786 \text{ or } \$7,645,786$$

Thus, the predicted average annual sales for a store with 4,000 square feet is 7,645.786 thousands of dollars, or $7,645,786.

Predictions in Regression Analysis: Interpolation versus Extrapolation

When using a regression model for prediction purposes, it is important that we consider only the **relevant range** of the independent variable in making our predictions. This relevant range encompasses all values from the smallest to the largest X used in developing the regression model. Hence, when predicting Y for a given value of X, we may *interpolate* within this relevant range of the X values, but we should not *extrapolate* beyond the range of X values. For example, when we use the square footage to predict annual sales, we note from Table 13.1 that the square footage varies from 1,102 to 5,841. Therefore, predictions of annual sales should be made only for stores that are between 1,102 and 5,841 square feet in size. Any prediction of annual sales for stores with size outside this range presumes that the fitted relationship holds outside the 1,102 to 5,841 range.

 **GENERATING SCATTER DIAGRAMS USING MICROSOFT EXCEL**

Overview

Both quick results users and developers should use this two-step procedure:

A. Use the Chart Wizard to generate a scatter diagram to plot the relationship between an X variable and a Y variable.

B. Modify the scatter diagram generated by the wizard to include a trend line, the regression equation, and the value of r^2.

13-2E.XLS   The 13-2E.XLS workbook file includes the scatter diagram for the Table 13.1 square footage and annual sales data for the site selection problem of section 13.2.

Details for All Users

◆ **A. Generating a Scatter Diagram Using the Chart Wizard**   To generate a scatter diagram to plot the relationship between an X variable and a Y variable, use the XY (Scatter) choice of the Chart Wizard. As an example, consider the site selection problem of section 13.2. To generate a scatter diagram that explores the relationship between the square footage of a store and the annual sales of the store, do the following:

❶ Open the Starting Point for Several Chapter 13 Sections workbook (STARTING POINT 13-SEVERAL.XLS) and click the Data sheet tab. Verify that store number, square feet, and annual sales data of Table 13.1 on page 775 have been entered into columns A, B, and C, respectively.

❷ Select Insert | Chart.

❸ In the Step 1 dialog box:
   a. Select the Standard Types tab and then select XY (Scatter) from the Chart type: list box. Select the first (top) choice from the Chart sub-types, which is identified as "Scatter. Compares pairs of values" when selected.
   b. Click the Next button.

❹ In the Step 2 dialog box:
   a. Select the Data Range tab. Enter B1:C15 in the Data range: edit box.
   b. Select the Columns option button in the Series in: group.
   c. Click the Next button.

❺ In the Step 3 dialog box:
   a. Select the Titles tab. Enter Regression Analysis for Site Selection in the Chart title: edit box, Square Feet in the Value (X) axis: edit box, and Annual Sales ($000) in the Value (Y) axis: edit box.
   b. Select, in turn, the Gridlines, Axes, Legend, and Data Table tabs and verify that their settings match those given in Table 2E.3 on page 88.
   c. Click the Next button.

❻ In the Step 4 dialog box:
   a. Select the As new sheet: option button and enter Trend in the edit box to the right of the option button.
   b. Click the Finish button.

The Chart Wizard inserts a chart sheet containing the scatter diagram for the data of Table 13.1 similar to one shown in Figure 13.3 on page 776.

◆ **Data Order and Scatter Diagrams**   When generating scatter diagrams, the Chart Wizard always assumes that the first column (or row) of data of the data range entered in the Step 2 dialog box contains values for the X variable (as it does in the preceding example). Had the X variable data been located in the second column of the data range, it would have been necessary to select the Series tab of the Step 2 dialog box and change the cell ranges in the X Values: and Y Values: edit boxes. Furthermore, because of a quirk in this wizard, the revised cell ranges must be entered as formulas that include sheet names, for example, =Data!B2:B15.

Using the simplified cell range form, for example, =B2:B15, would cause Microsoft Excel to display the misleading error message, "The formula you typed contains an error."

The contents of the Series tab for the preceding example, which did not need editing, is shown in Figure 13E.1. (See section S3.2 for a review of information about specifying worksheet locations.)

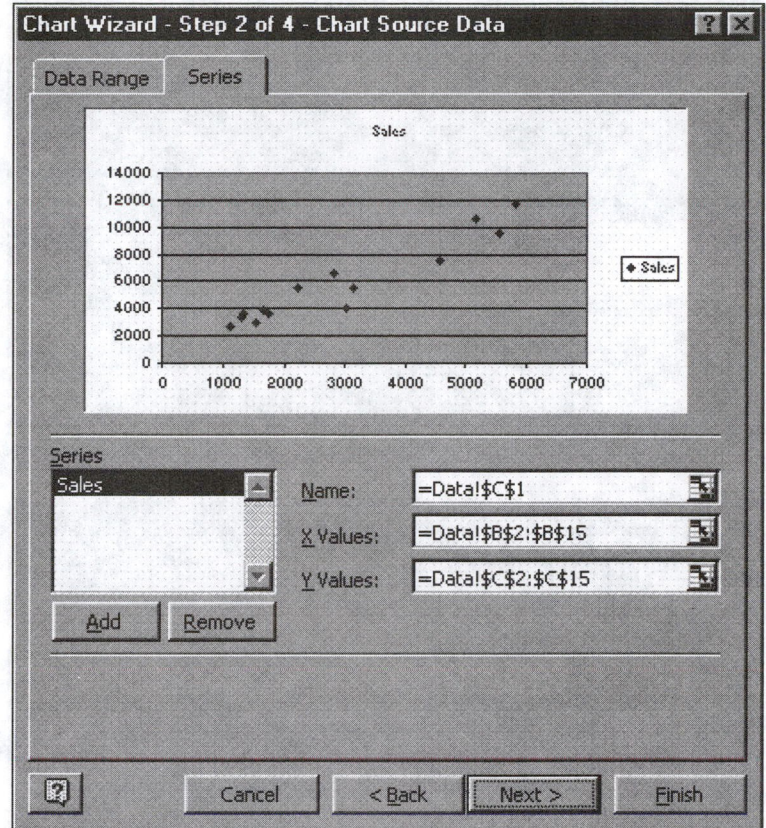

FIGURE 13E.1
Series tab of the Chart Wizard Step 2 dialog box for the *XY* (scatter) chart type for the site selection problem of section 13.2

◆ **B. Modifying the Scatter Diagram**   After the Chart Wizard generates a scatter diagram, we can modify the chart by adding a trend line, the regression equation, and the value of r^2. As an example, consider the chart produced in part A for the site selection problem of section 13.2. To add a trend line to this scatter diagram, do the following:

❶ Click the sheet tab of the Trend chart sheet generated in part A of this procedure.

❷ Select Chart | Add Trendline. (Note: the Chart choice appears on the Microsoft Excel menu bar only when a chart or chart sheet is selected.)

❸ In the Add Trendline dialog box:
a. Select the Type tab and select the Linear choice in the Trend/Regression type group (see Figure 13E.2, panel A).
b. Select the Options tab and select the Automatic option button and the Display equation on chart and Display R-squared value on chart check boxes (see Figure 13E.2, panel B).
c. Click the OK button.

FIGURE 13E.2
Panel A: Type tab of the Add
Trendline dialog box

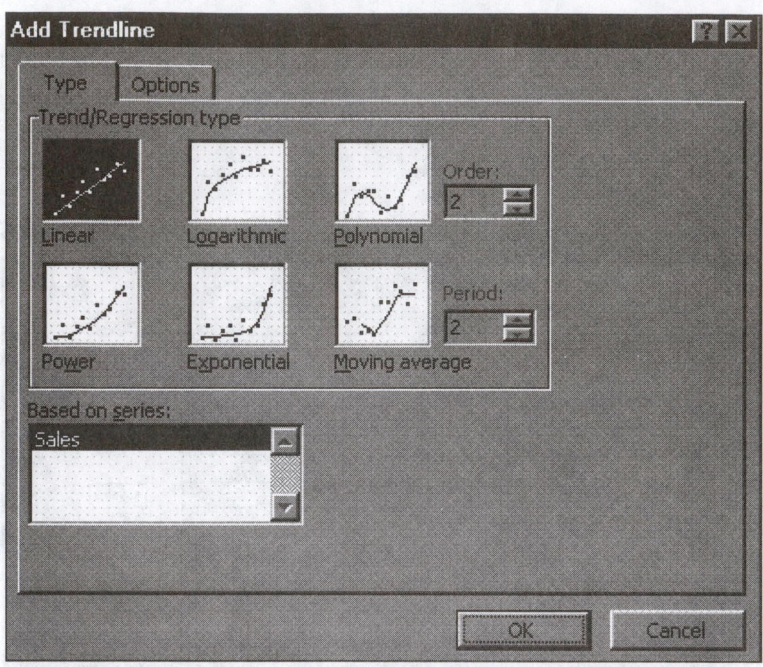

FIGURE 13E.2
Panel B: Options tab of the
Add Trendline dialog box

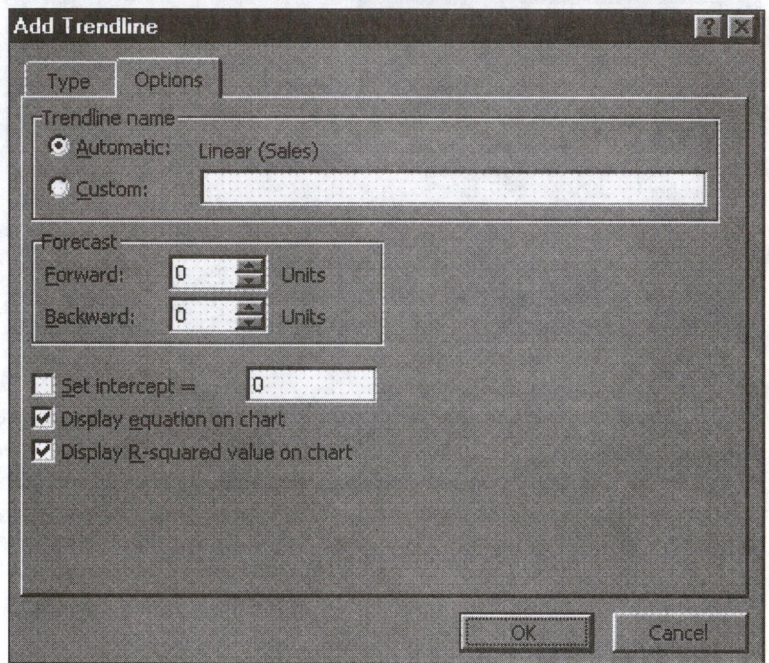

Microsoft Excel modifies the scatter diagram by adding a line of regression, the regression equation, and the value of the r^2 to the scatter diagram. The modified scatter diagram will be similar to the one shown in Figure 13.6 on page 791.

558

> **▲ WHAT IF EXAMPLE**
>
> We can change the values for the square footage or sales in the Data sheet to see their effects on the fit of the regression model as represented by the line of regression and regression equation on the scatter diagram. For example, if we change the value for the sales of store 1 from 3,681 to 4,500, we will observe on the scatter diagram that the Y intercept changes from 901.25 to 1,036.3, the slope changes from 1.6861 to 1.6599, and the value of r^2 changes from .9098 to .9041. To facilitate comparisons among many different alternative values, we can create scenarios using the Scenario Manager as discussed at the end of section 2.2E.

 **CALCULATING THE SIMPLE LINEAR REGRESSION COEFFICIENTS USING MICROSOFT EXCEL**

Overview

Both quick results users and developers should use the PHStat add-in to calculate the coefficients of the simple linear regression equation.

The 13-2E.XLS workbook file includes a table of coefficients of the simple linear regression equation for the Table 13.1 square footage and annual sales data for the site selection problem of section 13.2.

13-2E.XLS

Details for All Users

To calculate the coefficients of the simple linear regression equation, use the Regression | Simple Linear Regression choice of the PHStat add-in. This add-in modifies and extends the output generated by the Data Analysis Regression tool, which can also be used to calculate the regression coefficients. As an example, consider the site selection problem of section 13.2. To calculate the coefficients of the simple linear regression equation that represent the relationship between the square footage of a store and the annual sales of the store, do the following:

❶ Open the Starting Point for Several Chapter 13 Sections workbook (STARTING POINT 13-SEVERAL.XLS) and click the Data sheet tab. Verify that store number, square feet, and annual sales data of Table 13.1 on page 775 have been entered into columns A, B, and C, respectively.

❷ Select PHStat | Regression | Simple Linear Regression.

❸ In the Simple Linear Regression dialog box (see Figure 13E.3):
   a. Enter C1:C15 in the Y Variable Cell Range: edit box.
   b. Enter B1:B15 in the X Variable Cell Range: edit box.
   c. Select the First cells in both ranges contain label check box.
   d. Enter 95 in the Confidence Lvl. for regression coefficients: edit box.
   e. Select the Regression Statistics Table and the ANOVA and Coefficients Table check boxes.
   f. Enter Regression Analysis for Site Selection in the Output Title: edit box.
   g. Click the OK button.

The add-in inserts a worksheet that contains the regression coefficients and other summary information for the simple linear regression for the Table 13.1 data, similar to the worksheet shown in Figure 13.4 on page 778. This worksheet is *not* dynamically changeable, so any changes made to the underlying site selection data would require using the PHStat add-in a second time to produce updated results.

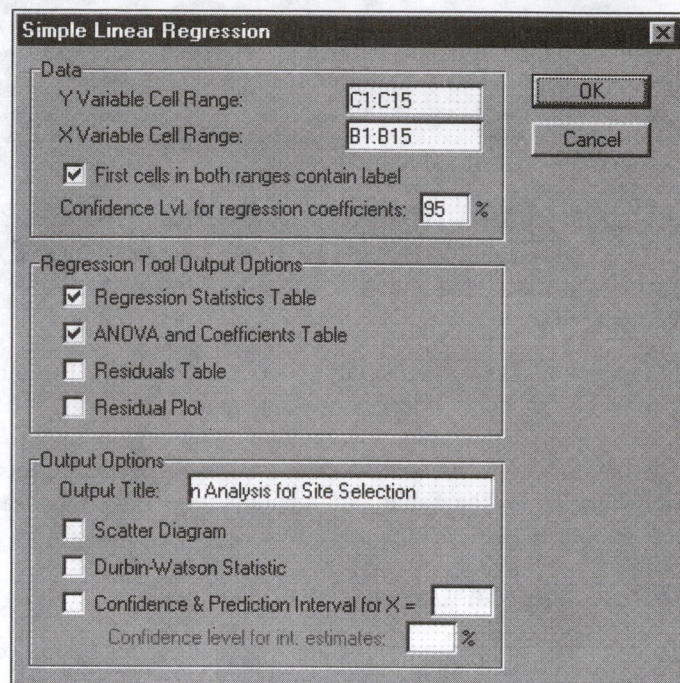

Problems for Section 13.2

Learning the Basics

• **13.1** Fitting a straight line to a set of data yields the following regression equation:

$$\hat{Y}_i = 2 + 5X_i$$

    (a) Interpret the meaning of the Y intercept b_0.
    (b) Interpret the meaning of the slope b_1.
    (c) Predict the average value of Y for X = 3.
    (d) If the values of X range from 2 to 25, should you use this model to predict the average value of Y when X equals

        (1) 3?   (4) 24?
        (2) −3?   (5) 26?
        (3) 0?

13.2 Fitting a straight line to a set of data yields the following regression equation

$$\hat{Y}_i = 16 - .5X_i$$

    (a) Interpret the meaning of the Y intercept b_0.
    (b) Interpret the meaning of the slope b_1.
    (c) Predict the average value of Y for X = 6.

Applying the Concepts

Note: Use Microsoft Excel to solve these problems.

● **13.3** The marketing manager of a large supermarket chain would like to determine the effect of shelf space on the sales of pet food. A random sample of 12 equal-sized stores is selected with the following results:

STORE	SHELF SPACE, X (FEET)	WEEKLY SALES, Y (HUNDREDS OF DOLLARS)	STORE	SHELF SPACE, X (FEET)	WEEKLY SALES, Y (HUNDREDS OF DOLLARS)
1	5	1.6	7	15	2.3
2	5	2.2	8	15	2.7
3	5	1.4	9	15	2.8
4	10	1.9	10	20	2.6
5	10	2.4	11	20	2.9
6	10	2.6	12	20	3.1

DATA FILE
PETFOOD.XLS

(a) Set up a scatter diagram.
(b) Assuming a linear relationship, use the least-squares method to find the regression coefficients b_0 and b_1.
(c) Interpret the meaning of the slope b_1 in this problem.
(d) Predict the average weekly sales (in hundreds of dollars) of pet food for stores with 8 feet of shelf space for pet food.
(e) Suppose that sales in store 12 are 2.6. Do parts (a)–(d) with this value and compare the results.
(f) What shelf space would you recommend that the marketing manager allocate to pet food? Explain.

WHAT IF?

13.4 Suppose that the management of a chain of package delivery stores would like to develop a model for predicting the weekly sales (in thousands of dollars) for individual stores based on the number of customers who made purchases. A random sample of 20 stores was selected from among all the stores in the chain with the following results.

STORES	CUSTOMERS	SALES ($000)	STORES	CUSTOMERS	SALES ($000)
1	907	11.20	11	679	7.63
2	926	11.05	12	872	9.43
3	506	6.84	13	924	9.46
4	741	9.21	14	607	7.64
5	789	9.42	15	452	6.92
6	889	10.08	16	729	8.95
7	874	9.45	17	794	9.33
8	510	6.73	18	844	10.23
9	529	7.24	19	1,010	11.77
10	420	6.12	20	621	7.41

DATA FILE
PACKAGE.XLS

(a) Set up a scatter diagram.
(b) Assuming a linear relationship, use the least-squares method to find the regression coefficients b_0 and b_1.
(c) Interpret the meaning of the slope b_1 in this problem.
(d) Predict the average weekly sales (in thousands of dollars) for stores that have 600 customers.

(e) Suppose that sales in store 19 are 14.77. Do parts (a)–(d) with this value and compare the results.

(f) What other factors besides the number of customers might affect sales?

• **13.5** A company manufacturing machine parts would like to develop a model to estimate the number of worker-hours required for production runs of varying lot size. A random sample of 14 production runs (2 each for lot sizes 20, 30, 40, 50, 60, 70, and 80) is selected with the following results:

LOT SIZE	WORKER-HOURS	LOT SIZE	WORKER-HOURS
20	50	50	112
20	55	60	128
30	73	60	135
30	67	70	148
40	87	70	160
40	95	80	170
50	108	80	162

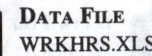

DATA FILE
WRKHRS.XLS

(a) Set up a scatter diagram.

(b) Assuming a linear relationship, use the least-squares method to find the regression coefficients b_0 and b_1.

(c) Interpret the meaning of the Y intercept b_0 and the slope b_1 in this problem.

(d) Predict the average number of worker-hours required for a production run with a lot size of 45.

(e) Why would it not be appropriate to predict the average number of worker-hours required for a production run with a lot size of 100? Explain.

(f) Suppose that the worker-hours for the lot size of 60 were 117 and 119. Do parts (a)–(d) with these values and compare the results.

WHAT IF?

13.6 A company that has the distribution rights to home video sales of previously released movies would like to be able to estimate the number of units that it can be expected to sell. Data are available for 30 movies that indicate the box office gross (in millions of dollars) and the number of units sold (in thousands) of home videos. The results are as follows:

MOVIE	BOX OFFICE GROSS ($ MILLIONS)	HOME VIDEO UNITS SOLD (000)	MOVIE	BOX OFFICE GROSS ($ MILLIONS)	HOME VIDEO UNITS SOLD (000)
1	1.10	57.18	16	9.36	190.80
2	1.13	26.17	17	9.89	121.57
3	1.18	92.79	18	12.66	183.30
4	1.25	61.60	19	15.35	204.72
5	1.44	46.50	20	17.55	112.47
6	1.53	85.06	21	17.91	162.95
7	1.53	103.52	22	18.25	109.20
8	1.69	30.88	23	23.13	280.79
9	1.74	49.29	24	27.62	229.51
10	1.77	24.14	25	37.09	277.68
11	2.42	115.31	26	40.73	226.73
12	5.34	87.04	27	45.55	365.14
13	5.70	128.45	28	46.62	218.64
14	6.43	126.64	29	54.70	286.31
15	8.59	107.28	30	58.51	254.58

DATA FILE
MOVIE.XLS

(a) Set up a scatter diagram.

(b) Use the least-squares method to find the regression coefficients b_0 and b_1.

(c) State the regression equation.

(d) Interpret the meaning of b_0 and b_1 in this problem.

(e) Predict the average video unit sales for a movie that had a box office gross of $20 million.

(f) What other factors in addition to box office gross might be useful in predicting video unit sales?

13.7 An agent for a residential real estate company in a large city would like to be able to predict the monthly rental costs for apartments based on the size of apartment as defined by square footage. A sample of 25 apartments in a particular residential neighborhood was selected and the information gathered revealed the following:

APARTMENT	MONTHLY RENT ($)	SIZE (SQUARE FEET)	APARTMENT	MONTHLY RENT ($)	SIZE (SQUARE FEET)
1	950	850	14	1,800	1,369
2	1,600	1,450	15	1,400	1,175
3	1,200	1,085	16	1,450	1,225
4	1,500	1,232	17	1,100	1,245
5	950	718	18	1,700	1,259
6	1,700	1,485	19	1,200	1,150
7	1,650	1,136	20	1,150	896
8	935	726	21	1,600	1,361
9	875	700	22	1,650	1,040
10	1,150	956	23	1,200	755
11	1,400	1,100	24	800	1,000
12	1,650	1,285	25	1,750	1,200
13	2,300	1,985			

DATA FILE
RENT.XLS

(a) Set up a scatter diagram.

(b) Use the least-squares method to find the regression coefficients b_0 and b_1?

(c) State the regression equation.

(d) Interpret the meaning of b_0 and b_1 in this problem.

(e) Predict the average monthly rental cost for an apartment that has 1,000 square feet.

(f) Why would it not be appropriate to use the model to predict the monthly rental for apartments that have 500 square feet?

(g) Your friends Jim and Jennifer are considering signing a lease for an apartment in this residential neighborhood. They are trying to decide between two apartments, one with 1,000 square feet for a monthly rent of $1,250 and the other with 1,200 square feet for a monthly rent of $1,425. What would you recommend to them? Why?

13.8 A limousine service operating from a suburban county wants to determine the length of time it would take to transport passengers from various locations to a major metropolitan airport during nonpeak hours. A sample of 12 trips on a particular day during nonpeak hours indicates the following

DISTANCE (MILES)	TIME (MINUTES)	DISTANCE (MILES)	TIME (MINUTES)
10.3	19.71	18.4	29.38
11.6	18.15	20.2	37.24
12.1	21.88	21.8	36.84
14.3	24.21	24.3	40.59
15.7	27.08	25.4	41.21
16.1	22.96	26.7	38.19

DATA FILE
LIMO.XLS

(a) Set up a scatter diagram.

(b) Assuming a linear relationship, use the least-squares method to find the regression coefficients of b_0 and b_1.

(c) Interpret the meaning of the Y intercept b_0 and the slope b_1 in this problem.

(d) Use the regression model developed in (b) to predict the average number of minutes to transport someone from a location that is 21 miles from the airport.

 **WHAT IF?**

(e) Suppose the distance for the last trip was 36.7 (instead of 26.7) miles and the time was 65 (instead of 38.19) minutes. Do parts (a)–(d) with these values and compare the results.

(f) What would the results of (e) lead you to think about the usefulness of the regression model?

13.3 MEASURES OF VARIATION

Obtaining the Sum of Squares

To examine how well the independent variable predicts the dependent variable in our statistical model, we need to develop several measures of variation. The first measure, the **total sum of squares (SST),** is a measure of variation of the Y_i values around their mean $\overline{Y}$. In a regression analysis the **total variation** or **sum of squares** can be subdivided into **explained variation** or **regression sum of squares (SSR)**, that which is attributable to the relationship between X and Y, and **unexplained variation** or **error sum of squares (SSE)**, that which is attributable to factors other than the relationship between X and Y. These different measures of variation can be seen in Figure 13.5.

FIGURE 13.5

Measures of variation in regression

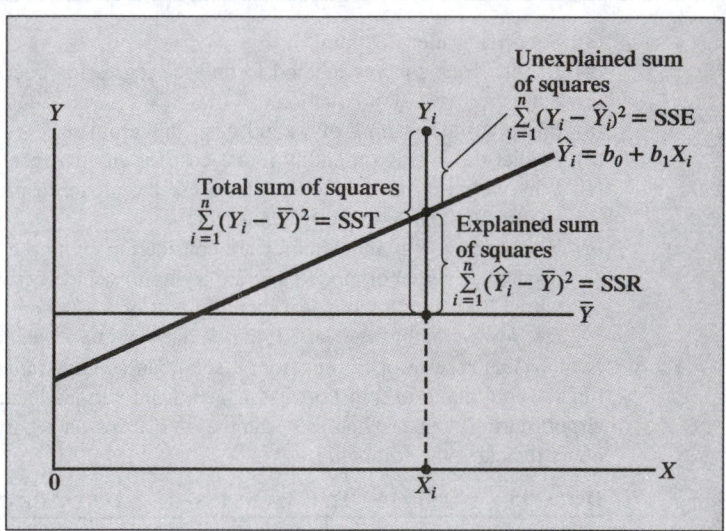

The regression sum of squares (SSR) represents the difference between $\hat{Y}_i$ (the value of Y that would be predicted from the regression relationship) and $\overline{Y}$ (the average value of Y). The error sum of squares (SSE) represents that part of the variation in Y that is not explained by the regression. It is based on the difference between Y_i and $\hat{Y}_i$. These measures of variation can be represented as follows:

Measures of Variation in Regression

Total sum of squares = regression sum of squares + error sum of squares

$$SST = SSR + SSE \qquad (13.4)$$

Total Sum of Squares (SST)

The total sum of squares (SST) is equal to the sum of the squared differences between each observed Y value and $\overline{Y}$, the average value of Y.

$$SST = \text{total variation or total sum of squares} = \sum_{i=1}^{n}(Y_i - \overline{Y})^2 \qquad (13.5)$$

Regression Sum of Squares (SSR)

The regression sum of squares (SSR) is equal to the sum of the squared differences between each predicted value of Y and the mean of Y.

$$SSR = \text{explained variation or regression sum of squares} \qquad (13.6)$$

$$= \sum_{i=1}^{n}(\hat{Y}_i - \overline{Y})^2$$

$$= SST - SSE$$

Error Sum of Squares (SSE)

The error sum of squares (SSE) is equal to the sum of the squared differences between each observed value of Y and each predicted value of Y.

$$SSE = \text{unexplained variation or error sum of squares} \qquad (13.7)$$

$$= \sum_{i=1}^{n}(Y_i - \hat{Y}_i)^2$$

Examining Figure 13.4 on page 778, we observe that

$$SSR = 106,208,120; \qquad SSE = 10,532,255; \qquad \text{and} \qquad SST = 116,740,375$$

We note also, from equation (13.4), that

$$SST = SSR + SSE$$

$$116,740,375 = 106,208,120 + 10,532,255$$

The total sum of squared differences around the average value of Y is equal to 116,740,375. This amount is subdivided into the sum of squares that is explained by the regression (SSR), equal to 106,208,120, and the sum of squares that is unexplained by the regression (the error sum of squares), equal to 10,532,255.

The Coefficient of Determination

By themselves, *SSR*, *SSE*, and *SST* provide little that can be directly interpreted. However, a simple ratio of the regression sum of squares (*SSR*) to the total sum of squares (*SST*) provides a measure of the usefulness of the regression equation. This ratio is called the **coefficient of determination** r^2 and is defined as

Coefficient of Determination

The coefficient of determination is equal to the regression sum of squares divided by the total sum of squares.

$$r^2 = \frac{\text{regression sum of squares}}{\text{total sum of squares}} = \frac{SSR}{SST} \qquad (13.8)$$

This coefficient of determination measures the proportion of variation in Y that is explained by the independent variable X in the regression model. For the site selection example, with $SSR = 106,208,120$; $SSE = 10,532,255$; and $SST = 116,740,375$

$$r^2 = \frac{106,208,120}{116,740,375} = .91$$

Therefore, 91% of the variation in annual sales can be explained by the variability in the size of the store as measured by the square footage. This is an example where there is a strong positive linear relationship between two variables because the use of a regression model has reduced the variability in predicting annual sales by 91%. Only 9% of the sample variability in annual sales can be explained by factors other than what is accounted for by the linear regression model that uses only square footage.

Standard Error of the Estimate

Although the least-squares method results in the line that fits the data with the minimum amount of variation, we have seen in the computation of the error sum of squares (*SSE*) that unless all the observed data points fall on the regression line, the regression equation is not a perfect predictor. Just as we do not expect all data values to be exactly equal to their arithmetic mean, neither can we expect all data points to fall exactly on the regression line. Therefore, we need to develop a statistic that measures the variability of the actual Y values from the predicted Y values, in the same way that we developed the standard deviation in chapter 3 as a measure of the variability of each observation around its mean. This standard deviation around the line of regression is called the **standard error of the estimate.**

The variability around the line of regression is illustrated in Figure 13.6 for the site selection data. We can see from Figure 13.6 that, although many of the actual values of Y fall near the predicted line of regression, there are several values above the line of regression as well as below the line of regression.

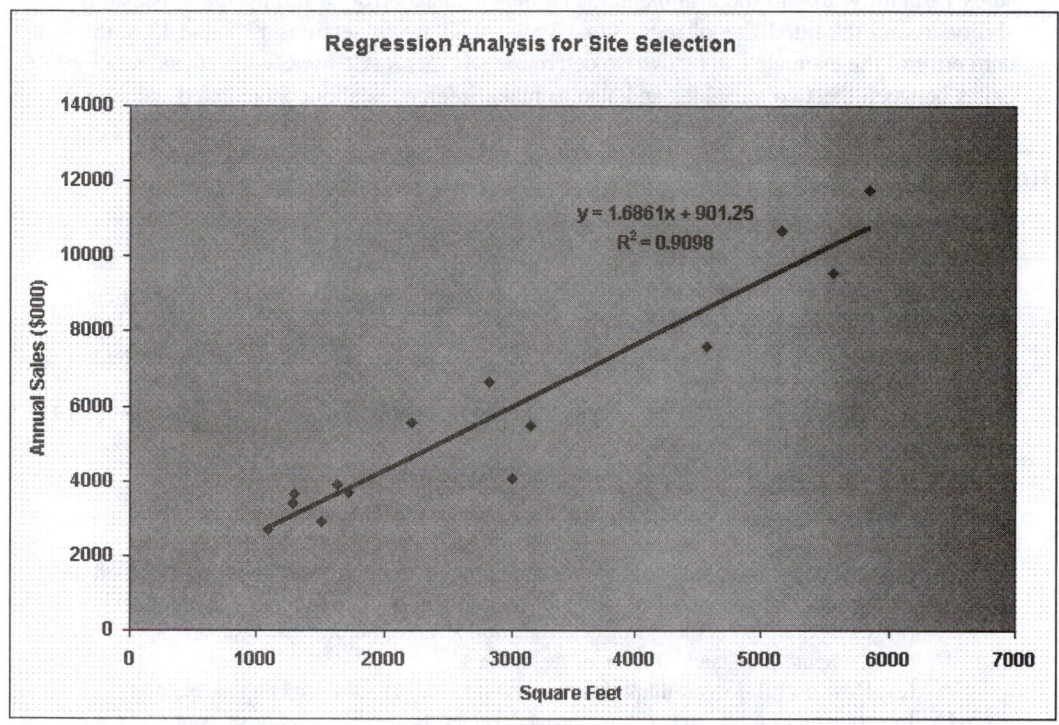

FIGURE 13.6

Scatter diagram and line of regression for site selection data obtained from Microsoft Excel

The standard error of the estimate, given by the symbol S_{YX}, is defined as

Standard Error of the Estimate

$$S_{YX} = \sqrt{\frac{SSE}{n-2}} = \sqrt{\frac{\sum_{i=1}^{n}(Y_i - \hat{Y}_i)^2}{n-2}} \qquad (13.9)$$

where

Y_i = actual value of Y for a given X_i

$\hat{Y}_i$ = predicted value of Y for a given X_i

SSE = error sum of squares

From equation (13.9), with $SSE = 10,532,255$, we have

$$S_{YX} = \sqrt{\frac{10,532,255}{14 - 2}}$$

$$S_{YX} = 936.85$$

This standard error of the estimate, equal to 936.85 thousands of dollars (i.e., \$936,850), is labeled Standard Error on the Microsoft Excel output of Figure 13.4 on page 778. The standard error of the estimate represents a measure of the variation around the fitted line of regression. It is measured in units of the dependent variable Y. The interpretation of the standard error of the estimate is similar to that of the standard deviation. Just as the standard deviation measures variability around the arithmetic mean, the standard error of the estimate measures variability around the fitted line of regression. As we shall see in sections 13.7 and 13.8, the standard error of the estimate can be used to determine whether a statistically significant relationship exists between the two variables and also to make inferences about a predicted value of Y.

Problems for Section 13.3

Learning the Basics

13.9 If the coefficient of determination r^2 is equal to .80, what does this mean?

• **13.10** If $SSR = 36$ and $SSE = 4$, find SST, then compute the coefficient of determination r^2 and interpret its meaning.

13.11 If $SSR = 66$ and $SST = 88$, compute the coefficient of determination r^2 and interpret its meaning.

13.12 If $SSE = 10$ and $SST = 30$, compute the coefficient of determination r^2 and interpret its meaning.

13.13 If $SSR = 120$, why is it impossible for SST to equal 110?

Applying the Concepts

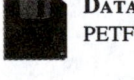
DATA FILE
PETFOOD.XLS

• **13.14** In Problem 13.3 on page 785 the marketing manager used shelf space for pet food to predict weekly sales. Use the computer output you obtained to solve that problem.
(a) Compute the coefficient of determination r^2 and interpret its meaning.
(b) Compute the standard error of the estimate.
(c) How useful do you think this regression model is for predicting sales?

DATA FILE
PACKAGE.XLS

13.15 In Problem 13.4 on page 785 a manager wanted to predict weekly sales at a chain of package delivery stores based on the number of customers who made purchases. Use the computer output you obtained to solve that problem.
(a) Compute the coefficient of determination r^2 and interpret its meaning.
(b) Compute the standard error of the estimate.
(c) How useful do you think this regression is for predicting sales?

DATA FILE
WORKHRS.XLS

• **13.16** In Problem 13.5 on page 786 a company wanted to predict the worker-hours required for production based on the lot size. Use the computer output you obtained to solve that problem.
(a) Compute the coefficient of determination r^2 and interpret its meaning.
(b) Compute the standard error of the estimate.
(c) How useful do you think this regression model is for predicting worker-hours?

DATA FILE
MOVIE.XLS

13.17 In Problem 13.6 on page 786 a company wanted to predict home video sales based on the box office gross of movies. Use the computer output you obtained to solve that problem.
(a) Compute the coefficient of determination r^2 and interpret its meaning.
(b) Compute the standard error of the estimate.
(c) How useful do you think this regression model is for predicting home video sales?

DATA FILE
RENT.XLS

13.18 In Problem 13.7 on page 787 an agent for a real estate company wanted to predict the monthly rent for apartments based on the size of the apartment. Use the computer output you obtained to solve that problem.

(a) Compute the coefficient of determination r^2 and interpret its meaning.

(b) Compute the standard error of the estimate.

(c) How useful do you think this regression model is for predicting monthly rent?

13.19 In Problem 13.8 on page 787 a limousine service wants to predict travel time to an airport based on the distance from the pickup location to the airport. Use the computer output you obtained to solve that problem.

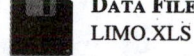

DATA FILE
LIMO.XLS

(a) Compute the coefficient of determination r^2 and interpret its meaning.

(b) Compute the standard error of the estimate.

(c) How useful do you think this regression model is for predicting travel time?

 **13.4 ASSUMPTIONS**

In our study of hypothesis testing and the analysis of variance, we have stated that the appropriate application of a particular statistical procedure is dependent on how well a set of assumptions for that procedure are met. The assumptions necessary for regression are analogous to those of the analysis of variance because they fall under the general heading of *linear models* (reference 6).

The three major **assumptions of regression** are listed in Exhibit 13.1.

Exhibit 13.1 Assumptions of Regression

✓ **1.** Normality of error

✓ **2.** Homoscedasticity

✓ **3.** Independence of errors

The first assumption, **normality,** requires that errors around the line of regression be normally distributed at each value of X (see Figure 13.7). Like the t test and the ANOVA F test, regression analysis is fairly robust against departures from the normality assumption. As long as the distribution of the errors around the line of regression at each level of X is not extremely different from a normal distribution, inferences about the line of regression and the regression coefficients will not be seriously affected.

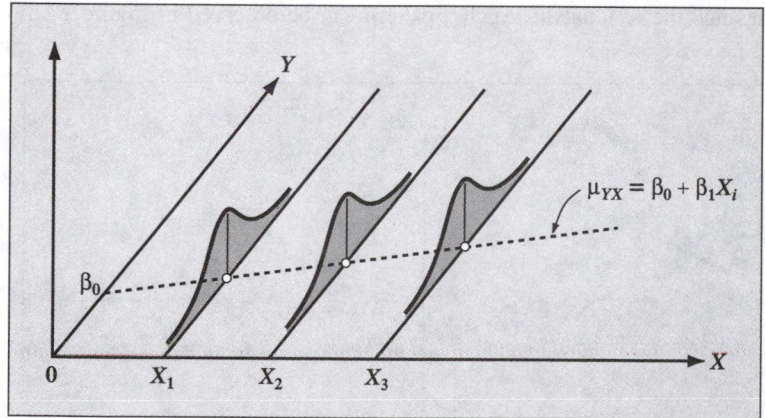

FIGURE 13.7
Assumptions of regression

The second assumption, **homoscedasticity,** requires that the variation around the line of regression be constant for all values of X. This means that the errors vary the same amount when X is a low value as when X is a high value (see Figure 13.7). The homoscedasticity assumption is important for using the least-squares method of determining the regression coefficients. If there are serious departures from this assumption, either data transformations (see section 14.8) or weighted least-squares methods (reference 6) can be applied.

The third assumption, **independence of errors,** requires that the errors be independent for each value of X. This assumption is particularly important when data are collected over a period of time. In such situations, the errors for a particular time period are often correlated with those of the previous time period.

13.5 RESIDUAL ANALYSIS

In the preceding discussion of the site selection data we have relied on a simple linear regression model in which the dependent variable is predicted based on a straight-line relationship with a single independent variable. In this section we use a graphical approach called **residual analysis** to evaluate the appropriateness of the regression model that has been fitted to the data. In addition, this approach also allows us to study potential violations of the assumptions of our regression model.

Evaluating the Aptness of the Fitted Model

The **residual** or estimated error value e_i is defined as the difference between the observed (Y_i) and predicted ($\hat{Y}_i$) values of the dependent variable for a given value of X_i. Thus the following definition applies:

> **The Residual**
>
> The residual equals the difference between the observed value of Y and the predicted value of Y.
>
> $$e_i = Y_i - \hat{Y}_i \tag{13.10}$$

We evaluate the aptness of the fitted regression model by plotting the residuals on the vertical axis against the corresponding X_i values of the independent variable on the horizontal axis. If the fitted model is appropriate for the data, there will be no apparent pattern in this plot of the residuals versus X_i. However, if the fitted model is not appropriate, there will be a relationship between the X_i values and the residuals e_i. Such a pattern can be observed in Figure 13.8.

FIGURE 13.8

Studying appropriateness of simple linear regression model

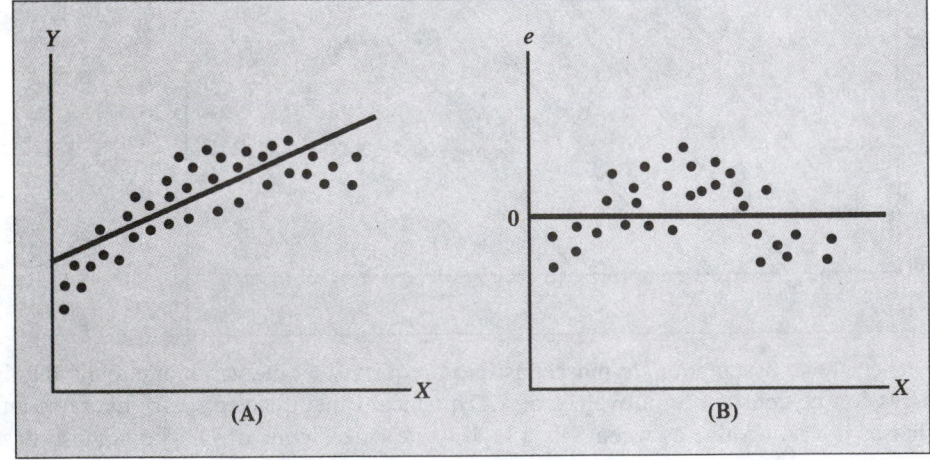

(A)   (B)

Panel (a) depicts a situation in which, although there is an increasing trend in Y as X increases, the relationship seems curvilinear because the upward trend decreases for increasing values of X. Thus a curvilinear model between the two variables seems more appropriate than a simple linear regression model. This curvilinear effect is highlighted in panel (b). Here there is a clear curvilinear effect between X_i and e_i. By plotting the residuals, we have filtered out or removed the linear trend of X with Y, thereby exposing the lack of fit in the simple linear model. Thus we conclude that the curvilinear model is a better fit and should be evaluated in place of the simple linear model (see section 14.6 for further discussion of fitting curvilinear models).

Having considered Figure 13.8, let us return to the evaluation of the site selection data. Figure 13.9 provides the output that includes the observed, predicted, residual, and a version of the Studentized residual values of the response variable (annual sales) in the simple linear model we have fitted.

	A	B	C	D
22	RESIDUAL OUTPUT			
23				
24	Observation	Predicted Sales	Residuals	Standard Residuals
25	1	3811.515529	-130.5155	-0.145001728
26	2	3669.880191	225.11981	0.250106339
27	3	5649.402647	1003.5974	1.114988776
28	4	10267.72633	-724.7263	-0.805165261
29	5	3079.732952	338.26705	0.375812033
30	6	4624.232585	938.76742	1.042963225
31	7	3115.141786	544.85821	0.605333196
32	8	2759.367307	-65.36731	-0.072622565
33	9	6214.257862	-746.2579	-0.829086623
34	10	3457.427185	-559.4272	-0.621519209
35	11	9603.389152	1070.6108	1.189440242
36	12	8601.82498	-1016.825	-1.129684566
37	13	10749.96093	1010.0391	1.122145471
38	14	5973.140561	-1888.141	-2.097709332

FIGURE 13.9

Residual statistics for site selection problem obtained from Microsoft Excel

Studentized residuals, expressed as equation (13.11), are the **standardized residuals** (the residual divided by its standard error) adjusted for the distance from the average X value. These Studentized residuals allow us to consider the magnitude of the residuals in units that reflect the standardized variation around the line of regression.

The Studentized Residual

$$\text{Studentized residual} = SR_i = \frac{e_i}{S_{YX}\sqrt{1 - h_i}} \qquad (13.11)$$

where, for observation i

$$h_i = \frac{1}{n} + \frac{(X_i - \overline{X})^2}{\sum_{i=1}^{n}(X_i - \overline{X})^2}$$

To determine whether the linear model is appropriate for these data, the residuals have been plotted against the independent variable (store size in square feet) in Figure 13.10. We observe that although there is widespread scatter in the residual plot, there is no apparent pattern or relationship between the residuals and X_i. The residuals appear to be evenly spread above and below 0 for the differing values of X. This result leads us to conclude that the fitted straight-line model is appropriate for the site selection sales data.

FIGURE 13.10

Plot of residuals against square footage of store for site selection problem obtained from Microsoft Excel

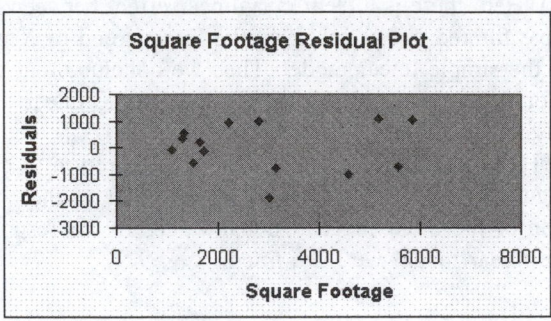

Evaluating the Assumptions

◆ *Homoscedasticity*   The assumption of homoscedasticity can also be evaluated from a plot of the residuals with X_i. For the site selection data of Figure 13.10 there do not appear to be major differences in the variability of the residuals for different X_i values as is the case in Figure 13.11. Thus, we may conclude that for our fitted model, there is no apparent violation in the assumption of equal variance at each level of X.

If we wish to observe a case in which the homoscedasticity assumption is violated, we should examine the hypothetical plot of SR_i with X_i in Figure 13.11. In this hypothetical plot there appears to be a *fanning* effect in which the variability of the residuals increases as X increases, demonstrating the lack of homogeneity in the variances of Y_i at each level of X.

FIGURE 13.11

Violations of homoscedasticity

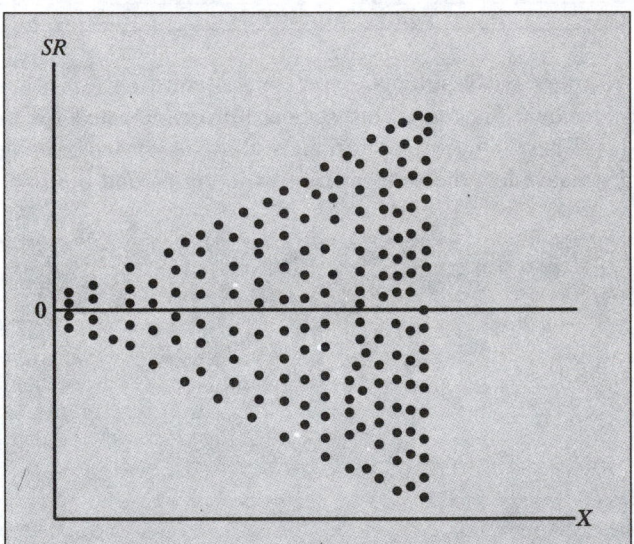

◆ *Normality*   The assumption of normality in the errors around the line of regression can be evaluated from a residual analysis by tallying the residuals or standardized residuals into a frequency distribution and displaying the results in a histogram (see chapter 2).

For the site selection data, the standardized residuals have been tallied into a frequency distribution as indicated in Table 13.2 with the results displayed in the histogram of Figure 13.12.

Table 13.2 *Frequency distribution of 14 standardized residual values for site selection data*

STANDARDIZED RESIDUALS	FREQUENCY
-2.25 but less than -1.75	1
-1.75 but less than -1.25	0
-1.25 but less than -0.75	3
-0.75 but less than -0.25	1
-0.25 but less than $+0.25$	2
$+0.25$ but less than $+0.75$	3
$+0.75$ but less than $+1.25$	2
$+1.25$ but less than $+1.75$	2
Total	14

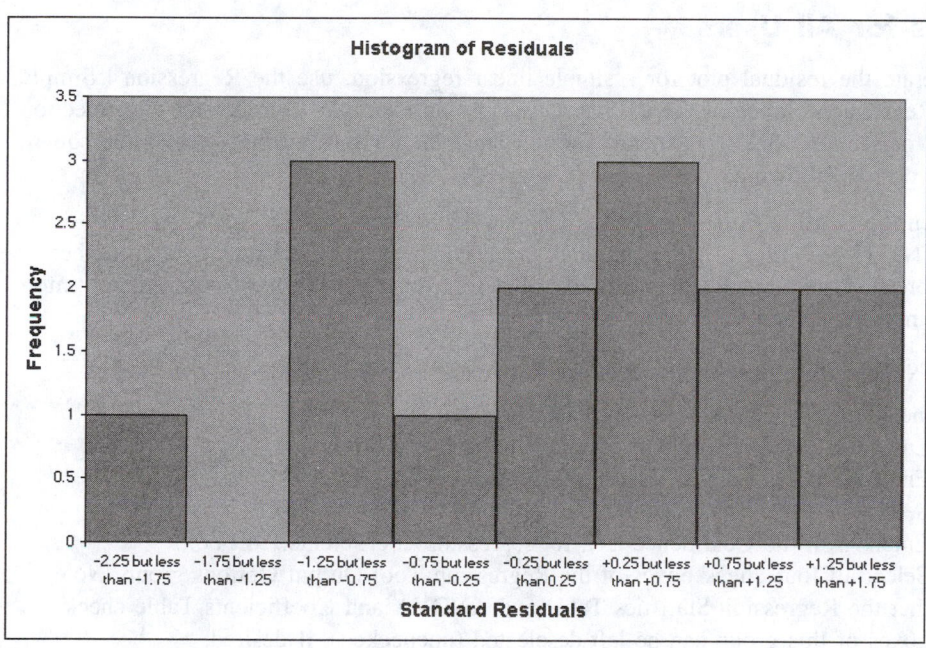

FIGURE 13.12
Histogram of standardized residuals for site selection data obtained from Microsoft Excel

It is difficult to evaluate the normality assumption for a sample of only 14 observations regardless of whether a histogram, stem-and-leaf display, box-and-whisker plot, or normal probability plot is obtained [formal test procedures are beyond the scope of this text (see reference 7)]. We can see from Figure 13.12 that although the data do not appear to be normally distributed, they are also not extremely skewed. The robustness of regression analysis to modest departures from normality, along with the small sample size, leads us to conclude that we should not be overly concerned about departures from this normality assumption in the site selection data.

◆ *Independence* The assumption of independence of the errors can be evaluated by plotting the residuals in the order or sequence in which the observed data were obtained. Data collected over periods of time sometimes exhibit an *autocorrelation* effect among successive observations. In these instances, there exists a relationship between consecutive residuals. Such a relationship, which violates the assumption of independence, is readily apparent in the plot of the residuals versus the time in which they were collected. This effect is measured by the Durbin-Watson statistic, which is the subject of section 13.6.

 GENERATING RESIDUAL PLOTS USING MICROSOFT EXCEL

Overview

Both quick results users and developers should use the PHStat add-in to generate the residual plot for a simple linear regression.

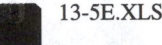
13-5E.XLS

The 13-5E.XLS workbook file includes a summary table and the residual plot for the simple linear regression equation for the Table 13.1 square footage and annual sales data for the site selection problem of section 13.2.

Details for All Users

To generate the residual plot for a simple linear regression, use the Regression | Simple Linear Regression choice of the PHStat add-in. As an example, consider the site selection problem of section 13.2. To generate the residual plot for the residuals versus the square footage, do the following:

❶ Open the Starting Point for Several Chapter 13 Sections workbook (STARTING POINT 13-SEVERAL.XLS) and click the Data sheet tab. Verify that store number, square feet, and annual sales data of Table 13.1 on page 775 have been entered into columns A, B, and C, respectively.

❷ Select PHStat | Regression | Simple Linear Regression.

❸ In the Simple Linear Regression dialog box:
   a. Enter C1:C15 in the Y Variable Cell Range: edit box.
   b. Enter B1:B15 in the X Variable Cell Range: edit box.
   c. Select the First cells in both ranges contain label check box.
   d. Enter 95 in the Confidence Lvl. for regression coefficients: edit box.
   e. Select all four check boxes of the Regression Tool Output Options group. Note that the Regression Statistics Table and ANOVA and Coefficients Table check boxes of this group can be left deselected (unchecked), if desired.
   f. Enter Regression Analysis for Site Selection in the Output Title: edit box.
   g. Click the OK button.

The add-in inserts a worksheet that contains a table of residuals and a residual plot for the simple linear regression for the Table 13.1 data. The table and plot will be similar to the ones shown in Figures 13.9 and 13.10 on pages 795 and 796, respectively. Because the inserted worksheet and chart are *not* dynamically changeable, any changes made to the underlying site selection data would require using the PHStat add-in a second time to produce an updated worksheet and chart.

Problems for Section 13.5

Learning the Basics

• **13.20** The following represents the residuals and X values obtained from a regression analysis along with the accompanying residual plot:

X	STANDARDIZED RESIDUALS	X	STANDARDIZED RESIDUALS
1	0.70	11	0.29
2	−0.78	12	−1.28
3	1.03	13	1.21
4	0.33	14	−0.37
5	2.39	15	1.02
6	−0.67	16	−0.16
7	0.16	17	1.42
8	1.65	18	−0.71
9	−1.19	19	−0.63
10	0.84	20	0.67

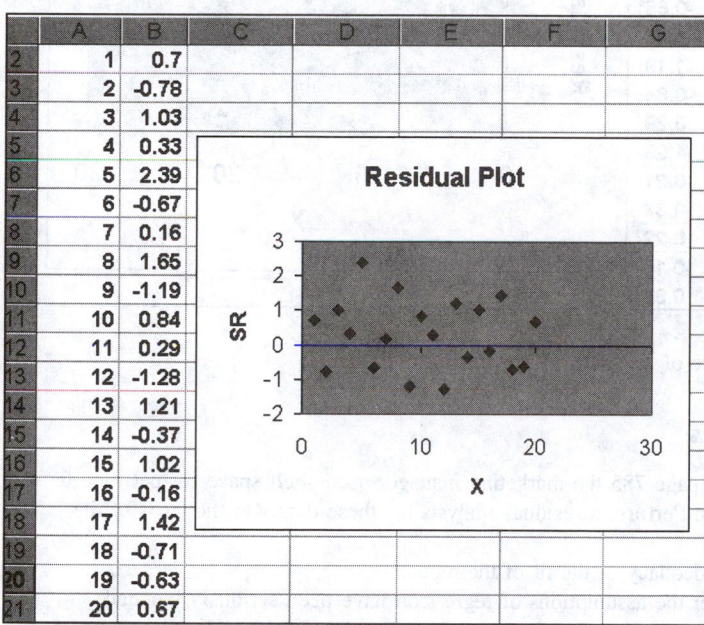

Is there any evidence of a pattern in the residuals? Explain.

13.21 The following represents the standardized residuals and X values obtained from a regression analysis along with the accompanying residual plot:

X	STANDARDIZED RESIDUALS	X	STANDARDIZED RESIDUALS
1	0.70	11	−0.29
2	1.58	12	−1.28
3	1.03	13	−0.21
4	0.33	14	−0.37
5	−0.39	15	0.22
6	−0.67	16	−0.16
7	−0.56	17	0.82
8	−1.65	18	0.41
9	−1.19	19	0.63
10	−0.84	20	0.67

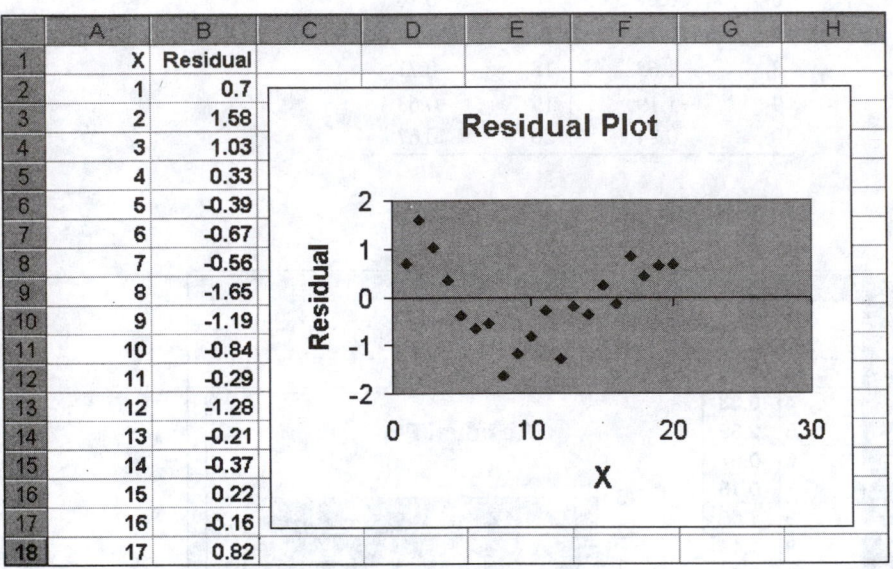

Is there any evidence of a pattern in the residuals? Explain.

Applying the Concepts

• **13.22** In Problem 13.3 on page 785 the marketing manager used shelf space for pet food to predict weekly sales. Perform a residual analysis for these data. On the basis of the results obtained,
(a) determine the adequacy of the fit of the model.
(b) evaluate whether the assumptions of regression have been seriously violated.

13.23 In Problem 13.4 on page 785 a manager wanted to predict weekly sales at a chain of package delivery stores based on the number of customers who made purchases. Perform a residual analysis for these data. On the basis of the results obtained,
(a) determine the adequacy of the fit of the model.
(b) evaluate whether the assumptions of regression have been seriously violated.

- **13.24** In Problem 13.5 on page 786 a company wanted to predict the worker-hours required for production based on the lot size. Perform a residual analysis for these data. On the basis of the results obtained,
  - (a) determine the adequacy of the fit of the model.
  - (b) evaluate whether the assumptions of regression have been seriously violated.

  DATA FILE
  WORKHRS.XLS

- **13.25** In Problem 13.6 on page 786 a company wanted to predict home video sales based on the box office gross. Perform a residual analysis for these data. On the basis of the results obtained,
  - (a) determine the adequacy of the fit of the model.
  - (b) evaluate whether the assumptions of regression have been seriously violated.

  DATA FILE
  MOVIE.XLS

- **13.26** In Problem 13.7 on page 787 an agent for a real estate company wanted to predict the monthly rent for apartments based on the size of the apartment. Perform a residual analysis for these data. On the basis of the results obtained,
  - (a) determine the adequacy of the fit of the model.
  - (b) evaluate whether the assumptions of regression have been seriously violated.

  DATA FILE
  RENT.XLS

- **13.27** In Problem 13.8 on page 787 a limousine service wanted to predict travel time to an airport based on the distance from the pickup location to the airport. Perform a residual analysis for these data. On the basis of the results obtained,
  - (a) determine the adequacy of the fit of the model.
  - (b) evaluate whether the assumptions of regression have been seriously violated.

  DATA FILE
  LIMO.XLS

13.6 MEASURING AUTOCORRELATION: THE DURBIN-WATSON STATISTIC

One of the basic assumptions of the regression model we have been considering is the independence of the errors. This assumption is often violated when data are collected over sequential periods of time because a residual at any one point in time may tend to be similar to residuals at adjacent points in time. Thus positive residuals would be more likely followed by positive residuals and negative residuals would be more likely followed by negative residuals. Such a pattern in the residuals is called **autocorrelation.** When substantial autocorrelation is present in a set of data, the validity of a fitted regression model may be in serious doubt.

Residual Plots to Detect Autocorrelation

As mentioned in section 13.5, the easiest way to detect autocorrelation in a set of data is to plot the residuals in time order. If a positive autocorrelation effect is present, clusters of residuals with the same sign will be present and an apparent pattern will be readily detected. To illustrate the autocorrelation effect, we consider the following example.

Suppose the manager of a package delivery store wants to predict weekly sales based on the number of customers making purchases for a period of 15 weeks. In this situation, because data are collected over a period of 15 consecutive weeks at the same store, we would need to be concerned with the autocorrelation effect of the residuals. The data for this store are summarized in Table 13.3. Figure 13.13 represents partial Excel output.

Table 13.3 *Customers and sales for period of 15 consecutive weeks*

WEEK	CUSTOMERS	SALES ($000)	WEEK	CUSTOMERS	SALES ($000)
1	794	9.33	9	880	12.07
2	799	8.26	10	905	12.55
3	837	7.48	11	886	11.92
4	855	9.08	12	843	10.27
5	845	9.83	13	904	11.80
6	844	10.09	14	950	12.15
7	863	11.01	15	841	9.64
8	875	11.49			

DATA FILE
COSTSALE.XLS

	A	B	C	D	E	F	G
1	Regression Analysis for Package Delivery Store						
2							
3	*Regression Statistics*						
4	Multiple R	0.810829997					
5	R Square	0.657445284					
6	Adjusted R Square	0.631094922					
7	Standard Error	0.936036681					
8	Observations	15					
9							
10	ANOVA						
11		*df*	*SS*	*MS*	*F*	*Significance F*	
12	Regression	1	21.860433	21.86043264	24.95014171	0.000245105	
13	Residual	13	11.390141	0.876164669			
14	Total	14	33.250573				
15							
16		*Coefficients*	*Standard Err*	*t Stat*	*P-value*	*Lower 95%*	*Upper 95%*
17	Intercept	-16.0321936	5.3101671	-3.019150493	0.009868641	-27.50410993	-4.560277262
18	Customers	0.030760228	0.0061582	4.995011683	0.000245105	0.017456271	0.044064185

(annotations: b_0 points to the Intercept row; b_1 points to the Customers row)

FIGURE 13.13 Microsoft Excel output for package delivery store data of Table 13.3

We note from Figure 13.13 that r^2 is .657, indicating that 65.7% of the variation in sales can be explained by variation in the number of customers. In addition, the Y intercept b_0 is -16.032 and the slope b_1 is .03076. However, before we can accept the validity of this model, we must undertake proper analyses of the residuals. Because the data have been collected over a consecutive period of 15 weeks, the residuals should be plotted over time to see whether a pattern exists. Figure 13.14 represents such a plot for the 15-week sales data. From Figure 13.14 we observe that the points tend to fluctuate up and down in a cyclical pattern. This cyclical pattern gives us strong cause for concern about the autocorrelation of the residuals and, hence, a violation of the assumption of independence of the residuals.

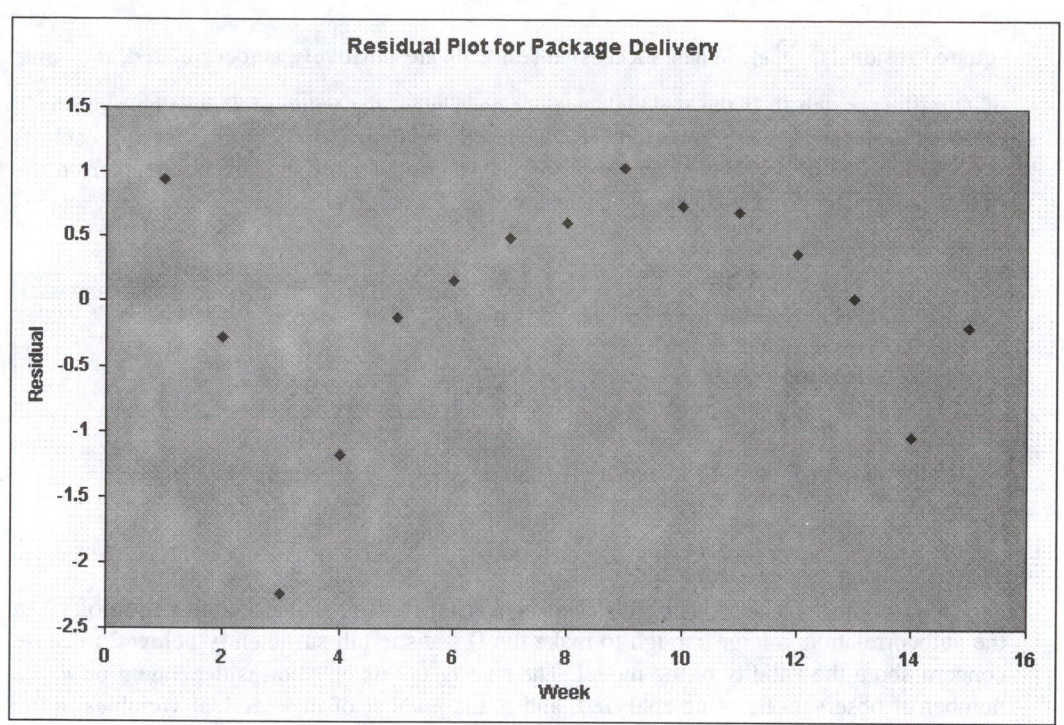

FIGURE 13.14 Microsoft Excel residual plot for package delivery store data of Table 13.3

The Durbin-Watson Statistic

In addition to residual plots, autocorrelation can also be detected and measured by using the **Durbin-Watson statistic.** This statistic measures the correlation between each residual and the residual for the time period immediately preceding the one of interest. The Durbin-Watson statistic D is defined as follows:

Durbin-Watson Statistic

$$D = \frac{\sum_{i=2}^{n}(e_i - e_{i-1})^2}{\sum_{i=1}^{n}e_i^2} \qquad (13.12)$$

where e_i = residual at the time period i.

To better understand what the Durbin-Watson statistic is measuring, we need to examine the composition of the D statistic presented in equation (13.12). The numerator $\sum_{i=2}^{n}(e_i - e_{i-1})^2$ represents the squared difference in two successive residuals, summed from the second observation to the nth observation. The denominator represents the sum of the squared residuals, $\sum_{i=1}^{n} e_i^2$. When successive residuals are positively autocorrelated, the value of D will approach 0. If the residuals are not correlated, the value of D will be close to 2. (If there is negative autocorrelation, D will be greater than 2 and could even approach its maximum value of 4.) The computation of the Durbin-Watson statistic obtained from the PHStat add-in for Microsoft Excel is illustrated in Figure 13.15.

FIGURE 13.15

PHStat add-in for Microsoft Excel output of the Durbin-Watson statistic for sales data

	A	B	C
1	Regression Analysis for Package Delivery Store		
2			
3	Squared Difference of Residuals	10.05752	
4	Squared Residuals	11.39014	
5	**Durbin-Watson Statistic**	**0.883003**	

For the data in Figure 13.15 we use equation (13.12) and obtain

$$D = \frac{10.058}{11.39} = .883$$

The crux of the issue in using the Durbin-Watson statistic is the determination of when the autocorrelation is large enough to make the D statistic fall sufficiently below 2 to cause concern about the validity of the model. The answer to this question is dependent on n, the number of observations being analyzed, and p, the number of independent variables in the model (in simple linear regression, $p = 1$). Table 13.4 has been extracted from appendix E, Table E.8, the table of the Durbin-Watson statistic.

Table 13.4 *Finding critical values of Durbin-Watson statistic*

					$\alpha = .05$					
	$p = 1$		$p = 2$		$p = 3$		$p = 4$		$p = 5$	
n	d_L	d_U	d_L	d_U	d_L	d_U	d_L	d_U	d_L	d_U
15	1.08	1.36	.95	1.54	.82	1.75	.69	1.97	.56	2.21
16	1.10	1.37	.98	1.54	.86	1.73	.74	1.93	.62	2.15
17	1.13	1.38	1.02	1.54	.90	1.71	.78	1.90	.67	2.10
18	1.16	1.39	1.05	1.53	.93	1.69	.82	1.87	.71	2.06

Source: Table E.8.

From Table 13.4 we observe that two values are shown in the table for each combination of α (level of significance), n (sample size), and p (number of independent variables in the model). The first value, d_L, represents the lower critical value when there is no autocorrelation in the data. If D is below d_L, we conclude that there is evidence of positive autocorrelation among the residuals. Under such a circumstance, the least-squares methods we have considered in this chapter are inappropriate and alternative methods need to be used (see reference 7). The second value, d_U, represents the upper critical value of D, above which we would conclude that there is no evidence of autocorrelation among the residuals. If D is between d_L and d_U, we are unable to arrive at a definite conclusion.

Thus, for our data concerning the package delivery stores, with one independent variable ($p = 1$) and 15 observations ($n = 15$), $d_L = 1.08$ and $d_U = 1.36$. Because $D = 0.883 < 1.08$, we conclude that there is autocorrelation among the residuals. Our least-squares regression analysis of the data of Figure 13.13 was inappropriate because of the presence of serious autocorrelation among the residuals. We need to consider the alternative approaches discussed in reference 6.

 ## 13.6E CALCULATING THE DURBIN-WATSON STATISTIC USING MICROSOFT EXCEL

Overview

Because the Durbin-Watson statistic requires the residuals generated by a regression analysis, both quick results users and developers should use the PHStat add-in to calculate this statistic.

The 13-6E.XLS workbook file includes the calculations for the Durbin-Watson statistic for the simple linear regression based on the Table 13.3 (page 802) weekly number of customers and weekly sales for the package delivery store problem of section 13.6.

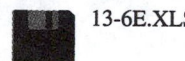 13-6E.XLS

Details for All Users

To calculate the Durbin-Watson statistic for a simple linear regression, use the Regression | Simple Linear Regression choice of the PHStat add-in. As an example, consider the package delivery store problem of section 13.6. To generate the Durbin-Watson statistic for the simple linear regression based on this problem, do the following:

❶ Open the Starting Point for Section 13.6E workbook (STARTING POINT 13-6E.XLS) and click the Data sheet tab. Verify that week number, weekly number of customers, and weekly sales data of Table 13.3 on page 802 have been entered into columns A, B, and C, respectively.

❷ Select PHStat | Regression | Simple Linear Regression.

❸ In the Simple Linear Regression dialog box:
   a. Enter C1:C16 in the Y Variable Cell Range: edit box.
   b. Enter B1:B16 in the X Variable Cell Range: edit box.
   c. Select the First cells in both ranges contain label check box.
   d. Enter 95 in the Confidence Lvl. for regression coefficients: edit box.
   e. Select the check boxes of the Regression Tool Output Options group as desired.
   f. Select Enter Regression Analysis for Package Delivery Store in the Output Title: edit box.
   g. Select the Durbin-Watson Statistic check box. (Selecting this check box will cause the residuals table to be produced, even if the Residuals Table check box was not selected in step 3e.)
   h. Click the OK button.

The add-in inserts two worksheets: one sheet contains a table of residuals (and other regression information, if selected in step 3e) and another sheet, similar to Figure 13.13 on page 802, contains the calculations for the Durbin-Watson statistic for the package delivery store problem of section 13.6. Because the worksheet containing the table of residuals is *not* dynamically changeable, any changes made to the underlying package delivery store data would require using the PHStat add-in a second time to produce updated Durbin-Watson statistics.

Developer Details

As discussed at the beginning of this section, because the Durbin-Watson statistic requires the residuals generated by a regression analysis, developers should use the PHStat add-in as explained in Details for All Users to calculate the Durbin-Watson statistic.

For developer reference, or for cases in which the residuals are already known, Table 13E.1 presents the design of the worksheet inserted by the PHStat add-in to calculate the Durbin-Watson statistic. This design assumes that the residuals are found in the cell range C25:C39 of a worksheet named Regression, as is the case for the workbook for the package delivery problem of this section.

This design uses the SUMXMY2 and SUMSQ worksheet functions to calculate the sum of squared differences between the set of e_i and e_{i-1} values, and the sum of the squared values for all residual values, respectively. The formats of these functions are:

$$\text{SUMXMY2}(X \text{ variable cell range}, Y \text{ variable cell range})$$

$$\text{SUMSQ}(\text{variable cell range})$$

With these values, the simple division formula in cell B5 calculates the statistic.

Table 13E.1 Calculations sheet design for the Durbin-Watson statistic inserted by the PHStat add-in for the package delivery store problem of section 13.6

	A	B
1	Regression Analysis for Package Delivery Store	
2		
3	Sum of Squared Differences of Residuals	=SUMXMY2(Regression!C26:C39, Regression!C25:C38)
4	Sum of Squared Residuals	=SUMSQ(Regression!C25:C39)
5	Durbin-Watson Statistic	=B3/B4

Problems for Section 13.6

Learning the Basics

• **13.28** Suppose the residuals for a set of data collected over 10 consecutive time periods are as follows:

TIME PERIOD	RESIDUAL	TIME PERIOD	RESIDUAL
1	−5	6	+1
2	−4	7	+2
3	−3	8	+3
4	−2	9	+4
5	−1	10	+5

(a) Plot the residuals over time. What conclusions can you reach about the pattern of the residuals over time?
(b) Compute the Durbin-Watson statistic.
(c) On the basis of (a) and (b), what conclusion can you reach about the autocorrelation of the residuals?

13.29 Suppose the residuals for a set of data collected over 15 consecutive time periods are as follows:

TIME PERIOD	RESIDUAL	TIME PERIOD	RESIDUAL
1	+4	9	+6
2	−6	10	−3
3	−1	11	+1
4	−5	12	+3
5	+2	13	0
6	+5	14	−4
7	−2	15	−7
8	+7		

(a) Plot the residuals over time. What conclusions can you reach about the pattern of the residuals over time?

(b) Compute the Durbin-Watson statistic. At the .05 level of significance, is there evidence of positive autocorrelation among the residuals?

(c) On the basis of (a) and (b), what conclusion can you reach about the autocorrelation of the residuals?

Applying the Concepts

• **13.30** In Problem 13.3 (pet food sales) on page 785 the marketing manager used shelf space for pet food to predict weekly sales.

(a) Is it necessary to compute the Durbin-Watson statistic? Explain.

(b) Under what circumstances would it be necessary to compute the Durbin-Watson statistic before proceeding with the least-squares method of regression analysis?

DATA FILE
PETFOOD.XLS

13.31 The owner of a single-family home in a suburban county in the northeastern United States would like to develop a model to predict electricity consumption in his "all electric" house (lights, fans, heat, appliances, and so on) based on outdoor atmospheric temperature (in degrees Fahrenheit). Monthly billing data and temperature information were available for a period of 24 consecutive months.

MONTH	KILOWATT USAGE	AVERAGE ATMOSPHERIC TEMPERATURE (°F)	MONTH	KILOWATT USAGE	AVERAGE ATMOSPHERIC TEMPERATURE (°F)
1	126	30	13	123	27
2	132	25	14	121	33
3	114	29	15	138	28
4	87	42	16	99	39
5	67	48	17	64	47
6	50	61	18	52	63
7	39	69	19	49	69
8	45	78	20	41	73
9	39	72	21	44	70
10	43	62	22	53	64
11	61	45	23	59	53
12	92	36	24	118	27

DATA FILE
ELECUSE.XLS

(a) Set up a scatter diagram.

(b) Assuming a linear relationship, use the least-squares method to find the regression coefficients b_0 and b_1.

(c) Interpret the meaning of the slope b_1 in this problem.

(d) Predict the average kilowatt usage when the average atmospheric temperature is 50 degrees F.

(e) Compute the coefficient of determination r^2 and interpret its meaning.

(f) Compute the standard error of the estimate.

(g) Plot the residuals versus the average atmospheric temperature.

(h) Plot the residuals versus the time period.

(i) Compute the Durbin-Watson statistic. At the .05 level of significance, is there evidence of positive autocorrelation among the residuals?

(j) On the basis of the results of (g)–(i), is there reason to question the validity of the model?

13.32 A mail-order catalog business selling personal computer supplies, software, and hardware maintains a centralized warehouse for the distribution of products ordered. Management is currently examining the process of distribution from the warehouse and is interested in studying the factors that affect warehouse distribution costs. Currently, a small handling fee is added to the order, regardless of the amount of the order. Data have been collected over the past 24 months indicating the warehouse distribution costs and the number of orders received. The results are as follows:

MONTH	DISTRIBUTION COST ($000)	NUMBER OF ORDERS	MONTH	DISTRIBUTION COST ($000)	NUMBER OF ORDERS
1	52.95	4,015	13	62.98	3,977
2	71.66	3,806	14	72.30	4,428
3	85.58	5,309	15	58.99	3,964
4	63.69	4,262	16	79.38	4,582
5	72.81	4,296	17	94.44	5,582
6	68.44	4,097	18	59.74	3,450
7	52.46	3,213	19	90.50	5,079
8	70.77	4,809	20	93.24	5,735
9	82.03	5,237	21	69.33	4,269
10	74.39	4,732	22	53.71	3,708
11	70.84	4,413	23	89.18	5,387
12	54.08	2,921	24	66.80	4,161

DATA FILE
WARECOST.XLS

(a) Set up a scatter diagram.

(b) Assuming a linear relationship, use the least-squares method to find the regression coefficients b_0 and b_1.

(c) Interpret the meaning of the slope b_1 in this problem.

(d) Predict the average monthly warehouse distribution costs when the number of orders is 4,500.

(e) Compute the coefficient of determination r^2 and interpret its meaning.

(f) Compute the standard error of the estimate.

(g) Plot the residuals versus the number of orders.

(h) Plot the residuals versus the time period.

(i) Compute the Durbin-Watson statistic. At the .05 level of significance, is there evidence of positive autocorrelation among the residuals?

(j) On the basis of the results of (g)–(i), is there reason to question the validity of the model?

13.33 The owner of a large chain of ice cream stores would like to study the effect of atmospheric temperature on sales during the summer season. A random sample of 21 days is selected with the results given as follows:

DAY	DAILY HIGH TEMPERATURE (°F)	SALES PER STORE ($000)	DAY	DAILY HIGH TEMPERATURE (°F)	SALES PER STORE ($000)
1	63	1.52	12	75	1.92
2	70	1.68	13	98	3.40
3	73	1.80	14	100	3.28
4	75	2.05	15	92	3.17
5	80	2.36	16	87	2.83
6	82	2.25	17	84	2.58
7	85	2.68	18	88	2.86
8	88	2.90	19	80	2.26
9	90	3.14	20	82	2.14
10	91	3.06	21	76	1.98
11	92	3.24			

DATA FILE
ICECREAM.XLS

Hint: Determine which are the independent and dependent variables.

(a) Set up a scatter diagram.

(b) Assuming a linear relationship, use the least-squares method to find the regression coefficients b_0 and b_1.

(c) Interpret the meaning of the slope b_1 in this problem.

(d) Predict the average sales per store for a day in which the temperature is 83°F.

(e) Compute the standard error of the estimate.

(f) Compute the coefficient of determination r^2 and interpret its meaning in this problem.

(g) Plot the residuals versus the temperature.

(h) Plot the residuals versus the time period.

(i) Compute the Durbin-Watson statistic. At the .05 level of significance, is there evidence of positive autocorrelation among the residuals?

(j) On the basis of the results of (g)–(i), is there reason to question the validity of the model?

(k) Suppose that the amount of sales on day 21 was 1.75. Do (a)–(j) and compare the differences in the results.

WHAT IF?

 ## 13.7 INFERENCES ABOUT THE SLOPE

In sections 13.1–13.4 we were concerned with the use of regression solely for the purpose of description. We used the least-squares method to determine the regression coefficients and to predict the value of Y from a given value of X. In addition, we discussed the standard error of the estimate along with the coefficient of determination.

Now that we have used residual analysis in section 13.5 to assure ourselves that the assumptions of the least-squares regression model have not been seriously violated and that the straight-line model is appropriate, we may concentrate on making inferences about the linear relationship between the variables in a population based on our sample results.

t Test for the Slope

We can determine the existence of a significant relationship between the X and Y variables by testing whether β_1 (the true slope) is equal to 0. If this hypothesis is rejected, we would conclude that there is evidence of a linear relationship. The null and alternative hypotheses are stated as follows:

$$H_0: \beta_1 = 0 \text{ (There is no linear relationship.)}$$
$$H_1: \beta_1 \neq 0 \text{ (There is a linear relationship.)}$$

and the test statistic is given by

Testing a Hypothesis for a Population Slope β_1 Using the t Test

The t statistic equals the difference between the sample slope and the hypothesized population slope divided by the standard error of the slope.

$$t = \frac{b_1 - \beta_1}{S_{b_1}} \qquad (13.13)$$

where

$$S_{b_1} = \frac{S_{YX}}{\sqrt{SSX}}$$

$$SSX = \sum_{i=1}^{n}(X_i - \bar{X})^2$$

and the test statistic t follows a t distribution with $n - 2$ degrees of freedom.

Returning to our site selection example, now we will test whether there is a significant relationship between the size of the store and the annual sales at the .05 level of significance. From the Microsoft Excel output of Figure 13.4 on page 778 we have[2]

$$b_1 = +1.686 \qquad n = 14 \qquad S_{b_1} = .1533$$

[2]More detailed computations of the t-test statistic are contained in section 13.10.

Therefore, to test the existence of a relationship at the .05 level of significance, we have

$$t = \frac{b_1}{S_{b_1}}$$

$$= \frac{1.686}{.1533} = 11.00$$

This t statistic is provided in the column titled t Stat by Microsoft Excel. Because $t = 11.00 > t_{12} = 2.1788$, we reject H_0. Using the p-value, we reject H_0 because the p-value is approximately 0. (It is labeled in scientific notation by Excel as approximately 1.27E-07, which is equal to .000000127 and is less than $\alpha = .05$.) Hence, we can conclude that there is a significant linear relationship between average annual sales and the size of the store (see Figure 13.16).

FIGURE 13.16
Testing hypothesis about population slope at .05 level of significance with 12 degrees of freedom

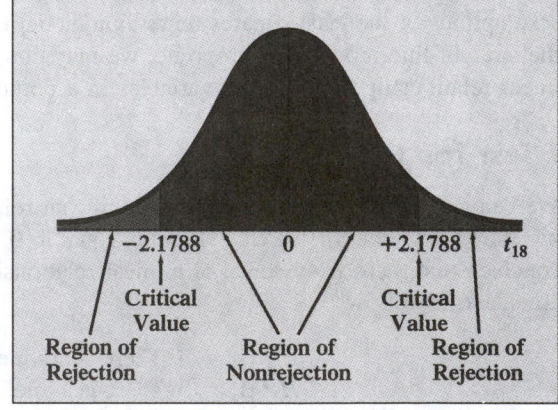

F Test for the Slope

An alternative approach for testing whether the slope in simple linear regression is statistically significant is to use an F test, as presented in Table 13.5. You may recall from section 9.2 that the F test is used to test the ratio of two variances. In testing for the significance of the slope, the measure of random error is the error variance (the error sum of squares divided by its degrees of freedom), so the F test is the ratio of the variance that is due to the regression (the regression sum of squares divided by the number of independent variables p) divided by the error variance as shown in equation (13.14).

Testing a Hypothesis for a Population Slope β_1 Using the F Test

The F statistic is equal to the regression mean square (MSR) divided by the error mean square (MSE).

$$F = \frac{MSR}{MSE} \qquad (13.14)$$

where

$$MSR = \frac{SSR}{p}$$

$$MSE = \frac{SSE}{n - p - 1}$$

p = number of explanatory variables in the regression model

F = test statistic from an F distribution with p and $n - p - 1$ degrees of freedom

Using a level of significance α, the decision rule is

Reject H_0 if $F > F_U$, the upper-tailed critical value from the F distribution with p and $n - p - 1$ degrees of freedom; otherwise do not reject H_0.

The complete set of results is organized into an analysis of variance (ANOVA) table as illustrated in Table 13.5.

Table 13.5 *ANOVA table for testing significance of regression coefficient*

SOURCE	d.f.	SUMS OF SQUARES	MEAN SQUARE (VARIANCE)	F
Regression	p	SSR	$MSR = \dfrac{SSR}{p}$	$F = \dfrac{MSR}{MSE}$
Error	$n - p - 1$	SSE	$MSE = \dfrac{SSE}{n - p - 1}$	
Total	$n - 1$	SST		

The completed ANOVA table is also available as part of the output from Excel (see Figure 13.4 on page 778). We observe from Figure 13.4 that the computed F statistic is 121.01 and the p-value is less than .001 (Excel computes the p-value as .000000127).

If a level of significance of .05 is chosen, we can determine from Table E.5 that the critical value on the F distribution (with 1 and 12 degrees of freedom) is 4.75, as depicted in Figure 13.17. From equation (13.14), because $F = 121.01 > 4.75$ or because the p-value $= .000000127 < .05$, we reject H_0 and conclude that the size of the store is significantly related to annual sales.

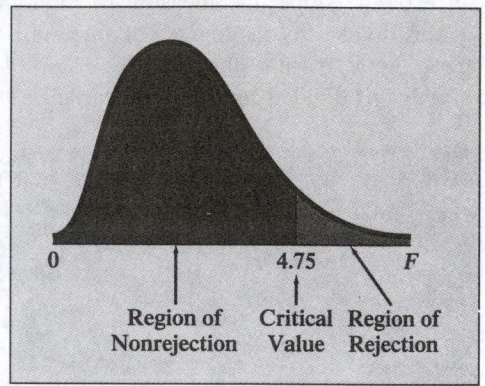

FIGURE 13.17
Testing for significance of slope at .05 level of significance with 1 and 12 degrees of freedom

Confidence Interval Estimate of the Slope (β_1)

An alternative to testing the existence of a linear relationship between the variables is to set up a confidence interval estimate of β_1 and to determine whether the hypothesized value ($\beta_1 = 0$) is included in the interval. The confidence interval estimate of β_1 is obtained as shown in equation (13.15).

Confidence Interval Estimate of the Slope

The confidence interval estimate of the slope is obtained by taking the sample slope b_1 and adding and subtracting the critical value of the t statistic multiplied by the standard error of the slope.

$$b_1 \pm t_{n-2}S_{b_1} \qquad (13.15)$$

From the Microsoft Excel output of Figure 13.4 on page 778 we have

$$b_1 = +1.686 \qquad n = 14 \qquad S_{b_1} = .1533$$

Thus,

$$b_1 \pm t_{n-2}S_{b_1} = +1.686 \pm (2.1788)(.1533)$$
$$= +1.686 \pm .334$$
$$+1.352 \le \beta_1 \le +2.02$$

From equation (13.15) the true slope is estimated with 95% confidence to be between $+1.352$ and $+2.02$ (i.e., \$1,352 to \$2,020). Because these values are above 0, we conclude that there is a significant linear relationship between annual sales and square footage size of the store. Had the interval included 0, the conclusion would have been that no relationship exists between the variables.

Problems for Section 13.7

Learning the Basics

- **13.34** Suppose you are testing the null hypothesis that the slope is not significant. From your sample of $n = 18$ you determine that

$$b_1 = +4.5 \qquad S_{b_1} = 1.5$$

  (a) What is the value of the t-test statistic?
  (b) At the $\alpha = .05$ level of significance, what are the critical values?
  (c) On the basis of your answers to (a) and (b), what statistical decision should be made?
  (d) Set up a 95% confidence interval estimate of the population slope β_1.

13.35 Suppose you are testing the null hypothesis that the slope is not significant. From your sample of $n = 20$, you determine that $SSR = 60$ and $SSE = 40$.
  (a) What is the value of the F-test statistic?
  (b) At the $\alpha = .05$ level of significance, what is the critical value?
  (c) On the basis of your answers to (a) and (b), what statistical decision should be made?

Applying the Concepts

- **13.36** In Problem 13.3 on page 785 the marketing manager used shelf space for pet food to predict weekly sales. Use the computer output you obtained to solve that problem.
  (a) At the .05 level of significance, is there evidence of a linear relationship between shelf space and sales?
  (b) Set up a 95% confidence interval estimate of the population slope β_1.

DATA FILE
PETFOOD.XLS

13.37 In Problem 13.4 on page 785 a manager wanted to predict weekly sales at a chain of package delivery stores based on the number of customers who made purchases. Use the computer output you obtained to solve that problem.
  (a) At the .05 level of significance, is there evidence of a linear relationship between the number of customers and sales?
  (b) Set up a 95% confidence interval estimate of the population slope β_1.

DATA FILE
PACKAGE.XLS

- **13.38** In Problem 13.5 on page 786 a company wanted to predict the worker-hours required for production based on the lot size. Use the computer output you obtained to solve that problem.
  (a) At the .05 level of significance, is there evidence of a linear relationship between lot size and worker-hours?
  (b) Set up a 95% confidence interval estimate of the population slope β_1.

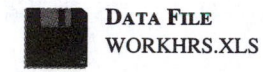

DATA FILE
WORKHRS.XLS

13.39 In Problem 13.6 on page 786 a company wanted to predict home video sales based on the box office gross of movies. Use the computer output you obtained to solve that problem.
  (a) At the .05 level of significance, is there evidence of a linear relationship between box office gross and home video sales?
  (b) Set up a 95% confidence interval estimate of the population slope β_1.

DATA FILE
MOVIE.XLS

13.40 In Problem 13.7 on page 787 an agent for a real estate company wanted to predict the monthly rent for apartments based on the size of the apartment. Use the computer output you obtained to solve that problem.
  (a) At the .05 level of significance, is there evidence of a linear relationship between the size of the apartment and the monthly rent?
  (b) Set up a 95% confidence interval estimate of the population slope β_1.

DATA FILE
RENT.XLS

13.41 In Problem 13.8 on page 787 a limousine service wanted to predict travel time to an airport based on the distance from the pickup location to the airport. Use the computer output you obtained to solve that problem.

DATA FILE
LIMO.XLS

(a) At the .05 level of significance, is there evidence of a relationship between distance and travel time?

(b) Set up a 95% confidence interval estimate of the population slope β_1.

 **13.8** **ESTIMATION OF PREDICTED VALUES**

In this section we will discuss methods of making predictive inferences about the mean of Y and an individual response value Y_I.

Obtaining the Confidence Interval Estimate

In the site selection example we predicted that the average yearly sales for stores with 4,000 square feet would be 7,645.786 (thousands of dollars). This estimate, however, is just a *point estimate* of the population average value. In chapter 7 we developed the concept of the confidence interval as an estimate of the population average. In a similar fashion a **confidence interval estimate for the mean response** can now be developed to make inferences about the predicted average value of Y.

Confidence Interval Estimate for μ_{YX}, the Mean of Y

$$\hat{Y}_i \pm t_{n-2} S_{YX} \sqrt{h_i} \qquad (13.16)$$

where

$$h_i = \frac{1}{n} + \frac{(X_i - \overline{X})^2}{\sum\limits_{i=1}^{n}(X_i - \overline{X})^2}$$

$\hat{Y}_i$ = predicted average value of Y: $\hat{Y}_i = b_0 + b_1 X_i$

S_{YX} = standard error of the estimate

n = sample size

X_i = given value of X

An examination of equation (13.16) indicates that the width of the confidence interval is dependent on several factors. For a given level of confidence, increased variation around the line of regression, as measured by the standard error of the estimate, results in a wider interval. However, as would be expected, increased sample size reduces the width of the interval. In addition, the width of the interval also varies at different values of X. When predicting Y for values of X close to $\overline{X}$, the interval is much narrower than for predictions for X values more distant from the mean. This effect can be seen from the square root portion of equation (13.16) and from Figure 13.18. As displayed in Figure 13.18, the interval estimate of the true mean of Y varies as a function of the closeness of the given X to $\overline{X}$. When predictions are to be made for X values that are distant from $\overline{X}$, a much wider interval will occur.

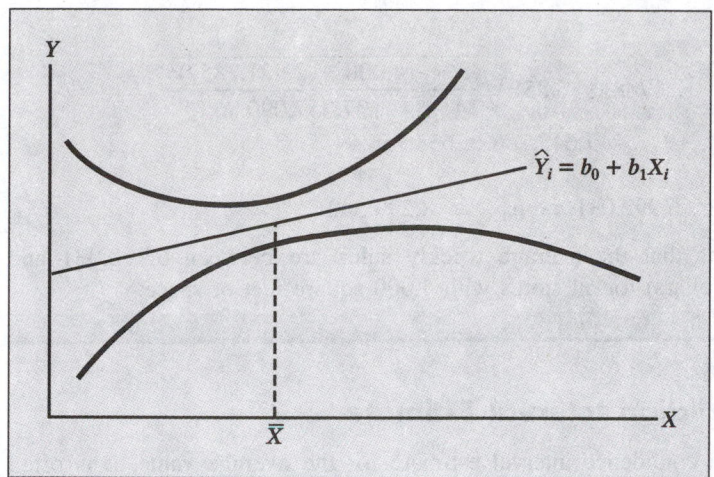

FIGURE 13.18
Interval estimates of μ_{YX}
for different values of X

$\hat{Y}_i = b_0 + b_1 X_i$

$\overline{X}$

Although the PHStat add-in for Microsoft Excel can be used to obtain a confidence interval estimate for the mean response, Example 13.3 illustrates equation (13.16) in the site selection example.

Example 13.3 *Setting Up a 95% Confidence Interval Estimate of the Mean Response* μ_{YX}

In the site selection example we obtained the simple linear regression model $\hat{Y}_i = 901.247 + 1.686\, X_i$. Set up a 95% confidence interval estimate of the true average annual sales for all stores that contain 4,000 square feet.

SOLUTION

Using our simple linear regression equation:

$$\hat{Y}_i = 901.247 + 1.686\, X_i$$

and for $X_i = 4,000$, we obtain

$$\hat{Y}_i = 901.247 + 1.686(4,000) = 7,645.786 \text{ (thousands of dollars)}$$

Also, given the following:

$$\overline{X} = 2{,}921.2857; \qquad S_{YX} = 936.85; \qquad \sum_{i=1}^{n}(X_i - \overline{X})^2 = 37{,}357{,}090.86$$

and from Table E.3, $t_{12} = 2.1788$. Thus,

$$\hat{Y}_i \pm t_{n-2} S_{YX}\sqrt{h_i}$$

where

$$h_i = \frac{1}{n} + \frac{(X_i - \overline{X})^2}{\displaystyle\sum_{i=1}^{n}(X_i - \overline{X})^2}$$

so that we have

$$\hat{Y}_i \pm t_{n-2} S_{YX}\sqrt{\frac{1}{n} + \frac{(X_i - \overline{X})^2}{\displaystyle\sum_{i=1}^{n}(X_i - \overline{X})^2}}$$

591

and

$$7{,}645.786 \pm (2.1788)(936.85) \sqrt{\frac{1}{14} + \frac{(4{,}000 - 2{,}921.2857)^2}{37{,}357{,}090.86}}$$
$$= 7{,}645{,}786 \pm 653.756$$

so

$$6{,}992.031 \leq \mu_{YX} \leq 8{,}299.542$$

Therefore, our estimate is that the average weekly sales are between 6,992.031 and 8,299.542 (thousands of dollars) for all stores with 4,000 square feet of space.

Obtaining the Prediction Interval Estimate

In addition to obtaining a confidence interval estimate for the average value, it is often important to be able to predict the response that would be obtained for an individual value. Although the form of the prediction interval estimate is similar to the confidence interval estimate of equation (13.16), the prediction interval is estimating an individual value, not a parameter. Thus, the **prediction interval for an individual response** Y_I at a particular value X_i is shown in equation (13.17).

Prediction Interval Estimate for an Individual Response Y_I

$$\hat{Y}_i \pm t_{n-2} S_{YX} \sqrt{1 + h_i} \tag{13.17}$$

where

h_i, $\hat{Y}_i$, S_{YX}, n, and X_i are defined as in equation (13.16) on page 814.

Although the PHStat add-in for Microsoft Excel can also be used to obtain a prediction interval estimate for the individual response, Example 13.4 illustrates equation (13.17) for the site selection data.

Example 13.4   *Setting Up a 95% Prediction Interval Estimate of the Individual Response Y_I*

In the site selection example we obtained the simple linear regression model $\hat{Y}_i = 901.247 + 1.686\,X_i$. Set up a 95% prediction interval estimate of the annual sales for an individual store that contains 4,000 square feet.

SOLUTION

Using our simple linear regression equation:

$$\hat{Y}_i = 901.247 + 1.686\,X_i$$

and for $X_i = 4{,}000$, we obtain

$$\hat{Y}_i = 901.247 + 1.686(4{,}000) = 7{,}645.786 \text{ (thousands of dollars)}$$

Also, given the following:

$$\overline{X} = 2{,}921.2857; \qquad S_{YX} = 936.85; \qquad \sum_{i=1}^{n}(X_i - \overline{X})^2 = 37{,}357{,}090.86$$

and from Table E.3, $t_{12} = 2.1788$. Thus,

$$\hat{Y}_i \pm t_{n-2} S_{YX} \sqrt{1 + h_i}$$

where

$$h_i = \frac{1}{n} + \frac{(X_i - \overline{X})^2}{\displaystyle\sum_{i=1}^{n}(X_i - \overline{X})^2}$$

so that we have

$$\hat{Y}_i \pm t_{n-2} S_{YX} \sqrt{1 + \frac{1}{n} + \frac{(X_i - \overline{X})^2}{\displaystyle\sum_{i=1}^{n}(X_i - \overline{X})^2}}$$

and

$$7{,}645.786 \pm (2.1788)(936.85) \sqrt{1 + \frac{1}{14} + \frac{(4{,}000 - 2{,}921.2857)^2}{37{,}357{,}090.86}}$$

$$= 7{,}645.786 \pm 2{,}143.357$$

so

$$5{,}502.43 \le Y_I \le 9{,}789.143$$

Therefore, our estimate is that the annual sales for an individual store with 4,000 square feet of space is between 5,502.43 and 9,789.143 (thousands of dollars).

If we compare the results of Examples 13.3 and 13.4, we observe that the width of the prediction interval for an individual store is much greater than the width of the confidence interval estimate for the average store. This is because there is much more variation in predicting an individual value than in predicting an average value.

 **ESTIMATING PREDICTED VALUES USING MICROSOFT EXCEL**

Overview

◆ *For Quick Results Users* Use the PHStat add-in to calculate the confidence interval estimate for the mean response and the prediction interval estimate for an individual response Y_I.

◆ *For Developers* Use either the Quick Results procedure or, if the standard error of the estimate S_{YX} is known, implement a worksheet that uses the TINV and TREND worksheet functions and arithmetic formulas to calculate the confidence interval and prediction interval estimates.

The 13-8E.XLS workbook file includes the calculations for the confidence interval estimate and the prediction interval estimate for the Table 13.1 square footage and annual sales data for the site selection problem of section 13.2.

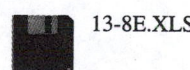 13-8E.XLS

Quick Results Details

To calculate the confidence interval estimate for the mean response and the prediction interval estimate for an individual response Y_I, use the Regression | Simple Linear Regression choice of the PHStat add-in. As an example, consider the site selection problem of section 13.8. To generate the confidence interval estimate of the average yearly sales for stores with 4,000 square feet and the prediction interval estimate of annual sales for a store with 4,000 square feet, do the following:

❶ Open the Starting Point for Several Chapter 13 Sections workbook (STARTING POINT 13-SEVERAL.XLS) and click the Data sheet tab. Verify that store number, square feet, and annual sales data of Table 13.1 on page 775 have been entered into columns A, B, and C, respectively.

❷ Select PHStat | Regression | Simple Linear Regression.

❸ In the Simple Linear Regression dialog box (see Figure 13E.4):
  a. Enter C1:C15 in the Y Variable Cell Range: edit box.
  b. Enter B1:B15 in the X Variable Cell Range: edit box.
  c. Select the First cells in both ranges contain label check box.
  d. Enter 95 in the Confidence Lvl. for regression coefficients: edit box.
  e. Select the Regression Statistics Table check box. (Other Regression Tool Output Options can be selected, if desired.)
  f. Enter Regression Analysis for Site Selection in the Output Title: edit box.
  g. Select the Confidence & Prediction Interval for X = check box and enter 4000 in its edit box. Enter 95 in the Confidence level for int. estimates: edit box.
  h. Click the OK button.

FIGURE 13E.4
PHStat Simple Linear Regression dialog box with confidence and prediction interval option selected

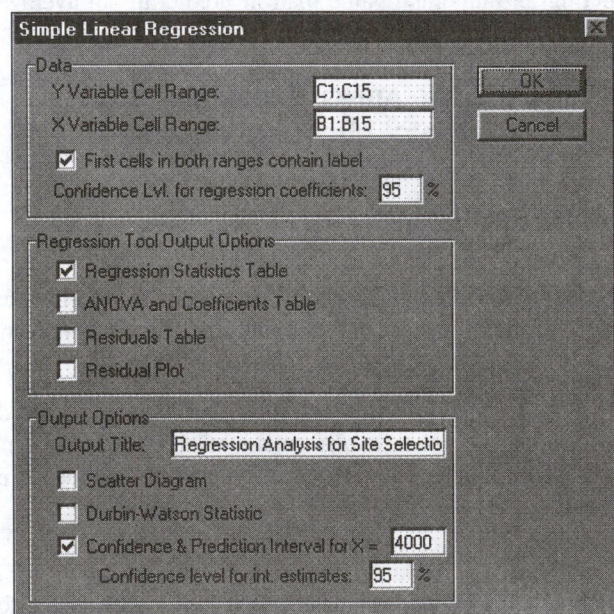

The add-in inserts two worksheets: one sheet contains a table of regression statistics; another sheet, similar to the one shown in Figure 13E.5, contains the calculations for the interval estimates for the site selection problem of section 13.8.

	A	B
1	Confidence Interval Estimate	
2		
3	X Value	4000
4	Confidence Level	95%
5	Sample Size	14
6	Degrees of Freedom	12
7	t Value	2.178812792
8	Sample Mean	2921.285714
9	Sum of Squared Difference	37357090.86
10	Standard Error of the Estimate	936.8500077
11	h Statistic	0.102577263
12	Average Predicted Y (YHat)	7645.786452
13		
14		
15	For Average Predicted Y (YHat)	
16	Interval Half Width	653.7557703
17	Confidence Interval Lower Limit	6992.030681
18	Confidence Interval Upper Limit	8299.542222
19		
20	For Individual Response Y	
21	Interval Half Width	2143.35692
22	Prediction Interval Lower Limit	5502.429532
23	Prediction Interval Upper Limit	9789.143371

FIGURE 13E.5
Confidence and prediction interval estimates for the site selection problem of section 13.8

▲ WHAT IF EXAMPLE

We can change the values for the X variable or the confidence level in the interval estimate worksheet inserted by the add-in to see their effects on the confidence interval estimate for the mean response and the prediction interval estimate for an individual response Y_I. For example, if we change the confidence level from 95% to 99%, we will observe that the confidence interval changes to between 6,729.27 and 8,562.30 and that the prediction interval changes to between 4,630.96 and 10,650.62. To facilitate comparisons among many different alternative values, we can create scenarios using the Scenario Manager as discussed at the end of section 2.2E.

Developer Details

If we know the standard error of the estimate S_{YX} for the regression, we can use the TINV (see section 7.2E) and TREND worksheet functions and arithmetic formulas as the basis for calculating confidence interval and prediction interval estimates. Table 13E.2 presents an estimate sheet design that calculates the interval estimates for the site selection problem of Section 13.8. This design uses the TINV function to return the t statistic value and the TREND function, the format of which is

TREND(Y *variable cell range*, X *variable cell range*, X-*value*)

to calculate the average predicted Y for a given X-value. The design assumes that the data of Table 13.1 have been placed in columns A through C of a worksheet named Data and that column D of that worksheet contains the squares of the differences for each X-value from its average value.

Table 13E.2 Estimate sheet design for confidence and prediction interval estimates for the site selection problem of section 13.8

	A	B
1	Regression Analysis for Site Selection	
2		
3	X Value	xxx
4	Confidence Level	.xx
5	Sample Size	=COUNT(Data!B:B)
6	Degrees of Freedom	=B5-2
7	t Value	=TINV(1-B4,B6)
8	Sample Mean	=AVERAGE(Data!B:B)
9	Sum of Squared Differences	=SUM(Data!D:D)
10	Standard Error of the Estimate	xxx
11	h Statistic	=1/B5+(B3-B8)^2/B9
12	Average Predicted Y (YHat)	=TREND(Data!C2:C15,Data!B2:B15,B3)
13		
14		
15	For Average Predicted Y (YHat)	
16	Interval Half Width	=B7*B10*SQRT(B11)
17	Confidence Interval Lower Limit	=B12-B16
18	Confidence Interval Upper Limit	=B12+B16
19		
20	For Individual Response Y	
21	Interval Half Width	=B7*B10*SQRT(1+B11)
22	Prediction Interval Lower Limit	=B12-B21
23	Prediction Interval Upper Limit	=B12+B21

To implement the Table 13E.2 design, do the following:

❶ Open the Starting Point for Several Chapter 13 Sections workbook (STARTING POINT 13-SEVERAL.XLS) and click the Data sheet tab. Verify that store number, square feet, and annual sales data of Table 13.1 on page 775 have been entered into columns A, B, and C, respectively.

❷ Select an unused worksheet (or select Insert | Worksheet if there are none) and rename the sheet Estimate.

❸ Enter the title, heading, and labels for column A as shown in Table 13E.2.

4 Enter the X Value, 4000, in cell B3, the confidence level, 95% (.95), in cell B4, and the standard error of the estimate, 936.8500077, in cell B10.

5 Enter the formulas for column B as shown in Table 13E.2. Do not worry about the #DIV/0! error messages that appear in many cells.

6 Click the Data sheet tab. Enter the heading (X-XBar)^2 in cell D1. Enter the formula =(B2-Estimate!$B$8)^2 in cell D2 and copy this formula down through row 15.

7 Click the Estimate sheet tab. Note that the #DIV/0! messages have been replaced by values.

8 Select the cell range A15:B15 and click the Merge and Center button on the formatting tool bar (see section S3.11). Repeat this step for the cell range A20:B20.

The completed worksheet will be similar to the one shown in Figure 13E.5 on page 819.

Problems for Section 13.8

Learning the Basics

13.42 Based on a sample of 20 observations, the least-squares method was used to obtain the following linear regression equation: $\hat{Y}_i = 5 + 3X_i$. In addition, $S_{YX} = 1.0$, $\overline{X} = 2$, and

$$\sum_{i=1}^{n}(X_i - \overline{X})^2 = 20.$$

(a) Set up a 95% confidence interval estimate of the true population average response for $X = 2$.
(b) Set up a 95% prediction interval estimate of the individual response for $X = 2$.

13.43 On the basis of a sample of 20 observations, the least-squares method was used to obtain the following linear regression equation: $\hat{Y}_i = 5 + 3X_i$. In addition, $S_{YX} = 1.0$, $\overline{X} = 2$, and

$$\sum_{i=1}^{n}(X_i - \overline{X})^2 = 20.$$

(a) Set up a 95% confidence interval estimate of the true population average response for $X = 4$.
(b) Set up a 95% prediction interval estimate of the individual response for $X = 4$.
(c) Compare the results of (a) and (b) with those of Problem 13.42 (a) and (b). Which interval is wider? Why?

Applying the Concepts

• **13.44** In Problem 13.3 on page 785 the marketing manager used shelf space for pet food to predict weekly sales. Use the computer output you obtained to solve that problem.
(a) Set up a 95% confidence interval estimate of the average weekly sales for all stores that have 8 feet of shelf space for pet food.

DATA FILE
PETFOOD.XLS

(b) Set up a 95% prediction interval of the weekly sales of an individual store that has 8 feet of shelf space for pet food.

(c) Explain the difference in the results obtained in (a) and (b).

DATA FILE
PACKAGE.XLS

13.45 In Problem 13.4 on page 785 a manager wanted to predict weekly sales at a chain of package delivery stores based on the number of customers who made purchases. Use the computer output you obtained to solve that problem.

(a) Set up a 95% confidence interval estimate of the average weekly sales for all stores that have 600 customers.

(b) Set up a 95% prediction interval of the weekly sales of an individual store that has 600 customers.

(c) Explain the difference in the results obtained in (a) and (b).

DATA FILE
WORKHRS.XLS

• **13.46** In Problem 13.5 on page 786 a company wanted to predict the worker-hours required for production based on the lot size. Use the computer output you obtained to solve that problem.

(a) Set up a 95% confidence interval estimate of the average worker-hours for a lot size of 45.

(b) Set up a 95% prediction interval of the worker-hours of an individual lot size of 45.

(c) Explain the difference in the results obtained in (a) and (b).

DATA FILE
MOVIE.XLS

13.47 In Problem 13.6 on page 786 a company wanted to predict home video sales based on the box office gross of movies. Use the computer output you obtained to solve that problem.

(a) Set up a 95% confidence interval estimate of the average video sales for all movies that gross $10 million at the box office.

(b) Set up a 95% prediction interval of the video sales of an individual movie that grosses $10 million at the box office.

(c) Explain the difference in the results obtained in (a) and (b).

DATA FILE
RENT.XLS

13.48 In Problem 13.7 on page 787 an agent for a real estate company wanted to predict the monthly rent for apartments based on the size of the apartment. Use the computer output you obtained to solve that problem.

(a) Set up a 95% confidence interval estimate of the average monthly rent for all apartments that are 1,000 square feet in size.

(b) Set up a 95% prediction interval of the monthly rent for an individual apartment that is 1,000 square feet in size.

(c) Explain the difference in the results obtained in (a) and (b).

DATA FILE
LIMO.XLS

13.49 In Problem 13.8 on page 787 a limousine service wanted to predict travel time to an airport based on the distance from the pickup location to the airport. Use the computer output you obtained to solve that problem.

(a) Set up a 95% confidence interval estimate of the average travel time for all trips with distances of 21 miles.

(b) Set up a 95% prediction interval of the travel time of an individual trip that has a distance of 21 miles.

(c) Explain the difference in the results obtained in (a) and (b).

13.9 PITFALLS IN REGRESSION AND ETHICAL ISSUES

Regression analysis is perhaps the most widely used and, unfortunately, the most widely misused statistical technique applied to business and economics. Some of the difficulties involved in using regression analysis are summarized in Exhibit 13.2.

Exhibit 13.2 Difficulties in Using Regression

✓ **1.** Lacking an awareness of the assumptions of least-squares regression

✓ **2.** Not knowing how to evaluate the assumptions of least-squares regression

✓ **3.** Not knowing what the alternatives to least-squares regression are if a particular assumption is violated

✓ **4.** Using a regression model without knowledge of the subject matter

The Pitfalls of Regression

The widespread availability of spreadsheet and statistical software has removed the computational block that prevented many users from applying regression analysis to situations that required forecasting. With this positive development of enhanced technology comes the realization that for many users, the access to powerful techniques has not been accompanied by an understanding of how to apply regression analysis properly. How can a user be expected to know what the alternatives to least-squares regression are if a particular assumption is violated, when he or she in many instances is not even aware of the assumptions of regression, let alone how the assumptions can be evaluated?

The necessity of going beyond the basic number crunching—of computing the Y intercept, the slope, and r^2—can be illustrated by referring to Table 13.6, a classical pedagogical piece of statistical literature that deals with the importance of observation through scatter plots and residual analysis.

Table 13.6 *Four sets of artificial data*

DATA SET A		DATA SET B		DATA SET C		DATA SET D	
X_i	Y_i	X_i	Y_i	X_i	Y_i	X_i	Y_i
10	8.04	10	9.14	10	7.46	8	6.58
14	9.96	14	8.10	14	8.84	8	5.76
5	5.68	5	4.74	5	5.73	8	7.71
8	6.95	8	8.14	8	6.77	8	8.84
9	8.81	9	8.77	9	7.11	8	8.47
12	10.84	12	9.13	12	8.15	8	7.04
4	4.26	4	3.10	4	5.39	8	5.25
7	4.82	7	7.26	7	6.42	19	12.50
11	8.33	11	9.26	11	7.81	8	5.56
13	7.58	13	8.74	13	12.74	8	7.91
6	7.24	6	6.13	6	6.08	8	6.89

Source: F. J. Anscombe, "Graphs in Statistical Analysis," by F.J. Ansombe in
AMERICAN STATISTICIAN *27 (1973); 17–21. Copyright © 1973 by American*
Statistical Assn. Reprinted by permission.

DATA FILE
ANSCOMBE.XLS

Anscombe (reference 1) showed that for the four data sets given in Table 13.6, the following results are obtained:

$$\hat{Y}_i = 3.0 + .5\,X_i$$
$$S_{YX} = 1.237$$
$$S_{b_1} = .118$$
$$r^2 = .667$$

$$SSR = \text{explained variation} = \sum_{i=1}^{n}(\hat{Y}_i - \overline{Y})^2 = 27.51$$

$$SSE = \text{unexplained variation} = \sum_{i=1}^{n}(Y_i - \hat{Y}_i)^2 = 13.763$$

$$SST = \text{total variation} = \sum_{i=1}^{n}(Y_i - \overline{Y})^2 = 41.273$$

Thus, with respect to these statistics associated with a simple linear regression, the four data sets are identical. Had we stopped our analysis at this point, valuable information in the data would be lost. This may be observed by examining Figure 13.19, which presents scatter diagrams for the four data sets, and Figure 13.20, which presents residual plots for the four data sets.

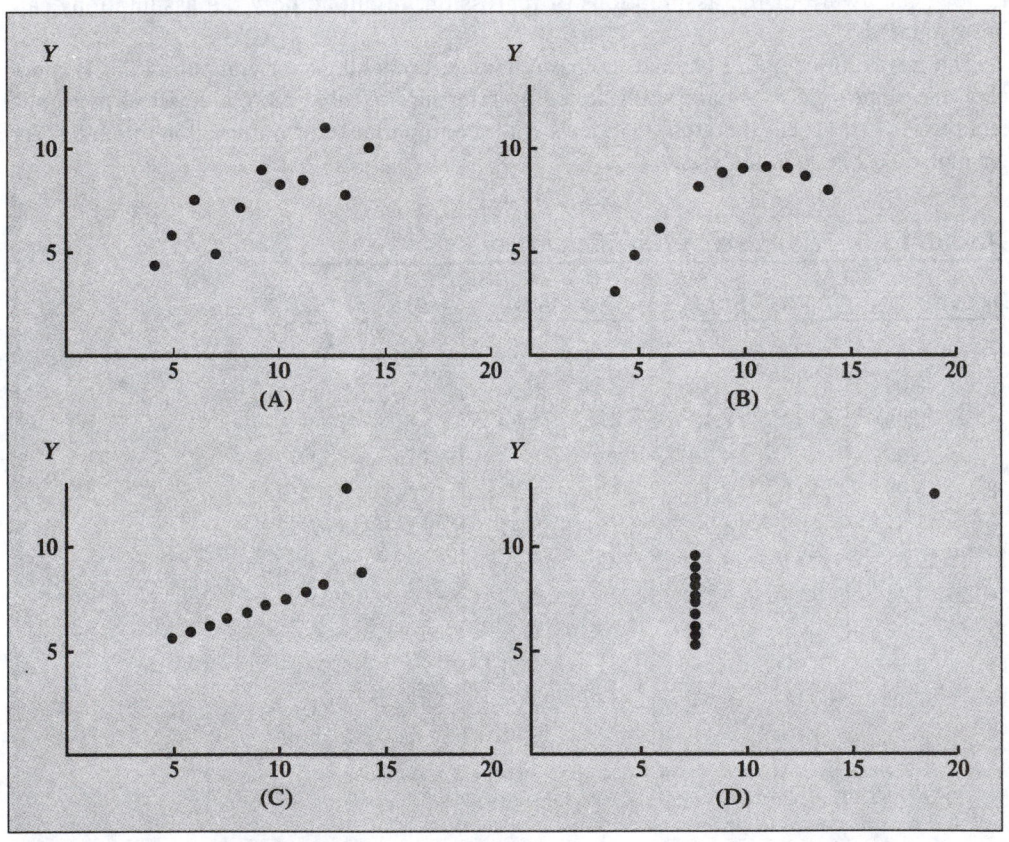

FIGURE 13.19   Scatter diagrams for four data sets

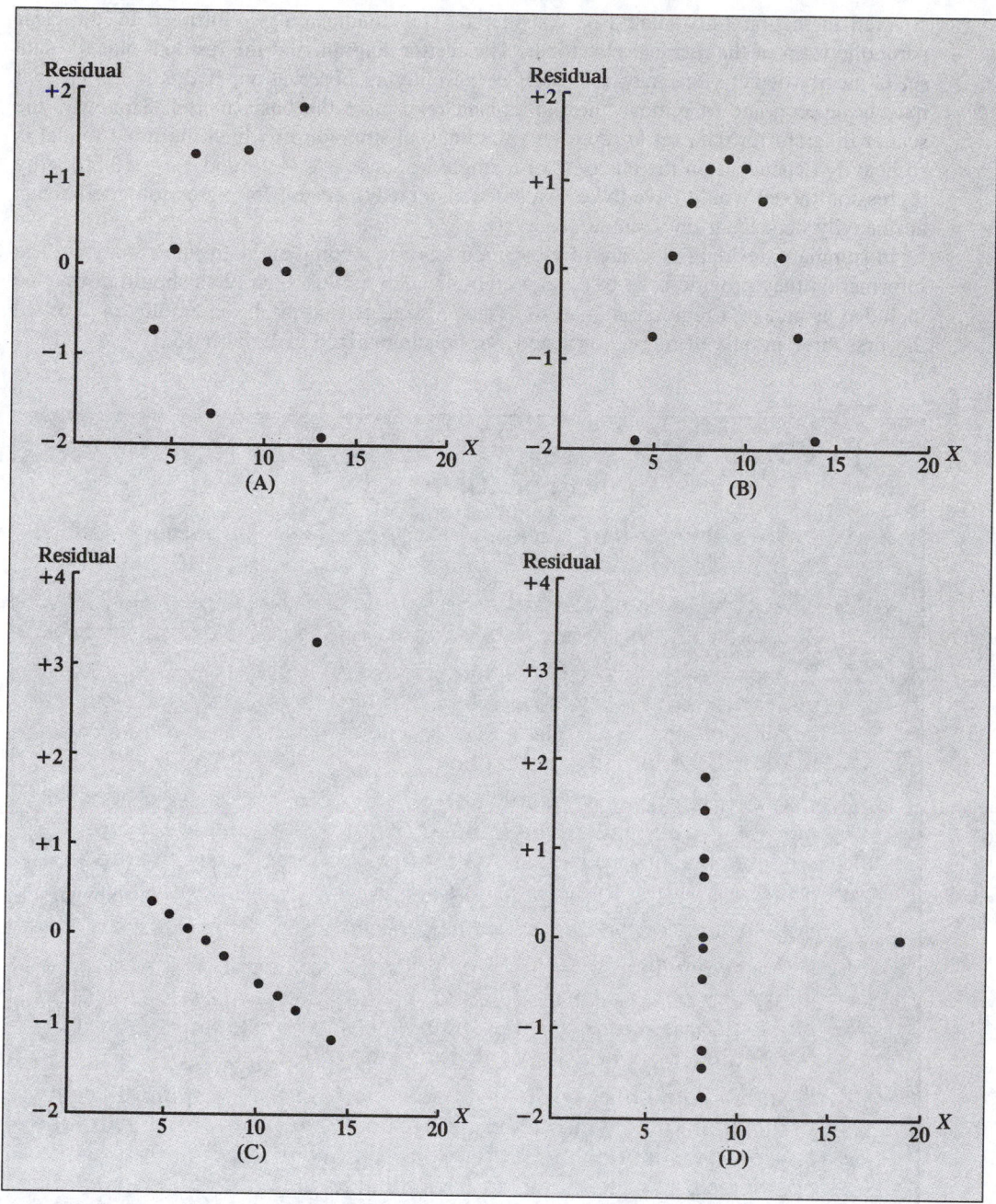

FIGURE 13.20   Residual plots for four data sets
Reprinted by permission of American Statistical Assn.

From the scatter diagrams of Figure 13.19 and the residual plots of Figure 13.20 we see how different the data sets are. The only data set that seems to follow an approximate straight line is data set A. The residual plot for data set A does not show any obvious patterns or outlying residuals. This is certainly not the case for data sets B, C, and D. The scatter plot for data set B seems to indicate that a curvilinear regression model (to be covered in section 14.6) should be considered. This conclusion is reinforced by the clear parabolic form of the residual plot for B. The scatter diagram and the residual plot for data set C clearly depict what may very well be an outlying observation. If this is the case, it may be appropriate to remove the outlier and reestimate the basic model. Similarly, the scatter diagram for data set D represents the unusual situation in which the fitted model is so heavily dependent on the outcome of a single response ($X_8 = 19$ and $Y_8 = 12.50$). Any regression model would have to be evaluated cautiously because its regression coefficients are heavily dependent on a single observation.

In summary, residual plots are of vital importance to a complete regression analysis. The information they provide is so basic to a credible analysis that such plots should always be included as part of a regression analysis. Thus, a strategy that might be employed to avoid the first three pitfalls of regression listed can be summarized in Exhibit 13.3.

Exhibit 13.3 Strategy for Avoiding the Pitfalls of Regression

✓ **1.** Always start with a scatter plot to observe the possible relationship between X and Y.

✓ **2.** Check the assumptions of regression after the regression model has been fit, before moving on to using the results of the model.

✓ **3.** Plot the residuals versus the independent variable. This will enable you to determine whether the model fit to the data is an appropriate one and will allow you to check visually for violations of the homoscedasticity assumption.

✓ **4.** Use a histogram, stem-and-leaf display, box-and-whisker plot, or normal probability plot of the residuals to graphically evaluate whether the normality assumption has been seriously violated.

✓ **5.** If the data have been collected in sequential order, plot the residuals in time order and compute the Durbin-Watson statistic.

✓ **6.** If the evaluation done in 3–5 indicates violations of the assumptions, use alternative methods to least-squares regression or alternative least-squares models (curvilinear or multiple regression), depending on what the evaluation has indicated.

✓ **7.** If the evaluation done in 3–5 does not indicate violations in the assumptions, then the inferential aspects of the regression analysis can be undertaken. Tests for the significance of the regression coefficients can be done and confidence and prediction intervals can be developed.

Ethical Considerations

Ethical considerations arise when a user wishing to develop forecasts manipulates the process of developing the regression model. The key here is intent. As listed in Exhibit 13.4, unethical behavior occurs in several ways in regression analysis.

Exhibit 13.4 Unethical Behavior in Regression Analysis

Unethical behavior occurs in regression analysis when one

✓ **1.** forecasts a response variable of interest with the willful intent of possibly excluding certain independent variables from inclusion in the model.

✓ **2.** deletes observations from the model to obtain a better model without giving reasons for deleting these observations.

✓ **3.** makes forecasts without providing an evaluation of the assumptions when one knows that the assumptions of least-squares regression have been violated.

All of these situations should make us realize even more the importance of following the steps given in Exhibit 13.3 and knowing the assumptions of regression, how to evaluate them, and what to do if any of them have been violated.

 **COMPUTATIONS IN SIMPLE LINEAR REGRESSION**

In our development of the simple linear regression model we have primarily focused on using the output of software such as Microsoft Excel. In this section we illustrate the computations that were involved in developing many of the statistics obtained for the simple linear regression model.

Computing the Y Intercept b_0 and the Slope b_1

Using the method of least-squares, equations (13.3a) and (13.3b) on page 777 need to be solved simultaneously to obtain the regression coefficients b_1 and b_0. Because there are two equations with two unknowns, the simultaneous solution to these two equations gives the following results:

Computational Formula for the Slope b_1

$$b_1 = \frac{SSXY}{SSX} \tag{13.18}$$

where

$$SSXY = \sum_{i=1}^{n}(X_i - \bar{X})(Y_i - \bar{Y}) = \sum_{i=1}^{n}X_i Y_i - \frac{\left(\sum_{i=1}^{n}X_i\right)\left(\sum_{i=1}^{n}Y_i\right)}{n}$$

$$SSX = \sum_{i=1}^{n}(X_i - \bar{X})^2 = \sum_{i=1}^{n}X_i^2 - \frac{\left(\sum_{i=1}^{n}X_i\right)^2}{n}$$

and

Examining equations (13.18) and (13.19), we see that there are five quantities that must be calculated to determine b_1 and b_0. These are n, the sample size; $\sum_{i=1}^{n} X_i$, the sum of the X values; $\sum_{i=1}^{n} Y_i$, the sum of the Y values; $\sum_{i=1}^{n} X_i^2$, the sum of the squared X values; and $\sum_{i=1}^{n} X_i Y_i$, the sum of the cross-products of X and Y. From the data in Table 13.1 on page 775 the number of square feet is used to predict the average annual sales in a store. The computation of the various sums needed [including $\sum_{i=1}^{n} Y_i^2$, the sum of the squared Y values that will be used to compute the sum of squares total (SST)] are presented in Table 13.7.

Table 13.7 *Computations for site selection problem*

STORE	SQUARE FEET X	SALES Y	X^2	Y^2	XY
1	1,726	3,681	2,979,076	13,549,761	6,353,406
2	1,642	3,895	2,696,164	15,171,025	6,395,590
3	2,816	6,653	7,929,856	44,262,409	18,734,848
4	5,555	9,543	30,858,025	91,068,849	53,011,365
5	1,292	3,418	1,669,264	11,682,724	4,416,056
6	2,208	5,563	4,875,264	30,946,969	12,283,104
7	1,313	3,660	1,723,969	13,395,600	4,805,580
8	1,102	2,694	1,214,404	7,257,636	2,968,788
9	3,151	5,468	9,928,801	29,899,024	17,229,668
10	1,516	2,898	2,298,256	8,398,404	4,393,368
11	5,161	10,674	26,635,921	113,934,276	55,088,514
12	4,567	7,585	20,857,489	57,532,225	34,640,695
13	5,841	11,760	34,117,281	138,297,600	68,690,160
14	3,008	4,085	9,048,064	16,687,225	12,287,680
Total	40,898	81,577	156,831,834	592,083,727	301,298,822

Using equations (13.18) and (13.19), we can compute the values of b_1 and b_0:

$$b_1 = \frac{SSXY}{SSX}$$

$$SSXY = \sum_{i=1}^{n}(X_i - \overline{X})(Y_i - \overline{Y}) = \sum_{i=1}^{n} X_i Y_i - \frac{\left(\sum_{i=1}^{n} X_i\right)\left(\sum_{i=1}^{n} Y_i\right)}{n}$$

$$= 301{,}298{,}822 - \frac{(81{,}577)(40{,}898)}{14}$$

$$= 301{,}298{,}822 - 238{,}309{,}724.71$$

$$= 62{,}989{,}097.29$$

$$SSX = \sum_{i=1}^{n}(X_i - \overline{X})^2 = \sum_{i=1}^{n} X_i^2 - \frac{\left(\sum_{i=1}^{n} X_i\right)^2}{n}$$

$$= 156{,}831{,}834 - \frac{(40{,}898)^2}{14}$$

$$= 156{,}831{,}834 - 119{,}474{,}743.1$$

$$= 37{,}357{,}090.86$$

so that

$$b_1 = \frac{62{,}989{,}097.29}{37{,}357{,}090.86}$$

$$= 1.68613$$

also

$$b_0 = \overline{Y} - b_1\overline{X}$$

and

$$\overline{Y} = \frac{\sum_{i=1}^{n} Y_i}{n} = \frac{81{,}577}{14} = 5{,}826.929$$

$$\overline{X} = \frac{\sum_{i=1}^{n} X_i}{n} = \frac{40{,}898}{14} = 2{,}921.2857$$

so that

$$b_0 = 5{,}826.929 - (1.68613)(2{,}921.2857)$$

$$= 901.2$$

Computing the Measures of Variation

Computational formulas can be developed to compute SST, SSR, and SSE, which were defined in equations (13.5), (13.6), and (13.7) on page 789.

Computational Formula for the Total Sum of Squares (SST)

$$SST = \text{total variation or total sum of squares}$$

$$= \sum_{i=1}^{n}(Y_i - \bar{Y})^2 \tag{13.20}$$

$$= \sum_{i=1}^{n}Y_i^2 - \frac{\left(\sum_{i=1}^{n}Y_i\right)^2}{n}$$

Computational Formula for the Regression Sum of Squares (SSR)

$$SSR = \text{explained variation or regression sum of squares}$$

$$= \sum_{i=1}^{n}(\hat{Y}_i - \bar{Y})^2 = b_0\sum_{i=1}^{n}Y_i + b_1\sum_{i=1}^{n}X_iY_i - \frac{\left(\sum_{i=1}^{n}Y_i\right)^2}{n} \tag{13.21}$$

Computational Formula for the Error Sum of Squares (SSE)

$$SSE = \text{unexplained variation or error sum of squares}$$

$$= \sum_{i=1}^{n}(Y_i - \hat{Y}_i)^2 = \sum_{i=1}^{n}Y_i^2 - b_0\sum_{i=1}^{n}Y_i - b_1\sum_{i=1}^{n}X_iY_i \tag{13.22}$$

Using the summary results from Table 13.7 on page 828,

$$SST = \text{total variation or total sum of squares} = \sum_{i=1}^{n}(Y_i - \bar{Y})^2 = \sum_{i=1}^{n}Y_i^2 - \frac{\left(\sum_{i=1}^{n}Y_i\right)^2}{n}$$

$$= 592{,}083{,}727 - \frac{(81{,}577)^2}{14}$$

$$= 592{,}083{,}727 - 475{,}343{,}352.1$$

$$= 116{,}740{,}375$$

$$SSR = \text{explained variation or regression sum of squares}$$

$$= \sum_{i=1}^{n}(\hat{Y}_i - \bar{Y})^2$$

$$= b_0\sum_{i=1}^{n}Y_i + b_1\sum_{i=1}^{n}X_iY_i - \frac{\left(\sum_{i=1}^{n}Y_i\right)^2}{n}$$

$$= (901.2)(81{,}577) + (1.68613)(301{,}298{,}822) - \frac{(81{,}577)^2}{14}$$

$$= 106{,}208{,}120$$

SSE = unexplained variation or error sum of squares

$$= \sum_{i=1}^{n} (Y_i - \hat{Y}_i)^2$$

$$= \sum_{i=1}^{n} Y_i^2 - b_0 \sum_{i=1}^{n} Y_i - b_1 \sum_{i=1}^{n} X_i Y_i$$

$$= 592{,}083{,}727 - (901.2)(81{,}577) - (1.68613)(301{,}298{,}822)$$

$$= 10{,}532{,}255.2$$

Computing the Standard Error of the Slope

In section 13.7 the standard error of the slope was used to test the existence of a relationship between the X and Y variables. The computational formula can be developed as follows:

$$S_{b_1} = \frac{S_{YX}}{\sqrt{SSX}}$$

$$SSX = \sum_{i=1}^{n} (X_i - \overline{X})^2$$

$$= \sum_{i=1}^{n} X_i^2 - \frac{\left(\sum_{i=1}^{n} X_i\right)^2}{n}$$

$$= 156{,}831{,}834 - \frac{(40{,}898)^2}{14}$$

$$= 37{,}357{,}090.86$$

$$S_{b_1} = \frac{936.85}{\sqrt{37{,}357{,}090.86}}$$

$$= .1533$$

For a second illustration of the computations involved in regression, we turn to Example 13.5.

Example 13.5   *Computing b_0, b_1, SST, SSR, SSE, and r^2*

In Table 13.6 on page 823, four data sets were considered. Data set A consisted of the following values:

X_i	Y_i
10	8.04
14	9.96
5	5.68
8	6.95
9	8.81
12	10.84
4	4.26
7	4.82
11	8.33
13	7.58
6	7.24

For these data, compute b_0, b_1, SST, SSR, SSE, and r^2.

SOLUTION

There are five quantities that must be calculated to determine b_1 and b_0. These are n, the sample size; $\sum_{i=1}^{n} X_i$, the sum of the X values; $\sum_{i=1}^{n} Y_i$, the sum of the Y values; $\sum_{i=1}^{n} X_i^2$, the sum of the squared X values; and $\sum_{i=1}^{n} X_i Y_i$, the sum of the cross products of X and Y. In addition, $\sum_{i=1}^{n} Y_i^2$, the sum of the squared Y values is needed to compute SST. The computation of these sums has been obtained from Microsoft Excel as presented in the accompanying figure.

Microsoft Excel computations for data set A

	A	B	C	D	E	F
1		X	Y	X^2	Y^2	XY
2		10	8.04	100	64.6416	80.4
3		14	9.96	196	99.2016	139.44
4		5	5.68	25	32.2624	28.4
5		8	6.95	64	48.3025	55.6
6		9	8.81	81	77.6161	79.29
7		12	10.84	144	117.5056	130.08
8		4	4.26	16	18.1476	17.04
9		7	4.82	49	23.2324	33.74
10		11	8.33	121	69.3889	91.63
11		13	7.58	169	57.4564	98.54
12		6	7.24	36	52.4176	43.44
13	Sums:	99	82.51	1001	660.1727	797.6

Using equations (13.18) and (13.19), we can compute the values of b_1 and b_0:

$$b_1 = \frac{SSXY}{SSX}$$

$$SSXY = \sum_{i=1}^{n}(X_i - \overline{X})(Y_i - \overline{Y}) = \sum_{i=1}^{n} X_i Y_i - \frac{\left(\sum_{i=1}^{n} X_i\right)\left(\sum_{i=1}^{n} Y_i\right)}{n}$$

$$= 797.6 - \frac{(99)(82.51)}{11}$$

$$= 797.6 - 742.59$$

$$= 55.01$$

$$SSX = \sum_{i=1}^{n}(X_i - \overline{X})^2 = \sum_{i=1}^{n} X_i^2 - \frac{\left(\sum_{i=1}^{n} X_i\right)^2}{n}$$

$$= 1{,}001 - \frac{(99)^2}{11}$$

$$= 1{,}001 - 891$$

$$= 110$$

so that

$$b_1 = \frac{55.01}{110}$$

$$= .5001$$

also

$$b_0 = \bar{Y} - b_1\bar{X}$$

and

$$\bar{Y} = \frac{\sum_{i=1}^{n} Y_i}{n} = \frac{82.51}{11} = 7.50$$

$$\bar{X} = \frac{\sum_{i=1}^{n} X_i}{n} = \frac{99}{11} = 9.0$$

so that

$$b_0 = 7.50 - (.5001)(9.0)$$

$$= 3.0$$

Using the summary results,

SST = total variation or total sum of squares

$$= \sum_{i=1}^{n}(Y_i - \bar{Y})^2 = \sum_{i=1}^{n} Y_i^2 - \frac{\left(\sum_{i=1}^{n} Y_i\right)^2}{n}$$

$$= 660.1727 - \frac{(82.51)^2}{11}$$

$$= 660.1727 - 618.9$$

$$= 41.2727$$

SSR = explained variation or regression sum of squares

$$= \sum_{i=1}^{n}(\hat{Y}_i - \bar{Y})^2 = b_0\sum_{i=1}^{n} Y_i + b_1\sum_{i=1}^{n} X_iY_i - \frac{\left(\sum_{i=1}^{n} Y_i\right)^2}{n}$$

$$= (3.0)(82.51) + (.5001)(797.6) - \frac{(82.51)^2}{11}$$

$$= 27.43$$

SSE = unexplained variation or error sum of shapes

$$= \sum_{i=1}^{n}(Y_i - \hat{Y}_i)^2 = \sum_{i=1}^{n} Y_i^2 - b_0\sum_{i=1}^{n} Y_i - b_1\sum_{i=1}^{n} X_iY_i$$

$$= 660.1727 - (3.0)(82.51) - (.5001)(797.6)$$
$$= 13.84$$

CORRELATION—MEASURING THE STRENGTH OF THE ASSOCIATION

The Correlation Coefficient

In our discussion of regression analysis we have been concerned with the prediction of the dependent variable Y based on the independent variable X. In contrast to this, in a correlation analysis our focus is on measuring the degree of association between two variables.

The strength of a relationship between two variables in a population is usually measured by the **coefficient of correlation** ρ, whose values range from -1 for perfect negative correlation up to $+1$ for perfect positive correlation. Figure 13.21 illustrates three different types of association between variables.

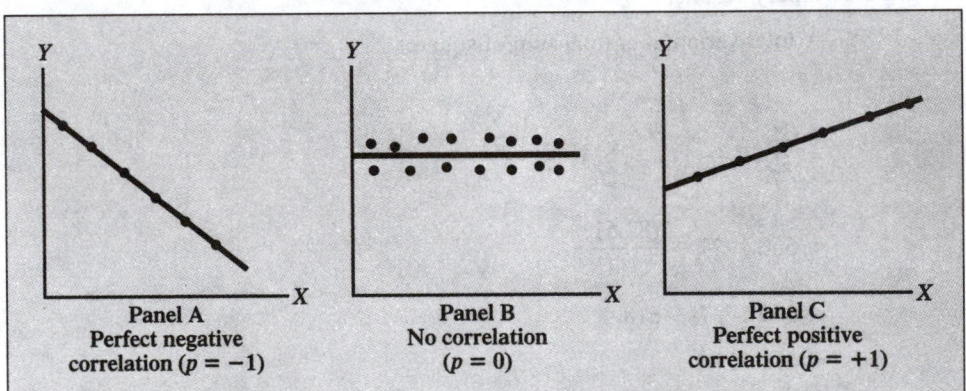

FIGURE 13.21   Types of association between variables

In Panel A of Figure 13.21 there is a perfect negative linear relationship between X and Y so that Y will decrease in a perfectly predictable manner as X increases. Panel B is an example in which there is no relationship between X and Y. As X increases, there is no change in Y, so there is no association between the values of X and the values of Y. Panel C depicts a perfect positive correlation between X and Y. In this case, Y increases in a perfectly predictable manner as X increases.

For situations in which our primary interest is regression analysis, the sample coefficient of correlation (r) is obtained from the coefficient of determination r^2.

$$r^2 = \frac{\text{regression sum of squares}}{\text{total sum of squares}} = \frac{SSR}{SST}$$

so that

In the site selection example, because $r^2 = .91$ and the slope b_1 is positive, the coefficient of correlation is computed as $+.954$. The closeness of the correlation coefficient to $+1.0$ implies a strong positive association between size of store and annual sales.

We have now computed and interpreted the correlation coefficient from a regression perspective. As we mentioned at the beginning of this chapter, regression and correlation are two separate techniques, with regression being concerned with prediction and correlation with association. In many applications we are concerned only with measuring association between variables, not with using one variable to predict another. If we are specifically interested in measuring correlation, the sample correlation coefficient r can be computed directly, using the following equation.

One application of the correlation coefficient occurs in finance where it is important to study the association between two investments over time. In section 4.8 we introduced the *covariance* that measured the covariation between two variables. One way of viewing the correlation coefficient is from the perspective that it is the covariance in standardized form so that its smallest possible value is -1 and its largest possible value is $+1$.

To illustrate an application of the correlation coefficient, suppose we want to study the association in the value of two currencies, the German mark and the Japanese yen, from 1988 to 1997. The results are summarized in Table 13.8.

Table 13.8 *Exchange rate of German mark and Japanese yen in U.S. dollars*

YEAR	GERMAN MARK	JAPANESE YEN
1988	1.76	128.17
1989	1.88	138.07
1990	1.62	145.00
1991	1.66	134.59
1992	1.56	126.78
1993	1.65	111.20
1994	1.62	102.21
1995	1.50	103.35
1996	1.54	115.87
1997	1.80	130.38

DATA FILE
MARKYEN.XLS

Source: Board of Governors of the Federal Reserve System, Table B-107.

From the data of Table 13.8 we use Microsoft Excel to compute SSX, SSY, SSXY, and r as displayed in Figure 13.22.

	A	B	C	D	E	F	G	H	I
1	Year	Mark(X)	Yen(Y)	(X-XBAR)^2	(Y-YBAR)^2	(X-XBAR)(Y-YBAR)			
2	1988	1.76	128.17	0.010201	21.233664	0.465408		Summary	
3	1989	1.88	138.07	0.048841	210.482064	3.206268		Xbar	1.659
4	1990	1.62	145	0.001521	459.587844	-0.836082		Ybar	123.562
5	1991	1.66	134.59	1E-06	121.616784	0.011028		SSXY	8.56242
6	1992	1.56	126.78	0.009801	10.355524	-0.318582		SSX	0.13129
7	1993	1.65	111.2	8.1E-05	152.819044	0.111258		SSY	1946.18
8	1994	1.62	102.21	0.001521	455.907904	0.832728		r	0.53566
9	1995	1.5	103.35	0.025281	408.524944	3.213708			
10	1996	1.54	115.87	0.014161	59.166864	0.915348			
11	1997	1.8	130.38	0.019881	46.485124	0.961338			

FIGURE 13.22 Summary computations for correlation of German mark and Japanese yen obtained from Microsoft Excel

To obtain the correlation coefficient *r*, we have

$$SSXY = \sum_{i=1}^{n}(X_i - \overline{X})(Y_i - \overline{Y}) = 8.56242$$

$$SSX = \sum_{i=1}^{n}(X_i - \overline{X})^2 = .13129$$

$$SSY = \sum_{i=1}^{n}(Y_i - \overline{Y})^2 = 1{,}946.17976$$

so that

$$r = \frac{SSXY}{\sqrt{SSX}\sqrt{SSY}} = \frac{8.56242}{\sqrt{.13129}\sqrt{1,946.17976}}$$

$$= .53566$$

The coefficient of correlation $r = +.536$ between the German mark and the Japanese yen indicates a moderate association. A higher price of the German mark is moderately associated with a higher price of the Japanese yen.

Now that we have computed the correlation coefficient r, we can use these sample results to determine whether there is any evidence of a statistically significant association between these variables. The population correlation coefficient ρ is hypothesized as equal to 0. Thus, the null and alternative hypotheses are

$$H_0: \rho = 0 \text{ (There is no correlation.)}$$
$$H_1: \rho \neq 0 \text{ (There is correlation.)}$$

The test statistic for determining the existence of a significant correlation is given by

Testing for the Existence of Correlation

$$t = \frac{r - \rho}{\sqrt{\dfrac{1 - r^2}{n - 2}}} \tag{13.25}$$

where the test statistic t follows a t distribution with $n - 2$ degrees of freedom.

For the currency data summarized in Figure 13.22, $r = +.53566$ and $n = 10$, so testing the null hypothesis we have

$$t = \frac{r}{\sqrt{\dfrac{1 - r^2}{n - 2}}}$$

$$= \frac{.53566}{\sqrt{\dfrac{1 - (.53566)^2}{10 - 2}}} = 1.794$$

Using the .05 level of significance, because $t = 1.794 < t_8 = 2.306$, we do not reject H_0. We conclude that there is no evidence of an association between the value of the German mark and the Japanese yen.

When we discussed inferences concerning the population slope, we used confidence intervals and tests of hypothesis interchangeably. However, the development of a confidence interval for the correlation coefficient is more complicated because the shape of the sampling distribution of the statistic r varies for different values of the true correlation coefficient. Methods for developing a confidence interval estimate for the correlation coefficient are presented in reference 6.

Overview

Both quick results users and developers should implement a formula using the CORREL function to calculate the correlation coefficient.

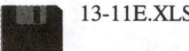
13-11E.XLS

The 13-11E.XLS workbook file contains the calculation of the correlation coefficient for German mark and Japanese yen data of Table 13.8 on page 836.

Details for All Users

We can use the CORREL worksheet function to calculate the correlation coefficient for two variables. The format of this function is:

CORREL(*X variable cell range, Y variable cell range*)

As an example, consider the study of the association in the value of two currencies problem of section 13.11. To implement the calculation for the correlation coefficient for this problem, do the following:

❶ Open the Starting Point for Section 13.11E workbook (STARTING POINT 13-11E.XLS) and click the Data sheet tab. Verify that the year, German mark, and Japanese yen data of Table 13.8 on page 836 have been entered into columns A, B, and C, respectively.

❷ Enter the title Two Currencies Analysis in cell E1. Select the cell range E1:F1 and click the Merge and Center button on the formatting toolbar (see section S3.11).

❸ Enter the label Correlation Coefficient in cell E3.

❹ Enter the formula =CORREL(B2:B11,C2:C11) in cell F3.

The resulting Data sheet will be similar to the one shown in Figure 13E.6.

FIGURE 13E.6

Correlation coefficient for the two currencies problem of section 13.11

	A	B	C	D	E	F
1	Year	Mark	Yen		Two Currencies Analysis	
2	1988	1.76	128.17			
3	1989	1.88	138.07		Correlation Coefficient	0.53566
4	1990	1.62	145.00			
5	1991	1.66	134.59			
6	1992	1.56	126.78			
7	1993	1.65	111.20			
8	1994	1.62	102.21			
9	1995	1.50	103.35			
10	1996	1.54	115.87			
11	1997	1.80	130.38			

Problems for Section 13.11

Learning the Basics

• **13.50** If $r^2 = .81$ and the slope of the fitted regression line is positive, find r.

13.51 If the coefficient of determination is .49 and the slope of the fitted regression line is -3, find the coefficient of correlation.

13.52 If $SSR = SST$ and the slope is a negative value, find r.

13.53 If $SSE = 0$, and the slope is a positive value, find r.

13.54 Given the following set of data from a sample of $n = 11$ items,

X	7	5	8	3	6	10	12	4	9	15	18
Y	21	15	24	9	18	30	36	12	27	45	54

(a) Compute the correlation coefficient r.

(b) At the .05 level of significance, is there evidence of a relationship between X and Y? Explain.

Applying the Concepts

13.55 The following data represent the approximate retail price (in $) and the energy cost per year (in $) of nine large side-by-side refrigerators.

BRAND	PRICE ($)	ENERGY COST PER YEAR ($)
KitchenAidSuperbaKSRS25QF	1,600	73
Kenmore(Sears)5757	1,200	73
WhirlpoolED25DQXD	1,550	78
AmanaSRD25S3	1,350	85
Kenmore(Sears)5647	1,700	93
GEProfileTPX24PRY	1,700	93
FrigidaireGalleryFRS26ZGE	1,500	95
MaytagRSW2400EA	1,400	96
GETFX25ZRY	1,200	94

DATA FILE
REFRIG.XLS

Source: "The Kings of Cool," Copyright © 1998 by Consumers Union of U.S., Inc. Adapted from CONSUMER REPORTS *(January 1998): 52, by permission of Consumers Union of U.S., Inc., Yonkers, NY 10703-1057. Although these data sets originally appeared in* CONSUMER REPORTS, *the selective adaptation and resulting conclusions presented are those of the authors and are not sanctioned or endorsed in any way by Consumers Union, the publisher of* CONSUMER REPORTS.

(a) Compute the correlation coefficient r.

(b) At the .05 level of significance, is there a relationship between X and Y? Explain.

(c) Would you expect the higher-priced refrigerators to have greater energy efficiency? Is this borne out by the data?

● **13.56** The following data represent the average charge (in dollars per minute) and the amount of minutes expended (in billions) for all telephone calls placed from the United States to 20 different countries during 1996.

COUNTRY	CHARGE PER MINUTE (IN DOLLARS)	MINUTES (IN BILLIONS)	COUNTRY	CHARGE PER MINUTE (IN DOLLARS)	MINUTES (IN BILLIONS)
Canada	0.34	3.049	India	1.38	0.287
Mexico	0.85	2.012	Brazil	0.96	0.284
Britain	0.73	1.025	Italy	1.00	0.279
Germany	0.88	0.662	Taiwan	0.97	0.273
Japan	1.00	0.576	Colombia	1.00	0.257
Dominican Republic	0.84	0.410	China	1.47	0.232
France	0.81	0.364	Israel	1.16	0.214

(continued)

615

COUNTRY	CHARGE PER MINUTE (IN DOLLARS)	MINUTES (IN BILLIONS)	COUNTRY	CHARGE PER MINUTE (IN DOLLARS)	MINUTES (IN BILLIONS)
South Korea	1.09	0.319	Australia	1.01	0.201
Hong Kong	0.90	0.317	Jamaica	1.03	0.188
Philippines	1.29	0.297	Netherlands	0.78	0.167

DATA FILE
INTPHONE.XLS

Source: The New York Times, *February 17, 1997, 46. Copyright by The New York Times Company. Reprinted by permission of* The New York Times.

(a) Compute the correlation coefficient r.

(b) At the .05 level of significance, is there a relationship between X and Y? Explain.

(c) One might expect that the higher the charge per minute, the lower the number of minutes that would be used. Does the correlation coefficient reflect this expected relationship? Explain.

13.57 The following data represent the retail price (in dollars) and the printing speed (in number of pages per minute of double-spaced black text with standard margins) for a sample of 19 computer printers:

BRAND	PRICE (IN DOLLARS)	TEXT SPEED (PAGES PER MINUTE)
Hewlett-PackardDeskJet855Cse	500	3.0
Hewlett-PackardDeskJet682C	300	2.5
Hewlett-PackardDeskJet600C	250	2.6
EpsonStylusColorII	230	2.5
Hewlett-PackardDeskWriter600	250	3.0
CanonBJC-610	430	1.3
CanonBJC-210	150	2.9
AppleColorStyleWriter1500	280	3.1
CanonBJC-4100	230	1.7
AppleColorStyleWriter2500	380	3.4
EpsonStylusColorIIs	190	0.7
Lexmark2070Jetprinter	350	2.1
Lexmark1020Jetprinter	150	1.3
NECSuperScript860	500	7.9
PanasonicKX-P6500	450	5.5
Hewlett-PackardLaserJet5L	480	4.2
TexasInstrumentsMicroLaserWin/4	380	4.2
CanonLBP-460	350	4.1
OkidataOL600e	400	3.9

DATA FILE
PRINTER.XLS

Source: "Computer Printers," Copyright ©1996 by Consumers Union of U.S., Inc. Adapted from CONSUMER REPORTS *(October 1996): 60–61, by permission of Consumers Union of U.S., Inc., Yonkers, NY 10703-1057. Although these data sets originally appeared in* CONSUMER REPORTS, *the selective adaptation and resulting conclusions presented are those of the authors and are not sanctioned or endorsed in any way by Consumers Union, the publisher of* CONSUMER REPORTS.

(a) Compute the correlation coefficient r.

(b) At the .05 level of significance, is there a relationship between X and Y? Explain.

(c) One might expect that the higher the price, the higher the text speed. Does the correlation coefficient reflect this expected relationship? Explain.

• **13.58** In Problem 13.3 on page 785 the marketing manager used shelf space for pet food to predict weekly sales. The coefficient of determination $r^2 = .684$.

(a) Compute the coefficient of correlation.
(b) Is there evidence of a significant correlation at the .05 level of significance?
(c) Compare the results of (b) to Problem 13.36(a) on page 813. What conclusion do you reach about the two tests?

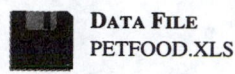

DATA FILE
PETFOOD.XLS

13.59 In Problem 13.4 on page 785 a manager wanted to predict weekly sales at a chain of package delivery stores based on the number of customers who made purchases. The coefficient of determination $r^2 = .913$.

(a) Compute the coefficient of correlation.
(b) Is there evidence of significant correlation at the .05 level of significance?
(c) Compare the results of (b) to Problem 13.37(a) on page 813. What conclusion do you reach about the two tests?

DATA FILE
PACKAGE.XLS

• **13.60** In Problem 13.5 on page 786 a company wanted to predict the worker-hours required for production based on the lot size. The coefficient of determination $r^2 = .9878$.

(a) Compute the coefficient of correlation.
(b) Is there evidence of a significant correlation at the .05 level of significance?
(c) Compare the results of (b) to Problem 13.38(a) on page 813. What conclusion do you reach about the two tests?

DATA FILE
WORKHRS.XLS

13.61 In Problem 13.6 on page 786 a company wanted to predict home video sales based on the box office gross of movies. The coefficient of determination $r^2 = .728$.

(a) Compute the coefficient of correlation.
(b) Is there evidence of a significant correlation at the .05 level of significance?
(c) Compare the results of (b) to Problem 13.39(a) on page 813. What conclusion do you reach about the two tests?

DATA FILE
MOVIE.XLS

13.62 In Problem 13.7 on page 787 an agent for a real estate company wanted to predict the monthly rent for apartments based on the size of the apartment. The coefficient of determination $r^2 = .723$.

(a) Compute the coefficient of correlation.
(b) Is there evidence of a significant correlation at the .05 level of significance?
(c) Compare the results of (b) to Problem 13.40(a) on page 813. What conclusion do you reach about the two tests?

DATA FILE
RENT.XLS

13.63 In Problem 13.8 on page 787 a limousine service wanted to predict travel time to an airport based on the distance from the pickup location to the airport. The coefficient of determination $r^2 = .918$.

(a) Compute the coefficient of correlation.
(b) Is there evidence of a significant correlation at the .05 level of significance?
(c) Compare the results of (b) to Problem 13.41(a) on page 813. What conclusion do you reach about the two tests?

DATA FILE
LIMO.XLS

◆ SUMMARY

As seen in the accompanying chapter summary chart, we developed the simple linear regression model, discussed the assumptions of the model, and showed how these assumptions could be evaluated. We then developed the t test for the significance of the slope and used the regression model for prediction. In addition, we studied the correlation coefficient and tested for its significance. In chapter 14, we will continue our discussion of regression analysis by considering a variety of multiple regression models.

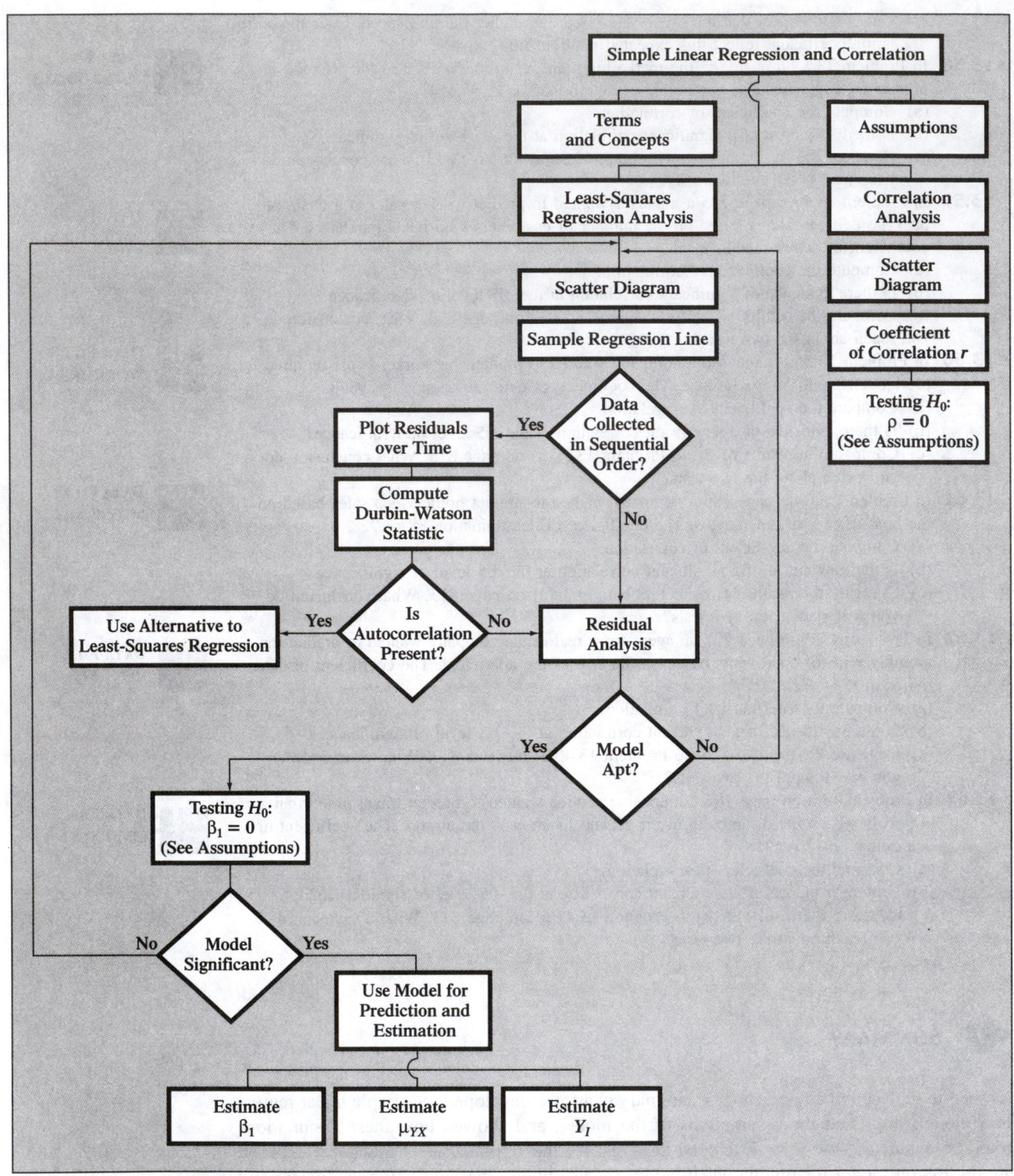

summary chart

Key Terms

assumptions of regression 793
autocorrelation 801
coefficient of correlation 834
coefficient of determination 790
confidence interval estimate for the mean response 814
correlation analysis 772
dependent variable 772
Durbin-Watson statistic 803
error sum of squares (SSE) 788
explained variation 788
explanatory variable 772
homoscedasticity 793

independence of error 794
independent variable 772
least-squares method 777
linear relationship 773
normality 793
prediction interval for an individual response 816
regression analysis 772
regression coefficient 777
regression sum of squares (SSR) 788
relevant range 779
residual analysis 794
residuals 794

response variable 772
scatter diagram 773
slope 773
standard error of the estimate 791
standardized residuals 795
Studentized residuals 795
total sum of squares (SST) 788
total variation 788
unexplained variation 788
Y intercept 773

Checking Your Understanding

13.64 What is the interpretation of the Y intercept and the slope in a regression model?

13.65 What is the interpretation of the coefficient of determination?

13.66 When will the unexplained variation or error sum of squares be equal to 0?

13.67 When will the explained variation or sum of squares that is due to regression be equal to 0?

13.68 Why should a residual analysis always be done as part of the development of a regression model?

13.69 What are the assumptions of regression analysis and how can they be evaluated?

13.70 What is the Durbin-Watson statistic and what does it measure?

13.71 Under what circumstances would it be important to compute the Durbin-Watson statistic? Explain.

13.72 What is the difference between a confidence interval estimate of the mean response μ_{YX} and a prediction interval estimate of Y_I?

Chapter Review Problems

13.73 Management of a soft-drink bottling company wished to develop a method for allocating delivery costs to customers. Although one aspect of cost clearly relates to travel time within a particular route, another type of cost reflects the time required to unload the cases of soft drink at the delivery point. A sample of 20 customers was selected from routes within a territory and the delivery time and the number of cases delivered were measured with the following results (at the top of page 844):

Assuming that we wanted to develop a model to predict delivery time based on the number of cases delivered:

(a) Set up a scatter diagram.
(b) Use the least-squares method to find the regression coefficients b_0 and b_1.
(c) State the regression equation.
(d) Interpret the meaning of b_0 and b_1 in this problem.
(e) Predict the average delivery time for a customer who is receiving 150 cases of soft drink.

CUSTOMER	NUMBER OF CASES	DELIVERY TIME (MINUTES)	CUSTOMER	NUMBER OF CASES	DELIVERY TIME (MINUTES)
1	52	32.1	11	161	43.0
2	64	34.8	12	184	49.4
3	73	36.2	13	202	57.2
4	85	37.8	14	218	56.8
5	95	37.8	15	243	60.6
6	103	39.7	16	254	61.2
7	116	38.5	17	267	58.2
8	121	41.9	18	275	63.1
9	143	44.2	19	287	65.6
10	157	47.1	20	298	67.3

DATA FILE
DELIVERY.XLS

(f) Would it be appropriate to use the model to predict the delivery time for a customer who is receiving 500 cases of soft drink? Why?

(g) Compute the coefficient of determination r^2 and explain its meaning in this problem.

(h) Compute the coefficient of correlation.

(i) Compute the standard error of the estimate.

(j) Perform a residual analysis using either the residuals or the Studentized residuals. Is there any evidence of a pattern in the residuals? Explain.

(k) At the .05 level of significance, is there evidence of a linear relationship between delivery time and the number of cases delivered?

(l) Set up a 95% confidence interval estimate of the average delivery time for customers that receive 150 cases of soft drink.

(m) Set up a 95% prediction interval estimate of the delivery time for an individual customer who is receiving 150 cases of soft drink.

(n) Set up a 95% confidence interval estimate of the population slope.

(o) Explain how the results obtained in (a)–(n) can help allocate delivery costs to customers.

13.74 A brokerage house would like to be able to predict the number of trade executions per day and has decided to use the number of incoming phone calls as a predictor variable. Data were collected over a period of 35 days with the following results (at the top of page 845):

(a) Set up a scatter diagram.

(b) Use the least-squares method to find the regression coefficients b_0 and b_1.

(c) State the regression equation.

(d) Interpret the meaning of b_0 and b_1 in this problem.

(e) Predict the average number of trades executed for a day in which the number of incoming calls is 2,000.

(f) Would it be appropriate to use the model to predict the average number of trades executed for a day in which the number of incoming calls is 5,000? Why?

(g) Compute the coefficient of determination r^2 and explain its meaning in this problem.

(h) Compute the coefficient of correlation.

(i) Compute the standard error of the estimate.

(j) Plot the residuals against the number of incoming calls and also against the days. Is there any evidence of a pattern in the residuals with either of these variables? Explain.

Day	Number of Incoming Calls	Trade Executions	Day	Number of Incoming Calls	Trade Executions
1	2,591	417	18	2,237	397
2	2,146	321	19	2,328	365
3	2,185	362	20	2,078	330
4	2,245	364	21	2,134	312
5	2,600	442	22	2,192	340
6	2,510	386	23	1,965	339
7	2,394	370	24	2,147	364
8	2,486	376	25	2,015	295
9	2,483	463	26	2,046	292
10	2,297	389	27	2,073	379
11	2,106	302	28	2,032	294
12	2,035	266	29	2,108	329
13	1,936	339	30	1,923	274
14	1,951	369	31	2,069	326
15	2,292	403	32	2,061	306
16	2,094	319	33	2,010	352
17	1,897	306	34	1,913	290
			35	1,904	283

DATA FILE
TRADES.XLS

(k) Compute the Durbin-Watson statistic for these data.

(l) On the basis of the results of (j) and (k), is there reason to question the validity of the fitted model? Explain.

(m) At the .05 level of significance, is there evidence of a linear relationship between the volume of trade executions and the number of incoming calls?

(n) Set up a 95% confidence interval estimate of the average number of trades executed for days in which the number of incoming calls is 2,000.

(o) Set up a 95% prediction interval estimate of the number of trades executed for a particular day in which the number of incoming calls is 2,000.

(p) Set up a 95% confidence interval estimate of the population slope.

(q) On the basis of the results of (a)–(p), do you think the brokerage house should focus on a strategy of increasing the total number of incoming calls or on a strategy that relies on trading by a small number of heavy traders? Explain.

13.75 Suppose we want to develop a model to predict selling price of houses based on assessed value. A sample of 30 recently sold single-family houses in a small western city is selected to study the relationship between selling price and assessed value (the houses in the city had been reassessed at full value 1 year prior to the study). The results are as shown at the top of page 846.

Hint: First determine which are the independent and dependent variables.

(a) Plot a scatter diagram and, assuming a linear relationship, use the least-squares method to find the regression coefficients b_0 and b_1.

(b) Interpret the meaning of the Y intercept b_0 and the slope b_1 in this problem.

(c) Use the regression model developed in (a) to predict the average selling price for a house whose assessed value is $70,000.

(d) Compute the standard error of the estimate.

OBSERVATION	ASSESSED VALUE ($000)	SELLING PRICE ($000)	OBSERVATION	ASSESSED VALUE ($000)	SELLING PRICE ($000)
1	78.17	94.10	16	84.36	106.70
2	80.24	101.90	17	72.94	81.50
3	74.03	88.65	18	76.50	94.50
4	86.31	115.50	19	66.28	69.00
5	75.22	87.50	20	79.74	96.90
6	65.54	72.00	21	72.78	86.50
7	72.43	91.50	22	77.90	97.90
8	85.61	113.90	23	74.31	83.00
9	60.80	69.34	24	79.85	97.30
10	81.88	96.90	25	84.78	100.80
11	79.11	96.00	26	81.61	97.90
12	59.93	61.90	27	74.92	90.50
13	75.27	93.00	28	79.98	97.00
14	85.88	109.50	29	77.96	92.00
15	76.64	93.75	30	79.07	95.90

DATA FILE
HOUSE1.XLS

(e) Compute the coefficient of determination r^2 and interpret its meaning in this problem.
(f) Compute the coefficient of correlation r.
(g) Perform a residual analysis on your results and determine the adequacy of the fit of the model.
(h) At the .05 level of significance, is there evidence of a linear relationship between selling price and assessed value?
(i) Set up a 95% confidence interval estimate of the average selling price for houses with an assessed value of $70,000.
(j) Set up a 95% prediction interval estimate of the selling price of an individual house with an assessed value of $70,000.
(k) Set up a 95% confidence interval estimate of the population slope.

• **13.76** Suppose we want to develop a model to predict assessed value based on heating area. A sample of 15 single-family houses is selected in a particular community. The assessed value (in thousands of dollars) and the heating area of the houses (in thousands of square feet) are recorded with the following results:

HOUSE	ASSESSED VALUE ($000)	HEATING AREA OF DWELLING (THOUSANDS OF SQUARE FEET)	HOUSE	ASSESSED VALUE ($000)	HEATING AREA OF DWELLING (THOUSANDS OF SQUARE FEET)
1	84.4	2.00	9	78.5	1.59
2	77.4	1.71	10	79.2	1.50
3	75.7	1.45	11	86.7	1.90
4	85.9	1.76	12	79.3	1.39
5	79.1	1.93	13	74.5	1.54
6	70.4	1.20	14	83.8	1.89
7	75.8	1.55	15	76.8	1.59
8	85.9	1.93			

DATA FILE
HOUSE2.XLS

Hint: First determine which are the independent and dependent variables.

(a) Plot a scatter diagram and, assuming a linear relationship, use the least-squares method to find the regression coefficients b_0 and b_1.

(b) Interpret the meaning of the Y intercept b_0 and the slope b_1 in this problem.

(c) Use the regression model developed in (a) to predict the average assessed value for a house whose heating area is 1,750 square feet.

(d) Compute the standard error of the estimate.

(e) Compute the coefficient of determination r^2 and interpret its meaning in this problem.

(f) Compute the coefficient of correlation r.

(g) Perform a residual analysis on your results and determine the adequacy of the fit of the model.

(h) At the .05 level of significance, is there evidence of a linear relationship between assessed value and heating area?

(i) Set up a 95% confidence interval estimate of the average assessed value for houses with a heating area of 1,750 square feet.

(j) Set up a 95% prediction interval estimate of the assessed value of an individual house with a heating area of 1,750 square feet.

(k) Set up a 95% confidence interval estimate of the population slope.

(l) Suppose that the assessed value for the fourth house was 79.7. Do (a)–(k) and compare the results.

WHAT IF?

• **13.77** The director of graduate studies at a large college of business would like to be able to predict the grade point index (GPI) of students in an MBA program based on the Graduate Management Aptitude Test (GMAT) score. A sample of 20 students who have completed 2 years in the program is selected; the results are as follows:

OBSERVATION	GMAT SCORE	GPI	OBSERVATION	GMAT SCORE	GPI
1	688	3.72	11	567	3.07
2	647	3.44	12	542	2.86
3	652	3.21	13	551	2.91
4	608	3.29	14	573	2.79
5	680	3.91	15	536	3.00
6	617	3.28	16	639	3.55
7	557	3.02	17	619	3.47
8	599	3.13	18	694	3.60
9	616	3.45	19	718	3.88
10	594	3.33	20	759	3.76

DATA FILE
GPIGMAT.XLS

Hint: First determine which are the independent and dependent variables.

(a) Plot a scatter diagram and, assuming a linear relationship, use the least-squares method to find the regression coefficients b_0 and b_1.

(b) Interpret the meaning of the Y intercept b_0 and the slope b_1 in this problem.

(c) Use the regression model developed in (a) to predict the average GPI for a student with a GMAT score of 600.

(d) Compute the standard error of the estimate.

(e) Compute the coefficient of determination r^2 and interpret its meaning in this problem.

(f) Compute the coefficient of correlation r.

(g) Perform a residual analysis on your results and determine the adequacy of the fit of the model.

(h) At the .05 level of significance, is there evidence of a linear relationship between GMAT score and GPI?

(i) Set up a 95% confidence interval estimate for the average GPI of students with a GMAT score of 600.

(j) Set up a 95% prediction interval estimate of the GPI for a particular student with a GMAT score of 600.

(k) Set up a 95% confidence interval estimate of the population slope.

(l) Suppose the GPIs of the 19th and 20th students were incorrectly entered. The GPI for student 19 should be 3.76, and the GPI for student 20 should be 3.88. Do (a)–(k) and compare the results.

13.78 The manager of the purchasing department of a large banking organization would like to develop a model to predict the amount of time it takes to process invoices. Data are collected from a sample of 30 days with the following results:

DAY	INVOICES PROCESSED	COMPLETION TIME (HOURS)	DAY	INVOICES PROCESSED	COMPLETION TIME (HOURS)
1	149	2.1	16	169	2.5
2	60	1.8	17	190	2.9
3	188	2.3	18	233	3.4
4	19	0.3	19	289	4.1
5	201	2.7	20	45	1.2
6	58	1.0	21	193	2.5
7	77	1.7	22	70	1.8
8	222	3.1	23	241	3.8
9	181	2.8	24	103	1.5
10	30	1.0	25	163	2.8
11	110	1.5	26	120	2.5
12	83	1.2	27	201	3.3
13	60	0.8	28	135	2.0
14	25	0.4	29	80	1.7
15	173	2.0	30	29	0.5

Hint: Determine which are the independent and dependent variables.

(a) Set up a scatter diagram.

(b) Assuming a linear relationship, use the least-squares method to find the regression coefficients b_0 and b_1.

(c) Interpret the meaning of the Y intercept b_0 and the slope b_1 in this problem.

(d) Use the regression model developed in (b) to predict the average amount of time it would take to process 150 invoices.

(e) Compute the standard error of the estimate.

(f) Compute the coefficient of determination r^2 and interpret its meaning.

(g) Compute the coefficient of correlation r.

(h) Plot the residuals against the number of invoices processed and also against time.

(i) Based on the plots in (h), does the model seem appropriate?

(j) Compute the Durbin-Watson statistic and, at the .05 level of significance, determine whether there is any autocorrelation in the residuals.

(k) On the basis of the results of (h)–(j), what conclusions can you reach concerning the validity of the model fit in (b)?

(l) At the .05 level of significance, is there evidence of a linear relationship between the amount of time and the number of invoices processed?

(m) Set up a 95% confidence interval estimate of the average amount of time taken to process 150 invoices.

(n) Set up a 95% prediction interval estimate of the amount of time it takes to process 150 invoices on a particular day.

13.79 Crazy Dave, a well-known baseball analyst, would like to study various team statistics for the 1997 baseball season to determine which variables might be useful in predicting the number of wins achieved by teams during the season. He has decided to begin by using the team earned run average (ERA) to predict the number of wins. The data for the 28 major league teams are as follows:

AMERICAN LEAGUE			NATIONAL LEAGUE		
TEAM	WINS	ERA	TEAM	WINS	ERA
Boston	78	4.85	Florida	92	3.83
Cleveland	86	4.73	Cincinnati	76	4.41
Kansas City	67	4.70	Chicago Cubs	68	4.44
Minnesota	68	5.01	San Francisco	90	4.39
Toronto	76	3.93	Los Angeles	88	3.62
Anaheim	84	4.52	Pittsburgh	79	4.28
Seattle	90	4.78	San Diego	76	4.98
Texas	77	4.69	New York Mets	88	3.95
Detroit	79	4.56	St. Louis	73	3.88
Chicago White Sox	80	4.73	Philadelphia	68	4.85
Milwaukee	78	4.22	Atlanta	101	3.18
Oakland	65	5.48	Montreal	78	4.14
Baltimore	98	3.91	Houston	84	3.66
New York Yankees	96	3.84	Colorado	83	5.29

DATA FILE
BB97.XLS

Hint: Determine which are the independent and dependent variables.
(a) Set up a scatter diagram.
(b) Assuming a linear relationship, use the least-squares method to find the regression coefficients b_0 and b_1.
(c) Interpret the meaning of the Y intercept b_0 and the slope b_1 in this problem.
(d) Use the regression model developed in (b) to predict the expected number of wins for a team with an ERA of 4.00.
(e) Compute the standard error of the estimate.
(f) Compute the coefficient of determination r^2 and interpret its meaning.
(g) Compute the coefficient of correlation.
(h) Perform a residual analysis on your results and determine the adequacy of the fit of the model.
(i) At the .05 level of significance, is there evidence of a linear relationship between the number of wins and the ERA?
(j) Set up a 95% confidence interval estimate of the average number of wins expected for teams with an ERA of 4.00.
(k) Set up a 95% prediction interval estimate of the number of wins for an individual team that has an ERA of 4.00.
(l) Set up a 95% confidence interval estimate of the slope.
(m) The 28 teams constitute a population. In order to use statistical inference [as in (i)–(l)], the data must be assumed to represent a random sample. What "population" would this sample be drawing conclusions about?
(n) What other independent variables might be considered for inclusion in the model?

Case Study — PREDICTING SUNDAY NEWSPAPER CIRCULATION

You are employed in the marketing department of a large nationwide newspaper chain. The parent company is interested in investigating the feasibility of beginning a Sunday edition for some of its newspapers. However, before proceeding with a final decision, it needs to estimate the amount of Sunday circulation that would be expected. In particular, it wishes to predict the Sunday circulation that would be obtained by newspapers (in three different cities) that have daily circulations of 200,000, 400,000, and 600,000, respectively.

You have been asked to develop a model that would enable you to make a prediction of the expected Sunday circulation and to write a report that presents your results and summarizes your findings. Toward this end, data collected from a sample of 32 newspapers are as follows:

| PAPER | CIRCULATION (IN 000) | | PAPER | CIRCULATION (IN 000) | |
	SUNDAY	DAILY		SUNDAY	DAILY
Des Moines Register	278,803	164,659	Long Island Newsday	646,446	559,233
Philadelphia Inquirer	865,989	422,829	San Diego Union		
New York Times	1,644,128	1,107,168	Tribune	456,494	383,263
New York News	974,034	619,032	Chicago Sun Times	438,337	491,143
Sacramento Bee	353,366	285,762	Minneapolis Star		
Los Angeles Times	1,361,988	1,068,812	Tribune	673,264	355,743
Boston Globe	751,377	466,317	Baltimore Sun	471,637	326,636
Cincinnati Enquirer	322,238	202,973	Pittsburgh Post Gazette	437,864	241,798
Miami Herald	492,235	362,184	Rocky Mountain News	415,962	326,189
Chicago Tribune	1,045,756	664,586	Boston Herald	193,462	285,930
Detroit News	789,666	236,246	New Orleans		
Houston Chronicle	740,952	549,856	Times–Picayune	304,991	265,820
Kansas City Star	415,918	278,394	Charlotte Observer	301,026	238,216
Omaha World			Hartford Courant	303,191	217,759
Herald	291,764	234,106	Rochester Democrat		
Denver Post	474,668	353,786	and Chronicle	246,520	147,331
St. Louis			St. Paul Pioneer Press	264,732	202,922
Post–Dispatch	539,421	318,994	Providence		
Portland Oregonian	440,096	353,745	Journal–Bulletin	243,643	168,368
Washington Post	1,123,305	818,231			

DATA FILE
NEWSCIRC.XLS

Source: From GALE DIRECTORY OF PUBLICATIONS, Circulation of 32 Newsmakers edited by Carolyn Fischer, 1998, 131th edition. Copyright © 1998 by Gale Research. All rights reserved. Reprinted by permission of Gale Research, 1-800-877-4253.

THE SPRINGVILLE HERALD CASE

In the implementation of the corporate strategic initiative of increasing home-delivery sales, the marketing department needs to work closely with the distribution department to accomplish a smooth initial delivery process for trial customers. This is of great importance in the effort to ensure that as many trial customers as possible are converted to long-term customers because a strong negative impression will be created by any problems that occur during the first week of newspaper delivery.

As part of its role in this process, it is important for the marketing department to be able to forecast the number of new subscribers in future months. A team consisting of managers from the marketing and distribution departments was convened to develop a better method of forecasting new subscriptions. Melissa Hogue, the marketing department head, asked Lauren Hall, who specializes in market forecasting, to provide some ideas about how the forecasting methods used could be improved. Lauren, who was recently hired by the company to provide special skills in quantitative forecasting methods, asked the team how the forecasting of new subscriptions had been done in the past. Al Baum, a member of the team, answered that usually after examining new sub-

scriptions in the previous 2 or 3 months, a group of three managers developed a consensus on what the final forecast should be. Lauren asked whether anyone had tried to determine what factors might be useful in helping predict monthly new subscriptions. Al replied that the forecasts in the last year had been particularly inaccurate because in some months a great deal of time had been spent on telemarketing and in other months less effort was made. Lauren suggested that data for the past 2 years be obtained from company records. She was particularly interested in obtaining data for the number of new subscriptions and the number of hours spent on telemarketing for new subscriptions for each month. The following table indicates the number of new subscriptions for the month and the number of hours spent on telemarketing for new subscriptions.

Number of new subscriptions and number of hours spent on telemarketing per month for 2-year time period

TIME PERIOD	TELEMARKETING HOURS	NEW SUBSCRIPTIONS
1	1,224	5,357
2	1,458	6,177
3	1,006	4,795
4	1,395	5,692
5	1,131	4,312
6	921	3,421
7	704	2,624
8	1,154	4,087
9	1,168	4,934
10	803	2,546
11	830	3,591
12	981	4,271
13	1,435	5,836
14	1,349	5,201
15	965	3,775
16	985	3,592
17	1,117	4,566
18	840	2,974
19	1,412	5,673
20	940	3,554
21	1,090	4,399
22	1,498	6,143
23	1,240	4,827
24	1,055	5,418

DATA FILE
SH13.XLS

Exercises

13.1 What criticism can you make concerning the method of forecasting that involved taking the new subscriptions for the last 3 months as the basis for future projections?

13.2 What factors other than number of telemarketing hours spent might be useful in predicting the number of new subscriptions? Explain.

13.3 **(a)** Analyze the data and develop a statistical model to predict the average number of new subscriptions for a month based on the number of hours spent on telemarketing for new subscriptions. Write a report giving detailed findings concerning the model that has been fit to the data.

(b) If there are expected to be 1,000 hours spent on telemarketing in the coming month, predict the average number of new subscriptions expected for the month. Indicate the assumptions upon which this prediction is based. Do you think these assumptions are valid? Explain.

(c) What would be the danger of predicting the average number of new subscriptions for a month in which 2,000 hours are spent on telemarketing? Explain.

References

1. Anscombe, F. J., "Graphs in Statistical Analysis," *American Statistician* 27 (1973): 17–21.

2. Hoaglin, D. C., and R. Welsch, "The Hat Matrix in Regression and ANOVA," *The American Statistician* 32 (1978): 17–22.

3. Hocking, R. R., "Developments in Linear Regression Methodology: 1959–1982," *Technometrics* 25 (1983): 219–250.

4. Hosmer, D., and S. Lemeshow, *Applied Logistic Regression* (New York: Wiley, 1989).

5. *Microsoft Excel 97* (Redmond, WA: Microsoft Corp., 1997).

6. Neter, J., M. H. Kutner, C. J. Nachtsheim, and W. Wasserman, *Applied Linear Statistical Models,* 4th ed. (Homewood, IL: Irwin, 1996).

7. Ramsey, P. P., and P. H. Ramsey, "Simple Tests of Normality in Small Samples," *Journal of Quality Technology* 22 (1990): 299–309.

Chapter 8

Student Solutions Manual

13.2 (a) When $X = 0$, the expected value of Y is 16.

(b) For increase in the value X by 1 unit, we can expect a decrease in 0.5 units in the value of Y.

(c) $\hat{Y} = 16 - 0.5X = 16 - 0.5(6) = 13$

13.4 (a)

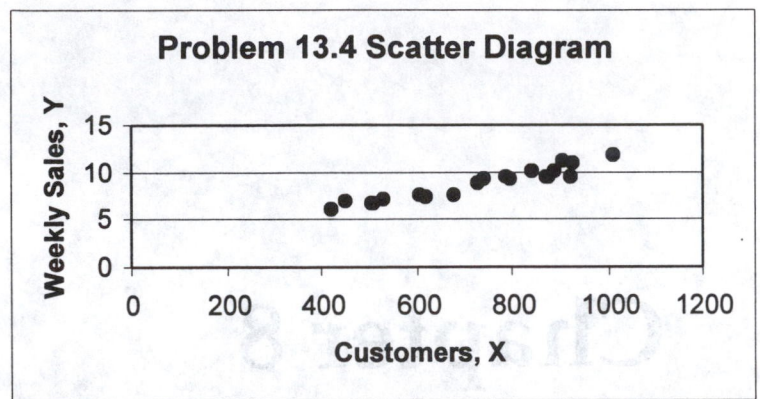

(c)

	Package Delivery Stores					
	Regression Statistics					
	Multiple R	0.9549132				
	R Square	0.91185922				
	Adjusted R Square	0.90696251				
	Standard Error	0.501495215				
	Observations	20				
	ANOVA					
		df	*SS*	*MS*	*F*	*Significance F*
	Regression	1	46.8335409	46.8335409	186.2187504	6.20621E-11
	Residual	18	4.526954104	0.25149745		
	Total	19	51.360495			
		Coefficients	*Standard Error*	*t Stat*	*P-value*	*Lower 95%*	*Upper 95%*
	Intercept	2.423044396	0.480964609	5.037885009	8.55388E-05	1.412574465	3.433514326
	Customers	0.008729338	0.00063969	13.64619912	6.20621E-11	0.007385398	0.010073278

For each increase of one additional customer, there is an expected increase in weekly sales of 0.00873 thousands of dollars, or $8.73.

(d) $\hat{Y} = 2.423 + 0.00873X = 2.423 + 0.00873(600) = 7.661$, or $7661

(e) $b_0 = 1.578$, $b_1 = 0.01009$

For each increase in shelf space of an additional foot, there is an expected increase in weekly sales of 0.01009 thousands of dollars, or $10.09.

$\hat{Y} = 1.578 + 0.01009X = 1.578 + 0.01009(600) = 7.632$, or $7632

13.6

Predicting Videos Sales						
Regression Statistics						
Multiple R	0.853088695					
R Square	0.727760321					
Adjusted R Square	0.718037476					
Standard Error	47.8667885					
Observations	30					
ANOVA						
	df	*SS*	*MS*	*F*	*Significance F*	
Regression	1	171499.778	171499.778	74.85054744	2.1259E-09	
Residual	28	64154.42435	2291.229441			
Total	29	235654.2023				
	Coefficients	*Standard Error*	*t Stat*	*P-value*	*Lower 95%*	*Upper 95%*
Intercept	76.53514239	11.83184172	6.468573886	5.23655E-07	52.29868609	100.7715987
Gross	4.333108105	0.500843491	8.651621088	2.1259E-09	3.30717557	5.359040641

(c) $\hat{Y} = 76.54 + 4.3331X$

(d)   For each increase of 1 million dollars in box office gross, expected home video units sold are estimated to increase by 4.3331 thousand, or 4333.1 units. 76.54 represents the portion of thousands of home video units that are not affected by box office gross.

(e) $\hat{Y} = 76.54 + 4.3331X = 76.54 + 4.3331(20) = 163.202$ or 163,202 units.

13.8

Predicting Travel Time						
Regression Statistics						
Multiple R	0.95793809					
R Square	0.917645385					
Adjusted R Square	0.909409923					
Standard Error	2.579237234					
Observations	12					
ANOVA						
	df	*SS*	*MS*	*F*	*Significance F*	
Regression	1	741.2582196	741.2582196	111.4261033	9.66059E-07	
Residual	10	66.52464708	6.652464708			
Total	11	807.7828667				
	Coefficients	*Standard Error*	*t Stat*	*P-value*	*Lower 95%*	*Upper 95%*
Intercept	3.374947947	2.610523714	1.292824091	0.225142125	-2.441662371	9.191558265
Distance	1.461229251	0.1384283	10.55585635	9.66059E-07	1.152791724	1.769666779

(c)   For each increase of one mile in distance, the expected travel time is estimated to increase by 1.46 minutes. The Y-intercept 3.375 represents the portion of the travel time that is not affected by distance.

(d) $\hat{Y} = 3.375 + 1.46X = 3.375 + 1.46(21) = 34.035$ minutes

(e) $b_0 = 11.497$, $b_1 = 0.9673$

For each increase of one mile in distance, the expected travel time is estimated to increase by 0.9673 minutes. The Y-intercept 11.497 represents the portion of the travel time that is not affected by distance.

$\hat{Y} = 11.497 + 0.9673X = 11.497 + 0.9673(21) = 31.81$ minutes

•13.10 $SST = 40$ and $r^2 = 0.90$. So, 90% of the variation in the dependent variable can be explained by the variation in the independent variable.

13.12 $r^2 = 0.333$. So, 33.3% of the variation in the dependent variable can be explained by the variation in the independent variable.

•13.14 (a) $r^2 = 0.684$. So, 68.4% of the variation in the dependent variable can be explained by the variation in the independent variable.
       (b) $s_{YX} = 0.308$
       (c)   Based on (a) and (b), the model should be very useful for predicting sales.

•13.16 (a) $r^2 = 0.988$. So, 98.8% of the variation in the dependent variable can be explained by the variation in the independent variable.
       (b) $s_{YX} = 4.71$
       (c)   Based on (a) and (b), the model should be very useful for predicting sales.

13.18 (a) $r^2 = 0.723$. So, 72.3% of the variation in the dependent variable can be explained by the variation in the independent variable.
       (b) $s_{YX} = 194.6$
       (c)   Based on (a) and (b), the model should be very useful for predicting monthly rent.

•13.20   A residual analysis of the data indicates no apparent pattern. The assumptions of regression appear to be met.

•13.22 (a)-(b) Based on a residual analysis, the model appears to be adequate.

•13.24 (a)-(b) Based on a residual analysis, the model appears to be adequate.

13.26 (a)-(b) Based on a residual analysis of the studentized residuals versus size, the model appears to be adequate.

•13.28 (a)   An increasing linear relationship exists.
       (b) $D = 0.109$
       (c)   There is strong positive autocorrelation among the residuals.

•13.30 (a)   No, since the data have been collected for a single period for a set of stores.
       (b)   If a single store was studied over a period of time and the amount of shelf space varied over time, computation of the Durbin-Watson statistic would be necessary.

13.32

Regression Analysis		Durbin-Watson Calculations				
Regression Statistics		Sum of Squared Difference of Residuals	1243.22442			
Multiple R	0.91880399	Sum of Squared Residuals	599.068347			
R Square	0.844200772	Durbin-Watson Statistic	2.07526308			
Adjusted R Square	0.837118989					
Standard Error	5.218273602					
Observations	24					
ANOVA						
	df	SS		MS	F	Significance F
Regression	1		3246.062049	3246.062049	119.2073751	2.38511E-10
Residual	22		599.0683465	27.23037939		
Total	23		3845.130396			
	Coefficients	Standard Error		t Stat	P-value	
Intercept	0.457625305	6.571882688		0.069633821	0.945114194	
Orders	0.016117564	0.001476209		10.918213	2.38511E-10	

(b) $b_0 = 0.458$, $b_1 = 0.0161$

(c)   For each increase of one order, the expected distribution cost is estimated to increase by 0.0161 thousand dollars, or \$16.10.

(d) $\hat{Y} = 0.458 + 0.0161X = 0.458 + 0.0161(4500) = 72.908$ or \$72,908

(e) $r^2 = 0.844$. So, 84.4% of the variation in distribution cost can be explained by the variation in the number of orders.

(f) $s_{YX} = 5.218$

(i) $D = 2.08 > 1.45$. There is no evidence of positive autocorrelation among the residuals.

(j)   Based on a residual analysis, the model appears to be adequate.

•13.34 (a) $t = b_1 / s_{b_1} = 4.5/1.5 = 3.00$

(b)   With n = 18, df = 18 − 2 = 16. $t_{16} = \pm 2.1199$

(c)   Reject H_0. There is evidence that the fitted linear regression model is useful.

(d) $b_0 - t_{16} s_{b_1} \le \beta_1 \le b_0 + t_{16} s_{b_1}$, $4.5 - 2.1199(1.5) \le \beta_1 \le 4.5 + 2.1199(1.5)$, $1.32 \le \beta_1 \le 7.68$

•13.36 (a) $t = 4.65 > t_{10} = 2.2281$ with 10 degrees of freedom for $\alpha = 0.05$. Reject H_0. There is evidence that the fitted linear regression model is useful.

(b) $0.0386 \le \beta_1 \le 0.1094$

•13.38 (a) $t = 31.15 > t_{12} = 2.1788$ with 12 degrees of freedom for $\alpha = 0.05$. Reject H_0. There is evidence that the fitted linear regression model is useful.

(b) $1.8236 \le \beta_1 \le 2.0978$

13.40 (a) $t = 7.74 > t_{23} = 2.0687$ with 23 degrees of freedom for $\alpha = 0.05$. Reject H_0. There is evidence that the fitted linear regression model is useful.

(b) $0.7805 \le \beta_1 \le 1.3497$

13.42　(a)　When $X = 2$, $\hat{Y} = 5 + 3X = 5 + 3(2) = 11$

$$h = \frac{1}{n} + \frac{(X_i - \bar{X})^2}{\sum\limits_{i=1}^{n}(X_i - \bar{X})^2} = \frac{1}{20} + \frac{(2-2)^2}{20} = 0.05$$

95% confidence interval: $\hat{Y} \pm t_{18} s_{YX} \sqrt{h} = 11 \pm 2.1009 \cdot 1 \cdot \sqrt{0.05}$

$10.53 \leq \mu_{YX} \leq 11.47$

　　　(b)　95% prediction interval: $\hat{Y} \pm t_{18} s_{YX} \sqrt{1+h} = 11 \pm 2.1009 \cdot 1 \cdot \sqrt{1.05}$

$8.847 \leq Y_I \leq 13.153$

•13.44　(a)　$1.7867 \leq \mu_{YX} \leq 2.2964$

　　　(b)　$1.3100 \leq Y_I \leq 2.7740$

　　　(c)　Part (b) provides an estimate for an individual response and Part (a) provides an estimate for an average predicted value.

•13.46　(a)　$98.08 \leq \mu_{YX} \leq 103.74$

　　　(b)　$90.27 \leq Y_I \leq 111.56$

　　　(c)　Part (b) provides an estimate for an individual response and Part (a) provides an estimate for an average predicted value.

13.48　(a)　$1153.0 \leq \mu_{YX} \leq 1331.5$

　　　(b)　$829.9 \leq Y_I \leq 1654.6$

　　　(c)　Part (b) provides an estimate for an individual response and Part (a) provides an estimate for an average predicted value.

•13.50　$r = +\sqrt{r^2} = +\sqrt{81} = +0.9$

13.52　$r^2 = SSR/SST = 1.\ r = -\sqrt{r^2} = -\sqrt{1} = -1$

13.54　(a)　$SSX = 217.64$, $SSY = 1958.73$, $SSXY = 652.91$.

$$r = \frac{SSXY}{\sqrt{SSX}\sqrt{SSY}} = \frac{652.91}{\sqrt{217.64}\sqrt{1958.73}} = +1.0$$

　　　(b)　$H_0: \rho = 0, H_1: \rho \neq 0$.

The test statistic t is positive infinity since $SSE = 0$ and $r = 1$. Reject H_0. There is evidence of an association between X and Y.

•13.56 (a) $r = -0.666$

(b) $H_0 : \rho = 0, H_1 : \rho \neq 0$.

$t = -3.78 < -t_{18} = -2.1009$. Reject H_0. There is evidence of an association between average charge per minute and minutes expended.

(c)   Yes. As suspected, the correlation in the data is negative.

•13.58 (a) $r = +0.827$

(b) $t = 4.65 > t_{10} = 2.2281$. Reject H_0. There is evidence of an association between the two variables.

(c)   Except for possible rounding error, the results of the two t tests are identical.

•13.60 (a) $r = +0.994$

(b) $t = 31.15 > t_{12} = 2.1788$. Reject H_0. There is evidence of an association between the two variables.

(c)   Except for possible rounding error, the results of the two t tests are identical.

13.62 (a) $r = +0.850$

(b) $t = 7.74 > t_{23} = 2.0687$. Reject H_0. There is evidence of an association between the two variables.

(c)   Except for possible rounding error, the results of the two t tests are identical.

13.74

Predicting Trade Executions		Durbin-Watson Calculations			
Regression Statistics		Sum of Squared Difference of Res	56000.13401		
Multiple R	0.793770534	Sum of Squared Residuals	28560.83537		
R Square	0.63007166	Durbin-Watson Statistic	1.960731655		
Adjusted R Square	0.618861711				
Standard Error	29.41903907				
Observations	35				
ANOVA					
	df	*SS*	*MS*	*F*	*Significance F*
Regression	1	48645.56463	48645.56463	56.20646637	1.27826E-08
Residual	33	28560.83537	865.4798597		
Total	34	77206.4			
	Coefficients	*Standard Error*	*t Stat*	*P-value*	
Intercept	-63.02045762	54.59736729	-1.154276493	0.256677777	
Calls	0.189005684	0.025210515	7.497097196	1.27826E-08	

(b) $b_0 = -63.02, b_1 = 0.189$

(c) $\hat{Y} = -63.02 + 0.189X$, where X is the number of incoming calls and $\hat{Y}$ is the estimated number of trade executions.

(d)   For each additional incoming call, the estimated number of trade executions increases by 0.189 minutes. $- 63.02$ is the portion of the estimated delivery time that is not affected by the number of incoming calls.

(e) $\hat{Y} = -63.02 + 0.189X = -63.02 + 0.189(2000) = 314.99$

(f)   No, 5000 incoming calls is outside the relevant range of the data used to fit the regression equation.

(g) $r^2 = 0.630$. So, 63.0% of the variation in trade executions can be explained by the variation in the number of incoming calls.

13.74 (h)   Since b_1 is positive, $r = +\sqrt{r^2} = +\sqrt{0.63} = +0.794$

cont.   (i) $s_{YX} = 29.42$

    (j)   Based on a visual inspection of the graphs of the distribution of studentized residuals and the residuals versus the number of cases, there is no pattern. The model appears to be adequate.

    (k) $D = 1.96$

    (l) $D = 1.96 > 1.52$. There is no evidence of positive autocorrelation. The model appears to be adequate.

    (m) $t = 7.50 > t_{33} = 2.0345$ with 33 degrees of freedom for $\alpha = 0.05$. Reject H_0. There is evidence that the fitted linear regression model is useful.

    (n), (o)

	A	B
1	Confidence Interval Estimate	
2		
3	X Value	2000
4	Confidence Level	95%
5	Sample Size	35
6	Degrees of Freedom	33
7	t Value	2.03451691
8	Sample Mean	2156.657143
9	Sum of Squared Difference	1361737.89
10	Standard Error of the Estimate	29.41903907
11	h Statistic	0.04659359
12	Average Predicted Y (YHat)	314.9909096
13		
14		
15	For Average Predicted Y (YHat)	
16	Interval Half Width	12.91971327
17	Confidence Interval Lower Limit	302.0711964
18	Confidence Interval Upper Limit	327.9106229
19		
20	For Individual Response Y	
21	Interval Half Width	61.23205321
22	Prediction Interval Lower Limit	253.7588564
23	Prediction Interval Upper Limit	376.2229628

    (p) $0.1377 \le \beta_1 \le 0.2403$

•13.76

	A	B	C	D	E	F	G
1	Predicting Assessed Value						
2							
3	Regression Statistics						
4	Multiple R	0.811995685					
5	R Square	0.659336993					
6	Adjusted R Square	0.633132146					
7	Standard Error	2.918927722					
8	Observations	15					
9							
10	ANOVA						
11		df	SS	MS	F	Significance F	
12	Regression	1	214.3741924	214.3741924	25.16087956	0.00023616	
13	Residual	13	110.7618076	8.520139046			
14	Total	14	325.136				
15							
16		Coefficients	Standard Error	t Stat	P-value	Lower 95%	Upper 95%
17	Intercept	51.91533994	5.562520886	9.333059777	3.98327E-07	39.89824648	63.9324334
18	Heating Area	16.63336947	3.316021403	5.016062157	0.00023616	9.469542149	23.79719679

•13.76
cont.

(a) $b_0 = 51.915$, $b_1 = 16.633$

(b)   For each additional 1000 square feet in heating area, the estimated assessed value increases by \$16,633. \$51,915 is the portion of the estimated assessed value that is not affected by heating area.

(c) $\hat{Y} = 51.915 + 16.633X = 51.915 + 16.633(1.75) = 81.024$ or \$81,024

(d) $s_{YX} = 2.919$

(e) $r^2 = 0.659$. 65.9% of the variation in assessed value can be explained by the variation in heating area.

(f)   Since b_1 is positive, $r = +\sqrt{r^2} = +\sqrt{0.659} = +0.812$

(g)   Based on a visual inspection of the graphs of the distribution of studentized residuals and the residuals versus the heating area, there is no pattern. The model appears to be adequate.

(h) $t = 5.02 > t_{13} = 2.1604$ with 13 degrees of freedom for $\alpha = 0.05$. Reject H_0. There is evidence that the fitted linear regression model is useful.

(i) $79.279 \le \mu_{YX} \le 82.769$

(j) $74.479 \le Y_I \le 87.569$

(k) $9.469 \le \beta_1 \le 23.797$

(l) $b_0 = 52.805$, $b_1 = 15.849$

For each additional 1000 square feet in heating area, the estimated assessed value increases by \$15,849. \$52,805 is the portion of the estimated assessed value that is not affected by heating area.

$\hat{Y} = 52.805 + 15.849X = 52.805 + 15.849(1.75) = 80.541$ or \$80,541

$s_{YX} = 2.598$

$r^2 = 0.689$. 68.9% of the variation in assessed value can be explained by the variation in heating area.

Since b_1 is positive, $r = +\sqrt{r^2} = +\sqrt{0.689} = +0.83$

Based on a visual inspection of the graphs of the distribution of studentized residuals and the residuals versus the heating area, there is no pattern. The model appears to be adequate.

$t = 5.37 > t_{13} = 2.1604$ with 13 degrees of freedom for $\alpha = 0.05$. Reject H_0. There is evidence that the fitted linear regression model is useful.

$78.987 \le \mu_{YX} \le 82.096$

$74.716 \le Y_I \le 86.367$

$9.471 \le \beta_1 \le 22.227$

13.78 (a)

	Regression Statistics		Durbin-Watson Calculations			
Predicting Invoice Completion Time			Durbin-Watson Calculations			
			Sum of Squared Difference of Residuals	5.57539132		
Multiple R	0.944668576		Sum of Squared Residuals	3.12818443		
R Square	0.892398719		Durbin-Watson Statistic	1.78230902		
Adjusted R Square	0.888555816					
Standard Error	0.334246724					
Observations	30					

ANOVA

	df	SS		MS	F	Significance F
Regression	1		25.94381557	25.94381557	232.2199511	4.3946E-15
Residual	28		3.128184432	0.111720873		
Total	29		29.072			

	Coefficients	Standard Error		t Stat	P-value	
Intercept	0.402374805		0.123582495	3.255920694	0.002954616	
Invoices Processed	0.012606814		0.000827286	15.23876475	4.3946E-15	

(b) $b_0 = 0.4024$, $b_1 = 0.012608$

(c)   For each additional invoice processed, the estimated completion time increases by 0.012608 hours. 0.4024 is the portion of the estimated completion time that is not affected by the number of invoices processed.

(d) $\hat{Y} = 0.4024 + 0.012608X = 0.4024 + 0.012608(150) = 2.2934$

(e) $s_{YX} = 0.3342$

(f) $r^2 = 0.892$. 89.2% of the variation in completion time can be explained by the variation in the number of invoices processed.

(g)   Since b_1 is positive, $r = +\sqrt{r^2} = +\sqrt{0.892} = +0.945$

(i)   Based on a visual inspection of the graphs of the distribution of studentized residuals and the residuals versus the number of invoices, there is no pattern. The model appears to be adequate.

(j) $D = 1.78$

(k) $D = 1.78 > 1.49$. There is no evidence of positive autocorrelation. The model appears to be adequate.

(l) $t = 15.24 > t_{28} = 2.0484$ with 28 degrees of freedom for $\alpha = 0.05$. Reject H_0. There is evidence that the fitted linear regression model is useful.

(m), (n)

Confidence Interval Estimate	
X Value	150
Confidence Level	95%
Sample Size	30
Degrees of Freedom	28
t Value	2.048409442
Sample Mean	129.9
Sum of Squared Difference	163238.70
Standard Error of the Estimate	0.334246724
h Statistic	0.035808298
Average Predicted Y (YHat)	2.29339697
For Average Predicted Y (YHat)	
Interval Half Width	0.12956144
Confidence Interval Lower Limit	2.16383553
Confidence Interval Upper Limit	2.422958411
For Individual Response Y	
Interval Half Width	0.696824836
Prediction Interval Lower Limit	1.596572134
Prediction Interval Upper Limit	2.990221807

Chapter 9
Multiple Regression Models

CHAPTER OBJECTIVES

✓ *To develop the multiple regression model as an extension of the simple linear regression model*
✓ *To evaluate the contribution of each independent variable to the regression model*
✓ *To measure the coefficient of partial determination*
✓ *To develop the curvilinear regression model*
✓ *To introduce categorical explanatory (dummy) variables into the regression model*
✓ *To examine regression models that involve transformations of a variable*
✓ *To show how stepwise regression or best-subsets regression can be used to build and select a regression model*

Introduction

In our discussion of simple regression in chapter 13, we focused on a model in which one independent or explanatory variable X was used to predict the value of a dependent or response variable Y. It is often the case that a better-fitting model can be developed if more than one explanatory variable is considered. Thus, in this chapter we extend our discussion to consider **multiple regression** models in which several explanatory variables can be used to predict the value of a dependent variable.

◆ **USING STATISTICS:** *Predicting Home Heating Oil Usage*

One of the important factors for determining the monthly costs of maintaining a single-family house in cold weather climates is the cost of heating fuel. In many parts of the United States the predominant fuel used for heating single-family houses is heating oil. Suppose that a developer wanted to investigate the factors involved in heating oil usage in an attempt to assist house purchasers in forecasting their monthly heating expenses. Certainly many factors could be considered. The most important factor is the average monthly atmospheric temperature where the house is located, but other factors could affect usage as well. Among these factors are the size of the house, the style of the house, the amount of insulation that the house has in the attic, the age of the oil burner, the temperature setting of the thermostatic control, and whether the thermostatic control can be changed automatically at various times during a 24-hour period.

In the interest of simplicity, suppose a real estate developer decides to collect data from similar houses that have been built in various locations and focuses on two factors, the average monthly atmospheric temperature and the amount of attic insulation.

14.1 DEVELOPING THE MULTIPLE REGRESSION MODEL

A sample of 15 houses with similar square footage built by a particular housing developer throughout the United States is selected and the heating oil consumption during the month of January is determined. Two independent variables are considered here—the average daily atmospheric temperature, as measured in degrees Fahrenheit, outside the house during that month (X_1) and the amount of insulation, as measured in inches, in the attic of the house (X_2). The results are presented in Table 14.1.

Table 14.1 *Monthly heating oil consumption, atmospheric temperature, and amount of attic insulation for random sample of 15 single-family houses*

HOUSE	MONTHLY CONSUMPTION OF HEATING OIL (GALLONS)	AVERAGE DAILY ATMOSPHERIC TEMPERATURE (°F)	AMOUNT OF ATTIC INSULATION (INCHES)	HOUSE	MONTHLY CONSUMPTION OF HEATING OIL (GALLONS)	AVERAGE DAILY ATMOSPHERIC TEMPERATURE (°F)	AMOUNT OF ATTIC INSULATION (INCHES)
1	275.3	40	3	9	237.8	23	10
2	363.8	27	3	10	121.4	63	3
3	164.3	40	10	11	31.4	65	10
4	40.8	73	6	12	203.5	41	6
5	94.3	64	6	13	441.1	21	3
6	230.9	34	6	14	323.0	38	3
7	366.7	9	6	15	52.5	58	10
8	300.6	8	10				

With two explanatory variables in the multiple regression model, a scatter diagram of the points can be plotted on a three-dimensional graph as shown in Figure 14.1.

DATA FILE
HTNGOIL.XLS

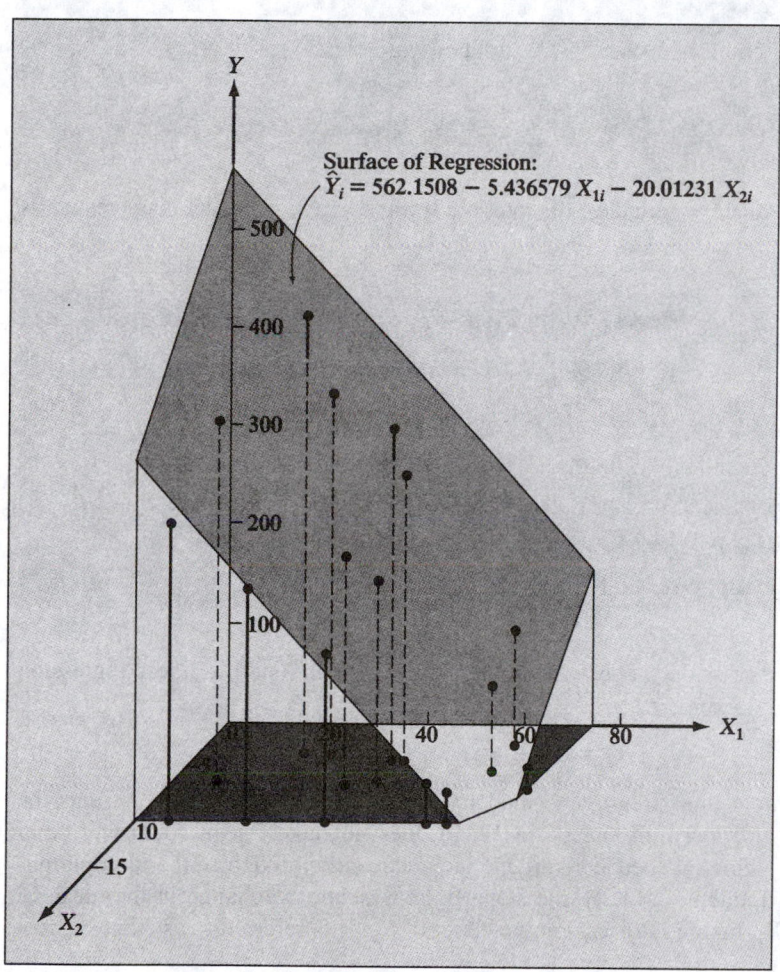

Surface of Regression:
$\hat{Y}_i = 562.1508 - 5.436579 X_{1i} - 20.01231 X_{2i}$

FIGURE 14.1
Scatter diagram of average daily atmospheric temperature X_1, amount of attic insulation X_2, and monthly heating oil consumption Y with indicated regression plane fitted by least-squares method

Interpreting the Regression Coefficients

When there are several explanatory variables present, the simple linear regression model of equation (13.1) on page 773 can be extended by assuming a linear relationship between each explanatory variable and the dependent variable. For example, with p explanatory variables, the multiple linear regression model is expressed as

---

Multiple Regression Model with p Independent Variables

$$Y_i = \beta_0 + \beta_1 X_{1i} + \beta_2 X_{2i} + \beta_3 X_{3i} + \cdots + \beta_p X_{pi} + \epsilon_i \qquad (14.1)$$

where

$\beta_0 = Y$ intercept

$\beta_1 = $ slope of Y with variable X_1 holding variables $X_2, X_3, \ldots, X_p$ constant

$\beta_2 = $ slope of Y with variable X_2 holding variables $X_1, X_3, \ldots, X_p$ constant

$\beta_3 = $ slope of Y with variable X_3 holding variables $X_1, X_2, \ldots, X_p$ constant

$\beta_p = $ slope of Y with variable X_p holding variables $X_1, X_2, X_3, \ldots, X_{p-1}$ constant

$\epsilon_i = $ random error in Y for observation i

For data with two explanatory variables, the multiple linear regression model is expressed as

---

Multiple Regression Model with Two Independent Variables

$$Y_i = \beta_0 + \beta_1 X_{1i} + \beta_2 X_{2i} + \epsilon_i \qquad (14.2)$$

where

$\beta_0 = Y$ intercept

$\beta_1 = $ slope of Y with variable X_1 holding variable X_2 constant

$\beta_2 = $ slope of Y with variable X_2 holding variable X_1 constant

$\epsilon_i = $ random error in Y for observation i

This multiple linear regression model can be compared with the simple linear regression model [equation (13.1)] expressed as

$$Y_i = \beta_0 + \beta_1 X_i + \epsilon_i$$

In the case of the simple linear regression model we note that the slope β_1 represents the change in the mean of Y per unit change in X and does not take into account any other variables besides the single independent variable included in the model. In the multiple linear regression model [equation (14.2)], the slope β_1 represents the change in the mean of Y per unit change in X_1, taking into account the effect of X_2. It is referred to as a **net regression coefficient.**

As in the case of simple linear regression, the sample regression coefficients (b_0, b_1, and b_2) are used as estimates of the true parameters (β_0, β_1, and β_2). Thus, the sample regression equation for a multiple linear regression model with two explanatory variables is

Multiple Linear Regression Equation with Two Independent Variables

$$\hat{Y}_i = b_0 + b_1X_{1i} + b_2X_{2i} \qquad (14.3)$$

Using the least-squares method, we obtain the values of the three sample regression coefficients from Microsoft Excel. Figure 14.2 presents partial output for the monthly heating oil consumption data from Microsoft Excel.

	A	B	C	D	E	F	G
1	SUMMARY OUTPUT						
2							
3	Regression Statistics						
4	Multiple R	0.982654757					
5	R Square	0.965610371					
6	Adjusted R Square	0.959878766					
7	Standard Error	26.01378323					
8	Observations	15	SSR				
9							
10	ANOVA	SST SSE					
11		df	SS	MS	F	Significance F	
12	Regression	2	228014.6263	114007.3132	168.4712028	1.65411E-09	
13	Residual	12	8120.603016	676.716918			
14	Total	14	236135.2293				
15							
16	b_1	Coefficients	Standard Error	t Stat	P-value	Lower 95%	Upper 95%
17	Intercept	562.1510092	21.09310433	26.65093769	4.77868E-12	516.1930837	608.1089348
18	Temperature b_2	-5.436580588	0.336216167	-16.16989642	1.64178E-09	-6.169132673	-4.704028503
19	Insulation	-20.01232067	2.342505227	-8.543127434	1.90731E-06	-25.11620102	-14.90844031

	A	B	C	D
23	RESIDUAL OUTPUT			
24				
25	Observation	Predicted Heating Oil	Residuals	Standard Residuals
26	1	284.6508237	-9.350823711	-0.388257405
27	2	355.3263714	8.473628646	0.351835215
28	3	144.564579	19.73542095	0.819438325
29	4	45.20670231	-4.406702309	-0.18297156
30	5	94.1359276	0.164072399	0.006812483
31	6	257.2333452	-26.33334524	-1.093392047
32	7	393.1478599	-26.44785994	-1.098146835
33	8	318.5351579	-17.93515786	-0.744689245
34	9	236.986449	0.813550957	0.033779611
35	10	159.6094702	-38.20947019	-1.586502985
36	11	8.650064348	22.74993565	0.944604587
37	12	219.1772811	-15.67728112	-0.650939497
38	13	387.9458549	53.15414512	2.207023795
39	14	295.5239849	27.47601511	1.14083707
40	15	46.70612846	5.793871536	0.240568488

FIGURE 14.2 Partial output obtained from Microsoft Excel for monthly heating oil consumption data

From Figure 14.2 we observe that the computed values of the regression coefficients are

$$b_0 = 562.151 \qquad b_1 = -5.43658 \qquad b_2 = -20.0123$$

Therefore, the multiple regression equation can be expressed as

$$\hat{Y}_i = 562.151 - 5.43658X_{1i} - 20.0123X_{2i}$$

where

$\hat{Y}_i$ = predicted average amount of heating oil consumed (gallons) during January for house i

X_{1i} = average daily atmospheric temperature (°F) during January for house i

X_{2i} = amount of attic insulation (inches) for house i

The Y intercept b_0, computed as 562.151, estimates the expected number of gallons of heating oil that would be consumed in January when the average daily atmospheric temperature is 0°F for a house that is not insulated (a house with 0 inches of attic insulation).

The slope of average daily atmospheric temperature with heating oil consumption (b_1, computed as -5.43658) means that, for a house with a given number of inches of attic insulation, the expected heating oil consumption is estimated to decrease by 5.43658 gallons per month for each 1°F increase in average daily atmospheric temperature. The slope of amount of attic insulation with heating oil consumption (b_2, computed as -20.0123) means that, for a month with a given average daily atmospheric temperature, the expected heating oil consumption is estimated to decrease by 20.0123 gallons for each additional inch of attic insulation.

COMMENT: *Interpreting the Slopes in Multiple Regression*

We have stated that the regression coefficients in multiple regression are net regression coefficients that measure the average change in Y per unit change in a particular X holding constant the effect of the other X variables. For example, in our study of heating oil consumption we have stated that, for a house with a given number of inches of attic insulation, the expected heating oil consumption is estimated to decrease by 5.43658 gallons per month for each 1°F increase in average daily atmospheric temperature. Another way to interpret this is to think of similar houses with an equal amount of attic insulation located in different geographic areas. For such houses (with the same attic insulation), expected heating oil consumption is predicted to decrease by 5.43658 gallons per month for each 1°F increase in average daily atmospheric temperature.

In a similar manner, the slope of heating oil consumption with attic insulation can be viewed from the perspective of two similar houses (except for the amount of attic insulation) located right next to each other. For these houses, the expected heating oil consumption is estimated to decrease by 20.0123 gallons for each additional inch of attic insulation. It is this conditional nature of the interpretation that is critical to understanding the magnitude of each slope. Otherwise, we could just compare the slopes.

Predicting the Dependent Variable Y

Now that the multiple regression model has been fitted to these data, we can predict the average monthly consumption of home heating oil and develop confidence and prediction interval estimates assuming that the regression model fitted is an appropriate one.

Suppose we want to predict the average number of gallons of heating oil consumed in a house that has 6 inches of attic insulation during a month in which the average daily atmospheric temperature is 30°F. Using our multiple regression equation

$$\hat{Y}_i = 562.151 - 5.43658X_{1i} - 20.0123X_{2i}$$

with $X_{1i} = 30$ and $X_{2i} = 6$, we have

$$\hat{Y}_i = 562.151 - (5.43658)(30) - (20.0123)(6)$$

and thus

$$\hat{Y}_i = 278.9798$$

Therefore, we estimate that an average of 278.98 gallons of heating oil would be used in houses with 6 inches of insulation when the average temperature is 30°F.

Coefficients of Multiple Determination

You may recall from section 13.3 that once a regression model has been developed, we can compute the coefficient of determination r^2. In multiple regression, because there are at least two explanatory variables, the **coefficient of multiple determination** represents the proportion of the variation in Y that is explained by the set of explanatory variables selected. For data with two explanatory variables, the coefficient of multiple determination $(r^2_{Y.12})$ is given by

The Coefficient of Multiple Determination

The coefficient of multiple determination is equal to the regression sum of squares divided by the total sum of squares.

$$r^2_{Y.12} = \frac{SSR}{SST} \tag{14.4}$$

where

$$SSR = \text{regression sum of squares}$$
$$SST = \text{total sum of squares}$$

In the heating oil consumption example, from Figure 14.2 on page 857, $SSR = 228,015$ and $SST = 236,135$ (rounded). Thus,

$$r^2_{Y.12} = \frac{SSR}{SST} = \frac{228,015}{236,135} = .9656$$

This coefficient of multiple determination, computed as .9656, means that 96.56% of the variation in home heating oil consumption can be explained by the variation in the average daily atmospheric temperature and the variation in the amount of attic insulation.

However, when dealing with multiple regression models, some researchers suggest that an **adjusted r^2** be computed to reflect both the number of explanatory variables in the model and the sample size. This is especially necessary when we are comparing two or more regression models that predict the same dependent variable but have different numbers of explanatory or predictor variables. Thus, the adjusted r^2 is

Adjusted r^2

$$r^2_{adj} = 1 - \left[(1 - r^2_{Y.12 \dots p}) \frac{n-1}{n-p-1} \right] \tag{14.5}$$

where p = number of explanatory variables in the regression equation.

Thus, for our monthly heating oil consumption data, because $r_{Y.12}^2 = .9656$, $n = 15$, and $p = 2$,

$$r_{adj}^2 = 1 - \left[(1 - r_{Y.12}^2) \frac{(15 - 1)}{(15 - 2 - 1)} \right]$$

$$= 1 - \left[(1 - .9656) \frac{14}{12} \right]$$

$$= 1 - .04$$

$$= .96$$

Hence, 96% of the variation in monthly heating oil consumption can be explained by our multiple regression model—adjusted for number of predictors and sample size.

Problems for Section 14.1

Learning the Basics

• **14.1** Suppose you have obtained the following multiple regression model:

$$\hat{Y}_i = 10 + 5X_{1i} + 3X_{2i} \quad \text{and} \quad r_{Y.12}^2 = .60$$

(a) Interpret the meaning of the slopes.
(b) Interpret the meaning of the Y intercept.
(c) Interpret the meaning of the coefficient of multiple determination $r_{Y.12}^2$.

14.2 Suppose you have obtained the following multiple regression model:

$$\hat{Y}_i = 50 - 2X_{1i} + 7X_{2i} \quad \text{and} \quad r_{Y.12}^2 = .40$$

(a) Interpret the meaning of the slopes.
(b) Interpret the meaning of the Y intercept.
(c) Interpret the meaning of the coefficient of multiple determination $r_{Y.12}^2$.

Applying the Concepts

• **14.3** A marketing analyst for a major shoe manufacturer is considering the development of a new brand of running shoes. The marketing analyst wishes to determine which variables can be used in predicting durability (or the effect of long-term impact). Two independent variables are to be considered, X_1 (FOREIMP), a measurement of the forefoot shock-absorbing capability, and X_2 (MIDSOLE), a measurement of the change in impact properties over time, along with the dependent variable Y (LTIMP), which is a measure of the long-term ability to absorb shock after a repeated impact test. A random sample of 15 types of currently manufactured running shoes was selected for testing. Using Microsoft Excel, we provide the following (partial) output:

ANOVA	DF	SS	MS	F	SIGNIFICANCE F
Regression	2	12.61020	6.30510	97.69	0.0001
Residual	12	0.77453	0.06454		
Total	14	13.38473			

VARIABLE	COEFFICIENTS	STANDARD ERROR	t STAT	p-VALUE
Intercept	−0.02686	.06905	−0.39	
Foreimp	0.79116	.06295	12.57	.0000
Midsole	0.60484	.07174	8.43	.0000

(a) Assuming that each independent variable is linearly related to long-term impact, state the multiple regression equation.

(b) Interpret the meaning of the slopes in this problem.
(c) Compute the coefficient of multiple determination $r_{Y.12}^2$ and interpret its meaning.
(d) Compute the adjusted r^2.

14.4 A mail-order catalog business selling personal computer supplies, software, and hardware maintains a centralized warehouse for the distribution of products ordered. Management is currently examining the process of distribution from the warehouse and is interested in studying the factors that affect warehouse distribution costs. Currently, a small handling fee is added to the order, regardless of the amount of the order. Data have been collected over the past 24 months indicating the warehouse distribution costs, the sales, and the number of orders received. The results are as follows:

Distribution cost data

MONTH	DISTRIBUTION COST ($000)	SALES ($000)	ORDERS	MONTH	DISTRIBUTION COST ($000)	SALES ($000)	ORDERS
1	52.95	386	4,015	13	62.98	372	3,977
2	71.66	446	3,806	14	72.30	328	4,428
3	85.58	512	5,309	15	58.99	408	3,964
4	63.69	401	4,262	16	79.38	491	4,582
5	72.81	457	4,296	17	94.44	527	5,582
6	68.44	458	4,097	18	59.74	444	3,450
7	52.46	301	3,213	19	90.50	623	5,079
8	70.77	484	4,809	20	93.24	596	5,735
9	82.03	517	5,237	21	69.33	463	4,269
10	74.39	503	4,732	22	53.71	389	3,708
11	70.84	535	4,413	23	89.18	547	5,387
12	54.08	353	2,921	24	66.80	415	4,161

DATA FILE
WARECOST.XLS

On the basis of the results obtained:
(a) State the multiple regression equation.
(b) Interpret the meaning of the slopes in this problem.
(c) Predict the average monthly warehouse distribution costs when sales are $400,000 and the number of orders is 4,500.
(d) Compute the coefficient of multiple determination $r_{Y.12}^2$ and interpret its meaning.
(e) Compute the adjusted r^2.

14.5 Suppose a consumer organization wanted to develop a model to predict gasoline mileage as measured by miles per gallon (MPG) based on the horsepower of the car's engine and the weight of the car. A sample of 50 recent car models was selected with the following results:

MPG	HORSEPOWER	WEIGHT	MPG	HORSEPOWER	WEIGHT
43.1	48	1,985	23.9	90	3,420
19.9	110	3,365	29.9	65	2,380
19.2	105	3,535	30.4	67	3,250
17.7	165	3,445	36.0	74	1,980
18.1	139	3,205	22.6	110	2,800
20.3	103	2,830	36.4	67	2,950
21.5	115	3,245	27.5	95	2,560
16.9	155	4,360	33.7	75	2,210
15.5	142	4,054	44.6	67	1,850

(continued)

MPG	HORSEPOWER	WEIGHT	MPG	HORSEPOWER	WEIGHT
18.5	150	3,940	32.9	100	2,615
27.2	71	3,190	38.0	67	1,965
41.5	76	2,144	24.2	120	2,930
46.6	65	2,110	38.1	60	1,968
23.7	100	2,420	39.4	70	2,070
27.2	84	2,490	25.4	116	2,900
39.1	58	1,755	31.3	75	2,542
28.0	88	2,605	34.1	68	1,985
24.0	92	2,865	34.0	88	2,395
20.2	139	3,570	31.0	82	2,720
20.5	95	3,155	27.4	80	2,670
28.0	90	2,678	22.3	88	2,890
34.7	63	2,215	28.0	79	2,625
36.1	66	1,800	17.6	85	3,465
35.7	80	1,915	34.4	65	3,465
20.2	85	2,965	20.6	105	3,380

DATA FILE
AUTO.XLS

On the basis of the results obtained:

(a) State the multiple regression equation.

(b) Interpret the meaning of the slopes in this problem.

(c) Predict the average miles per gallon for a car that has 60 horsepower and weighs 2,000 pounds.

(d) Compute the coefficient of multiple determination $r_{Y.12}^2$ and interpret its meaning.

(e) Compute the adjusted r^2.

• **14.6** Suppose a large consumer products company wants to measure the effectiveness of different types of advertising media in the promotion of its products. Specifically, two types of advertising media are to be considered: radio and television advertising and newspaper advertising (including the cost of discount coupons). A sample of 22 cities with approximately equal populations is selected for study during a test period of 1 month. Each city is allocated a specific expenditure level for both radio and television advertising and newspaper advertising. The sales of the product (in thousands of dollars) and also the levels of media expenditure during the test month are recorded with the following results:

CITY	SALES ($000)	RADIO AND TELEVISION ADVERTISING ($000)	NEWSPAPER ADVERTISING ($000)	CITY	SALES ($000)	RADIO AND TELEVISION ADVERTISING ($000)	NEWSPAPER ADVERTISING ($000)
1	973	0	40	12	1,577	45	45
2	1,119	0	40	13	1,044	50	0
3	875	25	25	14	914	50	0
4	625	25	25	15	1,329	55	25
5	910	30	30	16	1,330	55	25
6	971	30	30	17	1,405	60	30
7	931	35	35	18	1,436	60	30
8	1,177	35	35	19	1,521	65	35
9	882	40	25	20	1,741	65	35
10	982	40	25	21	1,866	70	40
11	1,628	45	45	22	1,717	70	40

DATA FILE
ADRADTV.XLS

On the basis of the results obtained:
(a) State the multiple regression equation.
(b) Interpret the meaning of the slopes in this problem.
(c) Predict the average sales for a city in which radio and television advertising is $20,000 and newspaper advertising is $20,000.
(d) Compute the coefficient of multiple determination $r^2_{Y.12}$ and interpret its meaning.
(e) Compute the adjusted r^2.

14.7 The director of broadcasting operations for a television station wants to study the issue of "standby hours," hours in which unionized graphic artists at the station are paid but are not actually involved in any activity. The variables to be considered are:

Standby hours (Y)—the total number of standby hours per week
Total staff present (X_1)—the weekly total of people-days over a 7-day week
Remote hours (X_2)—the total number of hours worked by employees at locations away from the central plant

The results for a period of 26 weeks are shown as follows:

WEEK	STANDBY HOURS	TOTAL STAFF PRESENT	REMOTE HOURS	WEEK	STANDBY HOURS	TOTAL STAFF PRESENT	REMOTE HOURS
1	245	338	414	14	161	307	402
2	177	333	598	15	274	322	151
3	271	358	656	16	245	335	228
4	211	372	631	17	201	350	271
5	196	339	528	18	183	339	440
6	135	289	409	19	237	327	475
7	195	334	382	20	175	328	347
8	118	293	399	21	152	319	449
9	116	325	343	22	188	325	336
10	147	311	338	23	188	322	267
11	154	304	353	24	197	317	235
12	146	312	289	25	261	315	164
13	115	283	388	26	232	331	270

DATA FILE
STANDBY.XLS

On the basis of the results obtained:
(a) State the multiple regression model.
(b) Interpret the meaning of the slopes in this problem.
(c) Predict the average standby hours for a week in which the total staff present is 310 people-days and the remote hours are 400.
(d) Compute the coefficient of multiple determination $r^2_{Y.12}$ and interpret its meaning.
(e) Compute the adjusted r^2.

14.2 RESIDUAL ANALYSIS FOR THE MULTIPLE REGRESSION MODEL

In section 13.5, we used residual analysis to evaluate whether the simple linear regression model is appropriate for the set of data being studied. In examining a multiple linear regression model with two explanatory variables, the residual plots listed in Exhibit 14.1 are of particular interest.

The first residual plot examines the pattern of residuals for the predicted values of Y. If the residuals show a pattern for different predicted values of Y, it provides evidence of a possible curvilinear effect in at least one explanatory variable and/or the need to transform the Y variable. The second and third residual plots involve the explanatory variables. Patterns in the plot of the residuals versus an explanatory variable may indicate the existence of a curvilinear effect and, therefore, lead to the possible transformation of that explanatory variable. The fourth type of plot is used to investigate patterns in the residuals when the data have been collected in time order. Associated with the residual plot versus time, as in section 13.6, the Durbin-Watson statistic can be computed and the existence of positive autocorrelation among the residuals can be determined.

The residual plots are available as part of the output of virtually all statistical and spreadsheet software. Figure 14.3 consists of the residual plots obtained from Microsoft Excel for the monthly heating oil consumption example. We can observe from Figure 14.3 that there appears to be very little or no pattern in the relationship between the residuals and the value of X_1 (temperature), the value of X_2 (attic insulation), or the predicted value of Y. Thus, we may conclude that the multiple linear regression model is appropriate for predicting heating oil usage.

PANEL A

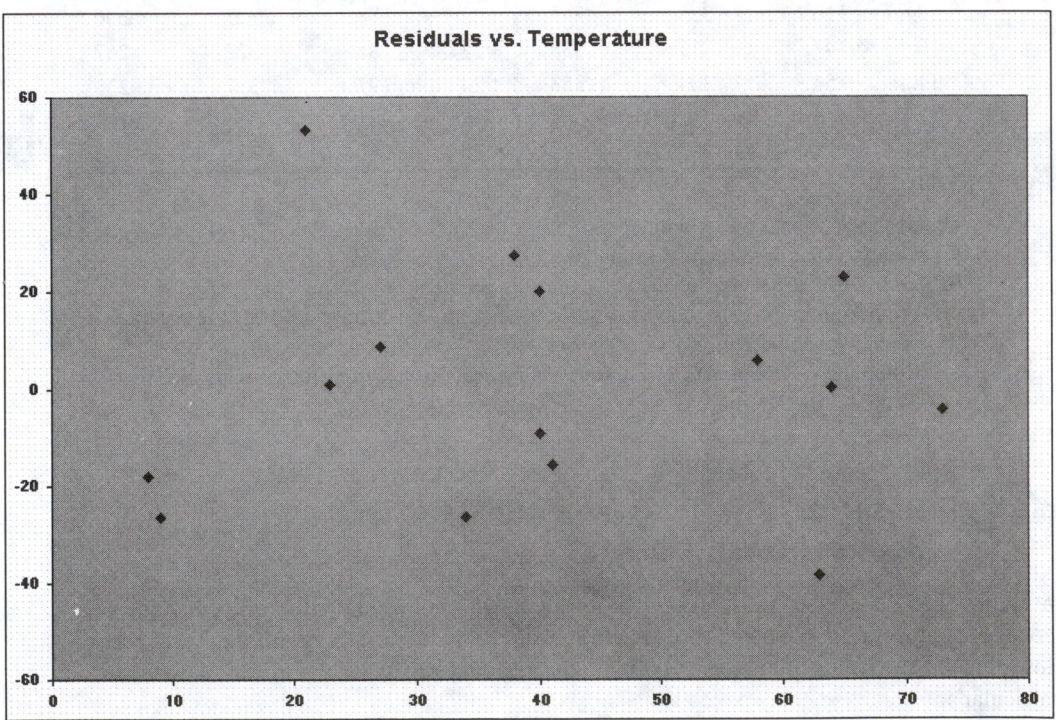

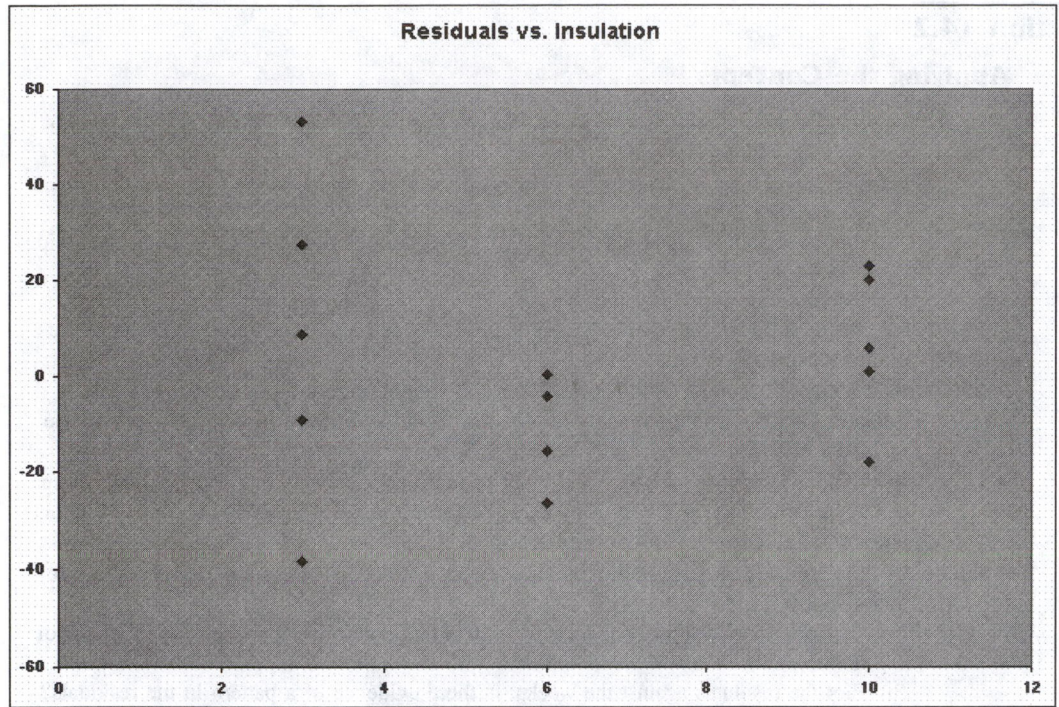

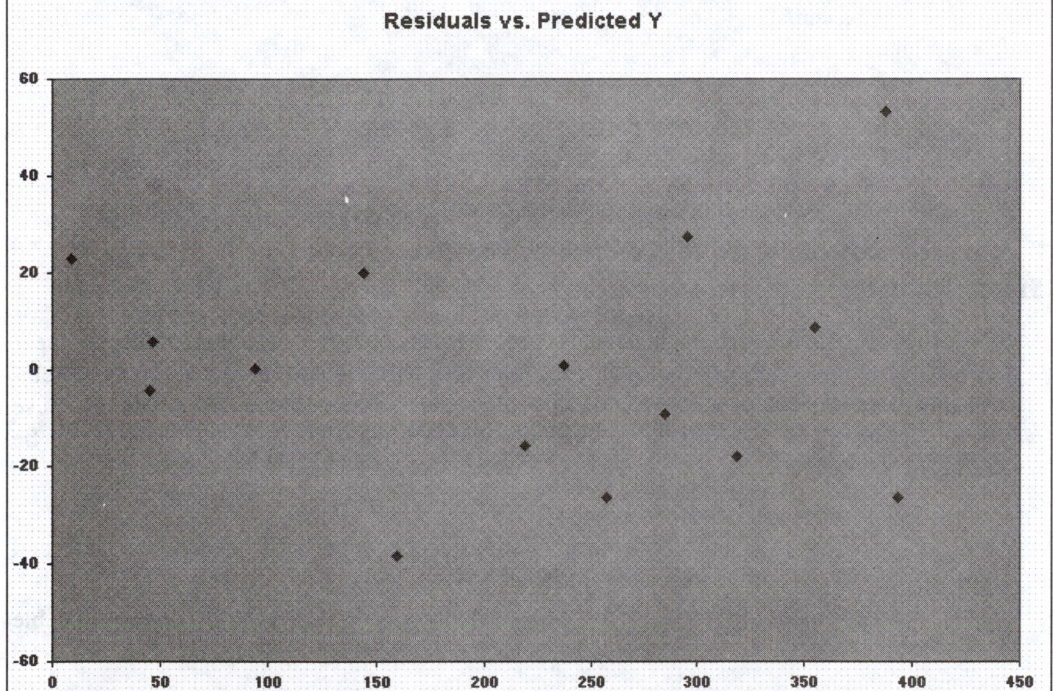

FIGURE 14.3 Residual plots for monthly heating oil consumption model obtained from Microsoft Excel: panel A, residuals versus temperature; panel B, residuals versus insulation; panel C, residuals versus predicted Y.

Problems for Section 14.2

Applying the Concepts

14.8 In Problem 14.4 on page 861 sales and number of orders were used to predict distribution cost at a mail-order catalog business.
 (a) Perform a residual analysis on your results and determine the adequacy of the fit of the model.
 (b) Plot the residuals against the months. Is there any evidence of a pattern in the residuals? Explain.
 (c) Compute the Durbin-Watson statistic.
 (d) At the .05 level of significance, is there evidence of positive autocorrelation in the residuals?

14.9 In Problem 14.5 on page 861 horsepower and weight were used to predict gasoline mileage. Perform a residual analysis on your results and determine the adequacy of the fit of the model.

 **14.10** In Problem 14.6 on page 862 the amount of radio and television advertising and newspaper advertising was used to predict sales. Perform a residual analysis on your results and determine the adequacy of the fit of the model.

14.11 In Problem 14.7 on page 863 the total staff present and remote hours were used to predict standby hours.
 (a) Perform a residual analysis on your results and determine the adequacy of the fit of the model.
 (b) Plot the residuals against the weeks. Is there evidence of a pattern in the residuals? Explain.
 (c) Compute the Durbin-Watson statistic.
 (d) At the .05 level of significance, is there evidence of positive autocorrelation in the residuals?

14.3 TESTING FOR THE SIGNIFICANCE OF THE MULTIPLE REGRESSION MODEL

Now that we have used residual analysis to assure ourselves that the multiple linear regression model is appropriate, we can determine whether there is a significant relationship between the dependent variable and the set of explanatory variables. Because there is more than one explanatory variable, the null and alternative hypotheses can be set up as follows:

H_0: $\beta_1 = \beta_2 = 0$ (There is no linear relationship between the dependent variable and the explanatory variables.)

H_1: At least one $\beta_j \neq 0$ (There is a linear relationship between the dependent variable and at least one of the explanatory variables.)

As was done in section 13.7 for simple linear regression, this null hypothesis is tested with an F test as summarized in Table 14.2.

652

The decision rule is

Reject H_0 if $F > F_U$,
the upper-tailed critical value of an F distribution with
p and $n - p - 1$ degrees of freedom;
otherwise do not reject H_0.

Table 14.2 *ANOVA table for testing significance of set of regression coefficients in multiple regression model with $p = 2$ explanatory variables*

SOURCE	d.f.	SUM OF SQUARES	MEAN SQUARE (VARIANCE)	F
Regression	p	SSR	$MSR = \dfrac{SSR}{p}$	$F = \dfrac{MSR}{MSE}$
Error	$n - p - 1$	SSE	$MSE = \dfrac{SSE}{n - p - 1}$	
Total	$n - 1$	SST		

The complete set of computations for our heating oil consumption example is shown in Figure 14.2 on page 857.

If a level of significance of .05 is chosen, we can determine from Table E.5 that the critical value on the F distribution (with 2 and 12 degrees of freedom) is 3.89, as depicted in Figure 14.4. Using equation (14.6), we can obtain the F statistic from Figure 14.2. Because $F = 168.47 > F_U = 3.89$ (see Figure 14.4) or because the p-value $= .000 < .05$, we can reject H_0 and conclude that at least one of the explanatory variables (temperature and/or insulation) is related to monthly heating oil consumption.

653

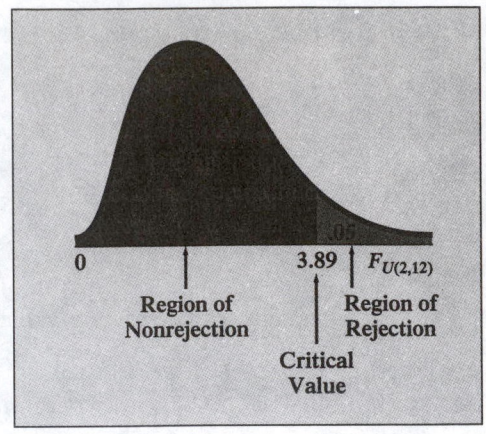

FIGURE 14.4
Testing for significance of set of regression coefficients at .05 level of significance with 2 and 12 degrees of freedom

Problems for Section 14.3

Learning the Basics

• **14.12** The following analysis of variance table was obtained from a multiple regression model with two independent variables.

SOURCE	DEGREES OF FREEDOM	SUM OF SQUARES	MEAN SQUARE	F
Regression	2	60		
Error	18	120		
Total	20	180		

(a) Determine the mean square that is due to regression and the mean square that is due to error.
(b) Determine the computed F statistic.
(c) Determine whether there is a significant relationship between Y and the two explanatory variables at the .05 level of significance.

14.13 The following analysis of variance table was obtained from a multiple regression model with two independent variables.

SOURCE	DEGREES OF FREEDOM	SUM OF SQUARES	MEAN SQUARE	F
Regression	2	30		
Error	10	120		
Total	12	150		

(a) Determine the mean square that is due to regression and the mean square that is due to error.
(b) Determine the computed F statistic.
(c) Determine whether there is a significant relationship between Y and the two explanatory variables at the .05 level of significance.

Applying the Concepts

• **14.14** In Problem 14.3 on page 860 the durability of a brand of running shoe was predicted based on a measurement of the forefoot shock-absorbing capability and a measurement of the change in impact properties over time. The following analysis of variance table was obtained:

ANOVA	DF	SS	MS	F	SIGNIFICANCE F
Regression	2	12.61020	6.30510	97.69	0.0001
Residual	12	0.77453	0.06454		
Total	14	13.38473			

(a) Determine whether there is a significant relationship between long-term impact and the two explanatory variables at the .05 level of significance.

(b) Interpret the meaning of the *p*-value.

14.15 In Problem 14.4 on page 861 sales and number of orders were used to predict distribution cost at a mail-order catalog business. Using the computer output you obtained to solve that problem,

(a) Determine whether there is a significant relationship between distribution cost and the two explanatory variables (sales and number of orders) at the .05 level of significance.

(b) Interpret the meaning of the *p*-value.

DATA FILE
WARECOST.XLS

14.16 In Problem 14.5 on page 861 horsepower and weight were used to predict gasoline mileage. Using the computer output you obtained to solve that problem,

(a) Determine whether there is a significant relationship between gasoline mileage and the two explanatory variables (horsepower and weight) at the .05 level of significance.

(b) Interpret the meaning of the *p*-value.

DATA FILE
AUTO.XLS

• **14.17** In Problem 14.6 on page 862 the amount of radio and television advertising and newspaper advertising was used to predict sales. Using the computer output you obtained to solve that problem,

(a) Determine whether there is a significant relationship between sales and the two explanatory variables (radio and television advertising and newspaper advertising) at the .05 level of significance.

(b) Interpret the meaning of the *p*-value.

DATA FILE
ADRADTV.XLS

14.18 In Problem 14.7 on page 863 the total staff present and remote hours were used to predict standby hours. Using the computer output you obtained to solve that problem,

(a) Determine whether there is a significant relationship between standby hours and the two explanatory variables (total staff present and remote hours) at the .05 level of significance.

(b) Interpret the meaning of the *p*-value.

DATA FILE
STANDBY.XLS

 **14.4** ## INFERENCES CONCERNING THE POPULATION REGRESSION COEFFICIENTS

In section 13.7 a test of hypothesis was performed on the slope in a simple linear regression model to determine the significance of the relationship between *X* and *Y*. In addition, a confidence interval was used to estimate the population slope. In this section these procedures will be extended to situations involving multiple regression.

Tests of Hypothesis

To test the hypothesis that the population slope β_1 was 0, we used equation (13.13):

$$t = \frac{b_1}{S_{b_1}}$$

However, this equation can be generalized for multiple regression as follows:

Testing for the Slope in Multiple Regression

$$t = \frac{b_k}{S_{b_k}} \qquad (14.7)$$

where

p = number of explanatory variables in the regression equation

b_k = slope of variable k with Y holding constant the effects of all other independent variables

S_{b_k} = standard error of the regression coefficient b_k

t = test statistic for a t distribution with $n - p - 1$ degrees of freedom

The results of this t test for each of the independent variables included in the regression model are provided as part of the output obtained in Figure 14.2 for Microsoft Excel on page 857.

Thus, if we wish to determine whether variable X_2 (amount of attic insulation) has a significant effect on the monthly consumption of home heating oil, taking into account the average daily atmospheric temperature, the null and alternative hypotheses would be

$$H_0: \beta_2 = 0$$
$$H_1: \beta_2 \neq 0$$

From equation (14.7), we have

$$t = \frac{b_2}{S_{b_2}}$$

and from the data of this example,

$$b_2 = -20.0123 \quad \text{and} \quad S_{b_2} = 2.3425$$

so that

$$t = \frac{-20.0123}{2.3425} = -8.5431$$

If a level of significance of .05 is selected, from Table E.3 we find that for 12 degrees of freedom, the critical values of t are -2.1788 and $+2.1788$ (see Figure 14.5). From Figure 14.2 on page 857 we observe that the p-value is .00000191 (or 1.91E-06 in scientific notation).

FIGURE 14.5

Testing for significance of regression coefficient at .05 level of significance with 12 degrees of freedom

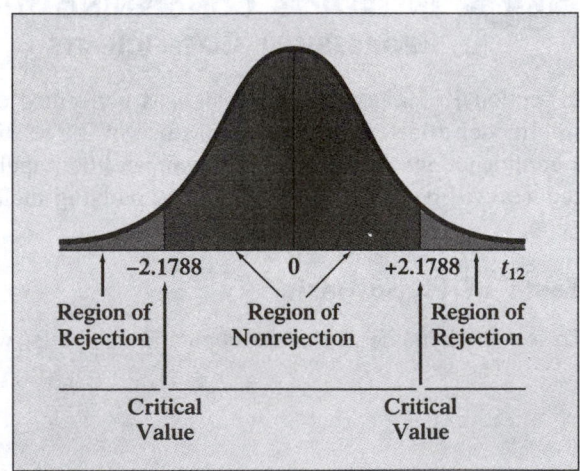

656

Because $t = -8.5431 < -t_{12} = -2.1788$ or the p-value of $.00000191 < .05$, we reject H_0 and conclude that there is a significant relationship between variable X_2 (amount of attic insulation) and heating oil consumption, taking into account the average daily atmospheric temperature X_1.

In a similar manner, in Example 14.1 we test for the significance of β_1, the slope of monthly consumption of heating oil with the atmospheric temperature.

Example 14.1 *Testing for the Significance of the Slope of Heating Oil Consumption and Atmospheric Temperature*

At the .05 level of significance, is there evidence that the slope of heating oil consumption with atmospheric temperature is different from zero?

SOLUTION

From Figure 14.2 on page 857, $t = -16.17 < -2.1788$ (the critical value for $\alpha = .05$) or the p-value $= .00000000164 < .05$. Therefore, there is a significant relationship between atmospheric temperature (X_1) and heating oil consumption, taking into account the attic insulation X_2.

As observed with each of these two X variables, the test of significance for a particular regression coefficient is actually a test for the significance of adding a particular variable into a regression model given that the other variable has been included. Therefore, the t test for the regression coefficient is equivalent to testing for the contribution of each explanatory variable.

Confidence Interval Estimation

Instead of testing the significance of a regression coefficient, we may be more concerned with estimating the population value of a regression coefficient. In multiple regression analysis, a confidence interval estimate for the population slope can be obtained from

Confidence Interval Estimate for the Slope

$$b_k \pm t_{n-p-1}S_{b_k} \tag{14.8}$$

For example, if we wish to obtain a 95% confidence interval estimate of the population slope β_1 (the effect of average daily temperature X_1 on monthly heating oil consumption Y, holding constant the effect of attic insulation X_2), from equation (14.8) and Figure 14.2 on page 857 we have

$$b_1 \pm t_{12}S_{b_1}$$

Because the critical value of t at the 95% confidence level with 12 degrees of freedom is 2.1788 (see Table E.3), we have

$$-5.43658 \pm (2.1788)(.33622)$$
$$-5.43658 \pm .732556$$
$$-6.169136 \leq \beta_1 \leq -4.704024$$

Thus, taking into account the effect of attic insulation, we estimate that the effect of average daily atmospheric temperature is to reduce the average consumption of heating oil by between approximately 4.7 and 6.17 gallons for each 1°F increase in temperature. We have 95% confidence that this interval correctly estimates the true relationship between these variables.

From a hypothesis-testing viewpoint, because this confidence interval does not include 0, we conclude that the regression coefficient β_1 has a significant effect. A confidence interval estimate for the slope of heating oil consumption with attic insulation is developed in Example 14.2.

Example 14.2 *Obtaining a Confidence Interval Estimate for the Slope of Heating Oil Consumption with Attic Insulation*

Set up a 95% confidence interval estimate of the population slope of heating oil consumption with attic insulation.

SOLUTION

Because the critical value of t at the 95% confidence level with 12 degrees of freedom is 2.1788 (see Table E.3), we have

$$-20.0123 \pm (2.1788)(2.3425)$$
$$-20.0123 \pm 5.1039$$
$$-25.1162 \leq \beta_2 \leq -14.9084$$

Thus, taking into account the effect of average daily atmospheric temperature, we estimate that the effect of attic insulation is to reduce the average consumption of heating oil by between approximately 14.9084 and 25.1162 gallons for each inch of attic insulation. We have 95% confidence that this interval correctly estimates the true relationship between these variables. From a hypothesis-testing viewpoint, because this confidence interval does not include 0, we conclude that the regression coefficient β_2 has a significant effect.

Problems for Section 14.4

Learning the Basics

• **14.19** Suppose you were given the following information from a multiple regression model

$$n = 25, \ b_1 = 5, \ b_2 = 10, \ S_{b_1} = 2, \ S_{b_2} = 8$$

(a) Which variable has a larger slope?
(b) Set up a 95% confidence interval estimate of the population slope for X_1.
(c) At the .05 level of significance, determine whether each explanatory variable makes a significant contribution to the regression model. On the basis of these results, indicate the independent variables that should be included in this model.

14.20 Suppose you were given the following information from a multiple regression model

$$n = 20, \ b_1 = 4, \ b_2 = 3, \ S_{b_1} = 1.2, \ S_{b_2} = 0.8$$

(a) Which variable has a larger slope?

(b) Set up a 95% confidence interval estimate of the population slope for X_1.

(c) At the .05 level of significance, determine whether each explanatory variable makes a significant contribution to the regression model. On the basis of these results, indicate the independent variables that should be included in this model.

Applying the Concepts

• **14.21** In Problem 14.3 on page 860 the durability of a brand of running shoe was predicted based on a measurement of the forefoot shock-absorbing capability and a measurement of the change in impact properties over time for a sample of 15 pairs of shoes. Use the following computer output:

VARIABLE	COEFFICIENTS	STANDARD ERROR	t STAT	p-VALUE
Intercept	−0.02686	.06905	−0.39	
Foreimp	0.79116	.06295	12.57	.0000
Midsole	0.60484	.07174	8.43	.0000

(a) Set up a 95% confidence interval estimate of the population slope between long-term impact and forefoot impact.

(b) At the .05 level of significance, determine whether each explanatory variable makes a significant contribution to the regression model. On the basis of these results, indicate the independent variables that should be included in this model.

14.22 In Problem 14.4 on page 861 sales and number of orders were used to predict distribution cost at a mail-order catalog business. Use the computer output you obtained to solve that problem.

(a) Set up a 95% confidence interval estimate of the population slope between distribution cost and sales.

(b) At the .05 level of significance, determine whether each explanatory variable makes a significant contribution to the regression model. On the basis of these results, indicate the independent variables that should be included in this model.

DATA FILE
WARECOST.XLS

14.23 In Problem 14.5 on page 861 horsepower and weight were used to predict gasoline mileage. Use the computer output you obtained to solve that problem.

(a) Set up a 95% confidence interval estimate of the population slope between gasoline mileage and horsepower.

(b) At the .05 level of significance, determine whether each explanatory variable makes a significant contribution to the regression model. On the basis of these results, indicate the independent variables that should be included in this model.

DATA FILE
AUTO.XLS

• **14.24** In Problem 14.6 on page 862 the amount of radio and television advertising and newspaper advertising was used to predict sales. Use the computer output you obtained to solve that problem.

(a) Set up a 95% confidence interval estimate of the population slope between sales and radio and television advertising.

(b) At the .05 level of significance, determine whether each explanatory variable makes a significant contribution to the regression model. On the basis of these results, indicate the independent variables that should be included in this model.

DATA FILE
ADRADTV.XLS

14.25 In Problem 14.7 on page 863 the total staff present and remote hours were used to predict standby hours. Use the computer output you obtained to solve that problem.

(a) Set up a 95% confidence interval estimate of the population slope between standby hours and total staff present.

(b) At the .05 level of significance, determine whether each explanatory variable makes a significant contribution to the regression model. On the basis of these results, indicate the independent variables that should be included in this model.

DATA FILE
STANDBY.XLS

In developing a multiple regression model, the objective is to use only those explanatory variables that are useful in predicting the value of a dependent variable. If an explanatory variable is not helpful in making this prediction, it could be deleted from the multiple regression model and a model with fewer explanatory variables could be used in its place.

An alternative method for determining the contribution of an explanatory variable is called the **partial *F*-test criterion.** It involves determining the contribution to the regression sum of squares made by each explanatory variable after all the other explanatory variables have been included in the model. The new explanatory variable is included only if it significantly improves the model.

To apply the partial *F*-test criterion in the home heating oil consumption example, we need to evaluate the contribution of the variable attic insulation (X_2) after average daily atmospheric temperature (X_1) has been included in the model and, conversely, we must also evaluate the contribution of the variable average daily atmospheric temperature (X_1) after attic insulation (X_2) has been included in the model.

In general, if there were several explanatory variables, the contribution of each explanatory variable to be included in the model can be determined by taking into account the regression sum of squares of a model that includes all explanatory variables except the one of interest, SSR (all variables except k). Thus, to determine the contribution of variable k, assuming that all other variables are already included, we would have

---

Determining the Contribution of an Independent Variable to the Regression Model

$SSR(X_k \mid$ all variables *except k*)

$$= SSR(\text{all variables } including\ k) - SSR(\text{all variables } except\ k) \qquad (14.9)$$

If, as in the monthly heating oil consumption example, there are two explanatory variables, the contribution of each can be determined from equations (14.10a) and (14.10b).

Determining the Contribution of X_1 and X_2 to a Regression Model

Contribution of variable X_1 given X_2 has been included:

$$SSR(X_1 \mid X_2) = SSR(X_1 \text{ and } X_2) - SSR(X_2) \qquad (14.10a)$$

Contribution of variable X_2 given X_1 has been included:

$$SSR(X_2 \mid X_1) = SSR(X_1 \text{ and } X_2) - SSR(X_1) \qquad (14.10b)$$

The terms $SSR(X_2)$ and $SSR(X_1)$, respectively, represent the sum of squares that is due to regression for a model that includes only the explanatory variable X_2 (amount of attic insulation) and only the explanatory variable X_1 (average daily atmospheric temperature). Output obtained from Microsoft Excel for these two models is presented in Figures 14.6 and 14.7.

	A	B	C	D	E	F	G
1	Regression Model: Heating Oil and Insulation						
2							
3	Regression Statistics						
4	Multiple R	0.465082527					
5	R Square	0.216301757					
6	Adjusted R Square	0.156017277					
7	Standard Error	119.3117327					
8	Observations	15					
9							
10	ANOVA		$SSR(X_2)$				
11		df	SS	MS	F	Significance F	
12	Regression	1	51076.46501	51076.46501	3.588017285	0.080660953	
13	Residual	13	185058.7643	14235.28956			
14	Total	14	236135.2293				
15							
16		Coefficients	Standard Error	t Stat	P-value	Lower 95%	Upper 95%
17	Intercept	345.3783784	74.69065911	4.624117426	0.000476363	184.0190506	506.7377061
18	Insulation	-20.35027027	10.743429	-1.894206241	0.080660953	-43.56003307	2.859492534

FIGURE 14.6 Partial output obtained from Microsoft Excel of simple linear regression model for amount of heating oil consumed and amount of attic insulation

	A	B	C	D	E	F	G
1	Regression Model: Heating Oil and Temperature						
2							
3	Regression Statistics						
4	Multiple R	0.86974117					
5	R Square	0.756449704					
6	Adjusted R Square	0.737715065					
7	Standard Error	66.51246564					
8	Observations	15					
9							
10	ANOVA		$SSR(X_1)$				
11		df	SS	MS	F	Significance F	
12	Regression	1	178624.4242	178624.4242	40.37706498	2.51847E-05	
13	Residual	13	57510.80511	4423.908086			
14	Total	14	236135.2293				
15							
16		Coefficients	Standard Error	t Stat	P-value	Lower 95%	Upper 95%
17	Intercept	436.4382299	38.63970893	11.29507033	4.30471E-08	352.9622299	519.9142299
18	Temperature	-5.462207697	0.859608768	-6.354295002	2.51847E-05	-7.319279177	-3.605136216

FIGURE 14.7 Partial output obtained from Microsoft Excel of simple linear regression model for amount of heating oil consumed and average daily atmospheric temperature

We can observe from Figure 14.6 that

$$SSR(X_2) = 51,076 \text{ (rounded)}$$

and, therefore, from equation (14.10a),

$$SSR(X_1 \mid X_2) = SSR(X_1 \text{ and } X_2) - SSR(X_2)$$

we have

$$SSR(X_1 \mid X_2) = 228,015 - 51,076 = 176,939$$

To determine whether X_1 significantly improves the model after X_2 has been included, we can now subdivide the regression sum of squares into two component parts as shown in Table 14.3.

Table 14.3 *ANOVA table dividing regression sum of squares into components to determine contribution of variable X_1*

SOURCE	d.f.	SUM OF SQUARES	MEAN SQUARE (VARIANCE)	F
Regression	2	228,015	114,007.5	
$\left\{ \begin{array}{c} X_2 \\ X_1 \mid X_2 \end{array} \right.$	$\left\{ \begin{array}{c} 1 \\ 1 \end{array} \right.$	$\left\{ \begin{array}{c} 51,076 \\ 176,939 \end{array} \right.$	176,939	261.47
Error	12	8,120	676.717	
Total	14	236,135		

The null and alternative hypotheses to test for the contribution of X_1 to the model are

H_0: Variable X_1 does not significantly improve the model once variable X_2 has been included.

H_1: Variable X_1 significantly improves the model once variable X_2 has been included.

The partial F-test criterion is expressed by

The Partial F-Test Criterion for Determining the Contribution of an Independent Variable

$$F = \frac{SSR(X_k \mid \text{all variables } except \ k)}{MSE} \tag{14.11}$$

In equation (14.11) F represents the F-test statistic that follows an F distribution with 1 and $n - p - 1$ degrees of freedom.

Thus, from Table 14.3 we have

$$F = \frac{176,939}{676.717} = 261.47$$

Because there are 1 and 12 degrees of freedom, respectively, if a level of significance of .05 is selected, we observe from Table E.5 that the critical value is 4.75 (see Figure 14.8).

FIGURE 14.8
Testing for contribution of regression coefficient to multiple regression model at .05 level of significance with 1 and 12 degrees of freedom

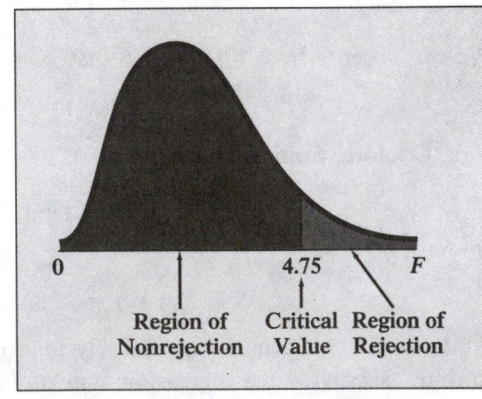

662

Because the computed F-value exceeds this critical F-value $(261.47 > 4.75)$, our decision is to reject H_0 and conclude that the addition of variable X_1 (average daily atmospheric temperature) significantly improves a regression model that already contains variable X_2 (attic insulation).

To evaluate the contribution of variable X_2 (attic insulation) to a model in which variable X_1 has been included, we need to use equation (14.10b):

$$SSR(X_2 \mid X_1) = SSR(X_1 \text{ and } X_2) - SSR(X_1)$$

From Figure 14.7 we determine that

$$SSR(X_1) = 178,624$$

Therefore, from Figures 14.2 and 14.7 on pages 857 and 875

$$SSR(X_2 \mid X_1) = 228,015 - 178,624 = 49,391 \text{ (rounded)}$$

Thus, to determine whether X_2 significantly improves a model after X_1 has been included, the regression sum of squares can be subdivided into two component parts as shown in Table 14.4. The null and alternative hypotheses to test for the contribution of X_2 to the model are

H_0: Variable X_2 does not significantly improve the model once variable X_1 has been included.

H_1: Variable X_2 significantly improves the model once variable X_1 has been included.

Using equation (14.11), we obtain

$$F = \frac{49,391}{676.717} = 72.99$$

as indicated in Table 14.4.

Table 14.4 *ANOVA table dividing the regression sum of squares into components to determine the contribution of variable X_2*

SOURCE	d.f.	SUM OF SQUARES	(VARIANCE) MEAN SQUARE	F
Regression	2	228,015	114,007.5	
$\left\{\begin{matrix} X_1 \\ X_2 \mid X_1 \end{matrix}\right.$	$\left\{\begin{matrix} 1 \\ 1 \end{matrix}\right.$	$\left\{\begin{matrix} 178,624 \\ 49,391 \end{matrix}\right.$	49,391	72.99
Error	12	8,120	676.717	
Total	14	236,135		

Because there are 1 and 12 degrees of freedom, respectively, if a .05 level of significance is selected, we again observe from Figure 14.8 that the critical value of F is 4.75.

Because the computed F-value exceeds this critical value $(72.99 > 4.75)$, our decision is to reject H_0 and conclude that the addition of variable X_2 (attic insulation) significantly improves the multiple regression model already containing X_1 (average daily atmospheric temperature).

Thus by testing for the contribution of each explanatory variable after the other has been included in the model, we determine that each of the two explanatory variables significantly improves the model. Therefore, our multiple regression model should include both average daily atmospheric temperature X_1 and the amount of attic insulation X_2 in predicting the monthly consumption of home heating oil.

Focusing on the interpretation of these conclusions, we note that there is a relationship between the value of the t-test statistic obtained from equation (14.7) and the partial F-test statistic [equation (14.11)] used to determine the contributions of X_1 and X_2 to the multiple regression model. The t-values were computed to be -16.17 and -8.5431, and the corresponding values of F were 261.47 and 72.99. This points up the following relationship[1] between t and F:

[1]*The relationship between t and F indicated in equation (14.12) holds when t is a two-tailed test.*

The Relationship between a t Statistic and an F Statistic

$$t_v^2 = F_{1,v} \qquad (14.12)$$

where v = number of degrees of freedom.

Coefficient of Partial Determination

In section 14.1 we discussed the coefficient of multiple determination ($r_{Y.12}^2$), which measured the proportion of the variation in Y that was explained by variation in the two explanatory variables. Now that we have examined ways in which the contribution of each explanatory variable to the multiple regression model can be evaluated, we can also compute the **coefficients of partial determination** ($r_{Y1.2}^2$ and $r_{Y2.1}^2$). The coefficients measure the proportion of the variation in the dependent variable that is explained by each explanatory variable while controlling for, or holding constant, the other explanatory variable(s). Thus, in a multiple regression model with two explanatory variables, we have

---

Coefficients of Partial Determination for a Two-Independent-Variable Model

$$r_{Y1.2}^2 = \frac{SSR(X_1 \mid X_2)}{SST - SSR(X_1 \text{ and } X_2) + SSR(X_1|X_2)} \qquad (14.13a)$$

and also

$$r_{Y2.1}^2 = \frac{SSR(X_2 \mid X_1)}{SST - SSR(X_1 \text{ and } X_2) + SSR(X_2|X_1)} \qquad (14.13b)$$

where

$SSR(X_1 \mid X_2)$ = sum of squares of the contribution of variable X_1 to the regression model given that variable X_2 has been included in the model

SST = total sum of squares for Y

$SSR(X_1 \text{ and } X_2)$ = regression sum of squares when both variables X_1 and X_2 are included in the multiple regression model

$SSR(X_2 \mid X_1)$ = sum of squares of the contribution of variable X_2 to the regression model given that variable X_1 has been included in the model

---

whereas in a multiple regression model containing several (p) explanatory variables, for the kth variable we have

For the monthly heating oil consumption example, we can compute

$$r^2_{Y1.2} = \frac{176{,}939}{236{,}135 - 228{,}015 + 176{,}939}$$
$$= 0.9561$$

and

$$r^2_{Y2.1} = \frac{49{,}391}{236{,}135 - 228{,}015 + 49{,}391}$$
$$= 0.8588$$

The coefficient of partial determination of variable Y with X_1 while holding X_2 constant ($r^2_{Y1.2}$) means that for a fixed (constant) amount of attic insulation, 95.61% of the variation in heating oil consumption in January can be explained by the variation in the average daily atmospheric temperature in that month. The coefficient of partial determination of variable Y with X_2 while holding X_1 constant ($r^2_{Y2.1}$) means that for a given (constant) average daily atmospheric temperature, 85.88% of the variation in heating oil consumption in January can be explained by variation in the amount of attic insulation.

 **14.5E   CALCULATING THE COEFFICIENTS OF PARTIAL DETERMINATION USING MICROSOFT EXCEL**

Overview

Because the coefficients of partial determination require values from several regression analyses, both quick results users and developers should use the PHStat add-in to calculate these coefficients.

The 14-5E.XLS workbook file includes the calculations for the coefficients of partial determination for the multiple regression based on the Table 14.1 data for the monthly heating oil example of section 14.5.

 14-5E.XLS

Details for All Users

To calculate the coefficients of partial determination for a multiple regression model that contains two independent variables, use the Regression | Multiple Regression choice of the PHStat add-in. As an example, consider the heating oil usage problem of section 14.1. To calculate the coefficients of partial determination for the multiple regression based on this problem, do the following:

❶ Open the Starting Point for Several Chapter 14 Sections workbook (STARTING POINT 14-SEVERAL.XLS) and click the Data sheet tab. Verify that the heating oil consumed, average temperature, and amount of attic insulation data of Table 14.1 on page 855 have been entered into columns A, B, and C, respectively.

❷ Select PHStat | Regression | Multiple Regression.

❸ In the Multiple Regression dialog box (see Figure 14E.1):
a. Enter A1:A16 in the Y Variable Cell Range: edit box.
b. Enter B1:C16 in the X Variables Cell Range: edit box.
c. Select the First cells in both ranges contain label check box.
d. Enter 95 in the Confidence Level for regr. coefficients: edit box.
e. Select the Regression Statistics Table, the ANOVA and Coefficients Table, and the Residuals Table check boxes.
f. Enter Regression Analysis for Heating Oil Usage in the Output Title: edit box.
g. Select the Coefficients of Partial Determination check box.
h. Click the OK button.

FIGURE 14E.1
PHStat Multiple Regression dialog box

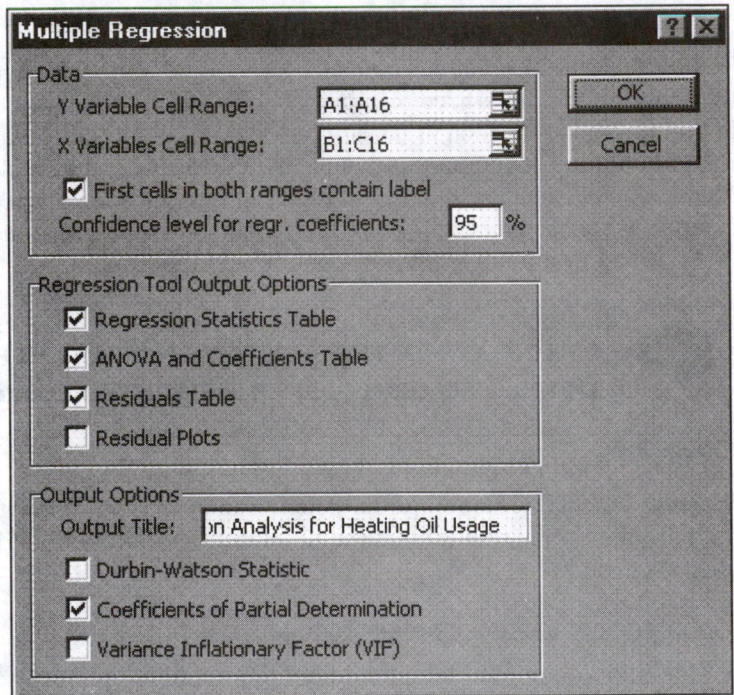

The add-in inserts four worksheets, three that contain the regression summaries—one for the multiple regression and one for each regression of one of the X variables and the Y variable—and a fourth that calculates the coefficients from the regression sums of squares found on the other three sheets. The fourth sheet will be similar to the worksheet shown in Figure 14E.2. Because the supporting regression worksheets are *not* dynamically changeable, any changes made to the underlying heating oil usage data would require using the PHStat add-in a second time to produce updated coefficients.

	A	B	C	D
1	Regression Analysis for Heating Oil Usage			
2	Coefficients of Partial Determination			
3				
4	SSR(X1,X2)	228014.6263		
5	SST	236135.2293		
6	SSR(X2)	51076.46501	SSR(X1 | X2)	176938.1613
7	SSR(X1)	178624.4242	SSR(X2 | X1)	49390.2021
8				
9	r2 Y1.2	0.956118787		
10	r2 Y2.1	0.858798655		

FIGURE 14E.2

Coefficients of partial determination for the heating oil usage problem of section 14.1

Developer Details

As discussed at the beginning of this section, because the coefficients of partial determination require values from several regression analyses, developers should use the PHStat add-in to calculate these coefficients.

For developer reference, Table 14E.1 presents the design of the worksheet inserted by the PHStat add-in to calculate the coefficients of partial determination for the heating oil problem of section 14.1. This design assumes the worksheets named MR, NOTX1, and NOTX2 contain the regression sum of squares values in cell C12 and that the MR sheet contains the total sum of squares in cell C14 (as would be the case with the regression sheets generated by the add-in). Note that the sum of squares and the total sum of squares values could be entered in cell range B4:B7 in lieu of the formulas shown and the design can be modified and extended for regressions involving more than two independent variables.

Table 14E.1 Calculations sheet design inserted by the PHStat add-in for the coefficients of partial determination for the heating oil problem of section 14.1

	A	B	C	D
1	Regression Analysis for Heating Oil Usage			
2	Coefficients of Partial Determination			
3				
4	SSR(X1,X2)	=MR!C12		
5	SST	=MR!C14		
6	SSR(X2)	=NOTX1!C12	SSR(X1,X2)	=B4-B6
7	SSR(X1)	=NOTX2!C12	SSR(X2,X1)	=B4-B7
8				
9	r2 Y1.2	=D6/(B5-B4+D6)		
10	r2 Y2.1	=D7/(B5-B4+D7)		

Problems for Section 14.5

Learning the Basics

• **14.26** The following analysis of variance table was obtained from a multiple regression model with two independent variables.

SOURCE	DEGREES OF FREEDOM	SUM OF SQUARES	MEAN SQUARE	F
Regression	2	60		
Error	18	120		
Total	20	180		

$SSR(X_1) = 45$ $SSR(X_2) = 25$

(a) Determine whether there is a significant relationship between Y and each of the explanatory variables at the .05 level of significance.

(b) Compute the coefficients of partial determination $r^2_{Y1.2}$ and $r^2_{Y2.1}$ and interpret their meaning.

14.27 The following analysis of variance table was obtained from a multiple regression model with two independent variables.

SOURCE	DEGREES OF FREEDOM	SUM OF SQUARES	MEAN SQUARE	F
Regression	2	30		
Error	10	120		
Total	12	150		

$SSR(X_1) = 20$ $SSR(X_2) = 15$

(a) Determine whether there is a significant relationship between Y and each of the explanatory variables at the .05 level of significance.

(b) Compute the coefficients of partial determination $r^2_{Y1.2}$ and $r^2_{Y2.1}$ and interpret their meaning.

Applying the Concepts

DATA FILE
WARECOST.XLS

14.28 In Problem 14.4 on page 861 sales and number of orders were used to predict distribution cost at a mail-order catalog business. Using the computer output you obtained to solve that problem,

(a) At the .05 level of significance, determine whether each explanatory variable makes a significant contribution to the regression model. On the basis of these results, indicate the regression model that should be used in the problem.

(b) Compute the coefficients of partial determination $r^2_{Y1.2}$ and $r^2_{Y2.1}$ and interpret their meaning.

DATA FILE
AUTO.XLS

14.29 In Problem 14.5 on page 861 horsepower and weight were used to predict gasoline mileage. Using the computer output you obtained to solve that problem,

(a) At the .05 level of significance, determine whether each explanatory variable makes a significant contribution to the regression model. On the basis of these results, indicate the regression model that should be used in the problem.

(b) Compute the coefficients of partial determination $r^2_{Y1.2}$ and $r^2_{Y2.1}$ and interpret their meaning.

DATA FILE
ADRADTV.XLS

● **14.30** In Problem 14.6 on page 862 the amount of radio and television advertising and newspaper advertising was used to predict sales. Using the computer output you obtained to solve that problem,

(a) At the .05 level of significance, determine whether each explanatory variable makes a significant contribution to the regression model. On the basis of these results, indicate the regression model that should be used in the problem.

(b) Compute the coefficients of partial determination $r^2_{Y1.2}$ and $r^2_{Y2.1}$ and interpret their meaning.

DATA FILE
STANDBY.XLS

14.31 In Problem 14.7 on page 863 the total staff present and remote hours were used to predict standby hours. Using the computer output you obtained to solve that problem,

(a) At the .05 level of significance, determine whether each explanatory variable makes a significant contribution to the regression model. On the basis of these results, indicate the regression model that should be utilized in the problem.

(b) Compute the coefficients of partial determination $r_{Y1.2}^2$ and $r_{Y2.1}^2$ and interpret their meaning.

 **14.6** THE CURVILINEAR REGRESSION MODEL

In our discussions of simple regression in chapter 13 and multiple regression thus far in this chapter we assumed that the relationship between Y and each explanatory variable is linear. However, in section 13.1 several different types of relationships between variables were introduced. One of the more common nonlinear relationships illustrated was a curvilinear relationship between two variables (see Figure 13.2, panels D–F, on page 774) in which Y increases (or decreases) at a changing rate for various values of X. This model of a curvilinear relationship between X and Y can be expressed as

Curvilinear Regression Model

$$Y_i = \beta_0 + \beta_1 X_{1i} + \beta_2 X_{1i}^2 + \epsilon_i \qquad (14.15)$$

where

$\beta_0 = Y$ intercept

$\beta_1 = $ coefficient of the linear effect on Y

$\beta_2 = $ coefficient of the curvilinear effect on Y

$\epsilon_i = $ random error in Y for observation i

This **curvilinear regression model** is similar to the multiple regression model with two explanatory variables [see equation (14.1) on page 856] except that the second explanatory variable in this instance is the square of the first explanatory variable.

As in the case of multiple linear regression, the sample regression coefficients (b_0, b_1, and b_2) are used as estimates of the population parameters (β_0, β_1, and β_2). Thus, the sample regression equation for the curvilinear model with one explanatory variable (X_1) and a dependent variable (Y) is

Sample Curvilinear Regression Equation

$$\hat{Y}_i = b_0 + b_1 X_{1i} + b_2 X_{1i}^2 \qquad (14.16)$$

In this equation, the first regression coefficient b_0 represents the Y intercept, the second regression coefficient b_1 represents the linear coefficient, and the third regression coefficient b_2 represents the quadratic or curvilinear effect.

Finding the Regression Coefficients and Predicting Y

To illustrate the curvilinear regression model, suppose the marketing department of a large supermarket chain wants to study the price elasticity for packages of disposable razors. A sample of 15 stores with equivalent store traffic and product placement (i.e., at the checkout counter) is selected. Five stores are randomly assigned to each of three price levels (79, 99, and 119 cents) for the packages of razors. The number of packages sold over a full week and the price at each store are presented in Table 14.5.

Table 14.5 *Sales and price of packages of disposable razors for sample of 15 stores*

SALES	PRICE (CENTS)	SALES	PRICE (CENTS)
142	79	115	99
151	79	126	99
163	79	77	119
168	79	86	119
176	79	95	119
91	99	100	119
100	99	106	119
107	99		

To help select the proper model for expressing the relationship between price and sales, a scatter diagram is plotted in Figure 14.9. An examination of Figure 14.9 indicates that as price increases sales decrease, which levels off with further increases in price. Sales for a price of 99 cents are substantially below sales at the 79 cent price, but sales for the $1.19 price are only slightly below sales at 99 cents. Therefore, it appears that a curvilinear model rather than a linear model may be the more appropriate choice to estimate sales based on price.

FIGURE 14.9
Scatter diagram of price (X) and sales (Y)

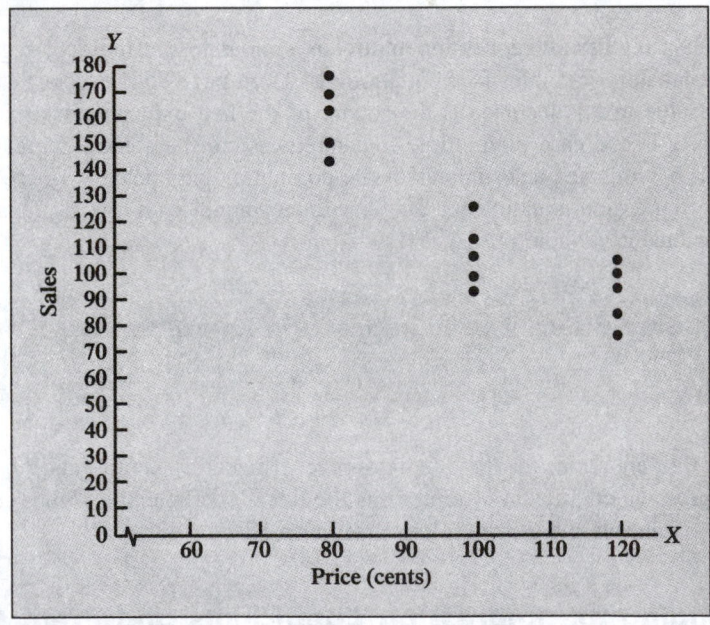

As in the case of multiple regression, the values of the three sample regression coefficients b_0, b_1, and b_2 are obtained from Microsoft Excel as illustrated in Figure 14.10. From Figure 14.10 we observe that

Regression Analysis for Price Elasticity

Regression Statistics	
Multiple R	0.928581176
R Square	0.862263
Adjusted R Square	0.839306834
Standard Error	12.86986143
Observations	15

ANOVA

	df	SS	MS	F	Significance F
Regression	2	12442.8	6221.4	37.56127994	6.82816E-06
Residual	12	1987.6	165.6333333		
Total	14	14430.4			

	Coefficients	Standard Error	t Stat	P-value	Lower 95%	Upper 95%
Intercept	729.8665	169.2575176	4.312165924	0.001009972	361.0860554	1098.646945
Price	-10.887	3.495239703	-3.114807831	0.008940633	-18.50247298	-3.271527022
Price^2	0.0465	0.017622784	2.638629696	0.0216284	0.008103254	0.084896746

b_0 b_1 b_2

FIGURE 14.10   Partial output from Microsoft Excel for razor sales data

$$b_0 = 729.8665 \qquad b_1 = -10.887 \qquad b_2 = .0465$$

Therefore, the sample curvilinear regression equation can be expressed as

$$\hat{Y}_i = 729.8665 - 10.887X_{1i} + .0465X_{1i}^2$$

where

$$\hat{Y}_i = \text{predicted average sales for store } i$$

$$X_{1i} = \text{price of disposable razors in store } i$$

As depicted in Figure 14.11, this curvilinear regression equation is plotted on the scatter diagram to see how well the selected regression model fits the original data.

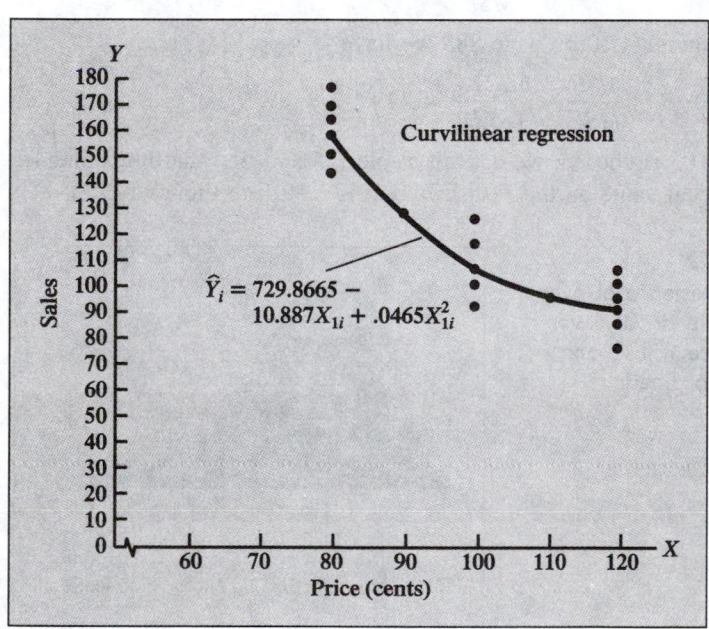

FIGURE 14.11

Scatter diagram expressing curvilinear relationship between price and sales for disposable razor data

From our curvilinear regression equation and Figure 14.11 the Y intercept (b_0, computed as 729.8665) has no direct interpretation for these data. It is simply a base or starting point. Similarly, the linear term b_1 also has no direct interpretation in our curvilinear regression equation. To interpret the coefficient b_2, we see from Figure 14.11 that sales decrease with increasing price. However, we can also observe that these decreases in sales level off or become reduced with increasing price. This can be demonstrated by predicting average sales for packages priced at 79, 99, and 119 cents. Using our curvilinear regression equation,

$$\hat{Y}_i = 729.8665 - 10.887X_{1i} + .0465X_{1i}^2$$

for $X_{1i} = 79$, we have

$$\hat{Y}_i = 729.8665 - 10.887(79) + .0465(79)^2 = 160$$

For $X_{1i} = 99$, we have

$$\hat{Y}_i = 729.8665 - 10.887(99) + .0465(99)^2 = 107.8$$

For $X_{1i} = 119$, we have

$$\hat{Y}_i = 729.8665 - 10.887(119) + .0465(119)^2 = 92.8$$

Thus, a store selling the razors for 79 cents is expected to sell 52.2 more packages than a store selling them for 99 cents, but a store selling them for 99 cents is expected to sell only 15 more packages than a store selling them for $1.19 (119 cents).

Testing for the Significance of the Curvilinear Model

Now that the curvilinear model has been fitted to the data, we can determine whether there is a significant overall relationship between sales Y and price X. In a manner similar to multiple regression (see section 14.3), the null and alternative hypotheses can be set up as follows:

H_0: $\beta_1 = \beta_2 = 0$ (There is no overall relationship between X_1 and Y.)

H_1: β_2 and/or $\beta_1 \neq 0$ (There is an overall relationship between X_1 and Y.)

The null hypothesis can be tested by using equation (14.6):

$$F = \frac{MSR}{MSE}$$

From the Excel output in Figure 14.10 on page 885 we have

$$F = \frac{MSR}{MSE} = \frac{6,221.4}{165.63} = 37.56$$

If a level of significance of .05 is chosen, we consult Table E.5 and find that for 2 and 12 degrees of freedom, the critical value on the F distribution is 3.89 (see Figure 14.12).

FIGURE 14.12
Testing for existence of overall relationship at .05 level
of significance with 2 and
12 degrees of freedom

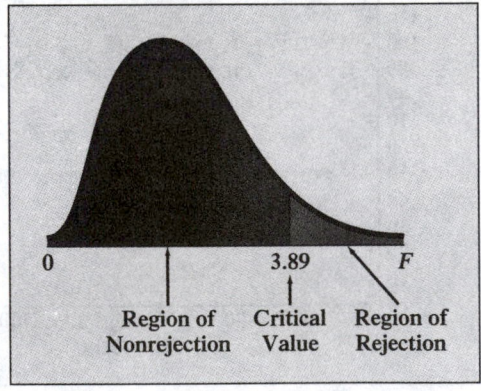

0   3.89 F

Region of   Critical   Region of
Nonrejection   Value   Rejection

Because $F = 37.56 > 3.89$ or because from Figure 14.10 the p-value $= .000006828 < .05$, we reject the null hypothesis H_0 and conclude that there is a significant overall relationship between sales and price of razors.

Testing the Curvilinear Effect

In using a regression model to examine a relationship between two variables, we would like to fit not only the most accurate model but also the simplest model expressing that relationship. Therefore, it becomes important to examine whether there is a significant difference between the curvilinear model

$$Y_i = \beta_0 + \beta_1 X_{1i} + \beta_2 X_{1i}^2 + \epsilon_i$$

and the linear model

$$Y_i = \beta_0 + \beta_1 X_{1i} + \epsilon_i$$

We can compare these two models by determining the regression effect of adding the curvilinear term, given that the linear term has already been included $[SSR(X_1^2 \mid X_1)]$.

You may recall that in section 14.4 we used the t test for the regression coefficient to determine whether each particular variable made a significant contribution to the regression model. Because the standard error of each regression coefficient and its corresponding t statistic are available as part of the Excel output (see Figure 14.10 on page 885), we test the significance of the contribution of the curvilinear effect with the following null and alternative hypotheses:

H_0: Including the curvilinear effect does not significantly improve the model ($\beta_2 = 0$)

H_1: Including the curvilinear effect significantly improves the model ($\beta_2 \neq 0$)

For our data

$$t = \frac{b_2}{S_{b_2}}$$

so that

$$t = \frac{.0465}{.01762} = 2.64$$

If a level of significance of .05 is selected, we use Table E.3 and find that with 12 degrees of freedom the critical values are -2.1788 and $+2.1788$ (see Figure 14.13).

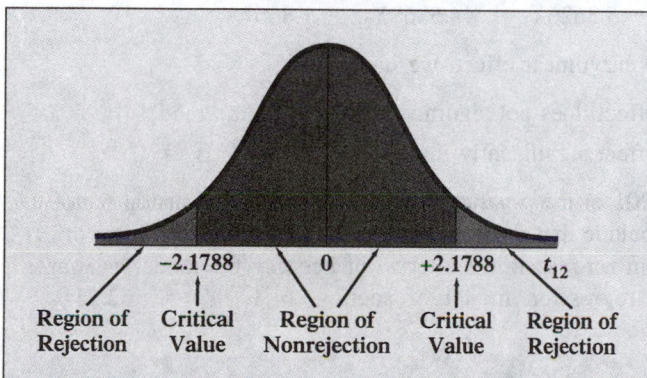

FIGURE 14.13

Testing for contribution of curvilinear effect to regression model at .05 level of significance with 12 degrees of freedom

Because $t = 2.64 > t_{12} = 2.1788$ or because the p-value $= .0216 < .05$, we reject H_0 and conclude that the curvilinear model is significantly better than the linear model in representing the relationship between sales and price.

As an additional illustration of a possible curvilinear effect, we turn to Example 14.3.

Example 14.3 Studying the Curvilinear Effect in a Multiple Regression Model

In section 14.1 a multiple regression model was fit to predict monthly heating oil consumption based upon atmospheric temperature and the amount of attic insulation. Examination of the residual plot for attic insulation showed some evidence of a curvilinear effect. Fit a multiple regression model that includes a curvilinear term for attic insulation. At the .05 level of significance, is there evidence of a curvilinear effect for attic insulation?

SOLUTION

Using Microsoft Excel, we obtain the following output:

	A	B	C	D	E	F	G
1	Regression Analysis: Curvilinear Effect for Attic Insulation?						
2							
3	Regression Statistics						
4	Multiple R	0.986157697					
5	R Square	0.972507004					
6	Adjusted R Square	0.965008914					
7	Standard Error	24.29377937					
8	Observations	15					
9							
10	ANOVA						
11		df	SS	MS	F	Significance F	
12	Regression	3	229643.1645	76547.72149	129.7006349	7.26403E-09	
13	Residual	11	6492.064875	590.1877159			
14	Total	14	236135.2293				
15							
16		Coefficients	Standard Error	t Stat	P-value	Lower 95%	Upper 95%
17	Intercept	624.5864209	42.43515952	14.71860664	1.39085E-08	531.1872173	717.9856245
18	Temperature	-5.362603095	0.317128467	-16.90987611	3.20817E-09	-6.060598498	-4.664607691
19	Insulation	-44.58678859	14.9546884	-2.981458884	0.012486902	-77.50185248	-11.67172469
20	Insulation^2	1.866704651	1.123755228	1.661131004	0.124891755	-0.606665181	4.340074482

Microsoft Excel output for multiple regression model with curvilinear term for attic insulation

The regression model is

$$\hat{Y}_i = 624.5864 - 5.3626X_{1i} - 44.5868X_{2i} + 1.8667X_{2i}^2$$

To test for the significance of the curvilinear effect, we have

H_0: Including the curvilinear effect does not significantly improve the model ($\beta_3 = 0$)

H_1: Including the curvilinear effect significantly improves the model ($\beta_3 \neq 0$)

From the output, $t = 1.661 < 2.201$ or the p-value $= .1249 > .05$. The decision is not to reject the null hypothesis. We conclude that there is no evidence that the curvilinear effect for attic insulation is different from zero, so in the interest of keeping the model as simple as possible, the multiple linear regression model of section 14.1, $\hat{Y}_i = 562.151 - 5.43658X_{1i} - 20.0123X_{2i}$, should be used.

Obtaining the Coefficient of Multiple Determination

In the multiple regression model, we computed the coefficient of multiple determination $r_{Y.12}^2$ (see section 14.1) to represent the proportion of variation in Y that is explained by variation in the explanatory variables. In curvilinear regression analysis, this coefficient can be computed from equation (14.4):

$$r_{Y.12}^2 = \frac{SSR}{SST}$$

From Figure 14.10,

$$SSR = 12{,}442.8 \qquad SST = 14{,}430.4$$

Thus,

$$r_{Y.12}^2 = \frac{SSR}{SST} = \frac{12{,}442.8}{14{,}430.4} = .862$$

This coefficient of multiple determination, computed as .862, means that 86.2% of the variation in sales can be explained by the curvilinear relationship between sales and price. An adjusted $r_{Y.12}^2$ can also be obtained that takes into account the number of explanatory variables and the degrees of freedom. In our curvilinear regression model, $p = 2$ because we have two explanatory variables, X_1, and its square, X_1^2. Thus, using equation (14.5) for the razor sales data, we have

$$
\begin{aligned}
r_{adj}^2 &= 1 - \left[(1 - r_{Y.12}^2) \frac{(15 - 1)}{(15 - 2 - 1)} \right] \\
&= 1 - \left[(1 - .862) \frac{14}{12} \right] \\
&= 1 - .161 \\
&= .839
\end{aligned}
$$

 **14.6E** CREATING CURVILINEAR REGRESSION MODELS USING MICROSOFT EXCEL

Overview

Both quick results users and developers should implement a formula that squares the values of an X variable to explore the curvilinear relationship between the X and the Y variable.

The 14-6E.XLS workbook file contains the calculations for the curvilinear regression model for the Table 14.5 disposable razors sales and price data for the price elasticity problem of section 14.6.

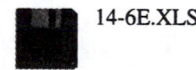 14-6E.XLS

Details for All Users

We can use simple formulas in the form =B2^2 to create the square of a first explanatory variable. As an example, consider the price elasticity problem of section 14.6. To implement the curvilinear model for these data, do the following:

❶ Open the Starting Point for Section 14.6E workbook (STARTING POINT 14-6E.XLS) and click the Data sheet tab. Verify that the sales and price data of Table 14.5 on page 884 have been entered into columns A and B, respectively.

❷ Enter the label Price^2 in cell C1 (of the Data sheet).

❸ Enter the formula =B2^2 in cell C2 and copy it down through row 16.

❹ Select PHStat | Regression | Multiple Regression.

❺ In the Multiple Regression dialog box:
  a. Enter A1:A16 in the Y Variable Cell Range: edit box.
  b. Enter B1:C16 in the X Variables Cell Range: edit box.
  c. Select the First cells in both ranges contain label check box.
  d. Enter 95 in the Confidence Level for regr. coefficients: edit box.
  e. Select the Regression Statistics Table and the ANOVA and Coefficients Table check boxes.
  f. Enter Regression Analysis for Price Elasticity in the Output Title: edit box.
  g. Click the OK button

The add-in inserts a worksheet similar to the one shown in Figure 14.10 on page 885. The cell range B17:B19 contains the three regression coefficients for the curvilinear regression equation.

Problems for Section 14.6

Learning the Basics

● **14.32** Suppose the following curvilinear regression model was fit for a sample of $n = 25$:

$$\hat{Y}_i = 5 + 3X_{1i} + 1.5X_{1i}^2$$

(a) Predict the average Y for $X = 2$.
(b) Suppose the t statistic for the curvilinear term is 2.35. At the .05 level of significance, is there evidence that the curvilinear model is better than the linear model?
(c) Suppose the t statistic for the curvilinear term is 1.17. At the .05 level of significance, is there evidence that the curvilinear model is better than the linear model?
(d) Suppose the regression coefficient for the linear effect was -3.0. Predict the average Y for $X = 2$.

Applying the Concepts

● **14.33** A researcher for a major oil company wishes to develop a model to predict miles per gallon based on highway speed. An experiment is designed in which a test car is driven at speeds ranging from 10 miles per hour to 75 miles per hour in 5-mile increments over two trial periods. The results are shown at the top of page 891.
  Assume a curvilinear relationship between speed and mileage based on the results obtained from Excel.
(a) Set up a scatter diagram between speed and miles per gallon.
(b) State the equation for the curvilinear model.
(c) Predict the average mileage obtained when the car is driven at 55 miles per hour.
(d) Perform a residual analysis on your results and determine the adequacy of the fit of the model.
(e) Determine whether there is a significant curvilinear relationship between mileage and speed at the .05 level of significance.

OBSERVATION	MILES PER GALLON	SPEED (MILES PER HOUR)	OBSERVATION	MILES PER GALLON	SPEED (MILES PER HOUR)
1	4.8	10	15	21.3	45
2	5.7	10	16	22.0	45
3	8.6	15	17	20.5	50
4	7.3	15	18	19.7	50
5	9.8	20	19	18.6	55
6	11.2	20	20	19.3	55
7	13.7	25	21	14.4	60
8	12.4	25	22	13.7	60
9	18.2	30	23	12.1	65
10	16.8	30	24	13.0	65
11	19.9	35	25	10.1	70
12	19.0	35	26	9.4	70
13	22.4	40	27	8.4	75
14	23.5	40	28	7.6	75

DATA FILE
SPEED.XLS

(f) At the .05 level of significance, determine whether the curvilinear model is a better fit than the linear regression model.

(g) Interpret the meaning of the coefficient of multiple determination $r^2_{Y.12}$.

(h) Compute the adjusted r^2.

14.34 An industrial psychologist would like to develop a model to predict the number of typing errors based on the amount of alcoholic consumption (in ounces). A random sample of 15 typists is selected with the following results:

TYPIST	ALCOHOLIC CONSUMPTION	NUMBER OF ERRORS	TYPIST	ALCOHOLIC CONSUMPTION	NUMBER OF ERRORS
1	0	2	9	2	9
2	0	6	10	3	13
3	0	3	11	3	18
4	1	7	12	3	16
5	1	5	13	4	24
6	1	9	14	4	30
7	2	12	15	4	22
8	2	7			

DATA FILE
ALCOHOL.XLS

Assuming a curvilinear relationship between alcoholic consumption and the number of errors and using Microsoft Excel:

(a) Set up a scatter diagram between alcoholic consumption X and number of errors Y.

(b) State the equation for the curvilinear model.

(c) Predict the average number of errors made by a typist who has consumed 2.5 ounces of alcohol.

(d) Perform a residual analysis on your results and determine the adequacy of the fit of the model.

(e) Determine whether there is a significant curvilinear relationship between alcoholic consumption and the number of errors made at the .05 level of significance.

677

(f) At the .05 level of significance, determine whether the curvilinear model is a better fit than the linear regression model.

(g) Interpret the meaning of the coefficient of multiple determination $r^2_{Y.12}$.

(h) Compute the adjusted r^2.

14.35 Suppose an agronomist wants to design a study in which a wide range of fertilizer levels (pounds per 1,000 square feet) is to be used to determine whether the relationship between the yield (in pounds) of tomatoes and amount of fertilizer can be fit by a curvilinear model. Six rates of application are to be used: 0, 20, 40, 60, 80, and 100 pounds per 1,000 square feet. These rates are then randomly assigned to plots of land with the following results:

PLOT	FERTILIZER APPLICATION RATE	YIELD	PLOT	FERTILIZER APPLICATION RATE	YIELD
1	0	6	7	60	46
2	0	9	8	60	50
3	20	19	9	80	48
4	20	24	10	80	54
5	40	32	11	100	52
6	40	38	12	100	58

Assuming a curvilinear relationship between the application rate and tomato yield and using Microsoft Excel:

(a) Set up a scatter diagram between application rate and yield.

(b) State the regression equation for the curvilinear model.

(c) Predict the average yield of tomatoes (in pounds) for a plot that has been fertilized with 70 pounds per 1,000 square feet.

(d) Perform a residual analysis on your results and determine the adequacy of the fit of the model.

(e) Determine from the curvilinear model whether there is a significant overall relationship between the application rate and tomato yield at the .05 level of significance.

(f) What is the p-value in (e)? Interpret its meaning.

(g) At the .05 level of significance, determine whether there is a significant curvilinear effect.

(h) What is the p-value in (g)? Interpret its meaning.

(i) Interpret the meaning of the coefficient of multiple determination $r^2_{Y.12}$.

(j) Compute the adjusted r^2.

14.36 An auditor for a county government would like to develop a model to predict the county taxes based on the age of single-family houses. A random sample of 19 single-family houses has been selected with the following results on page 893.

Assuming a curvilinear relationship between the age and county taxes and using Microsoft Excel:

(a) Set up a scatter diagram between age and county taxes.

(b) State the regression equation for the curvilinear model.

(c) Predict the average county taxes for a house that is 20 years old.

(d) Perform a residual analysis on your results and determine the adequacy of the fit of the model.

(e) Determine whether there is a significant overall relationship between age and county taxes at the .05 level of significance.

County Taxes ($)	Age (years)	County Taxes ($)	Age (years)
925	1	480	20
870	2	486	22
809	4	462	25
720	4	441	25
694	5	426	30
630	8	368	35
626	10	350	40
562	10	348	50
546	12	322	50
523	15		

DATA FILE
TAXES.XLS

(f) What is the p-value in (e)? Interpret its meaning.
(g) At the .05 level of significance, determine whether the curvilinear model is superior to the linear regression model.
(h) What is the p-value in (g)? Interpret its meaning.
(i) Interpret the meaning of the coefficient of multiple determination $r^2_{Y.12}$.
(j) Compute the adjusted r^2.

14.7 DUMMY-VARIABLE MODELS

In our discussion of multiple regression models we have assumed that each explanatory (or independent) variable is numerical. However, there are many occasions in which categorical variables need to be included in the model development process. For example, in section 14.1 we used the atmospheric temperature and the amount of attic insulation to predict the average monthly consumption of home heating oil. In addition to these numerical independent variables, we may want to include the effect of the style of the house (for example, ranch- or non-ranch-style houses) when developing a model to predict heating oil consumption.

The use of **dummy variables** is the vehicle that permits us to consider categorical explanatory variables as part of the regression model. If a given categorical explanatory variable has two categories, then only one dummy variable will be needed to represent the two categories. A particular dummy variable X_d is defined as

$$X_d = 0 \text{ if the observation is in category 1}$$
$$X_d = 1 \text{ if the observation is in category 2}$$

To illustrate the application of dummy variables in regression, we will examine a model for predicting the average assessed value from a sample of 15 houses based on the heating area (in thousands of square feet) and whether or not the house has a fireplace. The data are presented in Table 14.6. A dummy variable for fireplace (X_2) can be defined as

$$X_2 = 0 \text{ if the house does not have a fireplace}$$
$$X_2 = 1 \text{ if the house has a fireplace}$$

Table 14.6 Predicting assessed value based on heating area and presence of a fireplace

House	Assessed Value ($000)	Heating Area of Dwelling (thousands of square feet)	Fireplace	House	Assessed Value ($000)	Heating Area of Dwelling (thousands of square feet)	Fireplace
1	84.4	2.00	Yes	9	78.5	1.59	Yes
2	77.4	1.71	No	10	79.2	1.50	Yes
3	75.7	1.45	No	11	86.7	1.90	Yes
4	85.9	1.76	Yes	12	79.3	1.39	Yes
5	79.1	1.93	No	13	74.5	1.54	No
6	70.4	1.20	Yes	14	83.8	1.89	Yes
7	75.8	1.55	Yes	15	76.8	1.59	No
8	85.9	1.93	Yes				

DATA FILE
HOUSE3.XLS

Assuming that the slope of assessed value with heating area is the same for houses that have and do not have a fireplace, the regression model to be fitted is

$$Y_i = \beta_0 + \beta_1 X_{1i} + \beta_2 X_{2i} + \epsilon_i$$

where

Y_i = assessed value in thousands of dollars

β_0 = Y intercept

β_1 = slope of assessed value with heating area holding constant the effect of the presence of a fireplace

β_2 = incremental effect of the presence of a fireplace holding constant the effect of heating area

ϵ_i = random error in Y for house i

Figure 14.14 illustrates the output for this model obtained from Microsoft Excel.

	A	B	C	D	E	F	G
1	Regression Analysis for Fireplace Prediction						
2							
3	Regression Statistics						
4	Multiple R	0.900587177					
5	R Square	0.811057264					
6	Adjusted R Square	0.779566808					
7	Standard Error	2.262595954					
8	Observations	15					
9							
10	ANOVA						
11		df	SS	MS	F	Significance F	
12	Regression	2	263.7039146	131.8519573	25.75565321	4.54968E-05	
13	Residual	12	61.4320854	5.11934045			
14	Total	14	325.136				
15							
16		Coefficients	Standard Error	t Stat	P-value	Lower 95%	Upper 95%
17	Intercept	50.09048899	4.351657945	11.5106678	7.67943E-08	40.60904099	59.57193699
18	Heating	16.18583395	2.574441705	6.28712389	4.02437E-05	10.57660743	21.79506047
19	Fireplace	3.852982483	1.241222689	3.10418309	0.009118854	1.148590609	6.557374357

FIGURE 14.14 Microsoft Excel output for regression model that includes heating area and presence of fireplace

From this output, the sample regression equation may be stated as

$$\hat{Y}_i = 50.09 + 16.186X_{1i} + 3.853X_{2i}$$

For houses without a fireplace, this reduces to

$$\hat{Y}_i = 50.09 + 16.186X_{1i}$$

because $X_2 = 0$; whereas for houses with a fireplace, the regression equation is

$$\hat{Y}_i = 53.943 + 16.186X_{1i}$$

because $X_2 = 1$, so 3.853 is added to 50.09. In this model, we can interpret the slopes as follows:

1. Holding constant whether or not a house has a fireplace, for each increase of 1,000 square feet in heating area, the average assessed value is predicted to increase by 16.186 thousands of dollars (or $16,186).

2. Holding constant the heating area of the house, the presence of a fireplace is predicted to increase the assessed value of the house by an average of 3.853 thousand dollars (or $3,853).

We note from Figure 14.14 that the t statistic for the slope of heating area with assessed value is 6.29 and the p-value is approximately .000; the t statistic for presence of a fireplace is 3.10 and the p-value is .009. Thus, each of the two variables is making a significant contribution to the model at a level of significance of .01. In addition, 81.1% of the variation in assessed value is explained by variation in the heating area of the house and whether or not the house has a fireplace.

However, before we can use this model, we need to assure ourselves that the slope of assessed value with heating area is the same for houses with a fireplace as it is for houses without a fireplace. A hypothesis of equal slopes of a Y variable with X can be evaluated by defining an **interaction term** that consists of the product of the explanatory variable X_1 and the dummy variable X_2, and then testing whether this interaction variable makes a significant contribution to a regression model that contains the other X variables. If the interaction is significant, we cannot use our original model for prediction. For the data of Table 14.5,

$$X_3 = X_1 \times X_2$$

The Microsoft Excel output for this regression model, which includes the heating area of the house X_1, the presence of a fireplace X_2, and the interaction of X_1 and X_2 (which we have defined as X_3), is provided in Figure 14.15.

	A	B	C	D	E	F	G
1	Regression Analysis for Fireplace Prediction						
2							
3	Regression Statistics						
4	Multiple R	0.917906505					
5	R Square	0.842552352					
6	Adjusted R Square	0.799612084					
7	Standard Error	2.157268864					
8	Observations	15					
9							
10	ANOVA						
11		df	SS	MS	F	Significance F	
12	Regression	3	273.9441015	91.3147005	19.62149745	0.000100865	
13	Residual	11	51.19189849	4.653808954			
14	Total	14	325.136				
15							
16		Coefficients	Standard Error	t Stat	P-value	Lower 95%	Upper 95%
17	Intercept	62.9521815	9.612176928	6.54921169	4.13993E-05	41.79591203	84.10845098
18	Heating	8.362420012	5.817298426	1.437509201	0.178405557	-4.441373971	21.16621399
19	Fireplace	-11.84036371	10.64550326	-1.11224086	0.289751611	-35.27097026	11.59024285
20	X1*X2	9.518000219	6.416468169	1.483370597	0.166052624	-4.604558143	23.64055858

FIGURE 14.15 Microsoft Excel output for regression model that includes heating area, presence of fireplace, and interaction of heating area and presence of fireplace

To test the null hypothesis H_0: $\beta_3 = 0$ versus the alternative hypothesis H_1: $\beta_3 \neq 0$, from Figure 14.15 we observe that the t statistic for the interaction of heating area and presence of a fireplace is 1.48. Because the p-value $= .166 > .05$, the null hypothesis is not rejected. We conclude that the interaction term does not make a significant contribution to the model given that heating area and presence of a fireplace are already included.

Now that we have examined a regression model that includes a categorical explanatory variable, we turn to Example 14.4 to study a regression model in which there is more than one numerical explanatory variable along with a categorical variable.

Example 14.4 *Studying a Regression Model that Contains a Dummy Variable*

In the Using Statistics example (see pages 857–858), heating oil consumption was predicted based on atmospheric temperature X_1 and the amount of attic insulation X_2. Suppose that, of the 15 houses in that sample, houses 1, 4, 6, 7, 8, 10, and 12 are ranch-style houses. Fit the appropriate regression model based on these three independent variables.

SOLUTION

A dummy variable for ranch-style house X_3 is defined as

$$X_3 = 0 \text{ if the style is not ranch}$$
$$X_3 = 1 \text{ if the style is ranch}$$

Assuming that the slope between home heating oil consumption and atmospheric temperature X_1 and the amount of attic insulation X_2 is the same for both groups, the regression model is

$$Y_i = \beta_0 + \beta_1 X_{1i} + \beta_2 X_{2i} + \beta_3 X_{3i} + \epsilon_i$$

where

Y_i = monthly heating oil consumption in gallons

β_0 = Y intercept

β_1 = slope of heating oil consumption with atmospheric temperature holding constant the effect of attic insulation and the house style

β_2 = slope of heating oil consumption with attic insulation holding constant the effect of atmospheric temperature and the house style

β_3 = incremental effect of the presence of a ranch-style house holding constant the effect of atmospheric temperature and attic insulation

ϵ_i = random error in Y for house i

The following figure displays partial output obtained from Microsoft Excel.

	A	B	C	D	E	F	G
1	Regression Model with Ranch-Style						
2							
3	*Regression Statistics*						
4	Multiple R	0.994206187					
5	R Square	0.988445943					
6	Adjusted R Square	0.985294836					
7	Standard Error	15.7489393					
8	Observations	15					
9							
10	ANOVA						
11		*df*	*SS*	*MS*	*F*	*Significance F*	
12	Regression	3	233406.9094	77802.30312	313.6821708	6.21548E-11	
13	Residual	11	2728.319981	248.0290892			
14	Total	14	236135.2293				
15							
16		*Coefficients*	*Standard Error*	*t Stat*	*P-value*	*Lower 95%*	*Upper 95%*
17	Intercept	592.5401168	14.33698425	41.32948091	2.02317E-13	560.9846112	624.0956223
18	Temperature	-5.525100884	0.204431228	-27.0266971	2.07188E-11	-5.975051212	-5.075150557
19	Insulation	-21.37612794	1.448019304	-14.76232249	1.34816E-08	-24.56319855	-18.18905733
20	Ranch-Style	-38.97266608	8.358437237	-4.662673772	0.000690709	-57.3694717	-20.57586045

Excel output for regression model including dummy variable (style) for home heating oil data

From this output, the sample regression equation may be stated as

$$\hat{Y}_i = 592.5401 - 5.5251X_{1i} - 21.3761X_{2i} - 38.9726X_{3i}$$

For houses that are not ranch-style, this reduces to

$$\hat{Y}_i = 592.5401 - 5.5251X_{1i} - 21.3761X_{2i}$$

because $X_3 = 0$. For houses that are ranch-style, the sample regression equation reduces to

$$\hat{Y}_i = 553.5674 - 5.5251X_{1i} - 21.3761X_{2i}$$

because $X_3 = 1$. In addition, we note the following:

1. Holding constant the effect of attic insulation and the house style, for each additional 1°F increase in atmospheric temperature the average oil consumption is predicted to decrease by 5.525 gallons.

2. Holding constant the effect of atmospheric temperature and the house style, for each additional 1-inch increase in attic insulation the average oil consumption is predicted to decrease by 21.376 gallons.

3. b_3 measures the effect on oil consumption of having a ranch-style house ($X_3 = 1$) as compared with not having a ranch-style house ($X_3 = 0$). Thus, we estimate that with atmospheric temperature and attic insulation held constant, a ranch-style house is predicted to use 38.973 fewer gallons of heating oil per month than when the style is not ranch.

We note from the Excel output that the three t statistics representing the slopes for temperature, insulation, and presence of a ranch-style house are -27.03, -14.76, and -4.66. Each of the corresponding p-values is extremely small, all being less than .001. Thus, each of the three variables is making a significant contribution to the model. In addition, 98.8% of the variation in oil usage is explained by variation in the temperature, insulation, and whether or not the house is ranch-style.

Before we can use the model represented by the sample regression equation in Example 14.4 we need to find out whether the slope of oil usage with temperature is the same for ranch-style houses as it is for houses that are not ranch-style, and also whether the slope of oil usage with insulation is the same for ranch-style houses as it is for houses that are not ranch-style. To accomplish this, we develop Example 14.5.

Example 14.5 *Evaluating a Regression Model with Two Interactions*

For the data of Example 14.4 determine whether the two interaction effects make a significant contribution to the regression model.

SOLUTION

As we saw in our previous discussion of dummy variables, a hypothesis of equal slopes of an X variable with Y can be evaluated by defining an interaction term that consists of the product of the X variable and the dummy variable, and then testing whether this interaction variable makes a significant contribution to a regression model that contains the other X variables. In the case of two numerical X variables we evaluate the contribution of including both interaction terms, given that the three X variables are included in the model. For the oil usage example, we define

$$X_4 = X_1 \times X_3 \text{ and } X_5 = X_2 \times X_3$$

The Excel output for this regression model, which includes temperature X_1, insulation X_2, whether or not the house is ranch-style X_3, the interaction of X_1 and X_3 (which we have defined as X_4), and the interaction of X_2 and X_3 (which we have defined as X_5), is provided below.

	A	B	C	D	E	F	G
1	Regression Model withTwo Interactions						
2							
3	*Regression Statistics*						
4	Multiple R	0.994961389					
5	R Square	0.989948166					
6	Adjusted R Square	0.984363813					
7	Standard Error	16.23984198					
8	Observations	15					
9							
10	ANOVA						
11		*df*	*SS*	*MS*	*F*	*Significance F*	
12	Regression	5	233761.6371	46752.32743	177.2717932	1.04608E-08	
13	Residual	9	2373.592207	263.7324675			
14	Total	14	236135.2293				
15							
16		*Coefficients*	*Standard Error*	*t Stat*	*P-value*	*Lower 95%*	*Upper 95%*
17	Intercept	599.9782434	16.66954901	35.99247004	4.88276E-11	562.2690749	637.6874118
18	Temperature	-5.624853951	0.374378066	-15.02452857	1.11221E-07	-6.47175662	-4.777951282
19	Insulation	-21.84863672	1.955643913	-11.17209354	1.41221E-06	-26.27261397	-17.42465946
20	Ranch-Style	-76.85563866	33.87148579	-2.269036532	0.049440882	-153.4783213	-0.232956068
21	X1*X3	0.379785059	0.499241778	0.760723714	0.466284201	-0.749579165	1.509149284
22	X2*X3	3.924142467	3.954160702	0.992408443	0.346928526	-5.020797306	12.86908224

Excel output for regression model that includes temperature X_1, insulation X_2, whether or not the house is ranch-style X_3, the interaction of temperature and ranch-style X_4, and the interaction of insulation and ranch-style X_5

The null and alternative hypotheses are:

H_0: $\beta_4 = \beta_5 = 0$ (There is no interaction between X_1 or X_2 with X_3.)

H_1: β_4 and/or $\beta_5 \neq 0$ (X_1 and/or X_2 interacts with X_3.)

To test the null hypotheses we need to compare the regression sum of squares for the model that contains both interactions with the regression model that contains neither interaction. From the figure from Example 14.4 and the figure in this example, we have

$$SSR(X_1, X_2, X_3, X_4, X_5) = 233{,}761.6371 \text{ with 5 degrees of freedom}$$

and

$$SSR(X_1, X_2, X_3) = 233{,}406.9094 \text{ with 3 degrees of freedom}$$

Thus,

$$SSR(X_1, X_2, X_3, X_4, X_5) - SSR(X_1, X_2, X_3) = 233{,}761.6371 - 233{,}406.9094 = 354.7277$$

and the difference in degrees of freedom is $5 - 3 = 2$.

To test the null hypothesis, using an F test for the contribution of variables to a model based on Equation (14.11) on page 876 (because there is more than 1 degree of freedom in the numerator), we have

$$F = \frac{[SSR(X_1, X_2, X_3, X_4, X_5) - SSR(X_1, X_2, X_3)] / \text{difference in the regression d.f.}}{MSE(X_1, X_2, X_3, X_4, X_5)}$$

$$F = \frac{354.7277 / 2}{263.7325}$$

$$= 0.673$$

At the .05 level of significance, because $F = 0.673 < 4.26$, we conclude that neither interaction term makes a significant contribution to the model given that temperature X_1, insulation X_2, and whether or not the house is ranch-style X_3 are already included. Had we rejected this null hypothesis, we would need to test the contribution of each interaction separately in order to determine if both or only one interaction term should be included in the model.

14.7E CREATING DUMMY-VARIABLE MODELS USING MICROSOFT EXCEL

Overview

Both quick results users and developers should use the Excel find and replace feature to change a categorical explanatory variable into a dummy variable with the values 0 and 1.

The 14-7E.XLS workbook file contains the calculations for the dummy-variable regression model for the assessed value, heating area, and presence of a fireplace data of Table 14.6 used in the assessed value prediction problem of section 14.7.

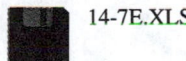

14-7E.XLS

Details for All Users

We can use the Excel find and replace feature to change a two-category categorical variable into a dummy variable with the values 0 and 1. As an example, consider the assessed value prediction problem of section 14.7. To implement a regression model that includes a dummy variable for the presence of a fireplace, do the following:

❶ Open the Starting Point for Section 14.7E workbook (STARTING POINT 14-7E.XLS) and click the Data sheet tab. Verify that the assessed value, heating area, and fireplace data of Table 14.6 on page 894 have been entered into columns A, B, and C, respectively.

❷ For later data verification, copy the contents of column C to column D.

❸ Select the range C2:C16 containing the categorical responses Yes and No, then select Edit | Replace.

❹ In the Replace dialog box (see Figure 14E.3):
   a. Enter Yes in the Find what: edit box.
   b. Enter 1 in the Replace with: edit box.
   c. Click the Replace All button.

❺ With the cell range C2:C16 still selected, select Edit | Replace a second time.

❻ In the Replace dialog box:
   a. Enter No in the Find what: edit box.
   b. Enter 0 in the Replace with: edit box.
   c. Click the Replace All button.

❼ Verify the changes by comparing the dummy-variable values in column C with their categorical equivalents in column D.

FIGURE 14E.3
Replace dialog box

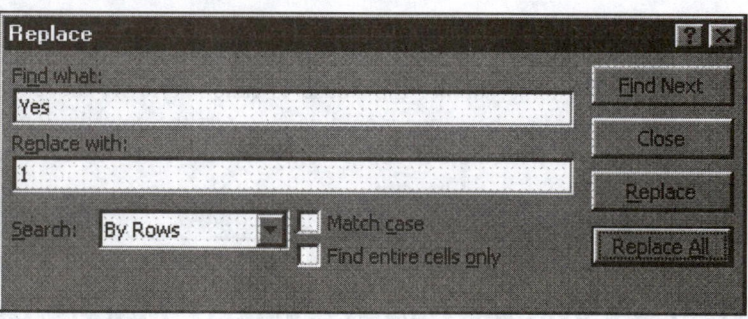

Once verified, the contents of columns A, B, and C can be used in a dummy-variable regression analysis. To generate this regression analysis, continue by doing the following:

❽ Click the Data sheet tab.

❾ Select PHStat | Regression | Multiple Regression.

❿ In the Multiple Regression dialog box:
   a. Enter A1:A16 in the Y Variable Cell Range: edit box.
   b. Enter B1:C16 in the X Variables Cell Range: edit box.
   c. Select the First cells in both ranges contain label check box.
   d. Enter 95 in the Confidence Level for regr. coefficients: edit box.

e. Select the Regression Statistics Table and the ANOVA and Coefficients Table check boxes.

f. Enter Regression Analysis for Fireplace Prediction in the Output Title: edit box.

g. Click the OK button.

The add-in inserts a worksheet similar to the one shown in Figure 14.14 on page 894.

Problems for Section 14.7

Learning the Basics

• **14.37** Suppose X_1 is a numerical variable and X_2 is a dummy variable. The following regression model has been fit for a sample of $n = 20$:

$$\hat{Y}_i = 6 + 4X_{1i} + 2X_{2i}$$

(a) Interpret the meaning of the slope for variable X_1.

(b) Interpret the meaning of the slope for variable X_2.

(c) Suppose that the t statistic for variable X_2 is 3.27. At the .05 level of significance, is there evidence that variable X_2 makes a significant contribution to the model?

Applying the Concepts

14.38 The chair of the accounting department of a large public university wishes to develop a regression model to predict the grade point average in accounting for students who are graduating as accounting majors, based on total SAT score for the student and whether or not the student received a grade of B or higher in the introductory statistics course (0 = no and 1 = yes).

(a) Explain the steps involved in developing a regression model for these data. Be sure to indicate the particular models that need to be evaluated and compared.

(b) If the regression coefficient for the variable of whether or not the student received a grade of B or higher in the introductory statistics course was +0.30, how would this be interpreted?

• **14.39** The marketing manager of a large supermarket chain would like to determine the effect of shelf space and whether the product was placed at the front or back of the aisle on the sales of pet food. A random sample of 12 equal-sized stores is selected with the following results:

STORE	SHELF SPACE, X (FEET)	LOCATION	WEEKLY SALES, Y (HUNDREDS OF DOLLARS)	STORE	SHELF SPACE, X (FEET)	LOCATION	WEEKLY SALES, Y (HUNDREDS OF DOLLARS)
1	5	Back	1.6	7	15	Back	2.3
2	5	Front	2.2	8	15	Back	2.7
3	5	Back	1.4	9	15	Front	2.8
4	10	Back	1.9	10	20	Back	2.6
5	10	Back	2.4	11	20	Back	2.9
6	10	Front	2.6	12	20	Front	3.1

DATA FILE
PETFOOD.XLS

Using Microsoft Excel:

(a) State the sample multiple regression equation.

(b) Interpret the meaning of the slopes in this problem.

(c) Predict the average weekly sales of pet food for a store with 8 feet of shelf space that is situated at the back of the aisle.

(d) Perform a residual analysis on your results and determine the adequacy of the fit of the model.

(e) Determine whether there is a significant relationship between sales and the two explanatory variables (shelf space and aisle position) at the .05 level of significance.

(f) At the .05 level of significance, determine whether each explanatory variable makes a contribution to the regression model. On the basis of these results, indicate the regression model that should be used in this problem.

(g) Set up 95% confidence interval estimates of the population slope for the relationship between sales and shelf space and between sales and aisle location.

(h) Compare the slope obtained in (b) with the slope for the simple linear regression model of Problem 13.3 on page 785. Explain the difference in the results.

(i) Interpret the meaning of the coefficient of multiple determination $r_{Y.12}^2$.

(j) Compute the adjusted r^2.

(k) Compare $r_{Y.12}^2$ with the r^2 value computed in Problem 13.14(a) on page 792.

(l) Compute the coefficients of partial determination and interpret their meaning.

(m) What assumption about the slope of shelf space with sales must be made in this problem?

(n) Include an interaction term in the model and at the .05 level of significance determine whether it makes a significant contribution to the model.

(o) On the basis of the results of (e) and (n), which model is more appropriate? Explain.

14.40 A bank would like to develop a model to predict the total sum of money customers withdraw from automatic teller machines (ATMs) on a weekend based on the median assessed value of houses in the vicinity of the ATM and the ATM's location (0 = not a shopping center; 1 = shopping center). A random sample of 15 ATMs is selected with the following results.

ATM NUMBER	WITHDRAWAL AMOUNT ($000)	MEDIAN ASSESSED VALUE OF HOUSES ($000)	LOCATION OF ATM
1	12.0	225	1
2	9.9	170	0
3	9.1	153	1
4	8.2	132	0
5	12.4	237	1
6	10.4	187	1
7	12.7	245	1
8	8.0	125	1
9	11.5	215	1
10	9.7	170	0
11	11.7	223	0
12	8.6	147	0
13	10.9	197	1
14	9.4	167	0
15	11.2	210	0

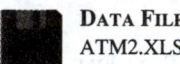
DATA FILE
ATM2.XLS

On the basis of the results obtained using Microsoft Excel:

(a) State the multiple regression equation.

(b) Interpret the meaning of the slopes in this problem.

(c) Predict the average withdrawal amount for a neighborhood in which the median assessed value of homes is $200,000 for an ATM located in a shopping center.

(d) Perform a residual analysis of your results and determine the adequacy of the model's fit.

(e) Determine whether there is a significant relationship between the withdrawal amount and the two explanatory variables (median assessed value of houses and the dummy variable of ATM location) at the .05 level of significance.

(f) At the .05 level of significance, determine whether each explanatory variable makes a contribution to the regression model. On the basis of these results, indicate the regression model that should be used in this problem.

(g) Set up 95% confidence interval estimates of the population slope for the relationship between the withdrawal amount and median assessed value of homes, and for the withdrawal amount and ATM location.

(h) Interpret the meaning of the coefficient of multiple determination $r^2_{Y.12}$.

(i) Compute the adjusted r^2.

(j) Compute and interpret the coefficients of partial determination.

(k) What assumption about the slope of withdrawal amount with median assessed value of homes must be made in this problem?

(l) Include an interaction term in the model and, at the .05 level of significance, determine whether it makes a significant contribution to the model.

(m) On the basis of the results of (f) and (l), which model is more appropriate? Explain.

• **14.41** A real estate association in a suburban community would like to study the relationship between the size of a single-family house (as measured by the number of rooms) and the selling price of the house. The study is to be carried out in two different neighborhoods, one on the east side of the community and the other on the west side. A random sample of 20 houses was selected with the following results:

SELLING PRICE	NUMBER OF ROOMS	NEIGHBORHOOD	SELLING PRICE	NUMBER OF ROOMS	NEIGHBORHOOD
109.6	7	East	108.5	6	West
107.4	8	East	181.3	13	West
140.3	9	East	137.4	10	West
146.5	12	East	146.2	10	West
98.2	6	East	142.4	9	West
137.8	9	East	123.7	8	West
124.1	10	East	129.6	8	West
113.2	8	East	143.6	9	West
127.8	9	East	160.7	11	West
125.3	8	East	148.3	9	West

DATA FILE
NEIGHBOR.XLS

Using Microsoft Excel:

(a) State the multiple regression equation.

(b) Interpret the meaning of the slopes in this problem.

(c) Predict the average selling price for a house with nine rooms that is located on the east side of the community.

(d) Perform a residual analysis on your results and determine the adequacy of the fit of the model.

(e) Determine whether there is a significant relationship between selling price and the two explanatory variables (rooms and neighborhood) at the .05 level of significance.

(f) At the .05 level of significance, determine whether each explanatory variable makes a contribution to the regression model. On the basis of these results, indicate the regression model that should be used in this problem.

(g) Set up 95% confidence interval estimates of the population slope for the relationship between selling price and number of rooms and between selling price and neighborhood.

(h) Interpret the meaning of the coefficient of multiple determination $r^2_{Y.12}$.

(i) Compute the adjusted r^2.

(j) Compute the coefficients of partial determination and interpret their meaning.

(k) What assumption about the slope of selling price with number of rooms must be made in this problem?

(l) Include an interaction term in the model and, at the .05 level of significance, determine whether it makes a significant contribution to the model.

(m) On the basis of the results of (f) and (l), which model is more appropriate? Explain.

14.42 The file UNIV&COL contains data on 80 colleges and universities. Among the variables included are the annual total cost (in thousands of dollars), the average total score on the Scholastic Aptitude Test (SAT), and whether the school is public or private (0 = public; 1 = private). Suppose we want to develop a model to predict the annual total cost based on SAT score and whether the school is public or private:

DATA FILE
UNIV&COL.XLS

(a) State the multiple regression equation.

(b) Interpret the meaning of the slopes in this problem.

(c) Predict the average total cost for a school with an average total SAT score of 1,000 that is a public institution.

(d) Perform a residual analysis on your results and determine the adequacy of the fit of the model.

(e) Determine whether there is a significant relationship between annual total cost and the two explanatory variables (total SAT score and whether the school is public or private) at the .05 level of significance.

(f) At the .05 level of significance, determine whether each explanatory variable makes a contribution to the regression model. On the basis of these results, indicate the regression model that should be used in this problem.

(g) Set up 95% confidence interval estimates of the population slope for the relationship between annual total cost and total SAT score and between annual total cost and whether the school is public or private.

(h) Interpret the meaning of the coefficient of multiple determination $r^2_{Y.12}$.

(i) Compute the adjusted r^2.

(j) Compute the coefficients of partial determination and interpret their meaning.

(k) What assumption about the slope of annual total cost with total SAT score must be made in this problem?

(l) Include an interaction term in the model and, at the .05 level of significance, determine whether it makes a significant contribution to the model.

(m) On the basis of the results of (f) and (l), which model is more appropriate? Explain.

14.8 USING TRANSFORMATIONS IN REGRESSION MODELS

In our discussion of multiple regression models we have thus far examined the multiple linear regression model [equation (14.1)], the curvilinear model [equation (14.15)], and a model containing the categorical explanatory (i.e., dummy) variable. In this section we discuss regression models in which the independent X variable, the dependent Y variable, or both, are transformed in order to either overcome violations of the assumptions of regression or make a model linear in its form. Among the many transformations available (see reference 5) are the square-root transformation and transformations involving the natural logarithm.[2]

[2]*The natural logarithm, usually abbreviated ln, is the logarithm to the base e, the mathematical constant approximately equal to 2.71828.*

The Square-Root Transformation

The **square-root transformation** is often used to overcome violations of the *homoscedasticity* assumption, as well as to transform a model that is not linear into one that is linear in form. If a square-root transformation were applied to the values of each of two explanatory variables, the multiple regression model would be

The use of a square-root transformation is illustrated in Example 14.6.

Example 14.6 *Using the Square-Root Transformation*

Given the following values for Y and X,

Y	X	Y	X
42.7	1	100.4	3
50 4	1	104.7	4
69.1	2	112.3	4
79.8	2	113.6	5
90.0	3	123.9	5

use a square-root transformation for the X variable and develop a scatter diagram.

SOLUTION

The first figure below displays the scatter diagram of X and Y; the second figure plots the square root of X and Y. These figures are obtained from Microsoft Excel.

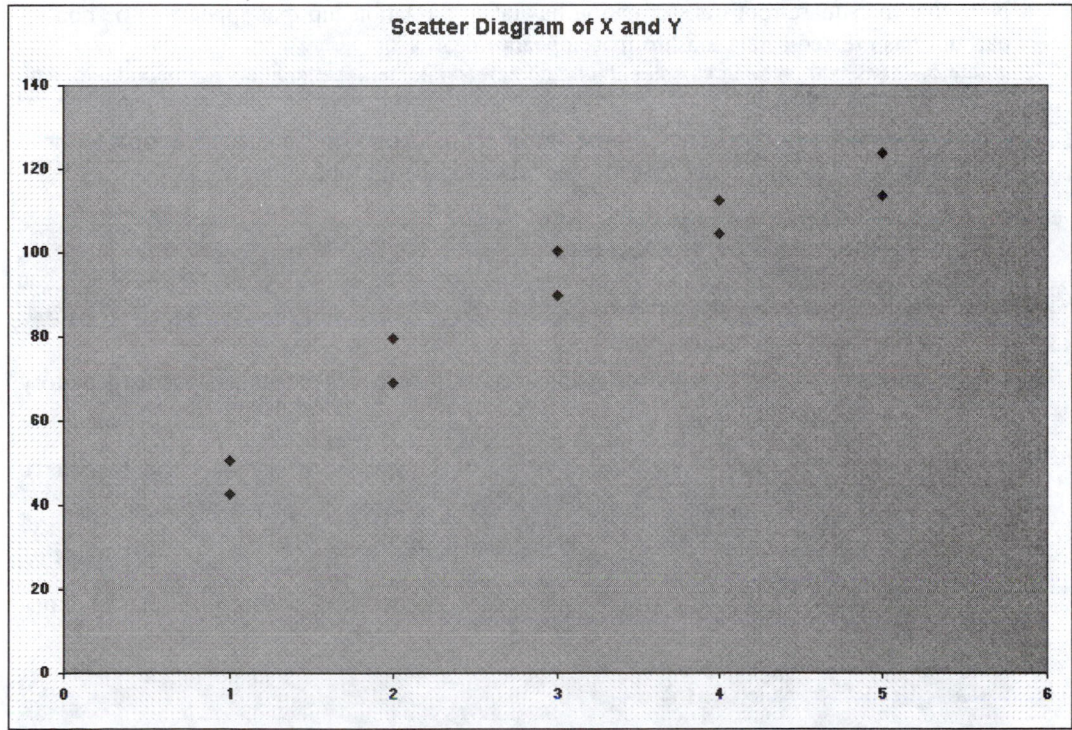

PANEL A Scatter diagram of X and Y

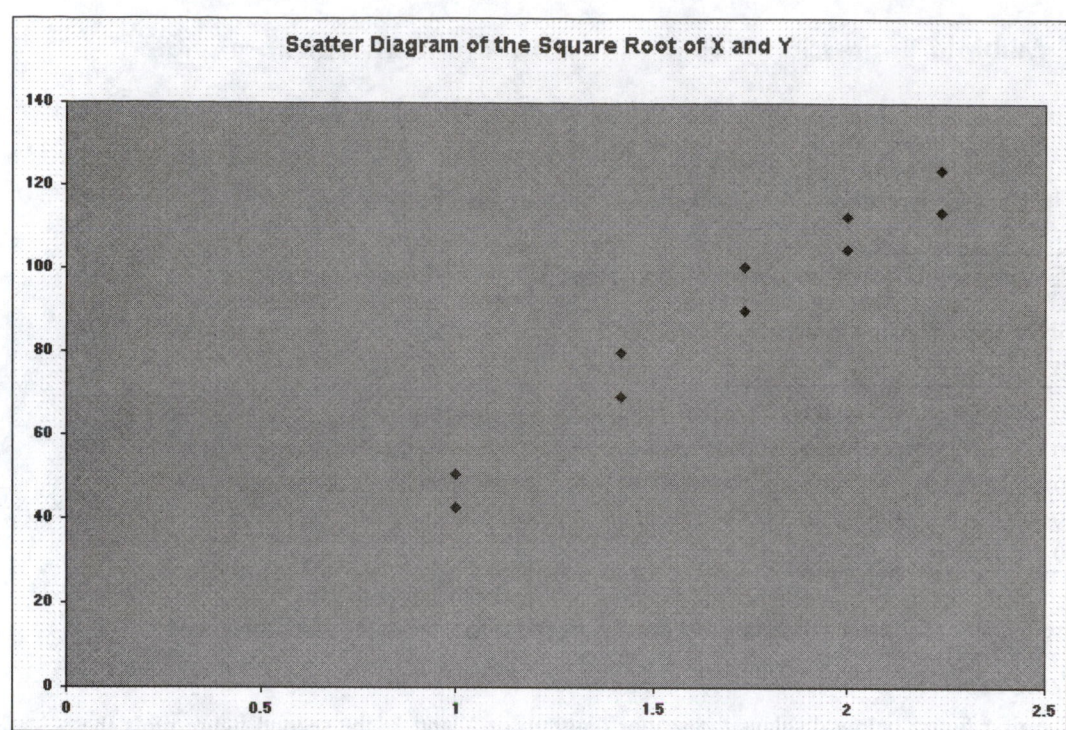

PANEL B Scatter diagram of square root of X and Y

Note that the square-root transformation has taken a relationship that appears to be nonlinear and has created a relationship that appears linear.

In some situations, the use of a **logarithmic transformation** can change a model whose form is nonlinear into a linear model. For example, the multiplicative model

Original Multiplicative Model

$$Y_i = \beta_0 X_{1i}^{\beta_1} X_{2i}^{\beta_2} \epsilon_i \qquad (14.18)$$

can be transformed (by taking natural logarithms of both the dependent and explanatory variables) to the model

Transformed Multiplicative Model

$$\ln Y_i = \ln \beta_0 + \beta_1 \ln X_{1i} + \beta_2 \ln X_{2i} + \ln \epsilon_i \qquad (14.19)$$

Hence, equation (14.19) is linear in the natural logarithms. In a similar fashion, the **exponential model**

Original Exponential Model

$$Y_i = e^{\beta_0 + \beta_1 X_{1i} + \beta_2 X_{2i}} \epsilon_i \qquad (14.20)$$

can also be transformed to linear form (by taking natural logarithms of both the dependent and explanatory variables). The resulting model is

Transformed Exponential Model

$$\ln Y_i = \beta_0 + \beta_1 X_{1i} + \beta_2 X_{2i} + \ln \epsilon_i \qquad (14.21)$$

Example 14.7 *Using the Natural Log Transformation*

Given the following values for Y and X,

Y	X	Y	X
19.5	1	11.8	3
18.8	1	9.7	4
15.3	2	9.6	4
14.9	2	7.6	5
12.2	3	7.4	5

use a natural log transformation for the Y variable and develop a scatter diagram.

SOLUTION

The first figure below displays the scatter diagram of X and Y; the second figure (on page 908) plots X versus the natural logarithm of Y. Note that the natural log transformation has taken a relationship that appears to be nonlinear and has created a relationship that appears linear.

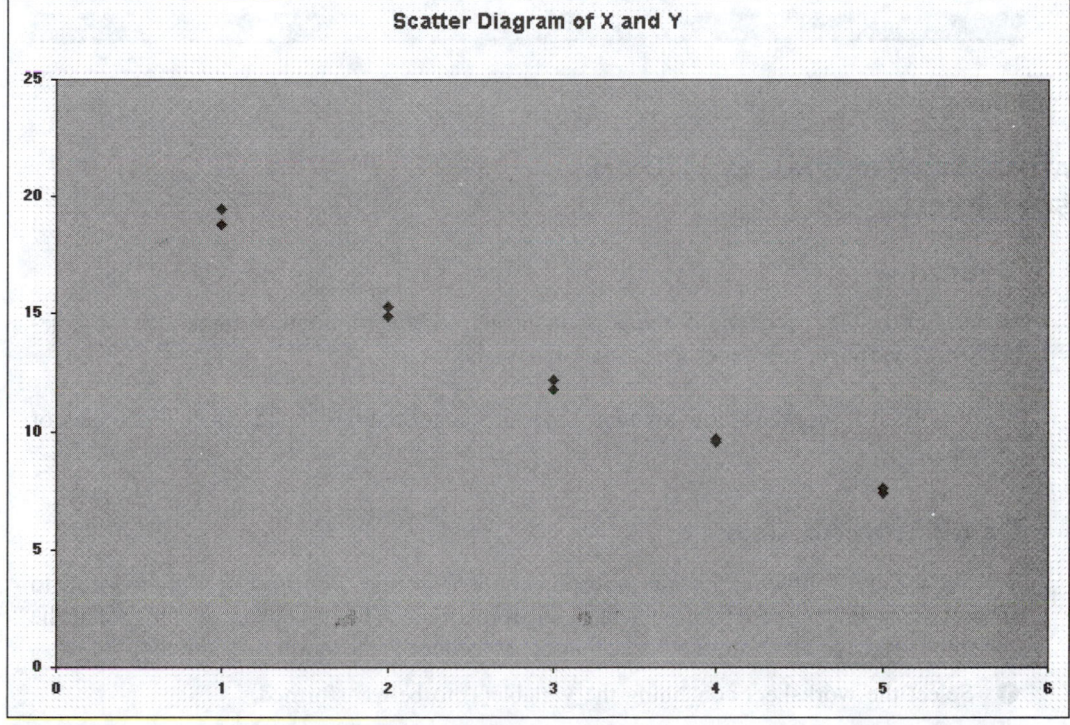

PANEL A
Scatter diagram of X and Y obtained from Microsoft Excel

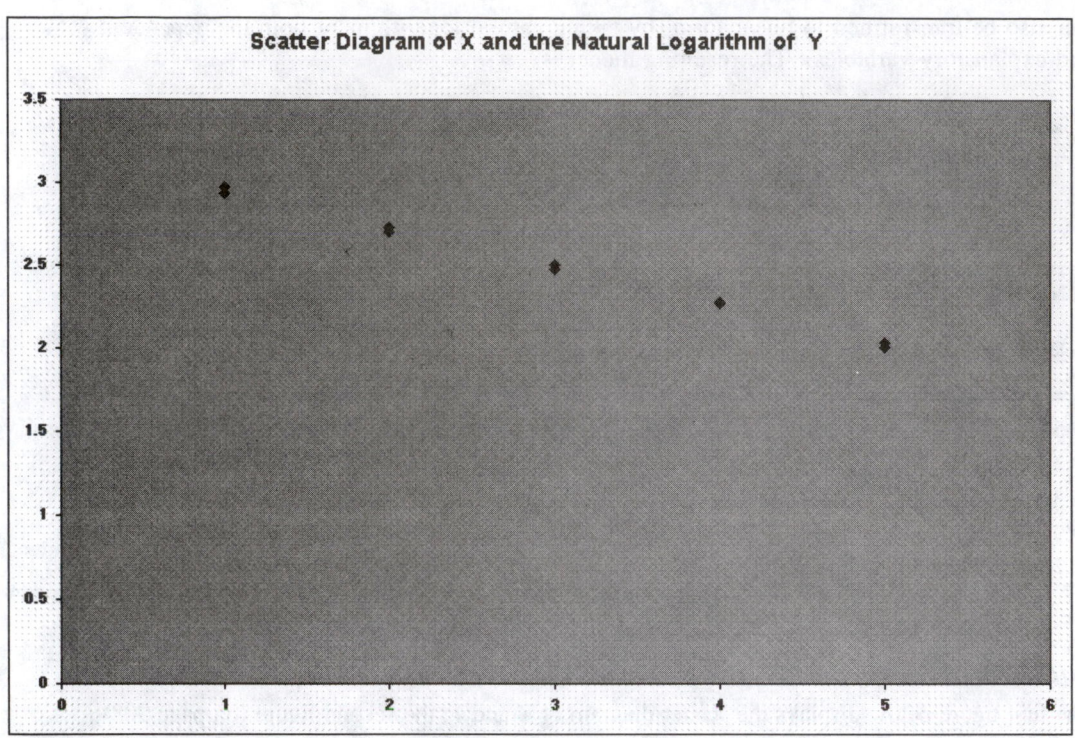

Scatter Diagram of X and the Natural Logarithm of Y

PANEL B

Scatter diagram of *X* and the natural logarithm of *Y* obtained from Microsoft Excel

14.8E GENERATING TRANSFORMATIONS USING MICROSOFT EXCEL

Overview

Both quick results users and developers should implement formulas to transform the independent *X* variable, the dependent *Y* variable, or both.

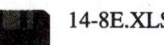 14-8E.XLS

The 14-8E.XLS workbook file contains several transformations based on the examples of section 14.8.

Details for All Users

We can use the following general procedure, similar to one discussed in section 14.6E to generate curvilinear models, to generate transformations. This procedure assumes that the variable data have been arranged in columns, as is done for examples in the text.

❶ Select the worksheet containing the variable(s) to be transformed.

❷ Select an unused column in the worksheet (or select Insert | Columns to insert a new, blank column).

③ Enter the formula to calculate the transformation in the row containing the first variable value and the column selected or inserted in step 2. Copy this formula down through all data rows.

④ Select PHStat | Regression | Multiple Regression.

⑤ In the Multiple Regression dialog box, make the appropriate entries and selections and click the OK button.

Use the SQRT function to calculate a square-root transformation. Use the LN function to calculate a natural logarithm transformation. Tables 14E.2 and 14E.3 show Data sheets after a square-root transformation has been applied to an X variable (Table 14E.2) and a natural logarithm transformation has been applied to the Y variable (Table 14E.3).

Table 14E.2 Data sheet after a square-root transformation

	A	B	C
1	Y	X	SQRT(X)
2	42.7	1	=SQRT(B2)
3	50.4	1	=SQRT(B3)
4	69.1	2	=SQRT(B4)
5	79.8	2	=SQRT(B5)
6	90	3	=SQRT(B6)
7	100.4	3	=SQRT(B7)
8	104.7	4	=SQRT(B8)
9	112.3	4	=SQRT(B9)
10	113.6	5	=SQRT(B10)
11	123.9	5	=SQRT(B11)

Table 14E.3 Data sheet after a natural logarithm transformation

	A	B	C
1	Y	LN(Y)	X
2	19.5	=LN(A2)	1
3	18.8	=LN(A3)	1
4	15.3	=LN(A4)	2
5	14.9	=LN(A5)	2
6	12.2	=LN(A6)	3
7	11.8	=LN(A7)	3
8	9.7	=LN(A8)	4
9	9.6	=LN(A9)	4
10	7.6	=LN(A10)	5
11	7.4	=LN(A11)	5

Problems for Section 14.8

Learning the Basics

14.43 Suppose the following model has been fit to a set of data

$$\ln \hat{Y}_i = 3.07 + .9 \ln X_{1i} + 1.41 \ln X_{2i}$$

(a) Predict the value of Y for $X_1 = 8.5$ and $X_2 = 5.2$.
(b) Interpret the meaning of the slopes.

● **14.44** Suppose the following model has been fit to a set of data

$$\ln \hat{Y}_i = 4.62 + .5 X_{1i} + .7 X_{2i}$$

(a) Predict the value of Y for $X_1 = 8.5$ and $X_2 = 5.2$.
(b) Interpret the meaning of the slopes.

Applying the Concepts

● **14.45** Referring to the data of Problem 14.33 on page 890 and using the file SPEED.XLS, perform a square-root transformation of the explanatory variable (speed).
(a) State the regression equation.
(b) Predict the average mileage obtained when the car is driven at 55 miles per hour.
(c) Perform a residual analysis of your results and determine the adequacy of the fit of the model.

(d) At the .05 level of significance, is there a significant relationship between mileage and the square root of speed?

(e) Interpret the meaning of the coefficient of determination r^2 in this problem.

(f) Compute the adjusted r^2.

(g) Compare your results with those obtained in Problem 14.33. Which model would you choose? Why?

• 14.46 Referring to the data of Problem 14.33 on page 890 and using the file SPEED.XLS, perform a natural logarithmic transformation of the response variable miles per gallon and reanalyze the data using this model. On the basis of your results:

(a) State the regression equation.

(b) Predict the average mileage obtained when the car is driven at 55 miles per hour.

(c) Perform a residual analysis of your results and determine the adequacy of the fit of the model.

(d) At the .05 level of significance, is there a significant relationship between the natural logarithm of miles per hour and speed?

(e) Interpret the meaning of the coefficient of determination r^2 in this problem.

(f) Compute the adjusted r^2.

(g) Compare your results with those obtained in Problems 14.33 and 14.45. Which model would you choose? Why?

14.47 Referring to the data of Problem 14.35 on page 892 and using the file TOMYLD2.XLS, perform a natural logarithmic transformation of the response variable yield and reanalyze the data. On the basis of your results:

(a) State the regression equation.

(b) Predict the average yield obtained when 55 pounds of fertilizer are applied per 1,000 square feet.

(c) Perform a residual analysis of your results and determine the adequacy of the fit of the model.

(d) At the .05 level of significance, is there a significant relationship between the natural logarithm of yield and the amount of fertilizer?

(e) Interpret the meaning of the coefficient of determination r^2 in this problem.

(f) Compute the adjusted r^2.

(g) Compare your results with those obtained in Problem 14.35. Which model would you choose? Why?

14.48 Referring to the data of Problem 14.35 on page 892 and using the file TOMYLD2.XLS, perform a square-root transformation of the explanatory variable amount of fertilizer and reanalyze the data. On the basis of your results:

(a) State the regression equation.

(b) Predict the average yield obtained when 55 pounds of fertilizer are applied per 1,000 square feet.

(c) Perform a residual analysis of your results and determine the adequacy of the fit of the model.

(d) At the .05 level of significance, is there a significant relationship between yield and the square root of the amount of fertilizer?

(e) Interpret the meaning of the coefficient of determination r^2 in this problem.

(f) Compute the adjusted r^2.

(g) Compare your results with those obtained in Problems 14.35 and 14.47. Which model would you choose? Why?

14.9 COLLINEARITY

One important problem in the application of multiple regression analysis involves the possible **collinearity** of the explanatory variables. This condition refers to situations in which some of the explanatory variables are highly correlated with each other. In such situations, collinear variables do not provide new information, and it becomes difficult to separate the effect of such variables on the dependent or response variable. In such cases, the values of the regression coefficients for the correlated variables may fluctuate drastically, depending on which independent variables are included in the model.

One method of measuring collinearity uses the **variance inflationary factor (*VIF*)** for each explanatory variable. VIF_j, the variance inflationary factor for variable j, is defined as in equation (14.22):

Variance Inflationary Factor

$$VIF_j = \frac{1}{1 - R_j^2} \qquad (14.22)$$

where R_j^2 is the coefficient of multiple determination of explanatory variable X_j with all other explanatory variables.

If there are only two explanatory variables, R_1^2 is just the coefficient of determination between X_1 and X_2. It would be identical to R_2^2, which is the coefficient of determination between X_2 and X_1. If, for example, there were three explanatory variables, then R_1^2 would be the coefficient of multiple determination of X_1 with X_2 and X_3; R_2^2 would be the coefficient of multiple determination of X_2 with X_1 and X_3; and R_3^2 would be the coefficient of multiple determination of X_3 with X_1 and X_2.

If a set of explanatory variables is uncorrelated, then VIF_j will be equal to 1. If the set is highly intercorrelated, then VIF_j might even exceed 10. Marquardt (see reference 3) suggests that if VIF_j is greater than 10, there is too much correlation between the variable X_j and the other explanatory variables. However, other researchers (see reference 6) suggest a more conservative criterion that would employ alternatives to least-squares regression if the maximum VIF_j exceeds 5.

If we reexamine the monthly heating oil consumption data of section 14.1, the correlation between the two explanatory variables, temperature and attic insulation, is computed as .00892. Therefore, because there are only two explanatory variables in the model, from equation (14.22):

$$VIF_1 = VIF_2 = \frac{1}{1 - (.00892)^2}$$

$$\cong 1.00$$

Thus, we may conclude that there is no reason to suspect any collinearity for the heating oil consumption data.

◆ 14.9E ◆ CALCULATING THE VARIANCE INFLATIONARY FACTOR USING MICROSOFT EXCEL

Overview

Because the variance inflationary factor (*VIF*) requires values from one or more regression analyses, both quick results users and developers should use the PHStat add-in to calculate these coefficients.

The 14-9E.XLS workbook file contains the calculations for the *VIF* (and other regression statistics) for the heating oil usage problem of section 14.1.

Details for All Users

To calculate the variance inflationary factor, use the Regression | Multiple Regression choice of the PHStat add-in. As an example, consider the heating oil usage problem of section 14.1. To calculate *VIF* for the multiple regression model based on this problem, do the following:

❶ Open the Starting Point for Several Chapter 14 Sections workbook (STARTING POINT 14-SEVERAL.XLS) and click the Data sheet tab. Verify that heating oil consumed, average temperature, and amount of attic insulation data of Table 14.1 on page 855 have been entered in columns A, B, and C, respectively.

❷ Select PHStat | Regression | Multiple Regression.

❸ In the Multiple Regression dialog box:
   a. Enter A1:A16 in the Y Variable Cell Range: edit box.
   b. Enter B1:C16 in the X Variables Cell Range: edit box.
   c. Select the First cells in both ranges contain label check box.
   d. Enter 95 in the Confidence Level for regr. coefficients: edit box.
   e. Select the Regression Statistics Table and the ANOVA and Coefficients Table check boxes.
   f. Enter Regression Analysis for Heating Oil Usage in the Output Title: edit box.
   g. Select the Variance Inflationary Factor (VIF) check box.
   h. Click the OK button.

The add-in inserts two worksheets, including a *VIF* worksheet that contains a variance inflationary factor similar to one shown in Figure 14E.4. Because this worksheet is *not* dynamically changeable, any changes made to the underlying heating oil usage data would require using the PHStat add-in a second time to produce the updated *VIF* value.

In this example, the add-in inserted only one *VIF* worksheet. More generally, the add-in will insert a *VIF* worksheet for each regression model pairing an explanatory variable X_j with all other X variables.

FIGURE 14E.4

VIF for the regression model for the heating oil usage problem of section 14.1

	A	B
1	Regression Model for X1 and X2	
2		
3	*Regression Statistics*	
4	Multiple R	0.008922039
5	R Square	7.96028E-05
6	Adjusted R Square	-0.076837351
7	Standard Error	21.45918623
8	Observations	15
9	VIF	1.000079609

Problems for Section 14.9

Learning the Basics

• **14.49** If the coefficient of determination between two independent variables is .20, what is the *VIF*?

14.50 If the coefficient of determination between two independent variables is .50, what is the *VIF*?

Applying the Concepts

• **14.51** Referring to Problem 14.4 on page 861 and doing multiple regression using the file WARECOST.XLS, determine the *VIF* for each explanatory variable in the model. Is there reason to suspect the existence of collinearity?

14.52 Referring to Problem 14.6 on page 862 and doing multiple regression using the file ADRADTV.XLS, determine the *VIF* for each explanatory variable in the model. Is there reason to suspect the existence of collinearity?

14.53 Referring to Problem 14.7 on page 863 and doing multiple regression using the file STANDBY.XLS, determine the *VIF* for each explanatory variable in the model. Is there reason to suspect the existence of collinearity?

◀14.10▶ MODEL BUILDING

In this chapter we developed the multiple linear regression model and subsequently discussed the curvilinear model, models involving dummy variables, and models involving transformations of variables. In this section we continue our discussion of regression by developing a model-building process that considers a set of several explanatory variables.

We start by referring to an example in which four explanatory variables (total staff present, remote hours, Dubner hours, and total labor hours) are to be considered in developing a regression model to predict standby hours of unionized graphic artists. The data are presented in Table 14.7.

Table 14.7 *Predicting standby hours based on total staff present, remote hours, Dubner hours, and total labor hours*

Week	Standby Hours	Total Staff Present	Remote Hours	Dubner Hours	Total Labor Hours
1	245	338	414	323	2,001
2	177	333	598	340	2,030
3	271	358	656	340	2,226
4	211	372	631	352	2,154
5	196	339	528	380	2,078
6	135	289	409	339	2,080
7	195	334	382	331	2,073
8	118	293	399	311	1,758
9	116	325	343	328	1,624
10	147	311	338	353	1,889
11	154	304	353	518	1,988
12	146	312	289	440	2,049
13	115	283	388	276	1,796
14	161	307	402	207	1,720
15	274	322	151	287	2,056
16	245	335	228	290	1,890
17	201	350	271	355	2,187
18	183	339	440	300	2,032
19	237	327	475	284	1,856
20	175	328	347	337	2,068

(continued)

Table 14.7 *Predicting standby hours based on total staff present, remote hours, Dubner hours, and total labor hours (continued)*

WEEK	STANDBY HOURS	TOTAL STAFF PRESENT	REMOTE HOURS	DUBNER HOURS	TOTAL LABOR HOURS
21	152	319	449	279	1,813
22	188	325	336	244	1,808
23	188	322	267	253	1,834
24	197	317	235	272	1,973
25	261	315	164	223	1,839
26	232	331	270	272	1,935

DATA FILE
STANDBY.XLS

Before we begin to develop a model to predict standby hours, we should keep in mind that a widely used criterion of model building is *parsimony*. This means that we wish to develop a regression model that includes the fewest number of explanatory variables that permit an adequate interpretation of the dependent variable of interest. Regression models with fewer explanatory variables are inherently easier to interpret, particularly because they are less likely to be affected by the problem of collinearity (described in section 14.9).

In addition, we should realize that the selection of an appropriate model when many explanatory variables are to be considered involves complexities that are not present for a model that contains only two explanatory variables. First, the evaluation of all possible regression models becomes more computationally complex. Second, although competing models can be quantitatively evaluated, there may not exist a *uniquely best* model but rather several *equally appropriate* models.

We begin our analysis of the standby hours data by first measuring the amount of collinearity that exists among the explanatory variables through the use of the variance inflationary factor [see equation (14.22) on page 911]. Figure 14.16 represents partial Microsoft Excel output of the variance inflationary factor values along with the fitted regression model.

PANEL A

	A	B
1	Model for X1 and all other X	
2		
3	*Regression Statistics*	
4	Multiple R	0.643681
5	R Square	0.414325
6	Adjusted R Square	0.334461
7	Standard Error	16.47151
8	Observations	26
9	VIF	1.707433

PANEL B

	A	B
1	Model for X2 and all other X	
2		
3	*Regression Statistics*	
4	Multiple R	0.434898
5	R Square	0.189136
6	Adjusted R Square	0.078564
7	Standard Error	124.9392
8	Observations	26
9	VIF	1.233253

PANEL C

	A	B
1	Model for X3 and all other X	
2		
3	*Regression Statistics*	
4	Multiple R	0.560992
5	R Square	0.314712
6	Adjusted R Square	0.221263
7	Standard Error	57.55254
8	Observations	26
9	VIF	1.45924

PANEL D

	A	B
1	Model for X4 and all other X	
2		
3	*Regression Statistics*	
4	Multiple R	0.70698
5	R Square	0.49982
6	Adjusted R Square	0.431614
7	Standard Error	114.4118
8	Observations	26
9	VIF	1.999281

	A	B	C	D	E	F	G
1	Regression Analysis for Standby Hours						
2							
3	Regression Statistics						
4	Multiple R	0.78935216					
5	R Square	0.623076833					
6	Adjusted R Square	0.551281944					
7	Standard Error	31.83500743					
8	Observations	26					
9							
10	ANOVA						
11		df	SS	MS	F	Significance F	
12	Regression	4	35181.79373	8795.448432	8.678568098	0.000268015	
13	Residual	21	21282.82166	1013.467698			
14	Total	25	56464.61538				
15							
16		Coefficients	Standard Error	t Stat	P-value	Lower 95%	Upper 95%
17	Intercept	-330.8318447	110.8953572	-2.983279491	0.007087351	-561.4514047	-100.2122847
18	Total Staff	1.245629161	0.412059739	3.022933435	0.006473169	0.388703875	2.102554447
19	Remote	-0.118417979	0.054324392	-2.179830739	0.040795357	-0.231391756	-0.005444202
20	Dubner	-0.297058588	0.1179313	-2.518912186	0.019945311	-0.542310195	-0.051806982
21	Total Labor	-0.130534912	0.059322942	2.200411993	0.039106915	0.00716608	0.253903744

(Coefficient labels circled: b_0, b_1, b_2, b_3, b_4)

FIGURE 14.16 Regression model obtained from Microsoft Excel to predict standby hours based on four explanatory variables

We observe that all the *VIF* values are relatively small, ranging from a high of 2.0 for the total labor hours to a low of 1.2 for remote hours. Thus, on the basis of the criteria developed by Snee (see reference 6), there is little evidence of collinearity among the set of explanatory variables.

The Stepwise Regression Approach to Model Building

We now continue our analysis of these data by attempting to determine the subset of all explanatory variables that yield an adequate and appropriate model without having to use the complete model. We begin by describing a widely used search procedure called **stepwise regression,** which attempts to find the "best" regression model without examining all possible regressions. Once a best model has been found, residual analysis is used to evaluate the aptness of the model.

Recall that in section 14.5 the partial *F*-test criterion was used to evaluate portions of a multiple regression model. Stepwise regression extends this partial *F*-test criterion to a model with any number of explanatory variables. An important feature of this stepwise process is that an explanatory variable that has entered into the model at an early stage may subsequently be removed once other explanatory variables are considered. That is, in stepwise regression, variables are either added to or deleted from the regression model at each step of the model-building process. The stepwise procedure terminates with the selection of a best-fitting model when no additional variables can be added to or deleted from the last model fitted.

This stepwise regression approach to model building was originally developed more than 30 years ago in an era in which regression analysis on mainframe computers involved the costly use of large amounts of processing time. Under such conditions, a search procedure such as stepwise regression, although providing a limited evaluation of alternative models, became widely used. In this current era of personal computers with extremely fast hardware, the evaluation of many different regression models can be done in very little time, at a very small cost. Thus we turn to a more general way of evaluating alternative regression models, the **best-subsets approach.**

The Best-Subsets Approach to Model Building

This approach evaluates either all possible regression models for a given set of independent variables or at least the best subsets of models for a given number of independent variables. Figure 14.17 represents partial output obtained from the PHStat add-in for Microsoft Excel in which the best three regression models for a given number of independent variables were provided according to two widely used criteria, the adjusted r^2 and the C_p statistic.

FIGURE 14.17

Best-subsets regression output obtained from the PHStat add-in for Microsoft Excel for standby hours data (Note: For extreme small r^2, adjusted r^2 can be negative)

	A	B	C	D	E	F	G
1	Best Subsets Analysis for Standby Hours Models						
2							
3	R2T	0.623077					
4	1 - R2T	0.376923					
5	n	26					
6	T	5					
7	n - T	21					
8							Consider
9		Cp	p+1	R Square	Adj. R Square	Std. Error	This Model?
10	X1	13.32152	2	0.366024	0.339608215	38.6206	No
11	X1X2	8.41933	3	0.489909	0.445553638	35.38734	No
12	X1X2X3	7.841813	4	0.536172	0.47292327	34.50286	No
13	X1X2X3X4	5	5	0.623077	0.551281944	31.83501	Yes
14	X1X2X4	9.344919	4	0.509194	0.442265513	35.49212	No
15	X1X3	10.64856	3	0.449898	0.402062546	36.74905	No
16	X1X3X4	7.751662	4	0.537791	0.474762012	34.44263	No
17	X1X4	14.79818	3	0.375417	0.321105605	39.15789	No
18	X2	33.20781	2	0.00909	-0.032197524	48.28359	No
19	X2X3	32.30673	3	0.061161	-0.020477044	48.00868	No
20	X2X3X4	12.13813	4	0.459059	0.385294475	37.26076	No
21	X2X4	23.24809	3	0.223752	0.156252135	43.65405	No
22	X3	30.38835	2	0.059696	0.020516668	47.03452	No
23	X3X4	11.82309	3	0.428816	0.379148026	37.44658	No
24	X4	24.1846	2	0.171045	0.136505715	44.16192	No

The first criterion that is often used is the adjusted r^2, which adjusts the r^2 of each model to account for the number of variables in the model as well as the sample size (see section 14.1). Because models with different numbers of independent variables are to be compared, the adjusted r^2 is more appropriate than r^2.

Referring to Figure 14.17, we observe that the adjusted r^2 reaches a maximum value of .551 when all four independent variables plus the intercept term (for a total of five estimated parameters) are included in the model.

A second criterion often used in the evaluation of competing models is based on the statistic developed by Mallows (see reference 5). This statistic, called C_p, measures the differences of a fitted regression model from a *true* model, along with random error. The **C_p statistic** is defined as

The C_p Statistic

$$C_p = \frac{(1 - R_p^2)(n - T)}{1 - R_T^2} - [n - 2(p + 1)] \qquad (14.23)$$

where

$p =$ number of independent variables included in a regression model

$T =$ total number of parameters (including the intercept) to be estimated in the full regression model

$R_p^2 =$ coefficient of multiple determination for a regression model that has p independent variables

$R_T^2 =$ coefficient of multiple determination for a full regression model that contains all T estimated parameters

Using equation (14.23) to compute C_p for the model containing total staff present and remote hours, we have

$$n = 26 \quad p = 2 \quad T = 4 + 1 = 5 \quad R_p^2 = .490 \quad R_T^2 = .623$$

so that

$$C_p = \frac{(1 - .49)(26 - 5)}{1 - .623} - [26 - 2(2 + 1)]$$

$$C_p = 8.42$$

When a regression model with p independent variables contains only random differences from a *true* model, the average value of C_p is $p + 1$, the number of parameters. Thus, in evaluating many alternative regression models, our goal is to find models whose C_p is close to or below $(p + 1)$.

From Figure 14.17 we observe that only the model with all four independent variables considered contains a C_p value equal to or below $p + 1$. Therefore this model should be chosen. Although it was not the case here, the C_p statistic often provides several alternative models for us to evaluate in greater depth using other criteria such as parsimony, interpretability, and departure from model assumptions (as evaluated by residual analysis).

Now that the explanatory variables to be included in the model have been selected, a residual analysis should be undertaken to evaluate the aptness of the fitted model. Figure 14.18 presents partial output obtained from Microsoft Excel for these purposes. We observe from Figure 14.18 that the plots of the standardized residuals versus the total staff, the remote hours, the Dubner hours, and the total labor hours all reveal no apparent pattern. In addition, a histogram of the standardized residuals (not shown here) indicates only moderate departure from normality.

Thus from Figure 14.16 on page 915, our sample regression equation can be expressed as

$$\hat{Y}_i = -330.83 + 1.2456X_{1i} - .1184X_{2i} - .2917X_{3i} + .1305X_{4i}$$

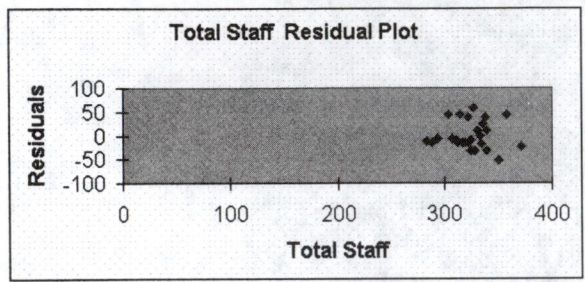

PANEL A

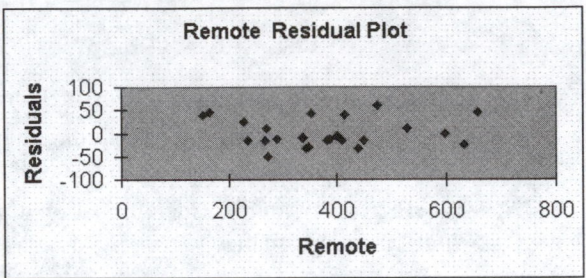

PANEL B

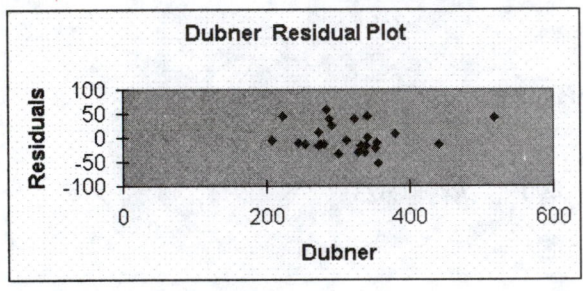

PANEL C

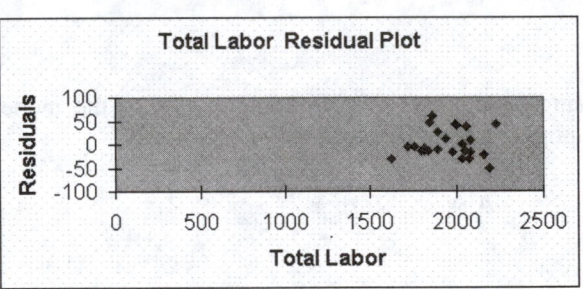

PANEL D

FIGURE 14.18 Residual plots for standby hours data obtained from Microsoft Excel

From this model we conclude that holding constant the effect of remote hours, Dubner hours, and total labor hours for each increase of one person in total staff, we predict that the average standby hours will increase by 1.2456 hours. Holding the total staff, Dubner hours, and total labor hours constant, we predict that the average standby hours will decrease by .1184 hour for each increase of 1 remote hour. Holding the total staff, remote hours, and total labor hours constant, we predict that the average standby hours will decrease by .2971 hour for each increase of 1 Dubner hour. Holding the total staff, remote hours, and Dubner hours constant, we predict that the average standby hours will increase by .1305 hour for each increase of 1 total labor hour.

To study a situation where there are several alternative models in which the C_p statistic is less than or equal to $(p + 1)$, we turn to Example 14.8.

Example 14.8 Choosing among Alternative Regression Models

Given the following output from a best-subsets regression analysis of a regression model with seven independent variables, determine which regression model you would choose as the *best* model.

VARIABLES	r^2	ADJ. r^2	C_p	
1	12.1	11.9	113.9	X_4
1	9.3	9.0	130.4	X_1
1	8.3	8.0	136.2	X_3
2	21.4	21.0	62.1	$X_3 X_4$
2	19.1	18.6	75.6	$X_1 X_3$
2	18.1	17.7	81.0	$X_1 X_4$
3	28.5	28.0	22.6	$X_1 X_3 X_4$
3	26.8	26.3	32.4	$X_3 X_4 X_5$
3	24.0	23.4	49.0	$X_2 X_3 X_4$
4	30.8	30.1	11.3	$X_1 X_2 X_3 X_4$
4	30.4	29.7	14.0	$X_1 X_3 X_4 X_6$
4	29.6	28.9	18.3	$X_1 X_3 X_4 X_5$
5	31.7	30.8	8.2	$X_1 X_2 X_3 X_4 X_5$
5	31.5	30.6	9.6	$X_1 X_2 X_3 X_4 X_6$
5	31.3	30.4	10.7	$X_1 X_3 X_4 X_5 X_6$
6	32.3	31.3	6.8	$X_1 X_2 X_3 X_4 X_5 X_6$
6	31.9	30.9	9.0	$X_1 X_2 X_3 X_4 X_5 X_7$
6	31.7	30.6	10.4	$X_1 X_2 X_3 X_4 X_6 X_7$
7	32.4	31.2	8.0	$X_1 X_2 X_3 X_4 X_5 X_6 X_7$

Partial output from best-subsets regression

SOLUTION

From this output, we determine which models have C_p values that are less than or equal to $(p + 1)$. Two models meet this criterion. The model with six independent variables (X_1, X_2, X_3, X_4, X_5, X_6) has a C_p value of 6.8, which is less than $p + 1 = 6 + 1 = 7$, and the full model with seven independent variables (X_1, X_2, X_3, X_4, X_5, X_6, X_7) has a C_p value of 8.0. One way to choose among models that meet these criteria is to determine whether the models contain a subset of variables that are common and then test whether the contribution of the additional variables is significant. In this case, because the models differ only by the inclusion of variable X_7 in the full model, we test whether variable X_7 made a significant contribution to the regression model given that variables X_1, X_2, X_3, X_4, X_5, and X_6 were already included in the model. If the contribution was statistically significant, then variable X_7 would be included in the regression model. If variable X_7 did not make a statistically significant contribution, variable X_7 would not be included in the model.

Exhibit 14.2 summarizes the steps involved in model building.

Exhibit 14.2 Steps Involved in Model Building

✓ **1.** Choose a set of independent variables to be considered for inclusion in the regression model.

✓ **2.** Fit a full regression model that includes all the independent variables to be considered so that the variance inflationary factor (*VIF*) for each independent variable can be determined.

✓ **3.** Determine whether any independent variables have a $VIF > 5$.

✓ **4.** There are three possible results that can occur.

   (a) None of the independent variables have a $VIF > 5$. If this is the case, proceed to step 5.

   (b) One of the independent variables has a $VIF > 5$. If this is the case, eliminate that independent variable and proceed to step 5.

   (c) More than one of the independent variables has a $VIF > 5$. If this is the case, eliminate the independent variable that has the highest *VIF* and go back to step 2.

✓ **5.** Perform a best-subsets regression with the remaining independent variables to obtain the best models (in terms of C_p) for a given number of independent variables.

✓ **6.** List all models that have $C_p \leq (p + 1)$.

✓ **7.** Among those models listed in step 6, choose a best model (as discussed in Example 14.8).

✓ **8.** Perform a complete analysis of the model chosen including residual analysis.

✓ **9.** Depending on the results of the residual analysis, add curvilinear terms and transform variables and reanalyze the data.

✓ **10.** Use the selected model for prediction.

The following figure represents a road map for these steps in model building.

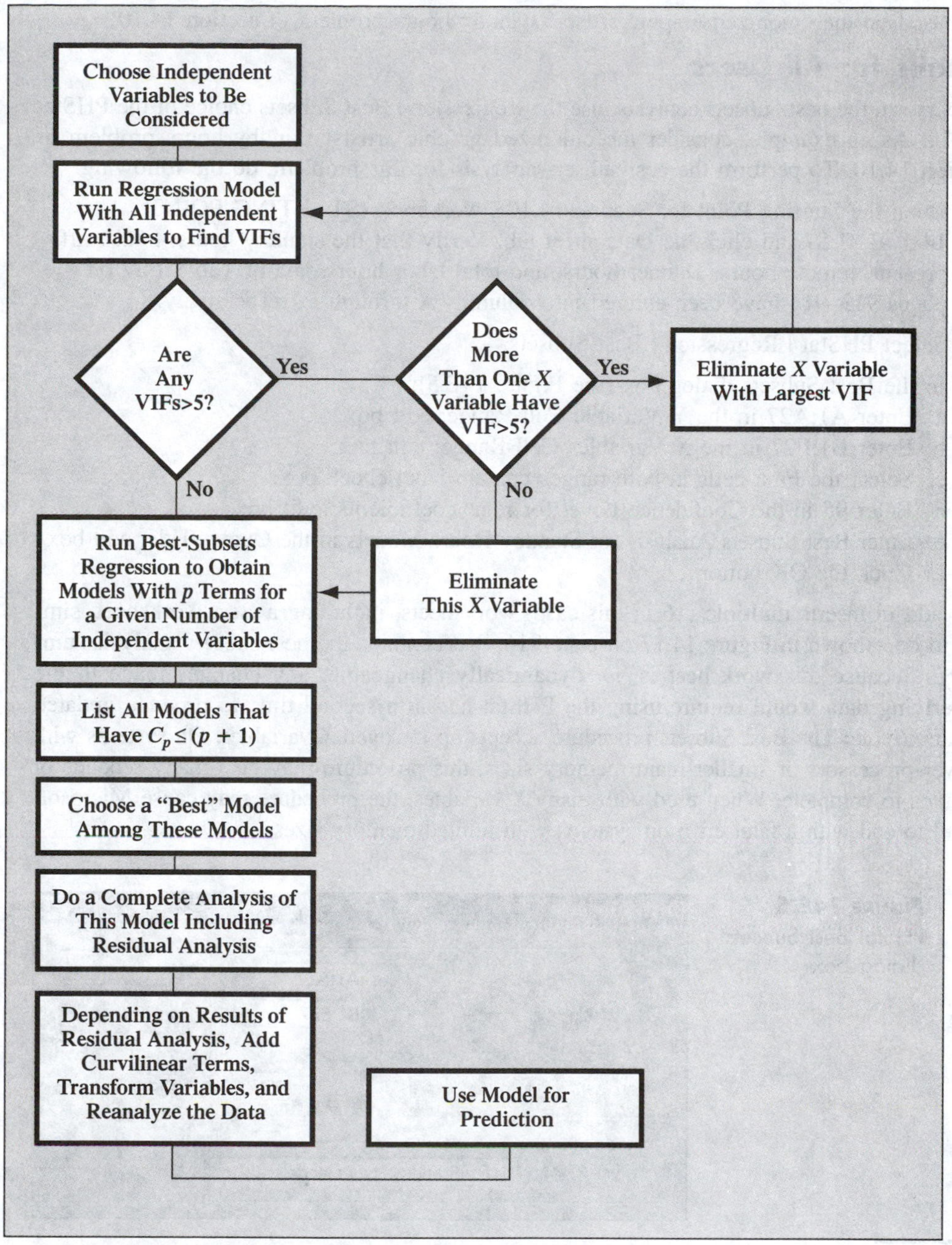

```
┌─────────────────────┐
│ Choose Independent  │
│ Variables to Be     │
│ Considered          │
└─────────────────────┘
          │
          ▼
┌─────────────────────┐
│ Run Regression Model│◄──────────────────────────────┐
│ With All Independent│                                │
│ Variables to Find   │                                │
│ VIFs                │                                │
└─────────────────────┘                                │
          │                                            │
          ▼                                            │
      ╱────────╲            ╱──────────╲        ┌──────────────────┐
     ╱   Are    ╲   Yes    ╱   Does     ╲  Yes  │ Eliminate X      │
    ╱    Any     ╲────────▶╱  More Than  ╲─────▶│ Variable With    │
    ╲  VIFs>5?   ╱         ╲  One X       ╱      │ Largest VIF      │
     ╲          ╱          ╲ Variable Have╱      └──────────────────┘
      ╲────────╱            ╲ VIF>5?     ╱
          │                  ╲──────────╱
          │ No                    │
          ▼                       │ No
┌─────────────────────┐          ▼
│ Run Best-Subsets    │   ┌──────────────────┐
│ Regression to Obtain│◄──│ Eliminate        │
│ Models With p Terms │   │ This X Variable  │
│ for a Given Number  │   └──────────────────┘
│ of Independent      │
│ Variables           │
└─────────────────────┘
          │
          ▼
┌─────────────────────┐
│ List All Models That│
│ Have Cp ≤ (p + 1)   │
└─────────────────────┘
          │
          ▼
┌─────────────────────┐
│ Choose a "Best"     │
│ Model Among These   │
│ Models              │
└─────────────────────┘
          │
          ▼
┌─────────────────────┐
│ Do a Complete       │
│ Analysis of This    │
│ Model Including     │
│ Residual Analysis   │
└─────────────────────┘
          │
          ▼
┌─────────────────────┐
│ Depending on Results│
│ of Residual Analysis│   ┌──────────────────┐
│ Add Curvilinear     │   │ Use Model for    │
│ Terms, Transform    │   │ Prediction       │
│ Variables, and      │   └──────────────────┘
│ Reanalyze the Data  │
└─────────────────────┘
```

List All Models That Have $C_p \leq (p + 1)$

Road map for model building

 **14.10E** **PERFORMING BEST-SUBSETS ANALYSIS USING MICROSOFT EXCEL**

Overview

Because the best-subsets analysis requires values from multiple regression analyses, both quick results users and developers should use the PHStat add-in to perform this analysis.

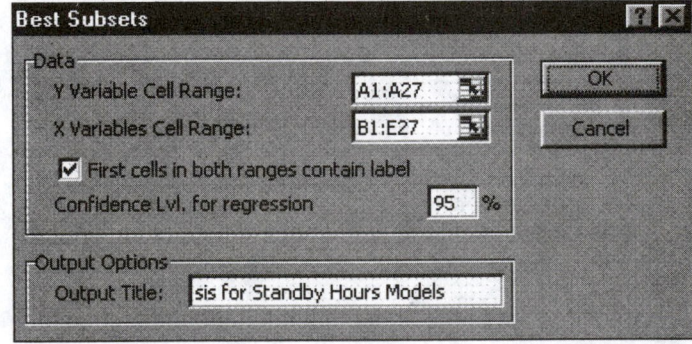

The 14-10E.XLS workbook file contains the best-subsets analysis for the regression models based on the unionized graphic artists' standby hours problem of section 14.10.

Details for All Users

To perform the best-subsets analysis, use the Regression | Best Subsets choice of the PHStat add-in. As an example, consider the unionized graphic artists' standby hours problem of section 14.10. To perform the best-subsets analysis for this problem, do the following:

❶ Open the Starting Point for Section 14.10E workbook (STARTING POINT 14-10E.XLS) and click the Data sheet tab. Verify that the standby hours, total staff present, remote hours, Dubner hours, and total labor hours data of Table 14.7 on pages 913–914 have been entered into columns A through E, respectively.

❷ Select PHStat | Regression | Best Subsets.

❸ In the Best Subsets dialog box (see Figure 14E.5):
 a. Enter A1:A27 in the Y Variable Cell Range: edit box.
 b. Enter B1:E27 in the X Variables Cell Range: edit box.
 c. Select the First cells in both ranges contain label check box.
 d. Enter 95 in the Confidence Level for regr. coefficients: edit box.
 e. Enter Best Subsets Analysis for Standby Hours Models in the Output Title: edit box.
 f. Click the OK button.

The add-in inserts multiple (16 in this case) worksheets, including a Best worksheet, similar to one shown in Figure 14.17 on page 916, that contains the best subsets analysis summary. Because this worksheet is *not* dynamically changeable, any changes made to the underlying data would require using the PHStat add-in a second time to produce updated results. (Note: The Best Subsets procedure accepts up to seven X variables. On systems with slower processors or smaller main memory sizes, this procedure may take many seconds or minutes to complete. When used with many X variables, the procedure may cause Microsoft Excel to end with a fatal error on systems with limited memory sizes.)

FIGURE 14E.5
PHStat Best Subsets dialog box

Problems for Section 14.10

Learning the Basics

• **14.54** Suppose that six independent variables are to be considered for inclusion in a regression model. A sample of 40 observations is selected with the following results:

$$n = 40 \quad p = 2 \quad T = 6 + 1 = 7 \quad R_p^2 = .274 \quad R_T^2 = .653$$

(a) Compute the C_p value for this two-independent-variable model.
(b) On the basis of the results of (a), does this model meet the criterion for further consideration as the best model to be selected? Explain.

14.55 Suppose that four independent variables are to be considered for inclusion in a regression model. A sample of 30 observations is selected with the following results:

The model that includes independent variables A and B has a C_p value equal to 4.6
The model that includes independent variables A and C has a C_p value equal to 2.4
The model that includes independent variables A, B, and C has a C_p value equal to 2.7

(a) On the basis of these results, which models meet the criterion for further consideration? Explain.
(b) How would you compare the model that contains independent variables A, B, and C to the model that contains independent variables A and B? Explain.

Applying the Concepts

• **14.56** Suppose we want to develop a model to predict the selling price of houses based on assessed value, time period in which a house was sold, and whether the house was new (0 = no; 1 = yes). A sample of 30 recently sold single-family houses in a small western city is selected to study the relationship between selling price and assessed value (the houses in the city had been reassessed at full value 1 year prior to the study). The results are as follows:

OBSERVATION	ASSESSED VALUE ($000)	SELLING PRICE ($000)	TIME	NEW	OBSERVATION	ASSESSED VALUE ($000)	SELLING PRICE ($000)	TIME	NEW
1	78.17	94.10	10	1	16	84.36	106.70	12	0
2	80.24	101.90	10	1	17	72.94	81.50	5	0
3	74.03	88.65	11	0	18	76.50	94.50	14	1
4	86.31	115.50	2	0	19	66.28	69.00	1	0
5	75.22	87.50	5	0	20	79.74	96.90	3	1
6	65.54	72.00	4	0	21	72.78	86.50	14	0
7	72.43	91.50	17	0	22	77.90	97.90	12	1
8	85.61	113.90	13	0	23	74.31	83.00	11	0
9	60.80	69.34	6	0	24	79.85	97.30	12	1
10	81.88	96.90	5	1	25	84.78	100.80	2	1
11	79.11	96.00	7	0	26	81.61	97.90	6	1
12	59.93	61.90	4	0	27	74.92	90.50	12	0
13	75.27	93.00	11	0	28	79.98	97.00	4	1
14	85.88	109.50	10	1	29	77.96	92.00	9	0
15	76.64	93.75	17	0	30	79.07	95.90	12	1

DATA FILE
HOUSE1.XLS

Develop the most appropriate multiple regression model to predict selling price. Be sure to perform a thorough residual analysis. In addition, provide a detailed explanation of your results.

14.57 The file UNIV&COL contains data on 80 colleges and universities. Among the variables included are the annual total cost (in thousands of dollars), the average total score on the Scholastic Aptitude Test (SAT), the room and board expenses (in thousands of dollars), whether the institution is public or private, and whether the TOEFL criterion is at least 550.

Develop the most appropriate multiple regression model to predict annual total cost. Be sure to perform a thorough residual analysis. In addition, provide a detailed explanation of your results.

DATA FILE
UNIV&COL.XLS

14.58 The file AUTO96 contains data on 89 automobile models from the year 1996. Among the variables included are gasoline mileage, weight, width, length of each automobile, and whether the car is front wheel drive or rear wheel drive. Develop the most appropriate multiple regression model to predict gasoline mileage. Be sure to perform a thorough residual analysis. In addition, provide a detailed explanation of your results.

DATA FILE
AUTO96.XLS

14.11 PITFALLS IN MULTIPLE REGRESSION AND ETHICAL ISSUES

Pitfalls in Multiple Regression

Model building is an art as well as a science. Different individuals may not always agree on the best multiple regression model. Nevertheless, we should use the process described in Exhibit 14.2 on page 920. In doing so, we must be aware of certain pitfalls that can interfere with the development of a useful model. In section 13.9 we discussed pitfalls in regression and ethical issues. Now that we have examined a variety of multiple regression models, we need to concern ourselves with some additional pitfalls related to the use of regression analysis. These are displayed in Exhibit 14.3.

> ### Exhibit 14.3 Additional Pitfalls in Multiple Regression
>
> ✓ **1.** The need to understand that the regression coefficient for a particular independent variable is interpreted from a perspective in which the values of all other independent variables are held constant.
>
> ✓ **2.** The need to evaluate residual plots for each independent variable.
>
> ✓ **3.** The need to evaluate interaction terms to determine whether the slope of other independent variables with the response variable is the same at each level of a dummy variable.
>
> ✓ **4.** The need to obtain the *VIF* for each independent variable before determining which independent variables should be included in the model.
>
> ✓ **5.** The need to examine several alternative models using best-subsets regression.

Ethical Considerations

Ethical considerations arise when a user wishing to make predictions manipulates the development process of the multiple regression model. The key here is intent. In addition to the situations discussed in section 13.9, unethical behavior occurs when someone uses multiple regression analysis and *willfully fails* to remove variables from consideration that exhibit a high collinearity with other independent variables or *willfully fails* to use methods other than least-squares regression when the assumptions necessary for least-squares regression have been seriously violated.

 ## SUMMARY

In this chapter, as illustrated in the accompanying summary chart, we developed the multiple regression model with two independent variables and determined the significance of the full model and tested whether each independent variable made a significant contribution to the model. In addition, the coefficients of partial determination were studied and we considered curvilinear regression, dummy variables, collinearity, transformations, and model building.

710

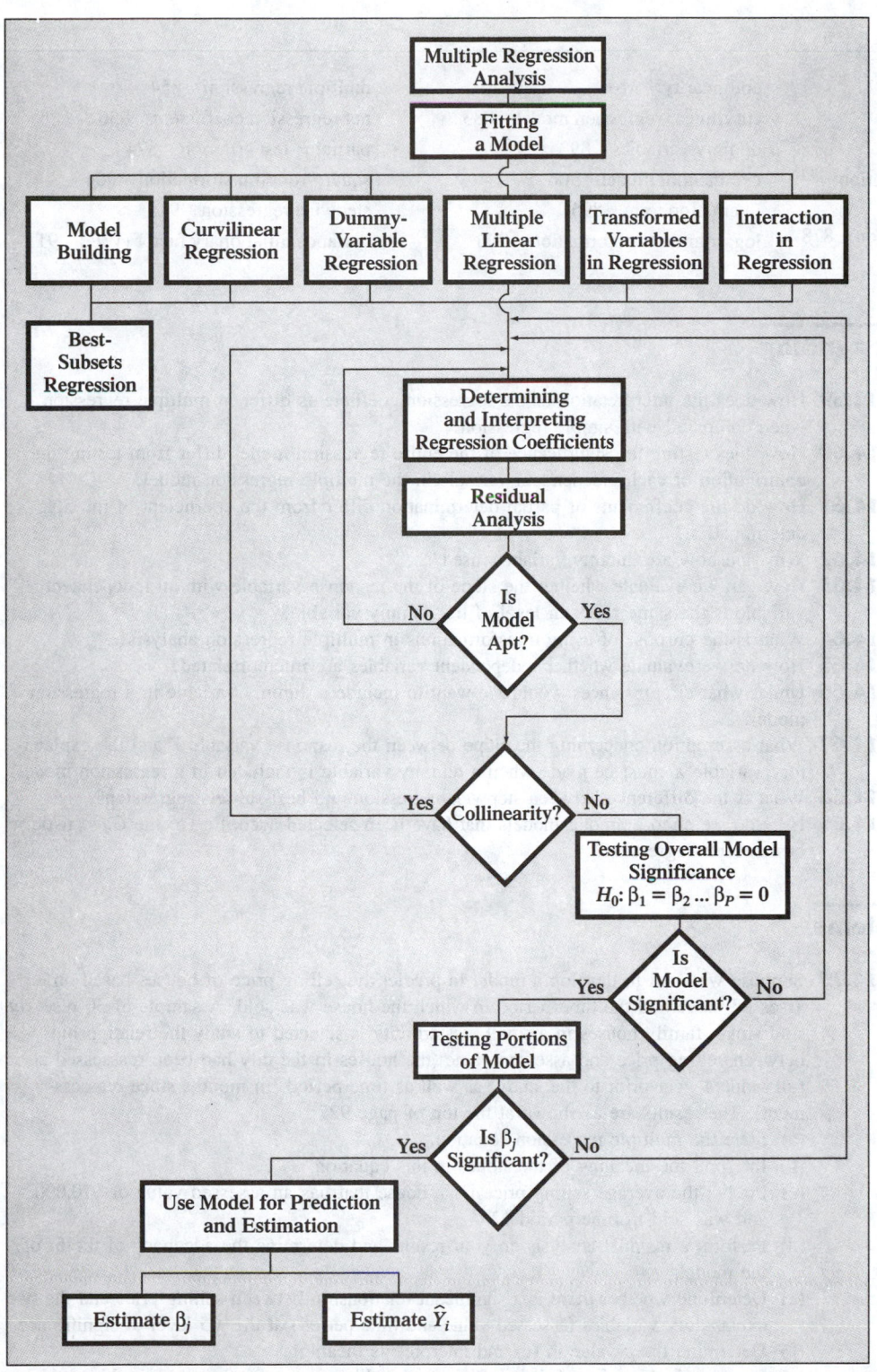

Chapter 14 summary chart

Key Terms

Checking Your Understanding

14.59 How does the interpretation of the regression coefficients differ in multiple regression when compared with simple regression?

14.60 How does testing the significance of the entire regression model differ from testing the contribution of each independent variable in the multiple regression model?

14.61 How do the coefficients of partial determination differ from the coefficient of multiple determination?

14.62 Why and how are dummy variables used?

14.63 How can we evaluate whether the slope of the response variable with an independent variable is the same for each level of the dummy variable?

14.64 What is the purpose of using transformations in multiple regression analysis?

14.65 How do we evaluate whether independent variables are intercorrelated?

14.66 Under what circumstances would we want to include a dummy variable in a regression model?

14.67 What assumption concerning the slope between the response variable Y and the explanatory variable X must be made when a dummy variable is included in a regression model?

14.68 What is the difference between stepwise regression and best-subsets regression?

14.69 How do we choose among models that have been selected according to the C_p statistic in best-subsets regression?

Chapter Review Problems

14.70 Suppose we want to develop a model to predict the selling price of houses based on assessed value and the time period in which the house was sold. A sample of 30 recently sold single-family houses in a small western city is selected to study the relationship between selling price and assessed value (the houses in the city had been reassessed at full value 1 year prior to the study) as well as time period (in months since reassessment). The results are as shown at the top of page 927:

(a) State the multiple regression equation.

(b) Interpret the meaning of the slopes in this equation.

(c) Predict the average selling price for a house that has an assessed value of $70,000 and was sold in time period 12.

(d) Perform a residual analysis on your results and determine the adequacy of the fit of the model.

(e) Determine whether there is a significant relationship between selling price and the two explanatory variables (assessed value and time period) at the .05 level of significance.

(f) Determine the p-value in (e) and interpret its meaning.

(g) Interpret the meaning of the coefficient of multiple determination $r^2_{Y.12}$ in this problem.

(h) Determine the adjusted r^2.

OBSERVATION	ASSESSED VALUE ($000)	SELLING PRICE ($000)	TIME PERIOD	OBSERVATION	ASSESSED VALUE ($000)	SELLING PRICE ($000)	TIME PERIOD
1	78.17	94.10	10	16	84.36	106.70	12
2	80.24	101.90	10	17	72.94	81.50	5
3	74.03	88.65	11	18	76.50	94.50	14
4	86.31	115.50	2	19	66.28	69.00	1
5	75.22	87.50	5	20	79.74	96.90	3
6	65.54	72.00	4	21	72.78	86.50	14
7	72.43	91.50	17	22	77.90	97.90	12
8	85.61	113.90	13	23	74.31	83.00	11
9	60.80	69.34	6	24	79.85	97.30	12
10	81.88	96.90	5	25	84.78	100.80	2
11	79.11	96.00	7	26	81.61	97.90	6
12	59.93	61.90	4	27	74.92	90.50	12
13	75.27	93.00	11	28	79.98	97.00	4
14	85.88	109.50	10	29	77.96	92.00	9
15	76.64	93.75	17	30	79.07	95.90	12

DATA FILE
HOUSE1.XLS

(i) At the .05 level of significance, determine whether each explanatory variable makes a significant contribution to the regression model. On the basis of these results, indicate the regression model that should be used in this problem.

(j) Determine the p-values in (i) and interpret their meaning.

(k) Set up a 95% confidence interval estimate of the true population slope between selling price and assessed value. How does the interpretation of the slope here differ from Problem 13.75 on page 845?

(l) Compute the coefficients of partial determination ($r^2_{Y1.2}$ and $r^2_{Y2.1}$) and interpret their meaning.

• **14.71** Suppose we want to develop a model to predict assessed value of single-family houses based on heating area and age. A sample of 15 single-family houses is selected. The assessed value (in thousands of dollars), the heating area of the houses (in thousands of square feet), and the age of the houses (in years) are recorded with the following results:

HOUSE	ASSESSED VALUE ($000)	HEATING AREA OF DWELLING (THOUSANDS OF SQUARE FEET)	AGE (YEARS)	HOUSE	ASSESSED VALUE ($000)	HEATING AREA OF DWELLING (THOUSANDS OF SQUARE FEET)	AGE (YEARS)
1	84.4	2.00	3.42	9	78.5	1.59	1.75
2	77.4	1.71	11.50	10	79.2	1.50	2.75
3	75.7	1.45	8.33	11	86.7	1.90	0.00
4	85.9	1.76	0.00	12	79.3	1.39	0.00
5	79.1	1.93	7.42	13	74.5	1.54	12.58
6	70.4	1.20	32.00	14	83.8	1.89	2.75
7	75.8	1.55	16.00	15	76.8	1.59	7.17
8	85.9	1.93	2.00				

DATA FILE
HOUSE2.XLS

(a) State the multiple regression equation.

(b) Interpret the meaning of the slopes in this equation.

(c) Predict the average assessed value for a house that has a heating area of 1,750 square feet and is 10 years old.

(d) Perform a residual analysis on your results and determine the adequacy of the fit of the model.

(e) Determine whether there is a significant relationship between assessed value and the two explanatory variables (heating area and age) at the .05 level of significance.

(f) Determine the p-value in (e) and interpret its meaning.

(g) Interpret the meaning of the coefficient of multiple determination $r_{Y.12}^2$ in this problem.

(h) Determine the adjusted r^2.

(i) At the .05 level of significance, determine whether each explanatory variable makes a significant contribution to the regression model. On the basis of these results, indicate the regression model that should be used in this problem.

(j) Determine the p-values in (i) and interpret their meaning.

(k) Set up a 95% confidence interval estimate of the true population slope between assessed value and heating area. How does the interpretation of the slope here differ from Problem 13.76 on page 846?

(l) Compute the coefficients of partial determination ($r_{Y1.2}^2$ and $r_{Y2.1}^2$) and interpret their meaning.

(m) The real estate assessor's office has been publicly quoted as saying that the age of a house has no bearing on its assessed value. On the basis of the results of (a)–(l), do you agree with this statement? Explain.

14.72 The file UNIV&COL.XLS contains data on 80 colleges and universities. Among the variables included are the annual total cost (in thousands of dollars), the average total score on the Scholastic Aptitude Test (SAT), and the room and board expenses (in thousands of dollars). Suppose we wanted to develop a model to predict the annual total cost based on SAT score and room and board expenses.

DATA FILE
UNIV&COL.XLS

(a) State the multiple regression equation.

(b) Interpret the meaning of the slopes in this equation.

(c) Predict the average annual total cost for a school that has an average SAT score of 1,100 and a room and board expense of $5,000.

(d) Perform a residual analysis on your results and determine the adequacy of the fit of the model.

(e) Determine whether there is a significant relationship between annual total cost and the two explanatory variables (SAT score and room and board expenses) at the .05 level of significance.

(f) Determine the p-value in (e) and interpret its meaning.

(g) Interpret the meaning of the coefficient of multiple determination $r_{Y.12}^2$ in this problem.

(h) Determine the adjusted r^2.

(i) At the .05 level of significance, determine whether each explanatory variable makes a significant contribution to the regression model. On the basis of these results, indicate the regression model that should be used in this problem.

(j) Determine the p-values in (i) and interpret their meaning.

(k) Set up a 95% confidence interval estimate of the true population slope between annual total cost and SAT score.

(l) Compute the coefficients of partial determination ($r_{Y1.2}^2$ and $r_{Y2.1}^2$) and interpret their meaning.

(m) Explain why the slope for total annual cost with room and board expenses seems substantially different from 1.0.

(n) What other factors that are not included in the model might account for the strong positive relationship between annual total cost and SAT score?

• **14.73** The file AUTO96.XLS contains data on 89 automobile models from the year 1996. Among the variables included are the gasoline mileage, the length (in inches), and the weight (in pounds), of each automobile. Suppose we wanted to develop a model to predict the gasoline mileage based on the length and weight of each automobile.

DATA FILE
AUTO96.XLS

(a) State the multiple regression equation.
(b) Interpret the meaning of the slopes in this equation.
(c) Predict the average gasoline mileage for an automobile that has a length of 195 inches and a weight of 3,000 pounds.
(d) Perform a residual analysis on your results and determine the adequacy of the fit of the model.
(e) Determine whether there is a significant relationship between gasoline mileage and the two explanatory variables (length and weight) at the .05 level of significance.
(f) Determine the p-value in (e) and interpret its meaning.
(g) Interpret the meaning of the coefficient of multiple determination $r^2_{Y.12}$ in this problem.
(h) Determine the adjusted r^2.
(i) At the .05 level of significance, determine whether each explanatory variable makes a significant contribution to the regression model. On the basis of these results, indicate the regression model that should be used in this problem.
(j) Determine the p-values in (i) and interpret their meaning.
(k) Set up a 95% confidence interval estimate of the true population slope between gasoline mileage and weight.
(l) Compute the coefficients of partial determination ($r^2_{Y1.2}$ and $r^2_{Y2.1}$) and interpret their meaning.

14.74 Crazy Dave, the well-known baseball analyst, has expanded his analysis of which variables are important in predicting a team's wins in a given season. He has collected the following data related to wins, ERA, and runs scored for a recent season (1997):

AMERICAN LEAGUE				NATIONAL LEAGUE			
TEAM	WINS	ERA	RUNS	TEAM	WINS	ERA	RUNS
Boston	78	4.85	851	Florida	92	3.83	740
Cleveland	86	4.73	868	Cincinnati	76	4.41	651
Kansas City	67	4.70	747	Chicago Cubs	68	4.44	687
Minnesota	68	5.01	772	San Francisco	90	4.39	784
Toronto	76	3.93	654	Los Angeles	88	3.62	742
Anaheim	84	4.52	829	Pittsburgh	79	4.28	725
Seattle	90	4.78	925	San Diego	76	4.98	795
Texas	77	4.69	807	New York Mets	88	3.95	777
Detroit	79	4.56	784	St. Louis	73	3.88	689
Chicago White Sox	80	4.73	779	Philadelphia	68	4.85	668
Milwaukee	78	4.22	681	Atlanta	101	3.18	791
Oakland	65	5.48	764	Montreal	78	4.14	691
Baltimore	98	3.91	812	Houston	84	3.66	777
New York Yankees	96	3.84	891	Colorado	83	5.29	923

DATA FILE
BB97.XLS

(a) State the multiple regression equation.
(b) Interpret the meaning of the slopes in this equation.
(c) Predict the average number of wins for a team that has an ERA of 4.00 and scored 750 runs.
(d) Perform a residual analysis on your results and determine the adequacy of the fit of the model.
(e) Determine whether there is a significant relationship between number of wins and the two explanatory variables (ERA and runs) at the .05 level of significance.
(f) Determine the p-value in (e) and interpret its meaning.
(g) Interpret the meaning of the coefficient of multiple determination $r^2_{Y.12}$ in this problem.

(h) Determine the adjusted r^2.

(i) At the .05 level of significance, determine whether each explanatory variable makes a significant contribution to the regression model. On the basis of these results, indicate the regression model that should be used in this problem.

(j) Determine the p-values in (i) and interpret their meaning.

(k) Set up a 95% confidence interval estimate of the true population slope between wins and ERA. Compare the results with those obtained in (l) of Problem 13.79 on page 849.

(l) Compute the coefficients of partial determination ($r^2_{Y1.2}$ and $r^2_{Y2.1}$) and interpret their meaning.

(m) Which seems to be more important in predicting wins, pitching as measured by ERA or offense as measured by runs scored? Explain.

14.75 Backpacks are commonly seen in many places, especially college campuses, shopping malls, airplanes, and hiking trails. In the August 1997 issue of *Consumer Reports* information was provided concerning different features of backpacks, including their prices ($), volume (cubic inches), and number of $5 \times 7\frac{3}{4}$-inch books the backpacks can hold. The results are summarized below:

PRICE	VOLUME	BOOKS	PRICE	VOLUME	BOOKS
48	2,200	59	40	1,950	46
45	1,670	49	40	1,810	44
50	2,200	48	45	1,910	48
42	1,700	52	27	1,875	42
29	1,875	52	25	1,450	42
50	1,500	49	35	1,519	43
48	1,874	50	95	1,102	73
38	1,586	47	40	1,316	55
33	1,910	53	40	1,760	43
40	1,500	49	25	1,844	42
35	1,950	49	50	2,150	52
32	1,385	45	35	1,810	50
40	1,700	38	50	2,180	46
35	2,000	51	35	1,635	40
28	1,500	46	15	1,245	47

DATA FILE
BACKPACK.XLS

Source: "Packs for Town and Country," Copyright © 1997 by Consumers Union of U.S., Inc. Adapted from CONSUMER REPORTS, *(August 1997): 20–21 by permission of Consumers Union of U.S., Inc., Yonkers, NY 10703-1057. Although these data sets originally appeared in* CONSUMER REPORTS, *the selective adaptation and resulting conclusions presented are those of the authors and are not sanctioned or endorsed in any way by Consumers Union, the publisher of* CONSUMER REPORTS.

Suppose we want to develop a multiple regression model to predict the price of a backpack on the basis of the volume and the number of books it can hold.

(a) State the multiple regression equation.

(b) Interpret the meaning of the slopes in this equation.

(c) Predict the average price for a backpack that has a volume of 2,000 cubic inches and can hold 50 books.

(d) Perform a residual analysis on your results and determine the adequacy of the fit of the model.

(e) Determine whether there is a significant relationship between price and the two explanatory variables (volume and number of books) at the .05 level of significance.

(f) Determine the p-value in (e) and interpret its meaning.

(g) Interpret the meaning of the coefficient of multiple determination $r^2_{Y.12}$ in this problem.

(h) Determine the adjusted r^2.

(i) At the .05 level of significance, determine whether each explanatory variable makes a significant contribution to the regression model. On the basis of these results, indicate the regression model that should be used in this problem.

(j) Determine the p-values in (i) and interpret their meaning.

(k) Set up a 95% confidence interval estimate of the true population slope between price and number of books.

(l) Compute the coefficients of partial determination ($r^2_{Y1.2}$ and $r^2_{Y2.1}$) and interpret their meaning.

(m) Do the results of (a)–(l) surprise you? Explain.

14.76 Referring to Problem 14.74 (relating wins to ERA) on page 929, suppose that in addition to using earned run average to predict the number of wins, Crazy Dave wants to include the league (American versus National) as an explanatory variable. On the basis of the results obtained using Microsoft Excel:

DATA FILE
BB97.XLS

(a) State the multiple regression equation.

(b) Interpret the meaning of the slopes in this problem.

(c) Predict the average number of wins for a team with an ERA of 4.00 in the American League.

(d) Perform a residual analysis on your results and determine the adequacy of the fit of the model.

(e) Determine whether there is a significant relationship between wins and the two explanatory variables (ERA and league) at the .05 level of significance.

(f) At the .05 level of significance, determine whether each explanatory variable makes a contribution to the regression model. On the basis of these results, indicate the regression model that should be used in this problem.

(g) Set up 95% confidence interval estimates of the population slope for the relationship between wins and ERA and between wins and league.

(h) Compare the slope obtained in (b) with the slope for the simple linear regression model of Problem 13.79 on page 849. Explain the difference in the results.

(i) Interpret the meaning of the coefficient of multiple determination $r^2_{Y.12}$.

(j) Determine the adjusted r^2.

(k) Compare $r^2_{Y.12}$ with the value computed in Problem 13.79(f). Explain the results.

(l) Compute the coefficients of partial determination and interpret their meaning.

(m) What assumption about the slope of wins with ERA must be made in this problem?

(n) Include an interaction term in the model and, at the .05 level of significance, determine whether it makes a significant contribution to the model.

(o) On the basis of the results of (f) and (n), which model is more appropriate? Explain.

14.77 Crazy Dave, the well-known baseball analyst, has expanded his analysis, presented in Problem 14.74 on page 929, of which variables are important in predicting a team's wins in a given season. He has collected data related to wins, ERA, saves, runs scored, hits allowed, walks allowed, and errors for the 1997 season (see the BB97.XLS file).

DATA FILE
BB97.XLS

(a) Develop the most appropriate multiple regression model to predict a team's wins. Be sure to include a thorough residual analysis. In addition, provide a detailed explanation of your results.

(b) Develop the most appropriate multiple regression model to predict a team's ERA on the basis of hits allowed, walks allowed, errors, and saves. Be sure to include a thorough residual analysis. In addition, provide a detailed explanation of your results.

14.78 The file UNIV&COL contains data on 80 colleges and universities. Among the variables included are the academic calendar type (1 = semester; 0 = other), annual total cost (in thousands of dollars), average total score on the Scholastic Aptitude Test (SAT), room and board expenses (in thousands of dollars), whether the institution is public or private, whether the TOEFL criterion is at least 550, and average indebtedness at graduation.

Develop the most appropriate multiple regression model to predict average indebtedness at graduation. Be sure to perform a thorough residual analysis. In addition, provide a detailed explanation of your results.

14.79 The file AUTO96 contains data on 89 automobile models from the year 1996. Among the variables included are the gasoline mileage, weight, width, length, wheel base, luggage capacity, and whether the car is front wheel drive (1) or rear wheel drive (0).

Develop the most appropriate multiple regression model to predict gasoline mileage. Be sure to perform a thorough residual analysis. In addition, provide a detailed explanation of your results.

14.80 Data are available for East Meadow, a suburban New York community, to predict appraised value and taxes of single-family houses based on lot size, number of bedrooms, number of bathrooms, age of the house, presence of an eat-in kitchen, and central air-conditioning.

(a) Develop the most appropriate multiple regression model to predict appraised value. Be sure to perform a thorough residual analysis. In addition, provide a detailed explanation of your results.

(b) Develop the most appropriate multiple regression model to predict taxes. Be sure to perform a thorough residual analysis. In addition, provide a detailed explanation of your results.

(c) Compare the results obtained in (a) and (b) with those of Problems 14.81–14.84.

14.81 Data similar to those in Problem 14.80 are available for Farmingdale, another suburban New York community.

(a) Perform an analysis similar to that of Problem 14.80 (a) and (b).

(b) Compare the results obtained in (a) with those of Problems 14.80 and 14.82–14.84.

14.82 Data similar to those in Problem 14.80 are available for Levittown, another suburban New York community.

(a) Perform an analysis similar to that of Problem 14.80 (a) and (b).

(b) Compare the results obtained in (a) with those of Problems 14.80, 14.81, and 14.83–14.84.

14.83 Data similar to those in Problem 14.80 are available for Islip, another suburban New York community.

(a) Perform an analysis similar to that of Problem 14.80 (a) and (b).

(b) Compare the results obtained in (a) with those of Problems 14.80–14.82 and 14.84.

14.84 Data similar to those in Problem 14.80 are available for Islip Terrace, another suburban New York community.

(a) Perform an analysis similar to that of Problem 14.80 (a) and (b).

(b) Compare the results obtained in (a) with those of Problems 14.80–14.83.

14.85 A headline on page 1 of *The New York Times* of March 4, 1990 read: "Wine Equation Puts Some Noses Out of Joint." The article proceeded to explain that Professor Orley Ashenfelter, a Princeton University economist, had developed a multiple regression model to predict the quality of French Bordeaux based on the amount of winter rain, the average temperature during the growing season, and the harvest rain. The equation developed was

$$Q = -12.145 + .00117WR + .6164TMP - .00386HR$$

where

Q = logarithmic index of quality where 1,961 equals 100
WR = winter rain (October through March) in millimeters
TMP = average temperature during the growing season (April through September) in degrees Celsius
HR = harvest rain (August to September) in millimeters

You are at a cocktail party, sipping a glass of wine, when one of your friends mentions to you that she has read the article. She asks you to explain the meaning of the coefficients in the equation and also asks you about analyses that might have been done and were not included in the article. You respond . . .

Case Study — EastWestSide Movers

The owner of an intracity moving company has typically used an estimator to determine the number of labor hours needed for a move. This has proved useful in the past, but he would like to be able to develop a more reliable estimate that would be more accurate in predicting the labor hours. In a prelimi- nary effort to provide a more accurate means of estimation, he has collected data for 36 moves in which the origin and destination were within the borough of Manhattan in New York City and the travel time was an insignificant portion of the hours worked. The results were as follows:

Observation	Labor Hours	Rooms	Cubic Feet Moved	Observation	Labor Hours	Rooms	Cubic Feet Moved
1	24.00	3.5	545	19	25.00	3.0	557
2	13.50	2.0	400	20	45.00	5.5	1,028
3	26.25	2.5	562	21	29.00	4.5	793
4	25.00	3.0	540	22	21.00	3.0	523
5	9.00	1.0	220	23	22.00	3.5	564
6	20.00	3.0	344	24	16.50	2.5	312
7	22.00	3.5	569	25	37.00	4.0	757
8	11.25	2.0	340	26	32.00	3.5	600
9	50.00	5.0	900	27	34.00	4.0	796
10	12.00	1.5	285	28	25.00	3.5	577
11	38.75	5.0	865	29	31.00	3.0	500
12	40.00	4.5	831	30	24.00	4.0	695
13	19.50	3.0	344	31	40.00	5.5	1,054
14	18.00	2.5	360	32	27.00	3.0	486
15	28.00	4.0	750	33	18.00	3.0	442
16	27.00	3.5	650	34	62.50	5.5	1,249
17	21.00	3.0	415	35	53.75	5.0	995
18	15.00	2.5	275	36	79.50	5.5	1,397

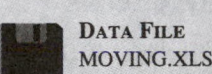

DATA FILE
MOVING.XLS

Develop a model to predict the labor hours based on the number of rooms and the number of cubic feet to be moved from the apartment of origin. Write an executive summary of no more than one page summarizing your conclusions. In addition, submit a technical appendix that provides and explains the statistical results.

Case Study — The Mountain States Potato Company

The Mountain States Potato Company is a potato-processing firm in eastern Idaho. A by-product of the process, called a filter cake, has been sold to area feedlots as cattle feed. Recently, one of the feedlot owners complained that the cattle were not gaining weight and believed that the problem was the filter cake they purchased from the Mountain States Potato Company.

Initially, all that was known of the filter cake system was that historical records showed that the percentage of solids had been running in the neighborhood of 11.5% in years past. At present,

719

the solids were running in the 8% to 9% range. Several additions had been made to the plant in the intervening years to significantly increase the water and solids volume and the clarifier temperature. What was actually affecting the solids was a mystery but, because the plant needed to get rid of its solid waste if it were going to run, something had to be done quickly. The only practical solution was to determine some way to get the solids content back up to the previous levels.

Individuals involved in the process were asked to identify variables that might be manipulated in order to affect the percentage of solids content. This review turned up six variables that would affect the percentage of solids. The variables are:

VARIABLE	COMMENTS
SOLIDS	Percent solids in filter cake.
PH	Acidity. This indicates bacterial action in the clarifier. As bacterial action progresses, organic acids are produced that can be measured using pH. This is controlled by the downtime of the system.
LOWER	Pressure of the vacuum line below the fluid line on the rotating drum.
UPPER	Pressure of the vacuum line above the fluid line on the rotating drum.
THICK	Cake thickness measured on the drum.
VARIDRIV	Setting used to control the drum speed. May differ from DRUMSPD because of mechanical inefficiencies.
DRUMSPD	Speed at which the drum was rotated when collecting filter cake. Measured with a stopwatch.

Data obtained by monitoring the process several times daily for 20 days are stored in the POTATO.XLS file. Develop a regression model to predict the percentage of solids. Write an executive summary of your findings to the president of the Mountain States Potato Company.

References

1. Hocking, R. R., "Developments in Linear Regression Methodology: 1959–1982," *Technometrics* 25 (1983): 219–250.

2. Hosmer, D., and S. Lemeshow, *Applied Logistic Regression* (New York: Wiley, 1989).

3. Marquardt, D. W., "You Should Standardize the Predictor Variables in Your Regression Models," discussion of "A Critique of Some Ridge Regression Methods," by G. Smith and F. Campbell, *Journal of the American Statistical Association* 75 (1980): 87–91.

4. *Microsoft Excel 97* (Redmond, WA: Microsoft Corp., 1997).

5. Neter, J., M. Kutner, C. Nachtsheim, and W. Wasserman, *Applied Linear Statistical Models* (Homewood, IL: Irwin, 1996).

6. Snee, R. D., "Some Aspects of Nonorthogonal Data Analysis, Part I. Developing Prediction Equations," *Journal of Quality Technology* 5 (1973): 67–79.

Chapter 9

Student Solutions Manual

14.2 (a) Holding constant the effect of X_2, for each additional unit of X_1 the response variable Y is expected to decrease on average by 2 units. Holding constant the effect of X_1, for each additional unit of X_2 the response variable Y is expected to increase on average by 7 units.

(b) The Y-intercept 50 represents the portion of the measurement of Y that is not affected by the factors measured by X_1 and X_2.

(c) 40% of the variation in Y can be explained or accounted for by the variation in X_1 and the variation in X_2.

14.4

Regression Analysis						
Regression Statistics						
Multiple R	0.93591442					
R Square	0.875935802					
Adjusted R Square	0.864120164					
Standard Error	4.766165573					
Observations	24					
ANOVA						
	df	*SS*	*MS*	*F*	*Significance F*	
Regression	2	3368.087376	1684.043688	74.1336022	3.0429E-10	
Residual	21	477.0430196	22.71633427			
Total	23	3845.130396				
	Coefficients	*Standard Error*	*t Stat*	*P-value*	*Lower 95%*	*Upper 95%*
Intercept	-2.728246583	6.157879754	-0.443049668	0.662260247	-15.53426079	10.07776763
Sales	0.047113872	0.02032792	2.317692762	0.030643769	0.004839642	0.089388103
Orders	0.011946926	0.002248569	5.313123092	2.87239E-05	0.007270769	0.016623083

(a) $\hat{Y} = -2.72825 + 0.47114X_1 + 0.011947X_2$

(b) For a given number of orders, each increase of $1000 in sales is expected to result in an average increase in distribution cost by $471.14. For a given amount of sales, each increase of one order is expected to result in the average increase in distribution cost by $11.95.

(c) $\hat{Y}_i = -2.72825 + 0.47114(400) + 0.011947(4500) = 69.878$ or $69,878

(d) $r_{Y.12}^2 = SSR / SST = 3368.087 / 3845.13 = 0.8759$. So, 87.59% of the variation in distribution cost can be explained by variation in sales and variation in number of orders.

(e) $r_{adj}^2 = 1 - \left[(1 - r_{Y.12}^2) \frac{n-1}{n-p-1} \right] = 1 - \left[(1 - 0.8759) \frac{24-1}{24-2-1} \right] = 0.8641$

	A	B	C	D	E	F	G
1	Regression Analysis of Sales						
2							
3	*Regression Statistics*						
4	Multiple R	0.899273236					
5	R Square	0.808692352					
6	Adjusted R Square	0.788554705					
7	Standard Error	158.9041256					
8	Observations	22					
9							
10	ANOVA						
11		*df*	*SS*	*MS*	*F*	*Significance F*	
12	Regression	2	2028032.69	1014016.345	40.15823435	1.50126E-07	
13	Residual	19	479759.9014	25250.52112			
14	Total	21	2507792.591				
15							
16		*Coefficients*	*Standard Error*	*t Stat*	*P-value*	*Lower 95%*	*Upper 95%*
17	Intercept	156.4304345	126.7578563	1.234088672	0.232217275	-108.8768901	421.7377592
18	RadioTV	13.08068096	1.759373685	7.434850861	4.88861E-07	9.398268369	16.76309354
19	Newspaper	16.79527808	2.963377915	5.667612623	1.83069E-05	10.59285489	22.99770127

(a)　　$\hat{Y} = 156.4 + 13.081 X_1 + 16.795 X_2$

(b)　　For a given amount of newspaper advertising, each increase by $1000 in radio and television advertising is expected to result in an average increase in sales by $13,081. For a given amount of radio and television advertising, each increase by $1000 in newspaper advertising is expected to result in the average increase in sales by $16,795.

(c)　　$\hat{Y}_i = 156.4 + 13.081(20) + 16.795(20) = 753.95$ or $753,950

(d)　　$r^2_{Y.12} = SSR/SST = 2028033/2507793 = 0.8087$. So, 80.87% of the variation in sales can be explained by variation in radio and television advertising and variation in newspaper advertising.

(e)　　$r^2_{adj} = 1 - \left[(1 - r^2_{Y.12}) \dfrac{n-1}{n-p-1} \right] = 1 - \left[(1 - 0.0.8087) \dfrac{22-1}{22-2-1} \right] = 0.7886$

14.8

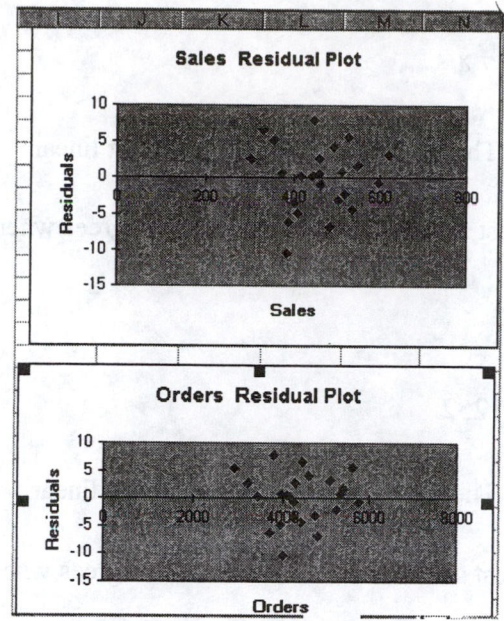

(a)   Based upon a residual analysis the model appears adequate.
(b)   There is no evidence of a pattern in the residuals versus time.
(c)   D = 2.26
(d)   D = 2.26 > 1.55. There is no evidence of positive autocorrelation in the residuals.

•14.10   There appears to be a curvilinear relationship in the plot of the residuals against both
            radio and television advertising. Thus, curvilinear terms for each of these explanatory
            models should be considered for inclusion in the model.

•14.12 (a) $MSR = SSR / p = 60 / 2 = 30$
             $MSE = SSE / (n - p - 1) = 120 / 18 = 6.67$
(b) $F = MSR / MSE = 30 / 6.67 = 4.5$
(c) $F = 4.5 > F_{U(2,21-2-1)} = 3.555$. Reject H_0. There is evidence of a significant linear
             relationship.

•14.14 (a)
             $F = 97.69 > F_{U(2,15-2-1)} = 3.89$. Reject H_0. There is evidence of a significant linear
             relationship with at least one of the independent variables.
(b)   The p value or probability of obtaining an F test statistic based on 2 and 12 degrees when
             H_0 is true is 0.0001.

14.16 (a) $MSR = SSR/p = 2451.974/2 = 1226.0$
$MSE = SSE/(n-p-1) = 819.8681/47 = 17.4$
$F = MSR/MSE = 1226.0/17.4 = 70.28$
$F = 70.28 > F_{U(2,50-2-1)} = 3.195$. Reject H_0. There is evidence of a significant linear relationship.

(b) The p value or probability of obtaining an F test statistic based on 2 and 47 degrees when H_0 is true is less than 0.001.

14.18 (a) $MSR = SSR/p = 27662.54/2 = 13831$
$MSE = SSE/(n-p-1) = 28802.07/23 = 1252$
$F = MSR/MSE = 13831/1252 = 11.05$
$F = 11.05 > F_{U(2,26-2-1)} = 3.422$. Reject H_0. There is evidence of a significant linear relationship.

(b) The p value or probability of obtaining an F test statistic based on 2 and 23 degrees when H_0 is true is less than 0.001.

14.20 (a) Variable X_1 has a larger slope than variable X_2.

(b) 95% confidence interval on β_1 : $b_1 \pm t_{n-p-1} s_{b_1}$, $4 \pm 2.1098 \cdot 1.2$
$1.46824 \le \beta_1 \le 6.53176$

(c) For X_1: $t = b_1/s_{b_1} = 4/1.2 = 3.33 > t_{17} = 2.1098$ with 17 degrees of freedom for $\alpha = 0.05$. Reject H_0. There is evidence that the variable X_1 contributes to a model already containing X_2.
For X_2: $t = b_2/s_{b_2} = 3/0.8 = 3.75 > t_{17} = 2.1098$ with 17 degrees of freedom for $\alpha = 0.05$. Reject H_0. There is evidence that the variable X_2 contributes to a model already containing X_1.
Both variables X_1 and X_2 should be included in the model.

14.22 (a) 95% confidence interval on β_1 : $b_1 \pm t_{n-p-1} s_{b_1}$, $0.471 \pm 2.0796 \cdot 0.0203$
$0.00488 \le \beta_1 \le 0.08932$

(b) For X_1: $t = b_1/s_{b_1} = 0.0471/0.0203 = 2.32 > t_{21} = 2.0796$ with 21 degrees of freedom for $\alpha = 0.05$. Reject H_0. There is evidence that the variable X_1 contributes to a model already containing X_2.
For X_2: $t = b_2/s_{b_2} = 0.01195/0.00225 = 5.31 > t_{21} = 2.0796$ with 21 degrees of freedom for $\alpha = 0.05$. Reject H_0. There is evidence that the variable X_2 contributes to a model already containing X_1.
Both variables X_1 and X_2 should be included in the model.

•14.24 (a) 95% confidence interval on β_1 : $b_1 \pm t_{n-p-1}s_{b_1}$, $13.0807 \pm 2.093 \cdot 1.7594$

$9.399 \le \beta_1 \le 16.763$

(b) For X_1: $t = b_1 / s_{b_1} = 13.0807/1.7594 = 7.43 > t_{19} = 2.093$ with 19 degrees of freedom for $\alpha = 0.05$. Reject H_0. There is evidence that the variable X_1 contributes to a model already containing X_2.

For X_2: $t = b_2 / s_{b_2} = 16.7953/2.9634 = 5.67 > t_{19} = 2.093$ with 19 degrees of freedom for $\alpha = 0.05$. Reject H_0. There is evidence that the variable X_2 contributes to a model already containing X_1.

Both variables X_1 and X_2 should be included in the model.

•14.26 (a) For X_1: $SSR(X_1|X_2) = SSR(X_1 \ and \ X_2) - SSR(X_2) = 60 - 25 = 35$

$$F = \frac{SSR(X_1|X_2)}{MSE} = \frac{35}{120/18} = 5.25 > F_{U(1,18)} = 4.41 \text{ with 1 and 18 degrees of freedom}$$

and $\alpha = 0.05$. Reject H_0. There is evidence that the variable X_1 contributes to a model already containing X_2.

For X_2: $SSR(X_2|X_1) = SSR(X_1 \ and \ X_2) - SSR(X_1) = 60 - 45 = 15$

$$F = \frac{SSR(X_2|X_1)}{MSE} = \frac{15}{120/18} = 2.25 < F_{U(1,18)} = 4.41 \text{ with 1 and 18 degrees of freedom}$$

and $\alpha = 0.05$. Do not reject H_0. There is not sufficient evidence that the variable X_2 contributes to a model already containing X_1.

Since variable X_2 does not significantly contribute to the model in the presence of X_1, only variable X_1 should be included and a simple linear regression model should be developed.

(b) $$r_{Y1.2} = \frac{SSR(X_1|X_2)}{SST - SSR(X_1 \ and \ X_2) + SSR(X_1|X_2)} = \frac{35}{180 - 60 + 35}$$

$= 0.2258$. Holding constant the effect of variable X_2, 22.58% of the variation in Y can be explained by the variation in variable X_1.

$$r_{Y2.1} = \frac{SSR(X_2|X_1)}{SST - SSR(X_1 \ and \ X_2) + SSR(X_2|X_1)} = \frac{15}{180 - 60 + 15}$$

$= 0.1111$. Holding constant the effect of variable X_1, 11.11% of the variation in Y can be explained by the variation in variable X_2.

14.28 (a) For X_1:

$SSR(X_1|X_2) = SSR(X_1 \ and \ X_2) - SSR(X_2) = 3368.087 - 3246.062 = 122.025$

$$F = \frac{SSR(X_1|X_2)}{MSE} = \frac{122.025}{477.043/21} = 5.37 > F_{U(1,21)} = 4.325 \text{ with 1 and 21 degrees of freedom}$$

and $\alpha = 0.05$. Reject H_0. There is evidence that the variable X_1 contributes to a model already containing X_2.

14.28 (a)
cont.

For X_2:

$$SSR(X_2|X_1) = SSR(X_1 \text{ and } X_2) - SSR(X_1) = 3368.087 - 2726.822 = 641.265$$

$$F = \frac{SSR(X_2|X_1)}{MSE} = \frac{641.265}{477.043/21} = 28.23 > F_{U(1,21)} = 4.325 \text{ with 1 and 21 degrees of freedom and}$$

$\alpha = 0.05$. Reject H_0. There is evidence that the variable X_2 contributes to a model already containing X_1.

Since each independent variable X_1 and X_2 makes a significant contribution to the model in the presence of the other variable, both variables should be included in the model.

(b)

$$r_{Y1.2} = \frac{SSR(X_1|X_2)}{SST - SSR(X_1 \text{ and } X_2) + SSR(X_1|X_2)}$$

$$= \frac{122.025}{3845.13 - 3368.087 + 122.025} = 0.2037.$$ Holding constant the effect of the number of orders, 20.37% of the variation in Y can be explained by the variation in sales.

$$r_{Y2.1} = \frac{SSR(X_2|X_1)}{SST - SSR(X_1 \text{ and } X_2) + SSR(X_2|X_1)}$$

$$= \frac{641.265}{3845.13 - 3368.087 + 641.265} = 0.5734.$$ Holding constant the effect of sales, 57.34% of the variation in Y can be explained by the variation in the number of orders.

•14.30 (a)

For X_1:

$$SSR(X_1|X_2) = SSR(X_1 \text{ and } X_2) - SSR(X_2) = 2028033 - 632259.4 = 1395773.6$$

$$F = \frac{SSR(X_1|X_2)}{MSE} = \frac{1395773.6}{479759.9/19} = 55.28 > F_{U(1,19)} = 4.381 \text{ with 1 and 19 degrees of}$$

freedom and $\alpha = 0.05$. Reject H_0. There is evidence that the variable X_1 contributes to a model already containing X_2.

For X_2:

$$SSR(X_2|X_1) = SSR(X_1 \text{ and } X_2) - SSR(X_1) = 2028033 - 1216940 = 811093$$

$$F = \frac{SSR(X_2|X_1)}{MSE} = \frac{811093}{479759.9/19} = 32.12 > F_{U(1,19)} = 4.381 \text{ with 1 and 19 degrees of freedom}$$

and $\alpha = 0.05$. Reject H_0. There is evidence that the variable X_2 contributes to a model already containing X_1.

Since each independent variable X_1 and X_2 makes a significant contribution to the model in the presence of the other variable, both variables should be included in the model.

•14.30 (b)

	A	B	C	D
1	Regression Analysis			
2	Coefficients of Partial Determination			
3				
4	SSR(X1,X2)	2028032.69		
5	SST	2507792.591		
6	SSR(X2)	632259.4483	SSR(X1 | X2)	1395773.241
7	SSR(X1)	1216939.671	SSR(X2 | X1)	811093.0188
8				
9	r2 Y1.2	0.744200787		
10	r2 Y2.1	0.628338834		

Holding constant the effect of newspaper advertising, 74.42% of the variation in Y can be explained by the variation in radio and television advertising.

Holding constant the effect of radio and television advertising, 62.83% of the variation in Y can be explained by the variation in newspaper advertising.

•14.32 (a) $\hat{Y} = 5 + 3X + 1.5X^2 = 5 + 3(2) + 1.5(2^2) = 17$

(b) $t = 2.35 > t_{22} = 2.0739$ with 22 degrees of freedom. Reject H_0. The curvilinear term is significant.

(c) $t = 1.17 < t_{22} = 2.0739$ with 22 degrees of freedom. Do not reject H_0. The curvilinear term is not significant.

(d) $\hat{Y} = 5 - 3X + 1.5X^2 = 5 - 3(2) + 1.5(2^2) = 5$

14.34 (b) $\hat{Y} = 4.181 + 0.438X + 1.1905X^2$

(c) $\hat{Y} = 4.181 + 0.438(2.5) + 1.1905(2.5^2) = 12.72$

(d) A residual analysis indicates no strong patterns.

(e) $F = 62.44 > F_{2,12} = 3.89$. Reject H_0. The overall model is significant. The p-value < 0.001.

(f) $t = 2.92 > t_{12} = 2.1788$. Reject H_0. The curvilinear effect is significant. The p-value = 0.013.

(g) $r_{Y.12}^2 = 0.912$. So, 91.2% of the variation in number of errors can be explained by the curvilinear relationship between number of errors and alcohol consumption.

(h) $r_{adj}^2 = 0.898$

Predicting County Taxes					
Regression Statistics					
Multiple R	0.967237191				
R Square	0.935547783				
Adjusted R Square	0.927491256				
Standard Error	48.40760315				
Observations	19				
ANOVA					
	df	SS	MS	F	Significance F
Regression	2	544220.9475	272110.4738	116.1229605	2.97785E-10
Residual	16	37492.73668	2343.296042		
Total	18	581713.6842			
	Coefficients	Standard Error	t Stat	P-value	Lower 95%	Upper 95%
Intercept	857.5883743	25.18944815	34.04554039	2.33404E-16	804.1891418	910.9876069
Age	-24.72208658	2.623757479	-9.422397754	6.24459E-08	-30.28420271	-19.15997045
Agesquared	0.293538926	0.051216368	5.731349866	3.09249E-05	0.1849651	0.402112752

(b) $\hat{Y} = 857.59 - 24.722X + 0.2935X^2$

(c) $\hat{Y} = 857.59 - 24.722(20) + 0.2935(20^2) = 480.6$

(d)   A residual analysis reveals patterns in the residuals vs. age, vs. the curvilinear variable (age squared), and vs. the fitted values.

(e) $F = 116.12 > F_{2,17} = 3.59$. Reject H_0. The overall model is significant.

(f)   The p-value < 0.001 indicates that the probability of having an F-test statistic of at least 116.12 when $\beta_1 = 0$ and $\beta_2 = 0$ is less than 0.001.

(g) $t = 5.73 > t_{17} = 2.1098$. Reject H_0. The curvilinear effect is significant.

(h)   The p-value < 0.001 indicates that the probability of having a t-test statistic with an absolute value of at least 5.73 when $\beta_2 = 0$ is less than 0.001.

(i) $r^2_{Y.12} = 0.936$. So, 93.6% of the variation in taxes can be explained by the curvilinear relationship between taxes and age of the house.

(j) $r^2_{adj} = 0.927$

14.38 (a)   First develop a multiple regression model using X_1 as the variable for the SAT score and X_2 a dummy variable with $X_2 = 1$ if a student had a grade of B or better in the introductory statistics course. If the dummy variable coefficient is significantly different than zero, you need to develop a model with the interaction term $X_1 X_2$ to make sure that the coefficient of X_1 is not significantly different if $X_2 = 0$ or $X_2 = 1$.

(b)   If a student received a grade of B or better in the introductory statistics course, the student would be expected to have a grade point average in accountancy that is 0.30 higher than a student who had the same SAT score, but did not get a grade of B or better in the introductory statistics course.

14.40 (a) $\hat{Y} = 2.9682 + 0.0393X_1 + 0.1228X_2$, where X_1 = median assessed value of homes and X_2 = ATM location.

(b)   Holding constant the effect of ATM location, for each additional thousand dollars of median assessed value, withdrawal amounts are expected to increase on average by 0.0393 thousands of dollars, or $39.30. For a given median assessed value of homes, a shopping center location is expected to increase withdrawal amounts on average by 0.1228 thousands of dollars, or $122.80.

14.40 (c) $\hat{Y} = 2.9682 + 0.0393(200) + 0.1228(1) = 10.95$ or $10,950

(d)   Based on a residual analysis, the model appears adequate.

(e) $F = 1642.59 > F_{2,12} = 3.89$. Reject H_0. There is evidence of a relationship between withdrawal amounts and the two dependent variables.

(f)   For X_1: $t = 53.52 > t_{12} = 2.1788$. Reject H_0. Median assessed value of homes makes a significant contribution and should be included in the model.
For X_2: $t = 2.25 > t_{12} = 2.1788$. Reject H_0. ATM location makes a significant contribution and should be included in the model.

(g) $0.0377 \le \beta_1 \le 0.0409$, $0.0039 \le \beta_2 \le 0.2418$

(h) $r_{Y.12}^2 = 0.996$. So, 99.6% of the variation in withdrawal amounts can be explained by variation in median assessed value of homes and variation in ATM location.

(i) $r_{adj}^2 = 0.996$

(j) $r_{Y1.2}^2 = 0.996$. Holding constant the effect of ATM location, 99.6% of the variation in withdrawal amounts can be explained by variation in median assessed value of homes.
$r_{Y2.1}^2 = 0.297$. Holding constant the effect of value of homes, 29.7% of the variation in withdrawal amounts can be explained by variation in ATM location.

(k)   The slope of withdrawal amount with median assessed value of home is the same regardless of whether the ATM is located in a shopping center or not.

(l) $\hat{Y} = 2.896 + 0.0392X_1 + 0.095X_2 + 0.00015X_1X_2$.
For $X_1 X_2$: the p-value is 0.926. Do not reject H_0. There is not evidence that the interaction term makes a contribution to the model.

(m)   The two-variable model in (a) should be used.

14.42

Predicting Total Cost

Regression Statistics

Multiple R	0.890548183
R Square	0.793076067
Adjusted R Square	0.787701419
Standard Error	2.715390278
Observations	80

ANOVA

	df	SS	MS	F	Significance F
Regression	2	2176.002359	1088.001179	147.5587096	4.55736E-27
Residual	77	567.747516	7.373344364		
Total	79	2743.749875			

	Coefficients	Standard Error	t Stat	P-value	Lower 95%	Upper 95%
Intercept	-3.467096383	3.083353539	-1.124456323	0.264312662	-9.60684507	2.672652303
Type of School	8.12843951	0.717002785	11.33669168	4.37476E-18	6.700702841	9.556176179
Average Total SAT	0.016080163	0.002776163	5.792219565	1.43648E-07	0.010552117	0.021608249

(a) $\hat{Y} = -3.467 + 0.01608X_1 + 8.12864X_2$, where X_1 = average total SAT score and X_2 = type of institution (public = 0, private = 1).

(b)   Holding constant the effect of type of institution, for each point increase on the average SAT, the total cost is expected to increase on average by 0.1608 thousands of dollars, or $160.80. For a given average SAT score, a private college or university is expected to increase total cost over a public institution by 8.12864 thousands of dollars, or $8,128.64.

cont. (c) $\hat{Y} = -3.467 + 0.01608(1000) + 8.12864(0) = 12.613$ or $12,613

(d)   Based on a residual analysis, the model appears adequate.

(e) $F = 147.56 > F_{2,77} = 3.1154$. Reject H_0. There is evidence of a relationship between total cost and the two dependent variables.

(f)   For X_1: $t = 5.79 > t_{77} = 1.9913$. Reject H_0. Average total SAT score makes a significant contribution and should be included in the model.

For X_2: $t = 11.34 > t_{77} = 1.9913$. Reject H_0. Type of institution makes a significant contribution and should be included in the model.

(g) $0.0106 \le \beta_1 \le 0.0216$, $6.7007 \le \beta_2 \le 9.5562$

(h) $r^2_{Y.12} = 0.793$. 79.3% of the variation in total cost can be explained by variation in average SAT score and variation in type of institution.

(i) $r^2_{adj} = 0.788$

(j)

	A	B	C	D
1	Predicting Total Cost			
2	Coefficients of Partial Determination			
3				
4	SSR(X1,X2)	2176.002359		
5	SST	2743.749875		
6	SSR(X2)	1228.375878	SSR(X1 | X2)	947.6264808
7	SSR(X1)	1928.628075	SSR(X2 | X1)	247.374284
8				
9	r2 Y1.2	0.625341653		
10	r2 Y2.1	0.303481374		

Holding constant the effect of average SAT score, 62.5% of the variation in total cost can be explained by variation in type of institution.

(k)   The slope of total cost with average SAT score is the same regardless of whether the institution is public or private.

(l)

	A	B	C	D	E	F	G
1	Interation Model for Predicting Total Cost						
2							
3	Regression Statistics						
4	Multiple R	0.895002968					
5	R Square	0.801030313					
6	Adjusted R Square	0.793176246					
7	Standard Error	2.680148892					
8	Observations	80					
9							
10	ANOVA						
11		df	SS	MS	F	Significance F	
12	Regression	3	2197.82682	732.6089402	101.9892437	1.42462E-26	
13	Residual	76	545.9230545	7.183198086			
14	Total	79	2743.749875				
15							
16		Coefficients	Standard Error	t Stat	P-value	Lower 95%	Upper 95%
17	Intercept	5.546801449	6.000353953	0.924412375	0.3581975	-6.403954486	17.49755738
18	Type of School	-4.27400243	7.150426227	-0.597726946	0.551799088	-18.51532875	9.967323887
19	Average Total SAT	0.007857324	0.005455545	1.44024545	0.153904874	-0.00300835	0.018722998
20	Type*SAT	0.010997125	0.006309084	1.743062048	0.085367166	-0.001568521	0.023562771

$\hat{Y} = 5.547 + 0.0079X_1 - 4.274X_2 + 0.011X_1X_2$.

For X_1X_2: the p-value is 0.085. Do not reject H_0. There is not evidence that the interaction term makes a contribution to the model.

•14.44 (a) $\ln \hat{Y} = 4.62 + 0.5(8.5) + 0.7(5.2) = 12.51$

            $\hat{Y} = e^{12.51} = 271{,}034.12$

      (b)   Holding constant the effects of X_2, for each additional unit of X_1 the natural logarithm of Y is expected to increase on average by 0.5. Holding constant the effects of X_1, for each additional unit of X_2 the natural logarithm of Y is expected to increase on average by 0.7.

•14.46 (a) $\ln \hat{Y} = 2.3882 + 0.004557 X_1$

      (b) $\ln \hat{Y} = 2.3882 + 0.004557(55) = 2.6388$

            $\hat{Y} = e^{2.6388} = 14.00$ miles per gallon.

      (c)   The residual analysis indicates a clear curvilinear pattern. The model does not adequately fit the data.

      (d) $t = 1.10 < t_{26} = 2.0555$. Do not reject H_0. The model does not provide a significant relationship.

      (e) $r^2_{Y.12} = 0.045$. Only 4.5% of the variation in the natural logarithm of miles per gallon can be explained by variation in the highway speed.

      (f) $r^2_{adj} = 0.008$

      (g)   The curvilinear model in Problem 14.33 is far superior to the inadequate models developed here and in Problem 14.45. The transformation of square root of highway speed or the natural logarithm of miles per gallon did virtually nothing to enhance the fit.

14.48 (a) $\hat{Y} = 4.666 + 5.0685 \sqrt{X_1}$

      (b) $\hat{Y} = 4.666 + 5.0685 \sqrt{55} = 42.255$ pounds

      (c)   The residual analysis does not indicate clear patterns.

      (d) $t = 12.86 > t_{10} = 2.2281$. Reject H_0. The model provides a significant relationship.

      (e) $r^2_{Y.12} = 0.943$. So, 94.3% of the variation in yield can be explained by variation in the square root of the amount of fertilizer applied.

      (f) $r^2_{adj} = 0.937$

      (g)   The curvilinear model in Problem 14.35 is slightly better than the model here. Both this model and the model in Problem 14.35 are better than the model in Problem 14.47.

14.50 $VIF = \dfrac{1}{1 - 0.5} = 2.0$

14.52 $R_1^2 = 0.008464$, $VIF_1 = \dfrac{1}{1 - 0.008464} = 1.009$

            $R_2^2 = 0.008464$, $VIF_2 = \dfrac{1}{1 - 0.008464} = 1.009$

       There is no reason to suspect the existence of collinearity.

•14.54 (a) $C_p = \dfrac{(1-R_p^2)(n-T)}{1-R_T^2} - [n-2(p+1)] = \dfrac{(1-0.274)(40-7)}{1-0.653} - [40-2(2+1)]$

$= 35.04$

(b) C_p overwhelmingly exceeds $p + 1 = 3$, the number of parameters (including the Y-intercept), so this model does not meet the criterion for further consideration as a best model.

•14.56 Let Y = selling price, X_1 = assessed value, X_2 = time period, and X_3 = whether house was new (0 = no, 1 = yes).

Based on a full regression model involving all of the variables, all of the VIF values (1.3, 1.0, and 1.3, respectively) are less than 5. There is no reason to suspect the existence of collinearity.

	A	B	C	D	E	F	G
1	Predicting Assessed Value						
2							
3	R2T	0.944827					
4	1 - R2T	0.055173					
5	n	30					
6	T	4					
7	n - T	26					
8							Consider
9		Cp	p+1	R Square	Adj. R Square	Std. Error	This Model?
10	X1	9.058101	2	0.925606	0.922948943	3.474934	No
11	X1X2	9.306584	3	0.929323	0.924087276	3.449169	No
12	X1X2X3	4	4	0.944827	0.93846131	3.105498	Yes
13	X1X3	2.848067	3	0.943028	0.938807597	3.096749	Yes
14	X2	371.6988	2	0.156074	0.125933484	11.70387	No
15	X2X3	336.853	3	0.234261	0.177540021	11.3531	No
16	X3	415.2237	2	0.063713	0.030273923	12.32769	No

Based on a best subsets regression and examination of the resulting C_p values, the best models appear to be a model with variables X_1 and X_2, which has $C_p = 2.8$, and the full regression model, which has $C_p = 4.0$. Based on a regression analysis with all original variables, variable X_3 fails to make a significant contribution to the model at the 0.05 level. Thus, the best model is the model using assessed value (X_1) and time (X_2) as the independent variables.

A residual analysis shows no strong patterns.

The final model is: $\hat{Y} = -44.988 + 1.7506X_1 + 0.368X_2$

$r_{Y.12}^2 = 0.943$, $r_{adj}^2 = 0.939$

Overall significance of the model: $F = 223.46, p < 0.001$
Each independent variable is significant at the 0.05 level.

14.58   Let Y = gasoline mileage, X_1 = weight, X_2 = width, X_3 = length, and X_4 = type of drive.

Based on a full regression model involving all of the variables:
$VIF_1 = 8.1$, $VIF_2 = 4.9$, $VIF_3 = 8.6$, $VIF_4 = 1.6$
Variable X_3 dropped from the model.

Based on regression model for remaining variables, all VIF values are less than 5.

A residual analysis shows no strong patterns. However, the p-values for the t-test statistics
Variables X_2 and X_4 are 0.445 and 0.126, respectively. X_2 dropped from the analysis since it is not
significant at the 0.05 level.

Based on a regression model using X_1 and X_4, the t-test statistic for variable X_4 has a p-value of
0.123 and should be dropped from the analysis.

The resulting model: $\hat{Y} = 47.885 - 0.0078538X_1$

$r^2 = 0.820$

Overall significance of the model: $F = 396.35$, $p < 0.001$

14.70   (a) $\hat{Y} = -44.988 + 1.7506X_1 + 0.368X_2$,
                  where X_1 = assessed value (in thousands of dollars) and X_2 = time period (in months).
        (b)   Holding constant the effects of time period, for each additional thousand dollars in assessed
                 value the selling price of the house is expected to increase on the average by 1.7506
                 thousands of dollars, or $1,750.60. Holding constant the effects of assessed value, for each
                 additional month the selling price of the house is expected to increase on the average by
                 0.368 thousands of dollars, or $368.
        (c) $\hat{Y} = -44.988 + 1.7506(70) + 0.368(12) = 81.969$ or $81,969
        (d)   All four residual plots indicate that the fitted model appears to be adequate.
        (e) $F = 301.27 > F_{U(2,26)} = 3.369$ with 2 and 26 degrees of freedom. Reject H_0. At least one of
                 the independent variables is linearly related to the dependent variable.
        (f)   The p value is less than 0.001. This means that the probability of obtaining an F test
                 statistic of 301.27 or greater if there were not relationship between the dependent variable
                 and independent variables is less than 0.001.
        (g) $r^2_{Y.12} = 0.959$. So, 95.9% of the variation in selling price can be explained
                 by the variation in assessed value and the variation in time period.
        (h) $r^2_{adj} = 0.955$
        (i)   For X_1: $t = 22.53 > t_{26} = 2.0555$ with 26 degrees of freedom. Reject H_0. There is evidence
                 that X_1 significantly contributes to a model already containing X_2. For X_2:
                 $t = 4.59 > t_{26} = 2.0555$ with 26 degrees of freedom. Reject H_0. There is evidence that X_2
                 significantly contributes to a model already containing X_1. Therefore, each independent
                 variable makes a significant contribution in the presence of the other variable, and both
                 variables should be included in the model.

(j) For X_1, the p value is less than 0.001. This means the probability of obtaining a t-test statistic which differs from zero by 22.53 or more (positively or negatively) when the null hypothesis that $\beta_1 = 0$ is true is less than 0.001. For X_2, the p value is less than 0.001. This means the probability of obtaining a t-test statistic which differs from zero by 4.59 or more (positively or negatively) when the null hypothesis that $\beta_2 = 0$ is true is less than 0.001.

(k) $1.509 \le \beta_1 \le 1.812$. This is a net regression coefficient. That is, taking into account the time period, this coefficient measures the expected average increase in selling price for each additional thousand dollars in assessed value. In Problem 13.75 the coefficient did not take into account (and hold constant) the effects of the time period.

(l) $r_{Y1.2} = 0.9513$. For a given time period, 95.13% of the variation in selling price can be explained by variation in assessed value. $r_{Y2.1} = 0.4481$. For a given assessed value, 44.81% of the variation in selling price can be explained by variation in time period.

14.72 (a) $\hat{Y} = -21.864 + 0.0239X_1 + 2.5191X_2$,
where X_1 = average SAT and X_2 = room and board expenses (in thousands of dollars).

(b) Holding constant the effects of room and board expenses, for each additional point of average SAT score the annual total cost is expected to increase on the average by 0.0239 thousands of dollars, or $23.90. Holding constant the effects of average SAT score, for each additional thousand dollars of room and board expense the annual total cost is expected to increase on the average by 2.5191 thousands of dollars, or $2519.10.

(c) $\hat{Y} = -21.864 + 0.0239(1100) + 2.5191(5.0) = 17.045$ or $17,045

(d) All four residual plots indicate that the fitted model appears to be adequate.

(e) $F = 100.15 > F_{U(2,77)} = 3.1154$ with 2 and 77 degrees of freedom. Reject H_0. At least one of the independent variables is linearly related to the dependent variable.

(f) The p value is less than 0.001. This means that the probability of obtaining an F test statistic of 100.15 or greater if there were not relationship between the dependent variable and independent variables is less than 0.001.

(g) $r^2_{Y.12} = 0.722$. 72.2% of the variation in annual cost can be explained by the variation in average SAT score and the variation in room and board expense.

(h) $r^2_{adj} = 0.715$

(i) For X_1: $t = 8.14 > t_{77} = 1.9913$ with 77 degrees of freedom. Reject H_0. There is evidence that X_1 significantly contributes to a model already containing X_2. For X_2: $t = 8.73 > t_{77} = 1.9913$ with 77 degrees of freedom. Reject H_0. There is evidence that X_2 significantly contributes to a model already containing X_1. Therefore, each independent variable makes a significant contribution in the presence of the other variable, and both variables should be included in the model.

(j) For X_1, the p value is less than 0.001. This means the probability of obtaining a t-test statistic which differs from zero by 8.14 or more (positively or negatively) when the null hypothesis that $\beta_1 = 0$ is true is less than 0.001. For X_2, the p value is less than 0.001. This means the probability of obtaining a t-test statistic which differs from zero by 8.73 or more (positively or negatively) when the null hypothesis that $\beta_2 = 0$ is true is less than 0.001.

(k) $0.0181 \le \beta_1 \le 0.0298$. This is a net regression coefficient. That is, taking into account the room and board expense, this coefficient measures the expected average increase in annual total cost for each point in average SAT score.

(l) $r_{Y1.2} = 0.4626$. For a given room and board expense, 46.26% of the variation in annual total cost can be explained by variation in average SAT score. $r_{Y2.1} = 0.4972$. For a given average SAT score, 49.72% of the variation in annual total cost can be explained by variation in room and board expense.

(m) Room and board expense is a minor component of annual total cost of attending a college or university. The remaining components tend to rise when room and board expenses rise.

14.74 (a) $\hat{Y} = 74.067 - 13.895X_1 + 0.0879X_2$, where X_1 = team E.R.A. and X_2 = runs scored.

(b) Holding constant the effects of runs scored, for each additional point of team E.R.A. the number of wins is expected to decrease on the average by 13.895 wins. Holding constant the effects of team E.R.A., for each additional run scored the number of wins is expected to increase on the average by 0.879 wins.

(c) $\hat{Y} = 74.067 - 13.895(4.00) + 0.0879(750) = 84.417$ wins

(d) All four residual plots indicate that the fitted model appears to be adequate.

(e) $F = 54.78 > F_{U(2,25)} = 3.3852$ with 2 and 25 degrees of freedom. Reject H_0. At least one of the independent variables is linearly related to the dependent variable.

(f) The p value is less than 0.001. This means that the probability of obtaining an F test statistic of 54.78 or greater if there were not relationship between the dependent variable and independent variables is less than 0.001.

(g) $r_{Y.12}^2 = 0.814$. 81.4% of the variation in team wins can be explained by the variation in team E.R.A. and the variation in team runs scored.

(h) $r_{adj}^2 = 0.799$

(i) For X_1: $t = -8.80 < -t_{25} = -2.0595$ with 25 degrees of freedom. Reject H_0. There is evidence that X_1 significantly contributes to a model already containing X_2. For X_2: $t = 7.79 > t_{25} = 2.0595$ with 25 degrees of freedom. Reject H_0. There is evidence that X_2 significantly contributes to a model already containing X_1. Therefore, each independent variable makes a significant contribution in the presence of the other variable, and both variables should be included in the model.

(j) For X_1, the p value is less than 0.001. This means the probability of obtaining a t-test statistic which differs from zero by 8.80 or more (positively or negatively) when the null hypothesis that $\beta_1 = 0$ is true is less than 0.001. For X_2, the p value is less than 0.001. This means the probability of obtaining a t-test statistic which differs from zero by 7.79 or more (positively or negatively) when the null hypothesis that $\beta_2 = 0$ is true is less than 0.001.

(k) $-17.1484 \le \beta_1 \le -10.642$. This is a net regression coefficient. That is, taking into account the number of runs scored, this coefficient measures the expected average increase in team wins for each point in team E.R.A.

14.74 (l)
cont.

$r_{Y1.2} = 0.7558$. For a given number of team wins, 75.58% of the variation in team wins can be explained by variation in team E.R.A.. $r_{Y2.1} = 0.7082$. For a given team E.R.A., 70.82% of the variation in team wins can be explained by variation in runs scored.

(m) Pitching as measured by team E.R.A. is slightly more important in predicting wins, based on the t-test statistics and coefficients of partial determination. However, both measures are important in predicting wins.

14.76 (a)
$\hat{Y} = 132.50 - 11.461X_1 - 2.563X_2$,
where X_1 = team E.R.A. and X_2 = league (American = 0, National = 1).

(b) Holding constant the effect of league, for each E.R.A. point, team wins are expected to decrease on average by 11.461. For a team E.R.A., being in the National League as opposed to the American League is expected to decrease team wins on average by 2.563.

(c) $\hat{Y} = 132.50 - 11.461(4.00) - 2.563(0) = 86.65$ wins

(d) Based on a residual analysis, the model appears adequate.

(e) $F = 7.64 > F_{2,25} = 3.39$. Reject H_0. There is evidence of a relationship between team wins and the two dependent variables.

(f) For X_1: $t = -3.87 < -t_{25} = -2.0595$. Reject H_0. Team E.R.A. makes a significant contribution and should be included in the model.
For X_2: $t = -0.81 > -t_{25} = -2.0595$. Do not reject H_0. League does not make a significant contribution and should not be included in the model.

(g) $-17.5548 \le \beta_1 \le -5.3679$, $-9.0676 \le \beta_2 \le 3.9419$

(h) The slope here takes into account the effect of the other predictor variable, league, while the solution for Problem 13.79 did not.

(i) $r_{Y.12}^2 = 0.379$. 37.9% of the variation in team wins can be explained by variation in team E.R.A. and variation in league.

(j) $r_{adj}^2 = 0.330$

(k) $r_{Y.12}^2 = 0.379$ while $r^2 = 0.363$. The inclusion of the league variable has resulted in the increase.

(l) $r_{Y1.2}^2 = 0.375$. Holding constant the effect of league, 37.5% of the variation in team wins can be explained by variation in team E.R.A..
$r_{Y2.1}^2 = 0.026$. Holding constant the effect of team E.R.A., 2.6% of the variation in team wins can be explained by variation in league.

(m) The slope of team wins with team E.R.A. is the same regardless of which the league the team is in.

(n) $\hat{Y} = 145.34 - 14.274X_1 - 22.86X_2 + 4.583X_1X_2$.
For the X_1X_2 coefficient: the p-value is 0.462. Do not reject H_0. There is not evidence that the interaction term makes a contribution to the model.

(o) A simple regression model with team E.R.A. as the independent variable should be used.

14.78   An analysis of the linear regression model with all possible independent variables reveals that one of the variables, annual total cost, has a VIF value (6.9) in excess of 5.0. Based on the procedure recommended in the text, this variable should be deleted from the model.

An analysis of the linear regression model with the remaining independent variables indicates none of the remaining variables have a VIF value that is 5.0 or larger. A best subsets regression produces several subsets of variables with C_p values at or below $p + 1$. The following variables belong to most of the subsets with better C_p values:

X_2 = Type of institution (0=public, 1=private)
X_3 = Average total SAT Score
X_4 = TOEFL Criterion at least 550 (0=no, 1=yes)

For a linear regression model fit to these three variables, there is not sufficient evidence that β_3 is significantly different from zero ($t = -0.86$, $p=0.392$). Removing X_3 and fitting a new model, there is not sufficient evidence that β_4 is significantly different from zero ($t = -1.38$, $p=0.172$).

The best model using best subsets regression appears to be the simple regression model:

$\hat{Y} = 10.9368 + 3.592X_2$. There is sufficient evidence that β_2 is significantly different from zero ($t = 3.90$, $p < 0.001$) and

$r^2 = 0.163$. A residual analysis shows no strong patterns.

14.80-14.84
Based on best subsets regression, the following were the best models for predicting appraised value using six measures available for the homes in the sample for each community – lot size, number of bedrooms, number of bathrooms, age of house, presence of an eat-in-kitchen (EIK), and presence of central air conditioning (CAC):

East Meadow:

$\hat{Y} = 137.68 + 41.969$ Bath $+ 13.581$ CAC

$r^2 = 0.503$, $r^2_{adj} = 0.489$. The C_p value $= 3.0$.

Farmingdale:

$\hat{Y} = 142.61 - 1.0451$Lotsize $- 5.103$Bedrooms $+ 31.958$Bathrooms $+ 22.14$EIK $+ 49.37$CAC

$r^2 = 0.437$, $r^2_{adj} = 0.385$. The C_p value $= 6.0$.

Levittown:

$\hat{Y} = 174.99 + 1.1906$ Lotsize $+ 14.341$ Bathrooms $- 1.1408$ Age $+8.718$ EIK

$r^2 = 0.359$, $r^2_{adj} = 0.331$. The C_p value $= 3.6$.

Islip:

$\hat{Y} = 100.99 + 0.7436$ Lotsize $+ 4.784$ Bedrooms $+ 11.904$ Bathrooms $- 0.1926$ Age $+ 15.09$ EIK $+ 27.09$ CAC

$r^2 = 0.366$, $r^2_{adj} = 0.317$. The C_p value $= 7.0$.

Islip Terrace:

$\hat{Y} = 98.87 + 11.141$ Bedrooms $- 0.7143$ Age $+ 35.70$ EIK $+ 16.811$ CAC

$r^2 = 0.590$, $r^2_{adj} = 0.549$. The C_p value $= 4.4$.

Based on best subsets regression, the following were the best models for predicting taxes using six measures available for the homes in the sample for each community – lot size, number of bedrooms, number of bathrooms, age of house, presence of an eat-in-kitchen (EIK), and presence of central air conditioning (CAC):

East Meadow:

$\hat{Y} = 175.0 + 153.5 \text{ Lotsize} + 178.37 \text{Bedrooms} + 784.3 \text{ Bathrooms}$

$r^2 = 0.598$, $r^2_{adj} = 0.581$. The C_p value = 4.0.

Farmingdale:

$\hat{Y} = 3168.8 + 499.5 \text{ Bathrooms} - 28.003 \text{ Age} + 562.9 \text{ CAC}$

$r^2 = 0.394$, $r^2_{adj} = 0.362$. The C_p value = 3.8.

Levittown:

$\hat{Y} = 2904.8 + 58.65 \text{ Lotsize} + 492.53 \text{ Bathrooms} - 35.47 \text{ Age} + 541.0 \text{ CAC}$

$r^2 = 0.504$, $r^2_{adj} = 0.483$. The C_p value = 4.7.

Islip:

$\hat{Y} = 2859.5 + 260.11 \text{ Bedrooms} + 441.6 \text{ Bathrooms} - 10.652 \text{ Age} + 800.2 \text{ CAC}$

$r^2 = 0.604$, $r^2_{adj} = 0.584$. The C_p value = 3.4.

Islip Terrace:

$\hat{Y} = 2835.8 + 591.2 \text{ Bathrooms}$

$r^2 = 0.257$, $r^2_{adj} = 0.240$. The C_p value = 1.0.

Lecture Notes

Statistics is Different

1. Technical Vocabulary
2. Highly Cumulative Course Content
3. Requires Precision in Procedures

What Can You Do?

- Learn the Language At Once
- Do NOT Get Behind
- Learn how to estimate an answer

Statistical Jargon–

What is Statistics?

A multi-use word

- ▸ It often is a summary value, people will refer to the statistic
- ▸ The Science of uncertainty
- ▸ the science of collecting, organizing, presenting, analyzing and interpreting numerical data for the purpose of decision making

- Our text implies that it is the science of understanding, managing and reducing variation.

Why study Statistics?

- Too much information
- Need help interpreting information
- Statistics regulates our lives
  - Insurance companies
  - EPA
  - Medical treatment decisions for example,

✓ <u>Salk Polio Vaccine</u>
  - ✓ 1954-400,000 children studied
  - ✓ well controlled
  - ✓ Polio is virtually unknown today
▸ Learn how to make "Good" data based decisions
  - A Bad One
  - 1986 The Challenger Shuttle blew up in 29 degree weather
  - NASA had failed to do a simple analysis of performance data at low temperatures.

Descriptive vs Inferential Statistics

- Descriptive statistics
  - organizing presenting and analyzing data in an informative way
  - describing past events
  - drawing pictures
  - Tools are tables, graphs and summary statistics
  - these are the statistics we think about and know most about

- **Inferential Statistics**
  - techniques used to "infer" or guess about a population based on a sample
  - making guesses about the future
  - deductive vs inductive reasoning
    - deductive reasons from the whole to draw conclusions about characteristics
    - inductive infers from observations (samples) to draw conclusions about the whole

- ▸ Do you ever perform inferential statistics?
  - ▸ testing temperature of water
  - ▸ sample size of new shampoo
  - ▸ drinking coffee/tea

- ▸ Variables- -things that can change
  - ▸ Most often we refer to the place holder for something that can change as a variable.
  - ▸ We will deal most often with Random Variables in this class.

Qualitative vs Quantitative Variables

Qualitative: non-numeric variables
- ▸ usually indicate the membership of an observation in a category
- ▸ we can count them,
- ▸ sometimes we put categories in order but not really anything else.

Quantitative: numeric variables which can be

- *Discrete:*
  - Such a variable can attain only certain values; gaps exist between possible values.
- *Continuous:*
  - Such a variable can attain any value within the range of the variable; no gaps exist between possible values.

You can see

→   Data can come in many forms
      and may be measured
      differently.

→   The data set dictates what
      procedures can be performed
      on that data set.

→   Data sets can be populations or
      samples.

Populations vs Samples

- *A Population*
  - a collection of all possible individuals, objects or measurements of interest.
  - students in this class
  - residents of Indiana
  - Citizens of the US
- *A Sample*
  - a part of a population
  - students in the class
  - students in the front row
  - students wearing blue

→ **Why use samples?**

- so we don't get burned
- so we don't get stuck with something we don't like
- cost of measuring, polling or counting the entire population is prohibitive
- observations may be destroyed or consumed by the test
- want an answer today (speed of response)
- Can't poll the entire population
- Sample results are no worse than those from populations

SAMPLING TECHNIQUES

→ **What kinds of samples are there?**

▸ probability: each item in the population has a known (non-zero) likelihood of being included in the sample

▸ non-probability: certain population items have a zero likelihood of being included in the sample

→ Results in BIASED samples

→ **What kinds of Probability Samples are there?**

▸ **The Simple Random Sample:**

- each item has an identical chance of being selected
- selection of one item does not influence selection of other items.
  - Draw items from a hat
  - The lottery
  - table of random numbers

- Could Build a "Sampling Frame"
  - list ever item in the population
  - Draw n items "randomly"
  - But, making the sampling frame may be
    - costly,
    - controversial
    - impossible

So, one uses one's knowledge of the population to design a sample

Types of Samples:

✓ **The Systematic Random Sample.**
  ▸ first item is chosen at random
  ▸ the remainder are drawn at intervals determined by the size of the sample
  ▸ not appropriate for situations where there is a pattern to the population

✓ **The Stratified Random Sample**
- ▸ **population divided into homogeneous subgroups**
- ▸ **sample from each of the groups.**

✓ **Cluster Sampling**
- ▸ **Divide the population into units**
- ▸ **select units at random**
- ▸ **measure all items in those units**

!! CAUTION !!

Statistical Results Depend on
   ✓ independence and
   ✓ lack of bias
in the simple random sample.

Other sampling methods results must be modified.

!! CAUTION #2 !!
Without Randomized Design,
there can be NO
Dependable Statistical Analysis

Problem Sampling:

1. The opportunity sample–Choose the first n units that come along.
2. Response Bias–Sample only those groups who have a vested interest in the results

STATISTICAL LIES - -

1. Improper Sized Graphs

2. Describing Data Using Emotionally Laden Words

3. Incomplete or Inadequate numerical representation

4. Post Hoc Fallacies

5. Claims Based on Very Small Samples

Reid's Table

Common Phrase	Absolute Value
One	1
A couple	2 to 4
Quite a few	3 to 6
Most (#)	4 to 6
A lot	7 to 11
A whole lot	8 to 17
Most (%)	10% to 20%
A few hundred	75 to 125
Half a million	90,000 to 125,000

The first act of reducing data to an organized form is to "order it", that is, put it in order.

  ✓ use the number line, order
     data from small to large.
  ✓ use the alphabet, order data
     from A to Z.
  ✓ use any other well-known
     organizational scheme, such as
     the common work day hours,
     and order data according to it.

Step Two:
understanding *frequencies*.

✓ the number of times an object or event of interest occurs in a given space or over a given time, a count.

✓ f_i

Other frequencies . . .

✓ relative frequency – the proportion of the whole each class represents.

✓ divide the class frequency by the total frequency or

$$\frac{f_i}{\sum_i f_i}$$

✓ cumulative frequency – each class frequency includes the counts from all previous classes.

✓ can be relative cumulative or absolute cumulative frequencies.

Step Three:
Organizing Frequencies

✓ a *frequency distribution* is one of the most common ways of presenting organized data.

✓ *frequency distribution* – divide the number line into intervals, then count and record the number of occurrences in each interval.

✓ *frequency distribution* – data grouped into mutually exclusive and exhaustive categories called *classes*.

Classy Terminology

✓ *class limits* – the upper and lower bound of the class.

✓ *class mark* – the midpoint of the class.

✓ *class interval* – the width of the class.

Class terms are interdependent.

→ If you know two, you can always get the third

Let's begin with limits. . . .What *ARE* limits, anyway?

- ✓ warning markers
- ✓ boundary lines
- ✓ edges of intervals
- ✓ a fixed value that a varying value can approach indefinitely.

Class limits are the upper and lower bound of the class   BUT

▸   the lower bound is closed, that is, the class includes the value of the lower bound.

▸ the upper bound is open, that is, the class goes up to but does NOT include the value of the upper bound.

TIP: The upper limit of one class is the lower limit of the next.

Class Mark

is the midpoint of the class, and is calculated by adding the upper and lower limit and dividing by two.

Class Interval

is the width of the class, and is calculated by subtracting the lower limit from the upper limit.

Class Intervals should be

- ✓ equal in width.
- ✓ can be estimated by subtracting the lowest number (minimum) from the highest number (maximum), and dividing by the number of classes you would like to have.
- ✓ may be dictated by previous experience.

Three Data Pictures . . .

1. **The Histogram**
   - ✓ bar chart of the frequency distribution
   - ✓ classes on the horizontal axis
   - ✓ frequencies on the vertical axis
   - ✓ the bars touch one another

2. The Frequency Polygon
✓ line graph
✓ connects class midpoints, class frequency points

3. **The Ogive**
   ✓ **line graph**
   ✓ **connects upper class limits, class cumulative frequency points**

Summary Statistics–
   ✓ they each reduce a data set to
      a single number

Types:

**Central Tendency–the center of the
data set or where we would look for
the data set if we had to find it**

**Dispersion–how spread out, how
different or diverse is the data**

PARAMETER vs STATISTIC

Population ➡ Parameter

Sample ➡ Statistic

Central Tendency Measures

Mean
- ✓ the arithmetic mean
- ✓ sum all observations and divide by the number of observations

Properties of the mean

→ All quantitative data have a mean

→ All observations are used

→ Each observation is equally important

→ A unique parameter/statistic per data set

→ The sum of the deviations from the mean is zero.

More Mean Properties:

→ It is the "balance point."

→ It is very sensitive to exceptionally large or small observations called outliers

→ Can't estimate in certain circumstances.

The "weighted" mean

✓ Technically, every mean is a weighted mean

✓ Data sets with unique entries have an implied weight of 1

✓ Means for data sets with multiple entries of the same value are often easier to calculate using weights

The Weighted Mean

$$\overline{X}_w = \frac{\sum (w_i * X_i)}{\sum w_i}$$

w_i = the weight on an observation

GPA calculations:

A C B B D C B A A C A B B B B

Grades (X_i)	# of grades (w_i)	w_iX_i
A=4	4	4*4=16
B=3	7	3*7=21
C=2	3	2*3=6
D=1	1	1*1=1
	Σw_i = 15	Σw_iX_i =44

GPA = 44/15=2.93

Median

✓ The "middle" of a data set

✓ The value below which are 50% of observations and above which are 50% of observations

✓ the $\frac{n+1}{2}$ observation in order

Properties of the Median:

→   Unique to a data set

→   Not sensitive to outliers

→   Any data that can be ordered can have one.

→   Can be used on open classes.

Mode

✓ **The most "ubiquitous" member of a data set**

✓ **That value that occurs most often**

Properties of the Mode:

→ All data may have modes

→ It is the only measure of center for categorical data

→ Not sensitive to outliers

→ Highly variable

→ Doesn't always exist

→ Sometimes there is more than one mode.

Centers of Grouped Data

The Grouped Data Mean

$$\bar{X} = \frac{\sum X_i f_i}{\sum f_i} = \frac{\sum X_i f_i}{n}$$

X_i is the midpoint of the class

(Compare to the weighted mean. . .)

The Median Class

the class with the $\frac{n}{2}$th entry.

The Modal Class

✓ the class with the largest frequency.

Auto Repair Cost for Minor Repairs

Costs $	Fre-quency			
0-100	10			
100-200	18			
200-300	60			
300-400	70			
400-500	42			

Summary Statistics and Graphs

Symmetric Data Distributions

$$\text{mean} = \text{median} = \text{mode}$$

Skewed Data Distributions

✓ Right or Positive Skewed Data

$$\text{mode} < \text{median} < \text{mean}$$

✓ Left or Negative Skewed Data

$$\text{mean} < \text{median} < \text{mode}$$

Some Important Distributions- -

Binomial- - -

Poisson- -

Hypergeometric- -

Uniform- -

Normal- - -

Exponential- - -

Descriptive Statistics- -

- **Measures of Center (think location)**

- **Measures of Dispersion (think area)**

    ▸ **these measures help us make judgments about how similar observations are to each other**

The simplest measure is the *range*

- ▸ how far apart the highest value and the lowest value in the set are.

- ▸ Range = (maximum-minimum)

- ▸ Used when reporting stock market movements

- ▸ Used to report temperatures over a day

Range Problems:

- doesn't account for actual distribution of the data between the high and the low point.

- sensitive to extreme values, just like the mean.

What are we looking for when we say dispersion?

✓ How far observations are from some fixed point.

✓ Why not use the mean as the fixed point?

✓ We defined deviations as the distance between an observation and the mean.

✓ Calculate the average deviation.

✓ Some data: the Doe Family Ages

- ▸ 2, 18, 34, 42

- ▸ $\mu = 96/4 = 24$

- ▸ Deviations

- ▸ The average deviation

The Variance

$$\sigma^2 = \frac{\sum(x_i - \mu)^2}{N}$$

It measures the "typical" squared distance from the mean.

Back to the Doe's:

▸ $\sigma^2 =$

- ▸ the typical squared difference from the average family age is 236 square years.

- ▸ 236 Square Years?

- ▸ What does *that* mean?

Problems with the Variance:

✓ It has excellent mathematical properties, but it is difficult to understand

✓ We will refer most often to its offspring, the standard deviation.

Standard Deviation:

- The square root of the variance

- $\sigma = \sqrt{\sigma^2}$

- It is no easier to calculate--you have to get a variance first--it is just easier to interpret.
- It is measured in the same units as the data is measured, so in this case, the standard deviation of the Doe family ages is 15.36 years.

These have been Parameters--

What about statistics?

The sample variance

$$s^2 = \frac{\sum (x_i - \bar{x})^2}{n-1}$$

Why is it different?

- Conceptually, a sample does not include all the information that a population does

- Samples tend to UNDER estimate the variability found in a population.

- If we divide by a slightly smaller number (n-1) we get a slightly larger number.

The sample standard deviation is just the square root of s^2.

$$s = \sqrt{s^2}$$

A Computational Formula

$$s^2 = \frac{\sum x^2 - \frac{(\sum x)^2}{n}}{n-1}$$

Compare to this for grouped data:

$$s^2 = \frac{\sum fx^2 - \frac{(\sum fx)^2}{n}}{n-1}$$

A hint of things to come:

Mean and Variance from a frequency distribution.

$$\mu = \frac{\sum f_i x_i}{\sum f_i} = \sum \frac{f_i}{N} x_i$$

$$\sigma^2 = \sum \frac{f_i}{N}(x_i - \mu)^2$$

Cookie Data Set				
X	f	fX	X^2	fX^2
2.50	4	10.00	6.25	25.00
2.63	6	15.75	6.89	41.34
2.75	10	27.50	7.56	75.63
2.88	19	54.63	8.27	157.05
3.00	22	66.00	9.00	198.00
3.13	14	43.75	9.77	136.72
3.25	13	42.25	10.56	137.31
3.38	8	27.00	11.39	91.13
3.50	1	3.50	12.25	12.25
3.63	1	3.63	13.14	13.14
	98	294		887.56
		3		

Using Measures of Dispersion

σ^2 or σ are often used for the same purposes--

    *measures of heterogeneity

- ▸ Intelligence test scores in a classroom
- ▸ Two cities with the same average income

*measures of consistency

▸ **the standard deviation of letter delivery times by the US post office**

▸ **standard deviation of major league batting averages**

→ An interesting question: the standard deviation of major league batting averages has declined from .049 to .031 over the last 100 years, but the mean has remained at .260.

→ What does this mean?

→ How likely is it that someone will break Ted Williams record average of .406 anytime soon?

Dispersion as Risk

Standard Deviation or Variance is often used as a measure of risk

→ Stockbrokers are concerned with assessing risk

→ Using standard deviation or variance is appropriate when the mean return on investments is the same.

What if they aren't?

→ When mean stock prices vary greatly, the size of the mean influences the size of the variance.

→ What if you want to compare dispersion for two items that are not measured in the same units?

Another Measure of Risk (A Better One)

The Coefficient of Variation

This measure of dispersion is free of units:

$$CV = \frac{s}{\bar{x}}(100)$$

Important Summary Statistic Behaviors

How do the mean and variance behave if we change the observations in a sample or population in a systematic way?

Summary Statistic Behavior		
X_i	$X_i + 5$	$3X_i$
1		
4		
10		
11		
14		
$\bar{x} = 8$	$\bar{x} =$	$\bar{x} =$
$s^2 = 28.5$	$s^2 =$	$s^2 =$

Using Standard Deviations to Describe Populations

✓ **Data sets with certain characteristics contain a predictable number of observations within certain limits.**

Two Ways to Predict How Many. . .

✓ **Chebyshev's Theorem**

✓ **The Empirical Rule**

Chebyshev's Theorem and Its Uses:

*Rule is very general.

*Applies regardless of the shape of the distribution.

*NOTHING is needed to be known about the distribution.

The Rule:

For any set of observations, the proportion of values that lie within k standard deviations of the mean is at least $(1-1/k^2)$, where k is a constant larger than 1.

817

How many major leaguers batting averages are found within 3 standard deviations of the mean of .260?

A More Limited but More Precise Rule- -

The Empirical or Normal Rule

For symmetric,
bell-shaped distributions:

- ✓ $\mu \pm 1\sigma$ contains about 68% of the observations.
- ✓ $\mu \pm 2\sigma$ contains about 95% of the observations.
- ✓ $\mu \pm 3\sigma$ contains about 99.7 % of the observations.

Using the Empirical Rule:
   Virtually all observations lie between ± 3σ so the range can be approximated.

The standard deviation of batting averages is .031. What is the approximate range of batting averages?

Comparing Chebyshev and the Empirical Rule:

*$\mu \pm 2\sigma$ contains about 95%

but

Chebyshev says $1 - 1/4 = 0.75$

*$\mu \pm 3\sigma$ contains about 99.7 %

but

Chebyshev says $1 - 1/9 = 0.89$

Pearson's First and Second Skewness Coefficients

First: $$S_k = \frac{(\bar{x} - mode)}{s}$$

Second: $$S_k = \frac{3(\bar{x} - median)}{s}$$

Truly: **The Third "Moment" of the Distribution Function**

Where are we?

Statistics

Descriptive

Centers
Dispersion
Pictures
The Past

Inferential

Probability
Guesses
The Future

Probability:

A value between zero and one, inclusive, describing the relative possibility (chance or likelihood) an event will occur.

Usually expressed as a decimal but can also be expressed as a fraction.

Certainty: When $P(A)=0$ or $P(A)=1$
Uncertainty: When $0<P(A)<1$

Probability is divided into Two Parts

Probability

Objective
- relies on theory or observation
- Classical
- Empirical

Subjective
- relies on opinion and beliefs

Classical Probability depends on theory or math and has Three Characteristics:

1. The assumption that all outcomes of an experiment are equally likely
2. The set of events is mutually exclusive
3. The set of events is collectively exhaustive

Examples:
   flipping coins and rolling dice

Important Result:

✓ IF the set of events is mutually exclusive and collectively exhaustive, the sum of the probabilities will equal ONE.

This also applies to . . .

Empirical Probability, which depends

on observations - -

experience - -

DATA

For both of these types of probability we use similar formal notation:

$$P(A) = \frac{\text{\# of favorable outcomes}}{\text{total number of possible outcomes}}$$

$$P(A) = \frac{\text{\# of occurrences from past}}{\text{total number of observations}}$$

Some Quick Reminder Definitions

Random Experiment: the process of observing the outcome of a chance event.

Elementary Outcomes: all the possible results of a random experiment.

Sample Space: the collection of all elementary outcomes.

Special Rule of Addition:

$$P(A \text{ or } B) = P(A) + P(B)$$

Two mutually exclusive events cannot happen at the same time.

General Rule of Addition:

$$P(A \text{ or } B) = P(A) + P(B) - P(A \text{ and } B)$$

Accounts for those events that may not be mutually exclusive, that is overlapping events, and corrects for them.

JOINT PROBABILITY

Events that over lap, that are NOT mutually exclusive, have joint probabilities.

If P(A and B) > O,

→   then A and B are not mutually exclusive events.

CONDITIONAL PROBABILITY

the probability that one event will happen, given that another already has.

$$P(A|B) = \frac{P(A \text{ and } B)}{P(B)}$$

General Rule of Multiplication

$$P(A \text{ and } B) = P(A|B)*P(B)$$

BUT IF

$$P(A \text{ and } B) = P(A)*P(B)$$

then A and B are INDEPENDENT.

This leads us to . . . .

Special Rule of Multiplication

which REQUIRES independence
of A and B.

$$P(A \text{ and } B) = P(A) * P(B)$$

Envisioning Joint and Conditional Probabilities.

The Contingency Table

-a cross classification of two categorized variables that can be used to calculate

♦ simple (marginal) probabilities

♦ joint probabilities

♦ conditional probabilities

A Contingency Table:

Earned Degrees by Level and Gender (in K's)					
	Level				
	BA/S	MA/S	Prof	PhD	TTL
F	379	119	18	10	526
M	325	105	27	16	474
TTL	704	224	46	26	1000

- ▸ Calculate the probabilities for each cell.

- ▸ Check the independence of
    - being female and earning a bachelor's.
    - being female and earning a professional degree.

- ▸ What is the probability of
    - being male given that one has earned a bachelor's degree?
    - earning a Master's degree given that one is male?

The Decision Tree Diagram

A contingency table built like the branches of a tree.

- The "large branches" are simple (marginal) probabilities.
- The "small branches" are conditional probabilities.
- The "leaves" are joint probabilities.

What is a *Random Variable (RV)?*

A process that assigns one numerical value (a probability) to each possible outcome in an experiment.

Notation: $X \equiv$ outcome

$P(X) \equiv$ probability of an outcome

Requirements for Random Variables:

1. An experiment can be conducted with outcomes determined by chance.
2. Outcomes constitute a sample space, where outcomes are mutually exclusive and exhaustive.
3. Outcomes have unique numerical values that occur with a probability distribution.
4. Values can be discrete or continuous.

Discrete versus Continuous Random Variables

Discrete:

1. can assume only a finite or countably infinite number of different values

2. "gaps" between values along the number line

3. It is possible to list all possible values of X and the associated P(X) in Probability Distributions

Continuous:

- an RV that can take on any value in an interval
- no "gaps" between values
- because there is an infinite number of X values, we cannot associate a probability with a single value, only a range of values.

Some Discrete Variables we treat as continuous:

- Family Income
- College GPA
- Corporate Taxes
- Sales Revenues
- Any monetary figure

    --dollar amounts are measured to the nearest penny, but are usually considered continuous.

Random Variables have Descriptive Statistics—called expected values!

$$\mu = E(x) = \sum x P(x)$$

$$\sigma^2 = E[(x - \mu)^2]$$
$$= \sum (x - \mu)^2 P(x)$$

$$\sigma = \sqrt{\sigma^2}$$

Bernoulli—

✓ **the simplest of all discrete distributions**

  ▸ **a single experiment MUST result in one of two mutually exclusive and exhaustive outcomes**

  ▸ **Outcomes are usually called "Success" and "Failure"**

  ▸ **P(success) = π**

  ▸ **P(failure) = $(1-\pi)$**

✓ examples:
  ▸ toss a coin- -heads or tails
  ▸ Select a product, defective or not
  ▸ select a person, employed or not
  ▸ select a shopper, purchase or not
  ▸ select an investment, profit or not

✓ an experiment like this is often referred to as a Bernoulli Trial

✓ The Bernoulli applies to an experiment that is performed only once.

What if we repeat our experiment more than once?

Characteristics of the Binomial

- ✓ Outcome of EACH TRIAL is one of two mutually exclusive categories.
- ✓ X is the result of counting the number of successes in a fixed number of trials, and is, thus, discrete.
- ✓ P(success) remains constant over all trials.
- ✓ Each trial is independent of the other.

Important information for the Binomial: Three Items

- ▸ **Sample Space of X**
- ▸ **Sample Size (n)**
- ▸ **P(success) for each trial, π**

Other Binomial Experiments:

- ▸ **Select 10 manufactured items, defective or not, Manufacturer claims 2% are defective.**

- Select 100 people, unemployed or not, unemployment rate is 35%

- Contact 300 people, reply or not, usually 5% of people reply

- Sample 60 purchasers of contact lenses after one year, still wearing or not, 25% of people stop wearing lenses in one year

Consider this problem and write down the probability distribution, as much as possible.

A large lot of fuses contains 10% defectives. Four fuses are drawn at random.

X	P(X)

Isn't there an easier way?

$$P(x) = \frac{n!}{x!(n-x)!}\pi^x(1-\pi)^{n-x}$$

Isn't there an even easier way?

=BINOMDIST(x,n,π,0)

BINOMIAL GRAPHS

→ Binomials are well-suited to histograms

→ Each value of X has a column above it

→ The column's height is P(X)

→ Columns are one unit wide, so the *AREA* of the column is P(X), too

How does varying the value of π change the look of the Binomial?

$\pi < 0.5$, Binomial is right skewed

$\pi > 0.5$, Binomial is left skewed

$\pi \approx 0.5$, Binomial is symmetric

How about changing the value of n?

▸ When n is small, the Binomial is determined by π.

▸ However, as n becomes large, the Binomial becomes more and more symmetric, regardless of π.

When is n large?

Rule of thumb is n is large when

$$n \geq 30$$

Final thoughts on the Bernoulli

- a one-parameter distribution
- π completely defines any Bernoulli
- $E(X) = \pi$ $V(X) = \pi(1-\pi)$

and the Binomial?

- the Binomial is the sum of n Bernoullis
- it is defined completely by n and π
- $E(x) = n\pi$ $V(X) = n\pi(1-\pi)$

Random Variables are Quite Flexible

We can add them,

multiply them by constants,

add constants to them,

In short, we can combine them into

LINEAR COMBINATIONS.

Consider a function

$$T = aX + bY + c.$$

✓ a, b and c are constants
✓ X and Y are some variables about which we know some things.
✓ Specifically we know:
  ▸ the mean of X is μ_x and the variance of X is σ^2_x
  ▸ the mean of Y is μ_y and the variance of Y is σ^2_y

Meet the

Expectations Operator, E which tells us to find the expected value of anything it is in front of.

Meet the

Variance Operator, V which tells us to find the variance of anything it is in front of.

A Rule About E:

▶ **The expected value of a constant is the constant.**

So $E(T) = E(aX + bY + c)$

$= E(aX) + E(bY) + E(c)$

$= E(a)E(X) + E(b)E(Y) + E(c)$

$= aE(X) + bE(Y) + c$

$= a\mu_x + b\mu_y + c$

Some Rules About V:

▸ **The variance of a constant is zero, if the constant is NOT attached to a variable.**

$$V(T) = V(aX + bY + c)$$

$$= V(aX) + V(bY) + V(c)$$

$$= V(aX) + V(bY) + 0$$

- The variance of a constant is the constant squared, if the constant is attached to a random variable.

- continuing from the previous page:

$$= V(aX) + V(bY) + 0$$

$$= V(a)V(X) + V(b)V(Y)$$

$$= a^2V(X) + b^2V(Y)$$

$$= a^2\sigma^2_x + b^2\sigma^2_y$$

HOW DO WE FIND

✓ EXPECTED VALUE OF X
✓ VARIANCE OF X

when X is described by a probability distribution?

$$\mu = E(X) = \Sigma[X * P(X)]$$

$$\sigma^2 = V(X) = \Sigma[(X-\mu)^2 * P(X)]$$

Pharmaceuticals Companies

Mature Corporation Current Value=$100		$X_1 = \Delta$ Value	
X_1	+$15	0	-$15
$P(X_1)$	0.40	0.50	0.10

$E(X_1)$

$V(X_1)$

$$\mu = E(X) = \Sigma[X * P(X)]$$

$$\sigma^2 = V(X) = \Sigma[(X-\mu)^2 * P(X)]$$

Startup Corporation Current Value=$100			$X_2 = \Delta$ Value
X_2	+$30	0	-$30
$P(X_2)$	0.40	0.50	0.10

$E(X_2)$

$V(X_2)$

A LINEAR COMBINATION OF BOTH STOCKS

✓ Invest 75% of $100 in Mature
✓ Invest 25% of $100 in Startup

✓ What is the expected return on a $100 investment?

$$E(\tfrac{3}{4}X_1 + \tfrac{1}{4}X_2) =$$

$$\tfrac{3}{4}E(X_1) + \tfrac{1}{4}E(X_2) =$$

$$\tfrac{3}{4}(\$4.50) + \tfrac{1}{4}(\$9.00) =$$

$$\$3.375 + \$2.25 = \$5.625$$

So, in general we would say

$$E(b_0 + b_1X_1 + b_2X_2 + b_3X_3 \ . \ . \ .) =$$

$$b_0 + b_1E(X_1) + b_2E(X_2) + b_3E(X_3) \ . \ .$$

when b_i is a constant.

☞   **What about the
Variance?**

$$V(\tfrac{3}{4}X_1 + \tfrac{1}{4}X_2) =$$

$$V(\tfrac{3}{4}X_1) + V(\tfrac{1}{4}X_2) =$$

$$(\tfrac{3}{4})^2 * V(X_1) + (\tfrac{1}{4})^2 * V(X_2) =$$

$$9/16 * 92.25 + 1/16 * 369 =$$

$$51.89 + 23.06 =$$

$$74.95$$

and in general we would say

$$V(b_0 + b_1 X_1 + b_2 X_2 + b_3 X_3 \ . \ . \ .) =$$

$$b_1^2 \sigma_1^2 + b_2^2 \sigma_2^2 + b_3^2 \sigma_3^2 \ . \ . \ .$$

when b_i is a constant.

BIG

DISCLAIMER

The calculations we have made are only accurate IF the two stocks are *INDEPENDENT* of one another.

How do we tell? We look at how the two variables

INTERACT.

Consider an *INTERACTION TERM*— the *COVARIANCE*

✓ Covariance is a measure of the strength of the linear relationship between two variables.

✓ Covariance answers the question, "As the observed values for one variable change, how do the observed values for the other variable change?"

Notation:

The covariance of X and Y

$$Cov(X, Y) = \sigma_{xy}$$

and it contributes to the variance of a linear combination generally this way:

$$\sigma^2_{ax+by} = a^2\sigma^2_x + b^2\sigma^2_y + 2ab\,\sigma_{xy}$$

→ If the variables do not interact in a systematic way, they are *INDEPENDENT* and the
$$Cov(X,Y) = 0$$

so, $2ab\sigma_{xy} = 0$

→ The variance of a linear combination of two independent variables is the sum of their variances; $Cov(X,Y)$ contributes nothing to the variance.

→ **If the variables DO interact in a systematic way, they are *DEPENDENT* and the *COVARIANCE* term contributes something to the calculation of the variance for a linear combination.**

→ **If $Cov(X,Y) > 0$, the variance is larger than when independent.**

→ **If $Cov(X,Y) < 0$, the variance is smaller than when independent.**

Covariance Formula for Probability Distributions:

$$\sigma_{xy} = \sum(X - E(X))(Y - E(Y))P(X,Y)$$

where $P(X,Y)$ is the

JOINT probability of X and Y

Where do we find joint probabilities?

→ in contingency tables
→ in decision trees

We have JOINT probabilities for our stock example. The probabilities are those associated with an

economic expansion, when
$$P[(\Delta X_1 > 0) \ \& \ (\Delta X_2 > 0)] = 0.40$$

economic stability, when
$$P[(\Delta X_1 = 0) \ \& \ (\Delta X_2 = 0)] = 0.50$$

or economic contraction, when
$$P[(\Delta X_1 < 0) \ \& \ (\Delta X_2 < 0)] = 0.10$$

$$Cov(X_1, X_2) = \sigma_{xy}$$

$$= \sum (X - E(X))(Y - E(Y))P(X,Y)$$

$$= [(10.50)*(21.00)*(.40)] +$$

$$[(-4.50)*(-9.00)*(.50)] +$$

$$[(-19.50)*(-39.00)*(.10)]$$

$$= 88.20 + 20.25 + 76.05$$

$$= 184.50$$

For our purposes, we are interested in the SIGN of the covariance.

✓ When the sign on the covariance is POSITIVE, then the variables have a *POSITIVE DEPENDENT RELATIONSHIP*, and covariance adds to total variance.

$$V(X + Y) = V(X) + V(Y) + 2Cov(XY)$$

✓ **When the sign on the covariance is NEGATIVE, then the variables have a *NEGATIVE DEPENDENT RELATIONSHIP*, and covariance subtracts from the total variance.**

$$V(X + Y) = V(X) + V(Y) - 2Cov(XY)$$

✓ When the covariance is ZERO, then the two variables are *INDEPENDENT* or *UNRELATED*.

$$V(X + Y) = V(X) + V(Y) + 0$$

So, the variance of a weighted sum of two variables includes this covariance term:

$$V(X + Y) = V(X) + V(Y) + 2Cov(XY)$$

or

$$\sigma^2_{ax+by} = a^2 \sigma^2_x + b^2 \sigma^2_y + 2ab\,\sigma_{xy}$$

For our stock example, what is the variance, if the two stocks are *INDEPENDENT?*

$$74.95$$

What is the variance, if the two stocks are *DEPENDENT?*

$$V(\tfrac{3}{4}X_1 + \tfrac{1}{4}X_2) =$$
$$(\tfrac{3}{4})^2 * V(X_1) + (\tfrac{1}{4})^2 * V(X_2) +$$
$$2(\tfrac{3}{4})(\tfrac{1}{4})Cov(X_1, X_2)$$
$$= 74.95 + (3/8)*184.50$$
$$= 74.95 + 69.19 = 144.14$$

How does varying the value of π change the look of the Binomial?

$\pi < 0.5$, Binomial is right skewed

$\pi > 0.5$, Binomial is left skewed

$\pi \approx 0.5$, Binomial is symmetric

How about changing the value of n?

▸ When n is small, the Binomial is determined by π.

▸ However, as n becomes large, the Binomial becomes more and more symmetric, regardless of π.

What happens when we increase n in the Binomial?

The binomial's asymmetry is overwhelmed as n gets large

☞ When n gets large enough, the binomial resembles a smooth, symmetric curve, that is bell-shaped- the NORMAL

Why is the NORMAL so important?

Data that are influenced by many small and unrelated random effects are approximately normally distributed.

The Normal is Everywhere:

- ▸ stock market fluctuations
- ▸ student weights
- ▸ yearly temperature averages
- ▸ SAT scores

For example, a student's SAT score is the result of genetics, nutrition, illness, last night's beer party, whether it was hard to find a parking spot the day the test was taken and cultural factors.

→ When you include ALL the factors, you get the normal!

Discrete versus Continuous Random Variables

→ We have just looked closely at a Discrete Random Variable, the Binomial.

→ We have just noted that it closely resembles the Normal under certain circumstances.

→ Let's consider Continuous Random Variables next.

Remember what we know about Discrete Random Variables:

- Can assume only a finite or countably infinite number of different values
- "Gaps" exist between values along the number line
- It is possible to list all possible values of X and the associated P(X)
- $\sum P(X) = 1$

Continuous Random Variables

- Random Variables that can take on any value in an interval

- No "gaps" between values

- Because there is an infinite number of X values, we cannot associate a probability with a single value, only a range of values.

A caveat--

Technically, all *measured* variables are discrete, because no continuous variable can be measured with exact precision.

Despite this fact, we treat many variables as continuous that are actually discrete.

- too tedious to list all X values
- differences between successive values of X are insignificant

CONTINUOUS LANGUAGE

Probability Density Function (pdf)

- the technical name of the curve

- often represented as f(x)
  (say "f of x")

- $f(x) \geq 0$ for all x
  (probabilities can't be negative)

- $\int f(x)\, dx = 1$
  (probabilities sum to 1)

Important Facts

For continuous distributions,

$$P(X=a) = 0$$

Only the probability that X falls in an *interval* can be measured, and is measured as an area under the curve.

$$P(a \leq X \leq b) = P(a < X < b)$$

The Most Important Distribution
The Normal Distribution

- a continuous distribution
- $f(x)$ = a really complicated formula
- the curve is bell shaped and symmetric about $X = \mu$
- the curve is defined
  from $-\infty$ to $+\infty$
- total area under the curve = 1
- $f(x) \geq 0$
- mean = median = mode = μ

Some Notation

→ The normal is a *two* parameter distribution, like the Binomial.

→ It is completely defined by the parameters μ and σ

$$X \sim N(\mu, \sigma)$$

If the values of X

✓ describe a "normal" curve,

✓ X is a "normal" RV and

✓ the population is said to be "normally distributed."

✓ The Normal is an example of a distribution that fits the Normal Rule.

The Normal is actually a family of distributions each characterized by its μ and σ.

The most important member is

$$X \equiv Z \sim N(0,1)$$

The Standard Normal

Any Normal can be transformed into the Standard Normal by the use of

$$Z = \frac{X - \mu}{\sigma}$$

the "Standardizing Transformation"

Z is a standardized score
or Z-score

→ Z is the number of standard deviations the related X is from its mean

→ Z is centered at 0

→ Z has a standard deviation of 1

→ $P(Z < a) = P(\dfrac{X - \mu}{\sigma} < \dfrac{a - \mu}{\sigma})$

CALCULATING NORMAL PROBABILITIES

→ **The Normal Formula is more complex than the Binomial**

→ **Statisticians in the past have had to rely on approximations from a table.**

→ **We are able to calculate them much more accurately**

THE NORMAL AND EXCEL

$$=\text{NORMDIST}(X, \mu, \sigma, 1)$$

→ the cumulative probability (the area under the curve) beginning at $-\infty$ and ending at X.

=NORMINV(π, μ, σ)

→ the X value associated with the probability (an area under the curve) assuming the probability is accumulated from -∞ on up.

1. EXCEL adds probability from negative infinity up.

2. You have to use your knowledge of the symmetry of the Normal distribution to find the areas between two possible values.

3. You do NOT have to translate your X values into Z-scores.

Do You Remember. . . .

All binomials turn into normals!

So. . . .?

   In general, computing the normal requires less time and resources.

Consider a Binomial

$$n = 25 \text{ and } \pi = 0.5$$

Find the probability that X is less than or equal to 14.

=BINOMDIST(14,25,0.5,1) = .7878

We can approximate this with the normal:

▸ $\mu = n\pi = (25)(.5) = 12.5$

▸ $\sigma = [n\pi(1-\pi)]^{\frac{1}{2}} = [(25)(.5)(.5)]^{\frac{1}{2}}$
   $= 2.5$

▸ =normdist(14,12.5,2.5,1)= .7257

Well, that's not very good!

   .7878-.7257=.0621 (about 8%)

In fact, we have not counted all the area we need to count!

A Continuity Correction Factor

In our example, we need to expand the area we are accumulating to include the "edge" of the column labeled 14 for the binomial, by adding 0.5 to 14.

=normdist(14.5,12.5,2.5,1)=.7881

A MUCH better approximation!

$$.7878 - .7881 = -.0003$$

The general rule for the continuity correction factor is:

$$P(a \leq X \leq b) \approx P(a - \tfrac{1}{2} < X < b + \tfrac{1}{2})$$

When is the normal approximation "Good Enough"?

Whenever n is large enough to make the number of expected successes and expected failures BOTH greater than 5

OR

when $n\pi \geq 5$ AND $n(1-\pi) \geq 5$

Another Continuous Distribution

Student's *t* distribution

- ▸ an approximation of the Standard Normal

- ▸ use it when we don't know σ

- ▸ we substitute s for σ in our standardization formula and get the *t*.

- ▸ looks like the Standard Normal, but it varies with n, the sample size

- t is flatter than a Normal because of the uncertainty added by using s

- t has "fat tails"

- as n increases, the t approaches the Standard Normal

The world is a big place

Collections of stuff are large

Its hard to get information that is "informative"

What to do when . . .

INQUIRING MINDS WANT TO KNOW:

- What percent in a voting population favors each candidate?

- What proportion of manufactured goods will be defective?

- What is the average length of pickles?

We talked earlier about why we don't want to measure every pickle in the world and calculate the average

MONEY!

TIME!

So, we SAMPLE

The question we MUST answer is

How big a sample do we need?

If n is the number of items in the sample, we will learn that all we need to know is controlled by

$$\frac{1}{\sqrt{n}}$$

!! CAUTION !!

Statistical Results Depend on
- ✓ independence and
- ✓ lack of bias

in the simple random sample.

Other sampling methods results must be modified.

!! CAUTION #2 !!
Without Randomized Design, there can be NO Dependable Statistical Analysis

The questions we listed earlier are questions about populations. We need to get good at making guesses about populations.

Let's start by guessing about the mean. . .

1. We take a sample and calculate the mean.

2. Then what? Is it the mean of the population? (Probably not!)

3. Is it close? Let's find out. . .

Introducing, *Sampling Error!!*

Sampling Error:

→ The difference between a statistic and its corresponding parameter.

Can we calculate it?

Theoretically. . .

Sampling Error = $(\bar{x} - \mu)$

But, with many important estimates, we have no idea of the size of sampling error, because we don't know the parameters.

What do we do?

→ Simulate Random Events and Study the Resulting Sampling Distributions.

Sampling Theory

Draw a lot of samples, all the same size, from a population.

Calculate the same statistic, say the mean, from each sample.

→ The sample means are random variables themselves(!)

→ They have means and variances like other Random Variables we have seen.

→ They have a distribution!

It is the
"Sampling Distribution"
of the Mean.

or

The Distribution
of the Sample Means

What do we see, after all this work?

→ The mean of the sampling distribution of $\overline{X}$ is identical to the population mean.

→ The variance of $\overline{X}$ is *smaller* than the variance of the population.

Calculate the variance of $\bar{x}$ using this formula:

$$\sigma_{\bar{x}}^2 = \frac{\sigma^2}{n}$$

Is this another of those approximation things?

Well, yes and no . . . .

What we have just seen is a simple example of the Central Limit Theorem.

"Central" refers to its importance to statistics.

The Theorem:

If $X \sim ?(\mu, \sigma^2)$ then $\bar{x} \sim N(\mu, \sigma^2/n)$, approximately, when n is large.

In English, please. . . .

We have a population and we don't know how it is distributed (what it looks like) but we DO know it has a mean = μ and variance = σ^2.

We create a random variable $\overline{X}$ (by drawing as many samples as we can and calculating their means).

SO . . .

$\overline{X}$ (the random variable) will be approximately normally distributed with mean = μ and variance = σ^2/n

(standard error = $\dfrac{\sigma}{\sqrt{n}}$)

IF

✓ n is large enough

✓ the sample is a simple random sample.

This theorem applies to other descriptive statistics as well. We could describe the distribution of s^2 the same way.

What does this mean?

Doesn't matter what X's distribution looks like, if n is big enough, the sample means will be close to normally distributed.

How large does n have to be?

$$n \geq 30$$

What if we know that X *IS* normally distributed?

Even Better! We know that the sample means will be EXACTLY normally distributed NO MATTER HOW BIG OR SMALL n IS!!!

What is the really big deal?

It doesn't matter what kind of a distribution your population is. If you draw a sufficiently large sample, you can make lots of good guesses about the parameters of the population, because you know how the sample statistics will be distributed.

Where were we . . .?

- ✓ Looking at a small population
- ✓ Calculating all possible sample means
- ✓ Realizing the means are a random variable themselves!
- ✓ Realizing that they must have a distribution of their own!
- ✓ How are the means distributed?

The distribution of the sample means is

   ✓ centered at the same place as the population (same mean)

   ✓ more compact (smaller variance)

We call the standard deviation of the sampling distribution the "standard error."

$$\sigma_{\bar{x}} = \frac{\sigma}{\sqrt{n}}$$

This is how it is calculated. So, the variance of the sampling distribution must be $\frac{\sigma^2}{n}$.

The Sample Proportion

$$p = \frac{x}{n}$$

when x is the number of successes in the sample and n is the size of the sample.

(What does this remind you of?)

p is the sample version of π

just like

$\bar{x}$ is the sample version of μ

The Sampling Distribution of the Sample Proportion,

$$p \sim N\left(p, \sqrt{\frac{p(1-p)}{n}}\right)$$

A certain type of rat is bred for use in lab research about breast cancer. This rat weighs 5 ounces on average with a standard deviation of 2 ounces.

What is the probability that a rat drawn at random weighs less than 4 ounces?

What is the probability that the mean rat weight is less than 4 ounces?

What is the difference between these two questions?

How do we use the CLT?

- ✓ to construct confidence intervals
- ✓ to perform hypothesis tests

A Point Estimate

- ✓ a single value used to estimate a population parameter, such as the sample mean, the sample standard deviation, the sample variance, the sample proportion, etc.

An Interval Estimate

✓ the range within which a population parameter *probably* lies.

What is a confidence interval?

✓ an interval estimate
✓ reported usually as the endpoints of the range
✓ centered over a point estimate

What is a Confidence Interval?

[24,600; 25,400]

-the upper and lower limit of a range within which a population parameter is expected to occur, with a chosen probability.

-the chosen probability is called the "level of confidence."

How do we construct Confidence Intervals (CIs)?

▸ a 95% CI for μ is given by

$$\overline{X} \pm 1.96\left(\frac{\sigma}{\sqrt{n}}\right)$$

▸ a 99% CI for μ is given by

$$\overline{X} \pm 2.58\left(\frac{\sigma}{\sqrt{n}}\right)$$

Formally,

$$\bar{X} \pm z_{\frac{\alpha}{2}}\left(\frac{\sigma}{\sqrt{n}}\right)$$

$\bar{X}$ is an estimator of μ

σ is the population standard deviation

n   is the sample size

α is a probability

About α:

▸ It is a number between 0 & 1

▸ It measures INcorrectness

▸ It is a probability

▸ There is always at least one z score associated with it.

▸ We call it Z_α sometimes

▸ Most often we call it $Z_{\alpha/2}$

Where does α come from?

- we pick it, depending on what we want.

- common α's are .01 or .05

- If α = .01, what is $\alpha/2$?

What does α do?

- it lets us calculate our level of confidence

- we decide how often we are willing to be wrong (α), then we figure out how often we will be right.

What is a level of confidence?

- ▸ a probability
- ▸ a statement of how likely it is that a population parameter will be contained in an interval.
- ▸ level of confidence $=(1-a)$.

If a = .01, what is the level of confidence?

So, a confidence interval for

$$\overline{X} \sim (\mu_{\overline{X}}, \sigma^2_{\overline{X}})$$

with level of confidence 100(1-a)%

$$\equiv \overline{X} \pm z_{\frac{\alpha}{2}} \left(\frac{\sigma}{\sqrt{n}}\right)$$

How about a picture?

A confidence interval for $\overline{X}$

What is the mean annual income of last year's graduates? A sample of 25 graduates was taken and
$\overline{X}$ = $19,500

σ = $2,000

What is a 95% confidence interval for the true mean?

$100(1-a) =$

so $a =$

and $a/2 =$

and $Z_{a/2} =$

and $CI =$

What about a 90% CI?

Margin of Error or
Acceptable Sampling Error

$$e = z\frac{\sigma}{\sqrt{n}}$$

Three Margin of Error Forms

$$z\left(\frac{\sigma}{\sqrt{n}}\right)$$

$$t\left(\frac{s}{\sqrt{n}}\right)$$

$$z\left(\sqrt{\frac{p(1-p)}{n}}\right)$$

In an election, we sample 400 voters and find that 203 of them favor a certain candidate. Based on our sample, what is a 99% confidence interval for the true π?

1. Test to see if we can use the normal approximation to the binomial.

2. Write down the sampling distribution of p, estimator for π.

3. Draw a picture

4. Write the CI formula

5. Write an Excel Command

6. Calculate CI.

7. What is the margin of error?

8. How would we interpret this CI?

$$[\ 44.3\%\ ,\ 57.2\%\ ]$$

or

$$50.8\% \pm 6.4\%$$

We are confident that 99% of similarly constructed confidence intervals contain the true population proportion.

THIS IS A STATISTICALLY ACCURATE INTERPRETATION OF THIS CONFIDENCE INTERVAL.

Another Accurate Interpretation:

99% of the sample proportions should be within 2.58 standard deviations of the population proportion.

Note: There is no longer any probability left in the interval we calculated,

→ so we CANNOT say that there is a 99% probability that the population proportion is between these limits–it either is or it isn't!

What can we usually say?

We usually say that we have 99% CONFIDENCE (we are 99% sure) that the population proportion lies within the interval between 44.3% and 57.2%.

Desirable Properties of CI's

1. The interval should have a high level of confidence, $(1-\alpha)$

2. The confidence interval should have a narrow width.

The width depends on three things:

1. $(1-\alpha)$
2. σ
3. n

Properties of ALL CI's

1. The larger $(1-\alpha)$, the wider the CI, ceteris paribus.

2. The smaller σ, the narrower the CI, ceteris paribus.

3. The larger n, the narrower the CI, ceteris paribus.

Using the Margin of Error- -

Suppose an economist wants to estimate the mean annual income of households in a certain district. σ is believed to be $4000. The economist wants the probability to be .95 that the sample mean will be within $500 of the true mean. How large a sample must be taken?

→ **TINV** *ALWAYS* assumes the probability you give it is divided between two tails.

For NORMSINV YOU have to divide alpha up.

→ **TINV** *ALWAYS* returns a *positive* T-value.

For NORMSINV YOU have to use the absolute value of the z-value you get.

=TINV(.05,60)	=TINV(.025,60)
=2.000297	=2.299048

FIND T-VALUES USING EXCEL:

→ TINV thinks for you.

→ You give it (1-confidence level)
and
→ degrees of freedom

→ it divides alpha in half and gives you the appropriate T-value.

=TINV((1-confidence level), df)

(1-confidence level = alpha)

→Compare this to NORMSINV

=ABS(NORMSINV((alpha)/2))

(alpha = 1-confidence level)

Hypothesis Testing

A formal procedure based on sample evidence and probability theory used to decide if a hypothesis-a guess-- is reasonable or unreasonable.

Sample means from an exam

$$56.54$$
$$58.41$$
$$57.96$$

$$\mu = 57.77 \text{ and } \sigma = 14.86$$

What might we like to know about this data?

Why perform Hypothesis Tests?

✓ in order to make a choice, to choose between two possible parameter values or ranges of values.

✓ to see if a value is truly the same as or different from another.

✓ to gauge the truth of a claim.

How are Hypotheses Tested?

- ◆ sample collected
- ◆ statistics are calculated
- ◆ systematically compare the sample to what is known or believed about the population.

If the true mean is 57.77, how likely is it we would get a sample mean of 56.54.

by chance?

To Perform a Hypothesis Test:

1. State all hypotheses.

2. Select a level of significance.

3. Identify the test statistic.

4. Write a decision rule.

5. Calculate and decide to fail to reject the null OR reject the null and conclude the alternate.

Some Definitions:

Hypothesis: a statement about a value or range of values a population parameter can take

- ▸ it is a target that we set up
- ▸ it is a guess and we use sample evidence to tell how good a guess.
- ▸ They come in pairs

$$H_0 \equiv \text{ the null}$$
$$H_1 \equiv \text{ the alternative}$$

The Null Hypothesis H_0:
 a statement about the value of a parameter

✓ "Null" means "nothing" and is used to imply "no change."

✓ It is the "maintained" hypothesis

✓ Can be Simple or Composite

- **Simple:** a single value
  Implies a 2 sided test
- **Composite:** a range
  Implies a 1 sided test.

- **Examples of nulls include:**

  - ★ The coin is "fair"

  - ★ The mean number of raisins in a box of raisin cereal is 7.

  - ★ The difference in effectiveness between 4 medications is due entirely to chance.

Alternative Hypothesis H_1:

A statement about the value of a population parameter that is DIFFERENT from the null

Some Alternative Hypotheses:

★ The coin is not "fair."

★ The mean number of raisins in a box of raisin cereal is not 7.

★ The difference in effectiveness between 4 medications is due to something other than chance.

Decision Rule:

a statement that specifies the set of values of the test statistic for which we would REJECT H_0 and those for which we would FAIL TO REJECT H_0.

Why this language?

If we FAIL TO REJECT H_0 that does NOT mean that H_0 is CERTAINLY true.

Level of Significance:

- ★ a probability

- ★ It is α!

- ★ So, the level of confidence plus the level of significance total up to 1.

- ★ α is the probability of rejecting the null when it is true, which is a mistake. We call this "Type I Error."

A Test Statistic:

✓ a random variable whose value is used to test a hypothesis, for example $\bar{x}$, p, Z score

✓ the one you use depends on the problem you have

✓ It is designed so that IF THE NULL IS TRUE, you know EXACTLY how the test statistic is distributed.

An Example:

Suppose we use Z as our test statistic. If our null hypothesis is true, then Z will have a standard normal distribution.

We calculate z = .878

Suppose we calculated z = 3

Rejection Region:
- ▸ or Critical Region
- ▸ all values where H_0 is rejected.

Acceptance Region:
- ▸ all values where we FTR H_0

Critical Value:

That value of the test statistic that separates the rejection region from the acceptance region.

More about Hypotheses. . .

The EPA estimated that 20% of auto emission systems are tampered with.

H_0:

H_1:

Kelloggs claims that breakfast cereal packages contain 32 oz, but customers have been complaining that they are underweight.

H_0:

H_1:

Currently, traffic light bulbs last 2000 hours, and are costly to change. There is a new bulb on the market. Should Bloomington buy the new bulbs?

H_0:

H_1:

The Food and Drug Administration approves new drugs for marketing. Which of the following should be the null hypothesis the FDA uses?

1. H_o: new drug is beneficial, so market it.

2. H_o: new drug is NOT beneficial, so don't market it.

Suppose FDA chose:

H_0: Drug is beneficial so market
H_1: Drug is not beneficial

Hypotheses are VERY important.

✓ everything is relative to H_0
✓ H_0 is "maintained" unless there
   is strong evidence to refute it.
✓ H_0 is "innocent"
✓ H_0 is rejected if there is no
   reasonable doubt it could be
   true.

We "reject the null" when our sample provides strong evidence that the null is probably false.

We "fail to reject the null" when our sample evidence is not strong enough to believe the null is probably false.

An example of a Hypothesis Test:

- ▸ 50% of eligible citizens were African American

- ▸ On an 80-person panel of possible jurors, only 4 were African American

- ▸ Could this be the result of pure chance?

☞ Assume juror selection was RANDOM.

☞ The number of jurors on a panel would be a Binomial random variable, n=80, p=.5

☞ =BINOMDIST(4,80,.5,1) =1.37889E-18 or .00000000000000014

☞ Since this is such a small number, it is evidence AGAINST our assumption.

Our 5 Step Process

1. State all hypotheses.

$H_0 : \pi = .50$

$H_1 : \pi \neq .50$

2. Select a level of significance.

$$\alpha = .000001$$

3. Identify the test statistic.

$$z = \frac{p - \pi}{\sigma_\pi}$$

4. Write a decision rule.

Reject H_0 if $| Z_{obs} | > | Z_{a/2} |$

5. Calculate and decide:

$Z_{obs} = -8.04984$

$Z_{.0000005} = -5.06639$

How many tails does *YOUR* test have?

Simple Null Hypothesis

- splits α up into two parts
- divides the Rejection Region up into two parts
- Acceptance Region is between the two Rejection Regions.
- So, it results in a "two-tailed test."
- Two-tailed tests have TWO critical values.

Composite Null Hypothesis

- α is all on one side or the other.
- the Rejection Region is all on one side or the other.
- the Acceptance Region is all on one side or the other.
- So, it results in a "one-tailed test."
- One-tailed tests have ONE critical value.

Restate our hypotheses and get a one-tailed test

$$H_0 : \pi \geq .50$$
$$H_1 : \pi < .50$$

A Few Words About Error . . .

In Hypothesis Testing we can have 4 possible outcomes: 2 correct and 2 incorrect.

		TRUTH	
		H_0 is TRUE	H_0 is FALSE
T E S T	FTR H_0		
	Reject H_0		

Type I Error: Rejecting the Null when we shouldn't (ouch!)

- ♦ US legal system assumes all defendants are innocent
- ♦ Type I Error is when an innocent person is found guilty.

$$\alpha = P(\text{Type I Error})$$
$$= P(\text{Reject } H_0 \mid H_0 \text{ is True})$$

or

how likely it is that the test statistic is in the Rejection Region when H_0 is True

Type II Error:

— NOT rejecting the Null when we SHOULD! (Ouch! Ouch!)

- ♦ when a jury finds a guilty person innocent!

$$\beta = P(\text{Type II Error})$$

or

how likely it is that the test statistic is in the Acceptance Region when H_0 is False

What is a p-value?
It is . . .

- ▸ a probability
- ▸ analogous to an a
- ▸ the *OBSERVED* level of significance. (Remember, a is the *CHOSEN* level of significance.)
- ▸ the area under the curve from the test statistic out.
- ▸ the probability that we would observe the sample statistic or a more extreme one if the null were true.

p-values are the NEW way to do hypothesis testing.

- a direct way of conducting a hypothesis test

- often interpreted as a measure of the strength of the evidence against H_0.

- the smaller the p-value, the stronger the evidence against H_0.

- p-values are compared to α

$$\text{p-value} > \alpha \rightarrow \text{FTR } H_0$$

$$\text{p-value} < \alpha \rightarrow \text{Reject } H_0$$

Calculating a P-value With Excel
Sigma Known

- ▸ for a ONE tailed test

  - ▸ $=1- \text{NORMSDIST}(\text{ABS}(z_{obs}))$

    or

  - ▸ $=1- \text{NORMDIST}(\text{abs}(\bar{x}, \mu, \sigma_{\bar{x}}, 1))$

▶ **for a TWO TAILED test**

$$=2*(1- \text{NORMSDIST}(\text{ABS}(z_{obs})))$$

or

$$=2*(1-\text{NORMDIST}(\text{ABS}(\bar{x},\mu,\sigma_{\bar{x}},1)))$$

M&M/Mars Company claims that 30% of the plain M&Ms in a king size package are brown. A king size package was purchased which contained 100 plain M&Ms, 37 of which were brown.

1.

2.

3.

4.

5.

Our M&Ms example and a TWO tailed test:

✓ H_0: $\pi = .30$

✓ H_1: $\pi \neq .30$

✓ test statistic = 1.53

✓ p-value
=2*(1-NORMDIST
(ABS(.37,.30,.046, 1))) =.1281

The ONE tailed test

✓ H_0: $\pi \geq .30$

✓ H_1: $\pi < .30$

✓ test statistic = 1.53

✓ p-value =

1-NORMDIST(ABS(.37,.30,.046,1))

= .064

Calculating a p-value with Excel
Sigma Unknown

▸ **TDIST requires**

  ▸ **a positive value in the first argument**

  ▸ **determines the p-value based on the number of tails.**

- ## TWO TAILED test

  $$\text{p-value} = \text{TDIST}(\text{ABS}(t_{obs}), df, 2)$$

- ## ONE TAILED test

  $$\text{p-value} = \text{TDIST}(\text{ABS}(t_{obs}), df, 1)$$

Pepperidge Farms claims that the average weight of its packages of Goldfish Crackers is at least 10 oz. A sample of 36 packages is selected and weighed. Local grocers will send back a shipment if the average weight is less than the claim.

$$\bar{x} = 9.94 \text{ oz}; \quad s = .22 \text{ oz}$$

Should this shipment be sent back?

1.

2.

3.

4.

5.

Goldfish Crackers test (ONE tail)

$$= TDIST(ABS(-1.636),35,1)$$

$$=0.055363$$

A TWO tailed version would be

H_0: $\boldsymbol{\mu} = 10$ oz

H_1: $\boldsymbol{\mu} \neq 10$ oz

$$=TDIST(ABS(-1.636),35,2)$$
$$=0.110726$$

Where are we?

- ✓ defined statistics
- ✓ created descriptions of samples and populations
- ✓ linked samples to probability via sampling distributions
- ✓ learned how to quantify confidence in an estimate
- ✓ learned how to test claims using a sample

Relationships Between Variables

We will:

1. Look at 2 or more variables
2. Study their relationship
3. Write an equation that explains their relationship
4. Use the equation to make predictions about one variable based on the other(s)

A Simple Model

- **Two Variables**
  - one dependent
    - what is being predicted
  - one independent
    - what is doing the predicting
- **Correlation Analysis**
  - Scatter Diagrams
- **Simple Linear Regression**

The Statistician Designs a Model

1. Select something you would like to explain or be able to predict.
2. Think about it.
   a. What affects it?
   b. What do you think is most important?
   c. How do you think the variables are related?
3. Gather Data
4. Estimate the Relationship

A statistical model breaks things up into

non-random, systematic behavior--
   that which is predictable or
   quantifiable

random behavior--
   that which is unpredictable

The random behavior is referred to as "error," and is due to a large number of minor, untraceable factors.

Correlation Analysis
— techniques to measure the strength of the association between two variables.

Step One: Scatter Diagrams
— a plot of paired observations

Some Data: Labor/Income Pairs		
8/512	12/579	16/853
20/944	24/725	28/1066
32/883	36/1200	40/914

Pearson's r

- ✓ coefficient of correlation
- ✓ measures the strength of a linear relationship between two variables
- ✓ ranges from -1 to +1
- ✓ -1 and +1 show *perfect* correlation
- ✓ $r = 0$ means no relationship
- ✓ r near 0 is a weak relationship
- ✓ r near ± 1 is a strong relationship

$$r = \frac{\sum(x_i - \bar{x})(y_i - \bar{y})}{\sqrt{\sum(x_i - \bar{x})^2 \sum(y_i - \bar{y})^2}}$$

Our Labor/Income data has

r = 0.759

What does this mean?

What is this Correlation Coefficient?

a *STANDARDIZED* Covariance

Remember. . .

✓ Covariance is a measure of the strength of the linear relationship between two variables.

✓ Covariance answers the question, "As the observed values for one variable change, how do the observed values for the other variable change?"

Why wasn't Covariance Good Enough?

✓ The units in which covariance is measured are a problem.

✓ We can't compare relative measures of covariance and get any meaning from them.

✓ All we could tell was if the relationship was positive or negative.

The Correlation Coefficient eliminates the units and so eliminates that problem.

ρ is the symbol for the *population* correlation coefficient

Some representative rho's . . .

WARNING!!!WARNING!!!WARNING!!!

Correlation is NOT Causation.
Large r's are interpreted as an
indication of a relationship between
two variables.

They may NOT be interpreted as "a
change in one causes a change in
the other."

Spurious Correlations: False,
counterfeit, bastard correlations.

In Correlation Analysis,

✓ both X and Y are random variables.

✓ correlation between X and Y is identical to correlation between Y and X.

✓ essentially, r gives us an indication of how well the variables move together in a straight-line fashion.

Can we do better?

CORRELATION COEFFICIENTS

▸ **don't tell us anything specific about the relationship between X and Y**

▸ **don't allow us to make predictions**

▸ **don't allow us to "control" any variable**

A Simple Linear Regression Model

$$Y = \beta_0 + \beta_1 X + \varepsilon$$

$$Y' = b_0 + b_1 X$$

✓ Y' is the *predicted* value of $Y|X$

✓ b_0 is the Y-intercept

✓ b_1 is the slope of the line

✓ X is any selected value of the independent variable

✓ ε is the "error" term

Linear Regression--
a Powerful Statistical Model

- models the relationship between X and Y as a straight line.

- values of X are assumed fixed, ie not random.

- The only randomness in the model comes from ε, the error term.

- $\varepsilon \sim N(0, \sigma)$

$$Y = \beta_0 + \beta_1 X + \varepsilon$$

(Our Model)

$$Y' = b_0 + b_1 X$$

(Our Estimated Line)

$$Y = \text{Observed Income}$$

$$Y' = \text{Predicted Income}$$

We would like Y and Y ' to be as close as possible to one another.

OR

We would like the *difference* between Y and Y ' to be as *small* as possible.

$$Y - Y' = e$$

In Fact, we want the TOTAL differences to be as small as possible.

And we call

$$\sum (y_i - y'_i)^2 = \sum e_i^2 \equiv SSE$$

The Sum of Squared Errors

This is an aggregate measure of how much the predicted Y's differ from the actual Y's.

When we do linear regression, this is the actual equation we minimize.

Yeah, . . .but, how?

We look for the "line of best fit."

It is the line that *minimizes* the sum of the squared difference between Y and Y'.

The "Least Squares Principle"

❑ uses calculus to estimate b_0 and b_1
❑ minimizes squared deviations

WE get to use EXCEL!!!

Tools==>Data Analysis==>Regression

Why "regression?"

Sir Francis Galton, cousin to Charles Darwin, noted that the heights of sons compared to their fathers showed a tendency to move toward the mean height of the population-- to regress or revert to the mean. Hence, regression to "mediocrity."

What does minimizing the error get us?

$$b_1 = \frac{\sum (x_i - \bar{x})(y_i - \bar{y})}{\sum (x_i - \bar{x})^2}$$

$$b_0 = \frac{\sum y_i}{n} - b_1 \frac{\sum x_i}{n} = \bar{y} - b\bar{x}$$

Note the prevalence of

$$\bar{y} \text{ and } \bar{x}$$

in the equations we have generated.

Variance, our measure of deviation we are familiar with, is the deviation from the mean of a variable. Let's look at our regression values in this context:

Variance: The "total" variation we are trying to explain.

(The sum of the squared deviations from the mean.)

$$SST(otal) = \sum (Y - \overline{Y})^2$$

Regression: The part of the variation we are able to explain

$$SSR(egression) = \sum (Y' - \overline{Y})^2$$

Error: The part of the variation we are NOT able to explain.

$$SSE(rror) = \sum (Y - Y')^2$$

$$SST = SSR + SSE$$

Look at the *PROPORTION* of SST we could explain:

The Coefficient of Determination

$$r^2 = \frac{SSR}{SST} = 1 - \frac{SSE}{SST}$$

- ✓ the proportion of the variance of Y in the sample that is explained by the regression
- ✓ always between 0 and 1
- ✓ the correlation coefficient squared
- ✓ a measure of the "fit" of the regression

SUMMARY OUTPUT

Regression Statistics	
Multiple R	0.76
R Square	0.58
Adjusted R Square	0.51
Standard Error	153.17
Observations	9

ANOVA

	df	SS	MS	F	Sig. F
Regression	1	222629	222629	9.49	0.02
Residual	7	164226	23461		
Total	8	386855			

	Coefficients	Standard Error	t Stat	P-value
Intercept	487.31	129.16	3.77	0.01
Hours	15.23	4.94	3.08	0.02

		Lower 95%	Upper 95%
Intercept	487.31	181.89	792.73
Hours	15.23	3.54	26.92

SUMMARY OUTPUT

Regression Statistics	
Multiple R	0.75
R Square	0.57
Adjusted R Square	0.56
Standard Error	136.59
Observations	92

ANOVA

	df	SS	MS	F	Sig. F
Regression	1	2197578	2197578	117.8	0.00
Residual	90	1679226	18658		
Total	91	3876804			

	Coefficients	Standard Error	t Stat	P-value
Intercept	494.11	36.07	13.70	0.00
Hours	15.07	1.39	10.85	0.00

	Coefficients	Upper 95%	Lower 95%
Intercept	494.11	422.44	565.77
Hours	15.07	12.31	17.83

"Standard Error of Coefficients"

✓ Found next to related coefficient.

✓ We would write it as s_b and it is like all the other standard errors we have seen before.

✓ It is used to perform hypothesis tests and construct confidence intervals for the coefficients.

"t-stat"

✓ A test statistic for a two-tailed hypothesis test of the coefficient it is related to.

✓ The null hypothesis for Excel is always $\beta_0 = 0$ or $\beta_1 = 0$. Excel always tests a null of "no effect."

✓ If you want to test another value, you must calculate the test statistic yourself.

"P-value"

✓ The p-value associated with the hypothesis test.

✓ Again, the null hypothesis for Excel is always $\beta_0 = 0$ or $\beta_1 = 0$.

✓ Again, if you test a different hypothesis, you must calculate the p-value yourself.

"Upper 95%" & "Lower 95%"

✓ The upper and lower bounds of the 95% confidence interval.

✓ You may request a second confidence interval if you wish when you run the regression.

Now for the ANOVA table

ANOVA is coined from ANanlysis Of VAriance

Regression: Statistics associated with the portion of the variation in Y explained by the model.

"Residual": would be better called "Error", the unexplained portion

Total: The sum of Regression and Residual.

df column:

✓ the number of degrees of freedom associated with each item in the table.
✓ Regression always has df = k, where k is the number of independent variables.
✓ "Residual" has the df equal to n-(k+1).
✓ Total has df = n-1

"SS"

✓ SS means Sum of Squares

✓ These are the sum of squares associated with SSR, SSE and SST we discussed last time.

✓ Note that SSR + SSE = SST

✓ They are used with df's to calculate MS

"MS"

✓ This is the "Mean Sum of Squares"

✓ It is the SS divided by the associated df.

✓ The only one we care about is that associated with the error (or residual.)

"Standard Error of the Estimate"

✓ Found under SUMMARY OUTPUT.

✓ A measure of goodness of fit of the model.

✓ It measures the "scatter" or dispersion of the data points around the regression line.

✓ A smaller value means a tighter fit.

✓ It is calculated as the square root of the Mean Squared Error, and is conceptually similar to a standard deviation.

✓ It is used to calculate confidence and prediction intervals for Y given X.

✓ It is not to be confused with the standard error of the coefficients.

A random sample of 12 companies was used to determine the following relationship between sales and earnings.

SUMMARY OUTPUT

Regression Statistics	
Multiple R	0.67
R Square	0.45
Adjusted R Square	0.40
Standard Error	2.52
Observations	12

Note about the Correlation Coefficient:

Multiple R is the same as Pearson's r
(the correlation coefficient)

Excel does NOT put the sign on Multiple R, so you have to.

The sign is the same as that of the slope coefficient, b_1.

	Coef	Stan Error	t Stat	P-value	Lower 95%	Upper 95%
Intercept	1.85	1.41	1.3	0.22	-1.30	5.00
Sales($M)	0.08	0.03	2.9	0.02	0.02	0.15

1. Write down the regression equation.

2. Interpret the goodness of fit of the equation.

3. Interpret the slope coefficient.

4. Is the independent variable significant?

ANOVA

	df	SS	MS	F	Sig F
Regression	1	52.62	52.6	8.3	0.02
Residual	10		6.34		
Total	11	116.01			

1. What is the value of SSE?

2. Calculate r^2.

1059

RESIDUAL OUTPUT

Obs	Predicted Earnings($M)	Residuals
1	9.31	-4.41
2	3.41	0.99
3	3.37	-2.07
4	7.84	0.16
5	6.75	-0.15
6	5.76	-1.66
7	3.31	-0.71
8	2.85	-1.15
9	3.49	0.01
10	6.13	2.07
11	4.24	1.76
12	7.64	5.16

What are the observed earnings for observation # 3?

What about for observation # 9?

Multiple Regression

✓ We can look at more complex, realistic models.

✓ more than one explanatory or independent variable.

✓ subscript "slope" coefficients and our X's when we write the equation.

✓ Allows us to "control" for the effect of other variables.

Assume a linear relationship exists between

Y – our dependent variable

and

X_1
X_2
X_3 our independent variables
.
.
.
X_k

We use the same technique as we did with the simple regression.

- ▸ make predictions
- ▸ subtract predictions from observations
- ▸ square the difference
- ▸ sum the squares

The set of predictions with the smallest sum is our "*PLANE* of best fit" for the multiple regression.

The Multiple Regression Equation

for 2 variables

$$Y' = b_0 + b_1X_1 + b_2X_2$$

for k variables

$$Y' = b_0 + b_1X_1 + b_2X_2 + \ldots + b_kX_k$$

The statistics we generate are analogous to those of the simple regression.

R^2

✓ is the coefficient of "*MULTIPLE*" determination.

✓ is the proportion of variation in Y explained by variation in all X's in the model.

✓ is SSR/SST

SSR, SSE and SST:

✓ calculated *exactly* the same way

✓ mean *exactly* the same thing

✓ SSR refers to multiple variables

Degrees of Freedom

✓ Regression: k, the number of independent variables.

✓ Residual: n-(k+1), the number of observations left over after subtracting one for the regression and one each for the independent variables.

✓ Total: n-1, the sum of above.

<u>*Coefficients:*</u>

✓ **the intercept: the value of the dependent variable when all of the independent variables are zero.**

✓ **the slopes: the change in Y for a one unit change in a selected X_i holding all the other X_i's constant.**
  ▸ **(net) regression coefficients**
  ▸ **a "conditional" slope**

Multiple R:

Correlation coefficients refer to two variables only, so for a multiple regression, this number has no meaning.

Can generate a "correlation matrix"
✓ a set of all correlations between all variables
✓ we won't do this in this class.

"Dummy Variables"

✓ categorical or indicator variables
✓ value of 0 or 1
✓ 1 indicates that a characteristic exists in this observation
✓ 0 indicates that a characteristic does not exist in this observation
✓ Regression can include these

An Example of a Multiple Regression

A district manager
✓ wishes to predict the number of cars sold at a dealership
✓ believes that the number of sales is a function of

▸ radio advertising (minutes)
▸ number of salespeople
▸ location of the dealership

A Dummy in Our Example. . .

Predict the number of cars sold at a dealership based on the location of the dealership

In City Limits? Yes/No

If yes = 1

If no = 0

Here is the complete regression: cars sold regressed on advertising, salespeople and location.

SUMMARY OUTPUT

Regression Statistics	
Multiple R	0.96
R Square	0.93
Adj R Square	0.90
Standard Error	7.25
Observations	12

ANOVA

	df	SS	MS	F	Sig F
Regression	3	5504.42	1834.81	34.93	0.00
Residual	8	420.25	52.53		
Total	11	5924.67			

	Coef	Stand Error	t Stat	P-value	Lower 95%	Upper 95%
In'cept	31.13	13.40	2.32	0.05	0.24	62.02
Adv	2.15	0.80	2.67	0.03	0.30	4.01
Sales	5.01	0.91	5.51	0.00	2.91	7.11
city	5.67	6.33	0.89	0.40	-8.94	20.27

RESIDUAL OUTPUT

Obs	Pred Cars	Residuals
1	125.67	1.33
2	138.62	-0.62
3	154.33	4.67
4	146.45	-2.45
5	127.88	11.12
6	131.39	-3.39
7	160.78	0.22
8	177.98	2.02
9	98.50	3.50
10	168.66	-5.66
11	120.00	-14.00
12	145.74	3.26

Interpretation of Output

R^2: 93% of the variation in monthly car sales is explained by the variation in minutes of radio advertising, number of salespeople and the location of the dealership.

Intercept: When

✓ there has been no radio advertising,
✓ there are no salespeople and
✓ the dealership is outside of the city,

about 31 cars are sold per month.

What if the dealership WAS in the city?

then 31.13 + 5.67 cars, on average are sold per month.

What if the dealership did 15 minutes of advertising, had 10 salespeople, but was not in the city?

Cars=31.13 + 2.15(15) + 5.01(10)

=113.48 cars sold on average

Testing Significance of Coefficients

▸ same rules as before

▸ same tests as before

▸ if p-value is smaller than your chosen α, reject the null of no effect and conclude the variable is important.

Confidence Intervals

▸ same construction as before

▸ same meaning as before

Important Concepts

▸ **Interpreting coefficients:**

    ▸ **(net) regression coefficients**

    ▸ **the effect of X_i on Y when everything else (all the other X_i's) is held constant.**

    ▸ **a "conditional" slope**

✓ **Interpreting the coefficient of determination.**

▸ R^2 **ALWAYS rises when you ADD another** X_i **to the regression.**

▸ **adj** R^2 **rises when the new** X_i **improves the models** _predictive_ **ability.**

✓ **If the purpose of the regression is to MAKE PREDICTIONS:**

▸ drop those variables that are not statistically significant

▸   **IF**

you are sure they do not add any predictive power to the model.

- ▸ rule of thumb is look at the size of the *t stat*

- ▸ if *t stat* $\geq$ 1 , it probably increases the predictive power of the model.

- ▸ REMEMBER: predictions based on values outside of your data ranges are very suspect.

✓ **If the purpose of the regression is to EXPLAIN Y:**

▸ THEORY (Yes, *YOUR* thoughts) guide which variables should be included in the model.

▸ NEVER use regression output to determine THEORY.

▸ Ideally, all the important X_i's will be included in the model and that will ensure that your ε_i's are random.

The Linear Regression Model

- ▸ **powerful**

- ▸ **flexible**

- ▸ **practical**

Also, it is

- ▸ **abused**
- ▸ **misused**
- ▸ **exploited**

In fact, most of what we have learned this semester

CAN BE AND IS

regularly used to

- ▸ baffle,
- ▸ obscure and
- ▸ suggest things that are not true.

Descriptive Statistics:

In 1984 the University of Virginia announced that its department of Rhetoric and Communications graduates *MEAN STARTING SALARY* was $55,000!

The MEDIAN salary was not reported, nor was the fact that NBA Center RALPH SAMPSON was a member of that class.

Many Statistics Are
Over Interpreted

BRITISH HE'S BATHE MORE
THAN SHE'S

	Winter	Summer
males:	1.7	2.1
females:	1.5	2.0

well, they SAY they do . . .

Population of a Region in China:

28 Million

Five years pass. . .

Population of SAME Region:

105 Million

(The first census was for tax purposes, the second for famine relief.)

Inferences:

ALL confidence intervals, hypothesis tests and regression analysis depends on the assumption of a

RANDOM SAMPLE.

You can perform a hypothesis test on the percentage of NRA members that favor gun-control, but your results are highly questionable in a more general context.

Extrapolation:

extend predictions beyond the range of data.

If you do this--and sometimes you may have to--be sure you qualify your predictions by saying they depend on nothing else changing!

Mark Twain did some calculations:

The Lower Mississippi has shortened itself at a rate of 1.33 miles per year.

Any calm person, who is not blind or idiotic . . . . can see that 742 years from now the Lower Mississippi will be only 1.75 miles long, and Cairo and New Orleans will have joined their streets together!

Appendix A

Sample Exams

Lecture Exam One

Lab Exam One

Lecture Exam Two

Lab Exam Two

Final Exam

1. Refer to Display #2. What percentage of the domestic passengers have flown 1300 miles or more?

   a. 6.70%
   b. 15.90%
   c. 65.24%
   d. 84.14%

2. A survey of top executives revealed that 35% of them regularly read Time magazine, 20% read Newsweek and 40% read U.S. News & World Report. Ten percent read both Time and U.S. News & World Report. What is the probability that a particular top executive reads either Time or U.S. News & World Report regularly?

   a. 0.088
   b. 0.85
   c. 0.75
   d. 0.65
   e. 0.06

3. Which approach to probability is exemplified by the following formula?

$$P(A) = \frac{\text{Number of times A occured in the past}}{\text{Total number of observations}}$$

   a. Classical approach
   b. Empirical approach
   c. Subjective approach
   d. Impassioned approach

4. What type of data is the number of robberies reported in Bloomington in 1998?

   a. Discrete
   b. Attribute
   c. Continuous
   d. Qualitative
   e. None of the above.

5.   Refer to Display #2. What is the mean value of the number of miles flown by domestic passengers?

   a.   28333.33
   b.   1236.59
   c.   1036.59
   d.   168.33
   e.   47.56

6.   The coefficient of variation for a set of annual incomes is 18%; the coefficient of variation for the length of service with the company is 29%. What does this indicate?

   a.   Dispersion in the two distributions of income and service length cannot be compared using percentages.
   b.   More dispersion in the distribution of the incomes compared with the dispersion of the length of service.
   c.   It is impossible to compare dispersions from different populations.
   d.   More dispersion in the lengths of service compared with incomes.
   e.   Dispersions are equal for the two distributions.

7.   You have decided to enter the stock market and are interested in purchasing two of your favorite stocks–Disney and Chiquita Bananas. You contact your broker and are told that the expected return on $100 in Disney is $200 with a standard deviation of $30. The expected return on $100 in Chiquita Bananas, however, is $500 with a standard deviation of $75. Which stock would represent the riskiest investment?

   a.   It is impossible to tell.
   b.   Disney
   c.   Chiquita Bananas
   d.   Both stocks have the same risk.

8.   Referring to Question #7, you decide to buy both stocks and make a total investment of $100, but that expected return on bananas has you salivating! So you decide to invest 2/3rds of your money in Chiquita and 1/3 in Disney.   What is the standard deviation of your portfolio investment, assuming these stocks are independent?

a. $32.02
b. $49.75
c. $50.99
d. $2475
e. $2600

9.   Referring to question #8, if the value of Chiquita and Disney were not independent, but the value of Chiquita fell at the same time Disney's value fell, which of the following statements is true?

a.   The standard deviation of the dependent portfolio is identical to that calculated in question #8.
b.   The standard deviation of the dependent portfolio is larger than that calculated in question #8.
c.   The standard deviation of the dependent portfolio is smaller than that calculated in question #8.
d.   The standard deviation of the dependent portfolio is impossible to determine without further information.

10. A child has been born into the Doe family every year for seven consecutive years. What is the variance of the ages of the Doe children?

a.   2
b.   1
c.   4
d.   49
e.   7

11. Three years have passed since you calculated the variance in question #10. How has the standard deviation of the ages of the Doe children changed?

a.   The standard deviation is unchanged.
b.   The standard deviation has increased by 3 years.
c.   The standard deviation has increased by 9 years.
d.   The standard deviation has decreased by an amount that depends on the actual ages of the Doe children.

12. Refer to Display #1. Which of the following statements about the skewness would be true, without calculating it?

    a. Skewness is large and negative.
    b. Skewness is large and positive.
    c. Skewness is small and negative.
    d. Skewness is small and positive.

13. Refer to Display #1. Calculate the sample variance, round to two significant digits.

    a. 4.17
    b. 14.14
    c. 121.00
    d. 200.00

14. Refer to Display #1. What is the sum of all the observations in the sample?

    a. 672
    b. 2064
    c. 2666
    d. 3168

15. If population A has a larger standard deviation than population B

    a. Population A will have a greater range than population B
    b. Population A will have a smaller range than population B
    c. Skewness will be increased
    d. Skewness will be decreased.
    e. We cannot say which has the greater range or skewness.

16. What measure of central tendency would be most appropriate for a measurement of salaries when there are a few people in the sample who make over one million dollars, but most of the employees sampled made under $50,000?

    a. Mean and mode
    b. Median and mode
    c. Mean and median
    d. Mean, median and mode

17. The mean weight of three gemstones is 12 grams. The weights of two of the stones are 9 grams and 11 grams. What is the weight of the third stone?

    a.   16 grams
    b.   10 grams
    c.   8 grams
    d.   14 grams
    e.   Insufficient information to determine.

18. A clothes store manager has sales data of trouser sizes for the last month's sales. Which measure of central tendency should the manager use, if the manager is interested in the most sellable size?

    a.   Mean
    b.   Median
    c.   Mode
    d.   Standard deviation
    e.   Interquartile range

19. For the following distribution of heights, what are the limits for the class with the greatest frequency?

Heights	60" up to 65"	65" up to 70"	70" up to 75"
Number	10	70	20

    a.   64 and up to 70
    b.   65 and 69
    c.   65 and 70
    d.   69.5 and 74.5
    e.   None of the above

20. The members of each basketball team wear numbers on the back of their jerseys. What scale of measurement are these numbers considered?

    a.   Nominal
    b.   Ordinal
    c.   Interval
    d.   Ratio
    e.   None of the above

United Airlines sampled members of their frequent flyer program about how many miles they had flown to the nearest 1000. The descriptive statistics, as calculated by Excel, are reproduced below.

Frequent Flyer Miles Flown	
Mean	55.54
Standard Error	2.041
Median	55.5
Mode	43
Standard Deviation	14.14
Sample Variance	
Kurtosis	-0.4096
Skewness	
Range	66
Minimum	
Maximum	88
Sum	
Count	48

Display #1

Miles Flown by a Sample of Domestic Passengers	
Miles Flown	Number of Passengers
100 up to 500	16
500 up to 900	41
900 up to 1300	81
1300 up to 1700	11
1700 up to 2100	9
2100 up to 2500	6

Display #2

E370 – Fall 1999
In Lab Examination One

Instructions

1. Answer the questions in the spaces provided on the exam. Do not provide computer output unless it is requested. Because this exam is multiple pages, be sure **your name is written on every page.**

2. You **MUST** show your work and explanations to receive full credit. You **MUST** write any EXCEL functions used, with arguments, as well as the numerical output. For example, NORMINV(0.5,0,1)=0. **Include four (4) decimal places in your answer.**

3. You may **ONLY USE EXCEL** for this exam.

4. Only the exam, pencils, erasers, and the formula sheet may be on your desk. Put all the rest of your belongings along the wall or at the front of the room.

5. **Remember, a student is to avoid even the appearance of cheating. Keep your eyes on your exam or on YOUR computer screen. ANY questionable behavior on your part is sufficient reason for me to confiscate your exam and ask you to leave the room.**

6. The value of each question is given by each question.

7. When you are finished, you may turn in all pages of the exam and leave the room as soon as you are able to **without disturbing your classmates.**

8. Stay calm and do your best!

1. The amount dispensed into bottles by a machine in a ketchup plant is supposed to be normally distributed with a mean of 10 ounces and a standard deviation of 0.5 ounce.

    a. (10 points) If the machine is working properly, what is the probability that a single bottle chosen at random from the assembly line will have more than 11 ounces or less than 9.5 ounces?

    b. (10 points) What would you think if a single bottle chosen at random held less than 8.5 ounces?

    c. (10 points) Between what two values would you expect to find 99% of the bottles filled by the machine, if it is operating properly?

2. As part of a new promotional campaign, the Pepsi Bottling Company is giving away small prizes to individuals finding bottle tops with Yoda inscribed on the inside. One-ninth of the bottles have the prizewinning tops, which are randomly distributed throughout Pepsi Bottling Company's output.

    a. (10 points) What is the probability that a randomly selected eight-pack will have more than one prizewinning bottle top?

b. (10 points) What is the probability that a randomly selected eight-pack will have no prizewinning bottle tops?

c. (10 points) The Amoco station on the corner of 3rd and Indiana received 50 eight-packs from Pepsi Bottling Company today. What is the probability that at least half of the eight-packs will include a winner?

d. (10 points) Verify your answer to (c.) using the normal distribution.

3. You own an electronics store and are thinking of expanding your inventory to include a new "high-end" (expensive) line of electronic gadgets. You decide to analyze the income of your customers. A random sample of customer incomes yields a mean income of $35,000 and a standard deviation of $4,421.

a. (10 points) What percentage of your customers would have an income above $38,000?

b.   (10 points) What is the likelihood that a customer selected at random
      would have an income between $29,000 and $36,000?

4.   (20 points) The following statistics were calculated by EXCEL from data
      recording the price of gasoline per gallon during the 1997 Memorial Day
      weekend in Detroit, Michigan. Fill in the missing entries. In the space to
      the right comment on the skewness of the distribution.

_____

Price of Gasoline per Gallon
_____

Mean	
Standard Error	0.010963142
Median	1.24
Mode	1.29
Standard Deviation	0.042460065
Sample Variance	
Kurtosis	-0.474296926
Skewness	
Range	
Minimum	1.15
Maximum	1.29
Sum	18.57
Count	15
Confidence Level(95.0%)	0.023513621

5. The following represent the amount of time (in seconds) to get from 0 to 60 from a road test of 12 German-made cars.

10	6.4	8.5	5.5	5.1	10.9
6.9	6.4	6	4.9	8.9	7.9

a. (15 points) Calculate all descriptive statistics using Excel. Print out the table of results and your data. **Be sure your name is on the sheet**.

b. (25 points) Create a probability histogram for your data using the following classes. Title and print the histogram. **Be sure your name is on the sheet**.

5 seconds or less; 5.1 to 6 seconds; 6.1 to 7 seconds; 7.1 to 8 seconds; 8.1 seconds or more.

Refer to the **Printout** at the back of this exam. Use this information to answer **questions 1 through 5** of this exam.

1.   We know nothing about the distribution from which this sample was drawn. We can say that the sampling distribution of the mean number of beds per facility is

   a.   approximately normal by the Central Limit Theorem.
   b.   normally distributed.
   c.   unknown by the Central Limit Theorem.
   d.   uniformly distributed by the Central Limit Theorem.

2.   What is the appropriate set of hypotheses if the New Mexico Director of Public Health wishes to determine if there is an insufficient number of beds available in licensed nursing facilities?

   a. H_0: $\mu \le 93.27$ H_1: $\mu > 93.27$
   b. H_0: $\mu < 100$ H_1: $\mu \ge 100$
   c. H_0: $\mu \ge 100$ H_1: $\mu < 100$
   d. H_0: $\mu = 93.27$ H_1: $\mu \ne 93.27$

3.   The 98% confidence interval for the population mean number of beds, calculated by EXCEL, is equal to

   a. $\mu \pm 40.85 \times$ NORMDIST(2.33, 0,1,1)
   b. $\mu \pm 40.85 \times$ NORMINV(.99, 0,1)
   c. $\mu \pm 5.67 \times$ NORMDIST(.98, 0,1,1)
   d. $\mu \pm 5.67 \times$ NORMINV(.99, 0, 1)

4.   If the 94% confidence interval for the population mean is [88.89, 97.65], the margin of error is

   a.   2.19
   b.   4.38
   c.   8.76
   d.   impossible to determine without more information.

5.   Compared to the 94% confidence interval, the 98% confidence interval is

   a.   wider.
   b.   narrower.
   c.   the same.
   d.   may be larger or smaller.

6.   The Central Limit Theorem applies to

a.   normal populations.
b.   right-skewed populations.
c.   uniform populations.
d.   all of the above.

7.   If Sample A with 34 items and Sample B with 68 items are taken from a population, then

a.   the relative variance of the sample means will depend on the population shape.
b.   the mean of Sample A has a smaller variance than the mean of Sample B.
c.   the mean of Sample A has a larger variance than the mean of Sample B.
d.   we cannot be sure which sample mean will have the greater variance.

8.   If the level of confidence is lowered from 95% to 90%, but the maximum allowable error and the standard deviation remain the same, what happens to the size of sample necessary to achieve that level of confidence?

a.   n increases.
b.   n decreases.
c.   n remains the same.
d.   n may increase, decrease or remain the same.

9.   Which of the following is characteristic of every binomial distribution?

a.   Each outcome is dependent on the previous outcome.
b.   The probability of success increases from trial to trial.
c.   Each outcome is mutually exclusive.
d.   The outcome of a trial depends on the number of trials.

10.   The probability of a person liking an advertisement is 0.6. Ten individuals are interviewed at random. What is the probability that at least one likes the advertisement?

a.   =BINOMDIST(0, 10, .6, 0)
b.   =BINOMDIST(1, 10, .6, 1)
c.   =1-BINOMDIST(1,10, .6, 1)
d.   =1-BINOMDIST(0, 10, .6, 0)

11. X is a normally distributed random variable with mean 100 and standard deviation of 20. Use the attached Standard Normal Table to determine that x for which P(110<X<x)= 0.10.

    a. 116
    b. 120
    c. 112
    d. 114

12. Fluctuations in the exchange rate of dollars against the pound sterling over a short time period were approximated by a normal distribution with a mean of 2.01 and a standard deviation of 0.13. Using the attached Standard Normal Table, what is the probability that the rate on a particular day was more than 1.90?

    a. 0.8461
    b. 0.3023
    c. 0.8023
    d. 0.3461

13. A 99% confidence interval can be interpreted as

    a. a 99% chance that the given interval includes the true value of the population parameter.
    b. Approximately 99 out of 100 similar intervals would include the true value of the population parameter.
    c. a 1% chance that the given interval does not include the true value of the population parameter.
    d. both a and c.

14. When the sample size increases, ceteris paribus, the width of a confidence interval for a population parameter will

    a. increase.
    b. decrease.
    c. remain unchanged.
    d. increase or decrease.

15. An increase in α, the level of significance, causes

    a. an increase in the probability of Type I error to occur.
    b. a decrease in the probability of Type I error to occur.
    c. no change in the probability of Type I or Type II error.
    d. the level of significance is not related to error of any type.

16. Which of the following is NOT a valid reason for selecting a sample instead of studying the whole population?

    a. The cost of studying an entire population can be too high.
    b. The population can be destroyed in the process of the study.
    c. The study of the population can be too time consuming.
    d. Sample results have more accuracy than population results.

17. If we reject the null hypothesis, what can we conclude subject to the α risk?

    a. The null hypothesis is false.
    b. The alternate hypothesis is false.
    c. The null hypothesis is true.
    d. Either or both the null and the alternative hypotheses are true.

18. A national manufacturer of unattached garages discovered that the distribution of the lengths of time it takes two construction workers to erect the Red Barn model is approximately normally distributed with a mean of 32 hours and a standard deviation of 2 hours. What percent of the garages take between 32 and 34 hours to erect?

    a. =1 - NORMDIST(32, 34, 2, 1)
    b. =NORMDIST(32, 32, 2, 0)
    c. =NORMDIST(34, 32, 2, 1) - NORMDIST(32, 32, 2, 1)
    d. =NORMDIST(32, 32, 2, 1) - 1 - NORMDIST(34, 32, 2, 1)

19. For the distribution of constructions times in question 18, the corresponding z value for the construction time to be at least 34 hours is

    a. .3413
    b. -.3413
    c. -1
    d. 1

20. Bottomline Ink, a forms management company, fills 100 orders a day with a 2% error rate in the completed orders. What is the standard deviation for this distribution of orders by Bottomline Ink?

    a. 0.02
    b. 1.4
    c. 2
    d. There is no standard deviation for a binomial distribution.

Printout

The following is an Excel generated list of descriptive statistics of the number of beds available for occupancy in a random sample of licensed nursing facilities in New Mexico. The sample was drawn in order to determine if, on average, the number of beds per facility was of sufficient size. A facility with fewer beds than 100 is considered insufficient.

Beds	
Mean	93.27
Standard Error	
Median	88
Mode	120
Standard Deviation	
Sample Variance	1668.95
Kurtosis	3.69
Skewness	1.41
Range	
Minimum	25
Maximum	244
Sum	4850
Count	

There are 20 parts to this test. Each part is worth 5 points. In every case appropriate EXCEL commands and work must be written and shown, along with the numerical results.

1. The following is a printout of some EXCEL-generated descriptive statistics for the total cost of a fixed market basket of goods purchased at a sample of Marsh grocery stores in Indianapolis.

Marsh	
Mean	131.17
Sample Variance	23.95
Range	19.30
Minimum	120.79
Sum	4590.82

a. Write a 95% confidence interval for the mean total cost of the market basket, based on the information above. Show how you obtained it using the information and the relevant EXCEL functions.

b. What is the margin of error if the level of confidence was 93%?

c. **From this point on in this question, assume that the population standard deviation of the fixed market basket is $4.98. Would the 95% confidence interval change? If it would change, write the new 95% confidence interval.**

d.   The Marsh stores claim that the national average for this fixed market basket is $132.69. Mr. Marsh would like to know if he can legitimately claim that Marsh prices are lower than the national average. He wishes to be very certain and likes a level of significance of .01. What set of hypotheses would he use?

e.   Calculate the test statistic.

f.   Calculate the appropriate critical value.

g.   Calculate the p-value.

h.   State your conclusion about the hypothesis test in terms of the original problem. (This means interpret your result!) At what level of significance would you change your mind?

2. During recent seasons, Major League Baseball has been criticized for the length of the games. A report indicated that the average game lasts 3 hours and 30 minutes. A sample of games revealed the following information, placed in a descriptive statistics format as generated by EXCEL. Note that the minutes were changed to fractions of hours, so that a game that lasted 2 hours and 24 minutes was reported at 2.40 hours.

Hours	
Standard Error	0.14
Standard Deviation	0.56
Sum	50.24
Confidence	0.29

a. MLB management would like to advertise that they were erroneously criticized for the extreme length of their games. What would be the appropriate set of hypotheses that the MLB would establish to test this claim?

b. Calculate the test statistic.

c. Calculate the p-value.

d.   State the conclusion of the hypothesis test in terms of the original problem at the 2.5% level.

3.   A typical college students drinks an average of 27 gallons of coffee each year, or 2.25 gallons per month. The following data is from a sample of students at Northwestern University from last month. At the 5% level, is there a significant difference between the average amount of coffee consumed at Northwestern University and the national claim?

Amount of Coffee in Gallons per Month Consumed by Northwestern University students.
1.75
1.96
1.57
1.82
1.85
1.82
2.43
2.65
2.6
2.24
1.69
2.66

a.   What is the appropriate set of hypotheses?

b.   Calculate the test statistic.

c.   Calculate the critical value.

d.   Calculate the p-value.

e.   State your conclusion to the hypothesis test in terms of the original
      problem.

4.   A random sample of credit card balances at Levenson Brothers Department
      Store generated the following descriptive statistics by EXCEL. Eighteen of
      the balances were paid in full the previous month. National statistics show
      that the average credit card holder carries a balance of $3,900 and only one
      third pay the entire balance each month.

      Balance in Dollars

Mean	1074.05
Standard Error	136.90
Standard Deviation	968.00
Sum	53702.37

a.	Levenson Brothers wishes to test to see if more of their credit card holders pay off their entire balance each month than the national claim. What would be the appropriate set of hypotheses to test this?

b.	Calculate the test statistic.

c.	Calculate the p-value.

d.	Evaluate this hypothesis at the 1% level of significance and state your conclusion in terms of the original problem.

1.　Refer to the **National Highway Data Set.** What is the sample size?
    a.　13
    b.　14
    c.　15
    d.　16

2.　Refer to the **National Highway Data Set.** Based on the results of the regression, what is the predicted size of the winning bid if eight different construction companies submit bids?
    a.　$3.99 M
    b.　$7.50 M
    c.　$10.77 M
    d.　$11.24 M

3.　Based on the results from the **National Highway Data Set**, if the National Highway Association was only going to go to Congress if there was evidence of a significant relationship between the number of bidders and the size of the bid, then at the 1% level of significance the NHA should
    a.　reject H_0 because the p-value is 0.003 and go to Congress.
    b.　fail to reject H_0 because the p-value is 0.003 and go to Congress.
    c.　reject H_0 because the p-value is 0.000 and not go to Congress.
    d.　fail to reject H_0 because the p-value is 0.000 and not go to Congress.

4.　In the **National Highway Data Set** the estimated linear relationship between BIDDERS and BIDSIZE is
    a.　BIDDERS = -0.47+11.24*BIDSIZE
    b.　BIDSIZE = 0.97 -0.13*BIDDERS
    c.　BIDSIZE = 11.24 - 0.47*BIDDERS
    d.　BIDDERS = -0.13 + 0.97 BIDSIZE

5.　According to the regression output from the **National Highway Data Set**, the correlation between BIDDERS and BIDSIZE is
    a.　positive and small.
    b.　positive and large.
    c.　negative and small.
    d.　negative and large.

6.　Calculate the r^2 for the regression in the **National Highway Data Set**.
    a.　0.29　　　c.　0.46
    b.　0.50　　　d.　0.71

7.　Calculate the standard error of the estimate in the **National Highway Data Set**.
    a.　0.08　　　c.　1.11
    b.　0.97　　　d.　4.01

Questions 8 and 9 refer to the following information. A student takes a mid-term and a final exam in a college biology course. From experience the mean of the midterm is known to be 72 with a standard deviation of 12 points. For the final the mean is known to be 68 with a standard deviation of 10 points. Assume that the two exams are independent.

8.　　What is the total number of points the average student can expect to earn in this course?
　　　a.　208 points
　　　b.　140 points
　　　c.　70 points
　　　d.　insufficient information to determine the number of points.

9.　　What is the standard deviation of the total number of points the average student can expect to earn in this course (to the nearest whole number)?
　　　a.　244
　　　b.　228
　　　c.　44
　　　d.　16

Questions 10, 11, and 12 refer to the following information. Bloomington Hospital hires RN's (Registered Nurses) and LPN's (Licensed Practical Nurses.) RN's are paid annually $30,000 on average, with a variance of $2,560,000. LPN's are paid annually $20,000 on average, with a variance of $422,500. The new hospital administrator is developing a plan for each hospital department that will assign 5 RN's and 8 LPN's to each department.

10.　　On average, how much can a given department at the hospital expect to pay for its nurses (RN's and LPN's)?
　　　a.　$290,000
　　　b.　$310,000
　　　c.　$325,000
　　　d.　$340,000

11.　　Assume that the salaries of RN's and LPN's are independent. What is the standard deviation of the salaries of nurses in a department to the nearest whole number?
　　　a.　9541　　　　　c.　16180000
　　　b.　81600　　　　d.　91040000

12.　　If the salaries of RNs and LPNs had a negative covariance, what would be the sign of the correlation coefficient for RN and LPN salaries?
　　　a.　indeterminate, since covariance has no relationship to correlation.
　　　b.　negative
　　　c.　positive
　　　d.　indeterminate without knowing which is independent and which is dependent.

13.   A regression of the square footage cleaned per night on the number of workers per janitorial firm yielded the following EXCEL output. Using this output, perform a test of the following hypotheses, where **α is the intercept and β is the slope coefficient:**

H_0: $\alpha = 0$
H_1: $\alpha \neq 0$

	Coefficients	Standard Error	t Stat	P-value
Intercept	0.55	0.624	0.88	0.3924
Employee	0.02	0.002	10.54	0.0000

At the 5% level of significance, you would

a.   Reject the null and conclude that the number of employees is significantly important to explaining the amount of square footage cleaned per night.

b.   Reject the null and conclude that the number of employees is completely irrelevant to explaining the amount of square footage cleaned per night.

c.   Fail to reject the null and conclude that the amount of square footage that is cleaned per night when there are zero employees working is zero.

d.   Fail to reject the null and conclude that the amount of square footage that is cleaned per night when there are zero employees working is significantly different from zero.

14.   Refer to question 13. Suppose you were testing the hypotheses:

H_0: $\beta = 0$
H_1: $\beta \neq 0$

At the 5% level of significance, you would

a.   Reject the null and conclude that the number of employees is significantly important to explaining the amount of square footage cleaned per night.

b.   Reject the null and conclude that the number of employees is completely irrelevant to explaining the amount of square footage cleaned per night.

c.   Fail to reject the null and conclude that the amount of square footage that is cleaned per night when there are zero employees working is zero.

d.   Fail to reject the null and conclude that the amount of square footage that is cleaned per night when there are zero employees working is significantly different from zero.

15. Refer to **Wishard Hospital Data Set.** The observed monthly salary for employee number 23 is (to the nearest dollar)
    a. $1544
    b. $1632
    c. $1719
    d. the answer requires further information.

16. According to the regression output in the **Wishard Hospital Data Set**, about how much does the salary of a technical employee differ from that of a clerical employee, ceteris paribus?
    a. $33 per month higher
    b. $33 per month lower
    c. $90 per month lower
    d. $90 per month higher

17. In the **Wishard Hospital Data Set** SSR is equal to
    a. 1066830
    b. 1398651
    c. 266708
    d. 2465481

18. According to the output in the **Wishard Hospital Data Set**, what is the predicted starting salary for a new clerical employee who is female and 18 years of age (to the nearest dollar)?
    a. 498
    b. 531
    c. 703
    d. 737

19. What is the coefficient of determination in the **Wishard Hospital Data Set**?
    a. 0.433
    b. 0.567
    c. 0.763
    d. 0.924

20. At the 5% level of significance, which of the independent variables in the **Wishard Hospital Data Set** would you find have a significant effect on the dependent variable?
    a. AGE, JOB
    b. SERV, AGE, SEX
    c. SERV, SEX
    d. SERV, AGE, JOB

21. If the purpose of the regression from the **Wishard Hospital Data Set** were to predict salaries only, which independent variables would you eliminate from the regression?
    a. AGE, JOB
    b. SERV, AGE, SEX
    c. SERV, SEX
    d. SERV, AGE, JOB

22. Based on the information in the **St. Luke's Hospital Data Set,** the appropriate test statistic to use to test the null hypothesis associated with the research question described in the printout is:

    a. $z = \dfrac{(\overline{X} - \mu)}{\sigma/\sqrt{n}}$

    b. $z = \dfrac{(\overline{X} - \mu)}{s/\sqrt{n}}$

    c. $t = \dfrac{(\overline{X} - \mu)}{s/\sqrt{n}}$

    d. $z = \dfrac{(p - \pi)}{\sqrt{\pi(1-\pi)/n}}$

23. Which of the following is the appropriate set of hypotheses for the test associated with the research question described in the **St. Luke's Hospital Data Set?**
    a. H_0: $\overline{x} \le 25$ H_1: $\overline{x} > 25$
    b. H_0: $\mu \le 26.07$ H_1: $\mu > 26.07$
    c. H_0: $\mu \le 25$ H_1: $\mu > 25$
    d. H_0: $\overline{x} \le 26.07$ H_1: $\overline{x} > 26.07$

24. The p-value associated with the above test of the **St. Luke's Hospital Data Set** is the probability of getting a sample of patient numbers with the sample mean
    a. less than that of the printout assuming the population mean is 25.
    b. greater to that of the printout assuming the population mean is 25.
    c. less than that of the printout assuming the population mean is 26.07.
    d. greater to that of the printout assuming the population mean is 26.07.

25. If the p-value calculated from the sample information in the **St. Luke's Hospital Data Set** is **.009** , the null hypothesis of the test would be
    a. rejected at α = .01 but not at α = .001.
    b. rejected at α = .001 but not at α = .01.
    c. rejected at both α = .01 and α = .001.
    d. Fail to reject the null at both α = .01 and α = .001.

26. Assume that the population of recovery room patients is normally distributed. We can say that the sampling distribution of the mean number of recovery room patients is
    a. approximately normal by the Central Limit Theorem.
    b. normally distributed.
    c. unknown by the Central Limit Theorem.
    d. student's t distributed.

27. If the population of recovery room patients from which the **St. Luke's Hospital Data Set** was drawn is normally distributed, the 99% confidence interval for the population mean calculated by EXCEL is equal to
    a. $\mu \pm 0.40 \times NORMDIST(2.58,0,1,1)$
    b. $\mu \pm 0.40 \times NORMINV(.995,0,1)$
    c. $\mu \pm 0.40 \times TINV(.01,14)$
    d. $\mu \pm 0.40 \times TDIST(2.33,14,1)$

28. The 90% confidence interval for the conditions in the previous question would be
    a. smaller than that of question 27.
    b. larger than that of question 27.
    c. the same as that of question 27.
    d. impossible to determine given the level of information.

29. Using the **St. Luke's Hospital Data Set** and your knowledge of measures of central tendency, select the best descriptor of the sample from below.
    a. symmetric
    b. left skewed
    c. right skewed
    d. unskewed

30. Sample means have
    a. the same variability as individual items in the population.
    b. more variability than individual items in the population.
    c. less variability than individual items in the population.
    d. indeterminate variability relative to individual items in the population.

31. Which statement is **NOT** correct regarding the level of significance?
    a. It denotes the probability of Type I error.
    b. It is usually indicated by the symbol α.
    c. If it is reduced, it is harder to reject the null hypothesis.
    d. It indicates the percent of the time a true null hypothesis is not rejected.

32. In a random sample of 200 IU students, 115 said that they would vote in favor of a proposal to install and enforce bike lanes on the IU campus. Calculate the test statistic associated with the alternative hypothesis that the proposal would pass in a general election.
    a. z=.0191
    b. z=.1804
    c. z=2.1429
    d. z=3.9226

33. Using EXCEL, at a significance level of 10%, the critical value against which the above test statistic would be compared would be calculated by
    a. =NORMSDIST(.1)
    b. =NORMSDIST(.95)
    c. =NORMSINV(.90)
    d. =NORMSINV(.95)

34. The mean number of travel days per year for the outside salespeople employed by a hardware distributor is to be estimated. The 0.90 degree of confidence is to be used. The mean of a small pilot study was 150 days, with a standard deviation of 14 days. If the population mean is to be estimated within two days, how many outside salespeople should be sampled? (NORMSINV(.95)=1.645; NORMSINV(.90)=1.28)
    a. 133
    b. 452
    c. 511
    d. 2100

35. Refer to the **Domino's Pizza Data Set.** What is the expected number of pizza orders per hour during Finals Week?
    a. 0.2 pizzas
    b. 6 pizzas
    c. 6.6 pizzas
    d. 30 pizzas

36. Refer to the **Domino's Pizza Data Set.** What is the standard deviation of the number of pizza orders per hour during Finals Week?
    a. 1.28
    b. 1.64
    c. 3
    d. 11.8

37. Refer to the **Domino's Pizza Data Set.** What is the probability of receiving fewer than seven orders per hour during Finals Week?
    a. 0.1
    b. 0.2
    c. 0.3
    d. 0.4

38. A new medicine has an 85% success rate. Twenty patients are treated with it. What is the probability that eighteen are cured with this new medicine?
    a.   =BINOMDIST(18,20,.85,0)
    b.   =BINOMDIST(18,20,.85,1)
    c.   =1-BINOMDIST(18,20,.85,0)
    d.   =1-BINOMDIST(17,20,.85,1)

39. A new medicine has an 85% success rate. Twenty patients are treated with it. What is the probability that at least eighteen are cured with this new medicine?
    a.   =BINOMDIST(18,20,.85,0)
    b.   =BINOMDIST(18,20,.85,1)
    c.   =1-BINOMDIST(18,20,.85,0)
    d.   =1-BINOMDIST(17,20,.85,1)

40. A new medicine has an 85% success rate. Twenty patients are treated with it. What is the mean of the distribution described by these parameters?
    a.   1.60
    b.   2.55
    c.   10
    d.   17

41. Refer to the **Executive Salaries Regression Equation**. The meaning of the coefficient on **Educ** is
    a.   ceteris paribus, for each additional dollar of the executive's salary the number of years the executive was in college rises by 1.45.
    b.   ceteris paribus, for each additional dollar of the executive's salary the number of years the executive was in college falls by 1.45.
    c.   ceteris paribus, for each additional year the executive was in college, the executive's salary rises by $1450.
    d.   ceteris paribus, for each additional year the executive was in college, the executive's salary falls by $1450.

42. Refer to the **Executive Salaries Regression Equation**. The meaning of the coefficient on **Level** is
    a.   ceteris paribus, the difference between the salary of an executive with low responsibility and one with high responsibility is $3,550.
    b.   ceteris paribus, an executive with low responsibility earns $3,550 more than one with high responsibility.
    c.   ceteris paribus, for each additional dollar of the executive's salary, the executive's responsibility rises by 3.55 levels.
    d.   ceteris paribus, for each additional dollar of the executive's salary, the executive's responsibility falls by 3.55 levels.

43. Which is indicative of an *inverse* relationship between X and Y?
    a.   a negative coefficient of determination.
    b.   a negative estimated intercept.
    c.   a negative p-value for the slope coefficient.
    d.   a negative coefficient of correlation.

44. In a right-tail test, the rejection region refers to
    a. the area to the left of the right-tail critical value.
    b. the area to the right of the left-tail critical value.
    c. the area to the right of the right-tail critical value.
    d. the area outside the left-tail and right-tail critical values.

45. Which of the following is **NOT** true of the *t* distribution?
    a. At the same level of confidence, confidence intervals calculated with the *t* are wider than those calculated with the **z**.
    b. The acceptance region for the same chosen level of significance will be wider with the *t* than with the **z**.
    c. It is more difficult to reject the null when using the *t* distribution than when using the **z**.
    d. The *t* distribution "family" consists of one member with mean = 0, and standard deviation = 1.

46. The mean age for the beer inventory at the corner grocery store is 10.2 days. Suppose the grocery store is closed for 5 and a half days by the health department. What is the mean age of the beer inventory when the store reopens?
    a. 4.7 days
    b. 10.2 days
    c. 15.7 days
    d. We cannot calculate the mean age of the beer without knowing the age of each can of beer.

National Highway Data Set

The National Highway Association is studying the relationship between the number of bidders on a highway project and the winning (that is, lowest) bid. The Association is considering asking for Congressional approval to require a minimum number of bids before a winner is selected. To gather evidence the Association took a random sample of 15 projects and regressed the size of the winning bid (BIDSIZE) in millions of dollars on the number of bidders (BIDDERS). The EXCEL output is given below.

SUMMARY OUTPUT

Regression Statistics	
Multiple R	0.71
R Square	
Adjusted R Square	0.46
Standard Error	
Observations	

ANOVA

	df	SS	MS	F
Regression	1	16.06	16.06	12.95
Residual	13	16.13	1.24	
Total	14	32.19		

	Coefficients	Standard Error	t Stat	P-value
Intercept	11.24	0.97	11.60	0.000
Bidders	-0.47	0.13	-3.60	0.003

Wishard Hospital Data Set

Wishard Hospital in Indianapolis was a large, county hospital and has recently become part of the IU Medical Center. Because Wishard was a county hospital, employee salaries were paid on a scale that may have been different from that of the rest of the IU Medical Center. The Hospital Administrator, John Roush, has asked for an analysis of the salaried employees so that the board of directors can take appropriate action to remedy any pay inequality. Mr. Roush has commissioned a random sample of 30 employees of the more than 1200 employees of Wishard Hospital. For each employee he has gathered information on monthly salary in dollars (PAY), length of service at Wishard in months (SERV), the employee's age (AGE), SEX (1=male, 0=female) and whether the employee has a clerical or technical job (JOB: 1=technical, 0=clerical) The regression output follows.

SUMMARY OUTPUT	
Regression Statistics	
Multiple R	0.66
R Square	
Adjusted R Square	0.34
Standard Error	236.53
Observations	

ANOVA				
	df	*SS*	*MS*	*F*
Regression	4		266707.60	4.77
Residual	25	1398650.98	55946.04	
Total	29	2465481.37		

	Coefficients	*Standard Error*	*t Stat*	*P-value*
Intercept	651.86	345.30	1.89	0.07
SERV	13.42	5.13	2.62	0.01
AGE	-6.71	6.35	-1.06	0.30
SEX	205.65	90.27	2.28	0.03
JOB	-33.45	89.55	-0.37	0.71

Wishard Hospital Data Set, Continued
RESIDUAL OUTPUT

Observation	Predicted PAY	Residuals
1	1823.91	-54.91
2	2031.94	-291.94
3	1938.10	2.90
4	2132.73	234.27
5	1938.09	528.91
6	1824.02	-184.02
7	1884.31	-128.31
8	1598.24	107.76
9	1940.40	-173.40
10	1443.87	-243.87
11	1645.23	60.77
12	1584.81	400.19
13	1692.10	-137.10
14	1544.44	204.56
15	1978.27	77.73
16	1766.02	-37.02
17	2246.81	-60.81
18	1658.63	199.37
19	1891.12	-72.12
20	1810.59	-460.59
21	1931.29	98.71
22	2112.50	437.50
23	1631.68	-87.68
24	1991.80	-225.80
25	1958.24	-21.24
26	1812.98	-121.98
27	1584.70	38.30
28	2000.90	-209.90
29	1897.82	103.18
30	1857.47	16.53

St. Luke's Hospital Data Set

The following is an EXCEL generated list of descriptive statistics of the number of patients in the recovery room at St. Luke's Hospital in Maumee, Ohio, over a random sample of days. The sample was drawn to determine if recent enlargements of the recovery room had increased the mean number of patients serviced there on a daily basis from 25 patients, the mean before the renovations.

Number of Patients	
Mean	
Standard Error	
Median	26
Mode	25
Standard Deviation	1.53
Sample Variance	2.35
Skewness	
Range	5
Minimum	24
Maximum	29
Sum	391
Count	15

Domino's Pizza Data Set

The number of pizza orders received in an hour at the Bloomington Domino's Pizza store during Finals Week is a random variable with the following probability distribution.

X	P(X)
4	0.1
5	0.1
6	0.2
7	0.3
8	0.3

Executive Salaries Regression Equation

Business Week ran a regression to explain the salary of executives in large corporations. The regression equation that resulted is:

Salary = 23.04 - .031(Age) + 1.45(Educ) + 3.55 (Level)

Salary is annual salary in thousands of dollars
Age is current age in years
Education (Educ) is the number of years the executive spent in college
Level is the level of responsibility the executive has in the corporation;
     where 1 = High and 0 = Low.

1133

Sample Exam Answers

Lecture Exams and Final

Lab Exam One

Lab Exam Two

Exam 1 – Pink Paper

1	B	5	C	9	B	13	D	17	A
2	D	6	D	10	C	14	C	18	C
3	B	7	D	11	A	15	E	19	C
4	A	8	C	12	D	16	B	20	A

Exam 2 – Blue Version

1	A	5	A	9	C	13	B	17	A
2	C	6	D	10	D	14	B	18	C
3	D	7	C	11	A	15	A	19	D
4	B	8	B	12	C	16	D	20	B

Final Exam – Blue Version
Answer Key

1	C	11	A	21	A	31	D	41	C
2	B	12	B	22	C	32	C	42	A
3	A	13	C	23	C	33	C	43	D
4	C	14	A	24	B	34	A	44	C
5	D	15	A	25	A	35	C	45	D
6	B	16	B	26	B	36	A	46	C
7	C	17	A	27	C	37	D		
8	B	18	B	28	A	38	A		
9	D	19	A	29	C	39	D		
10	B	20	C	30	C	40	D		

E370 – Fall 1999
In Lab Examination One
Answers

1.

    a.   =normdist(9.5,10,.5,1)=.1587 plus
                =1-normdist(11,10,.5,1)=.0228

    b.   The machine was not operating properly

    c.   =norminv(.005,10,.5) = 8.7121 and =norminv(.995,10,.5) = 11.2879

2.

    a.   Equals the sum of binomdist or 2 through 8 or 1-binomdist(1,8,1/9,1) = .2205

    b.   =binomdist(0,8,1/9,0) = .3897

    c.   Calculate the probability of getting at least one winner per 8-pack = 1-binomdist(0,8,1/9,0) =.6103. Calculate probability that at least half will have a winner =1-binomdist(24,50,.6103,1) = .9581

    d. μ=nπ = 50 * .6103 = 30.515
                σ^2=nπ(1-π) = 50 * .6103 * .3897
                σ=3.4484
                Command=1-normdist(24.5,30.515,3.4484,1)= .9545

3.

    a.   =1-normdist(38000,35000,4421,1) = .2487

    b.   =normdist(36000,35000,4421,1)-normdist(29000,35000,4421,1) =.5021

4.

Price of Gasoline per Gallon

Mean	**1.2380**
Standard Error	0.010963142
Median	1.24
Mode	1.29
Standard Deviation	0.042460065
Sample Variance	**.0018**
Kurtosis	-0.474296926
Skewness	**-.1412**
Range	**.14**
Minimum	1.15
Maximum	1.29
Sum	18.57
Count	15
Confidence Level(95.0%)	0.023513621

mean=1.2380
sample variance=.0018
skewness=-.1412
range=.14
Slight left skew, born out by mean < median < mode

5.

    a.

Seconds	
Mean	7.283333333
Standard Error	0.563516198
Median	6.65
Mode	6.4
Standard Deviation	1.952077371
Sample Variance	3.810606061
Kurtosis	-0.736732387
Skewness	0.593248004
Range	6
Minimum	4.9
Maximum	10.9
Sum	87.4
Count	12

E370 Sample Lab Exam #2
Fall 1999

ANSWERS

1.

a. $n = \text{sum/mean} = 4590.82/131.17 = 35$

$s_{\bar{x}} = \text{sqrt}(23.95/35) = .827216$

$CI = 131.17 +/- [\text{tinv}(.05,34)*.827216 = 2.03*.827216]$

$CI = 131.17 +/- 1.6792485$

$[129.4908, 132.8492]$

b. What is the margin of error if the level of confidence was 93%?

$E = \text{TINV}(.07,34)*.827216 = 1.870803*.827216 = 1.547558$

c. Yes.

$CI = 131.17 \pm [\text{normsinv}(.975) * (4.98/\text{sqrt}(35)) = 1.959961*.841774 = 1.649844]$

$[129.52, 132.82]$

d. H_0: $\mu \geq 132.69$

H_1: $\mu < 132.69$

e. $z = [(\bar{x}-\mu)/(\sigma/\text{sqrt}(n))] = (131.17-132.69) \div (4.98/\text{sqrt}(35)) = -1.80571$

f. $z_a = z_{.01} = \text{normsinv}(.01) = -2.32634$

g. $\text{p-value} = \text{normdist}(131.17, 132.69, .841774, 1) = 0.035482$

h. Fail to reject the hypothesis that the market basket average selling price is greater than or equal to $132.69. This result would be reversed at any chosen level of significance greater than .035482.

2.

a. H_0: $\mu \geq 3.5$

H_1: $\mu < 3.5$

b. $\text{se} = s/\text{sqrt}(n)$ so $.14 = .56/\text{sqrt}(n)$ so $n = 16$

$\text{mean} = \text{sum/n}$ so $50.24/16 = 3.14 = \bar{x}$

$t = [(\bar{x}-\mu)/\text{se}] = (3.14-3.5) \div .14 = -2.57143$

c. Critical value $= -\text{tinv}(2*.025, 15) = -2.131451$

d. $\text{p-value} = \text{tdist}(\text{abs}(-2.57143),15,1) = .010637$

e. Reject the null hypothesis and conclude that games are, on average, less than 3.5 hours and advertise that the MLB was erroneously criticized.

3. Problem 3 requires the assumption that coffee amounts are normally distributed. This wording would be written into the exam.

a. H_0: $\mu = 2.25$ gallons

H_1: $\mu \neq 2.25$ gallons

b. $t = ((\bar{x} - \mu)/se] = (2.0867 - 2.25) \div .116867792 = -1.39759$

c.   Critical Value $= \pm tinv(a, df) = \pm tinv(.05, 11) = \pm 2.200986$

d.   p-value $= tdist(abs(-1.397), 11, 2) = .189957$

e.   Fail to reject the null hypothesis that the average gallons of coffee drunk per month by a typical college student is different than 2.25.

4.   Test to see if it is appropriate to approximate the binomial with the normal: $np = .333 * 50 = 16.5 > 5$ and $n(1-p) = 50 * .667 = 33.5 > 5$

a. H_0: $\pi \leq .33$

       H_1: $\pi > .33$

b. $z = (p - \pi)/s_p = (.36 - .33)/.0665 = .45112782$ when $s_p = sqrt((.33 * .67)/50)$

c.   p-value $= 1 - normdist(.36, .33, .0665, 1) = .325949$

d.   At the 1% level of significance we fail to reject the null hypothesis that the proportion of Levenson Brothers cardholders that pays off balances monthly is the same as or smaller than the national average.

Appendix B

Correlation Guide for Student Solution Manual and Data Set CD ROM

Custom Chapter Number	Original Text Chapter Number
1	Text 1 / Soulutions Manual 1
2	Text 2 / Soulutions Manual 2
3	Text 3 / Soulutions Manual 3
4	Text 4 / Soulutions Manual 4
5	Text 6 / Soulutions Manual 6
6	Text 7 / Soulutions Manual 7
7	Text 8 / Soulutions Manual 8
8	Text 13 / Soulutions Manual 14
9	Text 14 / Soulutions Manual 14